PROGRESS IN BRAIN RESEARCH

VOLUME 175

NEUROTHERAPY: PROGRESS IN RESTORATIVE
NEUROSCIENCE AND NEUROLOGY

Other volumes in PROGRESS IN BRAIN RESEARCH

Volume 138: Plasticity in the Adult Brain: From Genes to Neurotherapy, by M.A. Hofman, G.J. Boer, A.J.G.D. Holtmaat, E.J.W. Van Someren, J. Verhaagen and D.F. Swaab (Eds.) – 2002, ISBN 0-444-50981-X.
Volume 139: Vasopressin and Oxytocin: From Genes to Clinical Applications, by D. Poulain, S. Oliet and D. Theodosis (Eds.) – 2002, ISBN 0-444-50982-8.
Volume 140: The Brain's Eye, by J. Hyönä, D.P. Munoz, W. Heide and R. Radach (Eds.) – 2002, ISBN 0-444-51097-4.
Volume 141: Gonadotropin-Releasing Hormone: Molecules and Receptors, by I.S. Parhar (Ed.) – 2002, ISBN 0-444-50979-8.
Volume 142: Neural Control of Space Coding, and Action Production, by C. Prablanc, D. Pélisson and Y. Rossetti (Eds.) – 2003, ISBN 0-444-509771.
Volume 143: Brain Mechanisms for the Integration of Posture and Movement, by S. Mori, D.G. Stuart and M. Wiesendanger (Eds.) – 2004, ISBN 0-444-513892.
Volume 144: The Roots of Visual Awareness, by C.A. Heywood, A.D. Milner and C. Blakemore (Eds.) – 2004, ISBN 0-444-50978-X.
Volume 145: Acetylcholine in the Cerebral Cortex, by L. Descarries, K. Krnjević and M. Steriade (Eds.) – 2004, ISBN 0-444-51125-3.
Volume 146: NGF and Related Molecules in Health and Disease, by L. Aloe and L. Calzà (Eds.) – 2004, ISBN 0-444-51472-4.
Volume 147: Development, Dynamics and Pathology of Neuronal Networks: From Molecules to Functional Circuits, by J. Van Pelt, M. Kamermans, C.N. Levelt, A. Van Ooyen, G.J.A. Ramakers and P.R. Roelfsema (Eds.) – 2005, ISBN 0-444-51663-8.
Volume 148: Creating Coordination in the Cerebellum, by C.I. De Zeeuw and F. Cicirata (Eds.) – 2005, ISBN 0-444-51754-5.
Volume 149: Cortical Function: A View from the Thalamus, by V.A. Casagrande, R.W. Guillery and S.M. Sherman (Eds.) – 2005, ISBN 0-444-51679-4.
Volume 150: The Boundaries of Consciousness: Neurobiology and Neuropathology, by Steven Laureys (Ed.) – 2005, ISBN 0-444-51851-7.
Volume 151: Neuroanatomy of the Oculomotor System, by J.A. Büttner-Ennever (Ed.) – 2006, ISBN 0-444-51696-4.
Volume 152: Autonomic Dysfunction after Spinal Cord Injury, by L.C. Weaver and C. Polosa (Eds.) – 2006, ISBN 0-444-51925-4.
Volume 153: Hypothalamic Integration of Energy Metabolism, by A. Kalsbeek, E. Fliers, M.A. Hofman, D.F. Swaab, E.J.W. Van Someren and R.M. Buijs (Eds.) – 2006, ISBN 978-0-444-52261-0.
Volume 154: Visual Perception, Part 1, Fundamentals of Vision: Low and Mid-Level Processes in Perception, by S. Martinez-Conde, S.L. Macknik, L.M. Martinez, J.M. Alonso and P.U. Tse (Eds.) – 2006, ISBN 978-0-444-52966-4.
Volume 155: Visual Perception, Part 2, Fundamentals of Awareness, Multi-Sensory Integration and High-Order Perception, by S. Martinez-Conde, S.L. Macknik, L.M. Martinez, J.M. Alonso and P.U. Tse (Eds.) – 2006, ISBN 978-0-444-51927-6.
Volume 156: Understanding Emotions, by S. Anders, G. Ende, M. Junghofer, J. Kissler and D. Wildgruber (Eds.) – 2006, ISBN 978-0-444-52182-8.
Volume 157: Reprogramming of the Brain, by A.R. Møller (Ed.) – 2006, ISBN 978-0-444-51602-2.
Volume 158: Functional Genomics and Proteomics in the Clinical Neurosciences, by S.E. Hemby and S. Bahn (Eds.) – 2006, ISBN 978-0-444-51853-8.
Volume 159: Event-Related Dynamics of Brain Oscillations, by C. Neuper and W. Klimesch (Eds.) – 2006, ISBN 978-0-444-52183-5.
Volume 160: GABA and the Basal Ganglia: From Molecules to Systems, by J.M. Tepper, E.D. Abercrombie and J.P. Bolam (Eds.) – 2007, ISBN 978-0-444-52184-2.
Volume 161: Neurotrauma: New Insights into Pathology and Treatment, by J.T. Weber and A.I.R. Maas (Eds.) – 2007, ISBN 978-0-444-53017-2.
Volume 162: Neurobiology of Hyperthermia, by H.S. Sharma (Ed.) – 2007, ISBN 978-0-444-51926-9.
Volume 163: The Dentate Gyrus: A Comprehensive Guide to Structure, Function, and Clinical Implications, by H.E. Scharfman (Ed.) – 2007, ISBN 978-0-444-53015-8.
Volume 164: From Action to Cognition, by C. von Hofsten and K. Rosander (Eds.) – 2007, ISBN 978-0-444-53016-5.
Volume 165: Computational Neuroscience: Theoretical Insights into Brain Function, by P. Cisek, T. Drew and J.F. Kalaska (Eds.) – 2007, ISBN 978-0-444-52823-0.
Volume 166: Tinnitus: Pathophysiology and Treatment, by B. Langguth, G. Hajak, T. Kleinjung, A. Cacace and A.R. Møller (Eds.) – 2007, ISBN 978-0-444-53167-4.
Volume 167: Stress Hormones and Post Traumatic Stress Disorder: Basic Studies and Clinical Perspectives, by E.R. de Kloet, M.S. Oitzl and E. Vermetten (Eds.) – 2008, ISBN 978-0-444-53140-7.
Volume 168: Models of Brain and Mind: Physical, Computational and Psychological Approaches, by R. Banerjee and B.K. Chakrabarti (Eds.) – 2008, ISBN 978-0-444-53050-9.
Volume 169: Essence of Memory, by W.S. Sossin, J.-C. Lacaille, V.F. Castellucci and S. Belleville (Eds.) – 2008, ISBN 978-0-444-53164-3.
Volume 170: Advances in Vasopressin and Oxytocin – From Genes to Behaviour to Disease, by I.D. Neumann and R. Landgraf (Eds.) – 2008, ISBN 978-0-444-53201-5.
Volume 171: Using Eye Movements as an Experimental Probe of Brain Function — A Symposium in Honor of Jean Büttner-Ennever, by Christopher Kennard and R. John Leigh (Eds.) – 2008, ISBN 978-0-444-53163-6.
Volume 172: Serotonin–Dopamine Interaction: Experimental Evidence and Therapeutic Relevance, by Giuseppe Di Giovanni, Vincenzo Di Matteo and Ennio Esposito (Eds.) – 2008, ISBN 978-0-444-53235-0.
Volume 173: Glaucoma: An Open Window to Neurodegeneration and Neuroprotection, by Carlo Nucci, Neville N. Osborne, Giacinto Bagetta and Luciano Cerulli (Eds.) – 2008, ISBN 978-0-444-53256-5.
Volume 174: Mind and Motion: The Bidirectional Link Between Thought and Action, by Markus Raab, Joseph G. Johnson and Hauke R. Heekeren (Eds.) – 2009, ISBN 978-0-444-53356-2.

PROGRESS IN BRAIN RESEARCH

VOLUME 175

# NEUROTHERAPY: PROGRESS IN RESTORATIVE NEUROSCIENCE AND NEUROLOGY

PROCEEDINGS OF THE 25TH INTERNATIONAL SUMMER SCHOOL OF BRAIN RESEARCH, HELD AT THE ROYAL NETHERLANDS ACADEMY OF ARTS AND SCIENCES, AMSTERDAM, THE NETHERLANDS, AUGUST 25–28, 2008

EDITED BY

JOOST VERHAAGEN
ELLY M. HOL
INGE HUITINGA
JAN WIJNHOLDS
ARTHUR A. BERGEN
GERARD J. BOER
DICK F. SWAAB

*Netherlands Institute for Neuroscience, Meibergdreef 47, 1105 BA Amsterdam, The Netherlands*

ELSEVIER

AMSTERDAM – BOSTON – HEIDELBERG – LONDON – NEW YORK – OXFORD
PARIS – SAN DIEGO – SAN FRANCISCO – SINGAPORE – SYDNEY – TOKYO

Elsevier
360 Park Avenue South, New York, NY 10010-1710
Linacre House, Jordan Hill, Oxford OX2 8DP, UK
Radarweg 29, PO Box 211, 1000 AE Amsterdam, The Netherlands

First edition 2009

Copyright © 2009 Elsevier B.V. All rights reserved

No part of this publication may be reproduced, stored in a retrieval system
or transmitted in any form or by any means electronic, mechanical, photocopying,
recording or otherwise without the prior written permission of the publisher

Permissions may be sought directly from Elsevier's Science & Technology Rights
Department in Oxford, UK: phone (+44) (0) 1865 843830; fax (+44) (0) 1865 853333;
email: permissions@elsevier.com. Alternatively you can submit your request online by
visiting the Elsevier web site at http://www.elsevier.com/locate/permissions, and selecting
*Obtaining permission to use Elsevier material*

Notice
No responsibility is assumed by the publisher for any injury and/or damage to persons
or property as a matter of products liability, negligence or otherwise, or from any use
or operation of any methods, products, instructions or ideas contained in the material
herein. Because of rapid advances in the medical sciences, in particular, independent
verification of diagnoses and drug dosages should be made

**British Library Cataloguing in Publication Data**
A catalogue record for this book is available from the British Library

**Library of Congress Cataloging-in-Publication Data**
A catalog record for this book is available from the Library of Congress

ISBN: 978-0-12-374511-8 (this volume)
ISSN: 0079-6123 (Series)

For information on all Elsevier publications
visit our website at elsevierdirect.com

Printed and bound in Great Britain

09 10 11 12 13    10 9 8 7 6 5 4 3 2 1

Working together to grow
libraries in developing countries

www.elsevier.com | www.bookaid.org | www.sabre.org

ELSEVIER    BOOK AID International    Sabre Foundation

# List of Contributors

F. Agosta, Neuroimaging Research Unit, Institute of Experimental Neurology, Division of Neuroscience, Scientific Institute and University Ospedale San Raffaele, Milan, Italy

J.K. Alexander, Neuroscience Graduate Studies Program, The Ohio State University College of Medicine, Columbus, OH, USA

R.R. Ali, Department of Genetics, UCL Institute of Ophthalmology, London, UK; Molecular Immunology Unit, UCL Institute of Child Health, London, UK

M. Azzouz, Academic Neurology Unit, University of Sheffield, Sheffield, UK

F. Baas, Neurogenetics Laboratory and Department of Neurology, Academic Medical Center, University of Amsterdam, Amsterdam, The Netherlands

K.S. Bankiewicz, Department of Neurological Surgery, University of California, San Francisco, CA, USA

A.L. Benabid, CEA-Minatec LETI and INSERM, Joseph Fourier University, Grenoble, France

F. Benedetti, Department of Neuroscience, University of Turin Medical School and National Institute of Neuroscience, Turin, Italy

G. Bernardi, Clinica Neurologica, Dipartimento di Neuroscienze, Università Tor Vergata, and Fondazione Santa Lucia, IRCCS, Rome, Italy

A. Björklund, Wallenberg Neuroscience Center, Lund University, Lund, Sweden

G.J. Boer, Laboratory for Neuroregeneration, Netherlands Institute for Neuroscience, an Institute of the Royal Academy of Arts and Sciences, Amsterdam, The Netherlands

N. Brazda, Molecular Neurobiology Laboratory, Department of Neurology, Heinrich Heine University of Düsseldorf, Düsseldorf, Germany

M. Bsibsi, Delta Crystallon BV, Leiden, The Netherlands

J. Burdick, Division of Engineering, Bioengineering, and Mechanical Engineering Options, California Institute of Technology, Pasadena, CA, USA

M. Carner, ENT Department, University of Verona, Verona, Italy

D. Centonze, Clinica Neurologica, Dipartimento di Neuroscienze, Università Tor Vergata, and Fondazione Santa Lucia, IRCCS, Rome, Italy

S. Chabardes, CEA-Minatec LETI and INSERM, Joseph Fourier University, Grenoble, France

G.J. Chader, Doheny Retina Institute, USC School of Medicine, Los Angeles, CA, USA

L. Colletti, ENT Department, University of Verona, Verona, Italy

V. Colletti, ENT Department, University of Verona, Verona, Italy

G. Courtine, Neurology Department, University of Zurich, Zurich, Switzerland

M.A. Curtis, Department of Anatomy with Radiology, Faculty of Medical and Health Sciences, University of Auckland, Auckland, New Zealand; Institute of Neuroscience and Physiology at Sahlgrenska Academy, Göteborg, Sweden

L.P. Deleyrolle, Department of Neurosurgery, McKnight Brain Institute, University of Florida, Gainesville, FL, USA

D. Denys, Department of Psychiatry, AMC, and the Netherlands Institute for Neuroscience, an Institute of the Royal Netherlands Academy of Arts and Sciences, Amsterdam, The Netherlands

V.R. Edgerton, Departments of Physiological Science and Neurobiology, and Brain Research Institute, University of California, Los Angeles, CA, USA

R. Eggers, Laboratory for Neuroregeneration, Netherlands Institute for Neuroscience, an Institute of the Royal Academy of Arts and Sciences, Amsterdam, The Netherlands

R.L.M. Faull, Department of Anatomy with Radiology, Faculty of Medical and Health Sciences, University of Auckland, Auckland, New Zealand

J. Fawcett, Cambridge University Centre for Brain Repair, Department of Clinical Neurosciences, Robinson Way, Cambridge, UK

M. Filippi, Neuroimaging Research Unit, Institute of Experimental Neurology, Division of Neuroscience, Scientific Institute and University Ospedale San Raffaele, Milan, Italy

A.J. Fong, Division of Engineering, Bioengineering, California Institute of Technology, Pasadena, CA, USA

J. Forsayeth, Department of Neurological Surgery, University of California, San Francisco, CA, USA

C. Foti, Medicina Fisica e Riabilitativa, Dipartimento di Sanità Pubblica e Biologia Cellulare, Università Tor Vergata, Rome, Italy

V. Fraix, CEA-Minatec LETI and INSERM, Joseph Fourier University, Grenoble, France

Y. Gerasimenko, Department of Physiological Science, University of California, Los Angeles, CA, USA and Pavlov Institute of Physiology, St. Petersburg, Russia

S.M. Gold, Multiple Sclerosis Program, Department of Neurology, and Cousins Center for Psychoneuroimmunology, Geffen School of Medicine, University of California Los Angeles, Los Angeles, CA, USA

P. Hadaczek, Department of Neurological Surgery, University of California, San Francisco, CA, USA

A.R. Harvey, School of Anatomy and Human Biology, The University of Western Australia, Crawley, WA, Australia

M. Hellström, School of Anatomy and Human Biology, The University of Western Australia, Crawley, WA, Australia

C. Henle, Laboratory for Biomedical Microtechnology, Department of Microsystems Engineering – IMTEK, University of Freiburg, Freiburg, Germany

S. Herwik, Microsystems Materials Laboratory, Department of Microsystems Engineering – IMTEK, University of Freiburg, Freiburg, Germany

M.S. Humayun, Doheny Retina Institute, USC School of Medicine, Los Angeles, CA, USA

R.M. Ichiyama, Institute of Membrane and Systems Biology, University of Leeds, Leeds, UK

M.O. Karl, Department of Biological Structure, Institute of Stem Cell and Regenerative Medicine, University of Washington, School of Medicine, Seattle, WA, USA

A.P. Kells, Department of Neurological Surgery, University of California, San Francisco, CA, USA

S. Kisban, Microsystems Materials Laboratory, Department of Microsystems Engineering – IMTEK, University of Freiburg, Freiburg, Germany

G. Koch, Clinica Neurologica, Dipartimento di Neuroscienze, Università Tor Vergata, and Fondazione Santa Lucia, IRCCS, Rome, Italy

J.H. Kordower, Department of Neurological Sciences, Rush University Medical Center, Chicago, IL, USA

P. Krack, CEA-Minatec LETI and INSERM, Joseph Fourier University, Grenoble, France

D.A. Lamba, Department of Biological Structure, Institute of Stem Cell and Regenerative Medicine, University of Washington, School of Medicine, Seattle, WA, USA

I. Lavrov, Department of Physiological Science, University of California, Los Angeles, CA, USA

C.A. Lemere, Brigham and Women's Hospital, Harvard Medical School, Boston, MA, USA

C. Lohmann, Department of Synapse and Network Development, Netherlands Institute for Neuroscience, Amsterdam, The Netherlands

C. Lubetzki, UMRS, Inserm 975, CR-Icm and Faculté of Médicine, Université Pierre & Marie Curie, Paris, France

R.E. MacLaren, Department of Genetics, UCL Institute of Ophthalmology, London, UK; Vitreoretinal Service, Moorfields Eye Hospital, London, UK

M.J.A. Malessy, Department of Neurosurgery, Leiden University Medical Center, Leiden, The Netherlands

M. Mantione, Department of Psychiatry, AMC, and the Netherlands Institute for Neuroscience, an Institute of the Royal Netherlands Academy of Arts and Sciences, Amsterdam, The Netherlands

H.J. Monzo, Department of Anatomy with Radiology, Faculty of Medical and Health Sciences, University of Auckland, Auckland, New Zealand

F. Mori, Clinica Neurologica, Dipartimento di Neuroscienze, Università Tor Vergata, and Fondazione Santa Lucia, IRCCS, Rome, Italy

H.W. Müller, Molecular Neurobiology Laboratory, Department of Neurology, Heinrich Heine University of Düsseldorf, Düsseldorf, Germany

A. Nanou, Academic Neurology Unit, University of Sheffield, Sheffield, UK

R.A. Pearson, Department of Genetics, UCL Institute of Ophthalmology, London, UK

B. Piallat, CEA-Minatec LETI and INSERM, Joseph Fourier University, Grenoble, France

G. Piaton, UMRS, Inserm 975, CR-Icm, Paris, France

P. Pollak, CEA-Minatec LETI and INSERM, Joseph Fourier University, Grenoble, France

A. Pollo, Department of Neuroscience, Faculty of Pharmacy, University of Turin and National Institute of Neuroscience, Turin, Italy

P.G. Popovich, Department of Molecular Virology, Immunology and Medical Genetics, Department of Neuroscience, Center for Brain and Spinal Cord Repair (CBSCR), The Institute for Behavioral Medicine Research, The Ohio State University College of Medicine, Columbus, OH, USA

V. Ramaglia, Neurogenetics Laboratory, Academic Medical Center, University of Amsterdam, Amsterdam, The Netherlands; Department of Infection, Immunity and Biochemistry, School of Medicine, Cardiff University, Cardiff, UK

S. Ramaswamy, Department of Neurological Sciences, Rush University Medical Center, Chicago, IL, USA

T.A. Reh, Department of Biological Structure, Institute of Stem Cell and Regenerative Medicine, University of Washington, School of Medicine, Seattle, WA, USA

B.A. Reynolds, Department of Neurosurgery, McKnight Brain Institute, University of Florida, Gainesville, FL, USA

J. Rodger, School of Animal Biology, The University of Western Australia, Crawley, WA, Australia

R.R. Roy, Departments of Physiological Science and Brain Research Institute, University of California, Los Angeles, CA, USA

B. Rubehn, Laboratory for Biomedical Microtechnology, Department of Microsystems Engineering – IMTEK, University of Freiburg, Freiburg, Germany

P. Ruther, Microsystems Materials Laboratory, Department of Microsystems Engineering – IMTEK, University of Freiburg, Freiburg, Germany

I. Sayeed, Brain Research Laboratory, Department of Emergency Medicine, Emory University, Atlanta, GA, USA

N.D. Schiff, Department of Neurology and Neuroscience, Weill Cornell Medical College, New York, NY, USA

R.O. Schlingemann, Medical Retina Unit and Ocular Angiogenesis Group, Department of Ophthalmology, University of Amsterdam, Academic Medical Centre, Amsterdam, The Netherlands

M. Schuettler, Laboratory for Biomedical Microtechnology, Department of Microsystems Engineering – IMTEK, University of Freiburg, Freiburg, Germany

D. Seilhean, Laboratoire de Neuropathologie Escourolle and Faculté of Médicine, Université Pierre & Marie Curie, Paris, France

R.V. Shannon, House Ear Institute, Los Angeles, CA, USA

W.M. Slocum, Department of Brain and Cognitive Sciences, Massachusetts Institute of Technology, Cambridge, MA, USA

S.M. Smirnakis, Department of Neuroscience, Baylor College of Medicine, Houston, TX, USA

K.E. Soderstrom, Department of Neurological Sciences, Rush University Medical Center, Chicago, IL, USA

J.C. Sowden, Developmental Biology Unit, UCL Institute of Child Health, London, UK

D.G. Stein, Brain Research Laboratory, Department of Emergency Medicine, Emory University, Atlanta, GA, USA

T. Stieglitz, Laboratory for Biomedical Microtechnology, Department of Microsystems Engineering – IMTEK, University of Freiburg, Freiburg, Germany

D.F. Swaab, Laboratory for Neuroregeneration, Netherlands Institute for Neuroscience, an Institute of the Royal Academy of Arts and Sciences, Amsterdam, The Netherlands

Y.C. Tai, Division of Engineering, Bioengineering, and Mechanical Engineering Options, California Institute of Technology, Pasadena, CA, USA

M.R. Tannemaat, Laboratory for Neuroregeneration, Netherlands Institute for Neuroscience, an Institute of the Royal Academy of Arts and Sciences, Amsterdam; Department of Neurology, Leiden University Medical Center, Leiden, The Netherlands

E.J. Tehovnik, Department of Brain and Cognitive Sciences, Massachusetts Institute of Technology, Cambridge, MA, USA

L.H. Thompson, Florey Neuroscience Institutes, University of Melbourne, Parkville, Victoria, Australia

A.S. Tolias, Department of Neuroscience, Baylor College of Medicine, Houston, TX, USA

N. Torres, CEA-Minatec LETI and INSERM, Joseph Fourier University, Grenoble, France

F. Valles, Department of Neurological Surgery, University of California, San Francisco, CA, USA

J.M. van Noort, Delta Crystallon BV, Leiden, The Netherlands

V. Varenika, Department of Neurological Surgery, University of California, San Francisco, CA, USA

J. Verhaagen, Laboratory for Neuroregeneration, Netherlands Institute for Neuroscience, an Institute of the Royal Academy of Arts and Sciences, Amsterdam, The Netherlands

S. Veronese, ENT Department, University of Verona, Verona, Italy

R.R. Voskuhl, Multiple Sclerosis Program, Department of Neurology, Geffen School of Medicine, University of California, Los Angeles, Los Angeles, CA, USA

H.U. Voss, Citigroup Biomedical Imaging Center, Weill Cornell Medical College, New York, NY, USA

J. Weiland, Doheny Retina Institute, USC School of Medicine, Los Angeles, CA, USA

E.L. West, Department of Genetics, UCL Institute of Ophthalmology, London, UK

A. Williams, MS Centre, Centre for Inflammation Research, QMRI, Edinburgh, UK

A.N. Witmer, Medical Retina Unit and Ocular Angiogenesis Group, Department of Ophthalmology, University of Amsterdam, Academic Medical Centre, Amsterdam, The Netherlands

# Preface

The 25th International Summer School of Brain Research was held in Amsterdam on August 25–28, 2008, at the Auditorium of the Royal Netherlands Academy of Arts and Sciences (KNAW). Since 1964, the Summer School has been a biennial event organized by the Netherlands Institute for Neuroscience, one of the life sciences institutes of the KNAW. The history of the Netherlands Institute for Neuroscience dates back to the beginning of the last century. At the meeting of the International Association of Academies held in Paris in 1901, the anatomist Wilhelm His proposed that research on the nervous system should be placed on an international footing. In 1904, this resulted in the formation of the International Academic Committee for Brain Research, which set itself the task of "organizing a network of institutions throughout the civilized world, dedicated to the study of the structure and functions of the central organ…." Several governments responded to this ambition by founding brain research institutes, among which was the Netherlands Central Institute for Brain Research, as it was called then, which opened its doors on June 8, 1909. Professor C. U. Ariëns Kappers (1877–1946) became the first director of the institute, a position he held until his death. In honor of its first director, the institute decided to create the C. U. Ariëns Kappers Award, which this year was presented to Dr. James Fawcett (Cambridge, UK), for his outstanding achievements in the field of molecular control of brain plasticity and regeneration.

The focus of this 25th Summer School was on exciting recent progress in restorative neurology and neuroscience and included lectures on major neurodegenerative disorders of the brain and the visual system, including Parkinson's disease, Alzheimer's disease, amyotrophic lateral sclerosis, Huntington's disease, macula degeneration, retinitis pigmentosa, glaucoma, peripheral nerve and spinal cord trauma, and multiple sclerosis. The primary goal of the 25th Summer School was to give an overview of new developments in translational research and in potential therapeutic strategies, including stem cell therapy, immunotherapy, gene therapy, pharmacotherapy, neuroprostheses, functional electrical stimulation, and "deep brain stimulation." Lectures covered all levels of biological organization, including novel molecular and cellular targets, electrophysiological, anatomical and behavioral substrates of neurodegeneration, and the application of whole brain in vivo imaging.

The enthusiasm with which a great number of scientists have agreed to come to Amsterdam and to contribute to this volume is very gratifying. We would like to acknowledge the generosity of both the Royal Netherlands Academy of Arts and Sciences and the Graduate School Neurosciences Amsterdam, under whose joint auspices this Summer School was being held, as well as many other generous financial supporters. Finally, we would like to express our special gratitude to Tini Eikelboom, Henk Stoffels, Wilma Verweij, and Wilma Top for their invaluable organizational and editorial assistance.

On behalf of the organizing committee
Joost Verhaagen

# Acknowledgments

The 25th International Summer School of Brain Research has been made possible by financial support from:

Abcam, UK
Alzheimer Nederland
Amsterdam Molecular Therapeutics (AMT)
Applied Biosystems
Bio-Connect BV
Boehringer Ingelheim
Bristol-Myers Squibb BV
Corning
Curatis Pharma GmbH
Eurogentec BV
Elsevier Science Publishers
Genzyme Europe BV
GlaxoSmithKline
Graduate School Neurosciences Amsterdam (ONWA)
Hersenstichting Nederland
Invitrogen
Leica Microsystems BV
Landelijke Stichting voor Blinden en Slechtzienden (LSBS)
Merck Research Labs
Royal Netherlands Academy of Arts and Sciences (KNAW)
Solvay Pharmaceuticals BV
Stichting Blindenhulp
Stichting Blindenpenning
Stichting Glaucoomfonds
Stichting MD Fonds
Stichting MS Research
Stichting Van den Houtenfonds
Tebu-bio
Uitgeverij Nieuwezijds
Zeiss
ZonMw

# Contents

List of Contributors.................................................................... v

Preface................................................................................ ix

Acknowledgments....................................................................... xi

**Section I. Stem Cells**

1. Cell transplantation strategies for retinal repair
   E.L. West, R.A. Pearson, R.E. MacLaren, J.C. Snowden and R.R. Ali (London, UK)        3

2. Strategies for retinal repair: cell replacement and regeneration
   D.A. Lamba, M.O. Karl and T.A. Reh (Seattle, WA, USA)...........................      23

3. The rostral migratory stream and olfactory system: smell, disease and slippery cells
   M.A. Curtis, H.J. Monzo and R.L.M. Faull
   (Auckland, New Zealand and Göteborg, Sweden)....................................      33

4. Identifying and enumerating neural stem cells: application to aging and cancer
   L.P. Delayrolle and B.A. Reynolds (Gainesville, FL, USA).........................     43

5. Transgenic reporter mice as tools for studies of transplantability and connectivity of dopamine neuron precursors in fetal tissue grafts
   L.H. Thompson and A. Björklund (Victoria, Australia and Lund, Sweden).........        53

**Section II. Immunotherapy and Vaccination Therapy**

6. Developing novel immunogens for a safe and effective Alzheimer's disease vaccine
   C.A. Lemere (Boston, MA, USA)..................................................       83

7. Innate immunity in the nervous system
   V. Ramaglia and F. Baas (Amsterdam, The Netherlands and Cardiff, UK).........         95

8. Neuroinflammation in spinal cord injury: therapeutic targets for neuroprotection and regeneration
   J.K. Alexander and P.G. Popovich (Columbus, OH, USA).............................    125

9. Toll-like receptors in the CNS: implications for neurodegeneration and repair
   J.M. van Noort and M. Bsibsi (Leiden, The Netherlands)..........................     139

**Section III. Gene Therapy**

10. Gene therapy and transplantation in the retinofugal pathway
    A.R. Harvey, M. Hellstrom and J. Rodger (Crawley, WA, Australia) . . . . . . . . . . . . 151

11. Controlled dissemination of AAV vectors in the primate brain
    V. Varenika, A.P. Kells, F. Valles, P. Hadaczek, J. Forsayeth and K.S. Bankiewicz
    (San Francisco, CA, USA) . . . . . . . . . . . . . . . . . . . . . . . . . . . . . . . . . . . . . . . . 163

12. From microsurgery to nanosurgery: how viral vectors may help repair the peripheral nerve
    M.R. Tannemaat, G.J. Boer, R. Eggers, M.J.A. Malessy and J. Verhaagen
    (Amsterdam, and Leiden, The Netherlands) . . . . . . . . . . . . . . . . . . . . . . . . . . . . 173

13. Gene therapy for neurodegenerative diseases based on lentiviral vectors
    A. Nanou and M. Azzouz (Sheffield, UK) . . . . . . . . . . . . . . . . . . . . . . . . . . . . . . 187

14. Trophic factors therapy in Parkinson's disease
    S. Ramaswamy, K.E. Soderstrom and J.H. Kordower (Chicago, IL, USA) . . . . . . . . 201

**Section IV. Pharmacotherapy**

15. Progesterone as a neuroprotective factor in traumatic and ischemic brain injury
    I. Sayeed and D.G. Stein (Atlanta, GA, USA) . . . . . . . . . . . . . . . . . . . . . . . . . . . 219

16. Estrogen and testosterone therapies in multiple sclerosis
    S.M. Gold and R.R. Voskuhl (Los Angeles, CA, USA) . . . . . . . . . . . . . . . . . . . . . 239

17. Treatment of retinal diseases with VEGF antagonists
    R.O. Schlingemann and A.N. Witmer (Amsterdam, The Netherlands) . . . . . . . . . . . 253

18. Pharmacological modification of the extracellular matrix to promote regeneration
    of the injured brain and spinal cord
    N. Brazda and H.W. Müller (Düsseldorf, Germany) . . . . . . . . . . . . . . . . . . . . . . . 269

19. The placebo response: neurobiological and clinical issues of neurological relevance
    A. Pollo and F. Benedetti (Turin, Italy) . . . . . . . . . . . . . . . . . . . . . . . . . . . . . . . . 283

**Section V. Neuroprostheses**

20. Brain—computer interfaces: an overview of the hardware to record neural signals
    from the cortex
    T. Stieglitz, B. Rubehn, C. Henle, S. Kisban, S. Herwik, P. Ruther and
    M. Schuettler (Freiburg, Germany) . . . . . . . . . . . . . . . . . . . . . . . . . . . . . . . . . . 297

21. Artificial vision: needs, functioning, and testing of a retinal electronic prosthesis
    G.J. Chader, J. Weiland and M.S. Humayun (Los Angeles, CA, USA) . . . . . . . . . . 317

22. Progress in restoration of hearing with the auditory brainstem implant
    V. Colletti, R.V. Shannon, M. Carner, S. Veronese and L. Colletti
    (Verona, Italy and Los Angeles, CA, USA) .................................. 333

23. Microstimulation of visual cortex to restore vision
    E.J. Tehovnik, W.M. Slocum, S.M. Smirnakis and A.S. Tolias
    (Cambridge, MA and Houston, TX, USA)..................................... 347

**Section VI. Deep Brain Stimulation, FES and TMS**

24. Functional neurosurgery for movement disorders: a historical perspective
    A.L. Benabid, S. Chabardes, N. Torres, B. Piallat, P. Krack, V. Fraix and
    P. Pollak (Grenoble, France)................................................ 379

25. Recovery of control of posture and locomotion after a spinal cord injury: solutions
    staring us in the face
    A.J. Fong, R.R. Roy, R.M. Ichiyama, I. Lavrov, G. Courtine, Y. Gerasimenko,
    Y.C. Tai, J. Burdick and V.R. Edgerton (Los Angeles, CA, USA, Leeds, UK,
    Zurich, Switzerland, St. Petersburg, Russia and Pasadena, CA, USA)............ 393

26. Deep brain stimulation in obsessive—compulsive disorder
    D. Denys and M. Mantione (Amsterdam, The Netherlands) .................... 419

27. The use of repetitive transcranial magnetic stimulation (rTMS) for the treatment of spasticity
    F. Mori, G. Koch, C. Foti, G. Bernardi and D. Centonze (Rome, Italy).......... 429

**Section VII. Mechanisms of Spontaneous Plasticity and Regeneration**

28. Calcium signaling and the development of specific neuronal connections
    C. Lohmann (Amsterdam, The Netherlands) ................................. 443

29. Remyelination in multiple sclerosis
    G. Piaton, A. Williams, D. Seilhean and C. Lubetzki (Paris, France and Edinburgh, UK) 453

30. Magnetic resonance techniques to quantify tissue damage, tissue repair, and functional cortical
    reorganization in multiple sclerosis
    M. Filippi and F. Agosta (Milan, Italy)..................................... 465

31. MRI of neuronal network structure, function, and plasticity
    H.U. Voss and N.D. Schiff (New York, NY, USA) ........................... 483

32. The eighteenth C.U. Ariëns Kappers lecture: an introduction
    J. Verhaagen and D.F. Swaab (Amsterdam, The Netherlands)................. 497

33. Molecular control of brain plasticity and repair
    J. Fawcett (Cambridge, UK).............................................. 501

Subject Index ................................................................ 511

**See Color Plate Section at the end of this book**

SECTION I

# Stem Cells

CHAPTER 1

# Cell transplantation strategies for retinal repair

E.L. West[1], R.A. Pearson[1], R.E. MacLaren[1,2], J.C. Sowden[3] and R.R. Ali[1,4],*

[1]Department of Genetics, UCL Institute of Ophthalmology, London, UK
[2]Vitreoretinal Service, Moorfields Eye Hospital, London, UK
[3]Developmental Biology Unit, UCL Institute of Child Health, London, UK
[4]Molecular Immunology Unit, UCL Institute of Child Health, London, UK

**Abstract:** Cell transplantation is a novel therapeutic strategy to restore visual responses to the degenerate adult neural retina and represents an exciting area of regenerative neurotherapy. So far, it has been shown that transplanted postmitotic photoreceptor precursors are able to functionally integrate into the adult mouse neural retina. In this review, we discuss the differentiation of photoreceptor cells from both adult and embryonic-derived stem cells and their potential for retinal cell transplantation. We also discuss the strategies used to overcome barriers present in the degenerate neural retina and improve retinal cell integration. Finally, we consider the future translation of retinal cell therapy as a therapeutic strategy to treat retinal degeneration.

**Keywords:** stem cell; progenitor cell; photoreceptor; retina; transplantation; degeneration

## Therapeutic strategies to restore the neural retina

Retinal degenerations, either inherited or age related, remain the largest cause of untreatable blindness in the developed world. While encompassing a range of causes, most have in common the loss of the sensory cells of the retina, the photoreceptors. Many therapeutic strategies aim to slow down the progression of retinal disease, as once photoreceptors are lost they will not regenerate. Stem cell therapy may have great therapeutic potential as a treatment for degenerative retinal disease, by providing the opportunity to replace the lost cells.

The most relevant clinical studies currently being conducted in patients with retinal degeneration are fetal retinal sheet transplants. This transplantation strategy relies on the immature retinal sheet extending cell processes and forming synaptic connections with the degenerate host retina. The rationale behind this is that the inner retinal neurons of the host remain intact and therefore only require synaptic connections with photoreceptors for visual function to be restored. To date, studies investigating retinal sheet transplantation in patients have shown some subjective visual improvement (Humayun et al., 2000; Berger et al., 2003; Kaplan et al., 1997; Radtke et al., 1999). A recent clinical study of retinitis pigmentosa and age-related macular degeneration patients who received fetal retinal sheet transplants (neural retina and retinal pigment epithelium, RPE), reported improvements in vision for 7 out of 10 patients, although the direct beneficial

*Corresponding author.
Tel.: +44 207 608 6817; Fax: +44 207 608 6991;
E-mail: r.ali@ucl.ac.uk

effects of the fetal retinal grafts are difficult to assess as all patients also received intraocular lens implants. Importantly, no overt immunological responses to the transplanted tissue were observed. However, the possibility of effector cell-mediated immune responses against the retinal grafts were not examined, and graft rejection cannot be completely discounted (Radtke et al., 2008).

Previous animal studies investigating retinal sheet transplantation have demonstrated increased visual responses localized to the region of the host neural retina overlaying the graft. Due to the lack of control animals with nonfunctional transplanted retinal sheets, it is difficult to determine whether this is the result of increased synaptic connectivity between the host and grafted retinal neurons, or a trophic response induced by the fetal retinal sheet on the remaining host photoreceptors (Mohand-Said et al., 2000; Arai et al., 2004; Liljekvist-Soltic et al., 2008; Seiler et al., 2005). In the former scenario, the intervening inner retinal layer of the graft forms a barrier to photoreceptor connectivity between the host inner retinal neurons and the graft photoreceptor layer. Therefore, synaptic connections made are unlikely to represent the normal principal retinal circuit, which comprises a single bipolar and ganglion cell, and may result in atypical visual responses (Fig. 1).

In summary, fetal retinal sheet transplants appear to offer limited potential for retinal repair but are currently one of few therapeutic options for most progressive retinal degenerations. Another therapeutic strategy currently under investigation is the transplantation of retinal cell suspensions. In theory, cell transplantation has the potential to not only maintain the diseased neural retina but also restore visual function and acuity. To date cell transplantation to restore the neural retina is still being investigated in animal models of photoreceptor degeneration, and clinical application is a distant prospect. However, the future therapeutic application of cell transplantation to human retinas must be considered and experimental strategies devised accordingly.

The brain and the neural retina are both derived from the neuroectoderm of the neural plate during embryonic development (Chow and Lang, 2001). Given that immature neurons and progenitor cells are intrinsically capable of migrating and differentiating during neural development, numerous studies have investigated the integration of brain-derived neural progenitors transplanted to the neural retina (Klassen et al., 2007b; Mellough et al., 2007; Mizumoto et al., 2003; Sakaguchi et al., 2003; Takahashi et al., 1998). However, cell transplantation to the adult retina has demonstrated limited cell integration of neural progenitor cells (Sakaguchi et al., 2005; Young et al., 2000). This was assumed to be due to the inhibitory environment present in the adult neural retina. Therefore, further studies have investigated the transplantation of neural precursor cells to the developing postnatal retina.

Promising results were observed after cell transplantation to the developing retina, and it was suggested that the age of the host tissue had a key role in determining the fate of transplanted precursor cells (Sakaguchi et al., 2003, 2004; Van Hoffelen et al., 2003; Chacko et al., 2000). Studies demonstrated well-integrated transplanted cells in all layers of the host retina. These cells exhibited retinal morphology for various cell types with extensive dendritic processes present in the plexiform layers, and all cells respecting the retinal architecture (Young et al., 2000; Takahashi et al., 1998). However, the integrated cells did not express any mature retinal cell markers, suggesting that their morphology was related to the retinal microenvironment in which they differentiated, rather than intrinsic signals (Marquardt and Gruss, 2002; Takahashi et al., 1998). Further studies using tissue-restricted reporter genes to demonstrate retinal cell fate determination also observed that integrated cells did not exhibit intrinsic features of mature retinal neurons (Sam et al., 2006).

The inability of neural progenitor cells to differentiate into photoreceptors, when transplanted into the developing eye, suggests the lineage restriction of these cells to brain-related cell types (Klassen et al., 2004a). Therefore, a more appropriate cell source for transplantation studies to the retina might be neural retinal progenitor cells. These cells develop in the retinal

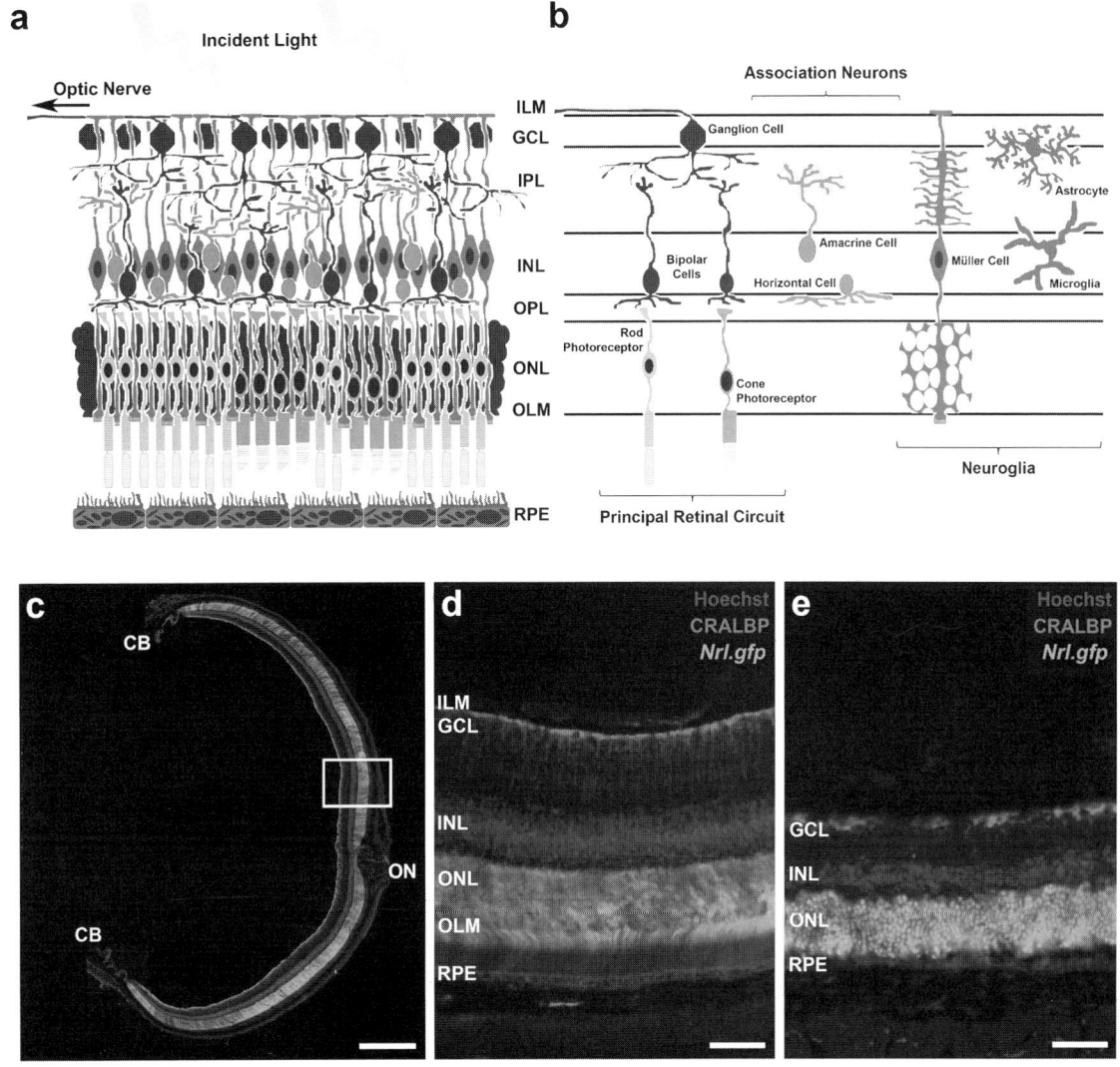

Fig. 1. The mammalian retina. (a) A schematic diagram illustrating the layers of the mammalian retina (green rod and purple cone photoreceptors; red Müller cells and RPE; blue nuclei). (b) A schematic diagram illustrating the position of the various cell types present in the adult neural retina. These cells are subdivided into (i) the principal retinal circuit, (ii) the association neurons, and (iii) the neuroglia. (c) A sagittal retinal section from an *Nrl.gfp* (green; rod photoreceptors) mouse. Scale bar, 200 μm. (d) A single fluorescence image of an adult *Nrl.gfp* retinal section stained for CRALBP (red), a protein present in Müller cells and the RPE. Scale bar, 40 μm. (e) A single fluorescence image of a degenerating retinal section stained for CRALBP (red), demonstrating the disorganization and loss of photoreceptor cells (*Nrl.gfp*; green). Scale bar, 40 μm. Nuclei were counterstained with Hoechst 33342 (blue). CB, ciliary body; ON, optic nerve; ILM, inner limiting membrane; GCL, ganglion cell layer; IPL, inner plexiform layer; INL, inner nuclear layer; OPL, outer plexiform layer; ONL, outer nuclear layer; OLM, outer limiting membrane; RPE, retinal pigment epithelium. (See Color Plate 1.1 in color plate section.)

microenvironment and may therefore have fewer inhibitory intrinsic signals enabling retinal-specific cell differentiation, compared with neural brain-derived progenitor cells. Retinal progenitors isolated from embryonic retinas have been transplanted into young (P17) dystrophic S334ter rats. The integration of these cells was observed in the form of neurite extensions into the host retina

(Qiu et al., 2005). Similar to studies using brain-derived neural progenitor cells, limited integration of retinal progenitor cells was observed after transplantation to adult retinas. Greater neurite extensions were observed following transplantation to young or developing postnatal retinas. However, the use of degenerate models with no remaining outer nuclear layer (ONL) makes it difficult to determine the extent of cell integration and mature retinal cell morphology of transplanted photoreceptors.

It was therefore assumed that the adult retina constituted an environment that inhibited retinal progenitor cell integration and differentiation possibly due to a lack of extrinsic cues that are present during development. However, recent studies have demonstrated morphological integration of early postnatal retinal precursor cells into the normal adult retina (MacLaren et al., 2006; Bartsch et al., 2008). The study by MacLaren et al. (2006) demonstrated that the integration of fully differentiated and functional photoreceptors can be achieved after transplantation into the adult retina, but only if the donor cells are postmitotic photoreceptor precursors. This was a surprising finding as it had been assumed that multipotent progenitor cells would be the best source of donor cells. Instead, these results suggest that the intrinsic nature of transplanted cells, rather than the extrinsic environment, is of greater importance for cell integration (Fig. 2). Unlike integrated neural progenitor cells, as well as mature photoreceptor morphology, the integrated photoreceptor precursor cells also demonstrated correctly localized mature retinal markers, such as rhodopsin (Rho) and peripherin-2 in the outer segments, and ribbon synapse proteins in the integrated spherules. They also demonstrated functional synaptic connectivity by increased light-induced pupil constriction following subretinal transplantation of functional compared with nonfunctional precursor cells, when transplanted into the $rho^{-/-}$ mouse (MacLaren et al., 2006). These results show that the ontogenetic stage of transplanted cells is crucial for the successful integration of retinal cells into the adult host ONL.

**Cell sources for retinal transplantation**

One fundamental problem for the application of photoreceptor cell transplantation for human retinal disease is that an appropriate source of the precursor cells is required. Postmitotic photoreceptor precursor cells can be derived from the P1-5 postnatal mouse retina. However, equivalent human retinal cells would have to be derived from second-trimester fetuses. Ethical considerations aside, such tissue is in very limited supply and may not provide a consistent source of cells for retinal cell transplantation. An expandable source of cells that could be cultured in vitro to the correct ontogenetic stage for transplantation may, therefore, be a more appropriate and reproducible source of photoreceptor precursor cells. Several potential such sources are discussed in the following text, including adult retinal stem-like (RS) cells, Müller stem-like (MS) cells, and embryonic stem (ES) cells.

Lower vertebrates such as fish and amphibians retain greater regenerative abilities than mammals. With regard to the eye, they continuously add new retinal neurons to the adult retina as they grow (Straznicky and Gaze, 1971; Johns, 1977; Johns and Easter, 1977). These new cells are added at the peripheral edge, at the ciliary margin zone (CMZ), in a manner that is thought to recapitulate embryonic retinal cell development (Harris and Perron, 1998). A similar zone of proliferating cells has been found in the chick that contributes to the postnatal growth of the retina (Fischer and Reh, 2000). However, the presence of a CMZ in the retina of the mouse has not been detected (Kubota et al., 2002). It has been speculated that the existence of a population of adult stem-like cells isolated from the ciliary body of the retina in mammals is the evolutionary equivalent of cells from the CMZ (Tropepe et al., 2000).

A population of quiescent cells from the ciliary body of the mammalian retina were discovered to proliferate in vitro, express immature retinal markers, and upon differentiation express markers of mature retinal cell types (Tropepe et al., 2000; Ahmad et al., 2000). These adult-derived

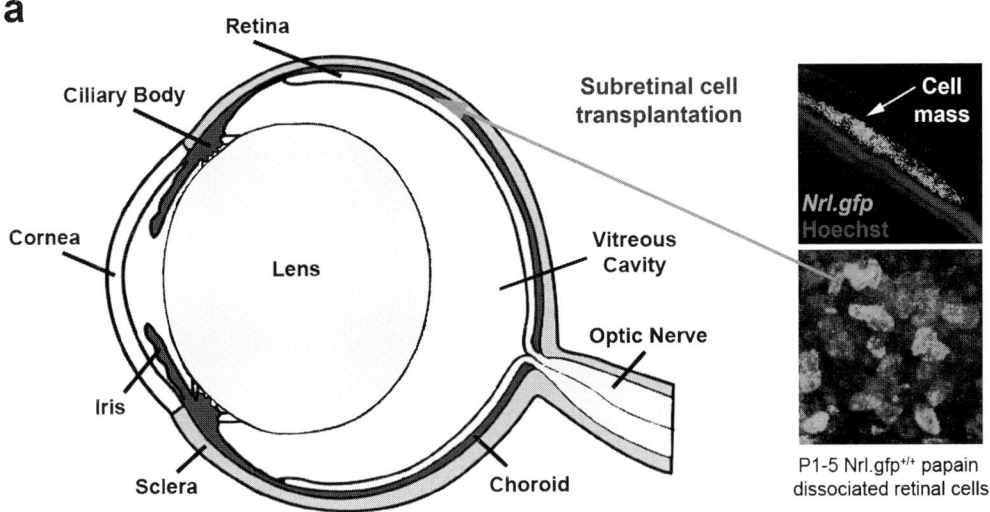

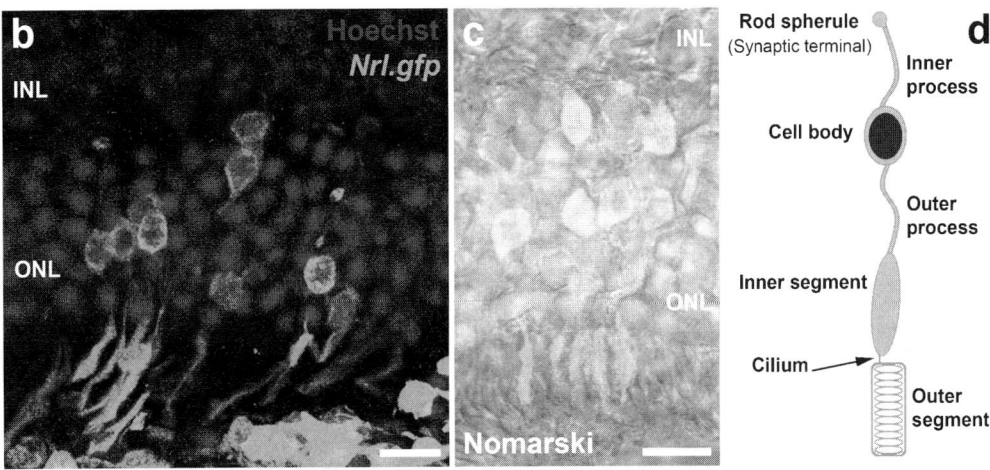

Fig. 2. Photoreceptor precursor cell transplantation into the adult eye. (a) A schematic diagram of a mouse eye illustrating the subretinal transplantation of *Nrl.gfp* precursor cells (green) and the resulting cell mass (inserts). (b) A confocal image of integrated *Nrl.gfp* rod photoreceptors, 21 days after transplantation to an adult recipient. (c) A Nomarski confocal image of integrated *Nrl.gfp* rod photoreceptors. (d) A schematic representation of the structure of a rod photoreceptor. Nuclei were counterstained with Hoechst 33342 (blue). Scale bars, 20 μm. INL, inner nuclear layer; ONL, outer nuclear layer. (See Color Plate 1.2 in color plate section.)

RS cells can be grown as neurospheres with epidermal growth factor (EGF) and fibroblast growth factor-2 (FGF2) and differentiated by culture on substrate-coated plates in a growth factor–free serum-containing medium, similar to adult neural stem (NS) cells (MacNeil et al., 2007; Ahmad et al., 2000). The addition of Wnt3a and FGF2 to adult RS cell neurosphere cultures has been shown to have an additive effect on cell proliferation, resulting in greater numbers of secondary neurospheres (Inoue et al., 2006). RS cell neurospheres have been derived from the iris, ciliary body, and pars plana, but not the anterior neural retina (Gu et al., 2007; Haruta et al., 2001; MacNeil et al., 2007). Further to the studies in rodent and porcine eyes, RS cells have also been isolated from adult human retinal tissue and shown to form neurospheres in vitro (Carter et al., 2007; Mayer et al., 2005).

Differentiated neurosphere cultures give rise to both neuronal and glial cell types, suggesting multipotentiality. However, the expression of a small number of mature retinal markers may not indicate completely differentiated and functionally mature retinal cell types (MacNeil et al., 2007; Kokkinopoulos et al., 2008). Several studies have investigated the induction of mature retinal phenotypes in RS cell cultures from both the adult ciliary body and iris by retroviral transduction of photoreceptor relevant transcription factors. The expression of *Crx* or *Otx2* in both cell types demonstrated the directed differentiation of cells positive for Rho, recoverin, and transducin protein expression (Akagi et al., 2004). In contrast, transduced mesencephalon-derived NS cells displayed little Rho immunoreactivity, suggesting that NS cells require greater manipulation to differentiate toward retinal cell lineages (Akagi et al., 2004; Haruta et al., 2001). In further studies, primate iris-derived cells were induced to differentiate into Rho-positive cells after transduction with a combination of both *Crx* and *NeuroD* retroviral vectors. Both rat and primate differentiated cells were shown to hyperpolarize after light stimulation, suggesting the generation of functional photoreceptor cell types (Akagi et al., 2005). Similar to this study, genetically modified mouse RS cells electroporated to express *Crx* have also been shown to induce differentiated cells that exhibit some functional properties of mature retinal photoreceptors (Jomary and Jones, 2008). This was in contrast to RS cells electroporated with a control plasmid, which differentiated to express mature photoreceptor cell markers but did not demonstrate light-sensitive properties (Jomary and Jones, 2008). These studies suggest that RS cells could be induced to differentiate into light-sensitive rod photoreceptor phenotypes; however, the expression of mature retinal markers by differentiated cells does not necessarily equate to functional photoreceptors (Bradford et al., 2005). Therefore, extensive in vitro characterization of differentiated cells is required prior to retinal cell transplantation. The potential of these cells to function in vivo and improve visual responses following transplantation into the degenerate neural retina has yet to be established (Akagi et al., 2005). RS cells isolated from the ciliary body or iris tissue have, to date, shown limited potential for cell integration after transplantation into adult wild-type or degenerate retinas (Akagi et al., 2003; Chacko et al., 2000; Klassen et al., 2007a; Canola et al., 2007). This is most likely due to reduced numbers of RS cell-derived photoreceptor precursors at the correct ontogenetic stage (MacLaren et al., 2006). Further investigation of homogeneous populations of transplanted cells at characterized stages of differentiation may enable RS cell-derived transplants to integrate with the host retina (Canola and Arsenijevic, 2007; Akagi et al., 2003).

To confirm that cultured cells can integrate into the neural retina, the transplantation of cultured retinal progenitor cells isolated from the embryonic retina has been investigated by a number of groups. The majority of studies have demonstrated the differentiation of transplanted progenitor cells into mature retinal phenotypes in the subretinal space. However, little integration into the host retina was observed following the subretinal transplantation of these cells (Akagi et al., 2003; Chacko et al., 2000). Other studies have observed very little differentiation and mature retinal cell marker expression, and concluded that progenitor cells required further differentiation in vitro prior to cell transplantation (Yang et al., 2002). Klassen et al. have examined the integration of cultured P1 retinal cells in the $rho^{-/-}$ mouse and demonstrated *gfp* (green fluorescent protein)-positive transplanted cells within the host neural retina. These integrated cells expressed mature retinal photoreceptor markers but lacked mature retinal cell morphology (Klassen et al., 2004b). A possible explanation for the differences observed in vivo following cultured retinal progenitor cell transplantation is the different culture conditions used. However, these investigations suggest that the culturing of cells in vitro prior to transplantation does not inhibit their migratory potential.

There has been some debate as to whether RS cells constitute a neural adult stem cell population like those found in the subventricular and subgranular zones of the brain, or whether these cells have limited self-renewal suggesting a progenitor-like phenotype (Xu et al., 2007; Inoue et al., 2005; Liu et al., 2005; Engelhardt et al., 2004; Kokkinopoulos et al., 2008; ). A growing number of investigations have found that RS cells can only be sustained in vitro for a limited period (Liu et al., 2005; Inoue et al., 2005; MacNeil et al., 2007). Due to the difficulty of propagating retinal cells individually, it is impossible to perform clonogenic analysis to establish if these cells divide asymmetrically. A study comparing the growth characteristics of adult rat-derived RS cells with those of NS cells demonstrated a lack of cell proliferation and self-renewal after 8 weeks in vitro for the former, while NS cells continued to proliferate in neurosphere cultures (Liu et al., 2005). It therefore remains to be determined whether RS cells can be sufficiently expanded in vitro for therapeutic purposes.

In the brain, the radial glial cells of the adult hippocampus proliferate and differentiate into neurons throughout life (Seri et al., 2001, 2004). A similar phenomenon has been observed in the adult neural retina of fish, with the generation of new neurons from the equivalent glial cell type in the retina, the Müller cells (Raymond et al., 2006; Bernardos et al., 2007). Several studies have demonstrated that Müller cells from the adult mammalian central retina also have some stem-like characteristics in vitro. This includes the formation of neurospheres and the expression of NS cell markers such as *Sox2*, *Pax6*, and *Chx10* (Lawrence et al., 2007; Das et al., 2006a; Nickerson et al., 2008). A spontaneously immortalized cell line of MS cells has been established from human retinal tissue, and their expansion did not appear to be limited like that of RS cells (Limb et al., 2002). Following differentiation, MS cells have been shown to express mature retinal cell markers, including peripherin, recoverin, and S-opsin (Lawrence et al., 2007; Das et al., 2006b). Of note, recent investigations in the $Chx10^{orJ/orJ}$ mouse have demonstrated a population of cells present in the central neural retina that exhibit properties similar to those of ciliary epithelium-derived RS cells (Dhomen et al., 2006; Kokkinopoulos et al., 2008). As the mutation of *Chx10* results in reduced retinal progenitor cell proliferation and microphthalmia, it has been suggested that these cells represent a dormant progenitor cell population that is maintained in the mutant central neural retina but not in wild-type retinas (Dhomen et al., 2006). When cultured in vitro, these cells express glial cell markers and may represent a similar Müller progenitor cell population. Similar to the differentiation of RS cells into mature retinal cell types, differentiated cells derived from MS cell cultures have yet to be functionally characterized to confirm that they represent fully differentiated retinal neurons.

So far, cultured MS cells have shown limited integration into host retinas following transplantation, similar to RS cells (Singhal et al., 2008; Lawrence et al., 2007; Bull et al., 2008). Increased integration was observed in degenerate retinas following chondroitinase ABC treatment at the time of cell transplantation, suggesting that chondroitin sulfate proteoglycans (CSPGs) form a significant barrier to cell migration and integration (Singhal et al., 2008). MS cell migration was enhanced further by substantial immune suppression, demonstrating a combinational effect (Singhal et al., 2008). As microglia can be activated by CSPGs and their breakdown products have been shown to exert anti-inflammatory effects, it is likely that an innate immune response against the transplanted MS cells is at least partially inhibiting successful cell integration (Jones and Tuszynski, 2002; Jones et al., 2002; Rolls et al., 2006). Further characterization of the developmental stage of the transplanted population may enable the functional integration of photoreceptor cells derived from this source. However, immune suppression would be required for the long-term integration of human-derived MS cells in models of retinal degeneration.

In contrast to adult-derived RS cells, which exhibit limited self-renewal, ES cells isolated from the inner cell mass of the blastocyst can be grown in culture for indefinite periods of time, after which they can be induced to differentiate into cell lineages of all three germ layers (Evans and Kaufman, 1981; Martin, 1981; Thomson et al., 1995, 1998; Suemori et al., 2001; Pera et al., 2000; Reubinoff et al., 2000). Therefore, established ES cell lines could provide an expandable source from which to derive photoreceptor precursor cells for retinal transplantation. Of concern for future clinical application is the culturing of ES cells with animal-derived reagents such as animal serum and animal-derived feeder layers, or by the use of animal cell culture conditioned medium. This is because the approval of therapeutic agents for use in humans requires them to be free of pathogens and animal contamination. Several studies have successfully cultured human ES cells in serum-free conditions and without feeder layers (Amit et al., 2004; Xu et al., 2001). A human ES cell line was recently established without the use of animal contaminated reagents, demonstrating that this should not be an issue for future clinical therapies (Ludwig et al., 2006a, b).

The differentiation of ES cells into neural progenitors that can produce the three main neural cell lineages of neurons, astrocytes, and oligodendrocytes has been well established (Joannides et al., 2007). Further to this, human ES cells have been shown to differentiate into various types of neurons, including dopaminergic neurons and oligodendrocytes (Yan et al., 2005; Perrier et al., 2004; Zhang et al., 2001; Nistor et al., 2005). Transplantation studies involving these differentiated cell types have demonstrated the potential of ES cells to produce differentiated neural cell populations that can be used for cell transplantation strategies (Nistor et al., 2005; Keirstead et al., 2005; Rodriguez-Gomez et al., 2007). The differentiation of ES cells into retinal cell lineages has not achieved the same progress as that seen for other neural cell types of the brain. However, recent advances in cell culture techniques have demonstrated the possibility of producing mature retinal cells from mouse, primate, and human ES cells (Osakada et al., 2008; Lamba et al., 2006). Previous studies have shown the differentiation of mouse ES cell-derived neural progenitors into photoreceptor-like cells after coculture with P1 or E6 retinal tissue. The differentiation of retinal cells was determined by immunohistochemistry and RT-PCR for photoreceptor-specific markers, including *Crx*, *Nrl*, *Rho kinase*, *arrestin*, and *interphotoreceptor retinoid-binding protein* (Zhao et al., 2002; Sugie et al., 2005). Despite the apparent generation of mature retinal phenotypes, the specific factors required to promote the differentiation of these cells were not established. Further studies using more defined culture conditions demonstrated the differentiation of mouse and human ES cells into immature retinal cells. However, coculture with retinal explants or cell suspensions was still required for the expression of mature photoreceptor markers such as *recoverin* (Ikeda et al., 2005; Lamba et al., 2006). Recently, Osakada et al. demonstrated the generation of mature rod and cone photoreceptors from ES cells, with the use of defined culture conditions. They found $17.2 \pm 1.8\%$ and $8.5 \pm 2.9\%$ of cells were Rho and recoverin positive, respectively, after the stepwise differentiation of mouse and human ES cells (Osakada et al., 2008). Despite the demonstration of gene expression for phototransduction components in these cells, further evidence of their function is still required. It will be of great interest to determine whether these cells, if differentiated to the correct ontogenetic stage, could functionally integrate into the adult neural retina.

**Optimization of transplanted cell integration**

Retinal disease has many different genetic and environmental causes, which result in a wide range of pathological conditions. A consistent outcome of these disorders is the degeneration and eventual loss of photoreceptors from the ONL. In order to replace these lost cells, transplanted photoreceptor precursors are required to migrate and integrate into the degenerated ONL. While the number of integrated photoreceptor precursor cells demonstrated in the adult neural retina is sufficient to restore the pupillary light reflex, only a relatively small number of transplanted cells integrate. Greater numbers of integrated cells would be required in order to improve visual acuity in degenerate models.

As photoreceptor precursor cells are intrinsically capable of migrating and differentiating into the adult neural retina, it follows that other barriers must be present that limit extensive cell integration (MacLaren et al., 2006). The ability of transplanted cells to integrate within the host opossum retina has been shown to decline with host maturation (Sakaguchi et al., 2003, 2004). This coincides with the maturation of glial elements, such as Müller cells, which form anatomical barriers within the host retina, including the outer limiting membrane (OLM).

The OLM has been shown to be a significant physical barrier to the migration and integration of photoreceptor precursor cells into the adult host ONL. OLM disruption, by the administration of the glial toxin alpha-aminoadipic acid (AAA), at the time of cell transplantation was shown to correspond with increased photoreceptor precursor cell integration (West et al., 2008). In mice with retinal dystrophy caused by defects in Crumbs homologue-1 (Crb1), a protein associated with adherens junction formation and stabilization, increased photoreceptor precursor cell integration has also been observed (Pearson et al., manuscript in preparation). However, OLM disruption has not been observed after retinal degeneration caused by other gene defects (Gouras and Tanabe, 2003; Sanyal and Hawkins, 1989). This suggests that the OLM would remain a significant barrier to transplanted photoreceptor cell integration in the majority of retinal degenerations (Fig. 3). The pharmacological induction of OLM disruption by AAA would not be suitable in degenerate retinas due to toxic effects on the supportive Müller glia (Pedersen and Karlsen, 1979; Ishikawa and Mine, 1983; Rich et al., 1995). An alternative method to induce transient OLM disruption is the use of small interfering ribonucleic acid (siRNA) to promote transcriptional gene silencing of relevant OLM-related proteins. Further investigation in degenerate models is required to establish the effect of OLM disruption on photoreceptor precursor cell integration in degenerate retinas.

A crucial difference between normal and degenerate retinas that may limit photoreceptor precursor cell integration is the presence of Müller cell activation in the latter. Following injury or degeneration of the neural retina, a process known as reactive gliosis occurs. This can vary in severity depending on the initiating insult and is indicated by the expression of glial fibrillary acidic protein (GFAP) by Müller cell processes (Lewis and Fisher, 2003). In contrast, in uninjured retinas, GFAP is only expressed by astrocytes present at the inner edge of the neural retina. In addition to the upregulation of GFAP and vimentin by Müller cells during reactive gliosis, Müller cell processes have also been shown to form glial barriers along the outer edge of the retina after retinal detachment (Fisher et al., 2005; Lewis and Fisher, 2000, 2003). This glial scarring constitutes a barrier to integrating transplanted cells and is a characteristic of many late-stage retinal disease models (Zhang et al., 2004; Ekstrom et al., 1988; Sheedlo et al., 1995; Iandiev et al., 2006; Fan et al., 1996; ).

Similar barriers to cell transplantation, such as the OLM and glial scarring, have been reported to limit the "integration" of retinal sheets with the host retina, as neurite extension does not occur in these regions (Zhang et al., 2003, 2004). Further to this, activated Müller cells and microglia are thought to produce increased extracellular matrix (ECM) components such as CSPGs, which have been shown to limit axon extension in the brain (Fawcett and Asher, 1999). Several studies have investigated the use of enzymes, such as chondroitinase ABC, neuraminidase X, and matrix metalloproteinase-2 (MMP-2), to break down these extracellular barriers in combination with cell transplantation and demonstrated encouraging results (Singhal et al., 2008; Suzuki et al., 2006, 2007; Zhang et al., 2007). It therefore seems that the investigation of techniques to reduce reactive gliosis and the subsequent glial scarring and ECM deposition will be important for successful photoreceptor precursor cell integration in late-stage retinal degeneration (Fig. 3).

One common feature of all retinal degenerations is cell death and the subsequent activation of the resident macrophage population, the microglia (Hughes et al., 2003; Hose et al., 2005; Zhang et al., 2005; Roque et al., 1996). This has also been demonstrated for injury-induced models of retinal

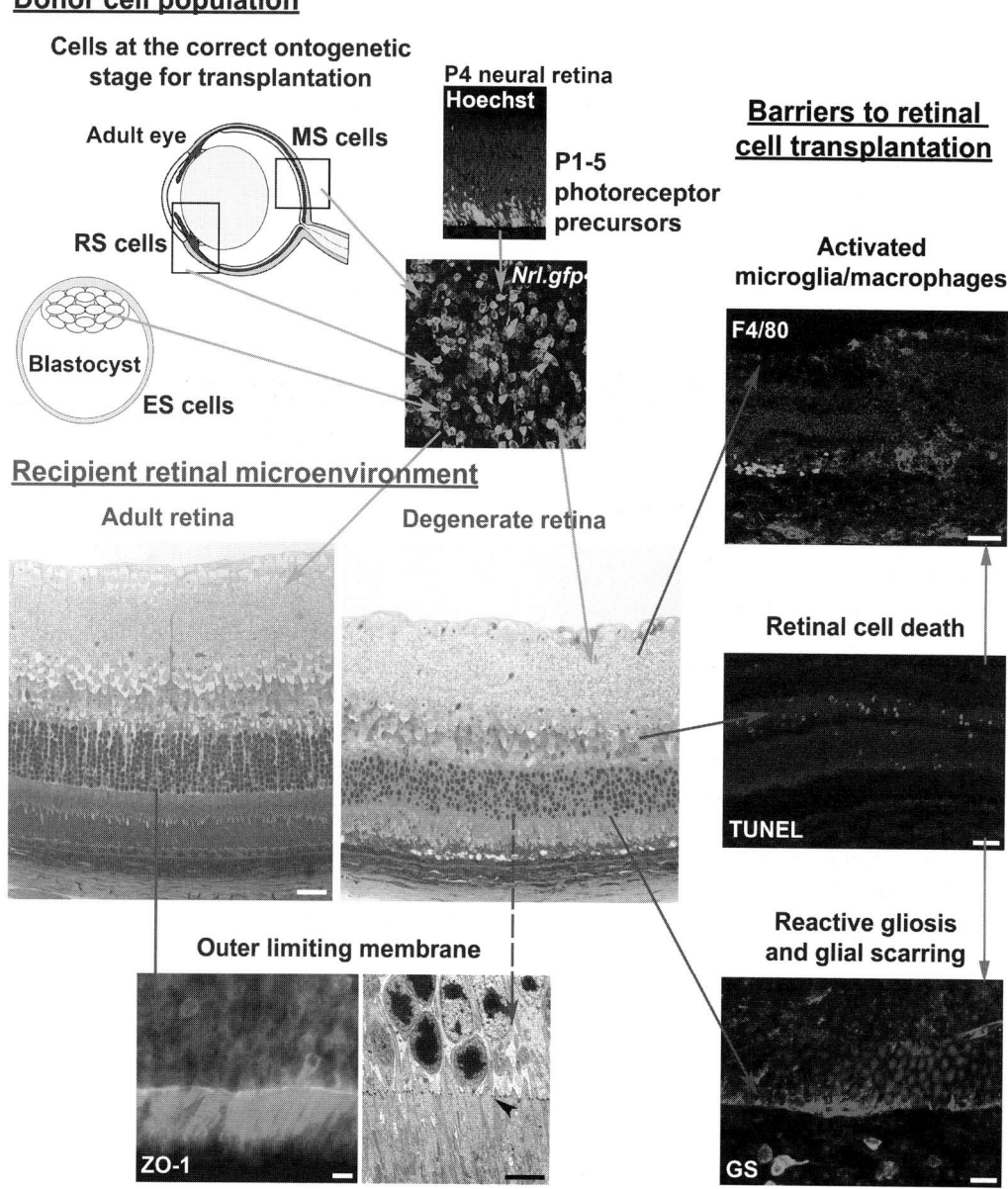

Fig. 3. A summary of retinal cell transplantation strategies. A diagram to summarize the various retinal cell transplantation strategies and the related barriers that may limit transplanted photoreceptor cell integration in the adult and degenerate neural retina, as discussed in the main text. The donor cell population (top; green) can be derived from a variety of cell sources, but must be differentiated to the correct ontogenetic stage (postmitotic rod precursors, *Nrl.gfp*; green) prior to transplantation to enable photoreceptor cell integration into the host adult retina (MacLaren et al., 2006). The recipient retinal microenvironment (middle; blue) may also limit photoreceptor cell integration if the relevant barriers are not modulated at the time of transplantation. Scale bar, 50 μm. The relevant barriers to retinal cell transplantation and integration (right; red) are indicated. The outer limiting membrane (indicated by the red or black arrow head) forms a barrier to increased cell integration in the adult retina and in some models of retinal degeneration. Scale bars, 10 μm and 5 μm. Other barriers, present predominantly in the degenerate retina, include retinal cell death and the resulting activated microglia/macrophages and reactive gliosis/glial scarring. Scale bars, 50, 100, and 20 μm, respectively. Nuclei were counterstained with Hoechst 33342 (blue). ES cells, embryonic stem cells; GS, glutamine synthetase; MS cells, Müller stem-like cells; RS cells, retinal stem-like cells; ZO-1, zonula occludens-1. (See Color Plate 1.3 in color plate section.)

degeneration (Harada et al., 2002). In our own investigations, we have noted that increased macrophage presence shortly after cell transplantation resulted in fewer integrated photoreceptors (unpublished results). It is not clear whether macrophages prevent precursor cell integration or cause the destruction of the integrated photoreceptors. However, the difference in inflammatory status between normal and degenerate retinas may be the cause of reduced photoreceptor cell integration observed in the latter. A recent study demonstrated increased numbers of sialoadhesin-expressing macrophages present in *rd1* and *rds* mouse models following precursor cell transplantation, and suggested that this may affect the survival of transplanted cells (Sancho-Pelluz et al., 2008). Previous studies have detected the presence of sialoadhesin-positive macrophages in untreated *rds* mice and a model of experimental autoimmune uveoretinitis (Jiang et al., 1999, 2006; Hughes et al., 2003). Sialoadhesin expression has been shown to contribute to the inflammatory response by promoting T cell and macrophage adhesion (Crocker et al., 1995; Jiang et al., 1999, 2006). Therefore, the increased inflammatory status of degenerate retinas may prompt the early rejection of transplanted cells, and initial innate immune suppression may be required to successfully transplant cells in these models of retinal degeneration (Fig. 3).

Immune rejection is a major problem in many transplantation paradigms. However, the brain and the eye are frequently described as immune-privileged sites, defined as sites that allow foreign grafts to survive for extended to indefinite periods of time. The eye contains several immune-privileged sites, namely, the anterior chamber, vitreous cavity, and subretinal space. Streilein et al. (2002) have performed extensive experiments examining the survival of neonatal retinal allografts in the eye. In combination with the eye maintaining an immune-privileged site, neonatal retinal tissue itself has been shown to be partially immune privileged when placed beneath the kidney capsule, a non-immune-privileged site. This is in contrast to skin grafts, a non-immune-privileged tissue type, which have been shown to be rejected by 12 days, and fully immune-privileged tissues, including the cornea and the RPE, which survived for indefinite periods of time (Ng et al., 2002; Hori et al., 2000; Wenkel and Streilein, 2000).

Of greatest relevance to photoreceptor precursor cell transplantation is that the subretinal space has been shown to elicit immune deviation after cell-associated or soluble antigen administration. The immune deviation of eye-derived antigens is a form of immune tolerance, a state of specific immunological unresponsiveness, mediated by antigen-specific T regulatory cells, also referred to as suppressor T cells (Streilein and Niederkorn, 1985; Wilbanks and Streilein, 1990). These cells are produced in the spleen and suppress delayed-type hypersensitivity immune reactions to alloantigens present in the eye. However, the immune deviation of alloantigens present in the subretinal space is lost if RPE cell viability is compromised or the outer blood-retinal barrier is disrupted (Wenkel and Streilein, 1998). Transplantation of neonatal retinal allografts to the subretinal space and vitreous cavity have been shown to induce immune deviation by 12 days, whereas transplantation to the subconjunctival space promoted antigen-specific delayed hypersensitivity (Jiang et al., 1993). However, neonatal retinal allografts eventually deteriorate in both the anterior chamber and the subretinal space by 35 days (5 weeks) post implantation. This appears to coincide with the loss of immune deviation and the onset of donor-specific delayed hypersensitivity (Jiang et al., 1995; Streilein et al., 2002).

The eye therefore represents a partially immune-privileged site and appears to eventually reject allogeneic cells transplanted to the subretinal space. This may be of concern for long-term retinal repair by cell transplantation. It remains to be seen whether a homogenous population of cultured photoreceptor precursor cells would elicit immune rejection following transplantation to the neural retina. Cultured neural progenitors have been shown to be less immunogeneic compared with freshly dissociated neural progenitors, the most likely explanation for this is the lack of donor-derived microglia in the cultured cell population (Hori et al., 2003; Ma and Streilein, 1998). Further investigation of cultured retinal progenitor cells transplanted to the subretinal space is required to establish the relevant issues of immune rejection for photoreceptor precursor cell transplantation.

**Future considerations for retinal cell therapy**

Studies have shown that precursor cells at the correct ontogenetic stage can migrate and integrate into the adult host ONL and form functional synaptic connections (MacLaren et al., 2006). Several studies have since demonstrated mature photoreceptor morphology of integrated precursor cells in adult retinas (MacLaren et al., 2006; West et al., 2008; Bartsch et al., 2008). Recent studies of ES cells have established defined culture conditions to differentiate ES cells into photoreceptors (Osakada et al., 2008). Despite the recent advances in the production of ES cell-derived retinal cells, these may not translate into successful cell transplantation strategies, namely, due to the foreign nature of these cells with regard to the host immune system. Classic immunosuppressive drug therapy could be used, or alternatively, a human ES cell bank of cell lines characterized by human leukocyte antigens (HLA) could be created to provide closely matched differentiated cells (Taylor et al., 2005). Several studies have investigated novel ways to promote prolonged immunological tolerance to transplanted alloantigens in the eye, such as the transplantation of retinal progenitors combined with immature dendritic cells or alpha-melanocyte-stimulating hormone–induced T regulatory cells to develop or transfer immune tolerance, respectively, against the alloantigens present. Such strategies appear to lead to enhanced transplanted cell survival (Ng et al., 2007; Oishi et al., 2007). It may therefore be possible to exploit the eye's natural immune deviation response to enable prolonged transplanted cell survival.

Barriers to photoreceptor precursor cell transplantation, such as the OLM and glial scarring, would still be present in the adult human retina. Intriguingly, however, cystoid macular edema (CME) is a condition seen in the end stages of many diseases of the outer retina, such as retinitis pigmentosa and diabetic maculopathy; microscopic examination of pathological specimens have shown that CME represents an intracytoplasmic swelling (edema) of Müller cells in the foveal region (Yanoff et al., 1984) which is similar to the effects of AAA described in previous studies (West et al., 2008; Ishikawa and Mine, 1983; Pedersen and Karlsen, 1979). Therefore, the diseased human fovea may have reduced OLM integrity and, as a result, constitute a particularly favorable site for future retinal cell transplantation strategies. This would be especially important if cone photoreceptor precursor cells are also able to integrate into the adult ONL, as observed for rod photoreceptor precursors (MacLaren et al., 2006). Other conditions that might be particularly suitable for cell replacement strategies include inherited retinal degenerations due to defects in Crb1, which have also been shown to result in reduced OLM integrity (Mehalow et al., 2003; van de Pavert et al., 2007).

A recent advance in stem cell biology has been the reprogramming of adult human fibroblasts by retroviral transduction to generate induced pluripotent stem (iPS) cells. Three independent studies used various combinations of four transcription factors, known to be required for pluripotency in ES cells (Friel et al., 2005), to induce adult cells to acquire pluripotent characteristics (Takahashi et al., 2007; Yu et al., 2007; Park et al., 2008). However, the use of retroviral transduction of transcription factors results in multiple random insertions of the transgene, which can also lead to oncogenesis in certain circumstances (Cattoglio et al., 2007). At present, very small numbers of human ES cell-like iPS cell colonies are produced (around 1 in 1000 cells). Therefore, further investigation of this cell population is required to improve the efficiency of the methods used and establish virus-free protocols of induction that would be less oncogeneic and have greater viability for therapeutic applications (Nakagawa et al., 2008; Kim et al., 2008; Okita et al., 2007). It will, however, be of significant interest to determine whether the current differentiation protocols for human ES cell-derived retinal cells also work for human iPS cells.

For retinal dystrophies caused by photoreceptor-specific gene mutations, autologous adult-derived cells do not initially appear to be the best source of new retinal neurons, as the genetic mutation will remain. However, by ex vivo gene therapy, they have the potential to replace and restore visual function in degenerate retinas. Future treatment for retinal degeneration due to photoreceptor cell loss may require a combination of gene and cell therapeutic strategies (Bainbridge et al., 2008; Maguire et al., 2008). An alternative to this is the use of allogeneic, but closely matched, adult donor cells from which photoreceptor precursor cells for transplantation can be generated. Similar to conventional organ transplantation, these cells could be derived from a close family member or HLA-matched donor tissue to reduce the possibility of transplanted cell rejection. However, for some retinal dystrophies that progress slowly, the integration of recently derived autologous photoreceptors may limit further degeneration, especially in diseases such as retinitis pigmentosa where the loss of peripheral rod photoreceptors leads to the secondary loss of cone photoreceptors vital for central vision. Therefore, the successful rescue of retinal degeneration via cell therapy is most likely to involve a combination of different strategies and methodologies, depending on the pathology of the retinal disease being treated.

In summary, the restoration of visual responses by photoreceptor precursor cell transplantation to the human retina remains a promising strategy for retinal repair. Many studies have demonstrated both the potential structural barriers to precursor cell transplantation present in the adult and degenerate retina, as well as the need for autologous cell transplantation to promote long-term survival of transplanted cells. Strategies to modulate these factors have highlighted some important considerations for future transplantation studies. The transplantation of photoreceptor precursor cells derived from the recently discovered iPS cells will be of great interest for future regenerative strategies of the neural retina. Since this review was written several papers of related interest have been published, these include Cicero et al. (2009) and Hirami et al. (2009).

## Abbreviations

| | |
|---|---|
| AAA | alpha-aminoadipic acid |
| CME | cystoid macular edema |
| CMZ | ciliary margin zone |
| Crb1 | Crumbs homologue-1 |
| CSPG | chondroitin sulfate proteoglycan |
| ECM | extracellular matrix |
| EGF | epidermal growth factor |
| ES cell | embryonic stem cell |
| FGF2 | fibroblast growth factor-2 |
| GFAP | glial fibrillary acidic protein |
| GFP | green fluorescent protein |
| HLA | human leukocyte antigen |
| iPS cell | induced pluripotent stem cell |
| ONL | outer nuclear layer |
| OLM | outer limiting membrane |
| MMP-2 | matrix metalloproteinase-2 |
| MS cell | Müller stem-like cell |
| NS cell | neural stem cell |
| Rho | rhodopsin |
| RPE | retinal pigment epithelium |
| RS cell | retinal stem-like cell |
| SiRNA | small interfering ribonucleic acid |

## Acknowledgments

The authors are generously supported by grants from the Wellcome Trust (082217); the Medical Research Council, UK (G03000341); Fight For Sight, UK; the Macula Vision Research Foundation; the Royal Blind Asylum and School; and the Scottish National Institution for the War Blinded. RAP is a Royal Society University Research Fellow. REM is a Health Foundation Clinician Scientist Fellow (N0141182824). RRA and REM were partially funded by the Department of Health's National Institute for Health Research Biomedical Research Centre at Moorfields Eye Hospital.

## References

Ahmad, I., Tang, L., & Pham, H. (2000). Identification of neural progenitors in the adult mammalian eye. *Biochemical and Biophysical Research Communications*, *270*, 517–521.

Akagi, T., Akita, J., Haruta, M., Suzuki, T., Honda, Y., Inoue, T., et al. (2005). Iris-derived cells from adult rodents and primates adopt photoreceptor-specific phenotypes. *Investigative Ophthalmology & Visual Science, 46*, 3411–3419.

Akagi, T., Haruta, M., Akita, J., Nishida, A., Honda, Y., & Takahashi, M. (2003). Different characteristics of rat retinal progenitor cells from different culture periods. *Neuroscience Letters, 341*, 213–216.

Akagi, T., Mandai, M., Ooto, S., Hirami, Y., Osakada, F., Kageyama, R., et al. (2004). Otx2 homeobox gene induces photoreceptor-specific phenotypes in cells derived from adult iris and ciliary tissue. *Investigative Ophthalmology & Visual Science, 45*, 4570–4575.

Amit, M., Shariki, C., Margulets, V., & Itskovitz-Eldor, J. (2004). Feeder layer- and serum-free culture of human embryonic stem cells. *Biology of Reproduction, 70*, 837–845.

Arai, S., Thomas, B. B., Seiler, M. J., Aramant, R. B., Qiu, G., Mui, C., et al. (2004). Restoration of visual responses following transplantation of intact retinal sheets in rd mice. *Experimental Eye Research, 79*, 331–341.

Bainbridge, J. W., Smith, A. J., Barker, S. S., Robbie, S., Henderson, R., Balaggan, K., et al. (2008). Effect of gene therapy on visual function in Leber's congenital amaurosis. *The New England Journal of Medicine, 358*, 2231–2239.

Bartsch, U., Oriyakhel, W., Kenna, P. F., Linke, S., Richard, G., Petrowitz, B., et al. (2008). Retinal cells integrate into the outer nuclear layer and differentiate into mature photoreceptors after subretinal transplantation into adult mice. *Experimental Eye Research, 86*, 691–700.

Berger, A. S., Tezel, T. H., Del Priore, L. V., & Kaplan, H. J. (2003). Photoreceptor transplantation in retinitis pigmentosa: short-term follow-up. *Ophthalmology, 110*, 383–391.

Bernardos, R. L., Barthel, L. K., Meyers, J. R., & Raymond, P. A. (2007). Late-stage neuronal progenitors in the retina are radial Muller glia that function as retinal stem cells. *The Journal of Neuroscience, 27*, 7028–7040.

Bradford, R. L., Wang, C., Zack, D. J., & Adler, R. (2005). Roles of cell-intrinsic and microenvironmental factors in photoreceptor cell differentiation. *Developmental Biology, 286*, 31–45.

Bull, N. D., Limb, G. A., & Martin, K. R. (2008). Human Muller stem cell (MIO-M1) transplantation in a rat model of glaucoma: survival, differentiation, and integration. *Investigative Ophthalmology & Visual Science, 49*, 3449–3456.

Canola, K., Angenieux, B., Tekaya, M., Quiambao, A., Naash, M. I., Munier, F. L., et al. (2007). Retinal stem cells transplanted into models of late stages of retinitis pigmentosa preferentially adopt a glial or a retinal ganglion cell fate. *Investigative Ophthalmology & Visual Science, 48*, 446–454.

Canola, K., & Arsenijevic, Y. (2007). Generation of cells committed towards the photoreceptor fate for retinal transplantation. *Neuroreport, 18*, 851–855.

Carter, D. A., Mayer, E. J., & Dick, A. D. (2007). The effect of postmortem time, donor age and sex on the generation of neurospheres from adult human retina. *The British Journal of Ophthalmology, 91*, 1216–1218.

Cattoglio, C., Facchini, G., Sartori, D., Antonelli, A., Miccio, A., Cassani, B., et al. (2007). Hot spots of retroviral integration in human CD34+ hematopoietic cells. *Blood, 110*, 1770–1778.

Chacko, D. M., Rogers, J. A., Turner, J. E., & Ahmad, I. (2000). Survival and differentiation of cultured retinal progenitors transplanted in the subretinal space of the rat. *Biochemical and Biophysical Research Communications, 268*, 842–846.

Chow, R. L., & Lang, R. A. (2001). Early eye development in vertebrates. *Annual Review of Cell and Developmental Biology, 17*, 255–296.

Cicero, S. A., Johnson, D., Reyntjens, S., Frase, S., Connell, S., Chow, L. M. L., et al. (2009). Cells previously identified as retinal stem cells are pigmented ciliary epithelial cells. *Proceedings of the National Academy of Sciences, 106*, 6685–6690.

Crocker, P. R., Freeman, S., Gordon, S., & Kelm, S. (1995). Sialoadhesin binds preferentially to cells of the granulocytic lineage. *The Journal of Clinical Investigation, 95*, 635–643.

Das, A. V., Mallya, K. B., Zhao, X., Ahmad, F., Bhattacharya, S., Thoreson, W. B., et al. (2006a). Neural stem cell properties of Muller glia in the mammalian retina: regulation by Notch and Wnt signaling. *Developmental Biology, 299*, 283–302.

Das, A. V., Zhao, X., James, J., Kim, M., Cowan, K. H., & Ahmad, I. (2006b). Neural stem cells in the adult ciliary epithelium express GFAP and are regulated by Wnt signaling. *Biochemical and Biophysical Research Communications, 339*, 708–716.

Dhomen, N. S., Balaggan, K. S., Pearson, R. A., Bainbridge, J. W., Levine, E. M., Ali, R. R., et al. (2006). Absence of chx10 causes neural progenitors to persist in the adult retina. *Investigative Ophthalmology & Visual Science, 47*, 386–396.

Ekstrom, P., Sanyal, S., Narfstrom, K., Chader, G. J., & van, V. T. (1988). Accumulation of glial fibrillary acidic protein in Muller radial glia during retinal degeneration. *Investigative Ophthalmology & Visual Science, 29*, 1363–1371.

Engelhardt, M., Wachs, F. P., Couillard-Despres, S., & Aigner, L. (2004). The neurogenic competence of progenitors from the postnatal rat retina in vitro. *Experimental Eye Research, 78*, 1025–1036.

Evans, M. J., & Kaufman, M. H. (1981). Establishment in culture of pluripotential cells from mouse embryos. *Nature, 292*, 154–156.

Fan, W., Lin, N., Sheedlo, H. J., & Turner, J. E. (1996). Muller and RPE cell response to photoreceptor cell degeneration in aging Fischer rats. *Experimental Eye Research, 63*, 9–18.

Fawcett, J. W., & Asher, R. A. (1999). The glial scar and central nervous system repair. *Brain Research Bulletin, 49*, 377–391.

Fischer, A. J., & Reh, T. A. (2000). Identification of a proliferating marginal zone of retinal progenitors in postnatal chickens. *Developmental Biology, 220*, 197–210.

Fisher, S. K., Lewis, G. P., Linberg, K. A., & Verardo, M. R. (2005). Cellular remodeling in mammalian retina: results from studies of experimental retinal detachment. *Progress in Retinal and Eye Research, 24*, 395–431.

Friel, R., van der Sar, S., & Mee, P. J. (2005). Embryonic stem cells: understanding their history, cell biology and signalling. *Advanced Drug Delivery Reviews, 57*, 1894–1903.

Gouras, P., & Tanabe, T. (2003). Ultrastructure of adult rd mouse retina. *Graefe's Archive for Clinical and Experimental Ophthalmology, 241*, 410–417.

Gu, P., Harwood, L. J., Zhang, X., Wylie, M., Curry, W. J., & Cogliati, T. (2007). Isolation of retinal progenitor and stem cells from the porcine eye. *Molecular Vision, 13*, 1045–1057.

Harada, T., Harada, C., Kohsaka, S., Wada, E., Yoshida, K., Ohno, S., et al. (2002). Microglia-Muller glia cell interactions control neurotrophic factor production during light-induced retinal degeneration. *The Journal of Neuroscience, 22*, 9228–9236.

Harris, W. A., & Perron, M. (1998). Molecular recapitulation: the growth of the vertebrate retina. *The International Journal of Developmental Biology, 42*, 299–304.

Haruta, M., Kosaka, M., Kanegae, Y., Saito, I., Inoue, T., Kageyama, R., et al. (2001). Induction of photoreceptor-specific phenotypes in adult mammalian iris tissue. *Nature Neuroscience, 4*, 1163–1164.

Hirami, Y., Osakada, F., Takahashi, K., Okita, K., Yamanaka, S., Ikeda, H., et al. (2009). Generation of retinal cells from mouse and human induced pluripotent stem cells. *Neuroscience Letters, 458*, 126–131.

Hori, J., Joyce, N., & Streilein, J. W. (2000). Epithelium-deficient corneal allografts display immune privilege beneath the kidney capsule. *Investigative Ophthalmology & Visual Science, 41*, 443–452.

Hori, J., Ng, T. F., Shatos, M., Klassen, H., Streilein, J. W., & Young, M. J. (2003). Neural progenitor cells lack immunogenicity and resist destruction as allografts. *Stem Cells, 21*, 405–416.

Hose, S., Zigler, J. S., Jr., & Sinha, D. (2005). A novel rat model to study the functions of macrophages during normal development and pathophysiology of the eye. *Immunology Letters, 96*, 299–302.

Hughes, E. H., Schlichtenbrede, F. C., Murphy, C. C., Sarra, G. M., Luthert, P. J., Ali, R. R., et al. (2003). Generation of activated sialoadhesin-positive microglia during retinal degeneration. *Investigative Ophthalmology & Visual Science, 44*, 2229–2234.

Humayun, M. S., de Juan, E., Jr., del Carro, M., Dagnelie, G., Radner, W., Sadda, S. R., et al. (2000). Human neural retinal transplantation. *Investigative Ophthalmology & Visual Science, 41*, 3100–3106.

Iandiev, I., Biedermann, B., Bringmann, A., Reichel, M. B., Reichenbach, A., & Pannicke, T. (2006). Atypical gliosis in Muller cells of the slowly degenerating rds mutant mouse retina. *Experimental Eye Research, 82*, 449–457.

Ikeda, H., Osakada, F., Watanabe, K., Mizuseki, K., Haraguchi, T., Miyoshi, H., et al. (2005). Generation of Rx+/Pax6+ neural retinal precursors from embryonic stem cells. *Proceedings of the National Academy of Sciences of the United States of America, 102*, 11331–11336.

Inoue, T., Kagawa, T., Fukushima, M., Shimizu, T., Yoshinaga, Y., Takada, S., et al. (2006). Activation of canonical Wnt pathway promotes proliferation of retinal stem cells derived from adult mouse ciliary margin. *Stem Cells, 24*, 95–104.

Inoue, Y., Yanagi, Y., Tamaki, Y., Uchida, S., Kawase, Y., Araie, M., et al. (2005). Clonogenic analysis of ciliary epithelial derived retinal progenitor cells in rabbits. *Experimental Eye Research, 81*, 437–445.

Ishikawa, Y., & Mine, S. (1983). Aminoadipic acid toxic effects on retinal glial cells. *Japanese Journal of Ophthalmology, 27*, 107–118.

Jiang, H. R., Hwenda, L., Makinen, K., Oetke, C., Crocker, P. R., & Forrester, J. V. (2006). Sialoadhesin promotes the inflammatory response in experimental autoimmune uveoretinitis. *Journal of Immunology, 177*, 2258–2264.

Jiang, H. R., Lumsden, L., & Forrester, J. V. (1999). Macrophages and dendritic cells in IRBP-induced experimental autoimmune uveoretinitis in B10RIII mice. *Investigative Ophthalmology & Visual Science, 40*, 3177–3185.

Jiang, L. Q., Jorquera, M., & Streilein, J. W. (1993). Subretinal space and vitreous cavity as immunologically privileged sites for retinal allografts. *Investigative Ophthalmology & Visual Science, 34*, 3347–3354.

Jiang, L. Q., Jorquera, M., Streilein, J. W., & Ishioka, M. (1995). Unconventional rejection of neural retinal allografts implanted into the immunologically privileged site of the eye. *Transplantation, 59*, 1201–1207.

Joannides, A. J., Fiore-Heriche, C., Battersby, A. A., Athauda-Arachchi, P., Bouhon, I. A., Williams, L., et al. (2007). A scaleable and defined system for generating neural stem cells from human embryonic stem cells. *Stem Cells, 25*, 731–737.

Johns, P. R. (1977). Growth of the adult goldfish eye. III. Source of the new retinal cells. *The Journal of Comparative Neurology, 176*, 343–357.

Johns, P. R., & Easter, S. S., Jr. (1977). Growth of the adult goldfish eye. II. Increase in retinal cell number. *The Journal of Comparative Neurology, 176*, 331–341.

Jomary, C., & Jones, S. E. (2008). Induction of functional photoreceptor phenotype by exogenous Crx expression in mouse retinal stem cells. *Investigative Ophthalmology & Visual Science, 49*, 429–437.

Jones, L. L., & Tuszynski, M. H. (2002). Spinal cord injury elicits expression of keratan sulfate proteoglycans by macrophages, reactive microglia, and oligodendrocyte progenitors. *The Journal of Neuroscience, 22*, 4611–4624.

Jones, L. L., Yamaguchi, Y., Stallcup, W. B., & Tuszynski, M. H. (2002). NG2 is a major chondroitin sulfate proteoglycan produced after spinal cord injury and is expressed by macrophages and oligodendrocyte progenitors. *The Journal of Neuroscience, 22*, 2792–2803.

Kaplan, H. J., Tezel, T. H., Berger, A. S., Wolf, M. L., & Del Priore, L. V. (1997). Human photoreceptor transplantation in retinitis pigmentosa: a safety study. *Archives of Ophthalmology, 115*, 1168–1172.

Keirstead, H. S., Nistor, G., Bernal, G., Totoiu, M., Cloutier, F., Sharp, K., et al. (2005). Human embryonic stem cell-derived oligodendrocyte progenitor cell transplants remyelinate and restore locomotion after spinal cord injury. *The Journal of Neuroscience, 25,* 4694–4705.

Kim, J. B., Zaehres, H., Wu, G., Gentile, L., Ko, K., Sebastiano, V., et al. (2008). Pluripotent stem cells induced from adult neural stem cells by reprogramming with two factors. *Nature, 454,* 646–650.

Klassen, H., Kiilgaard, J. F., Zahir, T., Ziaeian, B., Kirov, I., Scherfig, E., et al. (2007a). Progenitor cells from the porcine neural retina express photoreceptor markers after transplantation to the subretinal space of allorecipients. *Stem Cells, 25,* 1222–1230.

Klassen, H., Sakaguchi, D. S., & Young, M. J. (2004a). Stem cells and retinal repair. *Progress in Retinal and Eye Research, 23,* 149–181.

Klassen, H., Schwartz, P. H., Ziaeian, B., Nethercott, H., Young, M. J., Bragadottir, R., et al. (2007b). Neural precursors isolated from the developing cat brain show retinal integration following transplantation to the retina of the dystrophic cat. *Veterinary Ophthalmology, 10,* 245–253.

Klassen, H. J., Ng, T. F., Kurimoto, Y., Kirov, I., Shatos, M., Coffey, P., et al. (2004b). Multipotent retinal progenitors express developmental markers, differentiate into retinal neurons, and preserve light-mediated behavior. *Investigative Ophthalmology & Visual Science, 45,* 4167–4173.

Kokkinopoulos, I., Pearson, R. A., MacNeil, A., Dhomen, N. S., MacLaren, R. E., Ali, R. R., et al. (2008). Isolation and characterisation of neural progenitor cells from the adult Chx10(orJ/orJ) central neural retina. *Molecular and Cellular Neurosciences, 38,* 359–373.

Kubota, R., Hokoc, J. N., Moshiri, A., McGuire, C., & Reh, T. A. (2002). A comparative study of neurogenesis in the retinal ciliary marginal zone of homeothermic vertebrates. *Brain Research. Developmental Brain Research, 134,* 31–41.

Lamba, D. A., Karl, M. O., Ware, C. B., & Reh, T. A. (2006). Efficient generation of retinal progenitor cells from human embryonic stem cells. *Proceedings of the National Academy of Sciences of the United States of America, 103,* 12769–12774.

Lawrence, J. M., Singhal, S., Bhatia, B., Keegan, D. J., Reh, T. A., Luthert, P. J., et al. (2007). MIO-M1 cells and similar Muller glial cell lines derived from adult human retina exhibit neural stem cell characteristics. *Stem Cells, 25,* 2033–2043.

Lewis, G. P., & Fisher, S. K. (2000). Muller cell outgrowth after retinal detachment: association with cone photoreceptors. *Investigative Ophthalmology & Visual Science, 41,* 1542–1545.

Lewis, G. P., & Fisher, S. K. (2003). Up-regulation of glial fibrillary acidic protein in response to retinal injury: its potential role in glial remodeling and a comparison to vimentin expression. *International Review of Cytology, 230,* 263–290.

Liljekvist-Soltic, I., Olofsson, J., & Johansson, K. (2008). Progenitor cell-derived factors enhance photoreceptor survival in rat retinal explants. *Brain Research, 1227,* 226–233.

Limb, G. A., Salt, T. E., Munro, P. M., Moss, S. E., & Khaw, P. T. (2002). In vitro characterization of a spontaneously immortalized human Muller cell line (MIO-M1). *Investigative Ophthalmology & Visual Science, 43,* 864–869.

Liu, I. H., Chen, S. J., Ku, H. H., Kao, C. L., Tsai, F. T., Hsu, W. M., et al. (2005). Comparison of the proliferation and differentiation ability between adult rat retinal stem cells and cerebral cortex-derived neural stem cells. *Ophthalmologica, 219,* 171–176.

Ludwig, T. E., Bergendahl, V., Levenstein, M. E., Yu, J., Probasco, M. D., & Thomson, J. A. (2006a). Feeder-independent culture of human embryonic stem cells. *Nature Methods, 3,* 637–646.

Ludwig, T. E., Levenstein, M. E., Jones, J. M., Berggren, W. T., Mitchen, E. R., Frane, J. L., et al. (2006b). Derivation of human embryonic stem cells in defined conditions. *Nature Biotechnology, 24,* 185–187.

Ma, N., & Streilein, J. W. (1998). Contribution of microglia as passenger leukocytes to the fate of intraocular neuronal retinal grafts. *Investigative Ophthalmology & Visual Science, 39,* 2384–2393.

MacLaren, R. E., Pearson, R. A., MacNeil, A., Douglas, R. H., Salt, T. E., Akimoto, M., et al. (2006). Retinal repair by transplantation of photoreceptor precursors. *Nature, 444,* 203–207.

MacNeil, A., Pearson, R. A., MacLaren, R. E., Smith, A. J., Sowden, J. C., & Ali, R. R. (2007). Comparative analysis of progenitor cells isolated from the iris, pars plana, and ciliary body of the adult porcine eye. *Stem Cells, 25,* 2430–2438.

Maguire, A. M., Simonelli, F., Pierce, E. A., Pugh, E. N., Jr., Mingozzi, F., Bennicelli, J., et al. (2008). Safety and efficacy of gene transfer for Leber's congenital amaurosis. *The New England Journal of Medicine, 358,* 2240–2248.

Marquardt, T., & Gruss, P. (2002). Generating neuronal diversity in the retina: one for nearly all. *Trends in Neuroscience, 25,* 32–38.

Martin, G. R. (1981). Isolation of a pluripotent cell line from early mouse embryos cultured in medium conditioned by teratocarcinoma stem cells. *Proceedings of the National Academy of Sciences of the United States of America, 78,* 7634–7638.

Mayer, E. J., Carter, D. A., Ren, Y., Hughes, E. H., Rice, C. M., Halfpenny, C. A., et al. (2005). Neural progenitor cells from postmortem adult human retina. *The British Journal of Ophthalmology, 89,* 102–106.

Mehalow, A. K., Kameya, S., Smith, R. S., Hawes, N. L., Denegre, J. M., Young, J. A., et al. (2003). CRB1 is essential for external limiting membrane integrity and photoreceptor morphogenesis in the mammalian retina. *Human Molecular Genetics, 12,* 2179–2189.

Mellough, C. B., Cui, Q., & Harvey, A. R. (2007). Treatment of adult neural progenitor cells prior to transplantation affects graft survival and integration in a neonatal and adult rat model of selective retinal ganglion cell depletion. *Restorative Neurology and Neuroscience, 25,* 177–190.

Mizumoto, H., Mizumoto, K., Shatos, M. A., Klassen, H., & Young, M. J. (2003). Retinal transplantation of neural progenitor cells derived from the brain of GFP transgenic mice. *Vision Research, 43,* 1699–1708.

Mohand-Said, S., Hicks, D., Dreyfus, H., & Sahel, J. A. (2000). Selective transplantation of rods delays cone loss in a retinitis pigmentosa model. *Archives of Ophthalmology, 118,* 807–811.

Nakagawa, M., Koyanagi, M., Tanabe, K., Takahashi, K., Ichisaka, T., Aoi, T., et al. (2008). Generation of induced pluripotent stem cells without Myc from mouse and human fibroblasts. *Nature Biotechnology, 26,* 101–106.

Ng, T. F., Kitaichi, N., & Taylor, A. W. (2007). In vitro generated autoimmune regulatory T cells enhance intravitreous allogeneic retinal graft survival. *Investigative Ophthalmology & Visual Science, 48,* 5112–5117.

Ng, T. F., Osawa, H., Hori, J., Young, M. J., & Streilein, J. W. (2002). Allogeneic neonatal neuronal retina grafts display partial immune privilege in the subcapsular space of the kidney. *Journal of Immunology, 169,* 5601–5606.

Nickerson, P. E., Da, S. N., Myers, T., Stevens, K., & Clarke, D. B. (2008). Neural progenitor potential in cultured Muller glia: effects of passaging and exogenous growth factor exposure. *Brain Research, 1230,* 1–12.

Nistor, G. I., Totoiu, M. O., Haque, N., Carpenter, M. K., & Keirstead, H. S. (2005). Human embryonic stem cells differentiate into oligodendrocytes in high purity and myelinate after spinal cord transplantation. *Glia, 49,* 385–396.

Oishi, A., Nagai, T., Mandai, M., Takahashi, M., & Yoshimura, N. (2007). The effect of dendritic cells on the retinal cell transplantation. *Biochemical and Biophysical Research Communications, 363,* 292–296.

Okita, K., Ichisaka, T., & Yamanaka, S. (2007). Generation of germline-competent induced pluripotent stem cells. *Nature, 448,* 313–317.

Osakada, F., Ikeda, H., Mandai, M., Wataya, T., Watanabe, K., Yoshimura, N., et al. (2008). Toward the generation of rod and cone photoreceptors from mouse, monkey and human embryonic stem cells. *Nature Biotechnology, 26,* 215–224.

Park, I. H., Zhao, R., West, J. A., Yabuuchi, A., Huo, H., Ince, T. A., et al. (2008). Reprogramming of human somatic cells to pluripotency with defined factors. *Nature, 451,* 141–146.

Pedersen, O. O., & Karlsen, R. L. (1979). Destruction of Muller cells in the adult rat by intravitreal injection of D,L-alpha-aminoadipic acid. An electron microscopic study. *Experimental Eye Research, 28,* 569–575.

Pera, M. F., Reubinoff, B., & Trounson, A. (2000). Human embryonic stem cells. *Journal of Cell Science, 113*(Pt 1), 5–10.

Perrier, A. L., Tabar, V., Barberi, T., Rubio, M. E., Bruses, J., Topf, N., et al. (2004). Derivation of midbrain dopamine neurons from human embryonic stem cells. *Proceedings of the National Academy of Sciences of the United States of America, 101,* 12543–12548.

Qiu, G., Seiler, M. J., Mui, C., Arai, S., Aramant, R. B., de Juan, E., Jr., et al. (2005). Photoreceptor differentiation and integration of retinal progenitor cells transplanted into transgenic rats. *Experimental Eye Research, 80,* 515–525.

Radtke, N. D., Aramant, R. B., Petry, H. M., Green, P. T., Pidwell, D. J., & Seiler, M. J. (2008). Vision improvement in retinal degeneration patients by implantation of retina together with retinal pigment epithelium. *American Journal of Ophthalmology, 146,* 172–182.

Radtke, N. D., Aramant, R. B., Seiler, M., & Petry, H. M. (1999). Preliminary report: indications of improved visual function after retinal sheet transplantation in retinitis pigmentosa patients. *American Journal of Ophthalmology, 128,* 384–387.

Raymond, P. A., Barthel, L. K., Bernardos, R. L., & Perkowski, J. J. (2006). Molecular characterization of retinal stem cells and their niches in adult zebrafish. *BMC Developmental Biology, 6,* 36.

Reubinoff, B. E., Pera, M. F., Fong, C. Y., Trounson, A., & Bongso, A. (2000). Embryonic stem cell lines from human blastocysts: somatic differentiation in vitro. *Nature Biotechnology, 18,* 399–404.

Rich, K. A., Figueroa, S. L., Zhan, Y., & Blanks, J. C. (1995). Effects of Muller cell disruption on mouse photoreceptor cell development. *Experimental Eye Research, 61,* 235–248.

Rodriguez-Gomez, J. A., Lu, J. Q., Velasco, I., Rivera, S., Zoghbi, S. S., Liow, J. S., et al. (2007). Persistent dopamine functions of neurons derived from embryonic stem cells in a rodent model of Parkinson disease. *Stem Cells, 25,* 918–928.

Rolls, A., Cahalon, L., Bakalash, S., Avidan, H., Lider, O., & Schwartz, M. (2006). A sulfated disaccharide derived from chondroitin sulfate proteoglycan protects against inflammation-associated neurodegeneration. *The FASEB Journal, 20,* 547–549.

Roque, R. S., Imperial, C. J., & Caldwell, R. B. (1996). Microglial cells invade the outer retina as photoreceptors degenerate in Royal College of Surgeons rats. *Investigative Ophthalmology & Visual Science, 37,* 196–203.

Sakaguchi, D. S., Van Hoffelen, S. J., Grozdanic, S. D., Kwon, Y. H., Kardon, R. H., & Young, M. J. (2005). Neural progenitor cell transplants into the developing and mature central nervous system. *Annals of the New York Academy of Sciences, 1049,* 118–134.

Sakaguchi, D. S., Van Hoffelen, S. J., Theusch, E., Parker, E., Orasky, J., Harper, M. M., et al. (2004). Transplantation of neural progenitor cells into the developing retina of the Brazilian opossum: an in vivo system for studying stem/progenitor cell plasticity. *Developmental Neuroscience, 26,* 336–345.

Sakaguchi, D. S., Van Hoffelen, S. J., & Young, M. J. (2003). Differentiation and morphological integration of neural progenitor cells transplanted into the developing mammalian eye. *Annals of the New York Academy of Sciences, 995,* 127–139.

Sam, T. N., Xiao, J., Roehrich, H., Low, W. C., & Gregerson, D. S. (2006). Engrafted neural progenitor cells express a tissue-restricted reporter gene associated with differentiated retinal photoreceptor cells. *Cell Transplantation, 15,* 147–160.

Sancho-Pelluz, J., Wunderlich, K. A., Rauch, U., Romero, F. J., van Veen, T., Limb, G. A., et al. (2008). Sialoadhesin expression in intact degenerating retinas and following transplantation. *Investigative Ophthalmology & Visual Science*, 49, 5602–5610.

Sanyal, S., & Hawkins, R. K. (1989). Development and degeneration of retina in rds mutant mice: altered disc shedding pattern in the heterozygotes and its relation to ocular pigmentation. *Current Eye Research*, 8, 1093–1101.

Seiler, M. J., Sagdullaev, B. T., Woch, G., Thomas, B. B., & Aramant, R. B. (2005). Transsynaptic virus tracing from host brain to subretinal transplants. *The European Journal of Neuroscience*, 21, 161–172.

Seri, B., Garcia-Verdugo, J. M., Collado-Morente, L., McEwen, B. S., & Alvarez-Buylla, A. (2004). Cell types, lineage, and architecture of the germinal zone in the adult dentate gyrus. *The Journal of Comparative Neurology*, 478, 359–378.

Seri, B., Garcia-Verdugo, J. M., McEwen, B. S., & Alvarez-Buylla, A. (2001). Astrocytes give rise to new neurons in the adult mammalian hippocampus. *The Journal of Neuroscience*, 21, 7153–7160.

Sheedlo, H. J., Jaynes, D., Bolan, A. L., & Turner, J. E. (1995). Mullerian glia in dystrophic rodent retinas: an immunocytochemical analysis. *Brain Research. Developmental Brain Research*, 85, 171–180.

Singhal, S., Lawrence, J. M., Bhatia, B., Ellis, J. S., Kwan, A. S., MacNeil, A., et al. (2008). Chondroitin sulfate proteoglycans and microglia prevent migration and integration of grafted Muller stem cells into degenerating retina. *Stem Cells*, 26, 1074–1082.

Straznicky, K., & Gaze, R. M. (1971). The growth of the retina in *Xenopus laevis*: an autoradiographic study. *Journal of Embryology and Experimental Morphology*, 26, 67–79.

Streilein, J. W., Ma, N., Wenkel, H., Ng, T. F., & Zamiri, P. (2002). Immunobiology and privilege of neuronal retina and pigment epithelium transplants. *Vision Research*, 42, 487–495.

Streilein, J. W., & Niederkorn, J. Y. (1985). Characterization of the suppressor cell(s) responsible for anterior chamber-associated immune deviation (ACAID) induced in BALB/c mice by P815 cells. *Journal of Immunology*, 134, 1381–1387.

Suemori, H., Tada, T., Torii, R., Hosoi, Y., Kobayashi, K., Imahie, H., et al. (2001). Establishment of embryonic stem cell lines from cynomolgus monkey blastocysts produced by IVF or ICSI. *Developmental Dynamics*, 222, 273–279.

Sugie, Y., Yoshikawa, M., Ouji, Y., Saito, K., Moriya, K., Ishizaka, S., et al. (2005). Photoreceptor cells from mouse ES cells by co-culture with chick embryonic retina. *Biochemical and Biophysical Research Communications*, 332, 241–247.

Suzuki, T., Akimoto, M., Imai, H., Ueda, Y., Mandai, M., Yoshimura, N., et al. (2007). Chondroitinase ABC treatment enhances synaptogenesis between transplant and host neurons in model of retinal degeneration. *Cell Transplantation*, 16, 493–503.

Suzuki, T., Mandai, M., Akimoto, M., Yoshimura, N., & Takahashi, M. (2006). The simultaneous treatment of MMP-2 stimulants in retinal transplantation enhances grafted cell migration into the host retina. *Stem Cells*, 24, 2406–2411.

Takahashi, K., Tanabe, K., Ohnuki, M., Narita, M., Ichisaka, T., Tomoda, K., et al. (2007). Induction of pluripotent stem cells from adult human fibroblasts by defined factors. *Cell*, 131, 861–872.

Takahashi, M., Palmer, T. D., Takahashi, J., & Gage, F. H. (1998). Widespread integration and survival of adult-derived neural progenitor cells in the developing optic retina. *Molecular and Cellular Neurosciences*, 12, 340–348.

Taylor, C. J., Bolton, E. M., Pocock, S., Sharples, L. D., Pedersen, R. A., & Bradley, J. A. (2005). Banking on human embryonic stem cells: estimating the number of donor cell lines needed for HLA matching. *Lancet*, 366, 2019–2025.

Thomson, J. A., Itskovitz-Eldor, J., Shapiro, S. S., Waknitz, M. A., Swiergiel, J. J., Marshall, V. S., et al. (1998). Embryonic stem cell lines derived from human blastocysts. *Science*, 282, 1145–1147.

Thomson, J. A., Kalishman, J., Golos, T. G., Durning, M., Harris, C. P., Becker, R. A., et al. (1995). Isolation of a primate embryonic stem cell line. *Proceedings of the National Academy of Sciences of the United States of America*, 92, 7844–7848.

Tropepe, V., Coles, B. L., Chiasson, B. J., Horsford, D. J., Elia, A. J., McInnes, R. R., et al. (2000). Retinal stem cells in the adult mammalian eye. *Science*, 287, 2032–2036.

van de Pavert, S. A., Sanz, A. S., Aartsen, W. M., Vos, R. M., Versteeg, I., Beck, S. C., et al. (2007). Crb1 is a determinant of retinal apical Muller glia cell features. *Glia*, 55, 1486–1497.

Van Hoffelen, S. J., Young, M. J., Shatos, M. A., & Sakaguchi, D. S. (2003). Incorporation of murine brain progenitor cells into the developing mammalian retina. *Investigative Ophthalmology & Visual Science*, 44, 426–434.

Wenkel, H., & Streilein, J. W. (1998). Analysis of immune deviation elicited by antigens injected into the subretinal space. *Investigative Ophthalmology & Visual Science*, 39, 1823–1834.

Wenkel, H., & Streilein, J. W. (2000). Evidence that retinal pigment epithelium functions as an immune-privileged tissue. *Investigative Ophthalmology & Visual Science*, 41, 3467–3473.

West, E. L., Pearson, R. A., Tschernutter, M., Sowden, J. C., MacLaren, R. E., & Ali, R. R. (2008). Pharmacological disruption of the outer limiting membrane leads to increased retinal integration of transplanted photoreceptor precursors. *Experimental Eye Research*, 86, 601–611.

Wilbanks, G. A., & Streilein, J. W. (1990). Characterization of suppressor cells in anterior chamber-associated immune deviation (ACAID) induced by soluble antigen. Evidence of two functionally and phenotypically distinct T-suppressor cell populations. *Immunology*, 71, 383–389.

Xu, C., Inokuma, M. S., Denham, J., Golds, K., Kundu, P., Gold, J. D., et al. (2001). Feeder-free growth of undifferentiated human embryonic stem cells. *Nature Biotechnology*, 19, 971–974.

Xu, H., Sta Iglesia, D. D., Kielczewski, J. L., Valenta, D. F., Pease, M. E., Zack, D. J., et al. (2007). Characteristics of progenitor cells derived from adult ciliary body in mouse, rat, and human eyes. *Investigative Ophthalmology & Visual Science, 48,* 1674–1682.

Yan, Y., Yang, D., Zarnowska, E. D., Du, Z., Werbel, B., Valliere, C., et al. (2005). Directed differentiation of dopaminergic neuronal subtypes from human embryonic stem cells. *Stem Cells, 23,* 781–790.

Yang, P., Seiler, M. J., Aramant, R. B., & Whittemore, S. R. (2002). Differential lineage restriction of rat retinal progenitor cells in vitro and in vivo. *Journal of Neuroscience Research, 69,* 466–476.

Yanoff, M., Fine, B. S., Brucker, A. J., & Eagle, R. C., Jr. (1984). Pathology of human cystoid macular edema. *Survey of Ophthalmology, 28*(Suppl.), 505–511.

Young, M. J., Ray, J., Whiteley, S. J., Klassen, H., & Gage, F. H. (2000). Neuronal differentiation and morphological integration of hippocampal progenitor cells transplanted to the retina of immature and mature dystrophic rats. *Molecular and Cellular Neurosciences, 16,* 197–205.

Yu, J., Vodyanik, M. A., Smuga-Otto, K., Antosiewicz-Bourget, J., Frane, J. L., Tian, S., et al. (2007). Induced pluripotent stem cell lines derived from human somatic cells. *Science, 318,* 1917–1920.

Zhang, C., Lam, T. T., & Tso, M. O. (2005). Heterogeneous populations of microglia/macrophages in the retina and their activation after retinal ischemia and reperfusion injury. *Experimental Eye Research, 81,* 700–709.

Zhang, S. C., Wernig, M., Duncan, I. D., Brustle, O., & Thomson, J. A. (2001). In vitro differentiation of transplantable neural precursors from human embryonic stem cells. *Nature Biotechnology, 19,* 1129–1133.

Zhang, Y., Arner, K., Ehinger, B., & Perez, M. T. (2003). Limitation of anatomical integration between subretinal transplants and the host retina. *Investigative Ophthalmology & Visual Science, 44,* 324–331.

Zhang, Y., Kardaszewska, A. K., van Veen, T., Rauch, U., & Perez, M. T. (2004). Integration between abutting retinas: role of glial structures and associated molecules at the interface. *Investigative Ophthalmology & Visual Science, 45,* 4440–4449.

Zhang, Y., Klassen, H. J., Tucker, B. A., Perez, M. T., & Young, M. J. (2007). CNS progenitor cells promote a permissive environment for neurite outgrowth via a matrix metalloproteinase-2-dependent mechanism. *The Journal of Neuroscience, 27,* 4499–4506.

Zhao, X., Liu, J., & Ahmad, I. (2002). Differentiation of embryonic stem cells into retinal neurons. *Biochemical and Biophysical Research Communications, 297,* 177–184.

CHAPTER 2

# Strategies for retinal repair: cell replacement and regeneration

Deepak A. Lamba, Mike O. Karl and Thomas A. Reh*

*Department of Biological Structure, Institute of Stem Cell and Regenerative Medicine, University of Washington, School of Medicine, Seattle, WA, USA*

**Abstract:** The retina, like most other regions of the central nervous system, is subject to degeneration from both genetic and acquired causes. Once the photoreceptors or inner retinal neurons have degenerated, they are not spontaneously replaced in mammals. In this review, we provide an overview of retinal development and regeneration with emphasis on endogenous repair and replacement seen in lower vertebrates and recent work on induced mammalian retinal regeneration from Müller glia. Additionally, recent studies demonstrating the potential for cellular replacement using postmitotic photoreceptors and embryonic stem cells are also reviewed.

**Keywords:** photoreceptors; Müller glia; embryonic stem cells; transdifferentiation; development

## The retina and its degenerations

The retina, a thin sheet of neural tissue covering the inner surface of the posterior eye, contains the sensory cells and neurons necessary for the transduction of light to electrical signals responsible for our sense of sight. The retina contains a variety of neurons and glia. The rod and cone photoreceptors transduce the light to electrical potential changes, which are in turn relayed via synapses with bipolar cells to the retinal ganglion cells, which connect to brain nuclei via their axons in the optic nerve. In addition, the retina contains other types of interneurons, called horizontal cells and amacrine cells, which provide lateral interactions within the retinal circuit. Along with these various types of neurons, there are also three types of non-neuronal cells: Müller glia, pigmented epithelial cells, and astrocytes. The Müller glia span the retinal epithelium and are similar to astrocytes that are found in other regions of the CNS. The pigmented epithelial cells (RPE) form a layer adjacent to the photoreceptor layer; these cells are essential for the turnover of the rod and cone outer segments. A second type of glia, the astrocytes, line the inner surface of the retina and surround the blood vessels of the retina.

The retina, like many other regions of the nervous system, is subject to a variety of inherited and acquired degenerative conditions. One of the most common degenerations of the retina is Age-related Macular Degeneration (AMD), a disease characterized by the degeneration of the photoreceptors in the central region (i.e. foveal or macular). Prior to photoreceptor degeneration, the pigmented epithelial cells die most likely due to accumulations of insoluble Drusen in the

*Corresponding author.
Tel.: 206-543-8043; Fax: 206-543-1524;
E-mail: tomreh@u.washington.edu

DOI: 10.1016/S0079-6123(09)17502-7

adjacent Bruch's membrane. Once the RPE cells have degenerated, the cones soon follow. A more aggressive form of this disease, called the "wet" form, is characterized by new vessel growth (choroidal neovascularization) in the fovea, which causes a rapid loss in RPE cells and their associated cone photoreceptors. However, the inner retinal cells, the bipolar cells, amacrine cells, and ganglion cells survive this disease process for many years following the loss of the photoreceptors, and this has led to the possibility of photoreceptor replacement as a potential therapy. Another large class of degenerations, called Retinitis Pigmentosa, would also benefit from photoreceptor replacement. In these inherited diseases, rod photoreceptors degenerate initially, but cones frequently become involved as well, leading to total blindness.

Similar to other regions of the CNS, loss of photoreceptors in the mammalian retina does not lead to their spontaneous replacement. However, in many vertebrates, degeneration of retinal cells leads to efficient regeneration. This review outlines the current understanding of retinal regeneration in those species where it naturally occurs and describes recent attempts that have been made to activate retinal regeneration in mammalian models of retinal degeneration. In addition, we describe recent use of stem cell technologies to address problems of retinal repair.

**Retinal developmental biology**

The retina develops from a region of the neural plate, called the eye field, which is today identified by the expression of a set of transcription factors (EFTFs) that are necessary and in some cases sufficient for eye formation (Zuber et al., 2003). The single eye field is split into two, by a repression of the EFTFs in the midline, mediated by Shh signaling. The retinas arise as the distal parts of the optic vesicles, paired evaginations from the diencephalon. These vesicles undergo additional morphogenesis to form two-layered optic cups. The inner layer develops into the neural retina and the outer layer develops into the RPE.

The developing neural retina contains mitotically active retinal progenitors, cells that will give rise to all the different types of retinal neurons and the Müller glia. Richard Sidman discovered in thymidine birthdating studies in mouse retina that the different types of neurons are generated in a specific sequence, and subsequent analyses have confirmed that this sequence is conserved among vertebrates: ganglion cells, cone photoreceptors, horizontal cells, and most amacrine cells are generated during early stages of development (prenatally in mice), and most rod photoreceptors, bipolar cells, and Müller glia are generated in the latter half of the period of retinal histogenesis (Sidman, 1961). Retinal progenitors are similar to neural progenitors/stem cells found elsewhere in the central nervous system. They generate both neurons and glia, they respond to similar mitogenic factors (EGF, FGF, and Shh) (see Nelson et al., 2007 for review), they require many of the same proneural bHLH transcription factors (*Ascl1*, *Neurog2*, *NeuroD1*) (Akagi et al., 2004), and the Notch pathway is necessary for the maintenance of the retinal progenitor pool (see Nelson et al., 2007 for review).

**The ciliary marginal zone**

In many fish and amphibians, most of the retina of the adult is not generated during the period of embryonic development, but instead is produced by a specialized zone of stem and progenitor cells located at the junction between the neural retina and the non-neural ciliary epithelium. This zone of cells, that essentially forms a ring at the edge of the retina, is known as the CMZ, or ciliary marginal zone (also called the circumferential germinal zone, or CGZ). The cells in this region actively proliferate throughout life in fish and through metamorphosis in frogs, and produce all types of retinal neurons and the Müller glia, in a manner similar to embryonic retinal progenitors. The cells express both early and late progenitor genes, with cells in the earlier stages of development located more peripherally and later stage progenitors located centrally, adjacent to differentiated retina. It is thought that the most peripheral regions of the zone contain retinal "stem" cells and these produce progeny with progenitor characteristics. In many ways, this zone is reminiscent of the other zones of neural stem cells in the CNS, including the subventricular zone and the hippocampal progenitor zone, and many of the same factors that regulate proliferation in the CMZ are shared with other neural stem/progenitor zones (see Moshiri et al., 2004 for review).

The CMZ of fish and amphibians can contribute to regeneration after retinal damage. Several studies have shown that the production of neurons from the CMZ increases following retinal damage. In both fish and larval anurans, for example, neurotoxic damage to the retina causes an upregulation in proliferation in the CMZ, and an increase in the differentiation of the types of neurons destroyed by the toxins. Nevertheless, since this zone is located at the peripheral edge of the retina, the majority of the new neurons produced following retinal damage do not arise from the CMZ, but rather from the de-differentiation of either pigmented cells (in amphibians) or Müller glia (in fish) and therefore, these latter sources are more important in the regeneration process in poikilothermic vertebrates, and will be discussed more extensively in later sections.

Despite the lack of significant regeneration from the CMZ in fish and amphibians, it is still an important question as to whether this zone is present in homoeothermic vertebrates. Several years ago, we found that birds retain a rudiment of this zone, and that new neurons are generated for many weeks after hatching in chickens; in quails we found proliferating cells in this region even in year-old birds. In the posthatch chick, the zone produces all types of neurons, and the cells have a gene expression profile like that of the embryonic progenitors. It is not known whether this region contains retinal stem cells in birds, since neuronal production appears to decline as the birds mature; however, the overall organization of the zone and the factors that regulate and maintain neural stem/progenitor cell niches are also present in the bird CMZ.

Although opposums still show some proliferation at the retinal margin, the CMZ appears to be entirely in the eyes of eutherian mammals. Several groups have analyzed various species for evidence of ongoing proliferation in the margin of the retina, but no mitotic cells are present in normal mice, rats, or macaques (Ahmad et al., 2000; Kubota et al., 2002; Moshiri and Reh, 2004). Nevertheless, there is evidence that retinal stem or progenitor cells persist in the mammalian ciliary epithelium and that manipulation of critical signaling pathways may provide a way to reactivate them. Moshiri and Reh (2004) analyzed mice with a single functional allele of the patched gene, a negative regulator of Shh signaling, and found that proliferating cells are retained at the retinal margin into adulthood. Crossing the patched +/− mice with a photoreceptor degeneration mutant led to an increase in the proliferation of these cells, reminiscent of the response to retinal damage observed in the CMZ cells of lower vertebrates. Several attempts have also been made to stimulate proliferation of retinal stem/progenitor cells from this region by dissociated cell culture. Since cells from this region can give rise to spheres, several labs reported that they had identified retinal stem cells in mammals. However, not all that generates a sphere in low-density culture can be classified as a stem cell, and more recent investigations have failed to support the original observations. The spheres generated from the ciliary epithelium of mice retain characteristics of the non-neuronal ciliary epithelial cells and show only limited neuronal marker expression, even following transplantation (see below).

**Endogenous retinal repair mechanisms: Müller glia as a source of new neurons**

Glia cells are found in all parts of the CNS. In the mammalian retina, there are two types of macroglia, Müller glia and astrocytes. Out of these only Müller glia are derived from the retinal neuroepithelium, while astrocytes enter the retina through the optic nerve (Stone and Dreher, 1987; Watanabe and Raff, 1988). Microglial cells, of mesodermal origin (Chan et al., 2007) are also present in the retina. In mice and rats, Müller glia differentiate at the end of the first postnatal week, as one of the last cell types generated by the progenitors. The Müller glia span the retina radially, their mature cell processes surround the neurons and vasculature, provide various physiological functions for the retina including maintenance of the homeostasis of the retinal extracellular milieu (ions, water, neurotransmitter molecules, and pH) and support of the blood–retinal-barrier (Bringmann et al., 2000, 2006; Newman and Reichenbach, 1996).

It has been recognized for many years that damage to the retina causes Müller glial cells to undergo dramatic changes analogous to the phenomenon of reactive gliosis observed in astrocytes in other regions of the CNS. Depending on the species, Müller glia respond with hypertrophy, changes in gene expression and morphology, migration, and less often cell proliferation (Bringmann and Reichenbach, 2001). In mice and man, retinal damage induces Müller glia hypertrophy, which can lead to retinal disorganization and a scar-like subretinal or epiretinal "membrane." Similar to astrocytes in other areas of the CNS, reactive Müller glial cells are characterized by the upregulation in intermediate filament proteins (i.e. GFAP, vimentin, and nestin), cellular hypertrophy, and changes in gene expression (Eddleston and Mucke, 1993; Ridet et al., 1997). These changes in Müller glia have been proposed to contribute to neuronal survival after injury, since many cytokines and growth factors are upregulated during reactive gliosis.

In addition to this reactive gliotic response, neuronal damage in fish causes the Müller glia to undergo changes that lead to a reversal of their terminal differentiation, leading to an upregulation of genes normally present in retinal progenitors. In fish, Müller glia provide the primary source for regeneration for most, if not all, types of retinal neurons. After a variety of different types of retinal damage, fish Müller glia spontaneously re-enter the cell cycle, and after multiple rounds of division generate neurons and additional Müller glia. At least two of the genes known to be important for normal retinal progenitors, Ascl1 and Pax6, are also necessary for Müller glial cells to undergo their regenerative program. A similar response to damage is observed in the Müller glia of the posthatch chick. After NMDA-induced neuronal damage, the Müller glia upregulate the expression of many progenitor genes, including Pax6, Ascl1, Notch, Hes5, Pea3, Sox2, and FoxN4 (Fischer and Reh, 2001; Hayes et al., 2007). However, in contrast to the fish, the Müller glia that re-enter the mitotic cell cycle after damage in the posthatch chick retina only regenerate a subset of neuronal types: amacrine cells, bipolar cells, and ganglion cells. Photoreceptors are not produced by the Müller glia, at least under the conditions that have been tested to date.

The fact that mammalian Müller glia undergo a clear gliotic response rather than a regenerative response to damage has led to the suggestion that these two patterns of gene expression might be mutually exclusive. Observations in the chick retina support this idea. In the postnatal chick retina, the ability to activate a regenerative program is reduced in a central to peripheral fashion between P7 and P30. As the central retina loses its proliferative and regenerative response to injury, the Müller glia responds with an upregulation in GFAP expression. Therefore, it is possible that the de-differentiation required for regenerative neurogenesis is overall less advantageous than the quick fix of gliosis in limiting the damage in warm-blooded vertebrates. However, it is important to point out that in the uninjured fish retina, Müller glia normally express GFAP, as well as progenitor markers like Pax6 and Nestin, and at least some of the Müller glia are continuously proliferating at a low rate, generating rod photoreceptor precursors and other types of neurons (Bernardos et al., 2007; Raymond et al., 2006). In addition, in newts, a transient gliotic scar forms after damage, but it does not limit regeneration (Turner and Glaze, 1977). Clearly, we need to know a great deal more about the changes in gene expression following neuronal damage in the Müller glia of various species before we can make more specific hypotheses concerning the limits of regeneration in mammals.

Despite the lack of spontaneous neuronal regeneration in mammals following retinal damage (Chang et al., 2007; Close et al., 2006; Dyer and Cepko, 2000; Sahel et al., 1990; Zhao et al., 2005), several studies have tested various strategies designed to stimulate proliferation and neurogenesis from Müller glia in rats and mice. Ooto et al. (2004) used NMDA to induce damage in the adult rat retina and found that a small number of Müller glia incorporated BrdU, and even spontaneously differentiated into cells that expressed markers of bipolar cells and photoreceptors. Treatment of the damaged retina with retinoic acid (RA) or the mis-expressing basic helix-loop-helix and homeobox genes *Math3*, *NeuroD1*, and *Pax6* or *Crx* and *NeuroD1* led to further increases in the numbers of BrdU+ cells that expressed markers of retinal neurons, including bipolar, amacrine cells, or photoreceptors. More recent studies have also reported BrdU+ photoreceptors following either retinal damage or growth factor treatment in adult rodents, either in vitro or in vivo (Close et al., 2006; Das et al., 2006; Osakada et al., 2007; Wan et al., 2007, 2008). However, a major caveat in most of these experiments is that cytoplasmic markers, rather than nuclear markers, were used to identify the cells. Without careful optical sectioning and confocal microscopic analysis, it is impossible to exclude the possibility that the BrdU+ cells were misidentified in the densely packed retinal lamina. In addition, there have been no studies of functional regeneration in either the chick or the mammalian retina, and so it is still not known whether any Müller glial derived cells that express neuronal markers after damage in the mammalian retina are functionally integrated into the circuit.

Taken together, these studies show that there has been a progressive decline in the regenerative potential of Müller glia during vertebrate evolution. Nevertheless, several lines of evidence suggest that efforts to stimulate this process in mammals may ultimately be successful. The source of retinal regeneration in fish, the Müller glial cells, are present in all vertebrates, and thus could provide a source for new neurons in mammals. In addition, the pattern of gene expression in Müller glial following damage to the fish and avian retinas are remarkably similar, with the upregulation of many key progenitor genes, such as *Notch1*, *Pax6*, and *Ascl1*. Further studies will be required to determine which elements of the neurogenic program are re-expressed in mammalian Müller glia following retinal injury.

**Cell replacement strategies for retinal repair**

Since the mammalian retina has generally exhibited only very limited evidence for endogenous repair, a number of investigators have attempted to replace degenerated neurons with transplantation. A thorough review of this literature is beyond the scope of this Chapter; however, there are some key points that have emerged from these studies over the past 20 years. First, it has become clear that photoreceptors can be transplanted into the adult retina (either the normal intact retina or one that has lost photoreceptors) and that the transplanted cells can integrate with the host retinal neurons. Second, in some cases, the transplanted photoreceptors can mediate some light responses and therefore may be able to restore some function. Third, even the best cases of transplantation result in only a relatively small number of transplanted cells successfully integrating into the host retina. Fourth, the best examples of transplant integration and photoreceptor differentiation have come not from the transplant of progenitor cells or stem cells, but rather from the transplantation of newly postmitotic rod photoreceptors.

These previous studies thus support the overall possibility of retinal repair through cell replacement, but they also highlight a critical issue: since only relatively immature, postmitotic photoreceptors can integrate effectively, where can these cells be obtained? To extract these cells from fetal tissue at this late stage of development is not ethical or practical, several groups have attempted to produce retinal cells in vitro. While it has been known for some time that retinal progenitors from both rodents and humans can be expanded in dissociated cell cultures and retain their potential to differentiate into various types of retinal neurons, including rod and cone photoreceptors (Anchan et al., 1991; Kelley et al., 1995), long-term culture of fetal or embryonic retinal cells appears to inhibit the ability of the cells to differentiate into photoreceptors following transplantation. Neural progenitors isolated from other brain regions can also be transplanted into retina, but again, they fail to differentiate into photoreceptors in the host retina (Takahashi et al., 1998; Young et al., 2000). In addition, several laboratories have reported that neurospheres can be generated from adult pigmented iris and ciliary epithelium of rodents, humans, and pigs (Ahmad et al., 2000; Asami et al., 2007; Coles et al., 2004; Kohno et al., 2006; MacNeil et al., 2007; Sun et al., 2006; Tropepe et al., 2000). Since the presence of sphere-generating cells in the CNS has been associated with neural stem properties (Reynolds et al., 1992), the fact that pigmented cells from the iris and ciliary epithelium can generate spheres raises the possibility that they are latent retinal stem cells (Tropepe et al., 2000). However, despite considerable effort to promote neuronal differentiation of these cells, they appear to retain characteristics of the ciliary epithelium (Cicero et al., 2009) and do not efficiently differentiate into photoreceptors either in vivo following transplantation (Coles et al., 2004) or in vitro (Tropepe et al., 2000).

A number of groups have tried to use embryonic stem (ES) cells to derive retinal neurons for transplantation. These cells are particularly attractive, since they retain their pluripotentiality and ability to self-renew indefinitely. A number of reports, using either mouse or primate ES cells, have described different protocols to direct these cells to a retinal fate with variable success (Table 1). We therefore decided to test whether human ES cells might also be a source for replacement retinal neurons. These cells, first isolated from human embryos by James Thompson, can be maintained in an undifferentiated state and propagated for many generations. They have the potential to differentiate into any cell type in the body. By relying on the considerable amount of knowledge we have about retinal development, we devised a strategy to direct human ES cells to a retinal progenitor fate. The challenge was to be able to efficiently coax them to develop as retinal progenitors or preferably postmitotic photoreceptors.

Using a combination of a BMP inhibitor, Noggin, a *Wnt* inhibitor, Dkk1, and insulin-like growth factor (IGF-1) (Lamba et al., 2006) we developed a three-step protocol for generating retinal cells. We found that as many as 80% of all cells acquired a retinal progenitor fate as assessed by their expression of a set of transcription factors that define the "eye field" during embryonic development using RT-PCR or immunofluorescence. While many of the cells that are generated using this protocol express markers of retinal progenitor cells, many also express markers of differentiating retinal neurons; we found evidence for nearly all types of retinal neurons including ganglion cells (Brn3+), amacrine cells (HuC/D+), horizontal cells (Prox1+), bipolar cells (PKC-alpha+), and photoreceptors (*Crx*, *Nrl*, S-opsin, rhodopsin, and recoverin). More recently, we have used a lentivirus containing a photoreceptor-specific promoter (IRBP) to drive GFP to assess photoreceptor differentiation in live cultures; FACs purification of the IRBP-GFP expressing cells confirms that multiple markers of photoreceptors are consistently expressed in up to 10% of the total cells in the cultures. Overall, our results to date show that human photoreceptors, both rods and cones, can be generated from ES cells through a multistep approach and purified for further analysis and transplantation.

A similar protocol for directing human ES cells to a retinal fate has been taken by another group, using a combination of *Wnt* and *Nodal* antagonists (Osakada et al., 2008). While the approaches are similar in that antagonism of both the Wnt pathway and the BMP/nodal pathway appears to be required to direct ES cells to the retinal fate, there are some differences between the efficiency of photoreceptor differentiation, with the Osakada protocol yielding somewhat better expression of photoreceptor markers. The Osakada protocol involved very long-term culture of the cells following induction over the next 90–120 days. For photoreceptor differentiation, the 90–120-day-old retinally specified cells were then treated with a combination of RA and Taurine for another 30 days. Therefore, it is likely that refined protocols will ultimately show even more promise in production of human rods and cones from ES cells. Moreover, the possibility that ES cells may soon be derived without the need for embryos;

Table 1. Published protocols on generation of retinal cells from embryonic stem cells

| Cell line | Method | Reference |
| --- | --- | --- |
| *Mouse embryonic stem cells* | | |
| D3 mouse ES cell line | RA or ITSFn+FGF-2 followed by co-culture with P1 rat retina | Zhao et al. (2002) |
| D3 mouse ES cell line | Co-culture with PA6 stromal cell line+FGF-2+Dexamethasone+Cholera toxin | Hirano et al. (2003) |
| EB5 mouse ES cell line | RA followed by transplantation into retinal degeneration mouse | Meyer et al. (2004) |
| CCE mouse ES cell line | Transfection by electroporation with Rx gene followed by RA or PA6 cells and retinal explant co-culture | Tabata et al. (2004) |
| EB3 mouse ES cell line | RA or ITSFn then FGF-2+Laminin followed by co-culture with E6 chicken retina | Sugie et al. (2005) |
| EB5 mouse ES cell line | Dkk1+LeftyA with addition of FCS and activin after 3 days for 3 days | Ikeda et al. (2005), Osakada et al. (2008) |
| CCE and D3 mouse ES cell lines | Co-culture with PA6 stromal cell line+FGF-2+Dexamethasone+Cholera toxin followed by Wnt2b and transplant into E2 chicken embryos | Aoki et al. (2006) |
| *Primate embryonic stem cells* | | |
| CMK6 and CMK9 monkey ES cell lines | Co-culture with PA6 stromal cells | Kawasaki et al. (2002) |
| Cynomolgus monkey ES cell line | FGF-2 followed by co-culture with PA6 stromal cells | Ooto et al. (2003) |
| Cynomolgus monkey ES cell line | Co-culture with PA6 stromal cells followed by transplantation in vivo | Haruta et al. (2004) |
| CMK6 and CMK9 monkey ES cell lines | Dkk-1 and Lefty-A for 18 days followed by RA and Taurine for atleast 30 days | Osakada et al. (2008) |
| *Human embryonic stem cells* | | |
| H1, H7, and H9 human ES cell lines | Overgrowth under adherent conditions | Klimanskaya et al. (2004) |
| HES-1 human ES cell line | Noggin followed by FGF-2+EGF & transplant into adult and newborn rat retinas | Banin et al. (2006) |
| H-1 and Hsf-6 human ES cell lines | Dkk-1, Noggin and IGF-1 for 3 days followed by dkk-1, noggin, IGF-1 and bFGF for 3 weeks | Lamba et al. (2006) |
| khES-1 and khES-3 human ES cell lines | Dkk-1 and Lefty-A for 20 days followed by adherent culture for 90-120 days and then RA and taurine for another 30 days | Osakada et al. (2008) |

*Note*: RA, retinoic acid; ITSFn, combination of insulin+transferrin+selenium+fibronectin; Dkk-1, Dickkopf; FGF-2, fibroblast growth factor 2; FCS, fetal calf serum; IGF-1, insulin-like growth factor-1.

ES cell-like cells, known as induced pluripotent stem cells (iPS cells), have been derived by overexpressing a group of key pluripotency genes in adult fibroblasts (Takahashi et al., 2007; Yu et al., 2007). The production of photoreceptors from iPS cells would resolve some of the ethical concerns associated with the use of ES cells and also help in creating nonimmunogenic cells for transplants.

## Concluding remarks

The research reviewed in this Chapter highlights the significant progress that has been made in understanding the fundamental cellular mechanisms of regeneration in nonmammalian vertebrates, and the significant limits to these processes in mammals. The next five years should lead to a great deal of progress in our understanding of the molecular mechanisms underlying these phenomena in fish and amphibians and potentially some insight as to why they fail in mammals. However, a parallel, but potentially more direct path to retinal repair may lie in the use of ES cell derived retinal cells for transplantation. In either case, it is clear that recent advances have provided new hope for eventual treatment for a host of blinding degenerative diseases.

# References

Ahmad, I., Tang, L., & Pham, H. (2000). Identification of neural progenitors in the adult mammalian eye. *Biochemical and Biophysical Research Communications*, 270, 517–521.

Akagi, T., Inoue, T., Miyoshi, G., Bessho, Y., Takahashi, M., Lee, J. E., et al. (2004). Requirement of multiple basic helix-loop-helix genes for retinal neuronal subtype specification. *Journal of Biological Chemistry*, 279, 28492–28498.

Anchan, R. M., Reh, T. A., Angello, J., Balliet, A., & Walker, M. (1991). EGF and TGF-alpha stimulate retinal neuroepithelial cell proliferation in vitro. *Neuron*, 6, 923–936.

Aoki, H., Hara, A., Nakagawa, S., Motohashi, T., Hirano, M., Takahashi, Y., et al. (2006). Embryonic stem cells that differentiate into RPE cell precursors in vitro develop into RPE cell monolayers in vivo. *Experimental Eye Research*, 82, 265–274.

Asami, M., Sun, G., Yamaguchi, M., & Kosaka, M. (2007). Multipotent cells from mammalian iris pigment epithelium. *Developmental Biology*, 304, 433–446.

Banin, E., Obolensky, A., Idelson, M., Hemo, I., Reinhardtz, E., Pikarsky, E., et al. (2006). Retinal incorporation and differentiation of neural precursors derived from human embryonic stem cells. *Stem Cells*, 24, 246–257.

Bernardos, R. L., Barthel, L. K., Meyers, J. R., & Raymond, P. A. (2007). Late-stage neuronal progenitors in the retina are radial Müller glia that function as retinal stem cells. *Journal of Neuroscience*, 27, 7028–7040.

Bringmann, A., Pannicke, T., Grosche, J., Francke, M., Wiedemann, P., Skatchkov, S. N., et al. (2006). Müller cells in the healthy and diseased retina. *Progress in Retinal and Eye Research*, 25, 397–424.

Bringmann, A., & Reichenbach, A. (2001). Role of Müller cells in retinal degenerations. *Frontiers of Bioscience*, 6, E72–E92.

Bringmann, A., Skatchkov, S. N., Pannicke, T., Biedermann, B., Wolburg, H., Orkand, R. K., et al. (2000). Müller glial cells in anuran retina. *Microscopic Research Technology*, 50, 384–393.

Chan, W. Y., Kohsaka, S., & Rezaie, P. (2007). The origin and cell lineage of microglia: New concepts. *Brain Research Reviews*, 53, 344–354.

Chang, M. L., Wu, C. H., Jiang-Shieh, Y. F., Shieh, J. Y., & Wen, C. Y. (2007). Reactive changes of retinal astrocytes and Müller glial cells in kainate-induced neuroexcitotoxicity. *Journal of Anatomy*, 210, 54–65.

Cicero, S. A., Johnson, D., Reyntjens, S., Frase, S., Connell, S., Chow, L. M., et al. (2009). Cells previously identified as retinal stem cells are pigmented ciliary epithelial cells. *Proceedings of the National Academy of Sciences, USA*, 106(16), 6685–6690 (Epub 2009 Apr 3).

Close, J. L., Liu, J., Gumuscu, B., & Reh, T. A. (2006). Epidermal growth factor receptor expression regulates proliferation in the postnatal rat retina. *Glia*, 54, 94–104.

Coles, B. L., Angenieux, B., Inoue, T., Del Rio-Tsonis, K., Spence, J. R., McInnes, R. R., et al. (2004). Facile isolation and the characterization of human retinal stem cells. *Proceedings of the National Academy of Science USA*, 101, 15772–15777.

Das, A. V., Mallya, K. B., Zhao, X., Ahmad, F., Bhattacharya, S., Thoreson, W. B., et al. (2006). Neural stem cell properties of Müller glia in the mammalian retina: regulation by Notch and Wnt signaling. *Developmental Biology*, 299, 283–302.

Dyer, M. A., & Cepko, C. L. (2000). Control of Müller glial cell proliferation and activation following retinal injury. *Nature Neuroscience*, 3, 873–880.

Eddleston, M., & Mucke, L. (1993). Molecular profile of reactive astrocytes — implications for their role in neurologic disease. *Neuroscience*, 54, 15–36.

Fischer, A. J., & Reh, T. A. (2001). Müller glia are a potential source of neural regeneration in the postnatal chicken retina. *Nature Neuroscience*, 4, 247–252.

Haruta, M., Sasai, Y., Kawasaki, H., Amemiya, K., Ooto, S., Kitada, M., et al. (2004). In vitro and in vivo characterization of pigment epithelial cells differentiated from primate embryonic stem cells. *Investigative Ophthalmology and Visual Science*, 45, 1020–1025.

Hayes, S., Nelson, B. R., Buckingham, B., & Reh, T. A. (2007). Notch signaling regulates regeneration in the avian retina. *Developmental Biology*, 312, 300–311.

Hirano, M., Yamamoto, A., Yoshimura, N., Tokunaga, T., Motohashi, T., Ishizaki, K., et al. (2003). Generation of structures formed by lens and retinal cells differentiating from embryonic stem cells. *Developmental Dynamics*, 228, 664–671.

Ikeda, H., Osakada, F., Watanabe, K., Mizuseki, K., Haraguchi, T., Miyoshi, H., et al. (2005). Generation of Rx+/Pax6+ neural retinal precursors from embryonic stem cells. *Proceedings of National Academy of Science USA*, 102, 11331–11336.

Kawasaki, H., Suemori, H., Mizuseki, K., Watanabe, K., Urano, F., Ichinose, H., et al. (2002). Generation of dopaminergic neurons and pigmented epithelia from primate ES cells by stromal cell-derived inducing activity. *Proceedings of National Academy of Science USA*, 99, 1580–1585.

Kelley, M. W., Turner, J. K., & Reh, T. A. (1995). Regulation of proliferation and photoreceptor differentiation in fetal human retinal cell cultures. *Investigative Ophthalmology and Visual Science*, 36, 1280–1289.

Klimanskaya, I., Hipp, J., Rezai, K. A., West, M., Atala, A., & Lanza, R. (2004). Derivation and comparative assessment of retinal pigment epithelium from human embryonic stem cells using transcriptomics. *Cloning Stem Cells*, 6, 217–245.

Kohno, R., Ikeda, Y., Yonemitsu, Y., Hisatomi, T., Yamaguchi, M., Miyazaki, M., et al. (2006). Sphere formation of ocular epithelial cells in the ciliary body is a reprogramming system for neural differentiation. *Brain Research*, 1093, 54–70.

Kubota, R., Hokoc, J. N., Moshiri, A., McGuire, C., & Reh, T. A. (2002). A comparative study of neurogenesis in the retinal ciliary marginal zone of homeothermic vertebrates. *Brain Research and Developments in Brain Research*, 134, 31–41.

Lamba, D. A., Karl, M. O., Ware, C. B., & Reh, T. A. (2006). Efficient generation of retinal progenitor cells from human embryonic stem cells. *Proceedings of National Academy of Science USA*, 103, 12769–12774.

MacNeil, A., Pearson, R. A., MacLaren, R. E., Smith, A. J., Sowden, J. C., & Ali, R. R. (2007). Comparative analysis of progenitor cells isolated from the iris, pars plana, and ciliary body of the adult porcine eye. *Stem Cells, 25*, 2430–2438.

Meyer, J. S., Katz, M. L., Maruniak, J. A., & Kirk, M. D. (2004). Neural differentiation of mouse embryonic stem cells in vitro and after transplantation into eyes of mutant mice with rapid retinal degeneration. *Brain Research, 1014*, 131–144.

Moshiri, A., Close, J., & Reh, T. A. (2004). Retinal stem cells and regeneration. *International Journal of Developmental Biology, 48*, 1003–1014.

Moshiri, A., & Reh, T. A. (2004). Persistent progenitors at the retinal margin of ptc+/− mice. *Journal of Neuroscience, 24*, 229–237.

Nelson, B. R., Hartman, B. H., Georgi, S. A., Lan, M. S., & Reh, T. A. (2007). Transient inactivation of Notch signaling synchronizes differentiation of neural progenitor cells. *Developmental Biology, 304*, 479–498.

Newman, E., & Reichenbach, A. (1996). The Müller cell: A functional element of the retina. *Trends in Neurosciences, 19*, 307–312.

Ooto, S., Akagi, T., Kageyama, R., Akita, J., Mandai, M., Honda, Y., et al. (2004). Potential for neural regeneration after neurotoxic injury in the adult mammalian retina. *Proceedings of National Academy of Science USA, 101*, 13654–13659.

Ooto, S., Haruta, M., Honda, Y., Kawasaki, H., Sasai, Y., & Takahashi, M. (2003). Induction of the differentiation of lentoids from primate embryonic stem cells. *Investigative Ophthalmology and Visual Science, 44*, 2689–2693.

Osakada, F., Ikeda, H., Mandai, M., Wataya, T., Watanabe, K., Yoshimura, N., et al. (2008). Toward the generation of rod and cone photoreceptors from mouse, monkey and human embryonic stem cells. *Nature Biotechnology, 26*, 215–224.

Osakada, F., Ooto, S., Akagi, T., Mandai, M., Akaike, A., & Takahashi, M. (2007). Wnt signaling promotes regeneration in the retina of adult mammals. *Journal of Neuroscience, 27*, 4210–4219.

Raymond, P. A., Barthel, L. K., Bernardos, R. L., & Perkowski, J. J. (2006). Molecular characterization of retinal stem cells and their niches in adult zebrafish. *BMC Developmental Biology, 6*, 36.

Reynolds, B. A., Tetzlaff, W., & Weiss, S. (1992). A multipotent EGF-responsive striatal embryonic progenitor cell produces neurons and astrocytes. *Journal of Neuroscience, 12*, 4565–4574.

Ridet, J. L., Malhotra, S. K., Privat, A., & Gage, F. H. (1997). Reactive astrocytes: Cellular and molecular cues to biological function. *Trends in Neuroscience, 20*, 570–577.

Sahel, J. A., Albert, D. M., & Lessell, S. (1990). Proliferation of retinal glia and excitatory amino acids. *Ophtalmologie, 4*, 13–16.

Sidman, R. L. (1961). Histogenesis of the mouse retina. Studies with [3H] thymidine. In *The structure of the eye*. New York: Academic Press.

Stone, J., & Dreher, Z. (1987). Relationship between astrocytes, ganglion cells and vasculature of the retina. *Journal of Computational Neurology, 255*, 35–49.

Sugie, Y., Yoshikawa, M., Ouji, Y., Saito, K., Moriya, K., Ishizaka, S., et al. (2005). Photoreceptor cells from mouse ES cells by co-culture with chick embryonic retina. *Biochemical and Biophysical Research Communications, 332*, 241–247.

Sun, G., Asami, M., Ohta, H., Kosaka, J., & Kosaka, M. (2006). Retinal stem/progenitor properties of iris pigment epithelial cells. *Developmental Biology, 289*, 243–252.

Tabata, Y., Ouchi, Y., Kamiya, H., Manabe, T., Arai, K., & Watanabe, S. (2004). Specification of the retinal fate of mouse embryonic stem cells by ectopic expression of Rx/rax, a homeobox gene. *Molecular and Cellular Biology, 24*, 4513–4521.

Takahashi, K., Tanabe, K., Ohnuki, M., Narita, M., Ichisaka, T., Tomoda, K., et al. (2007). Induction of pluripotent stem cells from adult human fibroblasts by defined factors. *Cell, 131*, 861–872.

Takahashi, M., Palmer, T. D., Takahashi, J., & Gage, F. H. (1998). Widespread integration and survival of adult-derived neural progenitor cells in the developing optic retina. *Molecular and Cellular Neuroscience, 12*, 340–348.

Tropepe, V., Coles, B. L., Chiasson, B. J., Horsford, D. J., Elia, A. J., McInnes, R. R., et al. (2000). Retinal stem cells in the adult mammalian eye. *Science, 287*, 2032–2036.

Turner, J. E., & Glaze, K. A. (1977). Regenerative repair in the severed optic nerve of the newt (*Triturus viridescens*): Effect of nerve growth factor. *Experimental Neurology, 57*, 687–697.

Wan, J., Zheng, H., Chen, Z. L., Xiao, H. L., Shen, Z. J., & Zhou, G. M. (2008). Preferential regeneration of photoreceptor from Müller glia after retinal degeneration in adult rat. *Vision Research, 48*(2), 223–234.

Wan, J., Zheng, H., Xiao, H. L., She, Z. J., & Zhou, G. M. (2007). Sonic hedgehog promotes stem-cell potential of Müller glia in the mammalian retina. *Biochemical and Biophysical Research Communications, 363*, 347–354.

Watanabe, T., & Raff, M. C. (1988). Retinal astrocytes are immigrants from the optic nerve. *Nature, 332*, 834–837.

Young, M. J., Ray, J., Whiteley, S. J., Klassen, H., & Gage, F. H. (2000). Neuronal differentiation and morphological integration of hippocampal progenitor cells transplanted to the retina of immature and mature dystrophic rats. *Molecular and Cellular Neuroscience, 16*, 197–205.

Yu, J., Vodyanik, M. A., Smuga-Otto, K., Antosiewicz-Bourget, J., Frane, J. L., Tian, S., et al. (2007). Induced pluripotent stem cell lines derived from human somatic cells. *Science, 318*, 1917–1920.

Zhao, X., Das, A. V., Soto-Leon, F., & Ahmad, I. (2005). Growth factor-responsive progenitors in the postnatal mammalian retina. *Developmental Dynamics, 232*, 349–358.

Zhao, X., Liu, J., & Ahmad, I. (2002). Differentiation of embryonic stem cells into retinal neurons. *Biochemical and Biophysical Research Communications, 297*, 177–184.

Zuber, M. E., Gestri, G., Viczian, A. S., Barsacchi, G., & Harris, W. A. (2003). Specification of the vertebrate eye by a network of eye field transcription factors. *Development, 130*, 5155–5167.

CHAPTER 3

# The rostral migratory stream and olfactory system: smell, disease and slippery cells

Maurice A. Curtis[1,2,*], Hector J. Monzo[1] and Richard L.M. Faull[1]

[1]*Department of Anatomy with Radiology, Faculty of Medical and Health Sciences, University of Auckland, Auckland, New Zealand*
[2]*Institute of Neuroscience and Physiology at Sahlgrenska Academy, Göteborg, Sweden*

**Abstract:** In the mammalian brain, olfaction is an important sense that is used to detect odors of different kinds that can warn of off food, to produce a mothering instinct in a flock or group of animals, and to warn of danger such as fire or poison. The olfactory system is made up of a long-distance rostral migratory stream that arises from the subventricular zone in the wall of the lateral ventricle, mainly comprises neuroblasts, and stretches all the way through the basal forebrain to terminate in the olfactory bulb. The olfactory bulb receives a constant supply of new neurons that allow ongoing integration of new and different smells, and these are integrated into either the granule cell layer or the periglomerular layer. The continuous turnover of neurons in the olfactory bulb allows us to study the proliferation, migration, and differentiation of neurons and their application in therapies for neurodegenerative diseases. In this chapter, we will examine the notion that the olfactory system might be the route of entry for factors that cause or contribute to neurodegeneration in the central nervous system. We will also discuss the enzymes that may be involved in the addition of polysialic acid to neural cell adhesion molecule, which is vital for allowing the neuroblasts to move through the rostral migratory stream. Finally, we will discuss a possible role of endosialidases for removing polysialic acid from neural cell adhesion molecules, which causes neuroblasts to stop migrating and terminally differentiate into olfactory bulb interneurons

**Keywords:** rostral migratory stream; neuroblast; PSA-NCAM; progenitor migration

## The purpose and importance of olfaction

*To which organic sense do we owe the least and which appears to be the most dispensible? The sense of smell (Kant, 1978).*

*In one study of 750 consecutive patients with anosmia 68 per cent thought that a disordered sense of smell affected their quality of life and 56 per cent felt it altered their daily living (Deems et al., 1991; Jones and Rog, 1998).*

*Nothing is more memorable than a smell. One scent can be unexpected, momentary and fleeting, yet conjure up a childhood summer by a lake in the*

*Corresponding author.
Tel.: +64 9 3737599; Fax: +64 9 3737484;
E-mail: m.curtis@auckland.ac.nz

*mountains (Diane Ackerman: A Natural History of the Senses (Book))*.

The olfactory system is that part of the brain used for sensing smells. In most mammalian species, the olfactory system is divided into two discrete components called the main olfactory and accessory olfactory systems. The main olfactory system detects fluid signals, and the accessory olfactory system usually detects airborne and volatile stimuli in the form of pheromones. However, the focus of this chapter is on the main olfactory system, so the accessory system will not be considered further in this chapter.

The importance of olfaction in mammals is often thought minimal when considered in light of the small size of the olfactory system in humans. However, in other mammals, their existence relies on their ability to navigate, find appropriate mating partners, live nocturnally, and search for food, which requires an acute sense of smell. In the rodent brain, the size of the olfactory bulb (OB) reflects the importance of this sense and makes up approximately 20% of the weight of the brain (not including the olfactory tract or anterior olfactory cortex, AOC) as compared with humans where the OB is approximately 0.064% of the total brain weight (Curtis et al., 2007b). Humans use olfaction more than is readily attributed to them. Various experiments have revealed that when humans are asked to choose a potential mate based on the scent of the opposite sex, they are most likely to choose partners whose major histone compatibility components are complementary and therefore would likely produce offspring with a strong immune system (Kohl et al., 2001; Wedekind and Penn, 2000). Of course in humans, when looking for a partner, one's decision is usually based on factors other than just odor, but this may go part way to explaining attraction and certainly contributes since humans would seldom proceed in getting to know someone whose smell repulsed them. In addition, in one study of 750 consecutive patients with anosmia (the absence of the sense of smell), 68% thought that a disordered sense of smell affected their quality of life and 56% felt it altered their daily living (Deems et al., 1991; Jones and Rog, 1998).

One of the interesting hypotheses of recent times has been the olfactory vector hypothesis, in which it is proposed that neurodegenerative diseases such as Alzheimer's (AD) and Parkinson's disease (PD) arise from xenobiotic organisms that enter the brain via the nasal mucosa and then olfactory receptor neurons (Doty, 2003). The wealth of support for this hypothesis comes from the fact that in both of these diseases, and many others, 90% of cases are preempted by two to four years of anosmia before the motor or somatosensory symptoms are apparent in the individuals (Doty, 1991, 1994, 2001, 2003). According to postmortem studies, the earliest signs of AD and PD are seen in the olfactory bulb and tract, and they also affect the AOC and subsequently other mid- and hindbrain structures, suggesting that olfactory dysfunction may be a useful preclinical marker of ensuing neurodegeneration (Braak et al., 2003; Kovacs et al., 1999, 2001; Ohm and Braak, 1987). In one study of 1604 older nondemented people that tested positive for the apolipoprotein E4 gene and that were anosmic, had an odds ratio of 9.71 for developing cognitive decline within the following two years, whereas those with no olfactory dysfunction had an odds ratio of 1.91, indicating that anosmia is an accurate predictor of cognitive decline (Graves et al., 1999).

There are also many viruses that have a known route of entry to the brain via the nasal mucosa but have considerably less capacity to enter via other routes such as ingestion or inhalation. Poliovirus is one such virus that easily passes into the brain via nasal mucosa because there is an ineffective blood-brain barrier, and the olfactory receptor neurons readily pass the virus across the synapse via mass endocytosis and into glomeruli and then to the mitral cells that are transduced by the virus to the basal forebrain. This method of entry of the poliovirus was known in the 1930s, leading Canadian health authorities to cauterize the nasal passages of thousands of school-age children in the hope that this would eliminate this route of entry for the virus that was epidemic at the time (for review, see Doty, 2008). Whether there is a xenobiotic agent causing AD and PD remains to be proven, but there is considerable weight to the olfactory vector hypothesis, although concrete evidence remains to be determined one way or the other.

In summary, the olfactory system in humans and rodents is important for social interactions and food finding, and an impaired function may even be a useful biomarker for ensuing neurodegeneration. So then what are the basic components of the olfactory system in the mammalian brain?

**The rostral migratory stream (RMS) pathway: common anatomical features in vertebrates**

During development, neurons migrate from their place of birth in the subventricular zone (SVZ), differentiating en route and at their final destination where they become incorporated in to the functional circuitry of the forebrain. Rodent studies reveal that in the adult brain, neurogenesis continues in the SVZ throughout adult life from which cell replacement occurs, particularly in the OB (Altman, 1969; Benraiss et al., 2001; Lois et al., 1996; Luskin and Boone, 1994). The fate of the replacement cells is at least partly determined in the SVZ. Our studies (Curtis et al., 2003b, 2005a, b, 2006) and others' (Quinones-Hinojosa et al., 2006) on the biology of the SVZ in the normal and diseased human brain demonstrate that, as in the rodent, the SVZ retains a neurogenic potential in the adult human brain.

*The rodent RMS*

That the rodent RMS is organized around a tubular extension of the lateral ventricle reaching the OB would not seem strange if it was clearly this way in all species. In fact, the general organization of the RMS is a continuous tube in most mammalian species; however, in primates, the lumen of the tube becomes very narrow. In addition, the RMS takes a sigmoidal path rather than a direct rostral route. Many species possess an open tube between the lateral ventricle and the olfactory ventricle, and this allows the free flow of cerebrospinal fluid (CSF) as well as a continuous SVZ between the two regions (http://www.brainmuseum.org/Specimens/lagomorpha/domestic rabbit/sections/thumbnail.html; Rae, 1994). The critical features of the RMS in the rodent brain are that the cells within it form chains, which enables a tangential method of migration in which radial fibers are not needed, and these cells migrate through a glial tube network that offers the migrating cells cues about the direction and orientation (Lois and Alvarez-Buylla, 1994; Lois et al., 1996). The major difference between the SVZ in the mouse brain and the RMS in rodents is the ratio of glial cells to neuroblasts, which is approximately 3:2 and 4:1, respectively, and thus the SVZ is maintained as a niche for proliferation and the RMS for the migration of neuroblasts (Doetsch et al., 1997). The general anatomy of the rodent RMS is that from the rostrodorsal aspect of the striatum, the SVZ of the lateral ventricle gives rise to the RMS that proceeds first ventrally along the rostroventral aspect of the striatum and then courses rostrally into the olfactory tract, and the tract terminates in the granule cell layer of the OB (Lois and Alvarez-Buylla, 1994; Lois et al., 1996).

*The human RMS*

Recent studies in our laboratory have revealed that the human brain also has an RMS that is similar in its basic principles to that of the rodent; in particular, (1) the RMS contains numerous migratory progenitor cells, (2) it takes a route similar to that of rodents, and (3) the migratory progenitor cells differentiate in the OB to form neurons (Curtis et al., 2007c).

In the human brain, the RMS takes a course rostroventrally in relation to the striatum, and then the cells migrate forward in the olfactory tract to the OB. The human brain follows the basic structural organization of the mammalian brain; however, compared to the rodent, the human forebrain is extensively developed. Furthermore, the human OB, and hence the olfactory interneuron replacement system, is comparatively smaller than in rodents; thus the RMS has, until recently, remained illusive in the human brain.

The RMS begins as a cleft in the floor of the anterior horn of the lateral ventricle where large numbers of PCNA (proliferating cell nuclear antigen)-positive cells are present within the SVZ that overlies the caudate nucleus (CN).

The stream of cells first takes a caudal and ventral track along the undersurface of the CN, caudal to the genu of the corpus callosum and the frontal cortical white matter of the gyrus rectus (Fig. 1). When the descending limb reaches the ventral CN, the stream takes a rostral turn to form the "rostral limb" of the RMS; the rostral limb passes ventrally and rostrally to enter the AOC, and traverses the olfactory tract to enter the OB (Fig. 1). Near the SVZ, the mediolateral extent that contains proliferating cells covers 2.1 mm and forms the beginning of the "descending limb"; at the level of the rostral limb, proliferating cells only extend mediolaterally over approximatcly 0.6 mm. Thus, the overall mediolateral extent of the RMS from the SVZ to the olfactory tract is 2.7 mm, and the total length of the RMS pathway from the SVZ to the start of the olfactory tract is approximately 17 mm.

The RMS takes a path from medial, by the SVZ, to lateral, beneath the CN; close to the SVZ, the RMS is very wide. Thus, overall, the RMS is a funnel-shaped structure that is widest in the mediolateral extent at the SVZ and becomes narrower at the rostral limb.

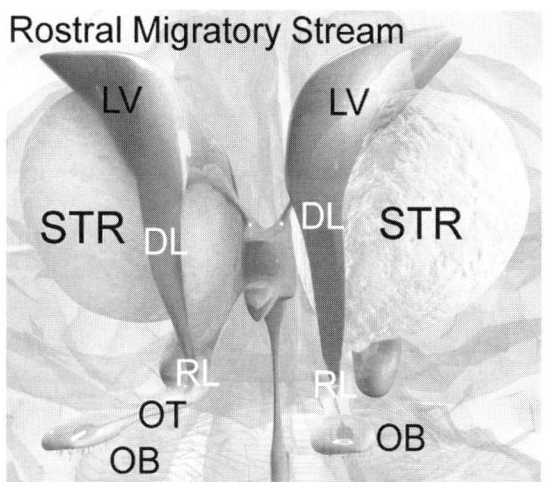

Fig. 1. The rostral migratory stream (RMS) in humans arises from a cleft in the floor of the lateral ventricles. The descending limb (DL) proceeds rostral to the striatum (STR) until it reaches the anterior olfactory cortex where the RMS turns rostrally to become the rostral limb (RL) then enters the olfactory tract (OT) and olfactory bulb (OB). The RMS is an extension of the lateral ventricle.

For the olfactory system, the underpinning driver of new cell production in the SVZ, their long-distance migration, and eventual maturation into granule cells is the loss of the granule cell population in the OB. However, in general, the OB must be intact for replacement to take place. In experiments where the OB was removed from the rodent brain, the SVZ reduced its production of progenitor cells over time (Kirschenbaum et al., 1999). When the OB tissue was replaced with "like" tissue, the SVZ upregulated progenitor cell production, whereas when the OB region was replaced with cortical tissue, the SVZ remained inactive. The replacement of olfactory interneurons in a lamina-specific way in the human brain after long-distance migration indicates the precise signaling in the ventriculo-olfactory neurogenic system (VONS) in the human brain (Curtis et al., 2007c; Kirschenbaum et al., 1999).

**Neuroblast migration in the RMS**

In order for olfactory neurogenesis to take place in the brain, long-distance migration of neuroblasts must occur from the SVZ surrounding the lateral ventricle to the OB via the RMS. However, progenitor cells that differentiate to form neurons and replace granule cells in the OB have been thought to arise locally within the OB in the human brain (Bedard and Parent, 2004). Our studies have demonstrated immunostaining for polysialylated neural cell adhesion molecule (PSA-NCAM), which is expressed by migrating cells in the rodent RMS (Bedard et al., 2002; Bonfanti, 2006), is present in abundance in the rodent and human SVZ, descending and rostral limbs of the RMS, the olfactory tract core and periphery, and the OB in the human brain (Curtis et al., 2007c). The orientation of the neuroblasts is in the long axis of the RMS and seems to be abundantly expressed in all the components of the RMS (Curtis et al., 2007c). However, the rodent brain differs from the human brain in that chains of cells use signals from one another and also from the surrounding glial cells for directional information and by migrating in a chainlike tangential manner, so there is no need for radial glia or radial fibers (Kornack and Rakic,

2001; Lois et al., 1996; Sanai et al., 2004). In the human RMS, neuroblasts seem to migrate in groups of up to six cells; the neuroblasts have a prominent leading process and a trailing process that is in contact with the leading process of the following cell.

In rodent brains, connexin 26 (cx26) and connexin 43 (cx43) are essential for the migration of neuroblasts in the cortex, and in particular, cx26 is responsible for signaling that orientates the leading process of the migrating cell and cx43 is responsible for signaling the intracellular machinery to engage somal translocation (Elias et al., 2007). Both forms of signaling are required for appropriate coordinated migration of neuroblasts toward the OB. In the human brain, it is not known if the same signaling occurs since the cells migrate in groups rather than long chains per se; however, there is an abundance of glial cells that form a network or mesh appearance throughout the RMS. Thus, since the cellular architecture is similar, it is likely that the same mechanism is present. The tangential mode of migration is unique from radial migration in that radial migration requires signaling from reelin (usually produced by Cajal–Retzius cells during development) that signals through its receptor VLDLR, but since radial fibers are absent during tangential migration, reelin is often not expressed and migrating cells require other signaling pathways for their cue to stop migrating. In adult brains, the start and stop signals are poorly understood.

## Polysialylated neural cell adhesion molecule: making the neuroblast slippery

Temporal and spatial regulation of cellular interactions is of paramount importance for the development and plasticity of the central nervous system (CNS). Several groups of cell adhesion molecules (CAMs) belonging to integrin, cadherin, immunoglobulin, and semaphorin superfamilies are involved in CNS morphogenesis and plasticity (for review, see Hortsch, 2003; Kruger et al., 2005); among these, the NCAM, a cell surface member of the immunoglobulin superfamily molecules, is widely expressed in the nervous system, and displays a tight association with the physiological and anatomical plasticity of the CNS.

Newborn neurons in the adult SVZ migrate a very long distance through the RMS, differentiating en route and at their final destination, the OB, where they are incorporated into the functional circuitry of the OB as granule neurons and periglomerular neurons (Carleton et al., 2003). Developmental stages in the course of adult neurogenesis such as neural stem cell self-renewal, progenitor cell-fate specification, and migration, differentiation, and survival/death of newborn neurons are spatially restricted to specific and permissive environments, and essentially directed by cellular interactions (Doetsch, 2003). Noteworthy effort has been made to the discovery of molecules mediating interactions between stem cells, immature neurons, and the local environment (for review, see Ninkovic and Gotz, 2007). To this end, isoforms of NCAM involved in direct cis/trans cell–cell and cell–extracellular matrix interactions as well as cytoplasmic signaling have emerged as key players in adult neurogenesis (for review, see Hinsby et al., 2004; Edelman, 1986; Rutishauser et al., 1982; Rutishauser and Landmesser, 1996).

The implication of NCAM in adult neurogenesis was driven by the finding of long chains of the alpha-2,8-linked polymer of sialic acid (polysialic acid, PSA) attached to the adhesion molecule, and the fact that this attachment resulted in a significant attenuation of adhesion forces and therefore reduction in the overall NCAM-mediated cell surface interactions (Johnson et al., 2005; Rutishauser and Landmesser, 1996). PSA exerts its function on the basis of being a spatially large molecule when hydrated and an unstable conformation that limits its ability to bind to a protein; both characteristics cause this molecule to occupy large spaces between cells, therefore impairing membrane–membrane contacts while lowering binding interactions with other cell surface molecules (i.e., PSA insulates the cells and imbues them with a "slippery" attribute) (for review, see Rutishauser, 2008).

The expression of NCAM and PSA-NCAM follows a characteristic pattern in the postnatal neurogenic niches. The adult mammal SVZ contains, among others, type A cells (identified as committed neuroblasts), type C cells (identified as transit-amplifying cells), and type B cells (identified as primary progenitors of new neurons) (for review, see Curtis et al., 2007a); whereas PSA attached to the NCAM is not detected on type B or C cells, type A cells display a strong signal when stained for this adhesion molecule (for review, see Ming and Song, 2005). A similar scenario is present in the SGZ, where radial astrocytes (identified as primary neural progenitor cells) do not show staining against PSA-NCAM, but D1 type cells (identified as transit-amplifying cells), and more committed precursor cells derived from D1 (types D2 and D3 cells), are positively immunolabeled for PSA-NCAM (Seri et al., 2004). However, available data indicate that the presence of PSA-NCAM on migratory neuroblasts is gradually downregulated as newborn neurons reach their final destination and dendritic growth and synaptic interaction steps are being completed (Rousselot et al., 1995). All together, these data establish PSA-NCAM as a key molecule for newborn neuroblasts to exit the neurogenic niche and migrate toward their final destination. Due to its insulative properties, the presence of PSA-NCAM on newborn neuroblasts appears to act by pausing their maturation process while migrating until they reach their final destination by avoiding any interaction with other cell types or the extracellular matrix.

Whereas most carbohydrates are attached to a variety of proteins, NCAMs are the preferred substrate for polysialylation. The polysialylation of NCAM takes place at the Golgi apparatus in a specific way: sialic acids are synthesized in the cytosol from UDP-*N*-acetylglucosamine in a four-step reaction catalyzed by the enzyme UDP-*N*-actylglucosamine 2-epimerase/*N*-acetylmannosamine kinase. Numerous activated sialic acid molecules are then consecutively linked to an asparaginyl-linked core carbohydrate and attached to the NCAM in a reaction mediated by two Golgi polysialyltransferase enzymes: PST (also known as ST8SiaIV) and STX (also known as ST8SiaII). Whereas addition of sialic acid monomers to the growing PSA polymer is actively mediated by the transferases, termination of the process seems to be related to a progressive diminution of the affinity of the transferases for PSA-NCAM as the polymer gets longer (for extended review, see Angata and Fukuda, 2003). Functional studies of PSA were greatly enhanced by the discovery and characterization of a phage-derived PSA-specific endoneuroaminidase (glycohydrolytic enzyme) that specifically degrades PSA (Vimr et al., 1984). Since then, several neuroaminidases (also known as sialidases) have been described in viruses, bacteria, protozoa, and vertebrate organisms, including four types in humans classified according to their subcellular location: NEU1 (intralysosomal), NEU2 (cytosolic), NEU3 (plasma membrane), NEU4 (mitochondrial membrane), and a recently discovered NEU4L (isoform of NEU4) specifically expressed in brain mitochondria (Hasegawa et al., 2007; Magesh et al., 2006). PSA-NCAMs have been postulated to act at different stages in the developing nervous system, including the promotion of neuronal progenitor migration toward their final destination (presence of PSA-NCAM) and to shift cells from migration to differentiation (cleavage of PSA from NCAMs). Other roles proposed to be directly related with the PSA-NCAM are the avoidance of premature synaptic establishments during axon projections (e.g., temporal insulation effect on growing axons until reaching the appropriate destination), correct timing for the synthesis of the neurotransmitter acetylcholine (e.g., in septal neurons, PSA blocks the access to brain-derived neurotrophic factor to the choline acetyltransferase enzyme, therefore avoiding the synthesis of the neurotransmitter), plasticity on input systems (e.g., establishment/dissolution of synaptic circuitries by PSA in order to allow the brain to appropriately process the inputs), and axon myelination (e.g., neural cell myelination has been proposed to be sensitive to PSA presence, and supporting evidence shows that prevention of PSA downregulation in growing axons and oligodendrocytes reduces normal myelination) (for review, see Rutishauser, 2008).

Finally, PSA-NCAM molecules are present within brain regions involved in learning and memory, such as the hippocampus and the neocortex.

The presence of PSA in neurons already integrated in adult circuitry may reflect a preestablished disposition in these regions for circuit plasticity and subsequent correlation with synaptic strengthening, adaptation to different environmental circumstances, memory acquisition, and memory consolidation (Becker et al., 1996). In addition, the level and distribution of PSA-NCAM have been shown to be altered in correspondence with neuropathological conditions in the adult mammal CNS. The number of PSA-NCAM-positive cells increases in the adult rodent SVZ after cortical lesions, and experimental evidences suggest that newly generated PSA-positive cells are mobilized from the neurogenic niche toward the lesion site within a few weeks after the insult (for review, see Bonfanti, 2006; Curtis et al., 2007b; Rutishauser, 2008).

To summarize, the importance of migratory molecules that enable the appropriate speed, direction, and timeliness of migration are critical. If the RMS neuroblasts overshoot their target or they differentiate in the RMS then the OB is not appropriately repopulated with new neurons. Therefore, tightly regulated migratory cues are required to ensure an ordered long-distance migratory system.

## The fate determinants for neuroblasts arriving at the olfactory bulb

When the olfactory-bound neuroblasts arrive at the OB, they have to become functional neurons, and their fate in the OB is determined either along their migratory route or once they get to the OB. It appears that a combination of these two is important for accurate placement of the neurons. First, there is the decision whether to become granule cells in the granule cell layer or whether to become periglomerular neurons. The granule cell layer accepts approximately 95% of the migrating neuroblasts, and the other 5% migrate further to the periglomerular layer and become GABAergic and dopaminergic neurons, respectively. En route to the OB, the stemness of neuroblasts is maintained by transcription factors such as olig2 that is normally involved in specifying an oligodendrocyte phenotype in precursor cells, but in the olfactory tract, it is abundant and likely acts to maintain cells in an undifferentiated state (Curtis et al., 2007c; Hack et al., 2005). However, once neuroblasts reach the OB, there is virtually no expression of olig2 and the level of Pax6 is upregulated. Studies have shown the importance of Pax6 as a specification cue for migrating progenitor cells to become periglomerular and granular interneurons upon reaching the OB (Hack et al., 2005). More specifically, Pax6 is one of the paired box gene family and is important in the developing and adult brain (Stoykova and Gruss, 1994). It is also one of the key factors for CNS patterning, and plays an essential role in the differentiation of cortical radial glia during development (Goetz, 1998; Simpson and Price, 2002). Hack et al. (2005) demonstrated in their studies that even overexpression of olig2 in the olfactory tract did not lead to a change in the production of oligodendrocytes, indicating that the specification of the cells to a neuronal fate had, at least in part, already taken place in the SVZ. It is for the transcription factors in the OB to specify periglomerular or granule cell neurons.

## Could the RMS be exploited for treating brain disorders?

The pathologies of Huntington's disease (HD), AD, and PD are evidenced by a loss of neurons in specific brain regions. In HD, the SVZ responds to the nearby striatal degeneration by upregulating the production of progenitor cells. In AD, there is a loss of cortical cholinergic cells, and the amyloid-beta protein is toxic to the SVZ progenitor cells. In PD, there is also a decrease in progenitor cell proliferation due to dopaminergic denervation of the SVZ progenitors by the substantia nigra. While the following is speculative, it would follow the theory that in AD and PD, the number of newborn replacement cells might not improve the numbers of cholinergic cortical neurons or newly born dopaminergic neurons if the disease itself is so pathological that the cells cannot survive in a stem cell-like state, which is the case in AD and in PD where they lack stimulation to divide (Höglinger et al., 2004; Ziabreva et al., 2006). Thus, it must be speculated that alleviation of the initial pathological events is required before redirecting neuroblasts from the RMS; it is for the science of the future to determine if this is in fact true. For HD, in theory, the possibilities seem less limited because the newborn cells are spared the accumulation of toxic substances, and there is a generalized increase in stem cell proliferation in the SVZ (Curtis et al., 2003a, b, 2005a, b, 2006). What happens to the RMS in HD remains a point of investigation. However, what is clear from the study of normal RMS in humans is that there are small pathways that appear to trail off from the main RMS motif. Where those trails lead to is also up for speculation, and how to improve the traffic over them will keep scientists challenged for some time to come.

## References

Ackerman, D. (1991). *A natural history of the senses*. Knopf Doubleday Publishing Group (ISBN-13: 9780679735663). New York, NY.

Altman, J. (1969). Autoradiographic and histological studies of postnatal neurogenesis. IV. Cell proliferation and migration in the anterior forebrain, with special reference to persisting neurogenesis in the olfactory bulb. *The Journal of Comparative Neurology*, *137*, 433–457.

Angata, K., & Fukuda, M. (2003). Polysialyltransferases: major players in polysialic acid synthesis on the neural cell adhesion molecule. *Biochimie*, *85*, 195–206.

Becker, C. G., Artola, A., Gerardy-Schahn, R., Becker, T., Welzl, H., & Schachner, M. (1996). The polysialic acid modification of the neural cell adhesion molecule is involved in spatial learning and hippocampal long-term potentiation. *Journal of Neuroscience Research*, *45*(2), 143–152.

Bedard, A., Levesque, M., Bernier, P. J., & Parent, A. (2002). The rostral migratory stream in adult squirrel monkeys: contribution of new neurons to the olfactory tubercle and involvement of the antiapoptotic protein Bcl-2. *The European Journal of Neuroscience*, *16*, 1917–1924.

Bedard, A., & Parent, A. (2004). Evidence of newly generated neurons in the human olfactory bulb. *Brain Research. Developmental Brain Research*, *151*, 159–168.

Benraiss, A., Chmielnicki, E., Lerner, K., Roh, D., & Goldman, S. A. (2001). Adenoviral brain-derived neurotrophic factor induces both neostriatal and olfactory neuronal recruitment from endogenous progenitor cells in the adult forebrain. *The Journal of Neuroscience*, *21*, 6718–6731.

Bonfanti, L. (2006). PSA-NCAM in mammalian structural plasticity and neurogenesis. *Progress in Neurobiology*, *80*, 129–164.

Braak, H., Del Tredici, K., Rub, U., de Vos, R. A., Jansen Steur, E. N., & Braak, E. (2003). Staging of brain pathology related to sporadic Parkinson's disease. *Neurobiology of Aging*, *24*, 197–211.

Carleton, A., Petreanu, L. T., Lansford, R., Alvarez-Buylla, A., & Lledo, P. M. (2003). Becoming a new neuron in the adult olfactory bulb. *Nature Neuroscience*, *6*, 507–518.

Curtis, M. A., Connor, B., & Faull, R. L. M. (2003a). Neurogenesis in the diseased adult human brain: new therapeutic strategies for neurodegenerative diseases. *Cell Cycle*, *2*, 428–430.

Curtis, M. A., Eriksson, P. S., & Faull, R. L. (2007a). Progenitor cells and adult neurogenesis in neurodegenerative diseases and injuries of the basal ganglia. *Clinical and Experimental Pharmacology and Physiology*, *34*(5–6), 528–532.

Curtis, M. A., Faull, R. L., & Eriksson, P. S. (2007b). The effect of neurodegenerative diseases on the subventricular zone. *Nature Reviews. Neuroscience*, *8*, 712–723.

Curtis, M. A., Faull, R. L., & Glass, M. (2006). A novel population of progenitor cells expressing cannabinoid receptors in the subependymal layer of the adult normal and Huntington's disease human brain. *Journal of Chemical Neuroanatomy*, *31*, 210–215.

Curtis, M. A., Kam, M., Nannmark, U., Anderson, M. F., Axell, M. Z., Wikkelso, C., et al. (2007c). Human neuroblasts migrate to the olfactory bulb via a lateral ventricular extension. *Science*, *315*, 1243–1249.

Curtis, M. A., Penney, E. B., Pearson, A. G., van Roon-Mom, W. M., Butterworth, N. J., Dragunow, M., et al. (2003b). Increased cell proliferation and neurogenesis in the adult human Huntington's disease brain. *Proceedings of the National Academy of Sciences of the United States of America*, *100*, 9023–9027.

Curtis, M. A., Penney, E. B., Pearson, J., Dragunow, M., Connor, B., & Faull, R. L. (2005a). The distribution of progenitor cells in the subependymal layer of the lateral ventricle in the normal and Huntington's disease human brain. *Neuroscience*, *132*, 777–788.

Curtis, M. A., Waldvogel, H. J., Synek, B., & Faull, R. L. (2005b). A histochemical and immunohistochemical analysis of the subependymal layer in the normal and Huntington's disease brain. *Journal of Chemical Neuroanatomy*, *30*, 55–66.

Deems, D. A., Doty, R. L., Settle, R. G., Moore-Gillon, V., Shaman, P., Mester, A. F., et al. (1991). Smell and taste disorders, a study of 750 patients from the University of Pennsylvania Smell and Taste Center. *Archives of Otolaryngology—Head & Neck Surgery*, *117*, 519–528.

Doetsch, F. (2003). The glial identity of neural stem cells. *Nature Neuroscience*, *6*, 1127–1134.

Doetsch, F., Garcia-Verdugo, J. M., & Alvarez-Buylla, A. (1997). Cellular composition and three-dimensional organization of the subventricular germinal zone in the adult mammalian brain. *The Journal of Neuroscience*, *17*, 5046–5061.

Doty, R. L. (1991). Olfactory capacities in aging and Alzheimer's disease. Psychophysical and anatomic considerations. *Annals of the New York Academy of Sciences, 640*, 20–27.

Doty, R. L. (1994). Olfaction and multiple chemical sensitivity. *Toxicology and Industrial Health, 10*, 359–368.

Doty, R. L. (2001). Olfactory deficit in Alzheimer's disease? *The American Journal of Psychiatry, 158*, 1533–1534. author reply 1534–1535

Doty, R. L. (2003). Odor perception in neurodegenerative diseases. *Handbook of olfaction and gustation* (2nd ed., pp. 479–502). New York: Marcel Dekker INC.

Doty, R. L. (2008). The olfactory vector hypothesis of neurodegenerative disease: is it viable? *Annals of Neurology, 63*, 7–15.

Edelman, G. M. (1986). Cell adhesion molecules in neural histogenesis. *Annual Review of Physiology, 48*, 417–430.

Elias, L. A., Wang, D. D., & Kriegstein, A. R. (2007). Gap junction adhesion is necessary for radial migration in the neocortex. *Nature, 448*, 901–907.

Goetz, M. (1998). How are neurons specified: master or positional control? *Trends in Neurosciences, 21*, 135–136.

Graves, A. B., Bowen, J. D., Rajaram, L., McCormick, W. C., McCurry, S. M., Schellenberg, G. D., et al. (1999). Impaired olfaction as a marker for cognitive decline: interaction with apolipoprotein E epsilon4 status. *Neurology, 53*, 1480–1487.

Hack, M. A., Saghatelyan, A., de Chevigny, A., Pfeifer, A., Ashery-Padan, R., Lledo, P., et al. (2005). Neuronal fate determinants of adult olfactory bulb neurogenesis. *Nature Neuroscience, 8*, 865–872.

Hasegawa, T., Sugeno, N., Takeda, A., Matsuzaki-Kobayashi, M., Kikuchi, A., Furukawa, K., et al. (2007). Role of Neu4L sialidase and its substrate ganglioside GD3 in neuronal apoptosis induced by catechol metabolites. *FEBS Letters, 581*, 406–412.

Hinsby, A. M., Berezin, V., & Bock, E. (2004). Molecular mechanisms of NCAM function. *Frontiers in Bioscience, 9*, 2227–2244.

Höglinger, G. U., Rizk, P., Muriel, M. P., Duyckaerts, C., Oertel, W. H., Caille, I., et al. (2004). Dopamine depletion impairs precursor cell proliferation in Parkinson disease. *Nature Neuroscience, 7*, 726–735.

Hortsch, M. (2003). Neural cell adhesion molecules—brain glue and much more!. *Frontiers in Bioscience, 8*, 357–359.

http://www.brainmuseum.org/Specimens/lagomorpha/domesticrabbit/sections/thumbnail.html.

Johnson, C. P., Fujimoto, I., Rutishauser, U., & Leckband, D. E. (2005). Direct evidence that neural cell adhesion molecule (NCAM) polysialylation increases intermembrane repulsion and abrogates adhesion. *The Journal of Biological Chemistry, 280*, 137–145.

Jones, N., & Rog, D. (1998). Olfaction: a review. *The Journal of Laryngology and Otology, 112*, 11–24.

Kant, I. (1978). *Anthropology from a pragmatic point of view* (V. L. Dowdell, Trans.). Carbondale, IL: Southern Illinois University Press (original work published 1798).

Kirschenbaum, B., Doetsch, F., Lois, C., & Alvarez-Buylla, A. (1999). Adult subventricular zone neuronal precursors continue to proliferate and migrate in the absence of the olfactory bulb. *The Journal of Neuroscience, 19*, 2171–2180.

Kohl, J. V., Atzmueller, M., Fink, B., & Grammer, K. (2001). Human pheromones: integrating neuroendocrinology and ethology. *Neuroendocrinology Letters, 22*, 309–321.

Kornack, D. R., & Rakic, P. (2001). The generation, migration, and differentiation of olfactory neurons in the adult primate brain. *Proceedings of the National Academy of Sciences of the United States of America, 98*, 4752–4757.

Kovacs, T., Cairns, N. J., & Lantos, P. L. (1999). Beta-amyloid deposition and neurofibrillary tangle formation in the olfactory bulb in ageing and Alzheimer's disease. *Neuropathology and Applied Neurobiology, 25*, 481–491.

Kovacs, T., Cairns, N. J., & Lantos, P. L. (2001). Olfactory centres in Alzheimer's disease: olfactory bulb is involved in early Braak's stages. *Neuroreport, 12*, 285–288.

Kruger, R. P., Aurandt, J., & Guan, K. L. (2005). Semaphorins command cells to move. *Nature Reviews. Molecular Cell Biology, 6*, 789–800.

Lois, C., & Alvarez-Buylla, A. (1994). Long-distance neuronal migration in the adult mammalian brain. *Science, 264*, 1145–1148.

Lois, C., Garcia-Verdugo, J. M., & Alvarez-Buylla, A. (1996). Chain migration of neuronal precursors. *Science, 271*, 978–981.

Luskin, M. B., & Boone, M. S. (1994). Rate and pattern of migration of lineally-related olfactory bulb interneurons generated postnatally in the subventricular zone of the rat. *Chemical Senses, 19*, 695–714.

Magesh, S., Suzuki, T., Miyagi, T., Ishida, H., & Kiso, M. (2006). Homology modeling of human sialidase enzymes NEU1, NEU3 and NEU4 based on the crystal structure of NEU2: hints for the design of selective NEU3 inhibitors. *Journal of Molecular Graphics & Modelling, 25*, 196–207.

Ming, G. L., & Song, H. (2005). Adult neurogenesis in the mammalian central nervous system. *Annual Review of Neuroscience, 28*, 223–250.

Ninkovic, J., & Gotz, M. (2007). Signaling in adult neurogenesis: from stem cell niche to neuronal networks. *Current Opinion in Neurobiology, 17*, 338–344.

Ohm, T. G., & Braak, H. (1987). Olfactory bulb changes in Alzheimer's disease. *Acta Neuropathologica, 73*, 365–369.

Quinones-Hinojosa, A., Sanai, N., Soriano-Navarro, M., Gonzalez-Perez, O., Mirzadeh, Z., Gil-Perotin, S., et al. (2006). Cellular composition and cytoarchitecture of the adult human subventricular zone: a niche of neural stem cells. *The Journal of Comparative Neurology, 494*, 415–434.

Rae, A. S. (1994). Nodules of cellular proliferation in sheep olfactory ventricle. *Clinical Neuropathology, 13*, 17–18.

Rousselot, P., Lois, C., & Alvarez-Buylla, A. (1995). Embryonic (PSA) N-CAM reveals chains of migrating neuroblasts between the lateral ventricle and the olfactory bulb of adult mice. *The Journal of Comparative Neurology, 351*, 51–61.

Rutishauser, U. (2008). Polysialic acid in the plasticity of the developing and adult vertebrate nervous system. *Nature Reviews. Neuroscience, 9*, 26–35.

Rutishauser, U., Hoffman, S., & Edelman, G. M. (1982). Binding properties of a cell adhesion molecule from neural tissue. *Proceedings of the National Academy of Sciences of the United States of America, 79*, 685–689.

Rutishauser, U., & Landmesser, L. (1996). Polysialic acid in the vertebrate nervous system: a promoter of plasticity in cell-cell interactions. *Trends in Neurosciences, 19*, 422–427.

Sanai, N., Tramontin, A. D., Quinones-Hinojosa, A., Barbaro, N. M., Gupta, N., Kunwar, S., et al. (2004). Unique astrocyte ribbon in adult human brain contains neural stem cells but lacks chain migration. *Nature, 427*, 740–744.

Seri, B., Garcia-Verdugo, J. M., Collado-Morente, L., McEwen, B. S., & Alvarez-Buylla, A. (2004). Cell types, lineage, and architecture of the germinal zone in the adult dentate gyrus. *The Journal of Comparative Neurology, 478*, 359–378.

Simpson, T. I., & Price, D. J. (2002). Pax6; a pleiotropic player in development. *Bioessays, 24*, 1041–1051.

Stoykova, A., & Gruss, P. (1994). Roles of Pax-genes in developing and adult brain as suggested by expression patterns. *The Journal of Neuroscience, 14*, 1395–1412.

Vimr, E. R., McCoy, R. D., Vollger, H. F., Wilkison, N. C., & Troy, F. A. (1984). Use of prokaryotic-derived probes to identify poly(sialic acid) in neonatal neuronal membranes. *Proceedings of the National Academy of Sciences of the United States of America, 81*, 1971–1975.

Wedekind, C., & Penn, D. (2000). MHC genes, body odours, and odour preferences. *Nephrology, Dialysis, Transplantation, 15*, 1269–1271.

Ziabreva, I., Perry, E., Perry, R., Minger, S. L., Ekonomou, A., Przyborski, S., et al. (2006). Altered neurogenesis in Alzheimer's disease. *Journal of Psychosomatic Research, 61*, 311–316.

CHAPTER 4

# Identifying and enumerating neural stem cells: application to aging and cancer

Loic P. Deleyrolle* and Brent A. Reynolds

*Department of Neurosurgery, McKnight Brain Institute, University of Florida, Gainesville, FL, USA*

**Abstract:** The discovery of stem cells in the adult central nervous system implied the potential for endogenous repair and exogenous cell-based therapeutics. The development of experimental protocols, like the neurosphere assay and the neural-colony forming cell assay, enable the accurate and meaningful investigation of neural stem cell properties and allow the exploration of mechanisms related to the role of neural stem cells in aging and cancer.

**Keywords:** neural stem cell; cancer; aging; neurosphere assay; neural-colony forming cell assay; mash modeling

## Neural stem cell discovery

The view of the brain being immutable dominated the field of neurobiology for most of the last century. The doctrine, which is often accredited to Ramon y Cajal, stated that in the adult brain "everything may die, nothing may be regenerated" (Ramon y Cajal, 1913). Therefore, the adult central nervous system (CNS) was thought not to exhibit a regenerative capacity, nor have the need for an endogenous stem cell population. This dogmatic view has faced several challenges, first initiated by Altman in the 1960s, Privat in the 1970s, and continued with Kaplan and Nottebohm in the 1980s (Altman, 1962, 1963; Altman and Das, 1965; Kaplan and Hinds, 1977; Privat, 1975; Privat and Leblond, 1972; Nottebohm, 1985). Altman demonstrated cell proliferation in the adult CNS, Privat showed generation of new glial cells in postnatal mammalian brain, and Kaplan revealed neuron genesis in the rat olfactory bulb and dentate gyrus of the hippocampus (Altman, 1962, 1963; Altman and Das, 1965; Kaplan and Hinds, 1977; Privat, 1975; Privat and Leblond, 1972). In the 1980s, Nottebohm published results that demonstrated neuronal replacement in birds during mating season, adding to the spattering of evidence of adult CNS plasticity and neurogenesis (Nottebohm, 1985). But the "no new cell genesis" dogma remained until the early 1990s when a Canadian group demonstrated that cells with the functional characteristics of stem cells could be isolated from the adult mammalian brain and expanded in vitro using defined conditions (Reynolds and Weiss, 1992). Since this original report, neural stem cells (NSCs) lining the ventricular system at all levels of the neuroaxis in the adult nervous system have been identified (Weiss et al., 1996). This resident population of NSCs generate functional neurons throughout the lifetime of the animal so as to replace cells

*Corresponding author.
Tel.: +352-273-8583; Fax: +352-392-8413;
E-mail: loic.deleyrolle@neurosurgery.ufl.edu

DOI: 10.1016/S0079-6123(09)17504-0

in at least two areas of the brain: the olfactory bulb and the hippocampus (Alvarez-Buylla and Garcia-Verdugo, 2002; Doetsch and Alvarez-Buylla, 1996; Gould and Gross, 2002; Rietze et al., 2000). While we do not fully understand the fundamental biological significance of adult neurogenesis, newly generated neurons are known to integrate into the existing circuitry of the region, and specific behaviors such as spatial learning are modulated by adult neurogenesis (Ming and Song, 2005). Indeed, exercise-induced increases in hippocampal neurogenesis are associated with augmented spatial learning (Kafri et al., 1999), while decreasing hippocampal neurogenesis leads to deficits in learning and memory (Shors et al., 2001). Of interest, the levels of neurogenesis in the dentate gyrus (Kuhn et al., 1996) and olfactory bulb (Enwere et al., 2004) are known to decline with increasing age. While the relationship between adult neurogenesis and learning and memory is somewhat controversial and unclear at this point (Shors, 2008), the reduction in proliferating precursors during aging and a correlating loss of cognitive ability is apparent. What is even more uncertain is the effect of aging on NSCs, which is due not only to the lack of studies that have directly addressed this issue but also an absence of markers and assays that are able to meaningfully identify and measure NSCs.

## Identifying neural stem cells using the neurosphere assay (NSA)

Due to the paucity of specific and definitive markers, together with poorly defined morphological characteristics, stem cells are defined by functional criterion. A common well-accepted definition of a stem cell currently exists and has been applied to a broad range of tissue-specific stem cells. Based on this functional description, a stem cell is defined as an undifferentiated cell, which retains the capacity to: (i) proliferate, (ii) exhibit extensive self-renewal, (iii) produce a large family of differentiated functional progeny, (iv) regenerate the tissue after injury, and (v) retain a flexible use of these alternatives (Potten and Loeffler, 1990; Reynolds and Rietze, 2005). Ideally, a cell is qualified as a stem cell when all of these criteria are met; however, due to technical or experimental limits, often only some of these criteria such as self-renewal and multipotency may be satisfied. These criteria are generally given greater importance than other attributes when considering whether such a cell is indeed a stem cell. In 1992, Reynolds and Weiss described a methodology (the Neurosphere Assay (NSA)) that allowed adult NSCs to be isolated, and their functional characteristics and developmental potential to be investigated (Fig. 1) (Reynolds and Weiss, 1992). This method uses minimalistic serum-free conditions in which the vast majority of the harvested primary CNS cells die, while the growth factor-responsive cells survive, divide, and generate clonally derived clusters of cells called neurospheres. In theory, these neurospheres can be indefinitely maintained in vitro, i.e. dissociated and re-plated to form secondary neurospheres, then tertiary, quaternary, and so on. When passed in this manner, the key in vitro elements of a stem cell can be demonstrated — extensive self-renewal, generation of a large number of progeny, and differentiation of the stem cell progeny into the three primary CNS phenotypes — neurons, astrocytes, and oligodendrocytes (Fig. 1a–b).

## Enumeration of neural stem cells using the neural-colony forming cell assay and a mathematical model

Due to its robustness and simplicity, the NSA remains the standard for determining the presence of NSCs. However, in light of observations that progenitor cells have the ability to form primary and secondary spheres (Bull and Bartlett, 2005; Louis et al., 2008; Reynolds and Rietze, 2005; Seaberg and van der Kooy, 2002), it is now clear that the majority of neurospheres (greater than 90%) are not derived from a stem cell. Consequently, the assumed one-to-one relationship between neurospheres and NSCs is incorrect and it would follow that neurosphere generation is not sufficient evidence of a NSC, nor can the enumeration of neurospheres be used as a means to detect changes in NSCs numbers. To overcome the absence of a meaningful assay for detecting specific changes in NSC activity, Louis et al. (2008) designed a methodology, the Neural-Colony Forming Cell Assay (N-CFCA). The method uses a serum-free medium and includes a semisolid matrix of collagen ensuring clonality of the colonies (Fig. 1c). Similar to the NSA, where the growth factor-responsive stem and progenitor cells proliferate, the design of the N-CFCA allows for cells to express their full proliferative capacity over a three to four week period. The read out of the assay is based on the observation that stem cells exhibit greater proliferative capability compared to progenitor cells, hence, the size of the clonally derived colonies (i.e. diameter) can be used to differentiate its founder cell type. Broad ranges of colony sizes are identified after several weeks in culture and validation of individual colonies reveal that only large colonies (i.e. >than 2 mm in diameter) exhibit stem cell characteristics (extensive self-renewal, generation of large number of progeny, and multipotency). Under these conditions the majority of the colonies (>90%) do not exhibit *in vitro* stem cell features and hence are classified as progenitor cells. Therefore, the N-CFCA assesses more stringently the potential for long-term proliferation and self-renewal than the NSA and represents a more meaningful method to enumerate NSCs.

## Neural stem cells: potential treatment for age-related CNS disorders

The persistence of NSC throughout the lifespan of the organism and their potential to generate large numbers of functional progeny provides a unique opportunity to repair cells lost to injury and disease. However, little is known about the effects of aging on this population and the ability of aged NSCs to respond to brain insults — this is not the case for all stem cell populations. Several studies have demonstrated that hematopoietic stem cells (HSCs) become functionally compromised as part of the aging process, leading to a reduced ability to maintain tissue homeostasis at youthful levels (Morrison et al., 1996; Rossi et al., 2005). This age-related decline in hematopoietic function appears not to be due to a reduction in the numbers of HSCs but rather a decline in their functional capacity. Similar to HSCs, mesenchymal stem cells (MSC) show alterations during aging such as changes in their frequency, proliferation, and differentiation potential, although the extend of this general decline in function is not as clear as that known for the HSCs, as a result of a lack of consensus as to the MSC phenotype (Roobrouck et al., 2008). Likewise, skeletal muscle stem cells (that have been identified as satellite cells), also display age-related decline in their function and regenerative abilities; interestingly, this appears to be largely determined by cell extrinsic factors (Conboy et al., 2003; Gopinath and Rando, 2008). Outside the hematopoietic system there is, in general, a lack of solid evidence on the effects of aging on stem cell activities and the mechanisms involved. This is due principally to the relatively recent identification of stem cells in many of the tissues of the body (i.e. brain, heart, muscle) and of markers and assays to measure their activity.

To date, our knowledge of the effects of aging on NSCs is scant due to a limited number of studies that have attempted to address this issue. Tropepe et al. (1997) reported a ~50% decline in the number of proliferating cells in the dorsolateral corner of the lateral ventricles in aged (2-year-old) versus young mice. This decline was accompanied by a 70% decline in the number of progeny migrating to the olfactory bulb and thus the number of olfactory neurons being generated. Somewhat surprisingly and inconsistent with this decline, was the preservation of the number of neurospheres that could be generated from young versus aged mice, suggesting that the stem cell population was unaffected. When Weiss and colleagues (Enwere et al., 2004) attempted to repeat this work, they also found a decline in dividing periventricular region (PVR) cells and progeny migrating to the olfactory bulb. However, unlike Tropepe et al., they reported a 60–80% reduction

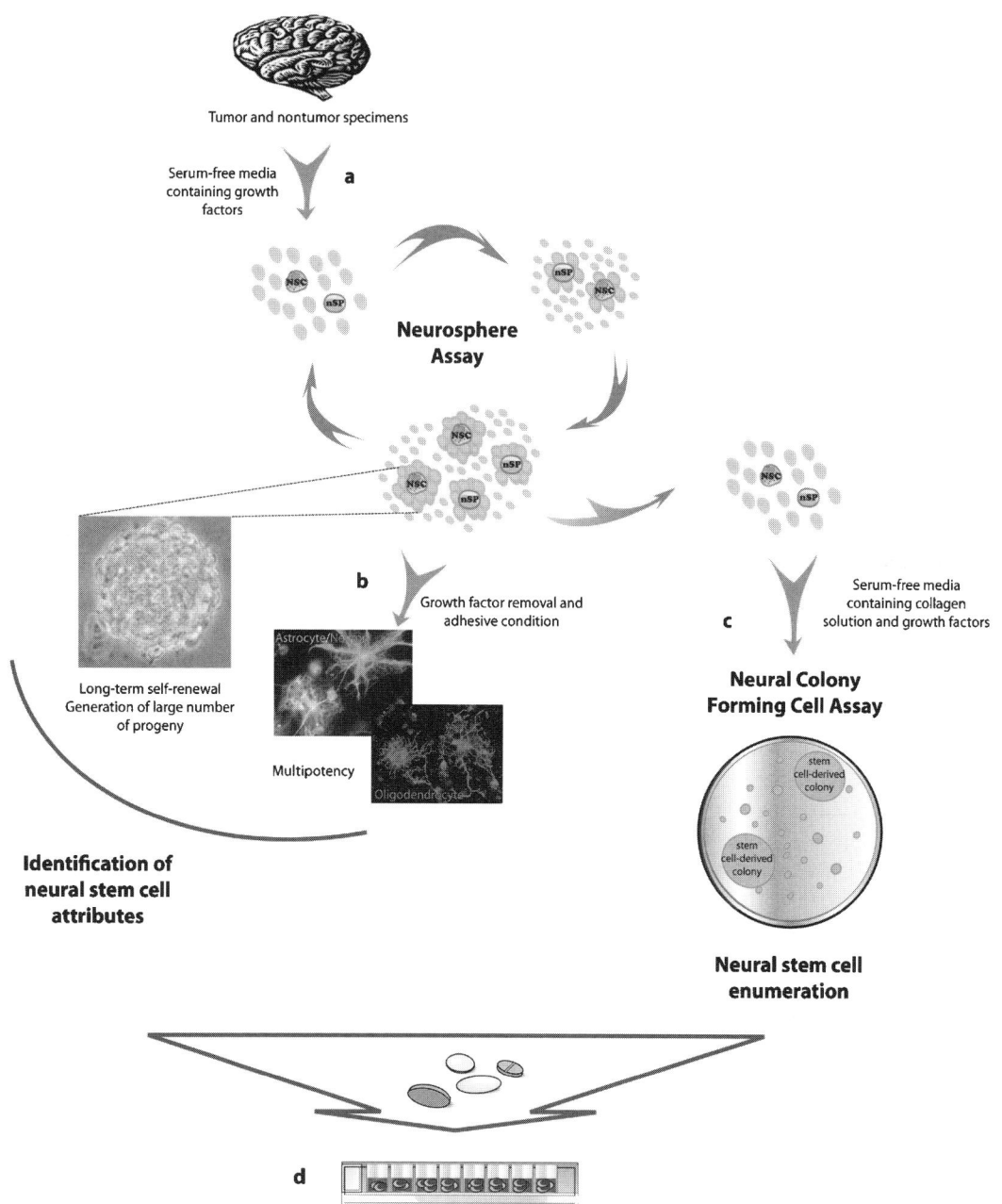

in the number of neurospheres generated in aged mice. A third study also reported a two-fold reduction in the number of putative NSCs (prospectively and equivocally identified on the basis of relative quiescence) and neurosphere-generating cells in aged (2-year-old) mice, suggesting that an age-related decline in NSCs does indeed occur (Maslov et al., 2004). Finally, the ability to mimic the age-related decline of putative NSCs (Kippin et al., 2005) by artificially increasing stem cell proliferation rates suggests that, akin to HSCs (Cheng et al., 2000), NSCs possess a limited regenerative capacity that is also contingent in part on mitotic activity. Very recently, Wu and colleagues looked at the effects of exercise on NSCs, neurite outgrowth, and survival of neuronal progenitor cells in the dentate gyrus of aged mice and reported an increase in NSC proliferation (Wu et al., 2008). However, like many such studies, their readout of NSC activity was based on overall BrdU incorporation, which does not equate to NSC activity or frequency (Marshall et al., 2007). In addition, the Morrison lab has made several significant contributions to the understanding of the genetic regulation of aging in NSCs (Molofsky et al., 2003; Nishino et al., 2008). While these studies begin to chip away at the genetic regulation that occurs during aging, the use of assays that are not selective for enumerating the frequency and function of NSCs limits its interpretation. Overall, while these studies suggest that an age-related functional impairment of stem cells is occurring in the brain, the limited research in this area together with appropriate assays, clearly demonstrates a significant deficiency in our most basic understanding of how ageing affects NSCs.

Providing opportunity to study and enumerate NSCs, the NSA and the N-CFCA, represent tools to investigate mechanisms of how the process of ageing impairs the prevalence and functional capacity of NSCs over time. Improving our understanding of how ageing affects NSCs in the brain coupled with the increasing evidence of the regenerative capacity of endogenous NSCs makes them an essential target for intervention in age-associated neurodegenerative disorders.

## Neural stem cells and brain tumors

One of the most prominent topics in the field of cancer biology is that tumor-initiating cells (TICs), which phenotypically mimic the cardinal properties of stem cells, are responsible for the origin and maintenance of solid tissue malignancies. The hypothesis that TICs exhibit stem cell characteristics was first postulated in the 1990s based on studies of acute myeloid leukemia (Bonnet and Dick, 1997; Lapidot et al., 1994), and has since been strengthened by findings related to breast (Al-Hajj et al., 2003; Ponti et al., 2005), prostate, lung, and mesenchymal tumors (Buzzeo et al., 2007). Owing to their long-term self-renewal capacity, NSCs may be ideal candidates for neoplastic transformation and constitute a possible source of brain tumor initiating cells. Deregulation of the control of

Fig. 1. (a) Tumor and nontumor brain tissue samples are harvested in single cell suspension and transfer in culture into the neurosphere assay (NSA) consisting of a serum-free media supplemented with growth factors. In these conditions, growth factor-responsive cells survive and proliferate to generate clonal cell aggregates called neurospheres. The neurospheres can be dissociated and replated to form secondary neurospheres. This process can be indefinitely repeated, resulting in an exponential increase in the number of cells and spheres that are generated, describing long-term self-renewal and proliferation properties. (b) Plating the cells or neurospheres on adhesive substrate coupled with growth factor withdrawal creates differentiative conditions leading to the generation of neurons, astrocytes, and oligodendrocytes (nuclei stained with DAPI, blue), demonstrating multipotency characteristic. Therefore, the NSA allows identification of stem cells characteristics. (c) Cells from primary tissue or from the NSA are dissociated into single cell suspension and transfer in the neural colony forming cell assay (N-CFCA) that consists of a serum-free media containing collagen solution and supplemented with growth factors. After 3 weeks of culture, the growth factor-responsive cells generate different ranges of colonies size (i.e. diameter). Quantification of the number of large colonies (>2 mm in diameter, exhibiting stem cell attributes) allows enumeration of actual neural stem cells. (d) The NSA and the N-CFCA represent meaningful tools to design stem cell-based treatment in the setting of age-related diseases and to devise rational drug discovery approaches to cure brain tumors. (See Color Plate 4.1 in color plate section.)

the NSCs self-renewal may therefore be involved in the genesis of cancer-initiating stem-like cells (Reya et al., 2001; Passegue et al., 2003). This concept has first been validated by the discovery of self-sustaining brain tumor stem cells (Ignatova et al., 2002). Ignatova et al. (2002) isolated neurosphere-forming cells from specimens of human glioblastoma multiforma (hGBM) that exhibit stem-like attributes. Existence of transformed, undifferentiated neural precursors responding to the same growth factors and possessing similar molecular features as adult somatic stem cells has been demonstrated in several brain tumors (Galli et al., 2004; Hemmati et al., 2003; Singh et al., 2003, 2004). Cancer stem-like cells, clonally derived from human brain cancer samples in the NSA, can initiate new tumors when transplanted into the brain of adult immunodeficient mice (Galli et al., 2004). These neo-tumors exhibit the in vivo features of hGBMs illustrating that the in vitro defined brain tumor stem cells can faithfully recapitulate the in vivo characteristics of primary tumor (Lee et al., 2006). As opposed to the stochastic model in which all tumor cells can function as TICs, these findings support the cancer stem cell hypothesis for solid tissue cancers. This hypothesis states that a subpopulation of cells is responsible for the growth and recurrence of the tumor, and it is there stem cell characteristics that make them resistant to conventional treatment. Although the specific identity of the origin of the brain tumor stem cells remains unresolved, several studies suggest that the mitogenic events leading to tumor formation seems to me more "efficient" in the mitotically active stem/precursor cells that represent though a preferential target of oncogenic transformation (Sanai et al., 2005; Holland et al., 1998; Holland et al., 2000; Vescovi et al., 2006). Therefore the tumor stem cells constitute a new target for cancer therapy. As mentioned above, the NSA is a useful assay for isolation, expansion, and identification of somatic NSCs and has become a method of choice to identify and study brain tumor stem cells as well. However, as in studying somatic precursor cells, both tumor stem and progenitor cells form spheres in the NSA, making enumeration of cancer stem cells inaccurate. Adaptation of the N-CFCA for studying tumor stem cells has allowed us (Louis et al., unpublished data) to distinguish between cancer cells with limited proliferative potential (i.e. progenitor cells) and those with an extensive proliferative potential (i.e. cancer stem cells) and the ability to initiate secondary tumor formation. In addition, we have used mathematical modeling to discover molecules that directly effect the proliferation of glioma cancer stem cells (Piccirillo et al., 2006).

The use of the N-CFCA to enumerate and appraise cancer stem cell can find application in the cancer field. Understand tumor stem cell regulation provides a basis for rational drug discovery to identify and selectively target cancer stem cells and their progeny. In addition, this suite of assays (NSA, N-CFCA, and the mathematical modeling) offer a means to screen molecules that specifically target distinct populations of cells (stem or progenitor cells) and to determine which populations are being effected. Further application may be found in using these assays as diagnostic tools to determine the presence, incidence, and response to treatment, for the purpose of designing better outcomes based on a more detailed understanding of the biology of CNS cancers.

## Conclusion

Demonstration of the persistence of NSCs in the CNS implies a therapeutic potential for these cells in brain regenerative medicine. Development of tools such as the NSA, mathematical modeling, and the neural-colony forming cell assay provide the opportunity to study and characterize NSCs and to comprehend the influences (genetic and epigenetic) that regulate their function. With an increased understanding of NSC biology comes the potential to harness their regenerative potential through strategies employed to repair and protect the injured and aging brain. Given the reality of an ageing population and the associated increase in neurodegenerative diseases, understanding mechanisms through which NSCs progressively lose their ability to maintain tissue homeostasis is of utmost importance for the future development of preventative or regenerative CNS therapies. The other side of the coin is the potential for NSCs and their progeny to be involved in tumor formation. A great deal of effort has recently gone into studying the stem cell characteristics of several types of CNS cancers. The ability to identify and culture brain tumor cells using the same growth conditions as is used for studying somatic NSCs, implies a similarity between these two cells types, possibly at a signaling or molecular level. The use of meaningful assays that allow one to dissect the contribution of particular phenotypes to tumor formation and to test agents that can regulate their proliferation and differentiation will be of benefit for developing the next generation of targeted therapies.

## References

Al-Hajj, M., Wicha, M. S., Benito-Hernandez, A., Morrison, S. J., & Clarke, M. F. (2003). Prospective identification of tumorigenic breast cancer cells. *Proceedings of the National Academy of Science USA*, *100*, 3983–3988.

Altman, J. (1962). Are new neurons formed in the brains of adult mammals? *Science*, *135*, 1127–1128.

Altman, J. (1963). Autoradiographic investigation of cell proliferation in the brains of rats and cats. *Anatomical Record*, *145*, 573–591.

Altman, J., & Das, G. D. (1965). Autoradiographic and histological evidence of postnatal hippocampal neurogenesis in rats. *Journal of Computational Neurology*, *124*, 319–335.

Alvarez-Buylla, A., & Garcia-Verdugo, J. M. (2002). Neurogenesis in adult subventricular zone. *Journal of Neuroscience*, *22*, 629–634.

Bonnet, D., & Dick, J. E. (1997). Human acute myeloid leukemia is organized as a hierarchy that originates from a primitive hematopoietic cell. *Nature Medicine*, *3*, 730–737.

Bull, N. D., & Bartlett, P. F. (2005). The adult mouse hippocampal progenitor is neurogenic but not a stem cell. *Journal of Neuroscience*, *25*, 10815–10821.

Buzzeo, M. P., Scott, E. W., & Cogle, C. R. (2007). The hunt for cancer-initiating cells: A history stemming from leukemia. *Leukemia*, *21*, 1619–1627.

Cheng, T., Rodrigues, N., Shen, H., Yang, Y., Dombkowski, D., Sykes, M., et al. (2000). Hematopoietic stem cell quiescence maintained by p21cip1/waf1. *Science*, *287*, 1804–1808.

Conboy, I. M., Conboy, M. J., Smythe, G. M., & Rando, T. A. (2003). Notch-mediated restoration of regenerative potential to aged muscle. *Science*, *302*, 1575–1577.

Doetsch, F., & Alvarez-Buylla, A. (1996). Network of tangential pathways for neuronal migration in adult mammalian brain. *Proceedings of the National Academy of Science USA*, *93*, 14895–14900.

Enwere, E., Shingo, T., Gregg, C., Fujikawa, H., Ohta, S., & Weiss, S. (2004). Aging results in reduced epidermal growth factor receptor signaling, diminished olfactory neurogenesis, and deficits in fine olfactory discrimination. *Journal of Neuroscience*, *24*, 8354–8365.

Galli, R., Binda, E., Orfanelli, U., Cipelletti, B., Gritti, A., De Vitis, S., et al. (2004). Isolation and characterization of tumorigenic, stem-like neural precursors from human glioblastoma. *Cancer Research*, *64*, 7011–7021.

Gopinath, S. D., & Rando, T. A. (2008). Stem cell review series: Aging of the skeletal muscle stem cell niche. *Aging Cell*, *7*, 590–598.

Gould, E., & Gross, C. G. (2002). Neurogenesis in adult mammals: Some progress and problems. *Journal of Neuroscience*, *22*, 619–623.

Hemmati, H. D., Nakano, I., Lazareff, J. A., Masterman-Smith, M., Geschwind, D. H., Bronner-Fraser, M., et al. (2003). Cancerous stem cells can arise from pediatric brain tumors. *Proceedings of the National Academy of Science USA*, *100*, 15178–15183.

Holland, E. C., Celestino, J., Dai, C., Schaefer, L., Sawaya, R. E., & Fuller, G. N. (2000). Combined activation of Ras and Akt in neural progenitors induces glioblastoma formation in mice. *Nature Genetics*, *25*, 55–57.

Holland, E. C., Hively, W. P., Gallo, V., & Varmus, H. E. (1998). Modeling mutations in the G1 arrest pathway in human gliomas: Overexpression of CDK4 but not loss of INK4a-ARF induces hyperploidy in cultured mouse astrocytes. *Genes Development*, *12*, 3644–3649.

Ignatova, T. N., Kukekov, V. G., Laywell, E. D., Suslov, O. N., Vrionis, F. D., & Steindler, D. A. (2002). Human cortical glial tumors contain neural stem-like cells expressing astroglial and neuronal markers in vitro. *Glia*, *39*, 193–206.

Kafri, T., van Praag, H., Ouyang, L., Gage, F. H., & Verma, I. M. (1999). A packaging cell line for lentivirus vectors. *Journal of Virology*, *73*, 576–584.

Kaplan, M. S., & Hinds, J. W. (1977). Neurogenesis in the adult rat: Electron microscopic analysis of light radioautographs. *Science*, *197*, 1092–1094.

Kippin, T. E., Martens, D. J., & van der Kooy, D. (2005). p21 loss compromises the relative quiescence of forebrain stem cell proliferation leading to exhaustion of their proliferation capacity. *Genes and Development*, *19*, 756–767.

Kuhn, H. G., Dickinson, A. H., & Gage, F. H. (1996). Neurogenesis in the dentate gyrus of the adult rat: Age-related decrease of neuronal progenitor proliferation. *Journal of Neuroscience*, *16*, 2027–2033.

Lapidot, T., Sirard, C., Vormoor, J., Murdoch, B., Hoang, T., Caceres-Cortes, J., et al. (1994). A cell initiating human acute myeloid leukaemia after transplantation into SCID mice. *Nature*, *367*, 645–648.

Lee, J., Kotliarova, S., Kotliarov, Y., Li, A., Su, Q., Donin, N. M., et al. (2006). Tumor stem cells derived from glioblastomas cultured in bFGF and EGF more closely mirror the phenotype and genotype of primary tumors than do serum-cultured cell lines. *Cancer Cell*, *9*, 391–403.

Louis, S. A., Rietze, R. L., Deleyrolle, L., Wagey, R. E., Thomas, T. E., Eaves, A. C., et al. (2008). Enumeration of neural stem and progenitor cells in the neural colony forming cell assay. *Stem Cells*, *26*, 988–996.

Marshall, G. P., 2nd., Reynolds, B. A., & Laywell, E. D. (2007). Using the neurosphere assay to quantify neural stem cells in vivo. *Current Pharmaceutical Biotechnology*, *8*, 141–145.

Maslov, A. Y., Barone, T. A., Plunkett, R. J., & Pruitt, S. C. (2004). Neural stem cell detection, characterization, and age-related changes in the subventricular zone of mice. *Journal of Neuroscience*, *24*, 1726–1733.

Ming, G. L., & Song, H. (2005). Adult neurogenesis in the mammalian central nervous system. *Annual Review of Neuroscience*, *28*, 223–250.

Molofsky, A. V., Pardal, R., Iwashita, T., Park, I. K., Clarke, M. F., & Morrison, S. J. (2003). Bmi-1 dependence distinguishes neural stem cell self-renewal from progenitor proliferation. *Nature*, *425*, 962–967.

Morrison, S. J., Wandycz, A. M., Akashi, K., Globerson, A., & Weissman, I. L. (1996). The aging of hematopoietic stem cells. *Nature Medicine*, *2*, 1011–1016.

Nishino, J., Kim, I., Chada, K., & Morrison, S. (2008). Hmga2 promotes neural stem cell self-renewal in young but not old mice by reducing p16Ink4a and p19Arf expression. *Cell*, *135*, 227–239.

Nottebohm, F. (1985). Neuronal replacement in adulthood. *Annals of the New York Academy of Science*, *457*, 143–161.

Passegue, E., Jamieson, C. H. M., Ailles, L. E., & Weissman, I. L. (2003). Normal and leukemic hematopoiesis: Are leukemias a stem cell disorder or a reacquisition of stem cell characteristics? *Proceedings of the National Academy of Science*, *100*, 11842–11849.

Piccirillo, S. G., Reynolds, B. A., Zanetti, N., Lamorte, G., Binda, E., Broggi, G., et al. (2006). Bone morphogenetic proteins inhibit the tumorigenic potential of human brain tumour-initiating cells. *Nature*, *444*, 761–765.

Ponti, D., Costa, A., Zaffaroni, N., Pratesi, G., Petrangolini, G., Coradini, D., et al. (2005). Isolation and in vitro propagation of tumorigenic breast cancer cells with stem/progenitor cell properties. *Cancer Research*, *65*, 5506–5511.

Potten, C. S., & Loeffler, M. (1990). Stem cells: attributes, cycles, spirals, pitfalls and uncertainties. Lessons for and from the crypt. *Development*, *110*, 1001–1020.

Privat, A. (1975). Postnatal gliogenesis in the mammalian brain. *International Review of Cytology*, *40*, 281–323.

Privat, A., & Leblond, C. P. (1972). The subependymal layer and neighboring region in the brain of the young rat. *Journal of Computational Neurology*, *146*, 277–302.

Ramon y Cajal, S. (1913). *Degeneration and regeneration of the nervous system*. Oxford University Press.

Reya, T., Morrison, S. J., Clarke, M. F., & Weissman, I. L. (2001). Stem cells, cancer, and cancer stem cells. *Nature*, *414*, 105–111.

Reynolds, B. A., & Rietze, R. L. (2005). Neural stem cells and neurospheres: Re-evaluating the relationship. *Nature Methods*, *2*, 333–336.

Reynolds, B. A., & Weiss, S. (1992). Generation of neurons and astrocytes from isolated cells of the adult mammalian central nervous system. *Science*, *255*, 1707–1710.

Rietze, R., Poulin, P., & Weiss, S. (2000). Mitotically active cells that generate neurons and astrocytes are present in multiple regions of the adult mouse hippocampus. *Journal of Computational Neurology*, *424*, 397–408.

Roobrouck, V. D., Ulloa-Montoya, F., & Verfaillie, C. M. (2008). Self-renewal and differentiation capacity of young and aged stem cells. *Experimental Cell Research*, *314*, 1937–1944.

Rossi, D. J., Bryder, D., Zahn, J. M., Ahlenius, H., Sonu, R., Wagers, A. J., et al. (2005). Cell intrinsic alterations underlie hematopoietic stem cell aging. *Proceedings of the National Academy of Science of USA*, *102*, 9194–9199.

Sanai, N., Alvarez-Buylla, A., & Berger, M. S. (2005). Neural stem cells and the origin of gliomas. *New England Journal of Medicine*, *353*, 811–822.

Seaberg, R. M., & van der Kooy, D. (2002). Adult rodent neurogenic regions: The ventricular subependyma contains neural stem cells, but the dentate gyrus contains restricted progenitors. *Journal of Neuroscience*, *22*, 1784–1793.

Shors, T. J. (2008). From stem cells to grandmother cells: How neurogenesis relates to learning and memory. *Cell Stem Cell*, *3*, 253–258.

Shors, T. J., Miesegaes, G., Beylin, A., Zhao, M., Rydel, T., & Gould, E. (2001). Neurogenesis in the adult is involved in the formation of trace memories. *Nature*, *410*, 372–376.

Singh, S. K., Clarke, I. D., Terasaki, M., Bonn, V. E., Hawkins, C., Squire, J., et al. (2003). Identification of a cancer stem cell in human brain tumors. *Cancer Research*, *63*, 5821–5828.

Singh, S. K., Hawkins, C., Clarke, I. D., Squire, J., Bayani, J., Hide, T., et al. (2004). Identification of human brain tumour initiating cells. *Nature*, *432*, 396–401.

Tropepe, V., Craig, C. G., Morshead, C. M., & van der Kooy, D. (1997). Transforming growth factor-alpha null and senescent mice show decreased neural progenitor cell proliferation in the forebrain subependyma. *Journal of Neuroscience*, *17*, 7850–7859.

Vescovi, A. L., Galli, R., & Reynolds, B. A. (2006). Brain tumour stem cells. *Nature Review Cancer, 6*, 425–436.

Weiss, S., Dunne, C., Hewson, J., Wohl, C., Wheatley, M., Peterson, A., et al. (1996). Multipotent CNS stem cells are present in the adult mammalian spinal cord and ventricular neuroaxis. *Journal Neuroscience, 16*, 7599–7609.

Wu, C. W., Chang, Y. T., Yu, L., Chen, H. I., Jen, C. J., Wu, S. Y., et al. (2008). Exercise enhances the proliferation of neural stem cells and neurite growth and survival of neuronal progenitor cells in dentate gyrus of middle-aged mice. *Journal of Applied Physiology, 105*, 1585–1594.

CHAPTER 5

# Transgenic reporter mice as tools for studies of transplantability and connectivity of dopamine neuron precursors in fetal tissue grafts

Lachlan H. Thompson[1,*] and Anders Björklund[2]

[1]*Florey Neuroscience Institutes, University of Melbourne, Parkville, Victoria, Australia*
[2]*Wallenberg Neuroscience Center, Lund University, Lund, Sweden*

**Abstract:** Cell therapy for Parkinson's disease (PD) is based on the idea that new midbrain dopamine (mDA) neurons, implanted directly into the brain of the patient, can structurally and functionally replace those lost to the disease. Clinical trials have provided proof-of-principle that the grafted mDA neurons can survive and function after implantation in order to provide sustained improvement in motor function for some patients. Nonetheless, there are a number of issues limiting the application of this approach as mainstream therapy, including: the use of human fetal tissue as the only safe and reliable source of transplantable mDA neurons, and variability in the therapeutic outcome. Here we review recent progress in this area from investigations using rodent models of PD, paying particular attention to the use of transgenic reporter mice as tools for neural transplantation studies. Cell type-specific expression of reporter genes, such as green fluorescent protein, affords valuable technical advantages in transplantation experiments, such as the ability to selectively isolate specific cell fractions from mixed populations prior to grafting, and the unambiguous visualization of graft-derived dopamine neuron fiber patterns after transplantation. The results from these investigations have given new insights into the transplantability of mDA precursors as well as their connectivity after grafting and have interesting implications for the development of stem cell based approaches for the treatment of PD.

**Keywords:** transplantation; regeneration; Parkinson's disease; ventral mesencephalon; GFP; cell sorting; cell therapy

## Introduction

The progressive and irreversible degeneration of midbrain dopamine (mDA) neurons is a major pathological hallmark of Parkinson's disease (PD) (German et al., 1989; Fearnley and Lees, 1991; McRitchie et al., 1997; Damier et al., 1999). The degeneration of the dopaminergic projections to the forebrain, the putamen and caudate nucleus, in particular, leads to the development of motor disturbances as one of the first detectable symptoms in patients. When loss of dopamine neurons reaches around 50% and the reduction in striatal dopamine reaches around 70%, the first signs of motor dysfunction become apparent including

*Corresponding author.
Tel.: +61 (0)3 8344 1888; Fax: +61 (0)3 9347 0446;
E-mail: lachlan.thompson@florey.edu.au

tremor at rest and difficulties in initiating and executing movements (Hornykiewicz, 1975; Fearnley and Lees, 1991). Untreated, these symptoms become progressively more severe as the disease continues. Current treatment strategies for PD aim to restore the level of dopaminergic signaling in the striatum in order to reinstate a normal pattern of information flow through the motor circuitry. Most commonly, this involves pharmacotherapy through administration of dopamine agonists or the dopamine precursor, levodopa (L-DOPA). Although this approach gives excellent results in many patients in the initial phase of treatment, prolonged treatment (>5 years) invariably leads to complications including a significant waning of the therapeutic effect and the appearance of dyskinesias related, at least in part, to the continued progression of the disease. Current efforts, therefore, are aimed at the development of restorative or neuroprotective treatments that will slow down disease progression and provide sustained recovery of function in the affected patients.

**Cell therapy for Parkinson's disease**

Cell replacement therapy for PD is based on the idea that implanted dopamine neurons may be able to substitute, structurally and functionally, for the lost nigro-striatal dopamine neurons. Experiments performed in rodent and primate models of PD have shown that transplanted mDA neurons can restore dopaminergic neurotransmission in the denervated host striatum and reverse some of the PD-like motor impairments induced by damage to the nigro-striatal system (for a comprehensive review, see Winkler et al., 2000). Clinical trials in patients with advanced PD have provided proof-of-principle that mDA neurons obtained from fetal human mesencephalic tissue can survive and function and provide long-lasting clinical benefits in some patients (for review, see Lindvall and Bjorklund, 2004). Cell therapy thus represents a promising alternative as a treatment with the potential to provide sustained symptomatic relief to PD patients over an extended period of time and without the complicating side effects associated with long-term drug treatment.

The cell therapy approach involves transplanting new mDA neurons directly into the brain in order to structurally and functionally replace those lost to the disease. At present, the only reliable source of transplantable mDA neurons are the developing dopamine neuroblasts contained in the embryonic ventral mesencephalon (VM; Figs. 1A and 4). When dissected from the developing embryo during the time of mDA neurogenesis [embryonic day (E) E10–E14 in rodents; 6–8 weeks of gestation in the human] VM cell preparations implanted into the striatum of the parkinsonian brain give rise to grafts that are rich in mDA neurons. The grafted dopamine neurons readily extend axons within the host striatum in order to establish a new and functional terminal network (Winkler et al., 2000; Kirik et al., 2001, Fig. 1E). In both animal models and in patients, this results in a significant degree of improvement in motor function that is comparable with that achieved during the early phase of L-DOPA therapy (Lindvall and Bjorklund, 2004; Olanow and Fahn, 2006). There are, however, a number of issues that limit the application of cell therapy as an optimal and mainstream therapeutic option for patients, including:

1. *Fetal tissue source.* In addition to ethical concerns, there are also considerable practical issues associated with the use of fetal VM as donor tissue in transplantation procedures. Not only is fetal tissue an unsustainable resource (with multiple fetal donors required to treat a single patient), but also is virtually impossible to standardize the preparations with respect to the number and kinds of cells present.
2. *Variability in therapeutic outcome.* The clinical outcome following intra-striatal transplantation of fetal VM has varied widely from patient to patient ranging from complete lack of therapeutic effect to substantial improvement in a range of motor functions, as assessed by the United Parkinson's Disease Rating Scale and timed motor

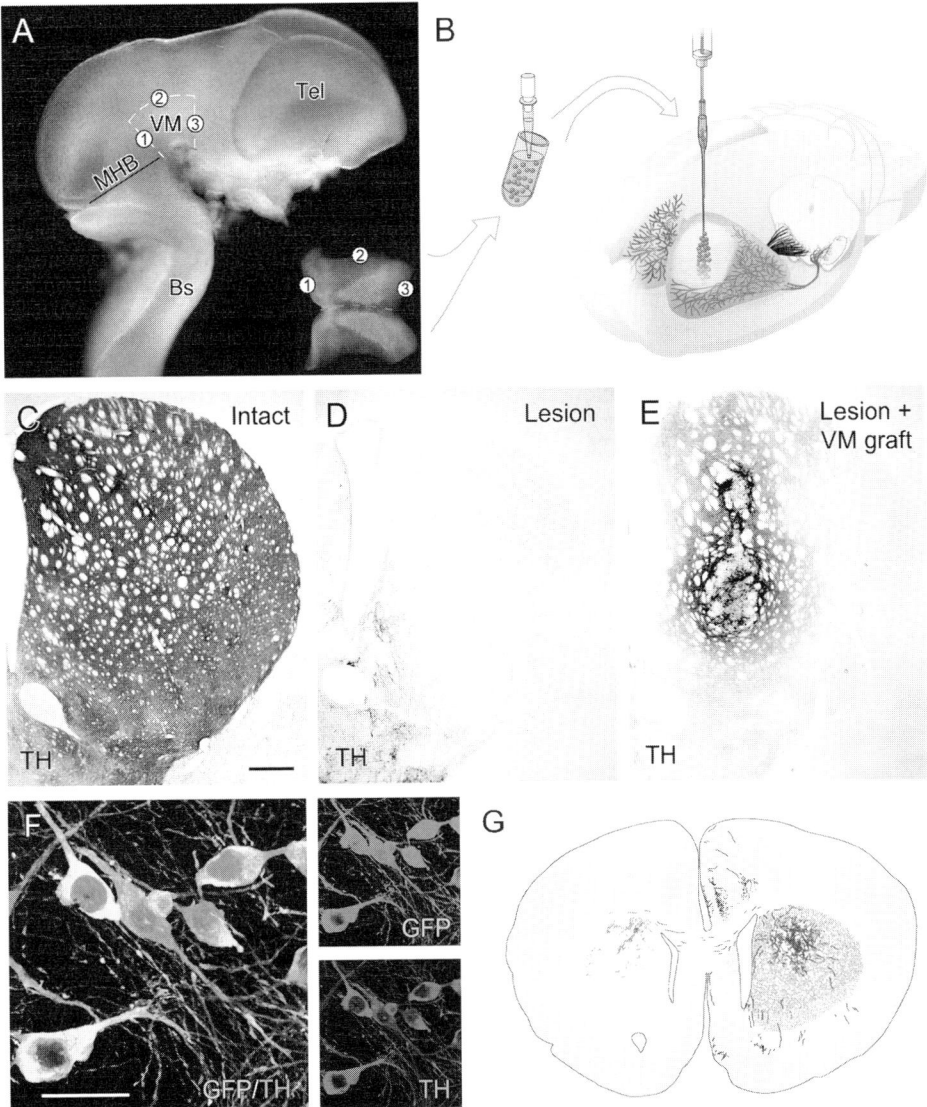

Fig. 1. Cell therapy for Parkinson's disease. (A) The developing mouse brain at embryonic day 12.5. The dashed lines indicate the approximate region of ventral mesencephalon (VM) dissected in order to generate cell preparations for grafting. The inset shows a piece of VM tissue dissected from a mouse in which all midbrain dopamine neurons express GFP (Pitx3-GFP mouse). The numbers indicated are for orientation relative to the intact brain. The red dashed line marks the midline. (B) A schematic overview of a typical transplantation procedure, whereby the dissected VM is prepared as a single cell suspension (through trypsin digestion and mechanical dissociation), and then cells are microinjected into the host brain. Image used here shows placement into the striatum. (C–E) TH immunohistochemistry in coronal sections through the adult rat brain. The dark staining of the striatum in the intact animal (C) represents the dense terminal network of TH-positive fibers originating from mDA neuronal projections. Lesioning of the mDA neurons through injection of 6-hydroxydopamine removes this TH-positive afferent innervation of the striatum (D). Panel (E) illustrates a 6-OHDA-lesioned animal 6 weeks after grafting of $1.0 \times 10^5$ E12.5 mouse VM cells into the striatum. The graft itself can be seen as a discrete teardrop-shaped deposit of darkly stained TH-positive cells, while the dark gray area surrounding the graft represents the new TH-positive innervation of the host striatum provided by the grafted mDA neurons. (F) Grafting of VM tissue from donor mice in which the dopamine neurons express GFP allows for the unequivocal identification of mDA neurons and their associated fiber outgrowth in the host brain. The images shown are from an intra-striatal graft of cells prepared from VM dissected from the TH-GFP mouse. (G) Schematic tracing of GFP immunoreactivity 6 weeks after grafting of E12.5 TH-GFP VM cells into the striatum of a neonatal rat reveals graft-specific patterns of mDA neuronal fiber outgrowth. Abbreviations: Bs, brainstem; MHB, mid-hindbrain boundary; Tel, telencephalon. Scale bars: C, 500 μm; F, 30 μm. (See Color Plate 5.1 in color plate section.)

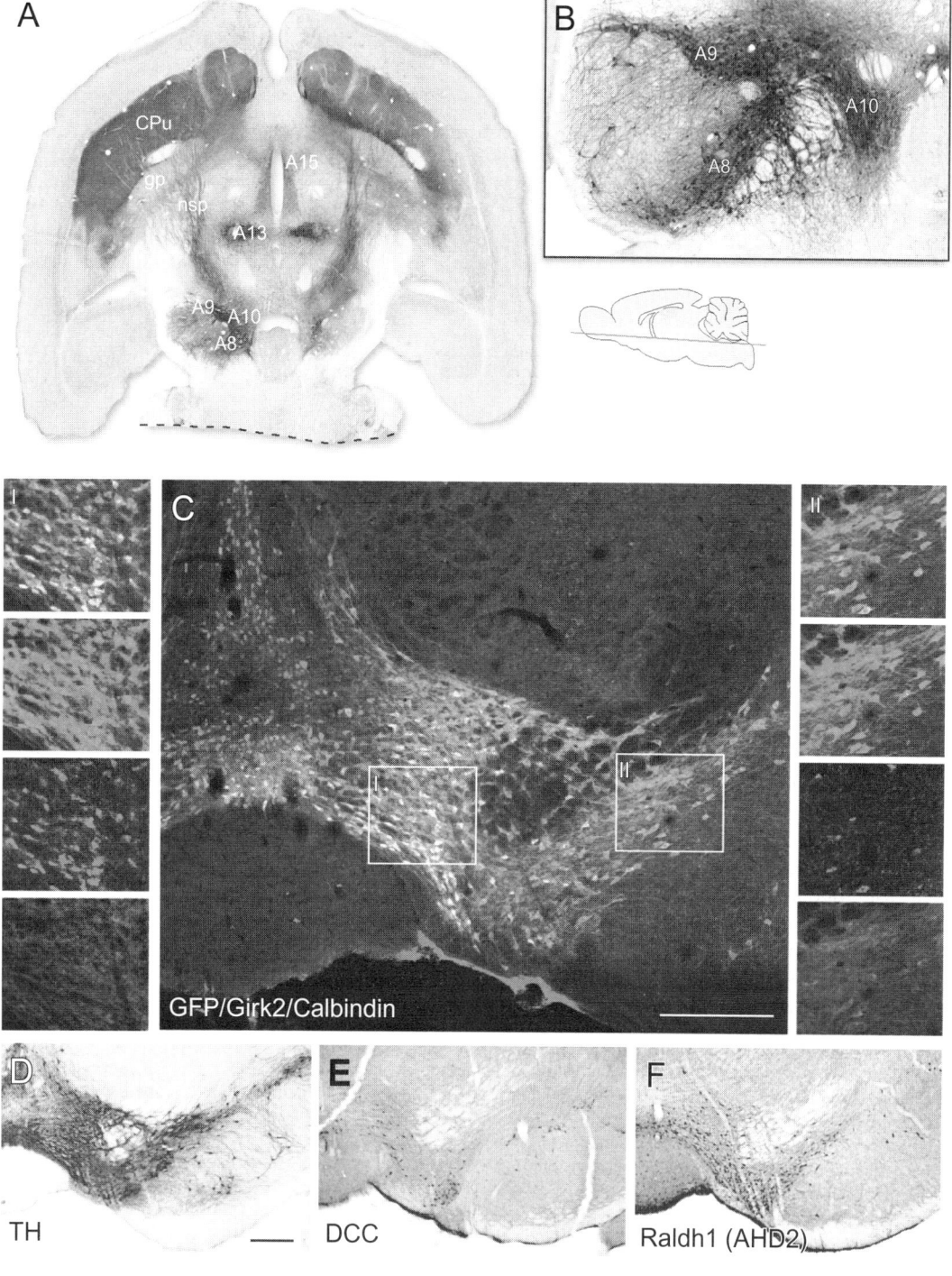

performance tests. In the most successful cases, patients have been able to completely withdraw from L-DOPA medication and have had sustained symptomatic relief for more than 10 years (Lindvall and Hagell, 2000; Dunnett, 2001). At the other end of the spectrum, some patients have not only shown lack of improvement, but have developed dyskinesias apparently associated with the graft (Freed et al., 2001; Olanow et al., 2003, 2009).

3. *Incomplete recovery*. Although intra-striatal VM grafts can facilitate recovery of simple/gross motor function, in both PD patients and in animal models, deficits in more complex aspects of fine motor skill and sensorimotor behaviors fail to improve based on current procedures (for review, see Winkler et al., 2000). While this is not necessarily a limiting issue for the application of cell therapy as a treatment option, it highlights there is significant scope for optimization of the therapeutic effect.

In this review, we discuss recent progress in the brain research field on these issues including: (a) the development of stem cell-based procedures for repair of the PD affected brain, (b) the relevance of mDA neuronal subtype for functional recovery, and (c) new insights into the potential for functional and anatomical reconstruction of the nigro-striatal pathway in the adult brain. We also highlight the important role of transgenic reporter mice as new tools for neural transplantation studies.

## Midbrain dopamine neurons

mDA neurons can be broadly divided into three distinct populations, including the retrorubral area (A8), substantia nigra (SN; A9), and ventral tegmental area (VTA; A10) cell groups (Fig. 2A and B). In the intact brain, they extend long axonal projections rostrally as an un-ramified fiber bundle, coursing through the medial forebrain bundle (MFB) and nigro-striatal pathway (Fig. 2A) in order to innervate various forebrain targets. The innervation of the striatum, which is derived predominately from the A9 neurons, forms a critical part of the basal ganglia circuitry that controls normal motor function (Bjorklund and Lindvall, 1984). The classification of mDA neurons as distinct populations (A8, A9, and A10; Fig. 2A and B), introduced by Dahlstrom and Fuxe (1964), is based on the cytoarchitectural arrangement and efferent projection patterns of these cell clusters, such that (a) *the A10 neurons* are located in a medial position spanning the midline and send projections to cortical and limbic structures including the nucleus accumbens, amygdala, hippocampus and the prefrontal and cingulate cortex to form the mesocorticolimbic pathway, (b) *the A9 neurons* form a compact layer of cells extending further laterally from the lateral border of the A10 group and send projections which predominately innervate the dorsolateral striatum to form the nigro-striatal pathway and to a lesser extent innervate extra-striatal areas including cortex, and (c) *the A8 neurons* lie caudal to the A9 cell group and innervate both limbic and striatal areas as well as providing a local innervation of A9 and A10 neurons

Fig. 2. Basic neuroanatomical features of the midbrain dopamine neuron projection system. (A) Immunohistochemistry for tyrosine hydroxylase in a horizontal section through the adult mouse brain shows the major midbrain dopaminergic cell groups (A8, A9, and A10) and their efferent projection patterns. The approximate section plane (red line) is indicated in the parasagittal diagram (note, this animal has received a partial 6-OHDA lesion on the right-hand side of the brain, reflected by a notable loss of TH-positive cell bodies on that side). (B) The spatial distribution of the A8, A9, and A10 cell groups shown at greater magnification. (C) Immunohistochemistry for GFP (green), Girk2 (blue), and calbindin (red) in a coronal section through the adult mouse midbrain. In this animal, GFP is expressed under control of the regulatory elements for Pitx3 and, therefore, in all dopamine neurons. Girk2 and calbindin broadly identify A9 and A10 dopamine neurons, respectively. The boxed areas show in greater detail cells in the VTA (I) and the substantia nigra pars compacta (II). Other proteins, including DCC (E) and Raldh1/AHD2 (F) identify ventral subsets of dopamine neurons in the midbrain. Tyrosine hydroxylase expression (D) is shown in an adjacent section as a point of reference. Abbreviations: CPu, caudate putamen unit; DCC, deleted in colorectal cancer; gp, globus pallidus; nsp, nigro-striatal pathway; TH, tyrosine hydroxylase. Scale bar: C, 200 μm (panels A, B, D–F, courtesy: S. Grealish). (See Color Plate 5.2 in color plate section.)

(Fallon and Moore, 1978; Swanson, 1982; Bjorklund and Lindvall, 1984; German and Manaye, 1993; Arts et al., 1996; for a recent review, see Bjorklund and Dunnett, 2007).

The anatomical division of dopamine-containing cell groups in the midbrain broadly defines the distribution of distinct subtypes of mDA neurons also on the basis of other criteria such as structural, functional, and molecular features as well as susceptibility to degeneration. The A10 group, for example, is predominately composed of small (10–15 μm in mouse) roughly spherical cells, many of which co-express calbindin and survive relatively well in PD pathology; while A9 neurons are larger (20–30 μm in mouse) with an angular morphology, the majority of which express the potassium channel subunit Girk2 (Fig. 2C) and these neurons are particularly vulnerable to degeneration in PD (McRitchie et al., 1996; Damier et al., 1999; Mendez et al., 2005; Thompson et al., 2005). However, the correlation between anatomical location across the A9/A10 cell groups and mDA neuronal subtype is by no means precise. Both calbindin- and Girk2-expressing neurons spill across the anatomical boundaries defining the VTA and SN regions, such that Girk2-positive neurons appear in the dorsolateral VTA and calbindin-positive mDA neurons are scattered throughout the dorsal part of the SN (Fig. 2C). Other proteins such as deleted in colorectal cancer (DCC) and Raldh1 define ventrally located subpopulations of mDA neurons that are distributed throughout both the VTA and SN (Fig. 2E and F; Osborne et al., 2005; Jacobs et al., 2007). Electrophysiological studies further highlight that the A9 and A10 groups are in themselves heterogeneous with respect to mDA subtype composition. For example, calbindin-positive and calbindin-negative neurons within either the VTA or SN define four subpopulations with distinct electrophysiological profiles (Neuhoff et al., 2002). Calbindin expression among mDA neurons has further significance in that it correlates with the cells that have the greatest capacity for survival in PD and a number of the associated animal models (German et al., 1992; Liang et al., 1996; Rodriguez et al., 2001; Maingay et al., 2006; Ekstrand et al., 2007). This aspect is evident when comparing the better survival of the calbindin-rich VTA population with the vulnerable SN populations, but also within the pars compacta layer of the SN (SNpc) itself. The SNpc mDA population can be further divided into dorsal and ventral tiers on the basis of both calbindin expression and vulnerability, whereby the dorsal tier contains virtually all calbindin-positive mDA neurons within the SNpc and is also more resistant to degeneration in PD relative to the ventral SNpc mDA neurons, which are among the most vulnerable (Damier et al., 1999).

## Isolation of transplantable midbrain dopamine neurons from stem cell-derived populations

The need for an alternative to fetal tissue as a source of transplantable mDA neurons has stimulated a great deal of research within the stem cell field. Of the many stem cell sources investigated to date, only those with pluripotent potential including embryonic stem (ES) cells and, more recently induced pluripotency stem (iPS) cells, have been shown to reliably generate mDA neurons (Lee et al., 2000; Perrier et al., 2004; Andersson et al., 2006b; Sonntag et al., 2007; Wernig et al., 2008). Unlike expanded populations of adult stem cells, which are restricted in the cell types they can generate, highly expanded populations of pluripotent ES cells can generate a vast array of differentiated cell types, including mDA neurons with A9 and A10 characteristics (Friling et al., 2009). In addition to servicing the need for an alternative to fetal tissue, stem cells are a promising cell source in that they present an opportunity to better standardize the transplantation procedure.

Although there has been substantial progress in the development of protocols that generate mDA neurons from ES cells, with yields of up to 60–80% of the total population of neurons in the culture dish (Chung et al., 2002; Andersson et al., 2006b), the translation of this success into an in vivo setting has been highly problematic. One of the most concerning issues has been the incidence of graft-derived tumors (Fig. 3). This likely reflects a lack of synchrony in the progressive patterning of ES cells into specified neural progenitors. In order to derive specialized neural cell types, ES cells are typically directed through a series of patterning steps that mimic corresponding events that occur during normal development, including a first phase of induction into primitive neuroectodermal cells, followed by differentiation into specific neural phenotypes (Fig. 3E and F). The transition at each step is often incomplete, however, so that at any point from the initial patterning of the undifferentiated ES cells, the cultures will contain a range of cells in various states of differentiation. This means at the time of transplantation, cultures that contain specified mDA neuronal progenitors, may also contain a residual component of earlier cell types capable of uncontrolled proliferation following implantation, including primitive neural stem cells such as neuroepithelial cells, or, in the worst case, undifferentiated ES cells.

Attempts to rid the cultures of these immature cell types through extended differentiation are complicated by the fact that this will also lead to continued differentiation and maturation of the mDA neurons, which will negatively impact on their ability to survive the implantation procedure, and also the potential for a small residual population of tumorogenic cell types to persist throughout the differentiation period. A promising alternative strategy is to separate the mDA progenitors from the mixed cell cultures through cell-sorting procedures prior to grafting, thereby leaving behind the tumorogenic cell types (Fig. 3F). ES cell lines genetically labeled with fluorescent reporter proteins are becoming increasingly common and allow for efficient separation of the labeled cells through fluorescent-activated cell sorting (FACS; Chung et al., 2006; Fukuda et al., 2006; Hedlund et al., 2008). Additionally, immunological labeling of cells using antibodies specifically directed against transmembrane proteins expressed by the target population is another effective means by which neural populations can be isolated from mixed cell preparations (Pruszak et al., 2007).

These cell-sorting procedures may involve both negative and positive sorting strategies, targeting either the tumor-forming population or the mDA neuronal progenitors, respectively, or a combination of both. In either case, the success of this strategy is dependent on: (1) knowing the identity of the target population, and (2) having a means to isolate this fraction. Targets for negative sorting include transmembrane proteins that are transiently expressed by ES cells and/or neural stem cells, including, for example, stage-specific embryonic antigens (SSEA1, 3, and 4) or CD133 (prominin). Specific antibodies raised against these proteins can be used to quite effectively remove most of the corresponding cell fractions from partially differentiated ES cell cultures (Pruszak et al., 2007). One concern with a reliance on negative sorting, however, is that the separation is rarely complete. Even with purity levels of 98–99%, which are common in these procedures, the small number of contaminating cells that avoid negative selection may be sufficient for tumor formation following implantation. Furthermore, given the heterogeneity of ES cell-derived populations, the possibility that multiple cell types can contribute to tumor formation is an important consideration. A number of recent investigations have instead chosen a positive selection strategy by, for example, isolating and grafting ES cell-derived cell fractions defined by Sox1 or PSA-NCAM expression, using a Sox1-GFP ES cell line (Chung et al., 2006; Fukuda et al., 2006) or an antibody against PSA-NCAM (Pruszak et al., 2007), respectively. Although the results have been promising with regard to avoiding tumor formation, both Sox1, which is expressed in neural stem and progenitor cells and PSA-NCAM, a transmembrane protein on immature neuroblasts, identify cell fractions which are significantly more broad than the mDA progenitor population. This means that the resulting graft will contain not only mDA neurons, but also many other neural cell types, some of which, may have unwanted effects. Serotonergic neurons, for example, which are often generated in ES cell differentiation procedures designed to yield mDA neurons, appear to play a significant role in the appearance of dyskinesias after grafting

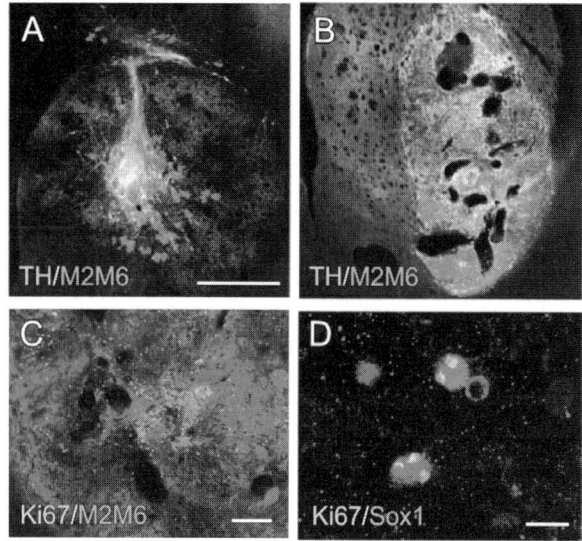

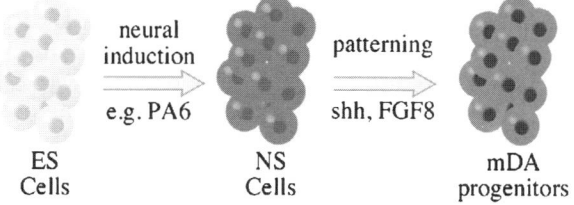

**E** Ideal differentiation of ES cells into mDA neuronal progenitors.

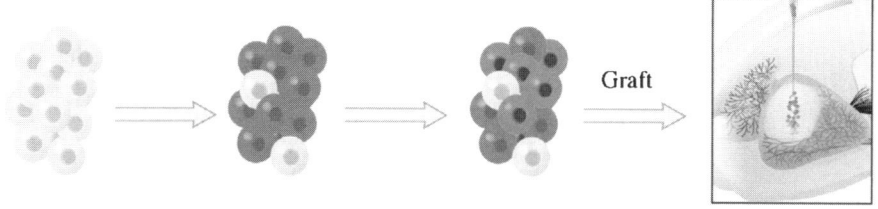

**F** In reality the differentiation process is often incomplete.

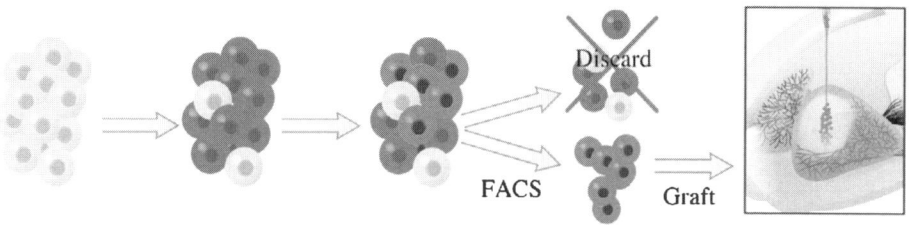

**G** Cell sorting offers a means to purify the mDA progenitor fraction prior to grafting.

(Carlsson et al., 2007). In an ideal scenario, it would be possible to selectively isolate only the mDA progenitors in order to standardize the cell preparations in terms of both the type and number of cells used for grafting. Unfortunately, this strategy is at present limited by a lack of information regarding the identity of the implantable mDA progenitor population in ES cell-derived cultures, and therefore, a means to isolate this fraction. Some recent insights, however, come from cell-sorting experiments using VM from transgenic reporter mice in which green fluorescent protein (GFP) is expressed in cell fractions corresponding to mDA progenitors at distinct stages of development.

## Reporter mice as tools in neural transplantation studies

Since their inception in the early 1980s, transgenic mice have become one of the most valuable and widely used tools in the brain research field. While the use of these animals has most commonly been associated with gain and loss of function studies, whereby specific genes are over-expressed or deleted, the development of so-called "reporter" mice has opened the way for other interesting applications. Reporter mice are typically engineered to express genes that encode nonmammalian proteins, such as, for example, GFP (originating from the jellyfish species *Aequorea victoria*), β-galactosidase (LacZ gene of *Escherichia coli*), or luciferase (*Photinus pyralis*). The first GFP-expressing mouse, reported by Okabe et al. (1997), was engineered to express GFP ubiquitously in all cell types by placing it under control of constitutively active promoter and enhancer elements. Since then, a raft of GFP mice have been developed in which GFP expression is restricted to specific cell populations by placing the GFP cDNA under the control of regulator elements that are active only in those cell types.

These GFP reporter mice have allowed for applications of particular interest to the neural transplantation field. For example, GFP-expressing cell fractions can be separated from mixed cell populations through FACS prior to transplantation. Additionally, the persistent expression of GFP following transplantation allows for the unambiguous identification of grafted cells and their processes within the host brain, with fine morphological detail. Here, we discuss specific examples of how these kinds of experiments have facilitated progress in the field of cell therapy for PD.

## Isolation of midbrain dopamine neuronal precursors from the developing midbrain

Dopamine neurons are generated in the mouse VM over a 3–4 day period between E9 and E13 (Bayer et al., 1995). In the early stages of development, at E9–E10, the mDA germinal zone consists of proliferating cells that express proteins characteristic of neural progenitor cells, including nestin and Sox2, and also genes involved in the intrinsic specification of mDA neurons, such as neurogenin2 (Ngn2) and Lmx1a (Andersson et al., 2006b; Kele et al., 2006; Thompson et al., 2006). Studies by Ono et al. (2007) have also

Fig. 3. Dopamine neuron-containing grafts derived from embryonic stem cells can give rise to tumors. (A, B) Grafts of mouse cells placed in the rat brain can be detected using antibodies against the mouse-specific antigens M2 and M6. Panels A and B illustrate the gross morphology of intra-striatal grafts derived from the same number ($1.0 \times 10^5$) of either E12.5 fetal VM cells (A) or partially differentiated ES cells (B), 6 weeks after grafting into neonatal rat hosts. Note the dramatically larger size of the ES cell-derived graft, along with pockets of necrosis (black) throughout the graft core. Both grafts also contain large numbers of TH-positive mDA neurons (green), which innervate the host striatum. (C) The ES cell-derived grafts contain a population of actively dividing (Ki67-positive, green) cells even 6 weeks after transplantation. (D) Many of the Ki67-positive cells are Sox1-positive and thus are likely to be primitive neural precursors. (E–G) Schematic representation of basic features of the differentiation procedures used to generate transplantable mDA neurons, and how a cell-sorting strategy might be used to avoid tumor formation after grafting. Abbreviations: FACS, fluorescence-activated cell sorting; FGF8, fibroblast growth factor 8; NS, neural stem; shh, sonic hedgehog. Scale bars: A–B, 1 mm; C, 200 μm; D, 20 μm. (See Color Plate 5.3 in color plate section.)

identified the transmembrane protein Corin as a maker of the floor plate cells which represent proliferative mDA progenitors. As development proceeds, these early mDA progenitors begin to leave the cell cycle and migrate out of the proliferative ventricular zone (VZ) to form an intermediate zone (IZ), populated by post-mitotic mDA neuroblasts that begin to express Nurr1. As these cells continue to differentiate, they will also begin to express Pitx3 and genes involved in the dopamine metabolic pathway, such as tyrosine hydroxylase (TH), and will begin to integrate into the developing host brain through the extension of axonal and dendritic processes. Thus, for much of the mDA neurogenic period the VM germinal zone contains mDA progenitors in various states of specification and differentiation, including: (a) a VZ comprised of proliferative mDA progenitors, (b) an IZ comprised of Nurr1-positive post-mitotic mDA neuroblasts, and (c) a mantle zone (MZ) containing the immature, TH-expressing neurons (Fig. 4).

Cell preparations of primary VM used for transplantation have typically been dissected midway through the neurogenic period (E12.5 in mouse, E14.5 in rat, and 6–8 weeks in human), and will therefore contain a mix of mDA cell types in these various states of differentiation. Historically, the immature TH-positive neurons have been viewed as the VM donor cells that give rise to a corresponding population of mDA neurons following grafting and, accordingly, the percentage yield of grafted mDA neurons has typically been estimated based on the TH-positive cell fraction contained in the VM at the time of dissection. Consistent with this idea, birth dating studies using E14.5 preparations of rat VM have shown that the mDA neurons in the resulting grafts are in fact post-mitotic at the time of grafting (Sinclair et al., 1999).

Experiments using donor tissue from earlier ages, and the use of transgenic mice in which GFP is driven by genes corresponding to distinct differentiation states within the mDA lineage, have shed new light on the identity of transplantable mDA progenitors across the neurogenic period. Transplantation studies using the Ngn2-GFP knock-in mouse have been particularly

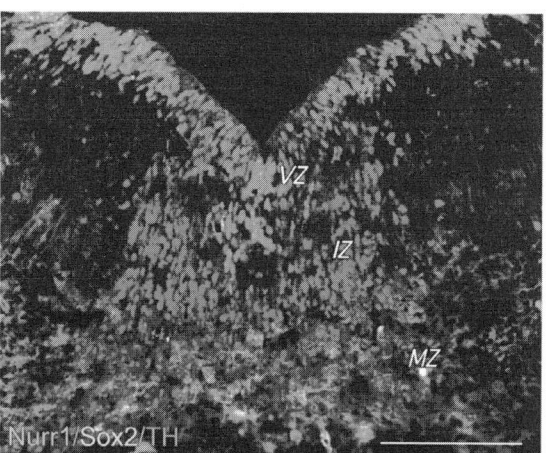

Fig. 4. Differentiation states of midbrain dopamine neuronal progenitors in the embryonic mouse brain. Immunohistochemistry for Sox2 (green), Nurr1 (red), and TH (blue) in a coronal section through the E12.5 mouse midbrain illustrates the distribution of neural progenitors in distinct states of differentiation. The ventricular zone (VZ) contains actively dividing Sox2-positive precursors. At this late stage of mDA neurogenesis (E12.5) most of the mDA progenitors have already exited the cell cycle and only very few of the VZ precursors will give rise to mDA neurons. Most of the transplantable mDA progenitors at E12.5 reside in the intermediate zone (IZ) as Nurr1-positive, post-mitotic neuroblasts. As these progenitors continue to differentiate, they move into the mantle zone (MZ) and begin to express TH. Very few of these TH-positive mDA neurons are able to survive the transplantation procedure. Scale bar: 200 μm. (See Color Plate 5.4 in color plate section.)

informative. In these mice GFP is driven by the regulatory elements for Ngn2 and therefore, its spatiotemporal expression pattern in the developing brain closely mimics that of Ngn2, which is involved in neuronal specification of neural progenitors in discrete regions, including the VM (Fig. 5A; Andersson et al., 2006a; Kele et al., 2006). At E12.5, endogenous Ngn2 is transiently expressed within a subset of progenitors in the VZ and is down-regulated as these cells exit the cell cycle and migrate into the IZ. The corresponding pattern of GFP expression lags slightly behind (likely due to differences in the rate at which GFP and Ngn2 are metabolized) such that it is expressed at highest levels in the Nurr1-positive/TH-negative post-mitotic neuroblasts (Fig. 5B). This GFP-positive population can be isolated at high purity by FACS (Fig. 5C), thus allowing for a comparison of the properties of the GFP-positive and GFP-negative populations. Experiments have shown that although transplantation of either fraction, isolated from E12.5 VM, yields neuron-rich grafts, virtually all of the mDA neurons are contained in grafts originating from the GFP-positive fraction (Thompson et al., 2006).

These data suggest that the identity of the transplantable mDA progenitor at E12.5 is a Nurr1-positive/TH-negative neuroblast at an intermediate state of differentiation and implies that other cell types contained in the GFP-

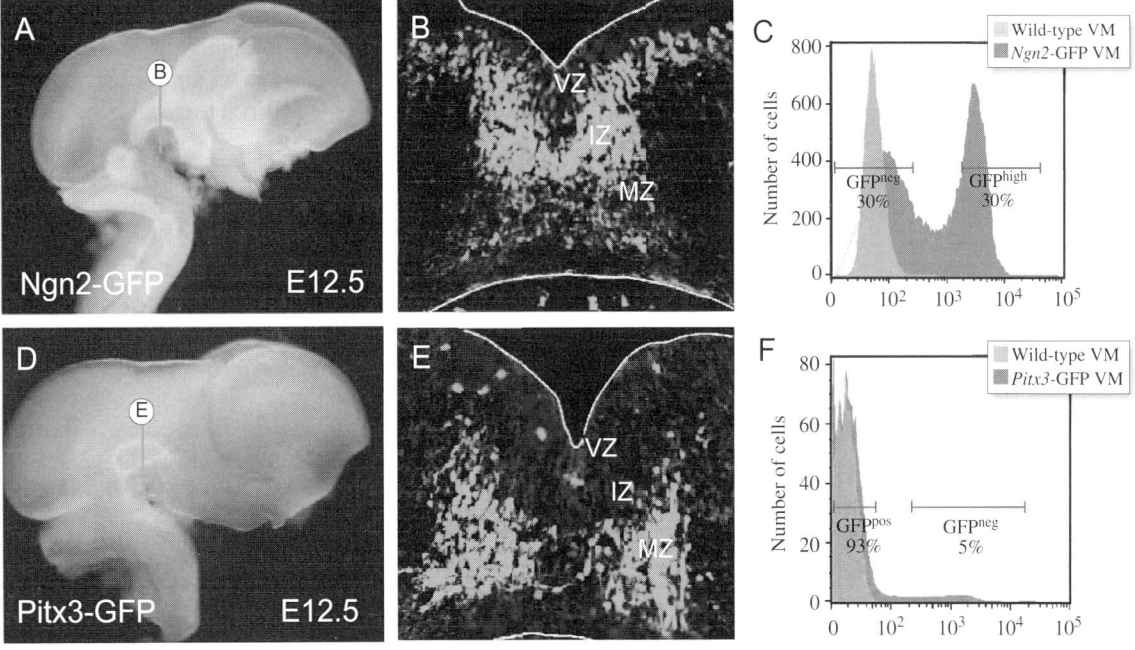

Fig. 5. Reporter mice can be used to isolate distinct progenitor populations from the embryonic midbrain. GFP fluorescence superimposed over brightfield photographs shows the regional distribution of GFP-expressing cells in the embryonic (E12.5) brains of the Ngn2-GFP (A) and Pitx3-GFP (D) mice. Immunohistochemistry for GFP in coronal sections through the midbrain of these mice illustrates the local distribution of GFP-positive cells. (B) In Ngn2-GFP mice, GFP expression identifies a population of newly post-mitotic neuroblasts in the intermediate zone (IZ). (E) In the Pitx3-GFP mice, the GFP-positive cells represent young mDA neurons in the mantle zone (MZ; the approximate section plane for the coronal images is indicated by the red line in A and D). These GFP-positive cell fractions can be selectively isolated from dissected VM tissue pieces through fluorescence-activated cell sorting (FACS). (C) FACS analysis of VM tissue pieces from E12.5 Ngn2-GFP mice identifies a distinct subpopulation of highly GFP-positive cells, which represents approximately 30% of all viable cells. (F) FACS analysis of E12.5 Pitx3-GFP VM identifies a subset of GFP-positive cells representing around 5% of the viable cell population. In order to establish the threshold for specific detection of GFP-positive cells, the gate settings on the FACS apparatus are determined using cell suspensions prepared from wild-type littermates (gray in C and F). Panel A is a modified reproduction from Thompson et al. (2006). (See Color Plate 5.5 in color plate section.)

negative fraction, including dividing progenitors in the VZ and immature TH-positive neurons in the MZ, do not contribute significantly to the mDA neuronal populations in grafts of E12.5 VM tissue. Supporting the lack of contribution from TH-positive MZ neurons, experiments using GFP-positive and GFP-negative fractions isolated by FACS from a Pitx3-GFP knock-in mouse (where GFP is expressed from the Pitx3 gene locus, Fig. 5D), show that the GFP-positive fraction (which will contain the young TH-positive neurons, Fig. 5E) gives rise to very few mDA neurons after grafting, and that the majority of the graftable mDA neurons actually come from the GFP-negative fraction, which will contain both of the earlier IZ and VZ cell types (Jonsson et al., 2009). An important conclusion from these findings is that even the most immature TH- and Pitx3-expressing neurons present in VM have an extremely limited capacity to survive the sorting and transplantation procedures.

There are at least two possible explanations for the apparent failure of E12.5-derived VZ progenitors to yield mDA neurons after grafting: (1) these cells lack the degree of intrinsic specification required to maintain a mDA differentiation program when removed from the developing brain and placed in an environment lacking instructive cues, or (2) the neurogenic program has shifted by E12.5, such that the VZ progenitors at this stage of development are now specified to produce alternative neural phenotypes. Results from grafting work utilizing VM tissue at earlier stages of development suggests the latter explanation to be more likely. In one study, the investigators have compared the ability of VM dissected from rats at various developmental stages between E11 and E14 to yield mDA neurons after grafting, and found that donor cells from E12 gave the greatest yield of mDA neurons as percentage of the total number of cells grafted (Torres et al., 2007). At this timepoint (corresponding to approximately E10 in the mouse) the VM consists almost exclusively of dividing VZ progenitors and thus, even at this early stage these cells are sufficiently specified to differentiate into mDA neurons after transplantation. Further evidence regarding the mDA developmental potential of VM progenitors at different ages comes from a comparison of the ability of VZ progenitors contained in E10.5 or E12.5 mouse VM to yield mDA neurons following grafting (Jonsson et al., 2009). In this study, the VZ progenitors were isolated from other cell types in the VM by virtue of the selective expression of the transmembrane protein Corin, which is transiently expressed by dividing VZ progenitors (Ono et al., 2007). FACS purification of the Corin-positive cell fractions, using a Corin-specific antibody, and subsequent grafting showed that the E10.5 VZ progenitors had a substantially greater capacity to generate mDA neurons relative to the corresponding fraction isolated from E12.5 VM.

*In summary,* the identity of the transplantable mDA progenitor population in fetal VM is dependent on the developmental stage of the embryo. At early ages, a significant proportion of the actively dividing progenitor cells within the Lmx1a-expressing ventral midline are already sufficiently specified to maintain a mDA terminal differentiation program following transplantation. At later stages of development there is a residual component of engraftable post-mitotic mDA neuroblasts generated from the active phase of mDA neurogenesis. Encouragingly, with regard to the implications for a purification strategy in an ES cell setting, both of these cell types are amenable to isolation and purification using genetic or immunological targeting strategies and also survive the FACS procedures reasonably well. This does not appear to be the case for the more mature TH/Pitx3-expressing neurons, which do not survive the dissociation and/or the FACS process, and suggests that the earlier mDA progenitors represent more optimal targets for sorting of ES cell-derived populations. Given the emphasis on avoiding tumor formation as a basis for adopting a cell-sorting approach, the early post-mitotic Nurr1-positive/TH-negative neuroblast would appear to be a good candidate. A remaining challenge is to identify an appropriate means by which to selectively isolate this fraction. GFP expression driven by Ngn2, for example, while relatively specific for mDA neuron specification in the VM, will be significantly broader in an ES cell context and thus, offer a more limited means for purification of the mDA component. Furthermore, in a clinical setting, immunological targeting of the mDA progenitors using specific antibodies is likely to be a more desirable approach than the engineering of cells to express fluorescent proteins. Accordingly, there may be considerable value in the identification of transmembrane proteins expressed selectively on transplantable mDA progenitors.

**Graft composition and its relevance for functional impact**

The developing midbrain is highly mixed with respect to the range of cell types present, containing progenitors for various glial and neuronal phenotypes. This is reflected in the composition of mature grafts of fetal VM, which are also highly heterogeneous. In addition to dopamine neurons, which in fact represent only a minor fraction of the grafted cells, the grafts will typically also contain serotonin-, GABA (γ-aminobutyric acid)-, enkephalin-, and substance P-containing neurons as well as many that cannot be readily identified based on neurochemical features (Bolam et al., 1987; Dunnett et al., 1988; Mahalik and Clayton, 1991; Thompson et al., 2008; Fig. 6). Glial cells with features consistent with an astrocytic or oligodendrocytic phenotype are also present. While the integration and functional impact of mDA neurons in fetal VM grafts has been intensely scrutinized, much less has been known about the properties of other neuronal phenotypes present, and possible consequences for the host. Studies using species-specific antibodies to characterize graft-derived fiber outgrowth following intra-striatal grafting of either porcine (Isacson and Deacon, 1996) or murine (Thompson et al., 2008) VM into the rat have shown, however, that the non-mDA neurons integrate extensively within the host, extending axons to innervate the surrounding striatum, and also over long distances to innervate extra-striatal structures including the thalamus, cortex, and midbrain (Fig. 6E). While any functional consequences of connections with targets outside the striatum, many of which appear to be GABAergic, remain unknown, there is now substantial evidence to suggest that graft-derived serotonergic innervation of the striatum can exacerbate or even cause dyskinetic behavior (Carlsson et al., 2007, 2009). This phenomenon appears to be caused by the storage and poorly controlled release of dopamine as a "false transmitter" by serotonergic neurons and will be most apparent when there are high numbers of grafted serotonergic neurons and relatively few dopaminergic fibers (of either host or graft origin) in the host striatum to buffer the resulting "overflow" of dopaminergic transmission (Carlsson et al., 2009). The ratio of mDA to serotonergic neurons in mature VM grafts is likely to be affected by various aspects of the transplantation procedure, including the dissection technique and age of the tissue (which can vary widely between different teams of investigators) and exemplifies the need to better standardize the composition of cell preparations used as a means to achieve a more predictable therapeutic outcome.

Grafts of fetal VM will also contain a mixture of different mDA neurons, including both the A9 and A10 subtypes (Mendez et al., 2005; Thompson et al., 2005). The large, Girk2-positive A9 neurons are typically distributed throughout the periphery of the graft, while the smaller, A10, calbindin-positive neurons tend to cluster in a more medial position in the center of the graft core (Fig. 7A and B). This predictable cytoarchitectural arrangement of mDA subtypes with the grafts is reminiscent of the medial to lateral arrangement of A10 (medial) and A9 (lateral) in the fully developed midbrain and suggests cell intrinsic signaling mechanisms within the graft milieu that confer positional instructions between different mDA subtypes. Remarkably, both the A9 and A10 mDA neurons are also capable of extending axonal projections over long distances in the host brain in order to innervate the normal developmental targets appropriate for each of these subtypes. Retrograde tracing experiments show that fluorescent microbeads injected into the dorsolateral striatum of rats with intra-striatal VM grafts will selectively identify the peripherally located Girk2-positive mDA neurons in the graft, while

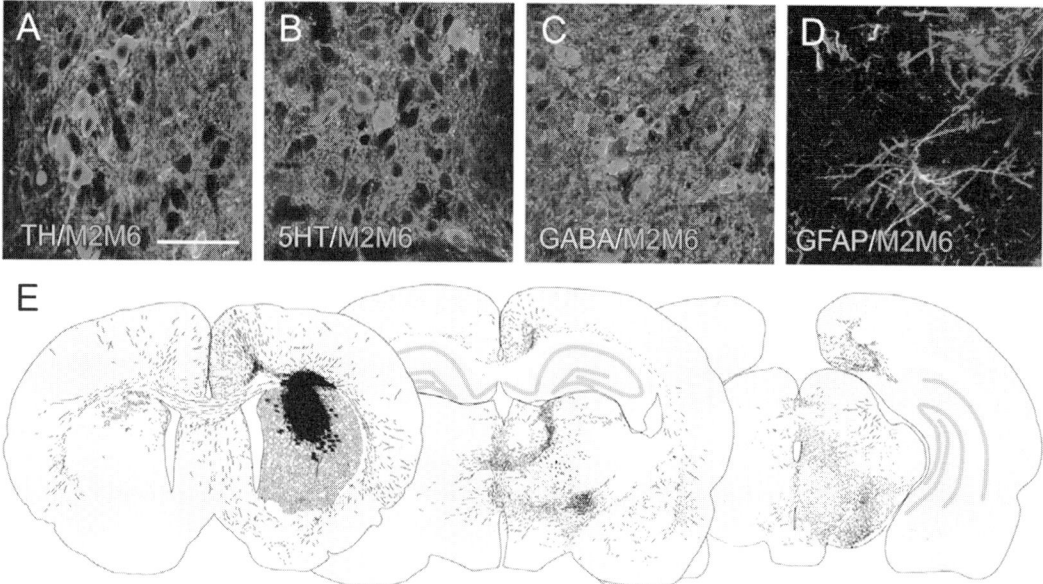

Fig. 6. Non-dopaminergic cell types in VM grafts. Grafts of VM will contain neurons corresponding to a variety of neurochemical phenotypes, and also various kinds of glial cells. Six weeks after transplantation of E12.5 mouse VM cells into the striatum of neonatal rats, immunohistochemistry for the mouse-specific M2 and M6 proteins allows for clear identification of the grafted cells. In addition to TH-positive mDA neurons (A), the grafts will also contain 5HT-positive serotonergic neurons (B), and a large number of γ-aminobutyric acid (GABA) containing neurons (C). The grafts also contain various glial subtypes, including those that are immunoreactive for glial fibrillary acidic protein (GFAP). The M2M6 antigens are expressed throughout the neuritic processes of certain classes of neurons, allowing for identification of patterns of fiber outgrowth. A schematic representation of immunohistochemistry for M2 and M6 in coronal sections 6 weeks after grafting of E12.5 mouse VM cells into the striatum of a neonatal rat, illustrates the pattern of graft-derived fiber outgrowth in the host brain. Double labeling of M2M6 and TH (not shown), indicates that while the vast majority of striatal M2M6-positive innervation is dopaminergic, most of the fibers found in structures outside the striatum are non-dopaminergic. Scale bar: A–D, 50 μm (images shown here are modified reproductions from Thompson et al., 2008). (See Color Plate 5.6 in color plate section.)

beads injected into the prefrontal cortex are transported predominately to the calbindin-positive mDA neurons in the center of the graft (Thompson et al., 2005). These findings suggest that the A9 and A10 mDA neuronal subtypes are intrinsically programmed for target-specific outgrowth already at the time they are taken for grafting (i.e., an early neuroblast stage at E12.5) and, importantly, that the adult brain retains the capacity to direct targeted outgrowth through appropriate interaction with membrane-bound growth and guidance proteins on the outgrowing mDA neuronal processes.

At a functional level these results also imply that the mDA neuronal subtype composition of intra-striatal VM grafts is an important consideration for therapeutic outcome. Specifically, the presence of A9 neurons that have the capacity to provide an extensive reinnervation of the denervated striatum appears to be an important requirement. Indeed, there is compelling evidence to support this from grafting experiments using VM tissue from the Pitx3-GFP mice, developed by Maxwell et al. (2005), in which GFP has been knocked into the gene locus for Pitx3. These mice can be bred as either heterozygous for GFP (Pitx3$^{+/GFP}$) or homozygous, so that GFP is knocked into both gene alleles (Pitx3$^{GFP/GFP}$). The Pitx3$^{GFP/GFP}$ mice are therefore Pitx3 knock-outs and display a phenotype consistent with the aphakia mice (which carry a loss of function mutation in the Pitx3 gene; Nunes et al., 2003) including a substantial loss of A9 mDA neurons in the adult midbrain and relative sparing of the A10

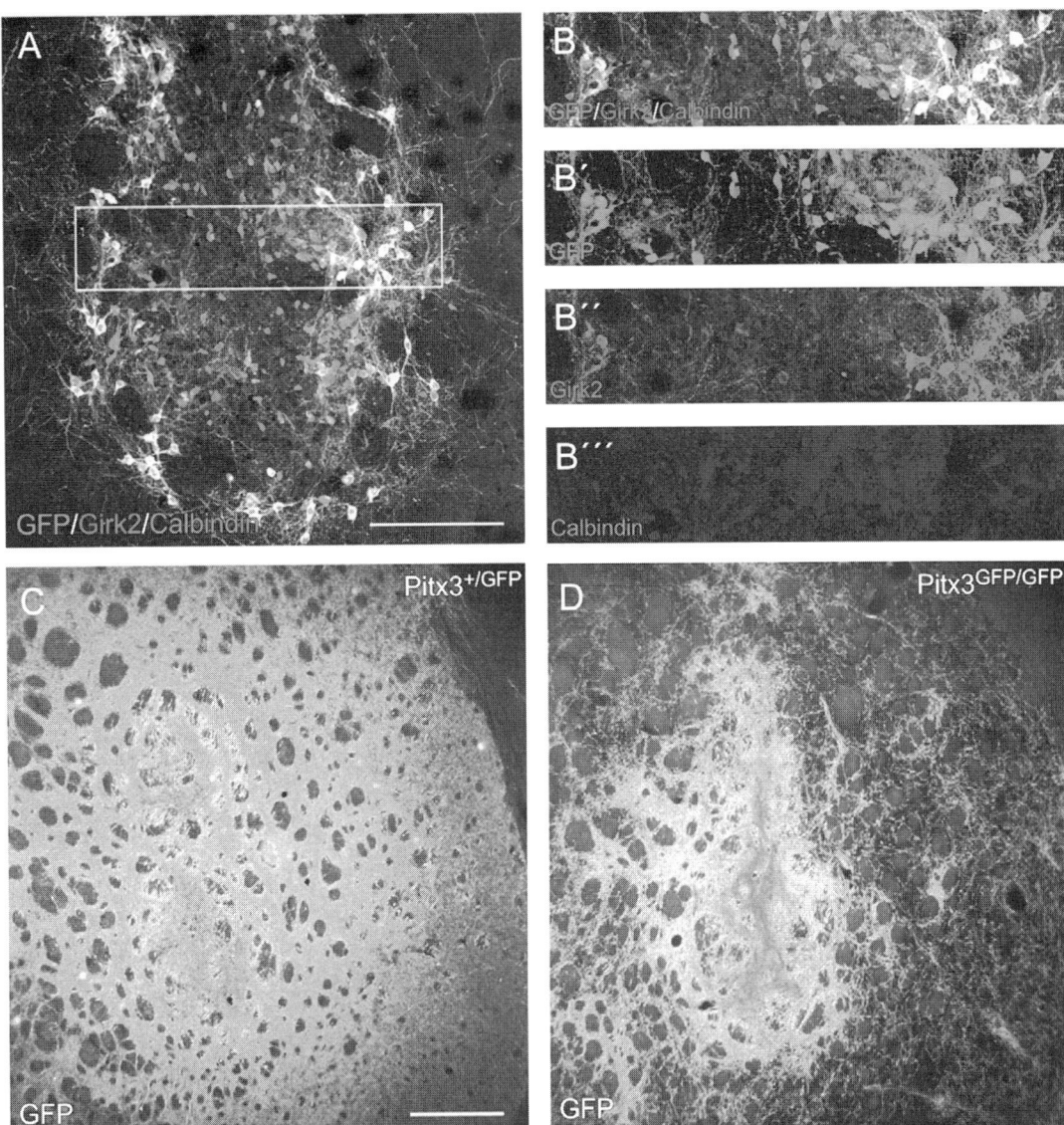

Fig. 7. Contribution of different midbrain dopamine neuronal subtypes in ventral mesencephalic grafts. The potassium channel protein, Girk2, and the calcium-binding protein, calbindin, can be used to broadly identify mDA neurons of the A9 and A10 subtype, respectively, in VM grafts. (A) Immunohistochemistry for Girk2 (red) and calbindin (blue) in coronal sections of a mouse that received an intra-striatal graft of E12.5 VM cells from Pitx3-GFP donor mice, reveals that the A9, Girk2-positive/GFP-positive neurons (yellow) are distributed throughout the periphery of the graft, while the A10, calbindin-positive/GFP-positive cells (aqua) are clustered mainly in the center of the graft. The boxed area, spanning peripheral and central aspects of the graft, is shown in higher magnification (B) and as individual color channels (B′–B′′′). The knock-in design of the Pitx3-GFP reporter mice means that VM cell suspensions can be prepared from mice either heterozygous ($Pitx3^{wt/GFP}$) or homozygous ($Pitx3^{GFP/GFP}$) for GFP, and therefore null for Pitx3 in the latter case. Darkfield images of immunohistochemistry for GFP 12 weeks after grafting of either $Pitx3^{wt/GFP}$ VM (C) or $Pitx3^{GFP/GFP}$ VM (D) shows that the $Pitx3^{GFP/GFP}$ grafts, which have a markedly reduced proportion of A9 neurons (not shown), also display a significantly reduced capacity to provide dopaminergic innervation of the host striatum. Scale bars: A, 200 μm; C, D, 500 μm. (See Color Plate 5.7 in color plate section.)

population (Maxwell et al., 2005). This is reflected in the mDA neuronal subtype composition of intra-striatal grafts of E12.5 VM taken from Pitx3$^{GFP/GFP}$ mice, which are rich in A10 mDA neurons but contain a substantially reduced number of A9 neurons. Importantly, these A9-deficient grafts provide only a poor level of innervation of the surrounding striatum (Fig. 7D and E) and fail to improve motor function in 6-OHDA-lesioned rats, as compared to similar sized grafts of Pitx3$^{+/GFP}$ (or wild type) VM with a normal complement of A9 and A10 mDA neuronal subtypes in roughly equal numbers. Similar results have been reported by Kuan et al. (2007) in a study demonstrating that the number of A9 mDA neurons in intra-striatal grafts positively correlates with the degree of functional recovery, and also with the ability of the grafts to ameliorate L-DOPA-induced dyskinetic behavior established prior to grafting.

Together, these findings highlight the importance of the presence of mDA neurons with the correct, nigral phenotype for functional recovery following intra-striatal grafting and underscore the previously held view that functional impact is dependent on robust innervation of the host striatum and integration at the synaptic level. Furthermore, in the likely case that procedural variations in the preparation and transplantation of fetal VM (donor age, dissection technique, etc.) will affect the relative numbers of A9 and A10 subtypes in the resulting grafts, this factor is likely to contribute to the variable outcome seen in VM-grafted PD patients.

The concept of graft composition and how this contributes to the overall functional effect of the graft is important to consider also in the context of stem cell-based approaches for cell therapy in PD. At present, even the most efficient procedures for the generation of mDA neurons from ES cells yield highly mixed populations which contain progenitors for a range of neuronal and glial phenotypes. This is reflected in the resulting grafts, which often contain high numbers of serotonin- and GABA-containing neurons (Bjorklund et al., 2002; Kim et al., 2002). Serotonergic neurons, which may cause unwanted side effects, are a particularly common "by-product" produced alongside dopamine neurons as a result of protocols used to pattern the ES cells. In the interest of eliminating unwanted cell types, and also in order to provide a higher degree of standardization of ES cell preparations, there is a need to improve on the currently available procedures for mDA-specific patterning. Along these lines, Andersson et al. (2006b) have found that forced expression of the intrinsic mDA determinant gene, *Lmx1a*, in ES cells at the onset of neural induction significantly increases the efficiency of mDA differentiation (up to 60% of all neurons) and also acts to suppress other neuronal fates including the generation of serotonergic neurons.

As discussed above, isolation of mDA progenitors from mixed cell populations prior to grafting may provide a valuable means to eliminate unwanted cell types. This approach might also enable greater control over the mDA neuronal subtype composition of ES cell-derived cell preparations used for grafting. Although there is limited information on the topic, current procedures for the mDA-specific patterning of ES cells, involving the application of sonic hedgehog and FGF8, appear to yield separate populations of mDA neurons with either A9 or A10 properties (Friling et al., 2009; Ferrari et al., 2006). Very little is currently known regarding either extrinsic patterning signals or intrinsic genetic determinants underlying the specification of different mDA neuronal subtypes during normal development. A rational basis for the development of mDA subtype-specific differentiation procedures will, therefore, depend on further progress in this area. Studies using fetal tissue, however, suggest that it may be possible to identify and isolate separate progenitor populations for A9 and A10 mDA neurons during the early phase of midbrain neurogenesis based on the differential expression of the transmembrane protein Corin (Jonsson et al., 2009). As described in the previous section, Corin is expressed in mice as early as E9 by radial glia within the proliferative Lmx1a-positive domain in the developing VM and appears to identify the early floor plate cells in this region (Ono et al., 2007). The expression is highest around the midline and becomes progressively lower toward the lateral parts of the Lmx1a-positive domain. Cell-sorting experiments, using an antibody directed against an extracellular part of Corin, have shown it is possible to target and separate the highly Corin-positive midline population from the weakly Corin-positive lateral populations through FACS procedures (Jonsson et al., 2009). Subsequent grafting of these isolated populations shows that the progenitors for A9 mDA neurons segregate to the lateral, weakly Corin-positive region at an early stage of mDA neurogenesis, and can be selectively isolated on this basis prior to grafting. While we will have to await further investigation to see if the same approach can be applied to differentiated cultures of ES cells, the results show that in principle it is possible to select for distinct mDA neuronal subtypes based on the targeting of transmembrane proteins differentially expressed by the corresponding progenitor populations. The ability to control not only the number but also the subtype of mDA neurons present in cell preparations used for grafting would represent an attractive prospect in the interest of standardizing graft composition as a means to achieve a more predictable therapeutic outcome.

## Reconstruction of the nigro-striatal pathway through intra-nigral grafting

Current transplantation procedures in PD involve the placement of VM cell preparations directly into the striatal target. In patients, this involves the stereotaxic injection of cells into the putamen and/or caudate nucleus, typically at multiple sites through 2–3 injection tracts. The resulting grafts are capable of providing an extensive, although partial, dopaminergic reinnervation of the host striatum along with significant improvements in motor function (Kordower et al., 1996; Mendez et al., 2005). Not all aspects of motor function are improved however. Despite substantial improvement in symptoms such as akinesia and rigidity, grafted patients will invariably continue to display deficits in motor function (Lindvall and Hagell, 2000). This is also the case in animal models of PD. While significant improvement in drug-induced rotational bias or simple motor tasks can be routinely achieved through intra-striatal grafting in experimental parkinsonian animals, more complex behaviors are only slightly improved or unaffected (for review, see Winkler et al., 2000). Even in the best cases of striatal reinnervation, using a micro-transplantation approach to spread cell deposits over multiple sites, behaviors such as skilled forelimb use and sensorimotor response, are only partially corrected (Nikkhah et al., 1993; Winkler et al., 1999; Kirik et al., 2001). While the underlying reason for the incomplete nature of functional recovery remains unclear, the placement of the cells not in their normal location in the midbrain, but in an ectopic position in the striatum may well be a contributing factor. Although this approach is indeed necessary in order to achieve a robust reinnervation of the host striatum, dopamine neurons placed in the striatum are likely to lack some of the key afferent inputs that may be required for optimal regulation of dopamine neuron function during ongoing behavior. In their appropriate midbrain location, mDA neurons receive afferent connections from a variety of nuclei including the locus coeruleus, medial raphe, striatum, cortex and subthalamic nucleus (Gerfen and Wilson, 1996). Furthermore, in addition to sending long axonal projections to the forebrain, nigral dopamine neurons integrate structurally and functionally also at the level of the midbrain through dendritic innervation of the underlying SN pars reticulata (Bjorklund and Lindvall, 1975; Geffen et al., 1976; Nieoullon et al., 1977; Robertson, et al., 1991a; Robertson et al., 1992). The release of dopamine by dendrites in the SNr facilitates striato-nigral transmission through activation of presynaptic D1 receptors located on the terminals of striato-nigral projections (Robertson, 1992; You et al., 1994; Rosales et al., 1997). This part of normal basal ganglia circuitry, lost during the disease process, is not in any way restored by placing new mDA neurons in the striatum. Thus, on the face of these apparent shortcomings related to the ectopic, striatal placement of new mDA neurons, there is a continuing interest in this field to achieve a more accurate reconstruction of the nigro-striatal pathway by grafting the cells back into their normal midbrain location.

Attempts to functionally reconstruct the nigro-striatal pathway in animal models of PD by placing grafts into the SN have met with limited success. The initial intra-nigral grafting experiments, using fetal rat VM allografted into rats with lesions of the intrinsic mDA system, reported that although the grafted mDA neurons survived in the midbrain location, the grafts had no detectable functional impact and failed to extend axons along the nigro-striatal pathway (Bjorklund et al., 1983; Dunnett et al., 1983). Subsequent studies, found some slight improvement in certain motor tests, including rotation induced by dopamine agonists and postural balance but again reported that the outgrowth of TH-expressing fibers from the grafts was limited to localized innervation within the midbrain (Nikkhah et al., 1994; Mendez et al., 1996; Bentlage et al., 1999; Winkler et al., 1999; Mukhida et al., 2001). This does suggest, however, that reinstatement of local release of dopamine at the level of the midbrain may indeed be an important consideration for functional repair. Some attempts have been made to improve on the behavioral recovery associated with intra-striatal grafting alone by performing "double grafts" where VM tissue is placed in both the striatum and SN of 6-OHDA-lesioned animals. The results have been mixed, however, with these studies reporting either no (Olsson et al., 1995; Robertson et al., 1991b) or a modest level of additional improvement (Mendez et al., 1996; Mukhida et al., 2001).

A more effective functional impact from intra-nigral grafts is most likely dependent on the axonal growth of the implanted mDA neurons along the nigro-striatal pathway and reinnervation of the host striatum. Earlier studies using so-called "bridge grafts" have demonstrated that mDA neurons grafted into the midbrain at least bare the intrinsic capacity for axonal extension over the long distances required to reach forebrain targets. In these experiments, growth-permissive substrates, such as peripheral nerve (Aguayo et al., 1984; Gage et al., 1985) or Schwann cells (Brecknell et al., 1996; Wilby et al., 1999), placed between the striatum and the intra-nigral grafts allow for the extension of TH-positive axons along the substrate which can then innervate the striatum. Other encouraging findings come from xenografting studies, which show that when pig or human VM tissue is grafted into the midbrain of 6-OHDA-lesioned adult rats the dopamine neurons can extend axons along the nigro-striatal pathway in order to innervate appropriate forebrain targets, including the striatum (Wictorin et al., 1992; Isacson et al., 1995). Furthermore, Bentlage et al. (1999) found that dopamine neurons in VM grafts placed in the SN of 6-OHDA-lesioned neonatal hosts (postnatal day 3 or 10) can also extend axons along the nigro-striatal pathway and provide a substantial level of innervation of the host striatum. The functional outcome as well as normalization of amphetamine- and apomorphine-induced c-Fos expression was comparable to that seen after intra-striatal grafting. As part of the same study, when the grafts were placed into the SN of slightly older animals (postnatal day 20) the functional improvement as well as the ability of mDA axons to reconnect with the striatum was lost.

These findings, along with the failure in earlier studies to detect a significant mDA outgrowth from intra-nigral VM grafts placed in the adult brain, lead to the view that the adult brain is incapable of supporting dopaminergic outgrowth along the nigro-striatal pathway. Subsequent studies, however, using donor cell preparations in which the mDA neurons express GFP, have shown that this is not the case (Thompson et al., 2009). In these experiments, the VM was dissected from TH-GFP mice at E12.5 and a micro-transplantation approach was used to deliver small deposits of cells to the midbrain of adult mice with partial lesions of the nigro-striatal pathway. The GFP expression allowed for unequivocal identification of graft-derived mDA-specific fiber outgrowth, even in the presence of remaining nigro-striatal projections within the host. Immunohistochemistry for GFP in horizontal sections through the host brain 16 weeks after grafting showed a remarkable degree of regrowth along the nigro-striatal pathway from the grafted, GFP-positive mDA neurons (Fig. 8). The pattern of outgrowth was quite specific and well matched to the normal structure of the intrinsic mDA projection system. Many GFP-positive fibers exited the rostral part of the graft in a highly polarized manner and ran parallel to the MFB as a loosely bundled collection of elongated fibers. On reaching the striatum, the GFP-positive fibers abruptly gave rise to a highly ramified terminal network to form a patchy innervation that was most prominently close to the palladial-striatal border. A number of the grafted mice also showed a complete normalization of amphetamine-induced turning behavior, indicating that the new innervation provided by the graft was indeed functional.

A notable feature of the graft-derived GFP-positive fiber outgrowth was the intermixing with remaining host fibers throughout the length of the nigro-striatal pathway. This raises the interesting possibility that the host fibers play a role in facilitating the growth of the graft fibers along the pathway. The concept of "pioneer axons" that first form connections with the target during development, when distances are relatively short, and act as a scaffold for the later growing axons, has been extensively demonstrated as a mechanism for axonal growth and guidance in the central nervous system (CNS) (Klose and Bentley, 1989; McConnell et al., 1989, 1994; Lin et al., 1995; Hidalgo and Brand, 1997; Molnar et al., 1998a; Pittman et al., 2008). Although closer inspection of the GFP-positive and host-derived mDA fiber patterns within the nigro-striatal pathway did not suggest a contact-mediated interaction, there is also the possibility that host mDA axons can support axonal growth and guidance through the release of diffusible chemoattractive or trophic factors, such as brain-derived neurotrophic factor (BDNF), which is known to be produced by mDA neurons (Bustos et al., 2004). If indeed this is the case, it may help to explain why previous intra-nigral grafting experiments have failed to detect a graft-derived dopaminergic outgrowth along the nigro-striatal pathway. In these earlier experiments, the lack of an independent marker for the grafted cells, such as GFP, required that the host system was completely removed prior to grafting in order that any TH-positive fiber outgrowth could reasonably be attributed to the grafted cells.

During normal development, the growth of mDA neuron axons rostrally along the MFB is determined by a combination of local directional cues, associated with the growth substratum, and more long distance chemoattractive influences present along the MFB and in the striatal primordium (Nakamura et al., 2000; Gates et al., 2004). It is unclear, however, to what extent the same mechanisms might operate in the adult brain, to guide axons from implanted mDA neurons. Nonetheless, there is certainly evidence to suggest that quite specific interactions exist between grafted mDA neuronal axons and the host environment that allow for target directed axonal outgrowth in the adult brain. When grafts are placed outside, but directly adjacent to the denervated striatum, for example, the outgrowing fibers will show a clear target preference for the nearby striatum (Bjorklund et al., 1983). As discussed in the previous section, the adult brain also retains the capacity to direct the projections from A9 and A10 mDA neuronal subtypes to their appropriate targets (Thompson et al., 2005). There is little information at present concerning the identity of molecules involved in the growth and guidance of mDA fibers in vivo. Expression analysis of known guidance molecules at the histological level suggests that DCC (Osborne et al., 2005) and Ephrin family proteins may be involved (Yue et al., 1999; Sieber et al., 2004). At the functional level, manipulations in vitro, using VM cell cultures, suggest that Netrin and Slit family proteins can modulate neurite outgrowth through their respective receptors, DCC and Robo1/2 (Lin et al., 2005; Lin and Isacson, 2006). There is also evidence to suggest that neurotrophic factors can play a role. Removal of the intrinsic innervation of the striatum has been shown to have a clear impact on the level of dopaminergic outgrowth from VM grafts. Specifically, grafts placed in the denervated striatum will innervate a larger striatal volume relative to those placed in the intact host (Doucet et al., 1989; Kirik et al., 2001; Thompson et al., 2005). Studies in 6-OHDA-lesioned rats suggest that the up-regulation of diffusible growth-promoting

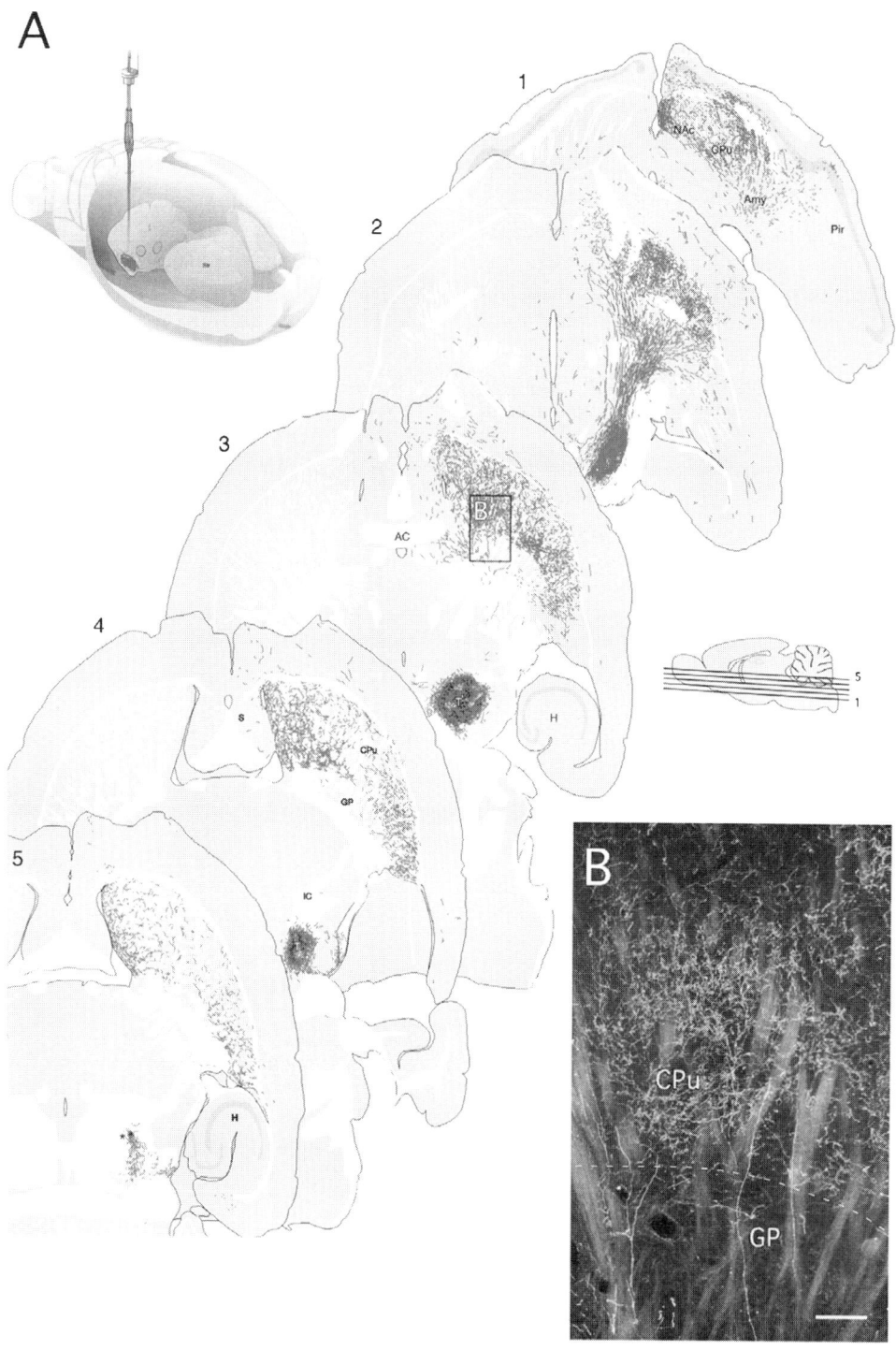

factors, including BDNF (Zhou et al., 1996) and glial cell line-derived neurotrophic factor (GDNF) (Yurek and Fletcher-Turner, 2001) may be responsible for this effect. GDNF, moreover, has been shown to act as an attractant for dopaminergic outgrowth from VM grafts in vivo (Rosenblad et al., 1996; Wilby et al., 1999).

An interesting feature of the fiber outgrowth in intra-nigral grafting experiments using GFP-positive donor cells is the close association between the outgrowing GFP-positive dopaminergic axons and the axons of the descending striato-nigral pathway. The GFP-positive mDA axons were seen to course through the nigro-striatal pathway in close apposition to DARPP-32-positive axon bundles, which belong to the medium-sized spiny (MSP) neurons within the striatum that form connections with the globus pallidus and the SN. The interaction between axonal projections that share a common pathway but with opposing trajectories has previously been described as a mechanism for axonal growth and guidance (Molnar and Blakemore, 1991; Molnar et al., 1998a, b; Canty and Murphy, 2008). According to this so-called "handshake" phenomenon, the pathways will converge during their initial development and will subsequently guide the outgrowing axons to their respective targets.

Whether this mechanism operates during the normal development of the nigro-striatal and striato-nigral projection systems is not currently known. Results from the TH-GFP grafting experiments suggest that it may be worth further investigation. Interestingly, the MSP neurons are known to be the major source of GDNF in the striatum of the adult brain (Trupp et al., 1997; Barroso-Chinea et al., 2005). Striatum-derived GDNF, transported and expressed along the striato-nigral pathway, may thus be able to create a growth-permissive environment for the graft-derived dopaminergic axons. GDNF is known to exert its growth-stimulating effect by interaction with the GDNF receptor, GRFα1, which is expressed by the developing dopamine neurons. Studies in other systems suggest that the GRFα1 receptor, when combined with GDNF, indeed can act as an attractant guidance molecule, as well as a cell adhesion factor, for axons that express the GDNF co-receptor Ret (Ledda et al., 2002; Paratcha and Ledda, 2008).

Thus, GDNF appears to be a strong candidate as a molecule involved in long distance growth of axons from transplanted mDA neurons along the nigro-striatal pathway. In further support of this, adeno-associated viral (AAV) vector-mediated over-expression of GDNF in the striatal target lead to a substantial increase in the level of GFP-positive outgrowth and striatal innervation from intra-nigral grafts. This may be in part due to an increase in the survival of the grafted mDA neurons through anterograde delivery of GDNF along the striato-nigral pathway to the SN (there were around 70% more mDA neurons in animals treated with GDNF). However, observations in mice where GDNF was also transported to the thalamus by anterograde transport from cortical projections suggest a specific chemoattractive effect is also involved. Only in these animals, were outgrowing GFP-positive axons seen to diverge from the nigro-striatal pathway in order

Fig. 8. Reconstruction of the nigro-striatal pathway through intra-nigral grafting. By using donor tissue from mice in which GFP expression is driven by the TH promoter, detection of GFP in the resulting grafts allows for highly sensitive and unambiguous characterization of graft-derived dopaminergic fiber patterns in the host brain. (A) Schematic reproduction of GFP immunoreactivity in horizontal sections 16 weeks after grafting of $1.5 \times 10^5$ E12.5 TH-GFP VM cells into the substantia nigra of a mouse that had previously received partial lesioning of the intrinsic dopamine neuron projection system. The approximate dorso-negativentral levels of the horizontal sections (1–5) are indicated in the parasagittal diagram. A schematic representation of the whole mouse brain illustrates the targeting of the substantia nigra using a micro-transplantation approach to inject the VM cells. (B) A darkfield photograph of immunohistochemistry for GFP shows the pattern of GFP-positive fibers coursing through the globus pallidus and forming a ramified terminal network in the striatum (from boxed area on Section 3 in panel A). The dashed line approximates the striato-pallidial border. Abbreviations: AC, anterior commissure; Amy, amygdala; CPu, caudate putamen; GP, globus pallidus; H, hippocampus; IC, internal capsule; NAc, nucleus accumbens; Pir, piriform cortex; S, septum; T, transplant. Scale bar: B, 200 μm. (See Color Plate 5.8 in color plate section.)

to innervate discrete areas of GDNF expression within the thalamus. The degree of GFP-positive terminal elaboration was also substantially greater in the GDNF compared to non-GDNF animals. Supply of exogenous GDNF has previously been shown to cause remodeling and sprouting of spared mDA axons in 6-OHDA-treated rats (Rosenblad et al., 1998; Brizard et al., 2006). It is likely that, in addition to attracting axons to innervate the striatum, GDNF can act locally to promote a more extensive ramification of the terminals and greater degree of striatal innervation. This improved innervation corresponded well to the more consistent degree of improvement in amphetamine-induced turning seen in GDNF-treated animals, as a group, compared to those with intra-nigral grafts but without AAV-GDNF injections.

The use of donor tissue from the TH-GFP reporter mouse has shed new light on the potential to reconstruct the nigro-striatal pathway in the adult brain. The technical advantage over earlier studies, provided by the GFP reporter, has been decisive by allowing for the unambiguous detection of graft derived, dopaminergic fiber patterns in the host brain, even in the presence of remaining mDA projections. The results show that dopamine neurons allografted into the adult, parkinsonian midbrain can structurally and functionally reinstate, at least partially, the nigro-striatal pathway. This finding is in line with earlier observations, made in other parts of the CNS (Bjorklund et al., 1986; Wictorin et al., 1991; Li and Raisman, 1993; Gaillard et al., 2007), showing that the adult brain retains a capacity for pathway reconstruction and repair that is much greater than previously realized, and that grafted fate-committed neuroblasts are capable of expressing their inherent capacity to grow long distance, target-specific projections also in the lesioned adult CNS. It also sets the scene for further work in this area aiming to explore the potential of intra-nigral grafting to yield functional recovery in behaviors that have remained poorly corrected through intra-striatal grafting. This will likely require manipulations that can facilitate an even greater degree of striatal innervation than has been observed in the experiments described here, in which the GFP-positive fiber coverage was still well below that which can be achieved with intra-striatal grafting. Furthermore, more extensive testing of the functional impact, involving tests of spontaneous and complex motor function, and utilizing rats, rather than mice, will also be important. Finally, the results set a new benchmark for criteria that can be used to assess the potential of transplantable mDA precursors derived from stem cells.

**Concluding remarks**

The use of transgenic reporter mice has had an important impact on our understanding of the key issues in the field of cell therapy for PD. Most notably, these tools have given valuable insight into the identity of the mDA progenitor most suitable for transplantation and have allowed for an unprecedented level of analysis of the structural integration of mDA neurons into the host brain following transplantation. Information on these principles of transplantability and connectivity will also be interesting to investigate in neural transplantation experiments involving the reconstruction of other systems in the CNS. Further work in this area will be greatly enhanced by the continued development of novel transgenic mouse lines driving reporter gene expression in specific cell types. Notably, a library of GFP reporter mice with gene-specific expression profiles has been established through the NIH-funded Gensat project (www.gensat.org), using bacterial artificial chromosome technology. This will undoubtedly have a positive impact for progress in the neural transplantation field.

**References**

Aguayo, A. J., Bjorklund, A., Stenevi, U., & Carlstedt, T. (1984). Fetal mesencephalic neurons survive and extend long axons across peripheral nervous system grafts inserted into the adult rat striatum. *Neuroscience Letters*, *45*, 53–58.

Andersson, E., Jensen, J. B., Parmar, M., Guillemot, F., & Bjorklund, A. (2006a). Development of the mesencephalic dopaminergic neuron system is compromised in the absence of neurogenin 2. *Development*, *133*, 507–516.

Andersson, E., Tryggvason, U., Deng, Q., Friling, S., Alekseenko, Z., Robert, B., et al. (2006b). Identification of intrinsic determinants of midbrain dopamine neurons. *Cell*, *124*, 393–405.

Arts, M. P., Groenewegen, H. J., Veening, J. G., & Cools, A. R. (1996). Efferent projections of the retrorubral nucleus to the substantia nigra and ventral tegmental area in cats as shown by anterograde tracing. *Brain Research Bulletin*, *40*, 219–228.

Barroso-Chinea, P., Cruz-Muros, I., Aymerich, M. S., Rodriguez-Diaz, M., Afonso-Oramas, D., Lanciego, J. L., et al. (2005). Striatal expression of GDNF and differential vulnerability of midbrain dopaminergic cells. *The European Journal of Neuroscience*, *21*, 1815–1827.

Bayer, S. A., Wills, K. V., Triarhou, L. C., & Ghetti, B. (1995). Time of neuron origin and gradients of neurogenesis in midbrain dopaminergic neurons in the mouse. *Experimental Brain Research*, *105*, 191–199.

Bentlage, C., Nikkhah, G., Cunningham, M. G., & Bjorklund, A. (1999). Reformation of the nigrostriatal pathway by fetal dopaminergic micrografts into the substantia nigra is critically dependent on the age of the host. *Experimental Neurology*, *159*, 177–190.

Bjorklund, A., & Dunnett, S. B. (2007). Dopamine neuron systems in the brain: an update. *Trends in Neuroscience*, *30*, 194–202.

Bjorklund, A., & Lindvall, O. (1975). Dopamine in dendrites of substantia nigra neurons: suggestions for a role in dendritic terminals. *Brain Research*, *83*, 531–537.

Bjorklund, A., & Lindvall, O. (1984). Dopamine-containing systems in the CNS. In A. Bjorklund & T. Hokfelt (Eds.), *Handbook of chemical neuroanatomy* (pp. 55–123). Amsterdam: Elsevier.

Bjorklund, A., Nornes, H., & Gage, F. H. (1986). Cell suspension grafts of noradrenergic locus coeruleus neurons in rat hippocampus and spinal cord: reinnervation and transmitter turnover. *Neuroscience*, *18*, 685–698.

Bjorklund, A., Stenevi, U., Schmidt, R. H., Dunnett, S. B., & Gage, F. H. (1983). Intracerebral grafting of neuronal cell suspensions. II. Survival and growth of nigral cell suspensions implanted in different brain sites. *Acta Physiologica Scandinavica. Supplementum*, *522*, 9–18.

Bjorklund, L. M., Sanchez-Pernaute, R., Chung, S., Andersson, T., Chen, I. Y., McNaught, K. S., et al. (2002). Embryonic stem cells develop into functional dopaminergic neurons after transplantation in a Parkinson rat model. *Proceedings of the National Academy of Sciences of the United States of America*, *99*, 2344–2349.

Bolam, J. P., Freund, T. F., Bjorklund, A., Dunnett, S. B., & Smith, A. D. (1987). Synaptic input and local output of dopaminergic neurons in grafts that functionally reinnervate the host neostriatum. *Experimental Brain Research*, *68*, 131–146.

Brecknell, J. E., Haque, N. S., Du, J. S., Muir, E. M., Fidler, P. S., Hlavin, M. L., et al. (1996). Functional and anatomical reconstruction of the 6-hydroxydopamine lesioned nigrostriatal system of the adult rat. *Neuroscience*, *71*, 913–925.

Brizard, M., Carcenac, C., Bemelmans, A. P., Feuerstein, C., Mallet, J., & Savasta, M. (2006). Functional reinnervation from remaining DA terminals induced by GDNF lentivirus in a rat model of early Parkinson's disease. *Neurobiology of Disease*, *21*, 90–101.

Bustos, G., Abarca, J., Campusano, J., Bustos, V., Noriega, V., & Aliaga, E. (2004). Functional interactions between somatodendritic dopamine release, glutamate receptors and brain-derived neurotrophic factor expression in mesencephalic structures of the brain. *Brain Research. Brain Research Reviews*, *47*, 126–144.

Canty, A. J., & Murphy, M. (2008). Molecular mechanisms of axon guidance in the developing corticospinal tract. *Progress in Neurobiology*, *85*, 214–235.

Carlsson, T., Carta, M., Munoz, A., Mattsson, B., Winkler, C., Kirik, D., et al. (2009). Impact of grafted serotonin and dopamine neurons on development of L-DOPA-induced dyskinesias in parkinsonian rats is determined by the extent of dopamine neuron degeneration. *Brain*, *132*, 319–335.

Carlsson, T., Carta, M., Winkler, C., Bjorklund, A., & Kirik, D. (2007). Serotonin neuron transplants exacerbate L-DOPA-induced dyskinesias in a rat model of Parkinson's disease. *The Journal of Neuroscience*, *27*, 8011–8022.

Chung, S., Shin, B. S., Hedlund, E., Pruszak, J., Ferree, A., Kang, U. J., et al. (2006). Genetic selection of sox1GFP-expressing neural precursors removes residual tumorigenic pluripotent stem cells and attenuates tumor formation after transplantation. *Journal of Neurochemistry*, *97*, 1467–1480.

Chung, S., Sonntag, K. C., Andersson, T., Bjorklund, L. M., Park, J. J., Kim, D. W., et al. (2002). Genetic engineering of mouse embryonic stem cells by Nurr1 enhances differentiation and maturation into dopaminergic neurons. *The European Journal of Neuroscience*, *16*, 1829–1838.

Dahlstrom, A., & Fuxe, K. (1964). Evidence for the existence of monoamine-containing neurons in the central nervous system. I. Demonstration of monoamines in the cell bodies of brain stem neurons. *Acta Physiologica Scandinavica. Supplementum*, *62*(232), 231–255.

Damier, P., Hirsch, E. C., Agid, Y., & Graybiel, A. M. (1999). The substantia nigra of the human brain. II. Patterns of loss of dopamine-containing neurons in Parkinson's disease. *Brain*, *122*(Pt 8), 1437–1448.

Doucet, G., Brundin, P., Seth, S., Murata, Y., Strecker, R. E., Triarhou, L. C., et al. (1989). Degeneration and graft-induced restoration of dopamine innervation in the weaver mouse neostriatum: a quantitative radioautographic study of [$^3$H]dopamine uptake. *Experimental Brain Research*, *77*, 552–568.

Dunnett, S. B., Bjorklund, A., & Lindvall, O. (2001). Cell therapy in Parkinson's disease: stop or go? *Nature Reviews Neuroscience*, *2*, 365–369.

Dunnett, S. B., Bjorklund, A., Schmidt, R. H., Stenevi, U., & Iversen, S. D. (1983). Intracerebral grafting of neuronal cell suspensions. IV. Behavioural recovery in rats with unilateral 6-OHDA lesions following implantation of nigral cell suspensions in different forebrain sites. *Acta Physiologica Scandinavica. Supplementum*, *522*, 29–37.

Dunnett, S. B., Hernandez, T. D., Summerfield, A., Jones, G. H., & Arbuthnott, G. (1988). Graft-derived recovery from 6-OHDA lesions: specificity of ventral mesencephalic graft tissues. *Experimental Brain Research, 71*, 411–424.

Ekstrand, M. I., Terzioglu, M., Galter, D., Zhu, S., Hofstetter, C., Lindqvist, E., et al. (2007). Progressive parkinsonism in mice with respiratory-chain-deficient dopamine neurons. *Proceedings of the National Academy of Sciences of the United States of America, 104*, 1325–1330.

Fallon, J. H., & Moore, R. Y. (1978). Catecholamine innervation of the basal forebrain. IV. Topography of the dopamine projection to the basal forebrain and neostriatum. *The Journal of Comparative Neurology, 180*, 545–580.

Fearnley, J. M., & Lees, A. J. (1991). Ageing and Parkinson's disease: substantia nigra regional selectivity. *Brain, 114*(Pt 5), 2283–2301.

Ferrari, D., Sanchez-Pernaute, R., Lee, H., Studer, L., & Isacson, O. (2006). Transplanted dopamine neurons derived from primate ES cells preferentially innervate DARPP-32 striatal progenitors within the graft. *The European Journal of Neuroscience, 24*, 1885–1896.

Freed, C. R., Greene, P. E., Breeze, R. E., Tsai, W. Y., DuMouchel, W., Kao, R., et al. (2001). Transplantation of embryonic dopamine neurons for severe Parkinson's disease. *The New England Journal of Medicine, 344*, 710–719.

Friling, S., Andersson, A., Thompson, L. H., Jonsson, M., Hebsgaard, J. B., and Nanou, E., et al. (2009). Efficient production of midbrain dopamine neurons by Lmx1a expression in embryonic stem cells. *PNAS, 106*(18), 7613–7618.

Fukuda, H., Takahashi, J., Watanabe, K., Hayashi, H., Morizane, A., Koyanagi, M., et al. (2006). Fluorescence-activated cell sorting-based purification of embryonic stem cell-derived neural precursors averts tumor formation after transplantation. *Stem Cells, 24*, 763–771.

Gage, F. H., Stenevi, U., Carlstedt, T., Foster, G., Bjorklund, A., & Aguayo, A. J. (1985). Anatomical and functional consequences of grafting mesencephalic neurons into a peripheral nerve "bridge" connected to the denervated striatum. *Experimental Brain Research, 60*, 584–589.

Gaillard, A., Prestoz, L., Dumartin, B., Cantereau, A., Morel, F., Roger, M., et al. (2007). Reestablishment of damaged adult motor pathways by grafted embryonic cortical neurons. *Nature Neuroscience, 10*, 1294–1299.

Gates, M. A., Coupe, V. M., Torres, E. M., Fricker-Gates, R. A., & Dunnett, S. B. (2004). Spatially and temporally restricted chemoattractive and chemorepulsive cues direct the formation of the nigro-striatal circuit. *The European Journal of Neuroscience, 19*, 831–844.

Geffen, L. B., Jessell, T. M., Cuello, A. C., & Iversen, L. L. (1976). Release of dopamine from dendrites in rat substantia nigra. *Nature, 260*, 258–260.

Gerfen, C. R., & Wilson, C. J. (1996). The basal ganglia. In A. Bjorklund, T. Hokfelt, & L. W. Swanson (Eds.), *Handbook of chemical neuroanatomy, intergrated systems of the CNS, part III* (pp. 371–468). Amsterdam: Elsevier.

German, D. C., Manaye, K., Smith, W. K., Woodward, D. J., & Saper, C. B. (1989). Midbrain dopaminergic cell loss in Parkinson's disease: computer visualization. *Annals of Neurology, 26*, 507–514.

German, D. C., & Manaye, K. F. (1993). Midbrain dopaminergic neurons (nuclei A8, A9, and A10): three-dimensional reconstruction in the rat. *The Journal of Comparative Neurology, 331*, 297–309.

German, D. C., Manaye, K. F., Sonsalla, P. K., & Brooks, B. A. (1992). Midbrain dopaminergic cell loss in Parkinson's disease and MPTP-induced parkinsonism: sparing of calbindin-D28k-containing cells. *Annals of the New York Academy of Sciences, 648*, 42–62.

Hedlund, E., Pruszak, J., Lardaro, T., Ludwig, W., Vinuela, A., Kim, K. S., et al. (2008). Embryonic stem cell-derived Pitx3-enhanced green fluorescent protein midbrain dopamine neurons survive enrichment by fluorescence-activated cell sorting and function in an animal model of Parkinson's disease. *Stem Cells, 26*, 1526–1536.

Hidalgo, A., & Brand, A. H. (1997). Targeted neuronal ablation: the role of pioneer neurons in guidance and fasciculation in the CNS of Drosophila. *Development, 124*, 3253–3262.

Hornykiewicz, O. (1975). Brain monoamines and parkinsonism. *National Institute on Drug Abuse Research Monograph Series*, 13–21.

Isacson, O., & Deacon, T. W. (1996). Specific axon guidance factors persist in the adult brain as demonstrated by pig neuroblasts transplanted to the rat. *Neuroscience, 75*, 827–837.

Isacson, O., Deacon, T. W., Pakzaban, P., Galpern, W. R., Dinsmore, J., & Burns, L. H. (1995). Transplanted xenogeneic neural cells in neurodegenerative disease models exhibit remarkable axonal target specificity and distinct growth patterns of glial and axonal fibres. *Nature Medicine, 1*, 1189–1194.

Jacobs, F. M., Smits, S. M., Noorlander, C. W., von Oerthel, L., van der Linden, A. J., Burbach, J. P., et al. (2007). Retinoic acid counteracts developmental defects in the substantia nigra caused by Pitx3 deficiency. *Development, 134*, 2673–2684.

Jonsson, M., Ono, Y., Bjorklund, A., and Thompson, L. H. (2009). Identification of transplantable dopamine neuron precursors at different stages of midbrain neurogenesis. *Experimental Neurology*, in press.

Kele, J., Simplicio, N., Ferri, A. L., Mira, H., Guillemot, F., Arenas, E., et al. (2006). Neurogenin 2 is required for the development of ventral midbrain dopaminergic neurons. *Development, 133*, 495–505.

Kim, J. H., Auerbach, J. M., Rodriguez-Gomez, J. A., Velasco, I., Gavin, D., Lumelsky, N., et al. (2002). Dopamine neurons derived from embryonic stem cells function in an animal model of Parkinson's disease. *Nature, 418*, 50–56.

Kirik, D., Winkler, C., & Bjorklund, A. (2001). Growth and functional efficacy of intrastriatal nigral transplants depend on the extent of nigrostriatal degeneration. *The Journal of Neuroscience, 21*, 2889–2896.

Klose, M., & Bentley, D. (1989). Transient pioneer neurons are essential for formation of an embryonic peripheral nerve. *Science, 245*, 982–984.

Kordower, J. H., Rosenstein, J. M., Collier, T. J., Burke, M. A., Chen, E. Y., Li, J. M., et al. (1996). Functional fetal nigral grafts in a patient with Parkinson's disease: chemoanatomic, ultrastructural, and metabolic studies. *The Journal of Comparative Neurology, 370*, 203–230.

Kuan, W. L., Lin, R., Tyers, P., & Barker, R. A. (2007). The importance of A9 dopaminergic neurons in mediating the functional benefits of fetal ventral mesencephalon transplants and levodopa-induced dyskinesias. *Neurobiology of Disease, 25*, 594–608.

Ledda, F., Paratcha, G., & Ibanez, C. F. (2002). Target-derived GFRalpha1 as an attractive guidance signal for developing sensory and sympathetic axons via activation of Cdk5. *Neuron, 36*, 387–401.

Lee, S. H., Lumelsky, N., Studer, L., Auerbach, J. M., & McKay, R. D. (2000). Efficient generation of midbrain and hindbrain neurons from mouse embryonic stem cells. *Nature Biotechnology, 18*, 675–679.

Li, Y., & Raisman, G. (1993). Long axon growth from embryonic neurons transplanted into myelinated tracts of the adult rat spinal cord. *Brain Research, 629*, 115–127.

Liang, C. L., Sinton, C. M., Sonsalla, P. K., & German, D. C. (1996). Midbrain dopaminergic neurons in the mouse that contain calbindin-D28k exhibit reduced vulnerability to MPTP-induced neurodegeneration. *Neurodegeneration, 5*, 313–318.

Lin, D. M., Auld, V. J., & Goodman, C. S. (1995). Targeted neuronal cell ablation in the Drosophila embryo: pathfinding by follower growth cones in the absence of pioneers. *Neuron, 14*, 707–715.

Lin, L., & Isacson, O. (2006). Axonal growth regulation of fetal and embryonic stem cell-derived dopaminergic neurons by Netrin-1 and Slits. *Stem Cells, 24*, 2504–2513.

Lin, L., Rao, Y., & Isacson, O. (2005). Netrin-1 and slit-2 regulate and direct neurite growth of ventral midbrain dopaminergic neurons. *Molecular and Cellular Neurosciences, 28*, 547–555.

Lindvall, O., & Bjorklund, A. (2004). Cell therapy in Parkinson's disease. *NeuroRx: The Journal of the American Society for Experimental NeuroTherapeutics, 1*, 382–393.

Lindvall, O., & Hagell, P. (2000). Clinical observations after neural transplantation in Parkinson's disease. *Progress in Brain Research, 127*, 299–320.

Mahalik, T. J., & Clayton, G. H. (1991). Specific outgrowth from neurons of ventral mesencephalic grafts to the catecholamine-depleted striatum of adult hosts. *Experimental Neurology, 113*, 18–27.

Maingay, M., Romero-Ramos, M., Carta, M., & Kirik, D. (2006). Ventral tegmental area dopamine neurons are resistant to human mutant alpha-synuclein overexpression. *Neurobiology of Disease, 23*, 522–532.

Maxwell, S. L., Ho, H. Y., Kuehner, E., Zhao, S., & Li, M. (2005). Pitx3 regulates tyrosine hydroxylase expression in the substantia nigra and identifies a subgroup of mesencephalic dopaminergic progenitor neurons during mouse development. *Developmental Biology, 282*, 467–479.

McConnell, S. K., Ghosh, A., & Shatz, C. J. (1989). Subplate neurons pioneer the first axon pathway from the cerebral cortex. *Science, 245*, 978–982.

McConnell, S. K., Ghosh, A., & Shatz, C. J. (1994). Subplate pioneers and the formation of descending connections from cerebral cortex. *The Journal of Neuroscience, 14*, 1892–1907.

McRitchie, D. A., Cartwright, H. R., & Halliday, G. M. (1997). Specific A10 dopaminergic nuclei in the midbrain degenerate in Parkinson's disease. *Experimental Neurology, 144*, 202–213.

McRitchie, D. A., Hardman, C. D., & Halliday, G. M. (1996). Cytoarchitectural distribution of calcium binding proteins in midbrain dopaminergic regions of rats and humans. *The Journal of Comparative Neurology, 364*, 121–150.

Mendez, I., Sadi, D., & Hong, M. (1996). Reconstruction of the nigrostriatal pathway by simultaneous intrastriatal and intranigral dopaminergic transplants. *The Journal of Neuroscience, 16*, 7216–7227.

Mendez, I., Sanchez-Pernaute, R., Cooper, O., Vinuela, A., Ferrari, D., Bjorklund, L., et al. (2005). Cell type analysis of functional fetal dopamine cell suspension transplants in the striatum and substantia nigra of patients with Parkinson's disease. *Brain, 128*, 1498–1510.

Molnar, Z., Adams, R., & Blakemore, C. (1998a). Mechanisms underlying the early establishment of thalamocortical connections in the rat. *The Journal of Neuroscience, 18*, 5723–5745.

Molnar, Z., Adams, R., Goffinet, A. M., & Blakemore, C. (1998b). The role of the first postmitotic cortical cells in the development of thalamocortical innervation in the reeler mouse. *The Journal of Neuroscience, 18*, 5746–5765.

Molnar, Z., & Blakemore, C. (1991). Lack of regional specificity for connections formed between thalamus and cortex in coculture. *Nature, 351*, 475–477.

Mukhida, K., Baker, K. A., Sadi, D., & Mendez, I. (2001). Enhancement of sensorimotor behavioral recovery in hemiparkinsonian rats with intrastriatal, intranigral, and intrasubthalamic nucleus dopaminergic transplants. *The Journal of Neuroscience, 21*, 3521–3530.

Nakamura, S., Ito, Y., Shirasaki, R., & Murakami, F. (2000). Local directional cues control growth polarity of dopaminergic axons along the rostrocaudal axis. *The Journal of Neuroscience, 20*, 4112–4119.

Neuhoff, H., Neu, A., Liss, B., & Roeper, J. (2002). I(h) channels contribute to the different functional properties of identified dopaminergic subpopulations in the midbrain. *The Journal of Neuroscience, 22*, 1290–1302.

Nieoullon, A., Cheramy, A., & Glowinski, J. (1977). Release of dopamine in vivo from cat substantia nigra. *Nature, 266*, 375–377.

Nikkhah, G., Bentlage, C., Cunningham, M. G., & Bjorklund, A. (1994). Intranigral fetal dopamine grafts induce behavioral compensation in the rat Parkinson model. *The Journal of Neuroscience, 14*, 3449–3461.

Nikkhah, G., Duan, W. M., Knappe, U., Jodicke, A., & Bjorklund, A. (1993). Restoration of complex sensorimotor behavior and skilled forelimb use by a modified nigral cell suspension transplantation approach in the rat Parkinson model. *Neuroscience, 56*, 33–43.

Nunes, I., Tovmasian, L. T., Silva, R. M., Burke, R. E., & Goff, S. P. (2003). Pitx3 is required for development of substantia nigra dopaminergic neurons. *Proceedings of the National Academy of Sciences of the United States of America, 100*, 4245–4250.

Okabe, M., Ikawa, M., Kominami, K., Nakanishi, T., & Nishimune, Y. (1997). 'Green mice' as a source of ubiquitous green cells. *FEBS Letters, 407*, 313–319.

Olanow, C. W., & Fahn, S. (2006). Fetal nigral transplantation as a therapy for Parkinson's disease. In C. W. Olanow & P. Brundin (Eds.), *Restorative therapies in Parkinson's disease* (pp. 93–118). New York: Springer Science and Business Media.

Olanow, C. W., Goetz, C. G., Kordower, J. H., Stoessl, A. J., Sossi, V., Brin, M. F., et al. (2003). A double-blind controlled trial of bilateral fetal nigral transplantation in Parkinson's disease. *Annals of Neurology, 54*, 403–414.

Olanow, C. W., Gracies, J. M., Goetz, C. G., Stoessl, A. J., Freeman, T., Kordower, J. H., et al. (2009). Clinical pattern and risk factors for dyskinesias following fetal nigral transplantation in Parkinson's disease: a double blind video-based analysis. *Movement Disorders, 24*, 336–343.

Olsson, M., Nikkhah, G., Bentlage, C., & Bjorklund, A. (1995). Forelimb akinesia in the rat Parkinson model: differential effects of dopamine agonists and nigral transplants as assessed by a new stepping test. *The Journal of Neuroscience, 15*, 3863–3875.

Ono, Y., Nakatani, T., Sakamoto, Y., Mizuhara, E., Minaki, Y., Kumai, M., et al. (2007). Differences in neurogenic potential in floor plate cells along an anteroposterior location: midbrain dopaminergic neurons originate from mesencephalic floor plate cells. *Development, 134*, 3213–3225.

Osborne, P. B., Halliday, G. M., Cooper, H. M., & Keast, J. R. (2005). Localization of immunoreactivity for deleted in colorectal cancer (DCC), the receptor for the guidance factor netrin-1, in ventral tier dopamine projection pathways in adult rodents. *Neuroscience, 131*, 671–681.

Paratcha, G., & Ledda, F. (2008). GDNF and GFRalpha: a versatile molecular complex for developing neurons. *Trends in Neuroscience, 31*, 384–391.

Perrier, A. L., Tabar, V., Barberi, T., Rubio, M. E., Bruses, J., Topf, N., et al. (2004). Derivation of midbrain dopamine neurons from human embryonic stem cells. *Proceedings of the National Academy of Sciences of the United States of America, 101*, 12543–12548.

Pittman, A. J., Law, M. Y., & Chien, C. B. (2008). Pathfinding in a large vertebrate axon tract: isotypic interactions guide retinotectal axons at multiple choice points. *Development, 135*, 2865–2871.

Pruszak, J., Sonntag, K. C., Aung, M. H., Sanchez-Pernaute, R., & Isacson, O. (2007). Markers and methods for cell sorting of human embryonic stem cell-derived neural cell populations. *Stem Cells, 25*, 2257–2268.

Robertson, G. S., Damsma, G., & Fibiger, H. C. (1991a). Characterization of dopamine release in the substantia nigra by in vivo microdialysis in freely moving rats. *The Journal of Neuroscience, 11*, 2209–2216.

Robertson, G. S., Fine, A., & Robertson, H. A. (1991b). Dopaminergic grafts in the striatum reduce D1 but not D2 receptor-mediated rotation in 6-OHDA-lesioned rats. *Brain Research, 539*, 304–311.

Robertson, H. A. (1992). Dopamine receptor interactions: some implications for the treatment of Parkinson's disease. *Trends in Neuroscience, 15*, 201–206.

Rodriguez, M., Barroso-Chinea, P., Abdala, P., Obeso, J., & Gonzalez-Hernandez, T. (2001). Dopamine cell degeneration induced by intraventricular administration of 6-hydroxydopamine in the rat: similarities with cell loss in parkinson's disease. *Experimental Neurology, 169*, 163–181.

Rosales, M. G., Martinez-Fong, D., Morales, R., Nunez, A., Flores, G., Gongora-Alfaro, J. L., et al. (1997). Reciprocal interaction between glutamate and dopamine in the pars reticulata of the rat substantia nigra: a microdialysis study. *Neuroscience, 80*, 803–810.

Rosenblad, C., Martinez-Serrano, A., & Bjorklund, A. (1996). Glial cell line-derived neurotrophic factor increases survival, growth and function of intrastriatal fetal nigral dopaminergic grafts. *Neuroscience, 75*, 979–985.

Rosenblad, C., Martinez-Serrano, A., & Bjorklund, A. (1998). Intrastriatal glial cell line-derived neurotrophic factor promotes sprouting of spared nigrostriatal dopaminergic afferents and induces recovery of function in a rat model of Parkinson's disease. *Neuroscience, 82*, 129–137.

Sieber, B. A., Kuzmin, A., Canals, J. M., Danielsson, A., Paratcha, G., Arenas, E., et al. (2004). Disruption of EphA/ephrin: a signaling in the nigrostriatal system reduces dopaminergic innervation and dissociates behavioral responses to amphetamine and cocaine. *Molecular and Cellular Neurosciences, 26*, 418–428.

Sinclair, S. R., Fawcett, J. W., & Dunnett, S. B. (1999). Dopamine cells in nigral grafts differentiate prior to implantation. *The European Journal of Neuroscience, 11*, 4341–4348.

Sonntag, K. C., Pruszak, J., Yoshizaki, T., van Arensbergen, J., Sanchez-Pernaute, R., & Isacson, O. (2007). Enhanced yield of neuroepithelial precursors and midbrain-like dopaminergic neurons from human embryonic stem cells using the bone morphogenic protein antagonist noggin. *Stem Cells, 25*, 411–418.

Swanson, L. W. (1982). The projections of the ventral tegmental area and adjacent regions: a combined fluorescent retrograde tracer and immunofluorescence study in the rat. *Brain Research Bulletin, 9*, 321–353.

Thompson, L., Barraud, P., Andersson, E., Kirik, D., & Bjorklund, A. (2005). Identification of dopaminergic neurons of nigral and ventral tegmental area subtypes in grafts of fetal ventral mesencephalon based on cell morphology, protein expression, and efferent projections. *The Journal of Neuroscience, 25*, 6467–6477.

Thompson, L. H., Andersson, E., Jensen, J. B., Barraud, P., Guillemot, F., Parmar, M., et al. (2006). Neurogenin2 identifies a transplantable dopamine neuron precursor in the developing ventral mesencephalon. *Experimental Neurology, 198,* 183–198.

Thompson, L. H., Grealish, S., Kirik, D., and Bjorklund, A. (2009). Re-construction of the nigro-striatal dopamine pathway in the adult mouse brain. *The European Journal of Neuroscience,* in press.

Thompson, L. H., Kirik, D., & Bjorklund, A. (2008). Non-dopaminergic neurons in ventral mesencephalic transplants make widespread axonal connections in the host brain. *Experimental Neurology, 213,* 220–228.

Torres, E. M., Monville, C., Gates, M. A., Bagga, V., & Dunnett, S. B. (2007). Improved survival of young donor age dopamine grafts in a rat model of Parkinson's disease. *Neuroscience, 146,* 1606–1617.

Trupp, M., Belluardo, N., Funakoshi, H., & Ibanez, C. F. (1997). Complementary and overlapping expression of glial cell line-derived neurotrophic factor (GDNF), c-ret proto-oncogene, and GDNF receptor-alpha indicates multiple mechanisms of trophic actions in the adult rat CNS. *The Journal of Neuroscience, 17,* 3554–3567.

Wernig, M., Zhao, J. P., Pruszak, J., Hedlund, E., Fu, D., Soldner, F., et al. (2008). Neurons derived from reprogrammed fibroblasts functionally integrate into the fetal brain and improve symptoms of rats with Parkinson's disease. *Proceedings of the National Academy of Sciences of the United States of America, 105,* 5856–5861.

Wictorin, K., Brundin, P., Sauer, H., Lindvall, O., & Bjorklund, A. (1992). Long distance directed axonal growth from human dopaminergic mesencephalic neuroblasts implanted along the nigrostriatal pathway in 6-hydroxydopamine lesioned adult rats. *The Journal of Comparative Neurology, 323,* 475–494.

Wictorin, K., Lagenaur, C. F., Lund, R. D., & Bjorklund, A. (1991). Efferent projections to the host brain from intrastriatal striatal mouse-to-rat grafts: time course and tissue-type specificity as revealed by a mouse specific neuronal marker. *The European Journal of Neuroscience, 3,* 86–101.

Wilby, M. J., Sinclair, S. R., Muir, E. M., Zietlow, R., Adcock, K. H., Horellou, P., et al. (1999). A glial cell line-derived neurotrophic factor-secreting clone of the Schwann cell line SCTM41 enhances survival and fiber outgrowth from embryonic nigral neurons grafted to the striatum and to the lesioned substantia nigra. *The Journal of Neuroscience, 19,* 2301–2312.

Winkler, C., Bentlage, C., Nikkhah, G., Samii, M., & Bjorklund, A. (1999). Intranigral transplants of GABA-rich striatal tissue induce behavioral recovery in the rat Parkinson model and promote the effects obtained by intrastriatal dopaminergic transplants. *Experimental Neurology, 155,* 165–186.

Winkler, C., Kirik, D., Bjorklund, A., & Dunnett, S. B. (2000). Transplantation in the rat model of Parkinson's disease: ectopic versus homotopic graft placement. *Progress in Brain Research, 127,* 233–265.

You, Z. B., Nylander, I., Herrera-Marschitz, M., O'Connor, W. T., Goiny, M., & Terenius, L. (1994). The striatonigral dynorphin pathway of the rat studied with in vivo microdialysis-I. Effects of K(+)-depolarization, lesions and peptidase inhibition. *Neuroscience, 63,* 415–425.

Yue, Y., Widmer, D. A., Halladay, A. K., Cerretti, D. P., Wagner, G. C., Dreyer, J. L., et al. (1999). Specification of distinct dopaminergic neural pathways: roles of the Eph family receptor EphB1 and ligand ephrin-B2. *The Journal of Neuroscience, 19,* 2090–2101.

Yurek, D. M., & Fletcher-Turner, A. (2001). Differential expression of GDNF, BDNF, and NT-3 in the aging nigrostriatal system following a neurotoxic lesion. *Brain Research, 891,* 228–235.

Zhou, J., Pliego-Rivero, B., Bradford, H. F., & Stern, G. M. (1996). The BDNF content of postnatal and adult rat brain: the effects of 6-hydroxydopamine lesions in adult brain. *Brain Research. Developmental Brain Research, 97,* 297–303.

# SECTION II

# Immunotherapy and Vaccination Therapy

CHAPTER 6

# Developing novel immunogens for a safe and effective Alzheimer's disease vaccine

Cynthia A. Lemere*

*Brigham and Women's Hospital, Harvard Medical School, Boston, MA, USA*

**Abstract:** Alzheimer's disease (AD) is the most prevalent form of neurodegeneration; however, therapies to prevent or treat AD are inadequate. Amyloid-beta (Aβ) protein accrues in cortical senile plaques, one of the key neuropathological hallmarks of AD, and is elevated in brains of early onset AD patients in a small number of families that bear certain genetic mutations, further implicating its role in this devastating neurological disease. In addition, soluble Aβ oligomers have been shown to be detrimental to neuronal function. Therapeutic strategies aimed at lowering cerebral Aβ levels are currently under development. One strategy is to immunize AD patients with Aβ peptides so that they will generate antibodies that bind to Aβ protein and enhance its clearance. As of 1999, Aβ immunotherapy, either through active immunization with Aβ peptides or through passive transfer of Aβ-specific antibodies, has been shown to reduce cerebral Aβ levels and improve cognitive deficits in AD mouse models and lower plaque load in nonhuman primates. However, a Phase II clinical trial of active immunization using full-length human Aβ1-42 peptide and a strong Th1-biased adjuvant, QS-21, ended prematurely in 2002 because of the onset of meningoencephalitis in ~6% of the AD patients enrolled in the study. It is possible that T cell recognition of the human full-length Aβ peptide as a self-protein may have induced an adverse autoimmune response in these patients. Although only ~20% of immunized patients generated anti-Aβ titers, responders showed some general slowing of cognitive decline. Focal cortical regions devoid of Aβ plaques were observed in brain tissues of several immunized patients who have since come to autopsy. In order to avoid a deleterious immune response, passive Aβ immunotherapy is under investigation by administering monthly intravenous injections of humanized Aβ monoclonal antibodies to AD patients. However, a safe and effective active Aβ vaccine would be more cost-effective and more readily available to a larger AD population. We have developed several novel short Aβ immunogens that target the Aβ N-terminus containing a strong B cell epitope while avoiding the Aβ mid-region and C-terminus containing T cell epitopes. These immunogens include dendrimeric Aβ1-15 (16 copies of Aβ1-15 on a lysine antigen tree), 2xAβ1-15 (a tandem repeat of two lysine-linked Aβ1-15 peptides), and 2xAβ1-15 with the addition of a three amino acid RGD motif (R-2xAβ1-15). Intranasal immunization with our short Aβ fragment immunogens and a mucosal adjuvant, mutant *Escherichia coli* heat-labile enterotoxin LT(R192G), resulted in reduced cerebral Aβ levels, plaque deposition, and gliosis, as well as increased plasma Aβ levels and

*Corresponding author.
Tel.: (617) 525-5214; Fax: (617) 525-5252;
E-mail: clemere@rics.bwh.harvard.edu

improved cognition in a transgenic mouse model of AD. Preclinical trials in nonhuman primates, and human clinical trials using similar Aβ immunogens, are now underway. Aβ immunotherapy looks promising but must be made safer and more effective at generating antibody titers in the elderly. It is hoped that these novel immunogens will enhance Aβ antibody generation across a broad population and avoid the adverse events seen in the earlier clinical trial.

**Keywords:** amyloid-beta; vaccine; Alzheimer's disease; immunotherapy; T cells; B cells; adjuvant; nonhuman primates; transgenic mice

## Introduction

Alzheimer's disease (AD) is the most common form of dementia, affecting ∼26 million people worldwide. Currently, there are no effective treatments and no known means to prevent this devastating neurological disease. While the major clinical symptoms include progressive memory loss, personality changes, language problems, and confusion, it is believed that the onset of two major pathological hallmarks of AD, extracellular amyloid-beta (Aβ) plaques, and intracellular neurofibrillary tangles containing hyperphosphorylated tau, precedes clinical symptoms by years. In addition, to plaques and tangles, AD brain is characterized by gliosis, inflammation, neuritic dystrophy, neuron loss, and changes in neurotransmitter levels (Dickson, 1997; Hardy and Selkoe, 2002). Aβ protein, a 40- to 42-amino acid protein, is generated by proteolytic cleavage from the beta-amyloid precursor protein (βAPP) by beta-secretase at its amino-terminus and by gamma-secretase at its carboxy-terminus (Wolfe, 2006). Aβ is now a major therapeutic target for AD because genetic mutations in APP and presenilin proteins 1 and 2 (PS1 and PS2), part of the gamma-secretase complex, are associated with AD in a small number of families; Aβ is deposited early in plaques and blood vessels in the brain; and Aβ oligomers and fibrillar aggregates are toxic to neurons (Selkoe, 2001; Klein, 2002; Walsh and Selkoe, 2004). Current therapeutic strategies aim to lower production of Aβ by inhibiting or modulating beta- or gamma-secretase, prevent formation of Aβ aggregates and/or dissolve preformed aggregates, and enhance clearance of Aβ from the brain. Aβ immunotherapy, one of the strategies under intense investigation, uses anti-Aβ antibodies by either active or passive vaccination to reduce Aβ deposition in the brain and enhance Aβ clearance.

## Aβ immunotherapy in rodents, monkeys, and humans

### Preclinical studies in rodents

In the mid-1990s, anti-Aβ antibodies were shown to dissolve Aβ aggregates and prevent monomers from aggregating in vitro (Solomon et al., 1996, 1997). In 1999, it was demonstrated for the first time that active vaccination with Aβ1-42 synthetic peptide resulted in anti-Aβ antibody generation and a significant reduction in plaque burden in the brains of APP transgenic mice in vivo (Schenk et al., 1999). Subsequently, we and many other groups confirmed and extended these studies in multiple transgenic AD mouse models using a variety of adjuvants and routes of administration, demonstrating that active Aβ immunization prevented or reduced plaque deposition (if given early enough) (Lemere et al., 2000; Weiner et al., 2000; Das et al., 2001; Sigurdsson et al., 2001; McLaurin et al., 2002), increased peripheral Aβ in blood (Lemere et al., 2003), and prevented or improved cognitive deficits (Janus et al., 2000; Morgan et al., 2000). The resulting antibodies recognized B cell epitopes within the first 15 amino acids of Aβ (Lemere et al., 2000, 2001; Town et al., 2001; McLaurin et al., 2002; Cribbs et al., 2003; Seabrook et al., 2004). T cell epitopes were mapped to the Aβ mid-region and carboxy-terminus (Monsonego et al., 2001; Cribbs et al., 2003).

Passive immunization was investigated by directly injecting Aβ monoclonal antibodies into AD transgenic mice, thereby bypassing a cellular immune response. Intraperitoneal injections of Aβ monoclonal antibodies decreased cerebral Aβ levels (Bard et al., 2000), increased peripheral Aβ levels in blood (DeMattos et al., 2001), and improved behavior (Kotilinek et al., 2002; Wilcock et al., 2004). In very old APP transgenic mice, passive Aβ vaccination improved cognitive performance within days, without reducing plaque burden, suggesting that removal of soluble oligomers may be enough to rescue cognition (Dodart et al., 2002). Direct application of Aβ monoclonal antibodies demonstrated plaque clearance and enhancement of microglial phagocytosis of Aβ using multiphoton imaging (Bacskai et al., 2001). Intrahippocampal injection of an Aβ monoclonal antibody into 3xTg-AD mice that develop both plaques and tangles cleared extracellular and intracellular Aβ as well as early tau aggregates but not hyperphosphorylated tau, a late-stage pathogenic marker (Oddo et al., 2004). Microhemorrhage was observed in very old APP transgenic mice with abundant vascular amyloid, although the mice still showed some cognitive benefit from the vaccine (Pfeifer et al., 2002; Wilcock et al., 2004; Racke et al., 2005). In addition, both active and passive Aβ vaccination was shown to protect against Aβ oligomer-induced long-term-potentiation (LTP) deficits in rats (Klyubin et al., 2005).

### Preclinical studies in nonhuman primates

Small Aβ immunization studies have been undertaken in nonhuman primate monkeys. For example, Gandy et al. (2004) (including our lab) immunized two 15-year-old rhesus monkeys with full-length Aβ peptide and adjuvant, resulting in anti-Aβ antibody generation and increased Aβ levels in blood. However, cerebral Aβ levels were unchanged, possibly because the animals were too young for plaque deposition in the brain and/or the trial was too short to show an effect. In another trial, we immunized four Caribbean vervets (African green monkeys from St. Kitts in the Eastern Caribbean) 16–25 years of age with full-length Aβ and complete and incomplete Freund's adjuvant (Lemere et al., 2004). Vervets naturally develop cerebral Aβ plaque pathology and vascular amyloid with aging, similar to other nonhuman primates. The immunized vervets generated anti-Aβ antibodies that recognized Aβ1-7 and bound monomeric, oligomeric, and fibrillar Aβ but not APP or APP C-terminal fragments. Plaque deposition was absent while insoluble Aβ$_{42}$ levels were significantly reduced in the brains of the 4 immunized vervets compared to 13 age-matched archived brain samples, as shown in Fig. 1. Plasma Aβ was elevated in the immunized animals. No adverse events were observed.

### Clinical Aβ immunotherapy trials in humans

The first clinical trial of Aβ immunotherapy, sponsored by ELAN/Wyeth, involved an active vaccine, AN1792, that contained Aβ1-42 synthetic peptide and a strong, T helper 1 (Th1)-biased adjuvant, QS-21. Although the Phase I and IIa trials were deemed safe, the multisite Phase IIb trial was halted in 2002 due to the occurrence of meningoencephalitis in ∼6% (18/300) of AD patients worldwide (Schenk, 2002). The cause of these adverse events is unknown, but many have speculated that they may have been due to an autoimmune-like T cell response to Aβ, a self-antigen (Orgogozo et al., 2003; Gilman et al., 2005). Other possibilities include reformulation of the vaccine in polysorbate 80 and a strong pro-inflammatory Th1 response due to the adjuvant, QS-21. Only 19% of the immunized patients generated an Aβ antibody response, perhaps due to the limited dosing (1–3 inoculations). Importantly, the occurrence of meningoencephalitis was independent of antibody response. The Aβ antibodies generated by AN1792 were shown to bind to the amino-terminus of Aβ (Lee et al., 2005) and to human Aβ plaques and vascular amyloid but not soluble Aβ$_{42}$ (Hock et al., 2002). Tau levels in CSF were lower in antibody responders (Gilman et al., 2005). Transient cortical shrinkage was observed by MRI in these same patients, although brain volumes resumed baseline levels after ∼1 year following vaccination (Fox et al., 2005). In most of the patients who have come to autopsy since the initiation of the trial, plaque

## i. Aß ELISA

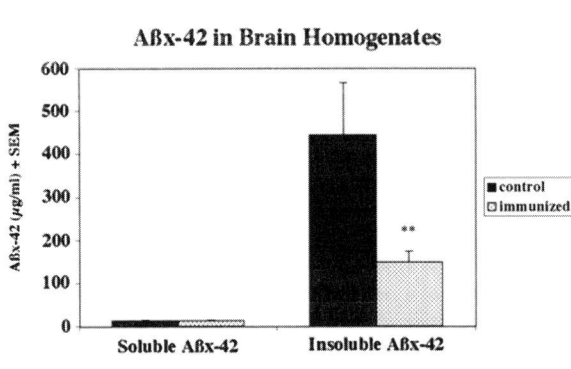

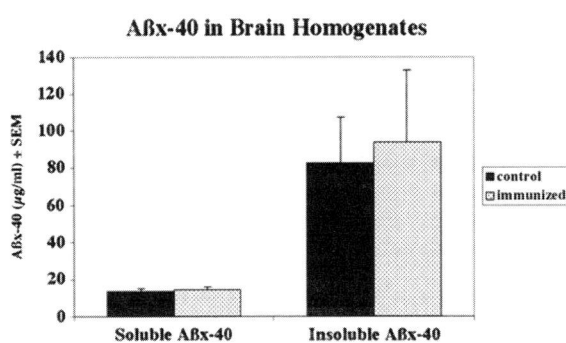

## ii. Aß Immunohistochemistry

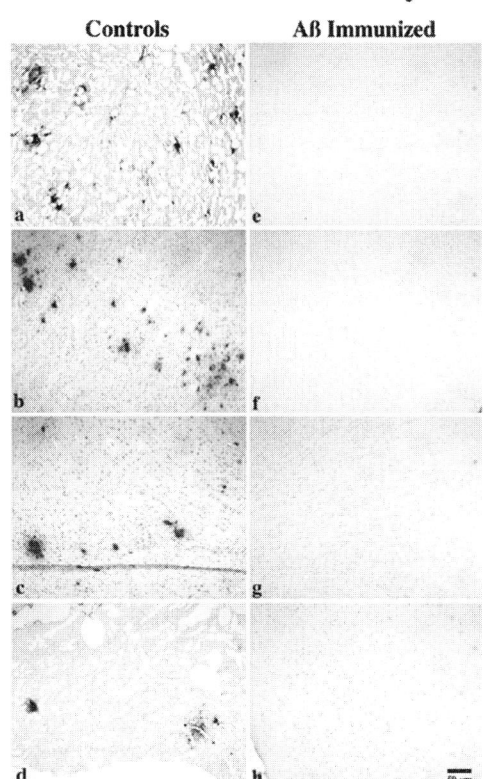

## iii. Aß Plaque Burden: Frontal Cortex

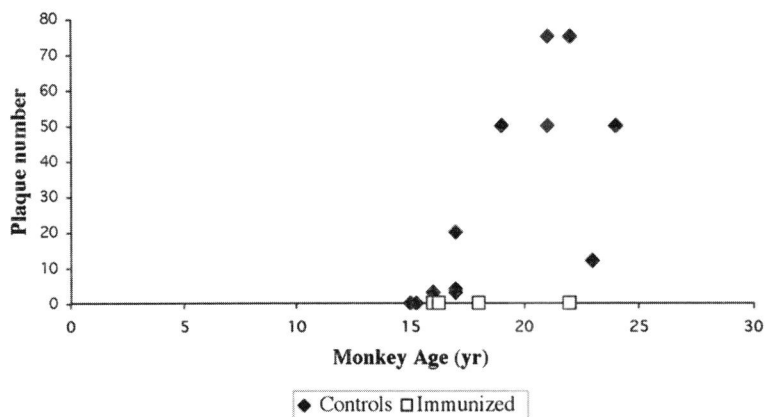

deposition was focally and regionally reduced (Nicoll et al., 2003, 2006; Ferrer et al., 2004; Masliah et al., 2005; Holmes et al., 2008). Some slowing of cognitive decline has been observed in antibody responders (Gilman et al., 2005; Hock et al., 2003; International Conference on Alzheimer's Disease, Chicago, IL, 2008). However, two patients from an earlier Phase I AN1792 trial who came to autopsy several years later, generated Aβ antibodies, had very few plaques (suggestive of plaque clearance, and yet had severe dementia (Holmes et al., 2008). A possible explanation for this finding may be that at the time of vaccination, the patients may have had substantial cerebral pathology, including neuron loss and neuritic plaques, which could not be reversed by the removal of Aβ plaques. Thus, the need for early intervention, possibly before the onset of symptoms, is critical in moving forward with Aβ-lowering treatments.

Several passive Aβ immunization trials are currently underway. The ELAN/Wyeth Phase II clinical trial results were reported at the 11th International Conference on Alzheimer's Disease (ICAD) in Chicago in July 2008. Intravenous administration of a humanized monoclonal antibody, Bapineuzimab, recognizing the amino-terminus of Aβ showed a nonsignificant trend for cognitive stabilization in mild-to-moderate AD patients. Post hoc analysis demonstrated significant cognitive benefits in multiple tests in Apo E4 noncarriers but only a trend in Apo E4 carriers, possibly due to accelerated pathogenesis in E4 carriers (ICAD, 2008). A Phase III trial is currently underway. Eli Lilly is currently testing a humanized monoclonal antibody that recognizes the mid-region of Aβ, and binds soluble Aβ protein.

## Possible mechanisms of Aβ clearance via immunotherapy

Several mechanisms have been proposed based on in vitro and in vivo studies. First, Aβ antibodies may prevent Aβ aggregation and/or dissolve preformed Aβ aggregates (Solomon et al., 1996, 1997). Second, upon binding of the Aβ antibodies to Aβ, the Fc portion of the antibodies may bind the Fc receptor on microglia, inducing phagocytosis of Aβ (Bard et al., 2000). This would require that Aβ antibodies cross the blood-brain barrier (BBB) and bind Aβ within the central nervous system (CNS). While some evidence has been reported to support this mechanism, two studies have demonstrated that Fc-mediated phagocytosis is not required for immunotherapy-induced Aβ clearance, as Fab fragments of Aβ antibodies (missing the Fc region) cleared Aβ when applied to the surface of APP transgenic mouse brain (Bacskai et al., 2002), and Aβ vaccination in APP transgenic mice lacking the FcR lowered cerebral Aβ (Das et al., 2003). A third mechanism proposes that the presence of Aβ antibodies in the periphery (i.e., blood) causes a shift in the gradient of Aβ transport across the BBB resulting in an increase in efflux from the brain to blood (DeMattos et al., 2001; Lemere et al., 2003). A fourth mechanism proposes that certain antibodies bind Aβ oligomers, thereby neutralizing the toxic effects of this Aβ species on synapses (Klyubin et al., 2005). These mechanisms are not mutually exclusive and may overlap under certain conditions. In addition, the mechanism of action may be disease state dependent. For example, a prevention vaccine

◀

Fig. 1. Cerebral Aβ levels. (i) Aβ ELISA was used to detect differences in soluble and insoluble Aβ levels in brain homogenates. No differences were observed between immunized (dotted) and control (solid) vervet soluble Aβx-42 or Aβx-40 cerebral levels. However, insoluble Aβx-42 was reduced 66% ($p<0.035$, 2-tailed Alternate Welch's T test) in the 4 immunized vervets compared to 13 aged age-matched controls. Insoluble Aβx-40 was much less abundant and was no significantly different between the two treatment groups. (ii) Aβ$_{42}$ immunohistochemistry using Mab 21F12 on paraffin frontal sections revealed plaque labeling in 11 of 13 age-matched control vervets (**a,** 22 years; **b,** 21 years; **c,** 23 years; **d,** 17 years). Aβ IR plaques were not detected in the frontal cortex of any of the four immunized vervets (**e,** 22 years; **f,** 18 years; **g,** 16 years; **h,** 16 years); five additional cortical regions per immunized vervet were also devoid of plaque labeling. Scale bar, 50 microns. (iii) Aβ deposition into cerebral plaques was quantified by visually counting the number of Aβ$_{42}$ (Mab 21F12-immunoreactive) plaques occupying four 4x fields ($\sim 2.4 \times 3$ mm) in the frontal cortex from each of the 4 immunized (open squares) and 13 control (solid diamonds) vervets. Although the two youngest controls (15 years each) did not show any Aβ plaque labeling, all of the older animals (aged 16–24 years) showed some plaque labeling. Adapted from Lemere et al. (2004) with permission from the American Society for Investigative Pathology.

may not require that the antibodies cross the BBB to induce Aβ phagocytosis, whereas a therapeutic vaccine (once plaque deposition is well underway) would likely benefit from the transport of Aβ antibodies into the CNS.

**Novel short Aβ immunogens for active vaccination**

Although the first clinical trial for an active Aβ vaccine ended prematurely due to adverse events, preclinical and, to some degree, clinical studies indicate potential for this therapeutic strategy to prevent AD or stop it in early stages. A modified active vaccine would be less costly to prepare than humanized monoclonal antibodies, would require fewer doctor visits, and would be more feasible than passive immunization for the large population of individuals with or at risk for AD. The effects of an active vaccine would be longer lasting than passive administration of anti-Aβ antibodies whose half-life is typically around 30 days. However, the cellular immune response to an active vaccine may be difficult to stop, should problems arise, once it is underway. Thus, going forward, many labs, including our own, have sought to develop second-generation active Aβ vaccines that target the B cell epitope in the Aβ amino-terminus and avoid an Aβ-specific T cell response directed at the mid-region or carboxyl-terminus of Aβ. Examples of three such vaccines being tested in our lab are described below. A more complete listing of novel immunogens under investigation for active Aβ immunization may be found in our recent review article (Lemere et al., 2007).

In an attempt to design a vaccine that would generate strong Aβ titers while avoiding an adverse T cell-mediated response, we first constructed a dendrimeric immunogen, dAβ1-15, consisting of 16 copies of Aβ1-15 peptide on a branched lysine tree (Seabrook et al., 2006). Weekly intranasal vaccination with dAβ1-15 with a mucosal adjuvant, mutant *Escherichia coli* heat-labile enterotoxin LT(R192G) (Dickinson and Clements, 1995), in J20 hAPP$_{FAD}$ transgenic mice (Mucke et al., 2000) 6–12 months of age resulted in a robust, predominantly Th2-biased (IgG2b) anti-Aβ antibody response and a significant reduction in cerebral plaque burden. Splenocytes from the immunized mice recognized and proliferated upon incubation with dAβ1-15 but only minimally upon incubation with full-length Aβ, providing evidence that this immunogen avoided an Aβ-specific cellular immune response.

Next, we tested two novel short Aβ immunogens by synthesizing two linear tandem-repeat copies of Aβ1-15 linked by two lysine residues, with or without an RGD motif at the amino-terminus of the peptide (Maier et al., 2006). RGD has been shown previously to immunogenicity of other antigens, such as *Streptococcus mutans* epitope (Yano et al., 2003), and may act as an adjuvant. In our study, intranasal immunization of wild-type mice (B6D2F1) with 2xAβ1-15 and R-2xAβ1-15 (with the RDG motif) and adjuvant LT(R192G) led to a strong humoral response, that is, high levels of anti-Aβ antibodies, primarily of Th2-biased IgG1 and IgG2b antibodies. Splenocytes from the immunized animals recognized and proliferated upon incubation with 2xAβ1-15 or R-2xAβ1-15 but not with full-length Aβ, suggesting that these vaccines avoided an Aβ-specific cellular immune response. The addition of the RGD motif did not provide sufficient adjuvant activity to induce a strong antibody response.

Each vaccine was tested in J20 hAPP$_{FAD}$ transgenic mice at an age when the mice first begin to show cerebral Aβ deposition. Weekly intranasal vaccination with R-2xAβ1-15 and adjuvant LT(R192G) in mice 6–12 months of age resulted in strong Aβ antibody production in the absence of an Aβ-specific cellular immune response. Aβ$_{42}$- and Aβ$_{40}$-immunoreactive plaques and thioflavin S fibrillar amyloid deposits were significantly reduced in the hippocampi of immunized mice (Fig. 2A–D), indicating an effect on both diffuse and compacted plaques. Insoluble Aβ levels were nonsignificantly reduced while soluble Aβ$_{40}$ and Aβ$_{42}$ were increased in brain homogenates, and Aβ levels were elevated in plasma (Fig. 2F). Weekly immunization with 2xAβ1-15 of J20 transgenic mice 4.5–12 months of age also resulted in high anti-Aβ titers but preferentially cleared Aβ$_{42}$-immunoreactive diffuse plaques and not Aβ$_{40}$-immunoreactive or

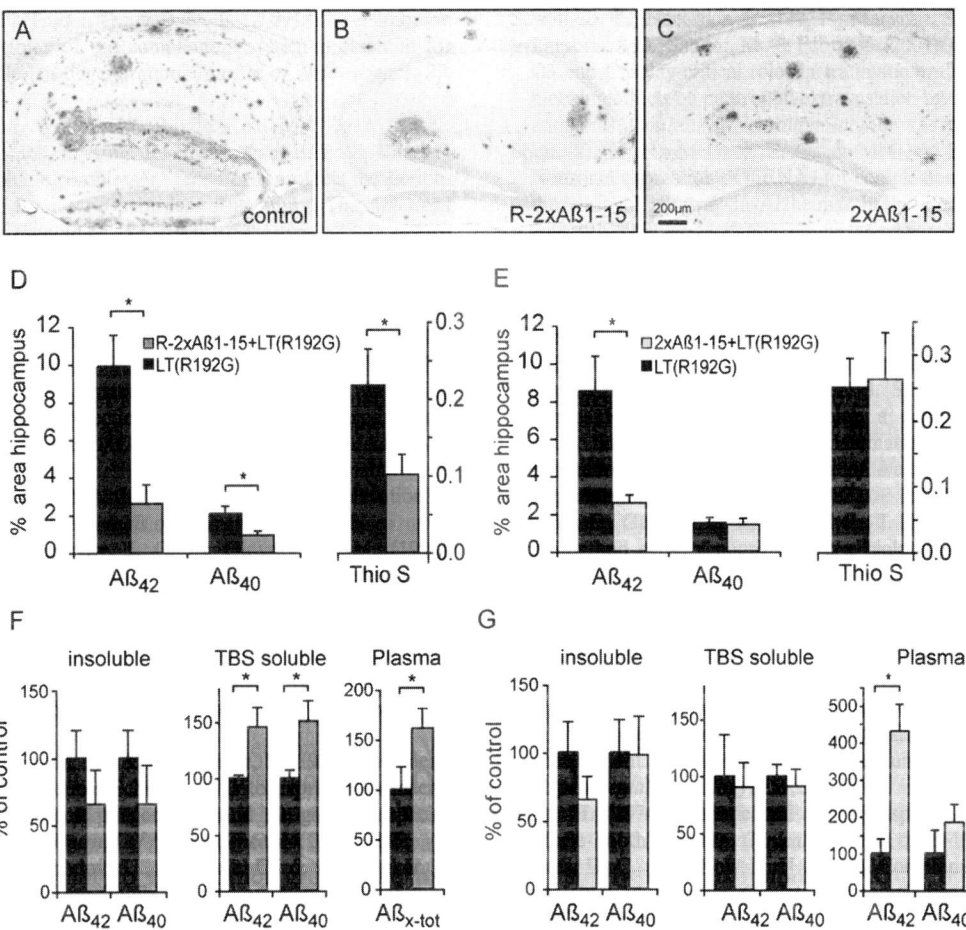

Fig. 2. Neuropathological and biochemical analysis of R-2xAβ1-15 (B, D) or 2x-Aβ1-15 (C, E) immunized animals compared to their corresponding group of adjuvant-treated control hAPP$_{FAD}$ animals (A, D, E). (A–C) Sections representing the median Aβ plaque load are shown for each group (A, B, C). (D and E) Quantitative image analysis of Aβ$_{42}$- and Aβ$_{40}$-specific immunoreactive and thioflavin S-positive plaque load. Aβ$_{42}$-, Aβ$_{40}$-, and thioflavin S-positive areas were significantly reduced in the hippocampus after immunization with R-2xAβ1-15+LT(R192G) (D, $p<0.05$, MWU). 2xAβ1-15+LT(R192G)-immunized mice showed a significant reduction of Aβ$_{42}$-specific immunoreactivity (E, $p<0.05$, MWU) compared to vehicle-treated controls. (F and G) Insoluble (guanidinium-soluble) brain Aβ, TBS-soluble brain Aβ, and plasma Aβ levels were analyzed by capture ELISA [absolute values of controls in R-2xAβ1-15 immunization experiment (F) were plasma Aβ$_{x-tot}$ $0.03\pm0.01$ pmol/ml, Aβ$_{42}$ insoluble $2052\pm417$ pmol/g, Aβ$_{40}$ insoluble $485\pm100$ pmol/g, Aβ$_{42}$ TBS soluble $1.6\pm0.1$ pmol/g, and Aβ$_{40}$ TBS soluble $0.3\pm0.1$ pmol/g; and in the 2xAβ1-15 immunization experiment (G, measured in a different ELISA run), Aβ$_{40}$ plasma $0.9\pm0.6$ pmol/ml, Aβ$_{42}$ plasma $0.3\pm0.1$ pmol/ml, Aβ$_{42}$ insoluble $3590\pm800$ pmol/g, Aβ$_{40}$ insoluble $132\pm32.5$ pmol/g, Aβ$_{42}$ TBS soluble $2.1\pm0.8$ pmol/g, and Aβ$_{40}$ TBS soluble $4.6\pm0.5$ pmol/g]. Asterisk indicates a significant difference (MWU, $p<0.05$). Adapted from Maier et al. (2006) with permission from the Society for Neuroscience.

compacted amyloid plaques (Fig. 2A–C and E). While there was a trend for reduced insoluble Aβ$_{42}$, soluble Aβ levels were unchanged and plasma Aβ$_{42}$ was elevated (Fig. 2G).

J20 mice immunized with 2xAβ-15 were subjected to cognitive evaluation using a reference memory test, the Morris water maze (MWM). After training, immunized mice were able to find the platform in the pool more efficiently than their vehicle control counterparts, indicating faster learning acquisition (Fig. 3A), and had better spatial memory retention of the platform location

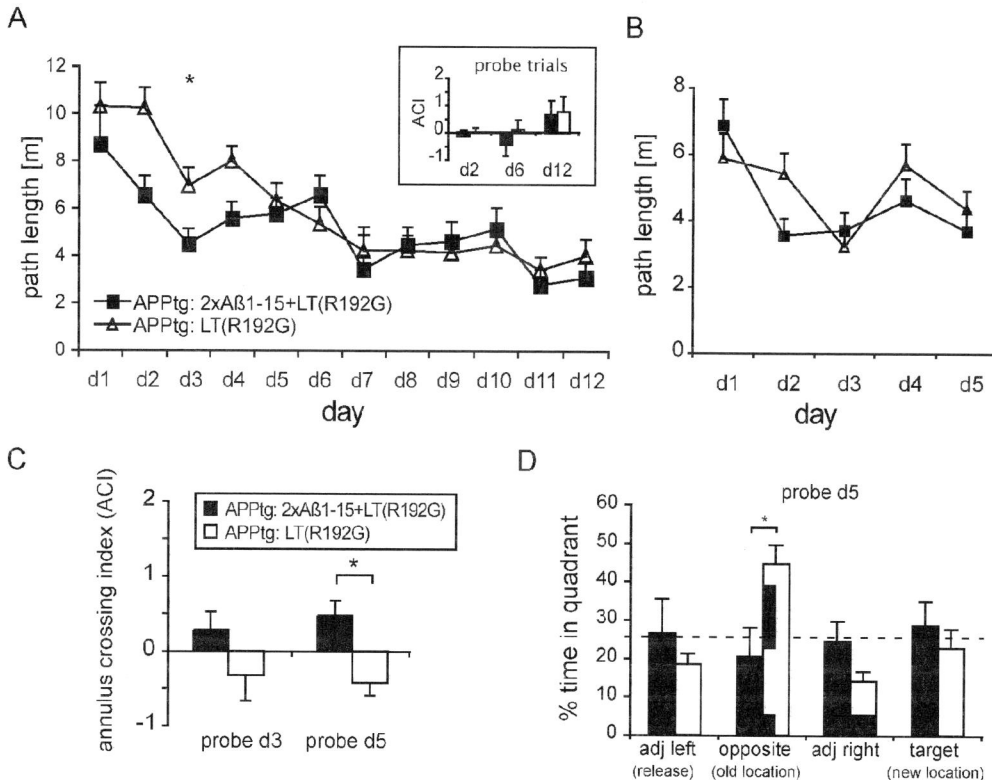

Fig. 3. The effect of immunization with 2xAβ1-15+LT(R192G) was assessed in a reference memory version of the Morris water maze (MWM). (A) 2xAβ1-15+LT(R192G)-immunized hAPP$_{FAD}$ mice ($n = 6$) showed significantly faster learning acquisition during the first four training sessions of a 12-day test as compared to adjuvant-only-treated control hAPP$_{FAD}$ animals ($n = 7$; $p < 0.05$, day 1 to day 5). Both cohorts of mice showed comparable spatial memory as evaluated by the annulus crossing index (ACI, defined as average frequency of swims over the platform site in the target quadrant minus average of swims over sites in other quadrants of the pool) at the end of training (inset, legend see C). (B) During the learning reversal task, 2xAβ1-15+LT(R192G)-immunized hAPP$_{FAD}$ mice showed a trend of faster initial acquisition of the new platform location as compared to control hAPP$_{FAD}$ mice during the first three training sessions ($p = 0.09$, day 1 to day 3). During the reversal stage (day 4 and 5), both cohorts of mice showed a comparable response to platform displacement. (C) ACI during probe trials administered 1 h after training on day 3 and day 5 in the platform reversal task. 2xAβ1-15+LT(R192G)-immunized mice (black bars) show a positive ACI for the platform location on day 3 or day 5, whereas control hAPP$_{FAD}$ mice (gray bars) show a significantly lower and negative ACI. (D) Quadrant dwell time of probe trial on day 5 indicates that control hAPP$_{FAD}$ mice (gray bars) persevered with their search in the original, previous location of the pool. Asterisks indicate significant results ($p < 0.05$). Adapted from Maier et al. (2006) with permission from the Society for Neuroscience.

when the platform was hidden (Fig. 3C and D). In general, the mice with the highest anti-Aβ titers performed the best in the MWM tasks.

## Conclusions

Aβ immunotherapy has been shown to lower cerebral Aβ levels, especially if given prior to or in the early stages of pathology, in AD-like transgenic mouse models, nonhuman primates, and human AD patients. Thus far, human clinical trials with active Aβ immunization have shown limited efficacy and resulted in adverse effects in 6% of immunized patients (AN1792 trial), possibly due to T cell-mediated inflammation in the brain. Passive immunotherapy trials involving the direct administration of humanized monoclonal antibodies should avoid T cell-mediated side effects but are costly and less feasible for a very large population. Several human trials are underway currently. Active Aβ immunotherapy remains under investigation, as it would be less costly and would provide a long-lasting immune response, thereby reducing travel to the doctor's office. Many active Aβ vaccines now target the amino-terminus of Aβ to generate a strong humoral response and avoid an Aβ-specific T cell response, thought to account for the adverse effects in the AN1792 trial. As more and more preclinical and clinical data are collected, it appears that Aβ immunotherapy, like other Aβ-lowering therapies, may have its best efficacy if given before or in the early stages of cognitive decline, prior to massive neuritic dystrophy and neuronal loss. Thus, identifying those at risk and in the earliest stages of the disease, through genetics, biomarkers, and/or imaging, is crucial to early intervention.

## Acknowedgments

This work was supported in part by NIH/NIA RO1 AG20159 (C.A.L.), and the generosity of the Brunozzi family.

## References

Bacskai, B., Kajdasz, S., McLellan, M., Games, D., Seubert, P., Schenk, D., et al. (2002). Non-Fc-mediated mechanisms are involved in clearance of amyloid-β *in vivo* by immunotherapy. *The Journal of Neuroscience, 22,* 7873–7878.

Bacskai, B. J., Kajdasz, S. T., Christie, R. H., Carter, C., Games, D., Seubert, P., et al. (2001). Imaging of amyloid-β deposits in brains of living mice permits direct observation of clearance of plaques with immunotherapy. *Nature Medicine, 7,* 369–372.

Bard, F., Cannon, C., Barbour, R., Burke, R. L., Games, D., Grajeda, H., et al. (2000). Peripherally administered antibodies against amyloid beta-peptide enter the central nervous system and reduce pathology in a mouse model of Alzheimer disease. *Nature Medicine, 6,* 916–919.

Cribbs, D. H., Ghochikyan, A., Vasilevko, V., Tran, M., Petrushina, I., Sadzikava, N., et al. (2003). Adjuvant-dependent modulation of Th1 and Th2 responses to immunization with beta-amyloid. *International Immunology, 15,* 505–514.

Das, P., Howard, V., Loosbrock, N., Dickson, D., Murphy, M. P., & Golde, T. E. (2003). Amyloid-beta immunization effectively reduces amyloid deposition in FcRgamma$^{-/-}$ knock-out mice. *The Journal of Neuroscience, 23,* 8532–8538.

Das, P., Murphy, M., Younkin, L., Younkin, S., & Golde, T. (2001). Reduced effectiveness of Aβ1-42 immunization in APP transgenic mice with significant amyloid deposition. *Neurobiology of Aging, 22,* 721–727.

DeMattos, R., Bales, K., Cummins, D., Dodart, J.-C., Paul, S., & Holtzman, D. (2001). Peripheral anti-Aβ antibody alters CNS and plasma clearance and decreases brain Aβ burden in a mouse model of Alzheimer's disease. *Proceedings of the National Academy of Sciences of the United States of America, 98,* 8850–8855.

Dickinson, B. L., & Clements, J. D. (1995). Dissociation of *Escherichia coli* heat-labile enterotoxin adjuvanticity from ADP-ribosyltransferase activity. *Infection and Immunity, 63,* 1617–1623.

Dickson, D. W. (1997). The pathogenesis of senile plaques. *Journal of Neuropathology and Experimental Neurology, 56,* 321–339.

Dodart, J.-C., Bales, K., Gannon, K., Greene, S., DeMattos, R., Mathis, C., et al. (2002). Immunization reverses memory deficits without reducing brain Aβ burden in Alzheimer's disease model. *Nature Neuroscience, 5,* 452–457.

Ferrer, I., Boada Rovira, M., Sanchez Guerra, M. L., Rey, M. J., & Costa-Jussa, F. (2004). Neuropathology and pathogenesis of encephalitis following amyloid-beta immunization in Alzheimer's disease. *Brain Pathology, 14,* 11–20.

Fox, N. C., Black, R. S., Gilman, S., Rossor, M. N., Griffith, S. G., Jenkins, L., et al. (2005). Effects of A{beta} immunization (AN1792) on MRI measures of cerebral volume in Alzheimer disease. *Neurology, 64,* 1563–1572.

Gandy, S., DeMattos, R. B., Lemere, C. A., Heppner, F. L., Leverone, J., Aguzzi, A., et al. (2004). Alzheimer's Abeta vaccination of rhesus monkeys (*Macaca mulatta*). *Mechanisms of Ageing and Development, 125,* 149–151.

Gilman, S., Koller, M., Black, R. S., Jenkins, L., Griffith, S. G., Fox, N. C., et al. (2005). Clinical effects of A{beta} immunization (AN1792) in patients with AD in an interrupted trial. *Neurology, 64,* 1553–1562.

Hardy, J., & Selkoe, D. J. (2002). The amyloid hypothesis of Alzheimer's disease: progress and problems on the road to therapeutics. *Science, 297,* 353–356.

Hock, C., Konietzko, U., Papassotiropoulos, A., Wollmer, A., Streffer, J., von Rotz, R. C., et al. (2002). Generation of antibodies for beta-amyloid by vaccination of patients with Alzheimer disease. *Nature Medicine, 8,* 1270–1275.

Hock, C., Konietzko, U., Streffer, J. R., Tracy, J., Signorell, A., Muller-Tillmanns, B., et al. (2003). Antibodies against beta-amyloid slow cognitive decline in Alzheimer's disease. *Neuron, 38,* 547–554.

Holmes, C., Boche, D., Wilkinson, D., Yadegarfar, G., Hopkins, V., Bayer, A., et al. (2008). Long-term effects of Aβ42 immunisation in Alzheimer's disease: follow-up of a randomised, placebo-controlled phase I trial. *Lancet, 372,* 216–223.

Janus, C., Pearson, J., McLaurin, J., Mathews, P. M., Jiang, Y., Schmidt, S. D., et al. (2000). A beta peptide immunization reduces behavioural impairment and plaques in a model of Alzheimer's disease. *Nature, 408*, 979–982.

Klein, W. (2002). Abeta toxicity in Alzheimer's disease: globular oligomers (ADDLs) as new vaccine and drug targets. *Neurochemistry International, 41*, 345–352.

Klyubin, I., Walsh, D. M., Lemere, C. A., Cullen, W. K., Shankar, G. M., Betts, V., et al. (2005). Amyloid beta protein immunotherapy neutralizes Abeta oligomers that disrupt synaptic plasticity in vivo. *Nature Medicine, 11*, 556–561.

Kotilinek, L. A., Bacskai, B., Westerman, M., Kawarabayashi, T., Younkin, L., Hyman, B. T., et al. (2002). Reversible memory loss in a transgenic model of Alzheimer's disease. *The Journal of Neuroscience, 22*, 6331–6335.

Lee, M., Bard, F., Johnson-Wood, K., Lee, C., Hu, K., Griffith, S. G., et al. (2005). Abeta42 immunization in Alzheimer's disease generates Abeta N-terminal antibodies. *Annals of Neurology, 58*, 430–435.

Lemere, C. A., Beierschmitt, A., Iglesias, M., Spooner, E. T., Bloom, J. K., Leverone, J. F., et al. (2004). Alzheimer's disease abeta vaccine reduces central nervous system abeta levels in a non-human primate, the Caribbean vervet. *The American Journal of Pathology, 165*, 283–297.

Lemere, C. A., Maier, M., Peng, Y., Jiang, L., & Seabrook, T. J. (2007). Novel Aβ immunogens: is shorter better? *Current Alzheimer Research, 4*, 427–436.

Lemere, C. A., Maron, R., Selkoe, D. J., & Weiner, H. L. (2001). Nasal vaccination with beta-amyloid peptide for the treatment of Alzheimer's disease. *DNA and Cell Biology, 20*, 705–711.

Lemere, C. A., Maron, R., Spooner, E. T., Grenfell, T. J., Mori, C., Desai, R., et al. (2000). Nasal A beta treatment induces anti-A beta antibody production and decreases cerebral amyloid burden in PD-APP mice. *Annals of the New York Academy of Sciences, 920*, 328–331.

Lemere, C. A., Spooner, E. T., LaFrancois, J., Malester, B., Mori, C., Leverone, J. F., et al. (2003). Evidence for peripheral clearance of cerebral Abeta protein following chronic, active Abeta immunization in PSAPP mice. *Neurobiology of Disease, 14*, 10–18.

Maier, M., Seabrook, T. J., Lazo, N. D., Jiang, L., Das, P., Janus, C., et al. (2006). Short amyloid-β (Aβ) immunogens reduce cerebral Aβ load and learning deficits in an Alzheimer's disease mouse model in the absence of an Aβ-specific cellular immune response. *The Journal of Neuroscience, 26*, 4717–4728.

Masliah, E., Hansen, L., Adame, A., Crews, L., Bard, F., Lee, C., et al. (2005). Abeta vaccination effects on plaque pathology in the absence of encephalitis in Alzheimer disease. *Neurology, 64*, 129–131.

McLaurin, J., Cecal, R., Kierstead, M., Tian, X., Phinney, A., Manea, M., et al. (2002). Therapeutically effective antibodies against amyloid-beta peptide target amyloid-beta residues 4–10 and inhibit cytotoxicity and fibrillogenesis. *Nature Medicine, 8*, 1263–1269.

Monsonego, A., Maron, R., Zota, V., Selkoe, D., & Weiner, H. (2001). Immune hyporesponsiveness to amyloid-β peptide in amyloid precursor protein transgenic mice: implications for the pathogenesis and treatment of Alzheimer's disease. *Proceedings of the National Academy of Sciences of the United States of America, 98*, 10273–10278.

Morgan, D., Diamond, D. M., Gottschall, P. E., Ugen, K. E., Dickey, C., Hardy, J., et al. (2000). Abeta peptide vaccination prevents memory loss in an animal model of Alzheimer's disease. *Nature, 408*, 982–985.

Mucke, L., Masliah, E., Yu, G.-Q., Mallory, M., Rockenstein, E. M., Tatsuno, G., et al. (2000). High-level neuronal expression of Aβ1-42 in wild-type human amyloid protein precursor transgenic mice: synaptotoxicity without plaque formation. *The Journal of Neuroscience, 20*, 4050–4058.

Nicoll, J. A., Wilkinson, D., Holmes, C., Steart, P., Markham, H., & Weller, R. O. (2003). Neuropathology of human Alzheimer disease after immunization with amyloid-beta peptide: a case report. *Nature Medicine, 9*, 448–452.

Nicoll, J. A. R., Barton, E., Boche, D., Neal, J. W., Ferrer, I., Thompson, P., et al. (2006). Aβ species removal after Aβ42 immunization. *Journal of Neuropathology and Experimental Neurology, 65*, 1040–1048.

Oddo, S., Billings, L., Kesslak, J. P., Cribbs, D. H., & LaFerla, F. M. (2004). Abeta immunotherapy leads to clearance of early, but not late, hyperphosphorylated tau aggregates via the proteasome. *Neuron, 43*, 321–332.

Orgogozo, J. M., Gilman, S., Dartigues, J. F., Laurent, B., Puel, M., Kirby, L. C., et al. (2003). Subacute meningoencephalitis in a subset of patients with AD after Abeta42 immunization. *Neurology, 61*, 46–54.

Pfeifer, M., Boncristiano, S., Bondolfi, L., Stalder, A., Deller, T., Staufenbiel, M., et al. (2002). Cerebral hemorrhage after passive anti-Abeta immunotherapy. *Science, 298*, 1379.

Racke, M. M., Boone, L. I., Hepburn, D. L., Parsadainian, M., Bryan, M. T., Ness, D. K., et al. (2005). Exacerbation of cerebral amyloid angiopathy-associated microhemorrhage in amyloid precursor protein transgenic mice by immunotherapy is dependent on antibody recognition of deposited forms of amyloid beta. *The Journal of Neuroscience, 25*, 629–636.

Schenk, D. (2002). Amyloid-β immunotherapy for Alzheimer's disease: the end of the beginning. *Nature, 3*, 824–828.

Schenk, D., Barbour, R., Dunn, W., Gordon, G., Grajeda, H., Guido, T., et al. (1999). Immunization with amyloid-β attenuates Alzheimer-disease-like pathology in the PDAPP mouse. *Nature, 400*, 173–177.

Seabrook, T. J., Iglesias, M., Bloom, J. K., Spooner, E. T., & Lemere, C. A. (2004). Differences in the immune response to long term Abeta vaccination in C57BL/6 and B6D2F1 mice. *Vaccine, 22*, 4075–4083.

Seabrook, T. J., Thomas, K., Jiang, L., Bloom, J., Spooner, E., Maier, M., et al. (2006). Dendrimeric Aβ1-15 is an effective immunogen in wildtype and APP tg mice. *Neurobiology of Aging, 28*, 813–823.

Selkoe, D. J. (2001). Alzheimer's disease: genes, proteins, and therapy. *Physiological Reviews, 81*, 741–766.

Sigurdsson, E. M., Scholtzova, H., Mehta, P. D., Frangione, B., & Wisniewski, T. (2001). Immunization with a nontoxic/nonfibrillar amyloid-beta homologous peptide reduces Alzheimer's disease-associated pathology in transgenic mice. *The American Journal of Pathology, 159*, 439–447.

Solomon, B., Koppel, R., Frenkel, D., & Hanan-Aharon, E. (1997). Disaggregation of Alzheimer β-amyloid by site-directed mAb. *Proceedings of the National Academy of Sciences of the United States of America, 94*, 4109–4112.

Solomon, B., Koppel, R., Hanan, E., & Katzav, T. (1996). Monoclonal antibodies inhibit in vitro fibrillar aggregation of the Alzheimer β-amyloid peptide. *Proceedings of the National Academy of Sciences of the United States of America, 93*, 452–455.

Town, T., Tan, J., Sansone, N., Obregon, D., Klein, T., & Mullan, M. (2001). Characterization of murine immunoglobulin G antibodies against human amyloid-β1-42. *Neuroscience Letters, 307*, 101–104.

Walsh, D. M., & Selkoe, D. J. (2004). Oligomers on the brain: the emerging role of soluble protein aggregates in neurodegeneration. *Protein and Peptide Letters, 11*, 213–228.

Weiner, H. L., Lemere, C. A., Maron, R., Spooner, E. T., Grenfell, T. J., Mori, C., et al. (2000). Nasal administration of amyloid-beta peptide decreases cerebral amyloid burden in a mouse model of Alzheimer's disease. *Annals of Neurology, 48*, 567–579.

Wilcock, D. M., Rojiani, A., Rosenthal, A., Subbarao, S., Freeman, M. J., Gordon, M. N., et al. (2004). Passive immunotherapy against Abeta in aged APP-transgenic mice reverses cognitive deficits and depletes parenchymal amyloid deposits in spite of increased vascular amyloid and microhemorrhage. *Journal of Neuroinflammation, 1*, 24.

Wolfe, M. (2006). Shutting down Alzheimer's. *Scientific American, 294*, 72–79.

Yano, A., Onozuka, A., Matin, K., Imai, S., Hanada, N., & Nisizawa, T. (2003). RGD motif enhances immunogenicity and adjuvanticity of peptide antigens following intranasal immunization. *Vaccine, 22*, 237–243.

CHAPTER 7

# Innate immunity in the nervous system

V. Ramaglia[1,2] and F. Baas[1,3],*

[1] Neurogenetics Laboratory, Academic Medical Center, University of Amsterdam, Amsterdam, The Netherlands
[2] Department of Infection, Immunity and Biochemistry, School of Medicine, Cardiff University, Cardiff, UK
[3] Department of Neurology, Academic Medical Center, University of Amsterdam, Amsterdam, The Netherlands

**Abstract:** The complement (C) system plays a central role in innate immunity and bridges innate and adaptive immune responses. A fine balance of C activation and regulation mediates the elimination of invading pathogens and the protection of the host from excessive C deposition on healthy tissues. If this delicate balance is disrupted, the C system may cause injury and contribute to the pathogenesis of various diseases, including neurodegenerative disorders and neuropathies. Here we review evidence indicating that C factors and regulators are locally synthesized in the nervous system and we discuss the evidence supporting the protective or detrimental role of C activation in health, injury, and disease of the nerve.

**Keywords:** complement; central nervous system; peripheral nervous system; degeneration; regeneration; neurodegenerative disorders; neuropathies

## The complement system

Complement was discovered in the 1890s (von Fodor, 1887; Nuttall, 1888; Buchner, 1889) as a heat-sensitive serum factor capable of lysing bacteria in the presence of the heat-stable antibody (Bordet, 1895). Molecular biology profoundly transformed our understanding of the complement system and from its original description as "complement" to humoral immunity (Ehrlich and Morgenroth, 1899), today it represents a key component of the innate immune system, defending the host against infections, bridging innate and adaptive immunity, and disposing of immune complexes and apoptotic cells (Walport, 2001a, b). Paradoxically, the same system responsible for such beneficial effects can be deleterious to the host. To prevent complement-mediated tissue injury, over 30 soluble and membrane-bound complement proteins are engaged in a fine coordination of activation and regulation. However, if the regulatory machinery fails, the complement system can contribute to tissue injury and the pathogenesis of various diseases.

The local synthesis of factors and regulators of the complement cascade in the brain (reviewed in Gasque et al., 2000) and the peripheral nerve (de Jonge et al., 2004) has been established but its role in nerve health, injury, and disease remains controversial. Complement activation in the brain could lead to cytolytic death of neurons and enhance a proinflammatory reaction contributing to the pathogenesis and progression of neurodegenerative diseases but could also be involved in brain tissue remodeling and repair by clearing toxic deposits, such as amyloid fibrils present in

*Corresponding author.
Tel.: +31 20 566 5998; Fax: +31 20 566 9312;
E-mail: f.baas@amc.uva.nl

neuritic plaques in Alzheimer's disease (AD), or by enhancing the expression of growth factors involved in the early processes of regeneration. Complement activation in the peripheral nerve could facilitate regeneration of injured axons by assisting in the efficient clearance of myelin debris, thought to be inhibitory for axon growth, but it could also exacerbate tissue damage during degeneration hampering the correct regeneration of the nerve. In addition, complement could contribute to the pathogenesis and progression of neuropathies and neurodegenerative diseases. Here we will review the evidence supporting the protective and detrimental role of the complement system in the nervous system with particular emphasis to peripheral nerves since this is the main research focus of the authors.

## Activation and regulation of the complement system

Activation of the complement system is rapid and efficient. Soluble complement components are present in the blood, body fluids, and tissues to readily trigger a defense reaction against external (i.e., pathogens) or internal (i.e., autoimmunity) danger signals (Kohl, 2006). Complement activation can occur via three routes: the classical, the lectin, and the alternative pathway. The classical pathway is activated by the recognition of an antigen–antibody complex by C1q. Upon binding, C1r cleaves C1s which in turn cleaves C2 and C4 into a small (C2b, C4a) and a large fragment (C2a, C4b). C2a and C4b together form the C3 convertase. The lectin pathway is triggered by binding of mannose-binding lectins (MBLs) to certain carbohydrates expressed on the pathogen surface. This activates the MBL-associated serine protease (MASP) 2, cleaving C4 and C2 (Fujita, 2002). The alternative pathway starts by spontaneous low-rate hydrolysis of C3 generating $C3(H_2O)$ which binds to factor B, permitting cleavage by factor D to form the fluid-phase C3 convertase $C3(H_2O)Bb$. This enzyme cleaves C3 and deposits C3b on surfaces where, in the absence of C inhibitors such as factor H, it binds and catalyses cleavage of factor B to form surface

bound C3 convertase C3bBb (Nauta et al., 2004). Irrespectively of the pathway involved, activation of the complement system leads to the cleavage of C3 and C5, generating the potent chemoattractants C3a and C5a as well as the C5b fragment. This initiates the assembly of the C5b-9 membrane attack complex (MAC), a lipophilic complex which forms pores in the pathogens cell membrane, leading to cell lysis (Nauta et al., 2004).

Recently, additional venues of complement activation have been identified. The so-called "C2 bypass" pathway consists of the direct cleavage of C3 by MASP-2 of the lectin pathway, bypassing the proteolytic cleavage of C2, thus formation of the C3 convertase (Atkinson and Frank, 2006). Direct cleavage of C3 and C5 by noncomplement proteins such as lysosomal enzymes released from neutrophils, kallikrein, part of the kinin and fibrinolysis systems, or trombin has also been shown and the route named "extrinsic protease" pathway (Markiewski and Lambris, 2007). Lastly, properdin stabilizes the C3 convertase on the pathogen surface but it also seems to act as a pattern recognition molecule and directly induce the C3 convertase on foreign surfaces (Hourcade, 2006; Spitzer et al., 2007) (Fig. 1A).

A sophisticated regulatory mechanism allows the complement system to rapidly attack invading pathogens while protecting host cells from its detrimental effects. This is achieved via the coordination of time, location, and molecular interactions. After activation, several thousand C3b molecules deposit every minute on a cell. In theory, this opsonization is not specifically directed against foreign cells. However, the thioster bound of C3b, which enables it to covalently bind to hydroxyl groups on nearby carbohydrates and protein-acceptor groups, has a short half-life, limiting its action to the site of activation. If the acceptor molecule is on the host cell, a set of soluble and membrane-bound negative regulators inactivate and degrade the activated complement components. On the other hand, positive regulators such as properdin, stabilize the convertase on foreign cells (Soares and Barlow, 2005).

These regulatory complement components either induce an accelerated decay of the convertase or act as cofactor for Factor I to degrade activated complement fragments. Decay accelerating factor (DAF/CD55) and C4-binding protein (C4BP) accelerate decay of the convertase; membrane cofactor protein (MCP/CD46) acts together with Factor I to degrade C3b to its inactive form iC3b; complement receptor 1 (CR1/CD35) and Factor H can do both. In addition, CD59 prevents formation of the MAC by inserting between the C8 and C9 subunits of the C5b-9 polymer (Soares and Barlow, 2005) (Fig. 1B).

## The role of complement in inflammation

For over 700 million years, the complement system has provided protection against microbial infections (Sunyer and Lambris, 1998), yet its function extends beyond a simple defense mechanism. Today it is clear that the complement system is a key regulator of various stages of an inflammatory reaction. These events are mediated by the potent complement anaphylatoxins C3a and C5a which propagate the immune reaction by binding to their receptors (C3aR, C5aR, C5L2) on the host cell. Macrophages, widely distributed in connective tissues and various organs (i.e., lung, liver, spleen), express complement receptors. The interactions between complement receptors on macrophages and opsonins on the target cell mediate activation of the phagocyte and secretion of cytokines and chemokines (reviewed by van Lookeren et al., 2007). Anaphylatoxins also activate mast cells to release histamine, tumor necrosis factor-α (TNF-α), cytokines, and chemokines, mediating vasodilation and leukocytes migration from the blood stream to the site of inflammation (Lee et al., 2002).

Complement is also involved in the disposal of immune complexes, necrotic, and apoptotic cells which are usually generated during an inflammatory reaction (Walport, 2001a). The clearance of immune complexes is facilitated by maintaining their solubility through the binding of the C1 complex, C4 and C3 to the antigen. This prevents an increase in the size of the opsonized complex which is easily recognized by phagocytes and readily removed to limit inflammation and prevent the propagation of injury on neighboring tissues. The removal of necrotic and apoptotic cells is critical for the termination of inflammation and prevention of autoimmunity. Dying cells undergo several changes to signal their removal to phagocytic cells. These changes involve modifications of the plasma membrane resulting in the exposure of self-antigens, normally sequestered within the viable cell, and the internalization of proteins normally expressed on the cell surface. The presentation of "eat me" signals and the downregulation of "don't eat me" signals trigger the binding of a number of opsonins to mediate the removal of dying cells. This is a key process in the maintenance of tissue homeostasis and normally does not require complement binding. However, during overwhelming apoptosis or impaired phagocytosis complement initiation factors can bind dying cells to ensure proper disposal of self-antigens (Trouw et al., 2008), avoiding the generation of an autoimmune reaction against the host.

## Complement in the central nervous system (CNS)

The primary site of synthesis of plasma complement proteins is liver. Extrahepatic complement biosynthesis occurs in many tissues and it may account for the rapid and efficient ability of the complement system to initiate and propagate an inflammatory response. Local production of complement proteins is especially important in tissues where the access to plasma proteins is hindered by a blood–tissue barrier (Morgan and Gasque, 1997; Laufer et al., 2001).

### *Local synthesis in the CNS*

The brain, shielded by the blood–brain barrier, has been for years considered an immune-privileged organ but local biosynthesis of complement occurs (Barnum, 1995; Gasque et al., 2000). Astrocytes, microglia, neurons, and oligodendrocytes are able to produce all complement components (Table 1). Membrane (CD59, MCP, DAF) and soluble [fH, fI, C1inh, clusterin (CLU)] complement inhibitors are expressed by human astrocytes, oligodendrocytes, and microglia whereas neurons express very

A

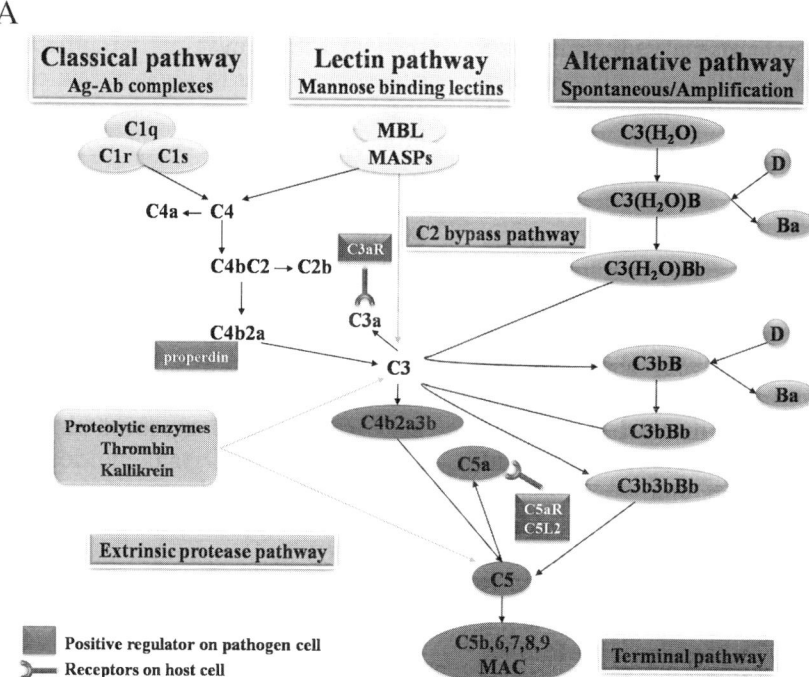

B

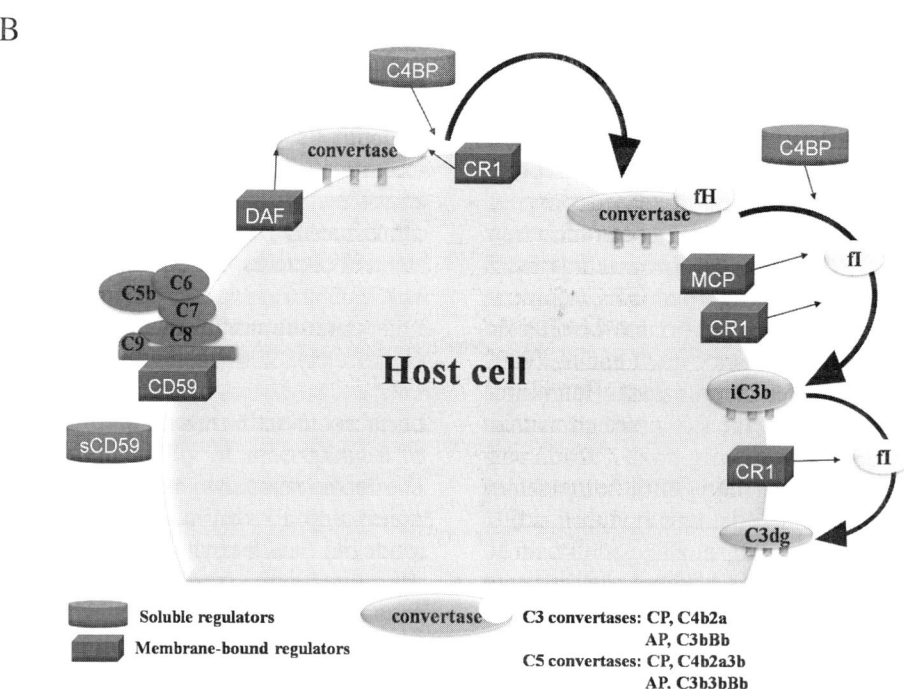

low levels of most complement inhibitors and lack DAF, rendering them extremely susceptible to complement-mediated lysis by the MAC (Gasque et al., 2002).

## Activation in CNS injury

Complement activation has been implicated in the development of secondary brain damage following traumatic brain injury (TBI) and axonal degeneration after spinal cord injury (SCI) (Table 2). Brain injury causes a primary mechanical damage derived from the impact and a secondary wave of neuroinflammatory events leading to delayed neuronal death. The delay in neuronal cell loss offers a window of opportunity for possible treatment.

The complement system is considered a key determinant of the post-traumatic neuroinflammation. Deposits of the terminal C9 have been found on injured neurons after brain trauma (Bellander et al., 1996) and high levels of fB, C3, and sC5b-9 have been detected in the cerebrospinal fluid (CSF) of patients with severe TBI during the first 10 days of trauma (Stahel et al., 2001; Kossmann et al., 1997). Mice deficient in the C3 or C5 complement components subjected to traumatic brain cryoinjury showed reduced neutrophils and secondary tissue damage compared to their wild-type littermates (Sewell et al., 2004). TBI on transgenic mice deficient in factor B ($fB^{-/-}$), necessary for activation of the alternative pathway of complement, showed reduced post-traumatic neuronal cell death, a strong upregulation of the anti-apoptotic mediator Bcl-2, and downregulation of the pro-apoptotic Fas receptor compared to the $fB^{+/+}$ littermates, implicating the alternative pathway of complement in the progression of the secondary neuronal death (Leinhase et al., 2006a). Neurological function after TBI was improved in transgenic mice with CNS-targeted overexpression of complement receptor type 1-related protein (Crry), a potent inhibitor of the C3 convertase (Rancan et al., 2003). Activation of the complement cascade also plays a role in the demyelination and axonal degeneration following SCI. High levels of complement factors (C1q, C4, fB, and C5b-9) and inhibitors (fH and CLU) have been found on neurons and oligodendrocytes of injured rat spinal cord up to 42 days post-injury (Anderson et al., 2004, 2005). Taken together these data implicate complement proteins in the secondary neurodegeneration after head injury and neuronal loss following SCI. Thus, limiting complement activation could represent a promising approach for treatment.

---

Fig. 1. Activation and regulation of the complement system. Pathways of complement activation are described in (A). The classical pathway is activated by the recognition of an antigen-antibody complex by C1q. Upon binding, C1r cleaves C1s which in turn cleaves C2 and C4 into a small (C2b, C4a) and a large fragment (C2a, C4b). C2a and C4b together form the C3 convertase. The lectin pathway is triggered by binding of mannose-binding lectins (MBLs) to certain carbohydrates expressed on the pathogen surface. This activates the MBL-associated serine protease (MASP) 2, cleaving C4 and C2. The alternative pathway starts by spontaneous low-rate hydrolysis of C3 generating $C3(H_2O)$ which binds to factor B, permitting cleavage by factor D to form the fluid-phase C3 convertase $C3(H_2O)Bb$. This enzyme cleaves C3 and deposits C3b on surfaces where, in the absence of C inhibitors such as factor H, it binds and catalyses cleavage of factor B to form surface bound C3 convertase C3bBb. Irrespectively of the pathway involved, activation of the complement system leads to the cleavage of C3 and C5, generating the potent chemoattractants C3a and C5a as well as the C5b fragment. This initiates the assembly of the C5b-9 membrane attack complex (MAC), a lipophilic complex which forms pores in the pathogens cell membrane, leading to cell lysis. The "C2 bypass" pathway consists of the direct cleavage of C3 by MASP-2 of the lectin pathway, bypassing the proteolytic cleavage of C2, thus formation of the C3 convertase. Direct cleavage of C3 and C5 by noncomplement proteins such as lysosomal enzymes released from neutrophils, kallikrein, part of the kinin and fibrinolysis systems, or trombin has also been shown and the route named "extrinsic protease" pathway. Lastly, properdin stabilizes the C3 convertase on the pathogen surface but it also seems to act as a pattern recognition molecule and directly induces the C3 convertase on foreign surfaces. Mechanisms of complement regulation are described in (B). Regulatory complement components either induce an accelerated decay of the convertase or act as cofactor for Factor I to degrade activated complement fragments. Decay accelerating factor (DAF/CD55) and C4-binding protein (C4BP) accelerate decay of the convertase; membrane cofactor protein (MCP/CD46) acts together with Factor I to degrade C3b to its inactive form iC3b; complement receptor 1 (CR1/CD35) and Factor H can do both. In addition, CD59 prevents formation of the MAC by inserting between the C8 and C9 subunits of the C5b-9 polymer.

Table 1. Expression of complement factors in the brain

| Complement factor | Brain cell/cell line | References | |
|---|---|---|---|
| C1q | a. Astroglioma<br>b. Microglia | a.<br>b. | Gasque et al. (1993)<br>Walker et al. (1995a) |
| C1r, C1s | a. Astroglioma<br>b. Neuroblastoma | a, b. | Gasque et al. (1993)<br>Veerhuis et al. (1996) |
| C2 | a. Astroglioma<br>b. Microglia | a.<br>b. | Gasque et al. (1993), Barnum et al. (1992a)<br>Walker et al. (1995a) |
| FD | a. Astroglioma | a. | Barnum et al. (1992a) |
| FB | a. Astroglioma and astrocytes | a. | Barnum et al. (1992a), Gasque et al. (1992) |
| Properdin | a. Astrocytes | a. | Avery et al. (1993) |
| C3 | a. Astroglioma and astrocytes<br>b. Oligodendrocytes<br>c. Microglia<br>d. Neuroblastoma | a.<br>b.<br>c.<br>d. | Barnum et al. (1992b, 1993), Gasque et al. (1992), Gordon et al. (1992)<br>Gasque and Morgan (1996)<br>Walker et al. (1995a)<br>Walker and McGeer (1993) |
| C4 | a. Astroglioma<br>b. Microglia<br>c. Neuroblastoma | a.<br>b.<br>c. | Gasque et al. (1993)<br>Walker et al. (1995a)<br>Walker and McGeer (1993) |
| C5 | a. Astroglioma | a. | Gasque et al. (1993) |
| C6, C7, C8, C9 | a. Astroglioma and astrocytes | a. | Gasque et al. (1995) |
| C1inh | a. Astroglioma and astrocytes<br>b. Neuroblastoma<br>c. Oligodendrocytes<br>d. Microglia | a, b.<br><br>c.<br>d. | Gasque et al. (1993, 1996)<br>Veerhuis et al. (1998)<br>Gasque and Morgan (1996)<br>Walker et al. (1995b) |
| C4BP | a. Oligodendrocytes | a. | Gasque and Morgan (1996) |
| FH | a. Astroglioma<br>b. Oligodendrocytes<br>c. Neuroblastoma | a.<br>b.<br>c. | Gasque et al. (1992)<br>Gasque and Morgan (1996)<br>Gasque et al. (1996) |
| FI | a. Astroglioma and astrocytes | a. | Gasque et al. (1992), Gordon et al. (1992) |
| S-protein, CLU | a. Astroglioma and astrocytes<br>b. Oligodendrocytes<br>c. Neuroblastoma | a.<br>b.<br>c. | Gasque et al. (1995)<br>Gasque and Morgan (1996)<br>Gasque et al. (1996) |
| MCP | a. Astroglioma and astrocytes<br>b. Oligodendrocytes<br>c. Neuroblastoma | a.<br>a, b.<br>c. | Yang et al. (1993)<br>Gasque and Morgan (1996)<br>Gasque et al. (1996) |
| CD59 | a. Astroglioma and astrocytes<br>b. Oligodendrocytes<br>c. Neuroblastoma and neurons | a.<br>a, b.<br>c. | Gordon et al. (1993), Yang et al., (1993)<br>Gasque and Morgan (1996)<br>Gasque et al. (1996), McGeer et al. (1991) |
| DAF | a. Astroglioma and astrocytes<br>b. Oligodendrocytes | a, b.<br>b. | Gasque and Morgan (1996)<br>Yang et al., (1993) |

Table 2. Complement implication in CNS injury

| Process | Evidence of C involvement | References |
|---|---|---|
| Traumatic brain injury | a. C9 deposits on neuronal membranes following experimental TBI in rats<br>b. C9 deposits in rat hippocampus following experimental perforant path transaction<br>c. Elevated fB, C3, and sC5b-9 levels in CSF of patients with TBI during the first 10 days of trauma<br>d. Reduced neuronal cell death in fB$^{-/-}$ mice following experimental TBI<br>e. Reduced neutrophils and secondary tissue damage in C3$^{-/-}$ and C5$^{-/-}$ mice following traumatic brain cryoinjury<br>f. Improved neurological function in transgenic mice overexpressing soluble Crry | a. Bellander et al. (1996)<br>b. Johnson et al. (1996)<br>c. Kossmann et al. (1997), Stahel et al. (2001),<br>d. Leinhase et al. (2006a)<br>e. Sewell et al. (2004)<br>f. Rancan et al. (2003) |
| Spinal cord injury | a. Elevated C1q, C4, fB, and C5b-9 deposits on neurons and oligodendrocytes of contusion-injured spinal cord at 1, 7, and 42 days post-injury<br>b. Upregulation of fH and clusterin on neurons and oligodendrocytes of contusion-injured spinal cord at 1, 7, and 42 days post-injury<br>c. Elevated C factors in sera of patients with spinal cord injury | a. Anderson et al. (2004)<br>b. Anderson et al. (2005)<br>c. Rebhun and Botvin (1980), Rebhun et al. (1991) |

## Activation in CNS disease

Complement activation is involved in a number of neurodegenerative disorders, including Alzheimer's, Huntington's, Parkinson's, Creutzfelt–Jakob disease and amyotrophic lateral sclerosis (ALS) (reviewed in Morgan et al., 1997; Bonifati and Kishore, 2007) and neuroinflammatory diseases such as multiple sclerosis (MS) (Ffrench-Constant, 1994) (Table 3).

### Alzheimer's disease

In his seminal observations, Alzheimer (1911) described a close association between amyloid plaques and brain phagocytes. Evidence accumulated over the past two decades suggested that inflammation may contribute to AD pathogenesis and implicates complement as a potential mediator of the inflammatory response (Eikelenboom et al., 1991; Emmerling et al., 2000). Complement transcripts and proteins are upregulated in AD brains, in same cases up to 80-fold (Yasojima et al., 1999a, b). Deposits of complement components including C1q, C1r, C1s, C2, C3, C4, C5, C6, C7, C8, and C9 have been found in pyramidal neurons (Shen et al., 1997) and plaques (Terai et al., 1997). Activated complement fragments have been described in close association with tangles and plaques, and MAC deposits have been found on dystrophic neuritis (McGeer and McGeer, 2002) whereas the complement regulatory protein CD59 is strongly decreased in the prefrontal cortex and hippocampus of AD patients (Zanjani et al., 2005). Both, the classical and alternative pathways of complement have been implicated in AD (Tacnet-Delorme et al., 2001; Strohmeyer et al., 2000) and seem to mediate neuronal injury via MAC-induced neurite disintegration (Yang et al., 2000) and increase in the level of reactive oxygen species (Luo et al., 2003). The role of complement in AD remains however controversial since evidence mainly from AD animal models treated with complement inhibitors have shown an increase in amyloid β accumulation and neuronal degeneration suggesting the possibility of a neuroprotective role of complement in AD pathology, probably by reducing accumulation and promoting the clearance of amyloid and degenerating neurons (Wyss-Coray et al., 2002). The question of whether the complement system plays a detrimental or neuroprotective role in AD still remains open.

Table 3. Examples of complement implication in neurodegenerative CNS diseases

| Process | Evidence of C involvement | References |
|---|---|---|
| Alzheimer's disease | a. C activation in and around neuritic plaques and neurofibrillary tangles in AD brain<br>b. C1q binds to fibrillar but not soluble β-amyloid activating the classical pathway in vitro<br>c. C1q binds to C reactive protein and serum amyloid P on neuritic amyloid plaques in vitro<br>d. C1q upregulation and colocalization with fibrillar Aβ in APP or presenilin mutant mice<br>e. Increased C mRNA level in AD brains (RT-PCR)<br>f. Decreased CD59 in AD frontal cortex and hippocampus<br>g. Increased expression of C receptors (C5aR, C3aR) on reactive glial cells in AD brains<br>h. C5b-9 deposits on dystrophic neuritis<br>i. Reduced level of activated glia surrounding Aβ plaques in APP/C1q$^{-/-}$ mice<br>j. C3 inhibition in APP mice overexpressing soluble complement receptor-related protein y (sCrry) in the brain show increased Aβ accumulation and neurodegeneration | a. Eikelenboom et al. (1989), Eikelenboom and Stam (1982, 1984), McGeer et al. (1989), Veerhuis et al. (1995), Webster et al. (1997a), McGeer and McGeer (2002)<br>b. Bradt et al. (1998), Jiang et al. (1994), Rogers et al. (1992), Webster et al. (1997b)<br>c. Akiyama et al. (1991), Gewurz et al. (1993)<br>d. Matsuoka et al. (2001)<br>e. May et al. (1990), Walker and McGeer (1992), Yasojima et al. (1999a, b)<br>f. Zanjani et al. (2005)<br>g. Gasque et al. (1997, 1998)<br>h. McGeer and McGeer (2002)<br>i. Fonseca et al. (2004)<br>j. Wyss-Coray et al. (2002) |
| Huntington's disease | a. Antibody-independent C activation on neurons in the caudate area<br>b. C activation products (C1q, C4, C3, iC3b, and C5b-9 neo) deposited on myelin and astrocytes<br>c. Increased C biosynthesis by microglia in HD caudate (ISH)<br>d. Increased expression of C receptors (C5aR, C3aR, C1qRp) in HD brains | a. Singhrao et al. (1999)<br>b. Singhrao et al. (1999)<br>c. Singhrao et al. (1999)<br>d. Gasque et al. (1997, 1998) |
| Pick's disease | a. Strong C immunoreactivity on neurons with Pick's bodies<br>b. Increased expression of C inhibitors (Sp, clusterin, CD59) by PiD neurons and astrocytes<br>c. Lack of expression of membrane C inhibitors (MCP, CR1, DAF) in PiD brains | a. Singhrao et al. (1996), Yasuhara et al. (1994)<br>b. Singhrao et al. (1996), Yasuhara et al. (1994)<br>c. Singhrao et al. (1996) |
| Amyotrophic lateral sclerosis | a. Increased C4d levels in CSF of ALS patients<br>b. C3 fragments in serum of ALS patients<br>c. C1qa, C1qb, and C1qc mRNAs upregulation in mutant SOD1-expressing motoneurons in mice at disease onset<br>d. C1q and C4 mRNAs upregulation in motoneurons of SOD1 mutant mice at onset and late stage of disease<br>e. C1q, C3c, and C3d in active microglia and reactive astrocytes | a. Tsuboi and Yamada (1994)<br>b. Goldknopf et al. (2006)<br>c. Lobsiger et al. (2007)<br>d. Ferraiuolo et al. (2007)<br>e. Our unpublished observations |

## Amyotrophic lateral sclerosis

ALS selectively kills brain and spinal cord motor neurons in both familial and sporadic cases via a mechanism which is yet to be understood. Gene array analysis of laser-microdissected motor neurons from mice with a mutation in the superoxide dismutase 1 (SOD1) gene, modeling the familial form of ALS, identified the induction of the three subunits ($\alpha$, $\beta$, and $\gamma$) of the C1q component of the classical pathway of complement at disease onset. Immunohistochemistry confirmed deposition of the C1q protein in a subset of lumbar ventral horn motor neurons (Lobsiger et al., 2007). Another group confirmed upregulation of classical pathway complement proteins, including C1q and C4, in motor neurons of SOD1 mutant mice both at disease onset and end-stage (Ferraiuolo et al., 2007). Our group investigated the expression and cellular distribution of complement components in spinal cord and motor cortex of both sporadic and familial ALS cases observing several components of the complement cascade, including C1q, C3c, and C3d in active microglia and reactive astrocytes (unpublished data). Complement subunits are present in the serum of ALS patients and are suggested as possible biomarkers of the disease (Goldknopf et al., 2006).

## Other diseases

Complement activation also occurs in the brain of MS patients (Yam et al., 1980; Storch et al., 1998) and a pathological role has been shown in the experimental autoimmune encephalomyelitis (EAE) model of MS (Piddlesden et al., 1994; Mead et al., 2002). Deposits of complement components and activated microglia have been reported in the substantia nigra of patients with sporadic and familial Parkinson's disease (Yamada et al., 1992). The striatum, neurons, myelin, and astrocytes of Huntington's disease patients and the extracellular deposits of the prion protein in Creutzfelt–Jakob disease are marked by deposits of activated complement fragments (Singhrao et al., 1999; Kovacs et al., 2004).

These data represent evidence for the involvement of the complement cascade in the pathogenesis of neurodegenerative disorders. Understanding the role of complement activation in neuron degeneration may have great importance in the development of new therapeutic strategies.

# Complement in the peripheral nervous system (PNS)

## Local synthesis in the PNS

We have recently shown endogenous synthesis and expression of components of the complement pathway in the healthy human peripheral nerve by serial analysis of gene expression (SAGE), a powerful tool of gene expression profiling (de Jonge et al., 2004). mRNAs encoding activating, inhibitory, and regulatory components of the complement system were highly represented in the SAGE (Table 4) and their expression was confirmed by northern blot and RT-PCR. Epithelial cells (Strunk et al., 1988), fibroblasts (Katz and Strunk, 1988), Schwann cells (Dashiell et al., 1997), and macrophages (McPhaden and Whaley, 1993) can synthesize complement in vitro but the high expression level, determined by SAGE, supports the hypothesis that Schwann cells are likely the main source of complement mRNA in the peripheral nerve trunk. Components of the classical (C1r, C1q, C1s, and C4), alternative (fD), and common pathway (C3) of complement together with regulatory members (CLU, C1inh, C4BP, MCP, DAF, and CD59) are produced in the peripheral nerve; the proteins are locally synthesized and differentially localized in various compartments of the nerve (see Table 4). The initial components of the classical and alternative pathway are localized in the axon whereas no complement regulators are expressed, leaving the axon without direct protection. The regulatory proteins CR1, MCP, DAF, and CD59 are expressed by Schwann cells. Surprisingly, the myelin sheath which extends from the Schwann cell plasma membrane, expresses only CD59, regulator of the C5b-9 complex, but lacks any

Table 4. Expression of complement factors in healthy human sciatic nerve

| Complement factor | RNA level (SAGE[a]) | Protein level (Western blot) | | | Protein localization (immunohistochemistry) | | | | | |
|---|---|---|---|---|---|---|---|---|---|---|
| | Nerve | Nerve | Brain | Liver | Axon | SC | Myelin | ED | EP | BV |
| C1q | 9 | ++ | + | + | − | − | − | + | − | + |
| C1r | 5 | + | + | ++ | + | + | − | − | + | − |
| C1s | 3 | ++ | + | + | + | + | − | − | − | − |
| FD | 16 | + | − | − | + | − | − | − | − | − |
| Properdin | − | + | + | + | n.d. | | | | | |
| C3 | 21 | + | + | + | − | − | − | + | − | + |
| C3c | | n.d. | | | − | − | − | − | − | − |
| C3d | | n.d. | | | − | − | − | − | − | − |
| C4 | 2 | n.d. | | | − | − | − | − | + | + |
| C5 | − | + | −/+ | + | − | − | − | − | − | − |
| C1inh | 9 | + | − | + | − | − | + | + | + | + |
| C4BP | − | ++ | −/+ | + | − | + | − | − | + | + |
| FH | 3 | ++ | + | ++ | n.d. | | | | | |
| FI | − | ++ | + | ++ | n.d. | | | | | |
| CLU | 22 | + | + | + | − | + | − | − | + | − |
| MCP | 1 | n.d. | n.d. | n.d. | − | − | − | − | − | − |
| CD59 | 2 | + | + | + | − | − | + | + | + | + |
| DAF | 1 | n.d. | n.d. | n.d. | − | + | − | + | − | − |
| MAC | | n.d. | n.d. | n.d. | − | − | − | − | − | + |

Source: Modified from de Jonge et al. (2004).
[a]Tags/10000; SC, Schwann cell; ED, endoneurium; EP, epineurium; BV, blood vessel; n.d., not determined; (−) negative staining; (+) positive staining; (++) very positive staining.

regulator of the C3 convertase (Koski, 1997). This differential distribution of complement regulatory proteins may protect the Schwann cell but not the myelin from deposition of C3b and iC3b fragments which mediate opsonization and phagocytosis by macrophages (Bruck and Friede, 1990).

Although the physiological function of the complement system in the healthy peripheral nerve is probably immune surveillance, an alternative function in the regulation of fatty acid homeostasis has been proposed (Chrast et al., 2004). Adipocytes, which constitute a large portion of the epineurial compartment of the nerve, express and secrete factors of the alternative complement pathway (FB, C3, and FD also known as adipsin) (Choy et al., 1992). These factors are sufficient to cleave C3 into C3a and C3b. Further, the N-terminal arginine of C3a can be cleaved by the carboxypeptidase B (CBP), an enzyme present in the endoneurial compartment of the nerve (Chrast et al., 2004). The resulting C3adesArg, also called acylation stimulating protein (ASP), is a potent stimulator of triglyceride synthesis in adipocytes and the inability to produce ASP in C3-deficient mice results in reduced fat mass even when placed on a high-fat diet (Cianflone et al., 1999). Locally produced triglycerides may represent a readily available energy source for peripheral axons which elongate far from their cell body, making energy supply from the soma difficult. Altogether, local synthesis of complement factors and ASP production are suggested as possible regulators of energy metabolism crass talk between various compartments of the nerve but more work aimed to elucidate this mechanism is ongoing.

### Activation in PNS injury

Transection or crush injury to peripheral nerves results in the disintegration of the axons in the distal stump of the traumatized nerve, a process known as Wallerian degeneration (WD) (Waller, 1850). The initial morphological changes are visible as early as 12 h after injury and include loss of axonal content. Approximately 2–3 days later, the first changes in myelin structure occur. Myelin collapses into ellipsoids in the distal stump and in the most distal end of the proximal stump up to the first intact node of Ranvier (King, 1999). At the same time, Schwann cells dedifferentiate, multiply within their basal lamina tubes, and downregulate myelin protein synthesis (LeBlanc and Poduslo, 1990). They break down the myelin ellipsoids into smaller ovoids and the myelin lipids eventually degenerate into neutral fat which are removed by macrophages. Starting at 2 days after injury, endoneurial macrophages proliferate, become activated, and participate in myelin removal; however, they cannot efficiently complete myelin clearance. Monocyte/macrophages recruited to the injured site from the blood stream are responsible for the rapid and efficient clearance of myelin debris (Kiefer et al., 2001).

Rapidly (within 1 h) after trauma, the complement cascade is activated locally at the site of injury (de Jonge et al., 2004) and, as a retrograde response, in the motor nuclei and sensory projections laying in the CNS (Liu et al., 1998; Svensson and Aldskogius, 1992; Svensson et al., 1995; Tornqvist et al., 1996; Mattsson et al., 1998) (see Table 5). The mechanical injury focally damages the nerve exposing axonal and myelin epitopes which are normally sequestered within the axolemma and in the myelin sheath. Myelin proteins can activate the classical and alternative pathways of complement in an antibody-independent manner as shown by in vitro studies of isolated rat or human myelin able to fix C1 and consume complement when incubated in normal and C2-deficient serum (Vanguri et al., 1982; Koski et al., 1985).

Activation of the complement system generates both opsonins and the MAC(C5b-9). The large split products of C3 and C5 (C3b and C5b, respectively) opsonize membranes, targeting them for disposal by phagocytes and for the anchoring of the C5b-9 complex. The small cleaved products, C3a and C5a, have a potent chemotactic function. Macrophage recruitment and activation is the most striking cellular response during WD after axonal injury and it is mediated by complement proteins. The initial demonstration of the importance of complement in the macrophage response during WD comes from early in vitro studies. In cocultures of macrophages and peripheral nerve segments, monoclonal antibodies blocking complement type 3 receptor (CR3), a receptor for iC3b, a product of the cleavage of C3 (Damoiseaux et al., 1989)

Table 5. Complement implication in PNS injury

| Process | Evidence of C involvement | References |
| --- | --- | --- |
| Wallerian degeneration | a. C3c deposits in crushed rat sciatic nerve 4 h post-injury<br>b. C1q, C1r, C1s upregulation in crushed rat sciatic nerve 2 days post-injury; C3c, C4c, MAC deposits at the crush site 3 days post-injury<br>c. Delayed WD in C3-depleted rats<br>d. Delayed WD in C5-deficient rats<br>e. C3-deficient serum blocks macrophage invasion of the degenerating nerve in vitro<br>f. CR3 is expressed by myelin-phagocytosing macrophages after nerve transection injury<br>g. mAb against CR3 prevent phagocytosis of opsonized myelin in vitro<br>h. Complement activation augment by twofold macrophage-mediated myelin phagocytosis via the CR3 receptor | a. de Jonge et al. (2004)<br>b. Ramaglia et al. (2007)<br>c. Dailey et al. (1998)<br>d. Liu et al. (1995)<br>e. Bruck and Friede (1991)<br>f. Reichert et al. (1994)<br>g. Bruck and Friede (1990)<br>h. Reichert and Rotshenker (2003) |
| Regeneration | a. Delayed in C3-depleted rats<br>b. Not altered in C5-deficient rats<br>c. Delayed in C57BL/Wlds mice | a. Dailey et al. (1998)<br>b. Liu et al. (1999)<br>c. Brown et al. (1992) |

expressed by macrophages during WD (Reichert et al., 1994), prevented the phagocytosis of opsonized myelin (Bruck and Friede, 1990). When C3-deficient serum was applied to the cocultures, macrophages were unable to invade the degenerating nerve whereas addition of C3-deficient serum after the macrophages had successfully entered the nerve impaired their ability to phagocytose myelin (Bruck and Friede, 1991). These data proved that myelin phagocytosis by macrophages is mediated by complement via CR3. Reichert and Rotshenker (2003) showed that scavenger receptor AI-II (SRAI/II) functions together with CR3 to mediate phagocytosis of myelin via the cyclic adenosine monophosphate (cAMP) cascade (Makranz et al., 2006).

The in vitro findings were followed by experiments in the in vivo models of WD. C3 depletion by intravenous administration of cobra venom factor (CVF) in Lewis rats reduced macrophage recruitment into the distal stump of the degenerating nerve. Macrophages failed to acquire the enlarged and vacuolated morphology, typical of the activated phenotype, and their capacity to clear myelin was diminished (Dailey et al., 1998). Mice congenitally deficient in the complement component C5 also showed delayed macrophage recruitment as well as axonal and myelin degradation from 1 to 21 days post-injury (Liu et al., 1999). These experiments point to a major role of complement in macrophage recruitment and activation but the contribution of the terminal MAC to WD is not pinpointed. C3 or C5 have multiple functions as generators of opsonins, chemoattractants, and anchor for the assembly of the MAC. Thus their depletion could affect different pathways. Whether C cleaved products alone or the entire complement cascade including the MAC is needed for WD is still a matter of debate (Barnum and Szalai, 2006). We recently showed that formation of the MAC is essential for rapid WD (Ramaglia et al., 2007). Crush injury of the sciatic nerve of rats deficient in the complement protein C6, necessary for the assembly of the MAC, resulted in delayed axonal and myelin degradation. Macrophage recruitment and activation was inhibited in the C6$^{def}$ rats and removal of myelin debris, which were visible in the C6$^{def}$ rats by 1 week post-injury, was less efficient and probably carried out by Schwann cells. Reconstitution of C6$^{def}$ rats with purified C6 protein could restore the wildtype phenotype. A model summarizing the involvement of complement in WD is proposed in Fig. 2. Following crush, C1q binds axonal and myelin epitopes exposed by the mechanical injury, activating the classical pathway. The damaged nerve is then opsonized by C4b, C3b, and C5b which act as opsonins for macrophages. In animals with a complete and functional complement system, formation of MAC creates pores on the axolemma allowing influx of calcium, into the axon. This activates calpains contributing to the rapid damage of the nerve. The MAC-induced neuronal debris is also targeted by C1q creating a positive feedback loop resulting in increased opsonization and macrophage activation and recruitment, leading to rapid degeneration. In the C6$^{def}$ rats the crush-derived debris is the only target for opsonization and not sufficient for efficient activation of macrophages to acquire the phagocytic phenotype. WD and myelin removal in the absence of C6 is carried out in a slow, opsonin-independent manner by proliferating Schwann cells. Thus, we showed that the entire complement cascade, including MAC deposition, is essential for rapid WD and efficient clearance of myelin after acute nerve trauma of the PNS.

### Effect of post-traumatic complement activation on peripheral nerve regeneration

Damaged peripheral axons often achieve a good morphological regeneration but regain function slowly and incompletely (Baker et al., 1994; Lundborg and Rosen, 2007). Peripheral nerve regeneration after injury requires axons to reenter the Schwann cell tubes injured at the lesion site. The search of axons for the appropriate Schwann cell tube is represented by the axonal branches emerging from the tip of the proximal undamaged nerve stump. Once in the distal stump, the axons need to re-navigate the paths followed before injury and generate specific synapses with exactly the same muscle fibers they had previously innervated. In this task they are guided by

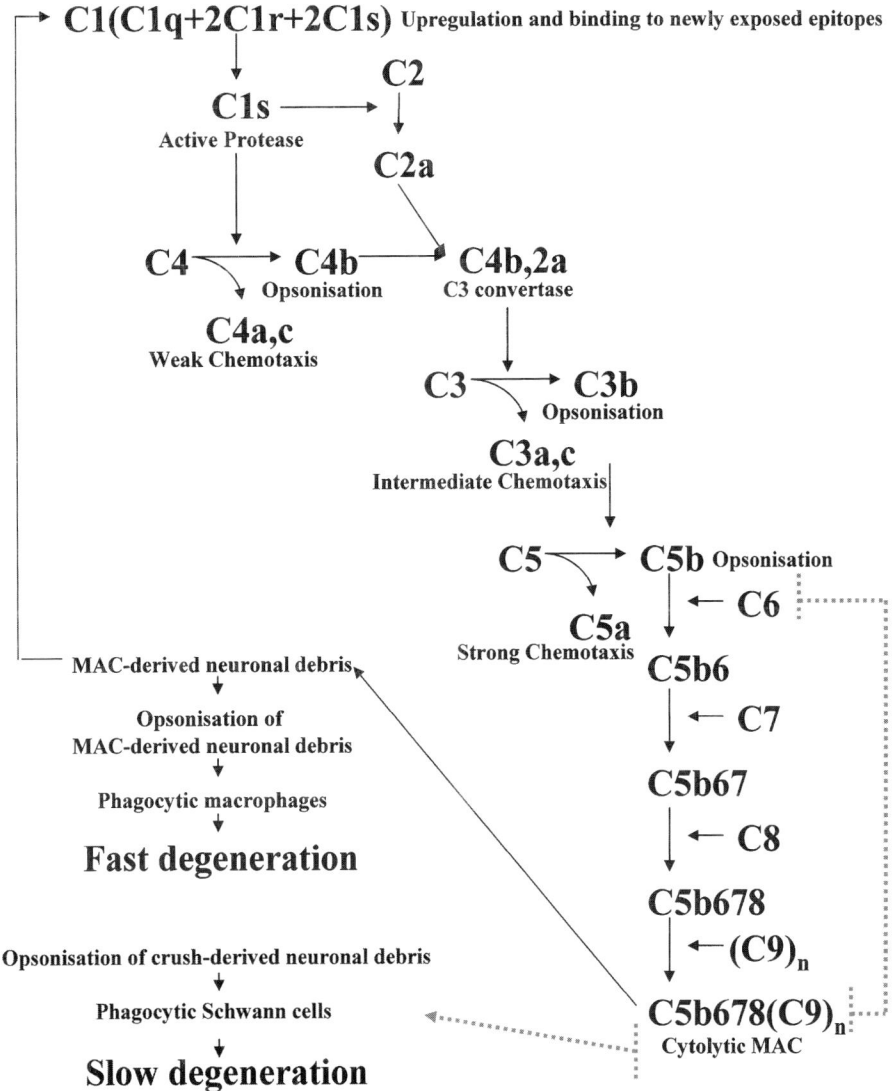

Fig. 2. Schematic diagram of complement-mediated damage during WD. Following injury, C1q binds to new epitopes exposed by the mechanical damage initiating the classical pathway of complement. The damaged nerve is then opsonized by C4b, C3b, and C5b while the respective cleaved products act as chemoattractants for macrophages. In WT animals, the presence of C6 leads to the formation of the cytolytic MAC. MAC makes pores in the axolemma allowing calcium into the axon. This activates proteases contributing to the damage of the nerve. The MAC-derived debris is recognized by C1q. This creates a positive feedback loop which results in an increased level of opsonin and chemoattractants. They lead to the recruitment of phagocytic macrophages thus to a fast degeneration. In the C6-deficient ($C6^{def}$) animals, the absence of C6 prevents MAC formation. The crush-derived debris is the only available substrate for opsonization which is not sufficient for efficient recruitment and activation of macrophages. An alternative opsonin-independent phagocytosis of the damaged nerve is initiated by Schwann cells. This results in a slow degeneration.

attractive and repulsive molecular cues (Tessier-Lavigne and Goodman, 1996; Yu and Bargmann, 2001) but physical factors also play a major role (Nguyen et al., 2002). It is important to note that excessive axonal branching at the injury site may impair the accuracy of target reinnervation. Since motor and sensory axons have equally little ability to identify Schwann cell tubes leading to their proper target, axons could be redistributed to wrong pathways (misdirection) and a single neuron could send axonal processes to multiple antagonistic muscles (hyperinnervation), impairing functional recovery. Thus, both a physiological control of axonal branching and the maintenance of intact endoneurial tubes are of high importance for regeneration of the adult peripheral nerve.

Histologically, regeneration is marked by the appearance of regenerative clusters of axons which are branches of the originally injured axon. Initially, these branches reside within a single Schwann cell but they are later separated by radial sorting. Once the 1:1 relationship between Schwann cell and axon is established, the pro-myelinating Schwann cell begins to ensheath the axon and starts to form myelin and the basal lamina tube. At this stage, regenerative clusters appear within adjacent Schwann cells as groups of small caliber, thinly myelinated axons. Continue axonal growth, at a rate of approximately 2.5 mm a day, eventually leads to the connection of one branch of the axon to its target. At this point, the rest of the branches are eliminated while the remaining axon thickens.

Complement activation during WD could be a "double-edged sword" for the subsequent regeneration of the nerve. On the one hand, it could be detrimental by perpetuating nonspecific tissue damage directly via the MAC and indirectly via the macrophages and their toxic mediators; on the other hand, the complement-mediated events could be beneficial to the termination of the inflammatory process and promote recovery via the breakdown of injured axons and myelin, early clearance of debris, and late secretion of anti-inflammatory cytokines by macrophages. The complement system has been suggested by us (de Jonge et al., 2004) and others (Dailey et al., 1998; DeJong and Smith, 1997; Reichert et al., 2001) as a key determinant of regeneration after axonal injury. Dailey et al. (1998) showed that axonal regeneration is delayed in C3 depleted animals whereas C5 deficiency did not alter regeneration after sciatic nerve injury (Liu et al., 1999). In the C57BL/Wlds mouse model with slow WD, regeneration is also delayed (Brown et al., 1992; Chen and Bisby, 1993). Components of the myelin sheath are generally considered inhibitors of axonal growth and their efficient clearance is considered a prerequisite for successful regeneration after axonal injury (Chen et al., 2000; McKerracher et al., 1994; Li et al., 1996). The delayed axonal regeneration of C57BL/Wlds mice seems to be partly dependent on the inhibitory effect of myelin-associated glycoprotein (MAG) on axonal regrowth since a cross of MAG-deficient mice and C57BL/Wlds mice showed an increase in the number of regrowing axons. However, no functional tests were performed (Schafer et al., 1996). Interestingly, a recent study demonstrated that local application of exogenous MAG for a short time (72 h) after injury reduces axonal branching without affecting axonal elongation and enhances the functional recovery after rat sciatic nerve transection, presumably by reducing hyperinnervation and misdirection (Tomita et al., 2007). Thus, lingering of myelin constituents in the degenerating nerve, as seen in the case of $C6^{def}$ animals in which myelin clearance after nerve injury is delayed (Ramaglia et al., 2007), can prevent branching and promote recovery.

We have recently shown that recovery of motor and sensory function after traumatic nerve injury is accelerated in $C6^{def}$ rats and that the same effect can be obtained by inhibition of post-traumatic complement activation (unpublished observations). This is probably due to the inhibition of destructive complement-mediated events during nerve degeneration which may interfere with the subsequent regeneration of the nerve. In addition, the possibility that the inhibitory effect of short-lasting myelin components in the degenerating nerve may control branching and promote correct reinnervation and faster recovery is not ruled out.

## Activation in PNS disease

Activation of the complement system occurs in both chronic and acute diseases of the PNS (Table 6).

### Neuroma/neurofibromatosis

Neuroma and neurofibromatosis are examples of chronic PNS disorders in which deposits of activated complement components have been detected. Neurofibromatosis is a genetic disorder which causes nerve tumors whereas neuromas occur in traumatized nerves as a result of impaired regeneration and it consists of axon sprouts, proliferating Schwann cells, perineurial cells, blood vessels, and connective tissue. Deposits of C3c and C4c were found in the perineurium of sprouting fibers and in proliferating Schwann cells of neuroma samples and in the perineurium of neurofibromatosis samples (de Jonge et al., 2004). This indicates that, although usually part of the acute phase response, complement activation can occur in chronic diseases probably triggered by active disease processes.

### Guillain–Barré syndrome

Complement activation is one of the earliest detectable events in acute acquired demyelinating neuropathies such as Guillain–Barré syndrome (GBS) (Hafer-Macko et al., 1996a). The C5b-9 complex is detected in peripheral nerve biopsy of

Table 6. Complement implication in PNS disease

| Process | Evidence of C involvement | References | |
|---|---|---|---|
| Neuroma | k. C3c and C3d deposits in the perineurium and Schwann cells | k. | de Jonge et al. (2004) |
| Neurofibromatosis | e. C3c and C3d deposits in the perineurium | e. | de Jonge et al. (2004) |
| GBS | a. C5b-9 deposits on degenerating myelin lamellae before macrophage invasion | d. | Hafer-Macko et al. (1996b) |
| | b. C9neo antigens on degenerating myelin sheaths in acute phase GBS | e. | Wanschitz et al. (2003) |
| | c. C5b-9 in cerebrospinal fluids, serum, and peripheral nerve biopsy | f. | Koski (1990) |
| | d. C5b-9 deposits on Schwann cell membranes in acute phase GBS | g. | Putzu et al. (2000) |
| | e. C-deficient serum rescue anti-GQ1b-mediated transmission block | e, f. | Plomp et al. (1999) |
| | f. C3c deposits on NMJs of mouse diaphragm treated with serum from MFS patient | g. | Paparounas et al. (1999) |
| | g. C3c binds node of Ranvier in desheated mouse sciatic nerve treated with ganglioside antisera or monoclonal anti-ganglioside antibodies | h. | Susuki et al. (2007) |
| | h. C3 and C5b-9 deposition at disrupted nodes of Ranvier in acute phase AMAN rabbits | i, j, k. | Halstead et al. (2004) |
| | i. C1q, C3c, C4, and MAC deposits at injured NMJs and pSCs of mouse diaphragm treated with anti-gangliosides antibodies | l. | Geleijns et al. (2006) |
| | j. C6 deficiency rescues NMJs and pSC injury | | |
| | k. CD59 deficiency exacerbates NMJs and pSC injury | | |
| | l. High MBL serum level and complex activity determined by the MBL2 haplotype is associated with severe GBS phenotype | | |
| EAN | f. C5b-9 deposits on Schwann cell surface and myelin before overt demyelination | f. | Stoll et al. (1991) |
| | g. CD59 upregulation on Schwann cells during demyelination and axonal degeneration | g. | Vedeler et al. (1999) |
| HMSNs | a. C3c, C4c, and C5b-9 deposits in sural nerve biopsies of HNPP and HSN patients | a, b. | Unpublished |
| | b. C4c deposition in sciatic nerves of C22 mice; C1q, C1r, fB, and C9 upregulation in sciatic nerve lysates of C22 mice | | |

patients with GBS as early as 3 days after onset of neurological symptoms and it is localized on the surface of Schwann cells and their myelin sheath (Hafer-Macko et al., 1996b; Putzu et al., 2000; Wanschitz et al., 2003). In the experimental autoimmune neuritis (EAN) model of GBS, deposition of C5b-9 complex occurs before the onset of clinical signs and short thereafter, and precedes demyelination (Stoll et al., 1991). In patients, vesicular degeneration of myelin lamellae is apparent before the invasion of macrophages, suggesting that complement activation could be responsible for the initial degenerative changes in the diseased nerve. C5b-9 is also detected in cerebrospinal fluids and serum of GBS patients (Koski, 1990).

More insight into the role of complement in acute acquired demyelinating diseases comes from studies of in vitro mouse hemidiaphragm preparations sensitized with anti-GQ1b antibodies which are associated with the regional variant of GBS, Miller Fisher syndrome (MFS) (Willison, 2005). The carbohydrate moieties of gangliosides are structurally similar to those of microbial glycans like the lipo-oligosaccharides expressed on the surface of the bacterium *Campylobacter jejuni*, highly associated with the development of GBS. This mimicry may explain the development of autoimmunity. The immune response, mediated by the anti-ganglioside antibody binding, not only targets the infection but also attacks the peripheral nerve (Willison and Yuki, 2002). Gangliosides are especially enriched in neuronal tissues, primarily localized at raft domains on the synaptic plasma membranes. Anti-GQ1b antibodies bind to motor nerve terminals (Ogawa-Goto and Abe, 1998) where they induce a massive release of acetylcholine and eventually block neuromuscular transmission, much similar to the effect of the pore-forming toxin, α-latrotoxin (Willison, 2005). Blockade of neuromuscular transmission corresponds to structural breakdown of the nerve terminals, loss of heavy neurofilament, and type III β-tubulin immunostaining. Presynaptic terminals appear disorganized, depleted of vescicles, and fragmented by Schwann cell processes. These lesions are completely rescued by application of complement-deficient serum, proving that the effects of anti-GQ1b antibodies on the neuromuscular junctions (NMJs) are entirely dependent on activation of the complement cascade (O'Hanlon et al., 2001). One mystery remains that the effects seen in the in vitro mouse model can only be obtained by using normal human serum (NHS) as a source of complement whereas mouse serum does not appear to be sufficiently activated by anti-ganglioside antibodies opsonized presynaptic membranes. How much NHS complement is regulated by mouse C regulators is also not clear. Willison (2005) proposed a model of the pathological processes leading to nerve terminal axonal injury and synaptic necrosis in the in vitro mouse model of MFS which is summarized in the schematic in Fig. 3.

Not only presynaptic neuronal membranes but also perisynaptic Schwann cells (pSCs) are the target of attack by anti-gangliosides antibodies which colocalize with deposits of C1q, C3c, C4, and MAC (Halstead et al., 2004). Hemidiaphragm preparations from $C6^{-/-}$ mice incubated with $C6^{-/-}$ serum failed to produce nerve terminal injury in the presence of anti-gangliosides antibodies showing that both presynaptic neuronal membranes and pSCs injury is caused by MAC-mediated lysis. Further, this process is controlled by the inhibitory complement regulator CD59 since neuronal and pSC injury is exacerbated in CD59-deficient mice with increased MAC formation (Halstead et al., 2004). The classical pathway has been suggested as initiator of complement activation in GBS but the involvement of the lectin pathway has not been ruled out (Halstead et al., 2004). High serum activity of MBL, which activates complement by the recognition of repetitive sugar groups on pathogens, is determined by the MBL2 haplotype and has been associated with the development and severity of GBS (Geleijns et al., 2006).

Whether the anti-ganglioside antibody-mediated nerve terminal degeneration also affects the proximal portion of the axon remains unclear. Anti-gangliosides antibodies bind nodes of Ranvier and can activate complement in desheated mouse sciatic nerve treated with sera from patients with autoimmune neuropathies supplemented with fresh human serum as a source of complement (Paparounas et al., 1999). In the in vivo rabbit

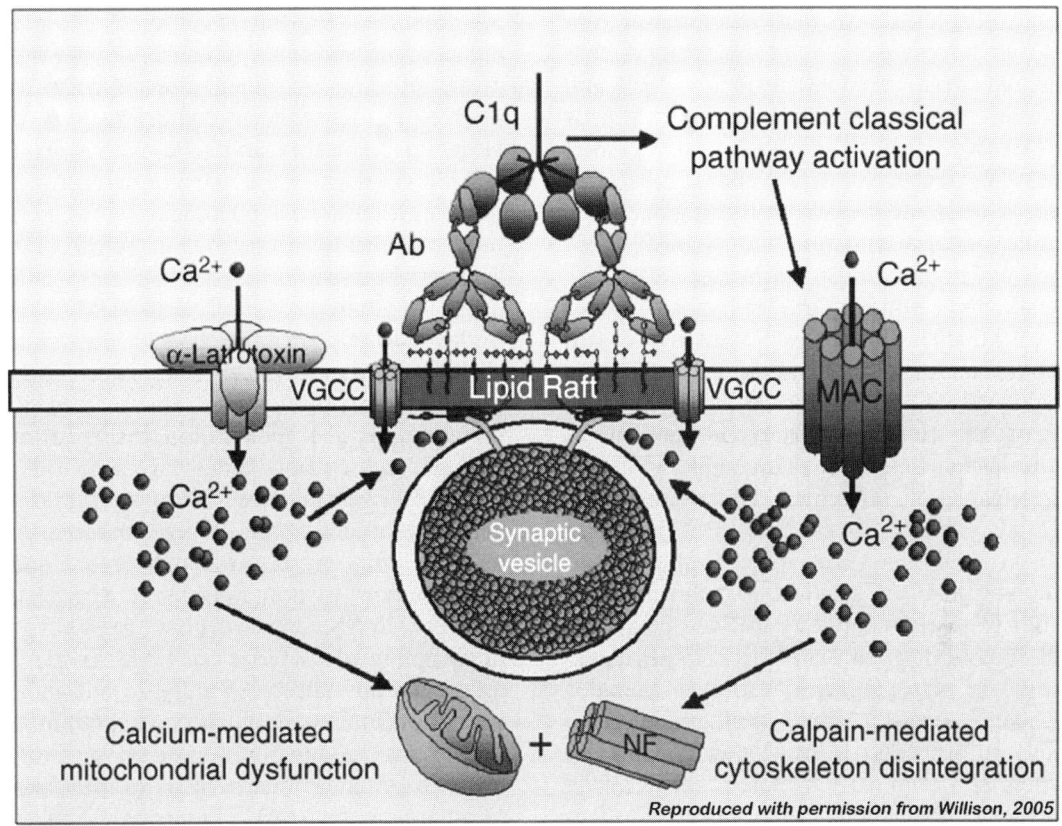

Fig. 3. Model of complement-mediated damage in inflammatory neuropathies. Antibodies bind to gangliosides in the extracellular layer of the plasma membrane of lipid raft domains at synapses. C1q binds to the antigen–antibody complex, initiating activation of the complement cascade which eventually results in the assembly and deposition of the MAC. MAC forms nonspecific pores in the presynaptic membrane leading to uncontrolled calcium influx. The entry of calcium is normally highly regulated by the voltage-gated calcium channels (VGCC). The level of intracellular calcium mediates the fusion of the synaptic vesicles to the intracellular leaflet of the presynaptic membrane and triggers neurotransmitter release in the synaptic cleft. The uncontrolled calcium influx leads to accumulation of intracellular calcium which triggers massive exocytosis and activates calcium-dependent proteases which degrade the cytoskeleton and trigger mitochondrial cell death, leading to structural and metabolic destruction of the nerve terminal. This mechanism is similar to that induced by the pore-forming toxin, α-latrotoxin.

model of GBS, resembling the acute motor axonal neuropathy (AMAN) variant (Susuki et al., 2004), the entire architecture of the nodes of Ranvier including the paranodal junctions, the nodal cytoskeleton, and the Schwann cell microvilli are disrupted during the acute phase of the disease accompanied by progressive limb weakness. Voltage-gated sodium ($Na_v$) channels, normally clustered at nodes of Ranvier, are disrupted or absent and colocalize with deposits of complement C3 and MAC (Susuki et al., 2007). Spontaneous recovery occurs in the AMAN rabbits 2 weeks after clinical onset (Susuki et al., 2004). During the early recovery phase, complement deposits at nodes of Ranvier decrease and $Na_v$ channels redistribute on both sides of affected nodes along with Schwann cells microvilli and paranodal myelin loops (Susuki et al., 2007). This architecture may represent the early stage of a later fusion of two heminodes to reconstitute new nodes of Ranvier in the later stages of the recovery phase.

The ability of (non-perisynaptic) Schwann cells to survive MAC deposition during inflammatory neuropathies is intriguing. This is likely influenced by the differential distribution of complement regulatory components on the Schwann cell surface (CR1, MCP, DAF, CD59) compared to its myelin sheath (only CD59) and the ability of cells to eliminate MAC channels by endocytosis and membrane shedding in the case of sublytic attack (Carney et al., 1986; Morgan et al., 1987). Increasing evidence points to a role of nonlethal amount of MAC deposition in the regulation of cell cycle and gene expression (Koski, 1997; Dashiell and Koski, 1999; Dashiell et al., 2000; Hila et al., 2001; David et al., 2006). MAC pores can reach a diameter of 100 Å (Ramm and Mayer, 1980). Multiple signaling molecules can cross these channels including calcium, diacylglycerol, cAMP, protein kinase C (PKC), and members of the mitogen-activated protein (MAP) kinase family which can influence cellular processes (Shin et al., 1996; Niculescu et al., 1997; Dashiell and Koski, 1999). In vitro studies demonstrated that sublytic MAC deposition on Schwann cells decreases the expression of P0 mRNA via activation of JNK1 (Dashiell and Koski, 1999; David et al., 2006). P0 protein constitutes a major component (>50%) of myelin membrane proteins and it is necessary for myelin compaction and integrity (Lemke, 1996; Giese et al., 1992; Quarles, 1997; Menichella et al., 2001). Thus it is conceivable that, by decreasing the expression of genes important in myelin compaction such as P0, sublytic complement activation could drive the Schwann cell toward a premyelinating phenotype possibly contributing to peripheral nerve demyelination (David et al., 2006). This is further supported by the observation that sublytic MAC levels rescue cultured Schwann cells from apoptosis via activation of PI-3 kinase, phosphorylation of BAD, and upregulation of Bcl-$x_L$ (Hila et al., 2001) and drive them into S phase, stimulating cell cycle and mitosis (Dashiell et al., 2000). The survival of Schwann cells and their shift from a myelinating to a proliferating phenotype could ultimately favor remyelination and recovery which occurs in most patients with GBS.

*Charcot–Marie–Tooth disease*

The immune system appears to be a determinant of disease severity also in the hereditary forms of neuropathies (Martini and Toyka, 2004). These are a clinically and genetically heterogeneous group of diseases which affect 1:2500–1:10,000 people in the western countries (Skre, 1974). Hereditary motor and sensory neuropathies (HMSNs) have been originally described on the basis of their clinical features by Charcot and Marie (1886) in France and Tooth (1886) in England from which they derived their traditional name of Charcot–Marie–Tooth (CMT) disease. Today mutations in more than 30 genes expressed by the Schwann cell or the axon have been identified as cause of the disease. Intriguingly, mutations in one gene can give rise to multiple diseases and multiple genes may be involved in a single disorder (reviewed in Shy et al., 2002). This suggests the existence of gene modifiers which influence the resulting phenotype. Although inflammation is not considered a criterion to classify HMSN, inflammatory foci have been described in nerve biopsy of HMSN patients and a few patients have responded to anti-inflammatory or immunomodulatory treatment which resulted in the amelioration of symptoms (Dyck, 1982; Watanabe et al., 2002; Vital et al., 2003; Ginsberg et al., 2004). These observations make the immune system an interesting candidate as modifier of inherited peripheral neuropathies.

Another argument for the role of the immune system in HMSNs comes from studies in mice models of the disease (reviewed in Martini and Toyka, 2004). The peripheral nerve of mice heterozygously deficient for P0 ($P0^{+/-}$) or lacking the gap junction protein connexin 32 ($Cx32^{-/-}$) or carrying four copies of the human peripheral myelin protein (PMP)22 gene (*C61*), models for HMSN1B, HMSN1X, and HMSN1A respectively, showed elevated number of CD8 lymphocytes and macrophages (Shy et al., 1997; Kobsar et al., 2002; Robertson et al., 2002). Interestingly, a quantitative analysis of immune cells in $P0^{+/-}$ mice of different age showed that the number of macrophages is high at age 4 months—typical disease onset—whereas the number of T-lymphocytes increases starting at age 6 months, indicating that the early increase in the number of macrophages does not result from T-cells infiltration. This is in contrast to the classical EAN model of immune-mediated neuropathies, in which T-lymphocytes are the first cell types to invade the nerve (Maurer et al., 2002b). The endoneurial macrophage population of $P0^{+/-}$ mice mainly consists of resident macrophages which proliferate and become activated within the nerve (Maurer et al., 2002a). The factors responsible for such increase and activation are not known. Monocyte chemoattractant protein-1 (MCP-1), released by Schwann cells, has been proposed as a possible candidate (Tofaris et al., 2002) but a possible involvement of the complement system has not yet been ruled out.

These immune cells seem to play a pathogenic role, at least in certain mouse models of HMSNs. When either $P0^{+/-}$ or $Cx32^{-/-}$ mice was crossbred with an immunodeficient strain of mice lacking the recombination activation gene (RAG) 1 ($Rag1^{-/-}$), the peripheral nerve of the double mutants showed no lymphocytes, reduced number of macrophages, and a significant amelioration of the demyelinating phenotype as documented by morphometry and electrophysiological recordings (Kobsar et al., 2003; Schmid et al., 2000). When $P0^{+/-}$ mice were crossbred with a mouse line lacking the macrophage colony-stimulating factor (M-CSF), critical for macrophage growth and differentiation, the $P0^{+/-}/M\text{-}CSF^{-/-}$ double mutants showed a decrease in the number of macrophage which resulted in the alleviation of the pathological abnormalities, proving an active role of macrophages in determining the demyelinating phenotype (Carenini et al., 2001).

We recently showed that the complement system is activated in HMSNs. We found deposition of activated C components in nerve biopsies of HMSN patients and transgenic mice with seven copies of the human peripheral myelin protein 22 (PMP22) gene (C22) (unpublished observations). In patients, MAC deposition correlated positively with myelin density. In young C22 mice, the amount of C4c deposition was higher than in adults, suggesting that the complement system is activated early in disease and that myelin could be the trigger of C activation. The role of complement in HMSNs is not clear but it offers the opportunity to investigate new specific drugs to alleviate the demyelinating phenotype and the progressive and disabling muscle atrophy derived from distal motor axon loss.

## Complement regulation in nerve injury and disease: a therapeutic approach

The implication of complement activation in the initiation, propagation, or exacerbation of PNS injury and disease, make it a potential target for therapeutic intervention. A number of complement inhibitors and modulators have been recently developed by Ricklin and Lambris (2007). Various steps of the proteolytic cascade and its (positive and negative) regulators have been targeted in CNS and PNS injury and disease and they are currently in the stage of preclinical development (Table 7).

The first anti-complement agent described was CVF (Gewurz et al., 1967; Muller-Eberhard and Fjellstrom, 1971). This venom protein binds to fB in plasma and, after cleavage of fB by fD, forms a stable C3 convertase (CVF-fB) which is resistant to fluid-phase regulators (Cooper, 1973). Decomplementation is achieved via the consumption of C3 by the CVF-fB convertase. Systemic complement depletion with CVF significantly reduces inflammation and demyelination in the EAN and adoptive transfer-EAN (AT-EAN) models of GBS (Vriesendorp et al., 1995). However, the massive C activation associated with decomplementation, makes CVF a potential threat for the development of iatrogenic shock syndrome as already observed in some animal models (Younger et al., 2001). The soluble form of complement receptor 1 (sCR1) has been a breakthrough in complement drug discovery and it has been successfully used to control C activation in animal models of GBS (Willison, 2005). Clinical signs of EAN, induced in rats by immunization with bovine peripheral nerve myelin, could be markedly suppressed by sCR1 treatment. Functional and structural lesions could also be prevented as shown by electrophysiological and morphological data (Jung et al., 1995; Vriesendorp et al., 1997). We have recently shown that systemic treatment with sCR1, and to a lesser extent with human C1inhibitor (Cetor), prevents

Table 7. Complement therapeutics in CNS and PNS injury and disease

| Agent | Activity | Model | Effect | References |
|---|---|---|---|---|
| CVF | Forms a C3 convertase able to escape regulation by FI and FH; C3 consumption | PNS<br>a. EAN<br>b. AT-EAN | a. Lowers clinical score, inflammation, and demyelination<br>b. Inhibits inflammation and demyelination | a. Vriesendorp et al. (1995)<br>b. Vriesendorp et al. (1997) |
| sCR1/TP10 | Extracellular region of CR1; inhibition of C3 convertase | PNS<br>a. Acute axonal injury<br>b. EAN<br>CNS<br>c. Spinal cord injury<br>d. TBI | a. Abrogates MAC formation, delays WD<br>b. Prevents demyelination and axonal degeneration; suppresses clinical sign of disease<br>c. Inhibits C3c formation and improves motor function at 3, 7, and 14 days post-injury<br>d. Inhibits neutrophil accumulation | a. Ramaglia et al. (2008)<br>b. Jung et al. (1995), Vriesendorp et al. (1997)<br>c. Li et al. (2005)<br>d. Kaczorowski et al. (1995) |
| C1inh/Cetor | Inhibition of C1r/C1s, kallikrein and other proteases | PNS<br>a. Acute axonal injury | a. Inhibits MAC formation, delays myelin degradation but not axonal damage | a. Ramaglia et al. (2008) |
| rhC1inh/Rhucin | Recombinant human C1inh produced in transgenic rabbits | PNS<br>a. Acute axonal injury | a. Inhibits C upregulation | a. Ramaglia et al. (2007) |
| APT070/Mirococept | sCR1 with lipopeptide membrane linker | PNS<br>a. MFS | a. Abrogates MAC formation and acute tissue injury | a. Halstead et al. (2005) |
| Eculizumab | Inhibition of C5a and C5b-9 | PNS<br>a. MFS | a. Abrogates MAC formation, structural and functional lesions in vitro; prevents respiratory paralysis and neuropathy phenotype | a. Halstead et al. (2008) |
| C5aR antagonist | Inhibition of C5a | CNS<br>a. TBI | a. Reduces neutrophils and secondary tissue damage in mice following traumatic brain cryoinjury | a. Sewell et al. (2004) |
| Crry-Ig | Recombinant Crry fused to the non-complement fixing mouse IgG1 Fc region; inhibition of C3 convertase | CNS<br>a. TBI | a. Attenuates neuroinflammation and secondary neurodegeneration in a closed-head injury model of TBI | a. Leinhase et al. (2006b) |
| mAb 1379 | Monoclonal anti-factor B antibody; inhibition of factor B | CNS<br>a. TBI | a. Inhibits C5a serum levels, attenuates cerebral tissue damage and neuronal cell death | a. Leinhase et al. (2007) |

early axon loss after axonal crush injury in rats (Ramaglia et al., 2008). The soluble complement regulator Mirococept/APT070 containing the region of CR1 which inhibits the C3/C5 convertase, can fully block MAC formation and abrogate acute tissue injury in ex vivo and in vivo mice immunized with anti-ganglioside antibody to mimic the MFS variant of GBS (Halstead et al., 2005). Similar results have been recently obtained by treatment with Eculizumab, the humanized antibody against C5, already in use in clinical trials for the treatment of paroxysmal nocturnal hemoglobinuria (Halstead et al., 2008). Eculizumab could prevent MAC deposition and terminal axonal neurofilament loss as well as block the large increase in the frequency of MEPPs and prevent the block in synaptic transmission.

Some of these compounds have been successfully used to control C activation in animal models of CNS injury and disease such as TBI (Leinhase et al., 2006b) and MS (Piddlesden et al., 1994). Pretreatment of rats with sCR1 reduced intracranial neutrophil infiltration following TBI (Kaczorowski et al., 1995). Post-traumatic systemic treatment of rats with a monoclonal anti-factor B antibody, inhibitor of the alternative pathway of complement, decreased tissue damage and neuronal cell death but it did not affect the neurological impairment (Leinhase et al., 2006a) whereas post-traumatic treatment with a recombinant Crry inhibited neuronal degeneration and improved neurological function after TBI (Leinhase et al., 2006b). In a rat model of MS, systemic treatment with sCR1 reduced the severity of clinical disease, inhibited inflammation, and demyelination (Piddlesden et al., 1994). These dramatic effects strongly implicate complement activation in the pathogenesis of neurodegenerative disease and secondary neurodegeneration after injury, offering an exciting prospect for therapy in these conditions.

Despite promising attempts, the design of the ideal complement drug meets various challenges. The diverse nature of acute (i.e., traumatic nerve injury) and chronic clinical conditions (i.e., neuropathies, neurodegenerative diseases) adds complexity to the challenge. The optimal target, the time, and extent of inhibition need to be critically defined for each condition. Our group observed that inhibition of the terminal pathway (i.e., MAC) of C is sufficient to promote post-traumatic nerve regeneration and recovery. This has a major advantage. The upstream factors of the C cascade have key functions in modulating an inflammatory reaction, essential to protect the body against infections and to maintain tissue homeostasis. Selective inhibition of the terminal pathway will target the "foe" side of the C cascade, leaving the upstream C factors, thus the "friend" side of the C cascade, undisturbed. This novel approach will limit the side effects which would derive from inhibition of the entire C cascade. Further, it will allow systemic administration of the selective inhibitor, circumventing the challenge of finding new venues to solely target local C activation.

**Abbreviations**

| | |
|---|---|
| AD | Alzheimer's disease |
| ALS | amyotrophic lateral sclerosis |
| AMAN | acute motor axonal neuropathy |
| ASP | acylation stimulating protein |
| AT-EAN | adoptive transfer-EAN |
| C | complement |
| cAMP | adenosine monophosphate |
| CBP | carboxypeptidase B |
| C4BP | C4-binding protein |
| CLU | clusterin |
| CMT | Charcot–Marie–Tooth |
| CNS | central nervous system |
| CSF | cerebrospinal fluid |
| CR1 | C receptor 1 |
| Crry | C receptor type 1-related protein |
| CVF | cobra venom factor |
| Cx32 | connexin 32 |
| DAF | decay accelerating factor |
| EAE | experimental autoimmune encephalomyelitis |
| EAN | experimental autoimmune neuritis |
| GBS | Guillain–Barré syndrome |
| HMSN | hereditary motor and sensory neuropathy |
| MAC | membrane attack complex |
| MAG | myelin-associated glycoprotein |

| | |
|---|---|
| MAP | mitogen-activated protein |
| MASP | MBL-associated serine protease |
| MCP | membrane cofactor protein |
| MCP-1 | monocyte chemoattractant protein-1 |
| M-CSF | macrophage colony-stimulating factor |
| MBL | mannose-binding lectin |
| MFS | Miller Fisher syndrome |
| MS | multiple sclerosis |
| NHS | normal human serum |
| NMJ | neuromuscular junction |
| PMP22 | peripheral myelin protein 22 |
| PKC | protein kinase C |
| pSC | perisynaptic Schwann cell |
| RAG | recombination activation gene |
| SAGE | serial analysis of gene expression |
| SCI | spinal cord injury |
| sCR1 | soluble C receptor 1 |
| SOD | superoxide dismutase |
| SRAI/II | scavenger receptor AI/II |
| TBI | traumatic brain injury |
| TNF-α | tumor necrosis factor-α |
| WD | Wallerian degeneration |

## Acknowledgment

We thank Prof. H. Willison for permission to reproduce Figure 3 and critically reading the manuscript.

## References

Akiyama, H., Yamada, T., Kawamata, T., & McGeer, P. L. (1991). Association of amyloid P component with complement proteins in neurologically diseased brain tissue. *Brain Research*, 548, 349–352.

Anderson, A. J., Najbauer, J., Huang, W., Young, W., & Robert, S. (2005). Upregulation of complement inhibitors in association with vulnerable cells following contusion-induced spinal cord injury. *Journal of Neurotrauma*, 22, 382–397.

Anderson, A. J., Robert, S., Huang, W., Young, W., & Cotman, C. W. (2004). Activation of complement pathways after contusion-induced spinal cord injury. *Journal of Neurotrauma*, 21, 1831–1846.

Atkinson, J. P., & Frank, M. M. (2006). Bypassing complement: evolutionary lessons and future implications. *The Journal of Clinical Investigation*, 116, 1215–1218.

Avery, V. M., Adrian, D. L., & Gordon, D. L. (1993). Detection of mosaic protein mRNA in human astrocytes. *Immunology and Cell Biology*, 71(Pt. 3), 215–219.

Baker, R. S., Stava, M. W., Nelson, K. R., May, P. J., Huffman, M. D., & Porter, J. D. (1994). Aberrant reinnervation of facial musculature in a subhuman primate: a correlative analysis of eyelid kinematics, muscle synkinesis, and motoneuron localization. *Neurology*, 44, 2165–2173.

Barnum, S. R. (1995). Complement biosynthesis in the central nervous system. *Critical Reviews in Oral Biology and Medicine*, 6, 132–146.

Barnum, S. R., Ishii, Y., Agrawal, A., & Volanakis, J. E. (1992a). Production and interferon-gamma-mediated regulation of complement component C2 and factors B and D by the astroglioma cell line U105-MG. *The Biochemical Journal*, 287(Pt. 2), 595–601.

Barnum, S. R., Jones, J. L., & Benveniste, E. N. (1992b). Interferon-gamma regulation of C3 gene expression in human astroglioma cells. *Journal of Neuroimmunology*, 38, 275–282.

Barnum, S. R., Jones, J. L., & Benveniste, E. N. (1993). Interleukin-1 and tumor necrosis factor-mediated regulation of C3 gene expression in human astroglioma cells. *Glia*, 7, 225–236.

Barnum, S. R., & Szalai, A. J. (2006). Complement and demyelinating disease: no MAC needed? *Brain Research Reviews*, 52, 58–68.

Bellander, B. M., von Holst, H., Fredman, P., & Svensson, M. (1996). Activation of the complement cascade and increase of clusterin in the brain following a cortical contusion in the adult rat. *Journal of Neurosurgery*, 85, 468–475.

Bonifati, D. M., & Kishore, U. (2007). Role of complement in neurodegeneration and neuroinflammation. *Molecular Immunology*, 44, 999–1010.

Bordet, J. (1895). Les leukocytes et les proprietes actives du serum chez les vaccines. *Annals of Institute Pasteur (Paris)*, 9, 462–506.

Bradt, B. M., Kolb, W. P., & Cooper, N. R. (1998). Complement-dependent proinflammatory properties of the Alzheimer's disease beta-peptide. *The Journal of Experimental Medicine*, 188, 431–438.

Brown, M. C., Lunn, E. R., & Perry, V. H. (1992). Consequences of slow Wallerian degeneration for regenerating motor and sensory axons. *Journal of Neurobiology*, 23, 521–536.

Bruck, W., & Friede, R. L. (1990). Anti-macrophage CR3 antibody blocks myelin phagocytosis by macrophages in vitro. *Acta Neuropathologica*, 80, 415–418.

Bruck, W., & Friede, R. L. (1991). The role of complement in myelin phagocytosis during PNS Wallerian degeneration. *Journal of the Neurological Sciences*, 103, 182–187.

Buchner, H. (1889). Uber die nahere Natur der bakterientodtenden Substanz in Blutserum. *Zentralblatt für Bakteriologie (Naturwissenschaften)*, 6, 561.

Carenini, S., Maurer, M., Werner, A., Blazyca, H., Toyka, K. V., Schmid, C. D., Raivich, G., & Martini, R. (2001). The role of macrophages in demyelinating peripheral nervous system of mice heterozygously deficient in P0. *Journal of Cell Biology*, 152, 301–308.

Carney, D. F., Hammer, C. H., & Shin, M. L. (1986). Elimination of terminal complement complexes in the plasma membrane of nucleated cells: influence of extracellular $Ca^{2+}$ and association with cellular $Ca^{2+}$. *Journal of Immunology*, *137*, 263–270.

Charcot, J., & Marie, P. (1886). Sue une forme particulaire d'atrophie musculaire progressive souvent familial debutant par les jamber et atteingnant plus tard les mains. *Reviews in Medicine*, *6*, 97–138.

Chen, M. S., Huber, A. B., van der Haar, M. E., Frank, M., Schnell, L., Spillmann, A. A., Christ, F., & Schwab, M. E. (2000). Nogo-A is a myelin-associated neurite outgrowth inhibitor and an antigen for monoclonal antibody IN-1. *Nature*, *403*, 434–439.

Chen, S., & Bisby, M. A. (1993). Long-term consequences of impaired regeneration on facial motoneurons in the C57BL/Ola mouse. *The Journal of Comparative Neurology*, *335*, 576–585.

Choy, L. N., Rosen, B. S., & Spiegelman, B. M. (1992). Adipsin and an endogenous pathway of complement from adipose cells. *The Journal of Biological Chemistry*, *267*, 12736–12741.

Chrast, R., Verheijen, M. H., & Lemke, G. (2004). Complement factors in adult peripheral nerve: a potential role in energy metabolism. *Neurochemistry International*, *45*, 353–359.

Cianflone, K., Maslowska, M., & Sniderman, A. D. (1999). Acylation stimulating protein (ASP), an adipocyte autocrine: new directions. *Seminars in Cell & Developmental Biology*, *10*, 31–41.

Cooper, N. R. (1973). Formation and function of a complex of the C3 proactivator with a protein from cobra venom. *Journal of Experimental Medicine*, *137*, 451–460.

Dailey, A. T., Avellino, A. M., Benthem, L., Silver, J., & Kliot, M. (1998). Complement depletion reduces macrophage infiltration and activation during Wallerian degeneration and axonal regeneration. *Journal of Neuroscience*, *18*, 6713–6722.

Damoiseaux, J. G., Dopp, E. A., Neefjes, J. J., Beelen, R. H., & Dijkstra, C. D. (1989). Heterogeneity of macrophages in the rat evidenced by variability in determinants: two new anti-rat macrophage antibodies against a heterodimer of 160 and 95 kd (CD11/CD18). *Journal of Leukocyte Biology*, *46*, 556–564.

Dashiell, S. M., & Koski, C. L. (1999). Sublytic terminal complement complexes decrease P0 Gene expression in Schwann cells. *Journal of Neurochemistry*, *73*, 2321–2330.

Dashiell, S. M., Rus, H., & Koski, C. L. (2000). Terminal complement complexes concomitantly stimulate proliferation and rescue of Schwann cells from apoptosis. *Glia*, *30*, 187–198.

Dashiell, S. M., Vanguri, P., & Koski, C. L. (1997). Dibutyryl cyclic AMP and inflammatory cytokines mediate C3 expression in Schwann cells. *Glia*, *20*, 308–321.

David, S., Hila, S., Fosbrink, M., Rus, H., & Koski, C. L. (2006). JNK1 activation mediates C5b-9-induced P0 mRNA instability and P0 gene expression in Schwann cells. *Journal of Peripheral Nervous System*, *11*, 77–87.

DeJong, B. A., & Smith, M. E. (1997). A role for complement in phagocytosis of myelin. *Neurochemical Research*, *22*, 491–498.

de Jonge, R. R., van Schaik, I. N., Vreijling, J. P., Troost, D., & Baas, F. (2004). Expression of complement components in the peripheral nervous system. *Human Molecular Genetics*, *13*, 295–302.

Dyck, P. J. (1982). Are motor neuropathies and motor neuron diseases separable? *Advances in Neurology*, *36*, 105–114.

Ehrlich, P., & Morgenroth, J. (1899). Zur Theorie der Lysenwirkung. *Berliner Klinische Wochenschrift*, *36*, 6.

Eikelenboom, P., Hack, C. E., Rozemuller, J. M., & Stam, F. C. (1989). Complement activation in amyloid plaques in Alzheimer's dementia. *Virchows Archive B: Cell Pathology Including Molecular Pathology*, *56*, 259–262.

Eikelenboom, P., Rozemuller, J. M., Kraal, G., Stam, F. C., McBride, P. A., Bruce, M. E., & Fraser, H. (1991). Cerebral amyloid plaques in Alzheimer's disease but not in scrapie-affected mice are closely associated with a local inflammatory process. *Virchows Archive B: Cell Pathology Including Molecular Pathology*, *60*, 329–336.

Eikelenboom, P., & Stam, F. C. (1982). Immunoglobulins and complement factors in senile plaques. An immunoperoxidase study. *Acta Neuropathologica*, *57*, 239–242.

Eikelenboom, P., & Stam, F. C. (1984). An immunohistochemical study on cerebral vascular and senile plaque amyloid in Alzheimer's dementia. *Virchows Archive B: Cell Pathology Including Molecular Pathology*, *47*, 17–25.

Emmerling, M. R., Watson, M. D., Raby, C. A., & Spiegel, K. (2000). The role of complement in Alzheimer's disease pathology. *Biochimica et Biophysica Acta*, *1502*, 158–171.

Ferraiuolo, L., Heath, P. R., Holden, H., Kasher, P., Kirby, J., & Shaw, P. J. (2007). Microarray analysis of the cellular pathways involved in the adaptation to and progression of motor neuron injury in the SOD1 G93A mouse model of familial ALS. *Journal of Neuroscience*, *27*, 9201–9219.

Ffrench-Constant, C. (1994). Pathogenesis of multiple sclerosis. *Lancet*, *343*, 271–275.

Fonseca, M. I., Zhou, J., Botto, M., & Tenner, A. J. (2004). Absence of C1q leads to less neuropathology in transgenic mouse models of Alzheimer's disease. *Journal of Neuroscience*, *24*, 6457–6465.

Fujita, T. (2002). Evolution of the lectin-complement pathway and its role in innate immunity. *Nature Reviews Immunology*, *2*, 346–353.

Gasque, P., Dean, Y. D., McGreal, E. P., VanBeek, J., & Morgan, B. P. (2000). Complement components of the innate immune system in health and disease in the CNS. *Immunopharmacology*, *49*, 171–186.

Gasque, P., Fontaine, M., & Morgan, B. P. (1995). Complement expression in human brain. Biosynthesis of terminal pathway components and regulators in human glial cells and cell lines. *Journal of Immunology*, *154*, 4726–4733.

Gasque, P., Ischenko, A., Legoedec, J., Mauger, C., Schouft, M. T., & Fontaine, M. (1993). Expression of the complement classical pathway by human glioma in culture. A model for complement expression by nerve cells. *Journal of Biological Chemistry*, *268*, 25068–25074.

Gasque, P., Julen, N., Ischenko, A. M., Picot, C., Mauger, C., Chauzy, C., et al. (1992). Expression of complement components of the alternative pathway by glioma cell lines. *Journal of Immunology, 149*, 1381–1387.

Gasque, P., & Morgan, B. P. (1996). Complement regulatory protein expression by a human oligodendrocyte cell line: cytokine regulation and comparison with astrocytes. *Immunology, 89*, 338–347.

Gasque, P., Neal, J. W., Singhrao, S. K., McGreal, E. P., Dean, Y. D., Van, B. J., & Morgan, B. P. (2002). Roles of the complement system in human neurodegenerative disorders: pro-inflammatory and tissue remodeling activities. *Molecular Neurobiology, 25*, 1–17.

Gasque, P., Singhrao, S. K., Neal, J. W., Gotze, O., & Morgan, B. P. (1997). Expression of the receptor for complement C5a (CD88) is up-regulated on reactive astrocytes, microglia, and endothelial cells in the inflamed human central nervous system. *The American Journal of Pathology, 150*, 31–41.

Gasque, P., Singhrao, S. K., Neal, J. W., Wang, P., Sayah, S., Fontaine, M., & Morgan, B. P. (1998). The receptor for complement anaphylatoxin C3a is expressed by myeloid cells and nonmyeloid cells in inflamed human central nervous system: analysis in multiple sclerosis and bacterial meningitis. *Journal of Immunology, 160*, 3543–3554.

Gasque, P., Thomas, A., Fontaine, M., & Morgan, B. P. (1996). Complement activation on human neuroblastoma cell lines in vitro: route of activation and expression of functional complement regulatory proteins. *Journal of Neuroimmunology, 66*, 29–40.

Geleijns, K., Roos, A., Houwing-Duistermaat, J. J., van Rijs, W., Tio-Gillen, A. P., Laman, J. D., van Doorn, P. A., & Jacobs, B. C. (2006). Mannose-binding lectin contributes to the severity of Guillain-Barre syndrome. *Journal of Immunology, 177*, 4211–4217.

Gewurz, H., Clark, D. S., Cooper, M. D., Varco, R. L., & Good, R. A. (1967). Effect of cobra venom-induced inhibition of complement activity on allograft and xenograft rejection reactions. *Transplantation, 5*, 1296–1303.

Gewurz, H., Ying, S. C., Jiang, H., & Lint, T. F. (1993). Nonimmune activation of the classical complement pathway. *Behring Institute Mitteilungen*, 138–147.

Giese, K. P., Martini, R., Lemke, G., Soriano, P., & Schachner, M. (1992). Mouse P0 gene disruption leads to hypomyelination, abnormal expression of recognition molecules, and degeneration of myelin and axons. *Cell, 71*, 565–576.

Ginsberg, L., Malik, O., Kenton, A. R., Sharp, D., Muddle, J. R., Davis, M. B., Winer, J. B., Orrell, R. W., & King, R. H. (2004). Coexistent hereditary and inflammatory neuropathy. *Brain, 127*, 193–202.

Goldknopf, I. L., Sheta, E. A., Bryson, J., Folsom, B., Wilson, C., Duty, J., Yen, A. A., & Appel, S. H. (2006). Complement C3c and related protein biomarkers in amyotrophic lateral sclerosis and Parkinson's disease. *Biochemical and Biophysics Research Communications, 342*, 1034–1039.

Gordon, D. L., Avery, V. M., Adrian, D. L., & Sadlon, T. A. (1992). Detection of complement protein mRNA in human astrocytes by the polymerase chain reaction. *Journal of Neuroscience Methods, 45*, 191–197.

Gordon, D. L., Sadlon, T., Hefford, C., & Adrian, D. (1993). Expression of CD59, a regulator of the membrane attack complex of complement, on human astrocytes. *Brain Research Molecular Brain Research, 18*, 335–338.

Hafer-Macko, C., Hsieh, S. T., Li, C. Y., Ho, T. W., Sheikh, K., Cornblath, D. R., McKhann, G. M., Asbury, A. K., & Griffin, J. W. (1996a). Acute motor axonal neuropathy: an antibody-mediated attack on axolemma. *Annals of Neurology, 40*, 635–644.

Hafer-Macko, C. E., Sheikh, K. A., Li, C. Y., Ho, T. W., Cornblath, D. R., McKhann, G. M., Asbury, A. K., & Griffin, J. W. (1996b). Immune attack on the Schwann cell surface in acute inflammatory demyelinating polyneuropathy. *Annals of Neurology, 39*, 625–635.

Halstead, S. K., Humphreys, P. D., Goodfellow, J. A., Wagner, E. R., Smith, R. A., & Willison, H. J. (2005). Complement inhibition abrogates nerve terminal injury in Miller Fisher syndrome. *Annals of Neurology, 58*, 203–210.

Halstead, S. K., O'Hanlon, G. M., Humphreys, P. D., Morrison, D. B., Morgan, B. P., Todd, A. J., Plomp, J. J., & Willison, H. J. (2004). Anti-disialoside antibodies kill perisynaptic Schwann cells and damage motor nerve terminals via membrane attack complex in a murine model of neuropathy. *Brain, 127*, 2109–2123.

Halstead, S. K., Zitman, F. M., Humphreys, P. D., Greenshields, K., Verschuuren, J. J., Jacobs, B. C., Rother, R. P., Plomp, J. J., & Willison, H. J. (2008). Eculizumab prevents anti-ganglioside antibody-mediated neuropathy in a murine model. *Brain, 131*(Pt. 5), 1168–1170.

Hila, S., Soane, L., & Koski, C. L. (2001). Sublytic C5b-9-stimulated Schwann cell survival through PI 3-kinase-mediated phosphorylation of BAD. *Glia, 36*, 58–67.

Hourcade, D. E. (2006). The role of properdin in the assembly of the alternative pathway C3 convertases of complement. *Journal of Biological Chemistry, 281*, 2128–2132.

Jiang, H., Burdick, D., Glabe, C. G., Cotman, C. W., & Tenner, A. J. (1994). Beta-amyloid activates complement by binding to a specific region of the collagen-like domain of the C1q A chain. *Journal of Immunology, 152*, 5050–5059.

Johnson, S., Young-Chan, C. S., Laping, N. J., & Finch, C. E. (1996). Perforant path transection induces complement C9 deposition in hippocampus. *Experimental Neurology, 138*, 198–205.

Jung, S., Toyka, K. V., & Hartung, H. P. (1995). Soluble complement receptor type 1 inhibits experimental autoimmune neuritis in Lewis rats. *Neuroscience Letters, 200*, 167–170.

Kaczorowski, S. L., Schiding, J. K., Toth, C. A., & Kochanek, P. M. (1995). Effect of soluble complement receptor-1 on neutrophil accumulation after traumatic brain injury in rats. *Journal of Cerebral Blood Flow and Metabolism, 15*, 860–864.

Katz, Y., & Strunk, R. C. (1988). Synthesis and regulation of complement protein factor H in human skin fibroblasts. *Journal of Immunology, 141*, 559–563.

Kiefer, R., Kieseier, B. C., Stoll, G., & Hartung, H. P. (2001). The role of macrophages in immune-mediated damage to the peripheral nervous system. *Progress in Neurobiology, 64*, 109–127.

King, R. H. M. (1999). *Atlas of peripheral nerve pathology*. New York: Oxford University Press Inc.

Kobsar, I., Berghoff, M., Samsam, M., Wessig, C., Maurer, M., Toyka, K. V., & Martini, R. (2003). Preserved myelin integrity and reduced axonopathy in connexin32-deficient mice lacking the recombination activating gene-1. *Brain, 126*, 804–813.

Kobsar, I., Maurer, M., Ott, T., & Martini, R. (2002). Macrophage-related demyelination in peripheral nerves of mice deficient in the gap junction protein connexin 32. *Neuroscience Letters, 320*, 17–20.

Kohl, J. (2006). The role of complement in danger sensing and transmission. *Immunologic Research, 34*, 157–176.

Koski, C. L. (1990). Characterization of complement-fixing antibodies to peripheral nerve myelin in Guillain-Barre syndrome. *Annals of Neurology, 27*(Suppl.), S44–S47.

Koski, C. L. (1997). Mechanisms of Schwann cell damage in inflammatory neuropathy. *The Journal of Infectious Diseases, 176*(Suppl. 2), S169–S172.

Koski, C. L., Vanguri, P., & Shin, M. L. (1985). Activation of the alternative pathway of complement by human peripheral nerve myelin. *Journal of Immunology, 134*, 1810–1814.

Kossmann, T., Stahel, P. F., Morganti-Kossmann, M. C., Jones, J. L., & Barnum, S. R. (1997). Elevated levels of the complement components C3 and factor B in ventricular cerebrospinal fluid of patients with traumatic brain injury. *Journal of Neuroimmunology, 73*, 63–69.

Kovacs, G. G., Kalev, O., & Budka, H. (2004). Contribution of neuropathology to the understanding of human prion disease. *Folia Neuropathologica, 42*(Suppl. A), 69–76.

Laufer, J., Katz, Y., & Passwell, J. H. (2001). Extrahepatic synthesis of complement proteins in inflammation. *Molecular Immunology, 38*, 221–229.

LeBlanc, A. C., & Poduslo, J. F. (1990). Axonal modulation of myelin gene expression in the peripheral nerve. *Journal of Neuroscience Research, 26*, 317–326.

Lee, D. M., Friend, D. S., Gurish, M. F., Benoist, C., Mathis, D., & Brenner, M. B. (2002). Mast cells: a cellular link between autoantibodies and inflammatory arthritis. *Science, 297*, 1689–1692.

Leinhase, I., Holers, V. M., Thurman, J. M., Harhausen, D., Schmidt, O. I., Pietzcker, M., Taha, M. E., Rittirsch, D., Huber-Lang, M., Smith, W. R., Ward, P. A., & Stahel, P. F. (2006a). Reduced neuronal cell death after experimental brain injury in mice lacking a functional alternative pathway of complement activation. *BMC Neuroscience, 7*, 55.

Leinhase, I., Rozanski, M., Harhausen, D., Thurman, J. M., Schmidt, O. I., Hossini, A. M., et al. (2007). Inhibition of the alternative complement activation pathway in traumatic brain injury by a monoclonal anti-factor B antibody: a randomized placebo-controlled study in mice. *Journal of Neuroinflammation, 2*(4), 13.

Leinhase, I., Schmidt, O. I., Thurman, J. M., Hossini, A. M., Rozanski, M., Taha, M. E., Scheffler, A., John, T., Smith, W. R., Holers, V. M., & Stahel, P. F. (2006b). Pharmacological complement inhibition at the C3 convertase level promotes neuronal survival, neuroprotective intracerebral gene expression, and neurological outcome after traumatic brain injury. *Experimental Neurology, 199*, 454–464.

Lemke, G. (1996). Unwrapping myelination. *Nature, 383*, 395–396.

Li, L. M., Zhu, Y., & Fan, G. Y. (2005). Effects of recombinant sCR1 on the immune inflammatory reaction in acute spinal cord injury tissue of rats. *Chinese Journal of Traumatology, 8*, 49–53.

Li, M., Shibata, A., Li, C., Braun, P. E., McKerracher, L., Roder, J., Kater, S. B., & David, S. (1996). Myelin-associated glycoprotein inhibits neurite/axon growth and causes growth cone collapse. *Journal of Neuroscience Research, 46*, 404–414.

Liu, L., Lioudyno, M., Tao, R., Eriksson, P., Svensson, M., & Aldskogius, H. (1999). Hereditary absence of complement C5 in adult mice influences Wallerian degeneration, but not retrograde responses, following injury to peripheral nerve. *Journal of Peripheral Nervous System, 4*, 123–133.

Liu, L., Persson, J. K., Svensson, M., & Aldskogius, H. (1998). Glial cell responses, complement, and clusterin in the central nervous system following dorsal root transection. *Glia, 23*, 221–238.

Lobsiger, C. S., Boillee, S., & Cleveland, D. W. (2007). Toxicity from different SOD1 mutants dysregulates the complement system and the neuronal regenerative response in ALS motor neurons. *Proceedings of the National Academy of Sciences of the United States of America, 104*, 7319–7326.

Lundborg, G., & Rosen, B. (2007). Hand function after nerve repair. *Acta Physiologica (Oxford), 189*, 207–217.

Luo, X., Weber, G. A., Zheng, J., Gendelman, H. E., & Ikezu, T. (2003). C1q-calreticulin induced oxidative neurotoxicity: relevance for the neuropathogenesis of Alzheimer's disease. *Journal of Neuroimmunology, 135*, 62–71.

Makranz, C., Cohen, G., Reichert, F., Kodama, T., & Rotshenker, S. (2006). cAMP cascade (PKA, Epac, adenylyl cyclase, Gi, and phosphodiesterases) regulates myelin phagocytosis mediated by complement receptor-3 and scavenger receptor-AI/II in microglia and macrophages. *Glia, 53*, 441–448.

Markiewski, M. M., & Lambris, J. D. (2007). The role of complement in inflammatory diseases from behind the scenes into the spotlight. *The American Journal of Pathology, 171*, 715–727.

Martini, R., & Toyka, K. V. (2004). Immune-mediated components of hereditary demyelinating neuropathies: lessons from animal models and patients. *Lancet Neurology, 3*, 457–465.

Matsuoka, Y., Picciano, M., Malester, B., LaFrancois, J., Zehr, C., Daeschner, J. M., Olschowka, J. A., Fonseca, M. I., O'Banion, M. K., Tenner, A. J., Lemere, C. A., & Duff, K.

(2001). Inflammatory responses to amyloidosis in a transgenic mouse model of Alzheimer's disease. *The American Journal of Pathology, 158,* 1345–1354.

Mattsson, P., Morgan, B. P., & Svensson, M. (1998). Complement activation and CD59 expression in the motor facial nucleus following intracranial transection of the facial nerve in the adult rat. *Journal of Neuroimmunology, 91,* 180–189.

Maurer, M., Kobsar, I., Berghoff, M., Schmid, C. D., Carenini, S., & Martini, R. (2002a). Role of immune cells in animal models for inherited neuropathies: facts and visions. *Journal of Anatomy, 200,* 405–414.

Maurer, M., Toyka, K. V., & Gold, R. (2002b). Cellular immunity in inflammatory autoimmune neuropathies. *Revista de Neurologia (Paris), 158,* S7–S15.

May, P. C., Lampert-Etchells, M., Johnson, S. A., Poirier, J., Masters, J. N., & Finch, C. E. (1990). Dynamics of gene expression for a hippocampal glycoprotein elevated in Alzheimer's disease and in response to experimental lesions in rat. *Neuron, 5,* 831–839.

McGeer, P. L., Akiyama, H., Itagaki, S., & McGeer, E. G. (1989). Activation of the classical complement pathway in brain tissue of Alzheimer patients. *Neuroscience Letters, 107,* 341–346.

McGeer, P. L., & McGeer, E. G. (2002). The possible role of complement activation in Alzheimer disease. *Trends in Molecular Medicine, 8,* 519–523.

McGeer, P. L., Walker, D. G., Akiyama, H., Kawamata, T., Guan, A. L., Parker, C. J., Okada, N., & McGeer, E. G. (1991). Detection of the membrane inhibitor of reactive lysis (CD59) in diseased neurons of Alzheimer brain. *Brain Research, 544,* 315–319.

McKerracher, L., David, S., Jackson, D. L., Kottis, V., Dunn, R. J., & Braun, P. E. (1994). Identification of myelin-associated glycoprotein as a major myelin-derived inhibitor of neurite growth. *Neuron, 13,* 805–811.

McPhaden, A. R., & Whaley, K. (1993). Complement biosynthesis by mononuclear phagocytes. *Immunologic Research, 12,* 213–232.

Mead, R. J., Singhrao, S. K., Neal, J. W., Lassmann, H., & Morgan, B. P. (2002). The membrane attack complex of complement causes severe demyelination associated with acute axonal injury. *Journal of Immunology, 168,* 458–465.

Menichella, D. M., Arroyo, E. J., Awatramani, R., Xu, T., Baron, P., Vallat, J. M., Balsamo, J., Lilien, J., Scarlato, G., Kamholz, J., Scherer, S. S., & Shy, M. E. (2001). Protein zero is necessary for E-cadherin-mediated adherens junction formation in Schwann cells. *Molecular and Cellular Neurosciences, 18,* 606–618.

Morgan, B. P., Dankert, J. R., & Esser, A. F. (1987). Recovery of human neutrophils from complement attack: removal of the membrane attack complex by endocytosis and exocytosis. *Journal of Immunology, 138,* 246–253.

Morgan, B. P., & Gasque, P. (1997). Extrahepatic complement biosynthesis: where, when and why? *Clinical and Experimental Immunology, 107,* 1–7.

Morgan, B. P., Gasque, P., Singhrao, S., & Piddlesden, S. J. (1997). The role of complement in disorders of the nervous system. *Immunopharmacology, 38,* 43–50.

Muller-Eberhard, H. J., & Fjellstrom, K. E. (1971). Isolation of the anticomplementary protein from cobra venom and its mode of action on C3. *Journal of Immunology, 107,* 1666–1672.

Nauta, A. J., Roos, A., & Daha, M. R. (2004). A regulatory role for complement in innate immunity and autoimmunity. *International Archives of Allergy and Immunology, 134,* 310–323.

Nguyen, Q. T., Sanes, J. R., & Lichtman, J. W. (2002). Pre-existing pathways promote precise projection patterns. *Nature Neuroscience, 5,* 861–867.

Niculescu, F., Rus, H., van Biesen, T., & Shin, M. L. (1997). Activation of Ras and mitogen-activated protein kinase pathway by terminal complement complexes is G protein dependent. *Journal of Immunology, 158,* 4405–4412.

Nuttall, G. (1888). Experimente uber die bacterienfeindlichen Einfluesse des thierischen Korpers. *Zeitschrift hygiene und infektionskranken, 4,* 353.

Ogawa-Goto, K., & Abe, T. (1998). Gangliosides and glycosphingolipids of peripheral nervous system myelins: a minireview. *Neurochemical Research, 23,* 305–310.

O'Hanlon, G. M., Plomp, J. J., Chakrabarti, M., Morrison, I., Wagner, E. R., Goodyear, C. S., Yin, X., Trapp, B. D., Conner, J., Molenaar, P. C., Stewart, S., Rowan, E. G., & Willison, H. J. (2001). Anti-GQ1b ganglioside antibodies mediate complement-dependent destruction of the motor nerve terminal. *Brain, 124,* 893–906.

Paparounas, K., O'Hanlon, G. M., O'Leary, C. P., Rowan, E. G., & Willison, H. J. (1999). Anti-ganglioside antibodies can bind peripheral nerve nodes of Ranvier and activate the complement cascade without inducing acute conduction block in vitro. *Brain, 122*(Pt. 5), 807–816.

Piddlesden, S. J., Storch, M. K., Hibbs, M., Freeman, A. M., Lassmann, H., & Morgan, B. P. (1994). Soluble recombinant complement receptor 1 inhibits inflammation and demyelination in antibody-mediated demyelinating experimental allergic encephalomyelitis. *Journal of Immunology, 152,* 5477–5484.

Plomp, J. J., Molenaar, P. C., O'Hanlon, G. M., Jacobs, B. C., Veitch, J., Daha, M. R., van Doorn, P. A., van der Meche, F. G., Vincent, A., Morgan, B. P., & Willison, H. J. (1999). Miller Fisher anti-GQ1b antibodies: alpha-latrotoxin-
like effects on motor end plates. *Annals of Neurology, 45,* 189–199.

Putzu, G. A., Figarella-Branger, D., Bouvier-Labit, C., Liprandi, A., Bianco, N., & Pellissier, J. F. (2000). Immunohistochemical localization of cytokines, C5b-9 and ICAM-1 in peripheral nerve of Guillain-Barre syndrome. *Journal of Neurological Sciences, 174,* 16–21.

Quarles, R. H. (1997). Glycoproteins of myelin sheaths. *Journal of Molecular Neuroscience, 8,* 1–12.

Ramaglia, V., King, R. H., Nourallah, M., Wolterman, R., de Jonge, R., Ramkema, M., Vigar, M. A., van der, W. S.,

Morgan, F., Troost, D., & Baas, F. (2007). The membrane attack complex of the complement system is essential for rapid Wallerian degeneration. *Journal of Neuroscience, 27,* 7663–7672.

Ramaglia, V., Wolterman, R., de Kok, M., Vigar, M. A., Wagenaar-Bos, I., King, R. H., Morgan, B. P., & Baas, F. (2008). Soluble complement receptor 1 protects the peripheral nerve from early axon loss after injury. *The American Journal of Pathology, 172,* 1043–1052.

Ramm, L. E., & Mayer, M. M. (1980). Life-span and size of the trans-membrane channel formed by large doses of complement. *Journal of Immunology, 124,* 2281–2287.

Rancan, M., Morganti-Kossmann, M. C., Barnum, S. R., Saft, S., Schmidt, O. I., Ertel, W., & Stahel, P. F. (2003). Central nervous system-targeted complement inhibition mediates neuroprotection after closed head injury in transgenic mice. *Journal of Cerebral Blood Flow and Metabolism, 23,* 1070–1074.

Rebhun, J., & Botvin, J. (1980). Complement elevation in spinal cord injury. *Annals of Allergy, 44,* 287–288.

Rebhun, J., Madorsky, J. G., & Glovsky, M. M. (1991). Proteins of the complement system and acute phase reactants in sera of patients with spinal cord injury. *Annals of Allergy, 66,* 335–338.

Reichert, F., & Rotshenker, S. (2003). Complement-receptor-3 and scavenger-receptor-AI/II mediated myelin phagocytosis in microglia and macrophages. *Neurobiology of Disease, 12,* 65–72.

Reichert, F., Saada, A., & Rotshenker, S. (1994). Peripheral nerve injury induces Schwann cells to express two macrophage phenotypes: phagocytosis and the galactose-specific lectin MAC-2. *Journal of Neuroscience, 14,* 3231–3245.

Reichert, F., Slobodov, U., Makranz, C., & Rotshenker, S. (2001). Modulation (inhibition and augmentation) of complement receptor-3-mediated myelin phagocytosis. *Neurobiology of Disease, 8,* 504–512.

Ricklin, D., & Lambris, J. D. (2007). Complement-targeted therapeutics. *Nature Biotechnology, 25,* 1265–1275.

Robertson, A. M., Perea, J., McGuigan, A., King, R. H., Muddle, J. R., Gabreels-Festen, A. A., Thomas, P. K., & Huxley, C. (2002). Comparison of a new pmp22 transgenic mouse line with other mouse models and human patients with CMT1A. *Journal of Anatomy, 200,* 377–390.

Rogers, J., Cooper, N. R., Webster, S., Schultz, J., McGeer, P. L., Styren, S. D., Civin, W. H., Brachova, L., Bradt, B., & Ward, P. (1992). Complement activation by beta-amyloid in Alzheimer disease. *Proceedings of the National Academy of Sciences of the United States of America, 89,* 10016–10020.

Schafer, M., Fruttiger, M., Montag, D., Schachner, M., & Martini, R. (1996). Disruption of the gene for the myelin-associated glycoprotein improves axonal regrowth along myelin in C57BL/Wlds mice. *Neuron, 16,* 1107–1113.

Schmid, C. D., Stienekemeier, M., Oehen, S., Bootz, F., Zielasek, J., Gold, R., Toyka, K. V., Schachner, M., & Martini, R. (2000). Immune deficiency in mouse models for inherited peripheral neuropathies leads to improved myelin maintenance. *Journal of Neuroscience, 20,* 729–735.

Sewell, D. L., Nacewicz, B., Liu, F., Macvilay, S., Erdei, A., Lambris, J. D., Sandor, M., & Fabry, Z. (2004). Complement C3 and C5 play critical roles in traumatic brain cryoinjury: blocking effects on neutrophil extravasation by C5a receptor antagonist. *Journal of Neuroimmunology, 155,* 55–63.

Shen, Y., Li, R., McGeer, E. G., & McGeer, P. L. (1997). Neuronal expression of mRNAs for complement proteins of the classical pathway in Alzheimer brain. *Brain Research, 769,* 391–395.

Shin, M. L., Rus, H. G., & Niculescu, F. I. (1996). Membrane attack by complement: assembly and biology of terminal complement complexes. In A. G. Lee (Ed.), *Biomembranes* (pp. 123–149). Greenwich, CT: JAI Press Inc.

Shy, M. E., Arroyo, E., Sladky, J., Menichella, D., Jiang, H., Xu, W., Kamholz, J., & Scherer, S. S. (1997). Heterozygous P0 knockout mice develop a peripheral neuropathy that resembles chronic inflammatory demyelinating polyneuropathy (CIDP). *Journal of Neuropathology and Experimental Neurology, 56,* 811–821.

Shy, M. E., Garbern, J. Y., & Kamholz, J. (2002). Hereditary motor and sensory neuropathies: a biological perspective. *Lancet Neurolgy, 1,* 110–118.

Singhrao, S. K., Neal, J. W., Gasque, P., Morgan, B. P., & Newman, G. R. (1996). Role of complement in the aetiology of Pick's disease? *Journal of Neuropathology and Experimental Neurology, 55,* 578–593.

Singhrao, S. K., Neal, J. W., Morgan, B. P., & Gasque, P. (1999). Increased complement biosynthesis by microglia and complement activation on neurons in Huntington's disease. *Experimental Neurology, 159,* 362–376.

Skre, H. (1974). Genetic and clinical aspects of Charcot-Marie-Tooth's disease. *Clinical Genetics, 6,* 98–118.

Soares, D. C., & Barlow, P. N. (2005). Complement control protein modules in the regulators of complement activation. In D. Morikis & J. D. Lambris (Eds.), *Structural biology of the complement system* (pp. 19–62). Boca Raton, FL: CRC Press.

Spitzer, D., Mitchell, L. M., Atkinson, J. P., & Hourcade, D. E. (2007). Properdin can initiate complement activation by binding specific target surfaces and providing a platform for de novo convertase assembly. *Journal of Immunology, 179,* 2600–2608.

Stahel, P. F., Morganti-Kossmann, M. C., Perez, D., Redaelli, C., Gloor, B., Trentz, O., & Kossmann, T. (2001). Intrathecal levels of complement-derived soluble membrane attack complex (sC5b-9) correlate with blood-brain barrier dysfunction in patients with traumatic brain injury. *Journal of Neurotrauma, 18,* 773–781.

Stoll, G., Schmidt, B., Jander, S., Toyka, K. V., & Hartung, H. P. (1991). Presence of the terminal complement complex (C5b-9) precedes myelin degradation in immune-mediated demyelination of the rat peripheral nervous system. *Annals of Neurology, 30,* 147–155.

Storch, M. K., Piddlesden, S., Haltia, M., Iivanainen, M., Morgan, P., & Lassmann, H. (1998). Multiple sclerosis: in situ evidence for antibody- and complement-mediated demyelination. *Annals of Neurology, 43,* 465–471.

Strohmeyer, R., Shen, Y., & Rogers, J. (2000). Detection of complement alternative pathway mRNA and proteins in the Alzheimer's disease brain. *Brain Research. Molecular Brain Research, 81*, 7–18.

Strunk, R. C., Eidlen, D. M., & Mason, R. J. (1988). Pulmonary alveolar type II epithelial cells synthesize and secrete proteins of the classical and alternative complement pathways. *Journal of Clinical Investigation, 81*, 1419–1426.

Sunyer, J. O., & Lambris, J. D. (1998). Evolution and diversity of the complement system of poikilothermic vertebrates. *Immunological Reviews, 166*, 39–57.

Susuki, K., Nishimoto, Y., Koga, M., Nagashima, T., Mori, I., Hirata, K., & Yuki, N. (2004). Various immunization protocols for an acute motor axonal neuropathy rabbit model compared. *Neuroscience Letters, 368*, 63–67.

Susuki, K., Rasband, M. N., Tohyama, K., Koibuchi, K., Okamoto, S., Funakoshi, K., Hirata, K., Baba, H., & Yuki, N. (2007). Anti-GM1 antibodies cause complement-mediated disruption of sodium channel clusters in peripheral motor nerve fibers. *Journal of Neuroscience, 27*, 3956–3967.

Svensson, M., & Aldskogius, H. (1992). Evidence for activation of the complement cascade in the hypoglossal nucleus following peripheral nerve injury. *Journal of Neuroimmunology, 40*, 99–109.

Svensson, M., Liu, L., Mattsson, P., Morgan, B. P., & Aldskogius, H. (1995). Evidence for activation of the terminal pathway of complement and upregulation of sulfated glycoprotein (SGP)-2 in the hypoglossal nucleus following peripheral nerve injury. *Molecular and Chemical Neuropathology, 24*, 53–68.

Tacnet-Delorme, P., Chevallier, S., & Arlaud, G. J. (2001). Beta-amyloid fibrils activate the C1 complex of complement under physiological conditions: evidence for a binding site for A beta on the C1q globular regions. *Journal of Immunology, 167*, 6374–6381.

Terai, K., Walker, D. G., McGeer, E. G., & McGeer, P. L. (1997). Neurons express proteins of the classical complement pathway in Alzheimer disease. *Brain Research, 769*, 385–390.

Tessier-Lavigne, M., & Goodman, C. S. (1996). The molecular biology of axon guidance. *Science, 274*, 1123–1133.

Tofaris, G. K., Patterson, P. H., Jessen, K. R., & Mirsky, R. (2002). Denervated Schwann cells attract macrophages by secretion of leukemia inhibitory factor (LIF) and monocyte chemoattractant protein-1 in a process regulated by interleukin-6 and LIF. *Journal of Neuroscience, 22*, 6696–6703.

Tomita, K., Kubo, T., Matsuda, K., Yano, K., Tohyama, M., & Hosokawa, K. (2007). Myelin-associated glycoprotein reduces axonal branching and enhances functional recovery after sciatic nerve transection in rats. *Glia, 55*, 1498–1507.

Tooth, H. (1886). *The peroneal type of progressive muscular atrophy*. London: Lewis.

Tornqvist, E., Liu, L., Aldskogius, H., Holst, H. V., & Svensson, M. (1996). Complement and clusterin in the injured nervous system. *Neurobiology of Aging, 17*, 695–705.

Trouw, L. A., Blom, A. M., & Gasque, P. (2008). Role of complement and complement regulators in the removal of apoptotic cells. *Molecular Immunology, 45*, 1199–1207.

Tsuboi, Y., & Yamada, T. (1994). Increased concentration of C4d complement protein in CSF in amyotrophic lateral sclerosis. *Journal of Neurology, Neurosurgery and Psychiatry, 57*, 859–861.

van Lookeren, C. M., Wiesmann, C., & Brown, E. J. (2007). Macrophage complement receptors and pathogen clearance. *Cellular Microbiology, 9*, 2095–2102.

Vanguri, P., Koski, C. L., Silverman, B., & Shin, M. L. (1982). Complement activation by isolated myelin: activation of the classical pathway in the absence of myelin-specific antibodies. *Proceedings of the National Academy of Sciences of the United States of America, 79*, 3290–3294.

Vedeler, C. A., Conti, G., Fujioka, T., Scarpini, E., & Rostami, A. (1999). The expression of CD59 in experimental allergic neuritis. *Journal of Neurological Sciences, 165*, 154–159.

Veerhuis, R., Janssen, I., Hack, C. E., & Eikelenboom, P. (1996). Early complement components in Alzheimer's disease brains. *Acta Neuropathologica, 91*, 53–60.

Veerhuis, R., Janssen, I., Hoozemans, J. J., De Groot, C. J., Hack, C. E., & Eikelenboom, P. (1998). Complement C1-inhibitor expression in Alzheimer's disease. *Acta Neuropathologica, 96*, 287–296.

Veerhuis, R., van der Valk, P., Janssen, I., Zhan, S. S., Van Nostrand, W. E., & Eikelenboom, P. (1995). Complement activation in amyloid plaques in Alzheimer's disease brains does not proceed further than C3. *Virchows Archive, 426*, 603–610.

Vital, A., Vital, C., Lagueny, A., Ferrer, X., Ribiere-Bachelier, C., Latour, P., & Petry, K. G. (2003). Inflammatory demyelination in a patient with CMT1A. *Muscle & Nerve, 28*, 373–376.

von Fodor, J. (1887). Die Faehigkeit des Bluts Bacterien zu vernichten. *Deutsche Medizinische Wochenschrift, 13*, 745.

Vriesendorp, F. J., Flynn, R. E., Pappola, M. A., & Koski, C. L. (1995). Complement depletion affects demyelination and inflammation in experimental allergic neuritis. *Journal of Neuroimmunology, 58*, 157–165.

Vriesendorp, F. J., Flynn, R. E., Pappola, M. A., & Koski, C. L. (1997). Soluble complement receptor 1 (sCR1) is not as effective as cobra venom factor in the treatment of experimental allergic neuritis. *International Journal of Neuroscience, 92*, 287–298.

Walker, D. G., Kim, S. U., & McGeer, P. L. (1995a). Complement and cytokine gene expression in cultured microglial derived from postmortem human brains. *Journal of Neuroscience Research, 40*, 478–493.

Walker, D. G., & McGeer, P. L. (1992). Complement gene expression in human brain: comparison between normal and Alzheimer disease cases. *Brain Research. Molecular Brain Research, 14*, 109–116.

Walker, D. G., & McGeer, P. L. (1993). Complement gene expression in neuroblastoma and astrocytoma cell lines of human origin. *Neuroscience Letters, 157*, 99–102.

Walker, D. G., Yasuhara, O., Patston, P. A., McGeer, E. G., & McGeer, P. L. (1995b). Complement C1 inhibitor is produced by brain tissue and is cleaved in Alzheimer disease. *Brain Research*, *675*, 75–82.

Waller, A. (1850). Experiments on the section of glossopharyngeal and hypoglossal nerves of the frog and observations on the alterations produced thereby in the structure of their primitive fibers. *Philosophical Transactions of the Royal Society of London B Bulletin*, *140*, 423–429.

Walport, M. J. (2001a). Complement. First of two parts. *The New England Journal of Medicine*, *344*, 1058–1066.

Walport, M. J. (2001b). Complement. Second of two parts. *The New England Journal of Medicine*, *344*, 1140–1144.

Wanschitz, J., Maier, H., Lassmann, H., Budka, H., & Berger, T. (2003). Distinct time pattern of complement activation and cytotoxic T cell response in Guillain-Barre syndrome. *Brain*, *126*, 2034–2042.

Watanabe, M., Yamamoto, N., Ohkoshi, N., Nagata, H., Kohno, Y., Hayashi, A., Tamaoka, A., & Shoji, S. (2002). Corticosteroid-responsive asymmetric neuropathy with a myelin protein zero gene mutation. *Neurology*, *59*, 767–769.

Webster, S., Bonnell, B., & Rogers, J. (1997a). Charge-based binding of complement component C1q to the Alzheimer amyloid beta-peptide. *The American Journal of Pathology*, *150*, 1531–1536.

Webster, S., Bradt, B., Rogers, J., & Cooper, N. (1997b). Aggregation state-dependent activation of the classical complement pathway by the amyloid beta peptide. *Journal of Neurochemistry*, *69*, 388–398.

Willison, H. J. (2005). The immunobiology of Guillain-Barre syndromes. *Journal of Peripheral Nervous System*, *10*, 94–112.

Willison, H. J., & Yuki, N. (2002). Peripheral neuropathies and anti-glycolipid antibodies. *Brain*, *125*, 2591–2625.

Wyss-Coray, T., Yan, F., Lin, A. H., Lambris, J. D., Alexander, J. J., Quigg, R. J., & Masliah, E. (2002). Prominent neurodegeneration and increased plaque formation in complement-inhibited Alzheimer's mice. *Proceedings of the National Academy of Sciences of the United States of America*, *99*, 10837–10842.

Yam, P., Petz, L. D., Tourtellotte, W. W., & Ma, B. I. (1980). Measurement of complement components in cerebral spinal fluid by radioimmunoassay in patients with multiple sclerosis. *Clinical Immunology and Immunopathology*, *17*, 492–505.

Yamada, T., McGeer, P. L., & McGeer, E. G. (1992). Lewy bodies in Parkinson's disease are recognized by antibodies to complement proteins. *Acta Neuropathologica*, *84*, 100–104.

Yang, C., Jones, J. L., & Barnum, S. R. (1993). Expression of decay-accelerating factor (CD55), membrane cofactor protein (CD46) and CD59 in the human astroglioma cell line, D54-MG, and primary rat astrocytes. *Journal of Neuroimmunology*, *47*, 123–132.

Yang, L. B., Li, R., Meri, S., Rogers, J., & Shen, Y. (2000). Deficiency of complement defense protein CD59 may contribute to neurodegeneration in Alzheimer's disease. *Journal of Neuroscience*, *20*, 7505–7509.

Yasojima, K., McGeer, E. G., & McGeer, P. L. (1999a). Complement regulators C1 inhibitor and CD59 do not significantly inhibit complement activation in Alzheimer disease. *Brain Research*, *833*, 297–301.

Yasojima, K., Schwab, C., McGeer, E. G., & McGeer, P. L. (1999b). Up-regulated production and activation of the complement system in Alzheimer's disease brain. *The American Journal of Pathology*, *154*, 927–936.

Yasuhara, O., Aimi, Y., McGeer, E. G., & McGeer, P. L. (1994). Expression of the complement membrane attack complex and its inhibitors in Pick disease brain. *Brain Research*, *652*, 346–349.

Younger, J. G., Sasaki, N., Waite, M. D., Murray, H. N., Saleh, E. F., Ravage, Z. B., Hirschl, R. B., Ward, P. A., & Till, G. O. (2001). Detrimental effects of complement activation in hemorrhagic shock. *Journal of Applied Physiology*, *90*, 441–446.

Yu, T. W., & Bargmann, C. I. (2001). Dynamic regulation of axon guidance. *Nature Neuroscience*, *4*(Suppl.), 1169–1176.

Zanjani, H., Finch, C. E., Kemper, C., Atkinson, J., McKeel, D., Morris, J. C., & Price, J. L. (2005). Complement activation in very early Alzheimer disease. *Alzheimer Disease and Associated Disorders*, *19*, 55–66.

CHAPTER 8

# Neuroinflammation in spinal cord injury: therapeutic targets for neuroprotection and regeneration

Jessica K. Alexander[1] and Phillip G. Popovich[2],*

[1]Neuroscience Graduate Studies Program, The Ohio State University College of Medicine, Columbus, OH, USA
[2]Department of Molecular Virology, Immunology and Medical Genetics, Department of Neuroscience, Center for Brain and Spinal Cord Repair (CBSCR), The Institute for Behavioral Medicine Research, The Ohio State University College of Medicine, Columbus, OH, USA

**Abstract:** Traumatic spinal cord injury triggers a complex local inflammatory reaction capable of enhancing repair and exacerbating pathology. The composition and effector potential of the post-injury cellular and molecular immune cascade changes as a function of time and distance from the lesion. Production along this time–space continuum of cytokines, proteases, and growth factors establishes dynamic environments that lead to the death, damage, repair or growth of affected neurons and glia. Microenvironmental cues, therefore, generated by the cells therein, may determine these distinct fates of repair versus pathology. To harness repair, it is necessary to manipulate the assembly and phenotype of cells that comprise the neuroinflammatory response to injury. Here, the potential of the neuroinflammatory response to cause outcomes such as pain, regeneration, and functional recovery is reviewed.

**Keywords:** spinal cord injury; neuroinflammation; inflammation; macrophage; microglia; regeneration

## Introduction to spinal cord injury (SCI) pathology: neuroinflammation in the pathogenesis of secondary injury

Traumatic SCI is generally sustained by the rapid displacement of an intervertebral disc onto anatomical structures that subserve motor, sensory, and autonomic function (Sekhon and Fehlings, 2001). Tissue that survives the initial trauma is susceptible to secondary mechanisms of neurodegeneration including ischemia (Fehlings et al., 1989; Hamamoto et al., 2007; Lee et al., 2005), edema (O'Carroll et al., 2008; Saadoun et al., 2008), glutamate excitotoxicity (Agrawal and Fehlings, 1997; Faden and Simon, 1988; Wrathall et al., 1997), and neuroinflammation (Donnelly and Popovich, 2008; Popovich and Longbrake, 2008).

Physical damage to the blood-spinal cord barrier (BSCB) during trauma initiates a feed-forward pathological cascade of cytotoxic edema (Simard et al., 2007) that may be treated using drugs that target ion flux across endothelia (Liang et al., 2007). Excess fluid accumulation in the spinal cord is exacerbated by cytokines and proteases that are released from activated microglia, astrocytes, and leukocytes (Aarabi et al., 2006; Armao et al., 1997; Koyanagi et al., 1989; Nagy et al., 1998).

As residents of the spinal cord, microglia and astrocytes are among the first cells to respond to

*Corresponding author.
Tel.: 614 688 8576; Fax: 614 292 7544;
E-mail: Phillip.Popovich@osumc.edu

DOI: 10.1016/S0079-6123(09)17508-8

tissue damage. While these cells are important for reestablishing tissue homeostasis (Bessis et al., 2007; Bush et al., 1999; Davalos et al., 2005; Elkabes et al., 1996; Faulkner et al., 2004; Koizumi et al., 2007; Myer et al., 2006; Okada et al., 2006; Pender and Rist, 2001; Persson et al., 2005), they also produce factors that hinder the ability of injured central nervous system (CNS) axons to regenerate or induce cellular reactions that cause neuronal hypersensitivity leading to neuropathic pain (Brambilla et al., 2005; Coull et al., 2005; DeLeo and Yezierski, 2001; Detloff et al., 2008; Guo et al., 2007; Horn et al., 2008; Wieseler-Frank et al., 2005). Reactive glia also release chemokines and cytokines that enhance recruitment of peripheral leukocytes (Bartholdi and Schwab, 1997; Rice et al., 2007; Yang et al., 2004). Once they enter the CNS, leukocytes produce cytokines, chemokines, proteases, and oxidative metabolites that enhance capillary permeability, upregulate endothelial adhesion molecules, and modify unctional adhesion molecules (Engelhardt and Ransohoff, 2005; Weber et al., 2007). Together, these changes provide a feed-forward mechanism for enhancing leukocyte adhesion and migration into the CNS (Pineau and Lacroix, 2008).

Neutrophils are the first circulating leukocytes to infiltrate sites of SCI (~2 h–3 days post-SCI) (Carlson et al., 1998; Fleming et al., 2006; Kigerl et al., 2006; Means and Anderson, 1983; Nguyen et al., 2008; Stirling and Yong, 2008; Taoka et al., 1997). Monocyte-derived macrophages infiltrate ~2 days after neutrophils and help clear apoptotic neutrophils from the lesion (Savill et al., 1989; Stirling and Yong, 2008). This custodial function of macrophages may be necessary for inducing a subset of functions that include release of resolvins and protectins to suppress further neutrophil recruitment (Nathan, 2006). Unlike neutrophils, macrophages persist in human and mouse SCI lesions as long as any study has examined—months in mice and years in humans (Chang, 2007; Fleming et al., 2006; Kigerl et al., 2006, 2007; Popovich et al., 2003). The nonspecific microbicidal activity of neutrophils and monocytes/macrophages can be destructive to host tissue after SCI; both cell types release proteases (e.g., matrix metalloproteases) and oxidative metabolites that can damage cells and compromise the BSCB (Noble et al., 2002; Scholz et al., 2007). Indeed, SCI pathology is reduced and spontaneous recovery of neurological function (motor, sensory, and autonomic) is improved when the recruitment and/or activation of blood-derived leukocytes is restricted (Blight, 1994; Giulian and Robertson, 1990; Gris et al., 2004; Popovich et al., 1999; Taoka et al., 1997). T and B lymphocytes also infiltrate the injured mammalian spinal cord, albeit in fewer numbers and at later times post-injury than monocytes (Ankeny et al., 2006; Popovich et al., 1997; Sroga et al., 2003). The functional significance of T and B cells in the injured spinal cord remains a point of controversy and is the subject of recent reviews (Ankeny and Popovich, 2009; Popovich and Longbrake, 2008).

## Manipulating neuroinflammation to improve recovery from SCI

Many therapies that have been, or are being, developed to treat SCI attempt to limit the neurotoxic potential of post-traumatic neuroinflammation, while other strategies attempt to "boost" the reparative functions of leukocytes (reviewed in Popovich and Longbrake, 2008). In this section, we review a range of pharamacotherapies and cell- or molecule-based therapies that target various aspects of post-traumatic neuroinflammation.

### *General immunosuppression*

Since the completion of the National Acute Spinal Cord Injury Studies (NASCIS) clinical trials, the glucocorticoid methylprednisolone (MP) has been widely used as the standard of care for treating acute SCI (Bracken et al., 1992, 1997). However, the efficacy of MP is controversial, and recent data indicate that the benefits of MP are small or inconsistent (Baptiste and Fehlings, 2007; Pointillart et al., 2000; Sayer et al., 2006). Moreover, the large doses of MP given to people with SCI appear to have adverse long-term consequences (Lee et al., 2007; Short et al., 2000; Suberviola et al., 2008). Consequently, the use of MP may be on the decline (Hugenholtz, 2003; Hurlbert and Hamilton, 2008).

Cyclosporine A (CsA), a calcineurin inhibitor, is a potent inhibitor of T-cell activation and has been widely used to prevent immunological rejection of organ transplants (Haddad et al., 2006). In models of SCI, CsA reduces neuroinflammation and improves recovery (Hayashi et al., 2005; Ibarra and Diaz-Ruiz, 2006), though lack of efficacy has also been observed (Rabchevsky et al., 2001). FK506 (tacrolimus) is another calcineurin inhibitor with potent immunosuppressive properties that has been shown to reduce measures of secondary injury after SCI (Guzmán-Lenis et al., 2008; Kaymaz et al., 2005; López-Vales et al., 2005; Nottingham et al., 2002). The benefits of CsA and FK506 are enhanced when used in combination with cell transplants (Hayashi et al., 2005; López-Vales et al., 2006).

Methotrexate, an antimetabolite immunosuppressant drug, has not yet been tested in experimental SCI; however, it is widely used in the treatment of rheumatoid arthritis (Ranganathan, 2008) and has been shown to ameliorate neuropathic pain in experimental models (Hashizume et al., 2000; Scholz et al., 2008). It does so in part by suppressing microglial activation (Scholz et al., 2008), an effect that has direct relevance to SCI. Indeed, post-traumatic activation of microglia remote from sites of injury is associated with the development of neuropathic pain (Detloff et al., 2008; Zhao et al., 2007). Despite the perceived benefits of these anti-inflammatory or immunosuppressive therapies, their effects are not restricted to cells of the immune system; neurons, astrocytes, and oligodendrocytes are also affected by these drugs (James et al., 2008; Kaminska et al., 2004; Lee et al., 2008; Sosa et al., 2005). Moreover, the systemic administration of a general immunosuppressant may be undesirable after SCI, as it could increase the incidence of sepsis and infection and limit the pro-regenerative and/or neuroprotective functions of immune cells (Schwartz and Yoles, 2006).

## *Cellular and molecular approaches to reducing neuroinflammation*

Therapies that target specific cell types or select intracellular signaling pathways may be more effective than immunosuppression in regulating post-traumatic neuroinflammation. For example, using different approaches in multiple species, selective depletion of neutrophils and/or monocytes has been shown to consistently improve neurological recovery and promote tissue sparing (Blight, 1994; Giulian and Robertson, 1990; Popovich et al., 1999; Taoka et al., 1997). Neuroprotection is also achieved by inhibiting the binding of leukocytes to endothelial adhesion molecules including selectins and integrins (Farooque et al., 1999, 2001; Gris et al., 2004; Mabon et al., 2000). Although the mechanisms responsible for leukocyte-mediated injury are unclear, reducing their entry into the lesion site limits oxidative stress in the spinal cord (Bao et al., 2004; Hamada et al., 1996; Taoka et al., 1997).

Inhibition of chemokines or cytokines is another approach for limiting leukocyte influx and reducing post-SCI neuroinflammation. Chemokines are chemotactic cytokines that recruit immune cells to injured tissues (Anthony et al., 2001; Cardona et al., 2008). After SCI, elevated levels of chemokines are found in serum (Davies et al., 2007; Liu et al., 2005), liver (Campbell et al., 2005), brain (Zhao et al., 2007), and spinal cord (Kigerl et al., 2007; Lee et al., 2000; McTigue et al., 1998; Rice et al., 2007). Importantly, treatment with the broad-spectrum chemokine receptor antagonist, vMIP-II, decreases leukocyte infiltration and astrogliosis while increasing axon and myelin sparing and neuronal survival (Ghirnikar et al., 2000, 2001). Functional recovery has not been evaluated in vMIP-II-treated rats.

CXCL10, a chemokine with broad chemoattractive potential, is increased by 6 h after SCI and remains elevated for at least 3 weeks post-injury in mice (Gonzalez et al., 2003; Jones et al., 2005). Anti-CXCL10 treatment decreases the number of macrophages, CD4+ and CD8+ T cells and B cells in injured mouse spinal cord (Gonzalez et al., 2007). Furthermore, CXCL10 neutralization decreases apoptosis, increases angiogenesis, and promotes axon sprouting and functional recovery (Glaser et al., 2006; Gonzalez et al., 2007).

Infliximab inhibits the pro-inflammatory cytokine TNF-α from binding to its receptor. In models of SCI, infliximab reduces inflammation and oxidative injury to levels achieved with high-dose glucocorticoids (Kurt et al., 2009). Genetic deletion of TNF-α receptors has also been shown to confer neuroprotection and ameliorate neuropathic pain after SCI (Kim et al., 2001; Vogel et al., 2006). IL-1 receptor antagonist (IL-1ra) is an endogenous regulatory cytokine that is increased after SCI in humans (Davies et al., 2007). In experimental SCI, exogenously administered IL-1ra reduces apoptosis and inhibits p38 activation (Nesic et al., 2001; Wang et al., 2005). However, like TNF-α, IL-1β may play an important role in coordinating various axes of CNS repair including remyelination (Mason et al., 2001; Temporin et al., 2008a, b).

Inhibition of select intracellular signaling pathways that are activated by pro-inflammatory cytokines are also feasible targets for regulating post-traumatic neuroinflammation. An example is p38, a member of the mitogen-activated protein kinase family (MAPK). After SCI, activated (phosphorylated) p38 is rapidly increased (minutes to hours) (Wang et al., 2005; Xu et al., 2006) then is sustained throughout the spinal cord for weeks or months post-injury (Crown et al., 2008; Detloff et al., 2008; Stirling and Yong, 2008). A single dose of SB203580, an inhibitor of p38 phosphorylation, reduces apoptosis (Wang et al., 2005; Xu et al., 2006), iNOS gene expression (Xu et al., 2006), and neuronal hyperexcitability and allodynia (Crown et al., 2008). Continuous intrathecal delivery of SB203580 has mixed effects. One study documented reduced apoptotic cell death and myelin pathology with improved recovery of motor function in SCI rats (Horiuchi et al., 2003), while a similar paradigm failed to improve motor recovery or tissue sparing (Stirling and Yong, 2008).

Minocycline is a broad-spectrum antibiotic that has anti-inflammatory and neuroprotective effects. When used in models of SCI, minocycline has produced mixed results. Some have demonstrated minocycline-mediated neuroprotection (Festoff et al., 2006; Lee et al., 2003; Stirling et al., 2004; Teng et al., 2004; Wells et al., 2003; Xu et al., 2004), while others have been unable to show any benefit of this drug (Pinzon et al., 2008; Zang and Cheema, 2003). The onset of below-level mechanical allodynia (a measure of neuropathic pain) in rodent models of SCI is associated with enhanced microglial activation (Detloff et al., 2008; Hains and Waxman, 2006; Zhao et al., 2007); indices of enhanced microglial activation and pain are reduced by minocycline (Hains and Waxman, 2006; Marchand et al., 2008; Zhao et al., 2007). The mechanisms by which minocycline confers its beneficial effects include inhibition of microglial-induced excitotoxicity (Baptiste et al., 2004; Tikka and Koistinaho, 2001) and cytokine signaling via p38 (Guo and Bhat, 2007; Henry et al., 2008; Yune et al., 2007). Intravenous minocycline is currently being tested in acute SCI as part of a Phase I/II clinical trial in Canada (Baptiste and Fehlings, 2007).

Collectively, the above data suggest that inhibition of acute inflammatory cascades will be neuroprotective and improve recovery from SCI. While this may be true, the efficacy of immune-based therapies will be enhanced further as we refine our understanding of the molecular signaling pathways that control neuroinflammation within the injured spinal cord. Indeed, inhibition of inflammatory cell function may not be categorically beneficial. For example, several injury- or "danger"-related endogenous molecules that are released at the sites of SCI (Matzinger, 2002; Pineau and Lacroix, 2008), including heat shock proteins (Lehnardt et al., 2008), extracellular matrix proteins (Schaefer et al., 2005; Termeer et al., 2002), and perhaps opioids (Hutchinson et al., 2007), engage toll-like receptor 2 and 4 (TLR2; TLR4) initiating a broad range of functions in microglia/macrophages (and perhaps astrocytes; Bowman et al., 2003). These functions are complex and some may be essential for coordinating tissue repair. For example, the absence of TLR2 and TLR4 has been shown to reduce intraspinal microglial activation and cytokine expression, and mitigate pain-like behavior after peripheral nerve injury (Kim et al., 2007;

Tanga et al., 2005). Similarly, in models of cerebral ischemia, infarct size and neurotoxicity are attenuated in the absence of TLR4 or when the brain is preconditioned with a TLR2 agonist (Hua et al., 2007, 2008); yet, pathology is exacerbated and spontaneous recovery is impaired in TLR4-deficient mice after traumatic SCI (Kigerl et al., 2006). These data suggest that under certain conditions, TLR4 signaling can down-regulate pro-inflammatory functions in innate immune cells and glia. In support of this view, neuroinflammation is enhanced in the absence of TLR4 signaling in mice with experimental autoimmune encephalomyelitis (EAE), a model of multiple sclerosis (Marta et al., 2008). As described above and in recent reviews, similarly divergent functions have been attributed to chemokines and cytokines.

Together, these data underscore the importance of understanding the dominant ligand/receptor pathways that are involved in signaling acute neuroinflammatory cascades after SCI and the microenvironment in which leukocytes become activated (Popovich and Longbrake, 2008). For example, the same population of microglia or macrophages will adopt unique morphological and functional phenotypes when exposed to distinct activation cues (Gordon, 2003; Stout et al., 2005). Importantly, these phenotypes are not static and may be dynamically influenced by changes in cytokines or other immune modulatory signals (Stout et al., 2005). After SCI, the spatiotemporal continuum of cell death and endogenous repair are dynamic processes requiring an array of cellular and molecular cascades. Consequently, the microenvironment and the cues that regulate glial and leukocyte function will change over time and with respect to location in the lesion, as will the effect that these cells have on the surrounding neuropil. Thus, it is not possible to categorically label post-traumatic inflammation as deleterious or reparative.

## Consequences of neuroinflammation on neuronal plasticity and regeneration

Microglia and macrophages have emerged as potentially invaluable participants in promoting neuron survival and axon growth. A number of studies have documented the ability of these cells to efficiently remove cellular debris and proteins that inhibit axon growth. These same cells can be activated to produce soluble factors that promote cell survival and/or direct the elongation of axons.

### *The role of macrophages in nerve regeneration and relevance to SCI*

Comparative analyses of axon regeneration in the CNS and peripheral nervous system (PNS) have revealed marked differences; PNS axons regenerate more efficiently than CNS axons. This is believed to be due, in part, to the increased efficiency by which macrophages remove cell, axon, and myelin debris—potent inhibitors of axon regeneration—in the distal portion of injured peripheral nerve (Vargas and Barres, 2007). Myelin phagocytosis and recovery of hind limb function is accelerated in nerve injured rats following local activation of macrophages with chemokines, cytokines, or TLR2 or TLR4 agonists (Boivin et al., 2007; Perrin et al., 2005). Conversely, inhibiting macrophage recruitment impairs myelin phagocytosis and delays the regenerative response (Barrette et al., 2008; Boivin et al., 2007; Vargas and Barres, 2007).

Based on these observations, efforts have been made to augment macrophage function at and distal to sites of CNS injury. Indeed, focal and systemic injections of TLR agonists accelerate macrophage/microglia-mediated clearance in spinal funiculi undergoing Wallerian degeneration (Perrin et al., 2005; Vallieres et al., 2006). When combined with overexpression of neurotrophic factors, systemic injection of lipopolysaccharide (LPS), a TLR4 agonist, can promote sprouting of intact axons into zones of Wallerian degeneration (Chen et al., 2008). Regeneration of the injured optic nerve is similarly enhanced when zymosan, a TLR2 agonist and potent activator of macrophages, is injected into the vitreous of the eye (Leon et al., 2000; Yin et al., 2003).

Augmenting the CNS macrophage response by local transplantation of ex vivo activated macrophages has also been shown to promote regeneration of injured optic nerve and spinal cord axons in rodents (Lazarov-Spiegler et al., 1996, 1998; Rapalino et al., 1998). However, the same approach was recently found to have no effect in a canine model of SCI (Assina et al., 2008). Based on these apparently conflicting data, it may be more useful to develop therapeutic strategies that enhance the quality rather than the quantity of the macrophage response that is elicited by SCI. Indeed, there is no shortage of activated macrophages in the injured mammalian spinal cord and markers of phagocytic activity are widespread (Birdsall Abrams et al., 2007; Fleming et al., 2006; Kigerl et al., 2006; Ritz and Hausmann, 2008; Sroga et al., 2003). This challenges the concept that CNS axon regeneration is impeded by a slow or diminutive macrophage response. Also, when compared to common strains of inbred mice (e.g., C57BL/6), 129 × 1/SvJ mice exhibit robust spontaneous axon growth coincident with a blunted inflammatory response at the site of injury (Ma et al., 2004). A similar relationship was recently described in the injured spinal cord of MRL/MpJ mice (Kostyk et al., 2008). These data can be interpreted to mean that acute inflammation impairs spontaneous axon growth. In support of this hypothesis, depletion or functional inactivation of innate (and adaptive, see above) immune cell function after acute SCI consistently confers neuroprotection and enhances spontaneous regrowth of axons (Blight, 1994; Ghirnikar et al., 2001; Glaser et al., 2006; Gris et al., 2004; Popovich et al., 1999).

## Growth promotion by neuroinflammation: a life or death decision?

Collectively, these data suggest that in order to promote the inherent ability of macrophages to promote axon regeneration, we must acquire a better understanding of the cues in the lesion microenvironment that regulate their activation, how these stimuli change over time, and ultimately, how activated macrophages influence neurons and glia in the context of these variables. Moreover, it is possible that regardless of how macrophages or microglia become activated, they uniformly develop the ability to simultaneously kill and promote repair/regeneration. Although these seemingly divergent functions are rarely studied in parallel, if they are triggered concurrently, then therapies that augment neuroinflammation in an effort to promote CNS repair may do so at a cost.

Recently, we monitored the axon growth promoting ability of macrophages and their ability to cause concurrent pathology (Gensel et al., 2009). Using an in vitro assay, we found that culture medium conditioned by zymosan-activated macrophages (ZAMs) promotes robust growth of dorsal root ganglia (DRG) axons; however, this growth was transient. With time (24–48 h in vitro), axon growth was reduced and accompanied by an increase in neuron toxicity. In contrast to zymosan, LPS-activated macrophages were only mildly neurotoxic and were unable to promote axon growth. Similar effects were observed in vivo for both TLR agonists.

When zymosan was microinjected into the intact spinal cord, it induced a robust but focal zone of macrophage activation (Popovich et al., 2002; Schonberg et al., 2007). At these sites, macrophages kill glia, and axons are damaged; however, this appears to be necessary for creating a microenvironment that is permissive for axon growth. Indeed, when enhanced green fluorescence protein (EGFP)-expressing DRG neurons were transplanted 4 mm rostral to the site of zymosan injection (but coincident with injection of zymosan), axon growth toward the ZAM foci was enhanced. In contrast, LPS did not cause focal spinal cord pathology nor was it able to promote DRG axon growth despite the induction of a florid macrophage response (Gensel et al., 2009). When macrophages were activated nearby (proximal) transplanted DRG somata, thereby mimicking the milieu created in our in vitro model (see above), axon growth was abolished and most transplanted DRG neurons were killed.

It is widely recognized that the secretory profile of macrophages is largely dictated by the ligands that activate them. Additional work is needed to determine the nature of the SCI-specific ligands found at sites of injury and the effect they have on macrophage function. For example, phagocytosis of apoptotic cells down-regulates inflammatory functions in macrophages (Henson et al., 2001), while opsonin-mediated phagocytosis of dead/

necrotic cells triggers production of pro-inflammatory mediators, many of which have been implicated as neurotoxic mediators (Meagher et al., 1992). Similarly, macrophages can be "programmed" to release cytokines that may enhance axon growth (e.g., IL-4), while simultaneously inhibiting their ability to produce neurotoxic mediators (e.g., nitric oxide) (Gordon, 2003). Again, these distinct functional profiles are dictated by the stimulus and environment in which macrophages are activated. The latter may be the most important factor in predicting whether effective nontoxic macrophage-mediated neuroprotection or regeneration is possible.

## Conclusions

SCI elicits a neuroinflammatory cascade that exerts complex and seemingly conflicting effects on neurons and glia. As part of this response, a plethora of cytokines, proteases, and growth factors are produced that can simultaneously kill, over-sensitize, or promote growth/repair of injured neurons and glia. These distinct fates are likely to be influenced by the composition and effector potential of the neuroinflammatory reaction, which will change in response to cues in the microenvironment that also change as a function of time and location within the lesion. Thus, even though generic immune suppressive therapies may promote acute neuroprotection, they may have little long-term benefit to SCI individuals. In fact, if used for too long, broad-spectrum immunosuppressants could impair the neuroinflammatory-mediated repair. Future research should work to define the molecular "fingerprint" of injurious and reparative neuroinflammation. By doing so, site-specific and time-dependent therapies may be developed.

## Acknowledgments

We are grateful to John Gensel for helpful discussions of this manuscript. Works described in this review were funded in part by NIH-NINDS grants NS047175 and NS37846 and the Craig H. Neilsen Foundation.

## References

Aarabi, B., Hesdorffer, D. C., Ahn, E. S., Aresco, C., Scalea, T. M., & Eisenberg, H. M. (2006). Outcome following decompressive craniectomy for malignant swelling due to severe head injury. *Journal of Neurosurgery, 104*, 469–479.

Agrawal, S. K., & Fehlings, M. G. (1997). Role of NMDA and non-NMDA ionotropic glutamate receptors in traumatic spinal cord axonal injury. *Journal of Neuroscience, 17*, 1055–1063.

Ankeny, D. P., Lucin, K. M., Sanders, V. M., McGaughy, V. M., & Popovich, P. G. (2006). Spinal cord injury triggers systemic autoimmunity: evidence for chronic B lymphocyte activation and lupus-like autoantibody synthesis. *Journal of Neurochemistry, 99*, 1073–1087.

Ankeny, D. P., & Popovich, P. G. (2009). Mechanisms and implications of adaptive immune responses after traumatic spinal cord injury. *Neuroscience, 158*(3), 1112–1121.

Anthony, D. C., Blond, D., Dempster, R., & Perry, V. H. (2001). Chemokine targets in acute brain injury and disease. *Progress in Brain Research, 132*, 507–524.

Armao, D., Kornfeld, M., Estrada, E. Y., Grossetete, M., & Rosenberg, G. A. (1997). Neutral proteases and disruption of the blood-brain barrier in rat. *Brain Research, 767*, 259–264.

Assina, R., Sankar, T., Theodore, N., Javedan, S. P., Gibson, A. R., Horn, K. M., et al. (2008). Activated autologous macrophage implantation in a large-animal model of spinal cord injury. *Neurosurgical Focus, 25*, E3.

Bao, F., Chen, Y., Dekaban, G. A., & Weaver, L. C. (2004). An anti-CD11d integrin antibody reduces cyclooxygenase-2 expression and protein and DNA oxidation after spinal cord injury in rats. *Journal of Neurochemistry, 90*, 1194–1204.

Baptiste, D. C., & Fehlings, M. G. (2007). Update on the treatment of spinal cord injury. *Progress in Brain Research, 161*, 217–233.

Baptiste, D. C., Hartwick, A. T., Jollimore, C. A., Baldridge, W. H., Seigel, G. M., & Kelly, M. E. (2004). An investigation of the neuroprotective effects of tetracycline deritvaives in experimental models of retinal cell death. *Molecular Pharmacology, 66*, 1113–1122.

Barrette, B., Hebert, M. A., Filali, M., Lafortune, K., Vallieres, N., Gowing, G., et al. (2008). Requirement of myeloid cells for axon regeneration. *Journal of Neuroscience, 28*, 9363–9376.

Bartholdi, D., & Schwab, M. E. (1997). Expression of pro-inflammatory cytokine and chemokine mRNA upon experimental spinal cord injury in mouse: an in situ hybridization study. *European Journal of Neuroscience, 9*, 1422–1438.

Bessis, A., Bechade, C., Bernard, D., & Roumier, A. (2007). Microglial control of neuronal death and synaptic properties. *Glia, 55*, 233–238.

Birdsall Abrams, M., Josephson, A., Dominguez, C., Oberg, J., Diez, M., Spenger, C., et al. (2007). Recovery from spinal cord injury differs between rat strains in a major histocompatibility complex-independent manner. *European Journal of Neuroscience, 26*, 1118–1127.

Blight, A. R. (1994). Effects of silica on the outcome from experimental spinal cord injury: implication of macrophages in secondary tissue damage. *Neuroscience, 60,* 263–273.

Boivin, A., Pineau, I., Barrette, B., Filali, M., Vallieres, N., Rivest, S., et al. (2007). Toll-like receptor signaling is critical for Wallerian degeneration and functional recovery after peripheral nerve injury. *Journal of Neuroscience, 27,* 12565–12576.

Bowman, C. C., Rasley, A., Tranguch, S. L., & Marriott, I. (2003). Cultured astrocytes express toll-like receptors for bacterial products. *Glia, 43,* 281–291.

Bracken, M. B., Shepard, M. J., Collins, W. F., Jr., Holford, T. R., Baskin, D. S., Eisenberg, H. M., et al. (1992). Methylprednisolone or naloxone treatment after acute spinal cord injury: 1-year follow-up data. Results of the Second National Acute Spinal Cord Injury Study. *Journal of Neurosurgery, 76,* 23–31.

Bracken, M. B., Shepard, M. J., Holford, T. R., Leo-Summers, L., Aldrich, E. F., Fazl, M., et al. (1997). Administration of methylprednisolone for 24 or 48 hours or tirilazad mesylate for 48 hours in the treatment of acute spinal cord injury. Results of the Third National Acute Spinal Cord Injury Randomized Controlled Trial. National Acute Spinal Cord Injury Study. *JAMA, 277,* 1597–1604.

Brambilla, R., Bracchi-Ricard, V., Hu, W. H., Frydel, B., Bramwell, A., Karmally, S., et al. (2005). Inhibition of astroglial nuclear factor kappaB reduces inflammation and improves functional recovery after spinal cord injury. *Journal of Experimental Medicine, 202,* 145–156.

Bush, T. G., Puvanachandra, N., Horner, C. H., Polito, A., Ostenfeld, T., Svendsen, C. N., et al. (1999). Leukocyte infiltration, neuronal degeneration, and neurite outgrowth after ablation of scar-forming, reactive astrocytes in adult transgenic mice. *Neuron, 23,* 297–308.

Campbell, S. J., Perry, V. H., Pitossi, F. J., Butchart, A. G., Chertoff, M., Waters, S., et al. (2005). Central nervous system injury triggers hepatic CC and CXC chemokine expression that is associated with leukocyte mobilization and recruitment to both the central nervous system and the liver. *American Journal of Pathology, 166,* 1487–1497.

Cardona, A. E., Li, M., Liu, L., Savarin, C., & Ransohoff, R. M. (2008). Chemokines in and out of the central nervous system: much more than chemotaxis and inflammation. *Journal of Leukocyte Biology, 84,* 587–594.

Carlson, S. L., Parrish, M. E., Springer, J. E., Doty, K., & Dossett, L. (1998). Acute inflammatory response in spinal cord following impact injury. *Experimental Neurology, 151,* 77–88.

Chang, H. T. (2007). Subacute human spinal cord contusion: few lymphocytes and many macrophages. *Spinal Cord, 45,* 174–182.

Chen, Q., Smith, G. M., & Shine, H. D. (2008). Immune activation is required for NT-3-induced axonal plasticity in chronic spinal cord injury. *Experimental Neurology, 209,* 497–509.

Coull, J. A., Beggs, S., Boudreau, D., Boivin, D., Tsuda, M., Inoue, K., et al. (2005). BDNF from microglia causes the shift in neuronal anion gradient underlying neuropathic pain. *Nature, 438,* 1017–1021.

Crown, E. D., Gwak, Y. S., Ye, Z., Johnson, K. M., & Hulsebosch, C. E. (2008). Activation of p38 MAP kinase is involved in central neuropathic pain following spinal cord injury. *Experimental Neurology, 213,* 257–267.

Davalos, D., Grutzendler, J., Yang, G., Kim, J. V., Zuo, Y., Jung, S., et al. (2005). ATP mediates rapid microglial response to local brain injury in vivo. *Nature Neuroscience, 8,* 752–758.

Davies, A. L., Hayes, K. C., & Dekaban, G. A. (2007). Clinical correlates of elevated serum concentrations of cytokines and autoantibodies in patients with spinal cord injury. *Archives of Physical Medicine and Rehabilitation, 88,* 1384–1393.

DeLeo, J. A., & Yezierski, R. P. (2001). The role of neuroinflammation and neuroimmune activation in persistent pain. *Pain, 90,* 1–6.

Detloff, M. R., Fisher, L. C., McGaughy, V., Longbrake, E. E., Popovich, P. G., & Basso, D. M. (2008). Remote activation of microglia and pro-inflammatory cytokines predict the onset and severity of below-level neuropathic pain after spinal cord injury in rats. *Experimental Neurology, 212,* 337–347.

Donnelly, D. J., & Popovich, P. G. (2008). Inflammation and its role in neuroprotection, axonal regeneration and functional recovery after spinal cord injury. *Experimental Neurology, 209,* 378–388.

Elkabes, S., DiCicco-Bloom, E. M., & Black, I. B. (1996). Brain microglia/macrophages express neurotrophins that selectively regulate microglial proliferation and function. *Journal of Neuroscience, 16,* 2508–2521.

Engelhardt, B., & Ransohoff, R. M. (2005). The ins and outs of T-lymphocyte trafficking to the CNS: anatomical sites and molecular mechanisms. *Trends in Immunology, 26,* 485–495.

Faden, A. I., & Simon, R. P. (1988). A potential role for excitotoxins in the pathophysiology of spinal cord injury. *Annals of Neurology, 23,* 623–626.

Farooque, M., Isaksson, J., & Olsson, Y. (1999). Improved recovery after spinal cord trauma in ICAM-1 and P-selectin knockout mice. *Neuroreport, 10,* 131–134.

Farooque, M., Isaksson, J., & Olsson, Y. (2001). White matter preservation after spinal cord injury in ICAM-1/P-selectin-deficient mice. *Acta Neuropathologica, 102,* 132–140.

Faulkner, J. R., Herrmann, J. E., Woo, M. J., Tansey, K. E., Doan, N. B., & Sofroniew, M. V. (2004). Reactive astrocytes protect tissue and preserve function after spinal cord injury. *Journal of Neuroscience, 24,* 2143–2155.

Fehlings, M. G., Tator, C. H., & Linden, R. D. (1989). The relationships among the severity of spinal cord injury, motor and somatosensory evoked potentials and spinal cord blood flow. *Electroencephalography and Clinical Neurophysiology, 74,* 241–259.

Festoff, B. W., Ameenuddin, S., Arnold, P. M., Wong, A., Santacruz, K. S., & Citron, B. A. (2006). Minocycline neuroprotects, reduces microgliosis, and inhibits caspase protease expression early after spinal cord injury. *Journal of Neurochemistry, 97*, 1314–1326.

Fleming, J. C., Norenberg, M. D., Ramsay, D. A., Dekaban, G. A., Marcillo, A. E., Saenz, A. D., et al. (2006). The cellular inflammatory response in human spinal cords after injury. *Brain, 129*, 3249–3269.

Gensel, J. C., Nakamura, S., Guan, Z., van Rooijen, N., Ankeny, D. P., & Popovich, P. G. (2009). Macrophages promote axon regeneration with concurrent neurotoxicity. *Journal of Neuroscience, 29*(12), 3956–3968.

Ghirnikar, R. S., Lee, Y. L., & Eng, L. F. (2000). Chemokine antagonist infusion attenuates cellular infiltration following spinal cord contusion injury in rat. *Journal of Neuroscience Research, 59*, 63–73.

Ghirnikar, R. S., Lee, Y. L., & Eng, L. F. (2001). Chemokine antagonist infusion promotes axonal sparing after spinal cord contusion injury in rat. *Journal of Neuroscience Research, 64*, 582–589.

Giulian, D., & Robertson, C. (1990). Inhibition of mononuclear phagocytes reduces ischemic injury in the spinal cord. *Annals of Neurology, 27*, 33–42.

Glaser, J., Gonzalez, R., Sadr, E., & Keirstead, H. S. (2006). Neutralization of the chemokine CXCL10 reduces apoptosis and increases axon sprouting after spinal cord injury. *Journal of Neuroscience Research, 84*, 724–734.

Gonzalez, R., Glaser, J., Liu, M. T., Lane, T. E., & Keirstead, H. S. (2003). Reducing inflammation decreases secondary degeneration and functional deficit after spinal cord injury. *Experimental Neurology, 184*, 456–463.

Gonzalez, R., Hickey, M. J., Espinosa, J. M., Nistor, G., Lane, T. E., & Keirstead, H. S. (2007). Therapeutic neutralization of CXCL10 decreases secondary degeneration and functional deficit after spinal cord injury in mice. *Regenerative Medicine, 2*, 771–783.

Gordon, S. (2003). Alternative activation of macrophages. *Nature Reviews Immunology, 3*, 23–35.

Gris, D., Marsh, D. R., Oatway, M. A., Chen, Y., Hamilton, E. F., Dekaban, G. A., et al. (2004). Transient blockade of the CD11d/CD18 integrin reduces secondary damage after spinal cord injury, improving sensory, autonomic, and motor function. *Journal of Neuroscience, 24*, 4043–4051.

Guo, G., & Bhat, N. R. (2007). p38alpha MAP kinase mediates hypoxia-induced motor neuron cell death: a potential target of minocycline's neuroprotective action. *Neurochemical Research, 32*, 2160–2166.

Guo, W., Wang, H., Watanabe, M., Shimizu, K., Zou, S., LaGraize, S. C., et al. (2007). Glial–cytokine–neuronal interactions underlying the mechanisms of persistent pain. *Journal of Neuroscience, 27*, 6006–6018.

Guzmán-Lenis, M. S., Vallejo, C., Navarro, X., & Casas, C. (2008). Analysis of FK506-mediated protection in an organotypic model of spinal cord damage: heat shock protein 70 levels are modulated in microglial cells. *Neuroscience, 155*, 104–113.

Haddad, E., McAlister, V., Renouf, E., Malthaner, R., Kjaer, M. S., & Gluud, L. L. (2006). Cyclosporin versus tacrolimus for liver transplanted patients. *Cochrane Database of Systematic Reviews*, Issue 4, Art. No. CD005161 (DOI:10.1002/14651858.CD005161.pub2).

Hains, B. C., & Waxman, S. G. (2006). Activated microglia contribute to the maintenance of chronic pain after spinal cord injury. *Journal of Neuroscience, 26*, 4308–4317.

Hamada, Y., Ikata, T., Katoh, S., Nakauchi, K., Niwa, M., Kawai, Y., et al. (1996). Involvement of an intercellular adhesion molecule 1-dependent pathway in the pathogenesis of secondary changes after spinal cord injury in rats. *Journal of Neurochemistry, 66*, 1525–1531.

Hamamoto, Y., Ogata, T., Morino, T., Hino, M., & Yamamoto, H. (2007). Real-time direct measurement of spinal cord blood flow at the site of compression: relationship between blood flow recovery and motor deficiency in spinal cord injury. *Spine, 32*, 1955–1962.

Hashizume, H., Rutkowski, M. D., Weinstein, J. N., & DeLeo, J. A. (2000). Central administration of methotrexate reduces mechanical allodynia in an animal model of radiculopathy/sciatica. *Pain, 87*, 159–169.

Hayashi, Y., Shumsky, J. S., Connors, T., Otsuka, T., Fischer, I., Tessler, A., et al. (2005). Immunosuppression with either cyclosporine a or FK506 supports survival of transplanted fibroblasts and promotes growth of host axons into the transplant after spinal cord injury. *Journal of Neurotrauma, 22*, 1267–1281.

Henry, C. J., Huang, Y., Wynne, A., Hanke, M., Himler, J., Bailey, M. T., et al. (2008). Minocycline attenuates lipopolysaccharide (LPS)-induced neuroinflammation, sickness behavior, and anhedonia. *Journal of Neuroinflammation, 5*, 15.

Henson, P. M., Bratton, D. L., & Fadok, V. A. (2001). The phosphatidylserine receptor: a crucial molecular switch? *Nature Reviews Molecular Cell Biology, 2*, 627–633.

Horiuchi, H., Ogata, T., Morino, T., Chuai, M., & Yamamoto, H. (2003). Continuous intrathecal infusion of SB203580, a selective inhibitor of p38 mitogen-activated protein kinase, reduces the damage of hind-limb function after thoracic spinal cord injury in rat. *Neuroscience Research, 47*, 209–217.

Horn, K. P., Busch, S. A., Hawthorne, A. L., van Rooijen, N., & Silver, J. (2008). Another barrier to regeneration in the CNS: activated macrophages induce extensive retraction of dystrophic axons through direct physical interactions. *Journal of Neuroscience, 28*, 9330–9341.

Hua, F., Ma, J., Ha, T., Kelley, J., Williams, D. L., Kao, R. L., et al. (2008). Preconditioning with a TLR2 specific ligand increases resistance to cerebral ischemia/reperfusion injury. *Journal of Neuroimmunology, 199*, 75–82.

Hua, F., Ma, J., Ha, T., Xia, Y., Kelley, J., Williams, D. L., et al. (2007). Activation of toll-like receptor 4 signaling contributes to hippocampal neuronal death following global cerebral ischemia/reperfusion. *Journal of Neuroimmunology, 190*, 101–111.

Hugenholtz, H. (2003). Methylprednisolone for acute spinal cord injury: not a standard of care. *CMAJ, 168*, 1145–1146.

Hurlbert, R. J., & Hamilton, M. G. (2008). Methylprednisolone for acute spinal cord injury: 5-year practice reversal. *Canadian Journal of Neurological Science, 35*, 41–45.

Hutchinson, M. R., Bland, S. T., Johnson, K. W., Rice, K. C., Maier, S. F., & Watkins, L. R. (2007). Opioid-induced glial activation: mechanisms of activation and implications for opioid analgesia, dependence, and reward. *ScientificWorld Journal*, 7, 98–111.

Ibarra, A., & Diaz-Ruiz, A. (2006). Protective effect of cyclosporin-A in spinal cord injury: an overview. *Current Medicinal Chemistry*, 13, 2703–2710.

James, S. E., Burden, H., Burgess, R., Xie, Y., Yang, T., Massa, S. M., et al. (2008). Anti-cancer drug induced neurotoxicity and identification of Rho pathway signaling modulators as potential neuroprotectants. *Neurotoxicology*, 29, 605–612.

Jones, T. B., Hart, R. P., & Popovich, P. G. (2005). Molecular control of physiological and pathological T-cell recruitment after mouse spinal cord injury. *Journal of Neuroscience*, 25, 6576–6583.

Kaminska, B., Gaweda-Walerych, K., & Zawadzka, M. (2004). Molecular mechanisms of neuroprotective action of immunosuppressants—facts and hypotheses. *Journal of Cellular and Molecular Medicine*, 8, 45–58.

Kaymaz, M., Emmez, H., Bukan, N., Dursun, A., Kurt, G., Pasaoglu, H., et al. (2005). Effectiveness of FK506 on lipid peroxidation in the spinal cord following experimental traumatic injury. *Spinal Cord*, 43, 22–26.

Kigerl, K. A., Lai, W., Rivest, S., Hart, R. P., Satoskar, A. R., & Popovich, P. G. (2007). Toll-like receptor (TLR)-2 and TLR-4 regulate inflammation, gliosis, and myelin sparing after spinal cord injury. *Journal of Neurochemistry*, 102, 37–50.

Kigerl, K. A., McGaughy, V. M., & Popovich, P. G. (2006). Comparative analysis of lesion development and intraspinal inflammation in four strains of mice following spinal contusion injury. *Journal of Comparative Neurology*, 494, 578–594.

Kim, D., Kim, M. A., Cho, I. H., Kim, M. S., Lee, S., Jo, E. K., et al. (2007). A critical role of toll-like receptor 2 in nerve injury-induced spinal cord glial cell activation and pain hypersensitivity. *Journal of Biological Chemistry*, 282, 14975–14983.

Kim, G. M., Xu, J., Xu, J., Song, S. K., Yan, P., Ku, G., et al. (2001). Tumor necrosis factor receptor deletion reduces nuclear factor-kappaB activation, cellular inhibitor of apoptosis protein 2 expression, and functional recovery after traumatic spinal cord injury. *Journal of Neuroscience*, 21, 6617–6625.

Koizumi, S., Shigemoto-Mogami, Y., Nasu-Tada, K., Shinozaki, Y., Ohsawa, K., Tsuda, M., et al. (2007). UDP acting at P2Y6 receptors is a mediator of microglial phagocytosis. *Nature*, 446, 1091–1095.

Kostyk, S. K., Popovich, P. G., Stokes, B. T., Wei, P., & Jakeman, L. B. (2008). Robust axonal growth and a blunted macrophage response are associated with impaired functional recovery after spinal cord injury in the MRL/MpJ mouse. *Neuroscience*, 156, 498–514.

Koyanagi, I., Iwasaki, Y., Isu, T., Akino, M., & Abe, H. (1989). Significance of spinal cord swelling in the prognosis of acute cervical spinal cord injury. *Paraplegia*, 27, 190–197.

Kurt, G., Ergun, E., Cemil, B., Borcek, A. O., Borcek, P., Gulbahar, O., et al. (2009). Neuroprotective effects of infliximab in experimental spinal cord injury. *Surgical Neurology*, 71(3), 332–336.

Lazarov-Spiegler, O., Solomon, A. S., & Schwartz, M. (1998). Peripheral nerve-stimulated macrophages simulate a peripheral nerve-like regenerative response in rat transected optic nerve. *Glia*, 24, 329–337.

Lazarov-Spiegler, O., Solomon, A. S., Zeev-Brann, A. B., Hirschberg, D. L., Lavie, V., & Schwartz, M. (1996). Transplantation of activated macrophages overcomes central nervous system regrowth failure. *FASEB Journal*, 10, 1296–1302.

Lee, H. C., Cho, D. Y., Lee, W. Y., & Chuang, H. C. (2007). Pitfalls in treatment of acute cervical spinal cord injury using high-dose methylprednisolone: a retrospect audit of 111 patients. *Surgical Neurology*, 68(Suppl. 1), S37–S41; discussion S41-42.

Lee, M., Lee, E. S., Kim, Y. S., Choi, B. H., Park, S. R., Park, H. S., et al. (2005). Ischemic injury-specific gene expression in the rat spinal cord injury model using hypoxia-inducible system. *Spine*, 30, 2729–2734.

Lee, J. M., Yan, P., Xiao, Q., Chen, S., Lee, K. Y., Hsu, C. Y., et al. (2008). Methylprednisolone protects oligodendrocytes but not neurons after spinal cord injury. *Journal of Neuroscience*, 28, 3141–3149.

Lee, Y. B., Yune, T. Y., Baik, S. Y., Shin, Y. H., Du, S., Rhim, H., et al. (2000). Role of tumor necrosis factor-alpha in neuronal and glial apoptosis after spinal cord injury. *Experimental Neurology*, 166, 190–195.

Lee, S. M., Yune, T. Y., Kim, S. J., Park, D. W., Lee, Y. K., Kim, Y. C., et al. (2003). Minocycline reduces cell death and improves functional recovery after traumatic spinal cord injury in the rat. *Journal of Neurotrauma*, 20, 1017–1027.

Lehnardt, S., Schott, E., Trimbuch, T., Laubisch, D., Krueger, C., Wulczyn, G., et al. (2008). A vicious cycle involving release of heat shock protein 60 from injured cells and activation of toll-like receptor 4 mediates neurodegeneration in the CNS. *Journal of Neuroscience*, 28, 2320–2331.

Leon, S., Yin, Y., Nguyen, J., Irwin, N., & Benowitz, L. I. (2000). Lens injury stimulates axon regeneration in the mature rat optic nerve. *Journal of Neuroscience*, 20, 4615–4626.

Liang, D., Bhatta, S., Gerzanich, V., & Simard, J. M. (2007). Cytotoxic edema: mechanisms of pathological cell swelling. *Neurosurgical Focus*, 22, E2.

Liu, S. Q., Ma, Y. G., Peng, H., & Fan, L. (2005). Monocyte chemoattractant protein-1 level in serum of patients with acute spinal cord injury. *Chinese Journal of Traumatology*, 8, 216–219.

López-Vales, R., Forés, J., Navarro, X., & Verdú, E. (2006). Olfactory ensheathing glia graft in combination with FK506 administration promote repair after spinal cord injury. *Neurobiology of Disease*, 24, 443–454.

López-Vales, R., Garcia-Alias, G., Fores, J., Udina, E., Gold, B. G., Navarro, X., et al. (2005). FK 506 reduces tissue damage and prevents functional deficit after spinal cord injury in the rat. *Journal of Neuroscience Research, 81,* 827–836.

Ma, M., Wei, P., Wei, T., Ransohoff, R. M., & Jakeman, L. B. (2004). Enhanced axonal growth into a spinal cord contusion injury site in a strain of mouse (129 × 1/SvJ) with a diminished inflammatory response. *Journal of Comparative Neurology, 474,* 469–486.

Mabon, P. J., Weaver, L. C., & Dekaban, G. A. (2000). Inhibition of monocyte/macrophage migration to a spinal cord injury site by an antibody to the integrin alphaD: a potential new anti-inflammatory treatment. *Experimental Neurology, 166,* 52–64.

Marchand, F., Tsantoulas, C., Singh, D., Grist, J., Clark, A. K., Bradbury, E. J., et al. (2008). Effects of etanercept and minocycline in a rat model of spinal cord injury. *European Journal of Pain, 12,* in press, doi:10.1016/j.ejpain.2008.08.001.

Marta, M., Andersson, A., Isaksson, M., Kampe, O., & Lobell, A. (2008). Unexpected regulatory roles of TLR4 and TLR9 in experimental autoimmune encephalomyelitis. *European Journal of Immunology, 38,* 565–575.

Mason, J. L., Suzuki, K., Chaplin, D. D., & Matsushima, G. K. (2001). Interleukin-1beta promotes repair of the CNS. *Journal of Neuroscience, 21,* 7046–7052.

Matzinger, P. (2002). The danger model: a renewed sense of self. *Science, 296,* 301–305.

McTigue, D. M., Tani, M., Krivacic, K., Chernosky, A., Kelner, G. S., Maciejewski, D., et al. (1998). Selective chemokine mRNA accumulation in the rat spinal cord after contusion injury. *Journal of Neuroscience Research, 53,* 368–376.

Meagher, L. C., Savill, J. S., Baker, A., Fuller, R. W., & Haslett, C. (1992). Phagocytosis of apoptotic neutrophils does not induce macrophage release of thromboxane B2. *Journal of Leukocyte Biology, 52,* 269–273.

Means, E. D., & Anderson, D. K. (1983). Neuronophagia by leukocytes in experimental spinal cord injury. *Journal of Neuropathology and Experimental Neurology, 42,* 707–719.

Myer, D. J., Gurkoff, G. G., Lee, S. M., Hovda, D. A., & Sofroniew, M. V. (2006). Essential protective roles of reactive astrocytes in traumatic brain injury. *Brain, 129,* 2761–2772.

Nagy, Z., Kolev, K., Csonka, E., Vastag, M., & Machovich, R. (1998). Perturbation of the integrity of the blood-brain barrier by fibrinolytic enzymes. *Blood Coagulation and Fibrinolysis, 9,* 471–478.

Nathan, C. (2006). Neutrophils and immunity: challenges and opportunities. *Nature Reviews Immunology, 6,* 173–182.

Nesic, O., Xu, G. Y., McAdoo, D., High, K. W., Hulsebosch, C., & Perez-Pol, R. (2001). IL-1 receptor antagonist prevents apoptosis and caspase-3 activation after spinal cord injury. *Journal of Neurotrauma, 18,* 947–956.

Nguyen, H. X., Galvan, M. D., & Anderson, A. J. (2008). Characterization of early and terminal complement proteins associated with polymorphonuclear leukocytes in vitro and in vivo after spinal cord injury. *Journal of Neuroinflammation, 5,* 26.

Noble, L. J., Donovan, F., Igarashi, T., Goussev, S., & Werb, Z. (2002). Matrix metalloproteinases limit functional recovery after spinal cord injury by modulation of early vascular events. *Journal of Neuroscience, 22,* 7526–7535.

Nottingham, S., Knapp, P., & Springer, J. (2002). FK506 treatment inhibits caspase-3 activation and promotes oligodendroglial survival following traumatic spinal cord injury. *Experimental Neurology, 177,* 242–251.

O'Carroll, S. J., Alkadhi, M., Nicholson, L. F., & Green, C. R. (2008). Connexin 43 mimetic peptides reduce swelling, astrogliosis, and neuronal cell death after spinal cord injury. *Cell Communication and Adhesion, 15,* 27–42.

Okada, S., Nakamura, M., Katoh, H., Miyao, T., Shimazaki, T., Ishii, K., et al. (2006). Conditional ablation of Stat3 or Socs3 discloses a dual role for reactive astrocytes after spinal cord injury. *Nature Medicine, 12,* 829–834.

Pender, M. P., & Rist, M. J. (2001). Apoptosis of inflammatory cells in immune control of the nervous system: role of glia. *Glia, 36,* 137–144.

Perrin, F. E., Lacroix, S., Aviles-Trigueros, M., & David, S. (2005). Involvement of monocyte chemoattractant protein-1, macrophage inflammatory protein-1alpha and interleukin-1beta in Wallerian degeneration. *Brain, 128,* 854–866.

Persson, M., Brantefjord, M., Hansson, E., & Ronnback, L. (2005). Lipopolysaccharide increases microglial GLT-1 expression and glutamate uptake capacity in vitro by a mechanism dependent on TNF-alpha. *Glia, 51,* 111–120.

Pineau, I., & Lacroix, S. (2008). Endogenous signals initiating inflammation in the injured nervous system. *Glia, 57,* 351–361.

Pinzon, A., Marcillo, A., Quintana, A., Stamler, S., Bunge, M. B., Bramlett, H. M., et al. (2008). A re-assessment of minocycline as a neuroprotective agent in a rat spinal cord contusion model. *Brain Research, 1243,* 146–151.

Pointillart, V., Petitjean, M. E., Wiart, L., Vital, J. M., Lassie, P., Thicoipe, M., et al. (2000). Pharmacological therapy of spinal cord injury during the acute phase. *Spinal Cord, 38,* 71–76.

Popovich, P. G., Guan, Z., McGaughy, V., Fisher, L., Hickey, W. F., & Basso, D. M. (2002). The neuropathological and behavioral consequences of intraspinal microglial/macrophage activation. *Journal of Neuropathology and Experimental Neurology, 61,* 623–633.

Popovich, P. G., Guan, Z., Wei, P., Huitinga, I., van Rooijen, N., & Stokes, B. T. (1999). Depletion of hematogenous macrophages promotes partial hindlimb recovery and neuroanatomical repair after experimental spinal cord injury. *Experimental Neurology, 158,* 351–365.

Popovich, P. G., & Longbrake, E. E. (2008). Can the immune system be harnessed to repair the CNS? *Nature Reviews Neuroscience, 9,* 481–493.

Popovich, P. G., van Rooijen, N., Hickey, W. F., Preidis, G., & McGaughy, V. (2003). Hematogenous macrophages express CD8 and distribute to regions of lesion cavitation after spinal cord injury. *Experimental Neurology, 182,* 275–287.

Popovich, P. G., Wei, P., & Stokes, B. T. (1997). Cellular inflammatory response after spinal cord injury in Sprague-Dawley and Lewis rats. *Journal of Comparative Neurology, 377*, 443–464.

Rabchevsky, A. G., Fugaccia, I., Sullivan, P. G., & Scheff, S. W. (2001). Cyclosporin A treatment following spinal cord injury to the rat: behavioral effects and stereological assessment of tissue sparing. *Journal of Neurotrauma, 18*, 513–522.

Ranganathan, P. (2008). An update on methotrexate pharmacogenetics in rheumatoid arthritis. *Pharmacogenomics, 9*, 439–451.

Rapalino, O., Lazarov-Spiegler, O., Agranov, E., Velan, G. J., Yoles, E., Fraidakis, M., et al. (1998). Implantation of stimulated homologous macrophages results in partial recovery of paraplegic rats. *Nature Medicine, 4*, 814–821.

Rice, T., Larsen, J., Rivest, S., & Yong, V. W. (2007). Characterization of the early neuroinflammation after spinal cord injury in mice. *Journal of Neuropathology and Experimental Neurology, 66*, 184–195.

Ritz, M. F., & Hausmann, O. N. (2008). Effect of 17beta-estradiol on functional outcome, release of cytokines, astrocyte reactivity and inflammatory spreading after spinal cord injury in male rats. *Brain Research, 1203*, 177–188.

Saadoun, S., Bell, B. A., Verkman, A. S., & Papadopoulos, M. C. (2008). Greatly improved neurological outcome after spinal cord compression injury in AQP4-deficient mice. *Brain, 131*, 1087–1098.

Savill, J. S., Wyllie, A. H., Henson, J. E., Walport, M. J., Henson, P. M., & Haslett, C. (1989). Macrophage phagocytosis of aging neutrophils in inflammation. Programmed cell death in the neutrophil leads to its recognition by macrophages. *Journal of Clinical Investigation, 83*, 865–875.

Sayer, F. T., Kronvall, E., & Nilsson, O. G. (2006). Methylprednisolone treatment in acute spinal cord injury: the myth challenged through a structured analysis of published literature. *Spine Journal, 6*, 335–343.

Schaefer, L., Babelova, A., Kiss, E., Hausser, H. J., Baliova, M., Krzyzankova, M., et al. (2005). The matrix component biglycan is proinflammatory and signals through Toll-like receptors 4 and 2 in macrophages. *Journal of Clinical Investigation, 115*, 2223–2233.

Scholz, J., Abele, A., Marian, C., Haussler, A., Herbert, T. A., Woolf, C. J., et al. (2008). Low-dose methotrexate reduces peripheral nerve injury-evoked spinal microglial activation and neuropathic pain behavior in rats. *Pain, 138*, 130–142.

Scholz, M., Cinatl, J., Schadel-Hopfner, M., & Windolf, J. (2007). Neutrophils and the blood-brain barrier dysfunction after trauma. *Medicinal Research Reviews, 27*, 401–416.

Schonberg, D. L., Popovich, P. G., & McTigue, D. M. (2007). Oligodendrocyte generation is differentially influenced by toll-like receptor (TLR) 2 and TLR4-mediated intraspinal macrophage activation. *Journal of Neuropathology and Experimental Neurology, 66*, 1124–1135.

Schwartz, M., & Yoles, E. (2006). Immune-based therapy for spinal cord repair: autologous macrophages and beyond. *Journal of Neurotrauma, 23*, 360–370.

Sekhon, L. H., & Fehlings, M. G. (2001). Epidemiology, demographics, and pathophysiology of acute spinal cord injury. *Spine, 26*, S2–S12.

Short, D. J., El Masry, W. S., & Jones, P. W. (2000). High dose methylprednisolone in the management of acute spinal cord injury—a systematic review from a clinical perspective. *Spinal Cord, 38*, 273–286.

Simard, J. M., Tsymbalyuk, O., Ivanov, A., Ivanova, S., Bhatta, S., Geng, Z., et al. (2007). Endothelial sulfonylurea receptor 1-regulated NC Ca-ATP channels mediate progressive hemorrhagic necrosis following spinal cord injury. *Journal of Clinical Investigation, 117*, 2105–2113.

Sosa, I., Reyes, O., & Kuffler, D. P. (2005). Immunosuppressants: neuroprotection and promoting neurological recovery following peripheral nerve and spinal cord lesions. *Experimental Neurology, 195*, 7–15.

Sroga, J. M., Jones, T. B., Kigerl, K. A., McGaughy, V. M., & Popovich, P. G. (2003). Rats and mice exhibit distinct inflammatory reactions after spinal cord injury. *Journal of Comparative Neurology, 462*, 223–240.

Stirling, D. P., Khodarahmi, K., Liu, J., McPhail, L. T., McBride, C. B., Steeves, J. D., et al. (2004). Minocycline treatment reduces delayed oligodendrocyte death, attenuates axonal dieback, and improves functional outcome after spinal cord injury. *Journal of Neuroscience, 24*, 2182–2190.

Stirling, D. P., & Yong, V. W. (2008). Dynamics of the inflammatory response after murine spinal cord injury revealed by flow cytometry. *Journal of Neuroscience Research, 86*, 1944–1958.

Stout, R. D., Jiang, C., Matta, B., Tietzel, I., Watkins, S. K., & Suttles, J. (2005). Macrophages sequentially change their functional phenotype in response to changes in microenvironmental influences. *Journal of Immunology, 175*, 342–349.

Suberviola, B., Gonzalez-Castro, A., Llorca, J., Ortiz-Melon, F., & Minambres, E. (2008). Early complications of high-dose methylprednisolone in acute spinal cord injury patients. *Injury, 39*, 748–752.

Tanga, F. Y., Nutile-McMenemy, N., & DeLeo, J. A. (2005). The CNS role of toll-like receptor 4 in innate neuroimmunity and painful neuropathy. *Proceedings of the National Academy of Sciences of the United States of America, 102*, 5856–5861.

Taoka, Y., Okajima, K., Uchiba, M., Murakami, K., Kushimoto, S., Johno, M., et al. (1997). Role of neutrophils in spinal cord injury in the rat. *Neuroscience, 79*, 1177–1182.

Temporin, K., Tanaka, H., Kuroda, Y., Okada, K., Yachi, K., Moritomo, H., et al. (2008a). IL-1beta promotes neurite outgrowth by deactivating RhoA via p38 MAPK pathway. *Biochemical and Biophysical Research Communications, 365*, 375–380.

Temporin, K., Tanaka, H., Kuroda, Y., Okada, K., Yachi, K., Moritomo, H., et al. (2008b). Interleukin-1 beta promotes sensory nerve regeneration after sciatic nerve injury. *Neuroscience Letters*, *440*, 130–133.

Teng, Y. D., Choi, H., Onario, R. C., Zhu, S., Desilets, F. C., Lan, S., et al. (2004). Minocycline inhibits contusion-triggered mitochondrial cytochrome c release and mitigates functional deficits after spinal cord injury. *Proceedings of the National Academy of Sciences of the United States of America*, *101*, 3071–3076.

Termeer, C., Benedix, F., Sleeman, J., Fieber, C., Voith, U., Ahrens, T., et al. (2002). Oligosaccharides of hyaluronan activate dendritic cells via toll-like receptor 4. *Journal of Experimental Medicine*, *195*, 99–111.

Tikka, T. M., & Koistinaho, J. E. (2001). Minocycline provides neuroprotection against N-methyl-d-aspartate neurotoxicity by inhibiting microglia. *Journal of Immunology*, *166*, 7527–7533.

Vallieres, N., Berard, J. L., David, S., & Lacroix, S. (2006). Systemic injections of lipopolysaccharide accelerates myelin phagocytosis during Wallerian degeneration in the injured mouse spinal cord. *Glia*, *53*, 103–113.

Vargas, M. E., & Barres, B. A. (2007). Why is Wallerian degeneration in the CNS so slow? *Annual Review of Neuroscience*, *30*, 153–179.

Vogel, C., Stallforth, S., & Sommer, C. (2006). Altered pain behavior and regeneration after nerve injury in TNF receptor deficient mice. *Journal of the Peripheral Nervous System*, *11*, 294–303.

Wang, X. J., Kong, K. M., Qi, W. L., Ye, W. L., & Song, P. S. (2005). Interleukin-1 beta induction of neuron apoptosis depends on p38 mitogen-activated protein kinase activity after spinal cord injury. *Acta Pharmacologica Sinica*, *26*, 934–942.

Weber, C., Fraemohs, L., & Dejana, E. (2007). The role of junctional adhesion molecules in vascular inflammation. *Nature Reviews Immunology*, *7*, 467–477.

Wells, J. E., Hurlbert, R. J., Fehlings, M. G., & Yong, V. W. (2003). Neuroprotection by minocycline facilitates significant recovery from spinal cord injury in mice. *Brain*, *126*, 1628–1637.

Wieseler-Frank, J., Maier, S. F., & Watkins, L. R. (2005). Immune-to-brain communication dynamically modulates pain: physiological and pathological consequences. *Brain, Behavior and Immunity*, *19*, 104–111.

Wrathall, J. R., Teng, Y. D., & Marriott, R. (1997). Delayed antagonism of AMPA/kainate receptors reduces long-term functional deficits resulting from spinal cord trauma. *Experimental Neurology*, *145*, 565–573.

Xu, L., Fagan, S. C., Waller, J. L., Edwards, D., Borlongan, C. V., Zheng, J., et al. (2004). Low dose intravenous minocycline is neuroprotective after middle cerebral artery occlusion-reperfusion in rats. *BMC Neurology*, *4*, 7.

Xu, Z., Wang, B.-R., Wang, X., Kuang, F., Duan, X.-L., Jiao, X.-Y., et al. (2006). ERK1/2 and p38 mitogen-activated protein kinase mediate iNOS-induced spinal neuron degeneration after acute traumatic spinal cord injury. *Life Sciences*, *79*, 1895–1905.

Yang, L., Blumbergs, P. C., Jones, N. R., Manavis, J., Sarvestani, G. T., & Ghabriel, M. N. (2004). Early expression and cellular localization of proinflammatory cytokines interleukin-1beta, interleukin-6, and tumor necrosis factor-alpha in human traumatic spinal cord injury. *Spine*, *29*, 966–971.

Yin, Y., Cui, Q., Li, Y., Irwin, N., Fischer, D., Harvey, A. R., et al. (2003). Macrophage-derived factors stimulate optic nerve regeneration. *Journal of Neuroscience*, *23*, 2284–2293.

Yune, T. Y., Lee, J. Y., Jung, G. Y., Kim, S. J., Jiang, M. H., Kim, Y. C., et al. (2007). Minocycline alleviates death of oligodendrocytes by inhibiting pro-nerve growth factor production in microglia after spinal cord injury. *Journal of Neuroscience*, *27*, 7751–7761.

Zang, D. W., & Cheema, S. S. (2003). Leukemia inhibitory factor promotes recovery of locomotor function following spinal cord injury in the mouse. *Journal of Neurotrauma*, *20*, 1215–1222.

Zhao, P., Waxman, S. G., & Hains, B. C. (2007). Modulation of thalamic nociceptive processing after spinal cord injury through remote activation of thalamic microglia by cysteine cysteine chemokine ligand 21. *Journal of Neuroscience*, *27*, 8893–8902.

CHAPTER 9

# Toll-like receptors in the CNS: implications for neurodegeneration and repair

Johannes M. van Noort* and Malika Bsibsi

*Delta Crystallon BV, Leiden, The Netherlands*

**Abstract:** The role of Toll-like receptors (TLRs) in the CNS is only starting to be uncovered. As in peripheral organs, multiple TLRs are dynamically expressed. They are involved in mounting a host-defense response against microbial invasion of the CNS. The many different TLRs expressed on microglia are likely the most important first line of defense in this respect. Intriguingly, microglial TLR tend to trigger a very standard cytokine and chemokine response, irrespective of the type of TLR agonist they meet. The main purpose of this standardized response by microglia may be to recruit the assistance by other cells rather than to immediately mount a destructive response toward invaders. As is generally the case for microglial responses, TLR-mediated responses can also work out in either beneficial or detrimental ways, depending on the strength and timing of the activating signal.

Yet, the role of TLRs in the CNS extends well beyond controlling host-defense responses alone. Other cells in the CNS, including astrocytes, neurons, and oligodendrocytes, can also express multiple functional TLRs upon activation. These play important roles in tissue development, cellular migration, and differentiation; in limiting inflammation; and in mounting repair processes following trauma. The TLR-mediated reactions of these other neural cells to TLR agonists is highly cell specific and does not necessarily resemble that of microglia at all. It appears likely that endogenous agonists for TLRs are particularly relevant to activate these endogenous TLR functions on neural cells, also during development when microbial invaders have not yet entered the stage. In this chapter, current data are reviewed to highlight the emerging variety of functional roles of TLRs in the CNS.

**Keywords:** toll-like receptors; microglia; astrocytes; neurons; cytokines

## Introduction and scope of this chapter

TLRs comprise a family of at least 11 conserved pattern-recognition receptors that mediate diverse cellular responses to a large variety of structures, often of microbial origin. Upon activation by typically microbial structures including lipopolysaccharides, teichoic acids, peptidoglycans, and various forms of nucleic acids, TLR family members trigger innate inflammatory responses that are usually dominated by the NF-κB-mediated production of cytokines such as TNF-α and IL-1β, and chemokines including IL-8 and CCL5. It is no surprise, therefore, that the

*Corresponding author.
Tel.: +31 71 518 1541; Fax: +31 71 518 1901;
E-mail: hans.vannoort@tno.nl

DOI: 10.1016/S0079-6123(09)17509-X

TLR-mediated cellular response — also in the CNS — is often understood to be a pro-inflammatory host-defense response, primarily designed to eliminate invading pathogens. Yet, TLRs activate many more types of cellular responses than only host-defense responses. TLRs also appear on neural cells under conditions where no foreign invaders are obvious, for example, in the CNS of people with Alzheimer's disease (Walter et al., 2007) or multiple sclerosis (Bsibsi et al., 2002). Evidence is accumulating that especially in response to endogenous molecular ligands, TLRs also play a role in tissue development, cellular migration and differentiation, and repair processes. Especially these more subtle and beneficial roles of TLR functions may well be very relevant to the CNS. After all, even in neural cell types including neurons and oligodendrocytes, cells that do not play an active role in destructive inflammatory responses, functional TLRs, are indeed expressed. Even on astrocytes, TLRs can activate production of a wide range of neuroprotective and anti-inflammatory mediators rather than merely stimulating pro-inflammatory factors. A one-dimensional view of the role of TLRs in the CNS would therefore be underestimating their clearly multifaceted functional roles.

In this chapter, current evidence for the expression and functional roles of TLRs on major cell types in the CNS will be reviewed. It will be separately discussed what is currently known about the role of TLRs in microglia, astrocytes, and neurons. This will illustrate how variable TLR-mediated responses can be depending on the cell type. It must be kept in mind, however, that research on the role of TLRs in the CNS has only just begun, and many questions are still unanswered. Oligodendrocytes, for example, are known to express TLRs at least in cell culture (Bsibsi et al., 2002) but functional data on oligodendroglial TLRs remain lacking. Yet, the evidence that has become available over the past years begins to paint an exciting picture of TLRs as a dynamically expressed family of receptors that play important roles not only in defending the CNS against pathogens but also in shaping the CNS itself during development, and repairing it after damage.

## General features of TLRs

The family of TLRs share structural properties not only in the extracellular leucine-rich repeat structures designed to register the presence of multiple ligands but also in their intracellular domains which interact with intracellular adaptor proteins that relay the agonist engagement signal. Currently, five of such adaptors are known. The dominant and founding member of the family of these adaptors is MyD88 (or MyD88-1), which relays the signal for most TLR family members and tends to predominantly induce NF-κB-mediated activation of genes, including those encoding TNF-α, CCL5 (RANTES), IL-1β, and CXCL8. This dominant pathway has been particularly well characterized in myelomonocytic cells, which explains why TLR-mediated responses are generally portrayed as pro-inflammatory. Yet, signaling by TLR is much more complex than this. The intracellular domain of TLR3, for example, is slightly different from that of the others and unable to interact with MyD88-1. It can only bind to a variant member of the adaptor family called MyD88-3, also referred to as TRIF/TICAM-1. Activation of this signaling route by TLR3 leads to the dominant induction of interferon response factors instead of NF-κB and subsequent activation of genes encoding type-I interferons including interferon-α and -β. Also TLR4 can interact with MyD88-3, but TLR3 is particularly dominant in doing so. The way TLRs interact with each of the five different adaptors is not just controlled by structural preferences embedded in the structure of the intracellular domain of TLRs, but also by the mere availability of adaptors. Not every cell expresses the same set. For example, it is rapidly becoming clear that selective expression of the less frequently used adaptor MyD88-5 in neurons renders these cells uniquely sensitive to TLR-mediated activation of the JNK pathway to apoptosis, instead of the NF-κB pathway to inflammatory mediators. In this way, selective expression of adaptors strongly influences the quality of the response mounted by different types of cells to a given TLR agonist.

Very often, activation of TLR-mediated signaling by various agonists does not involve a straightforward key-and-lock mode of ligand–receptor binding. That extracellular domains of all TLRs share important structural features, yet mediate responses to widely different agonists, already clarifies that more complex interactions are involved. Recognition of double-stranded RNA by TLR3 is relatively simple, involving two separate TLR3 extracellular domains acting in concert to bind one double-stranded RNA helix. In general, however, many different additional proteins are required for the activation of TLR-mediated signaling by their agonists, including co-receptors and docking molecules on the cell surface and binding catalysts, such as CD14 and heat shock proteins, that promote certain interactions (Bsibsi et al., 2007; Jou et al., 2006; Lehnardt et al., 2008). This interplay with several additional proteins is probably crucial in allowing structurally conserved TLRs to respond to myriad different agonists that share no obvious structural similarity. In addition to the essential contribution by co-receptors and accessory interaction partners, most TLRs also operate as homo- or heterodimers. In short, activation of TLR-mediated signaling is far more complex than simply having one agonist molecule bound to the extracellular domain of one TLR. For many of the currently known TLR-signaling scenarios, different partners have already been identified. For many other scenarios, details remain to be clarified.

**Expression and function of TLRs in microglia**

Much like other macrophage-like cells, microglia can express essentially all different TLR family members. While TLR expression is hardly detectable in resting microglia in a healthy CNS, multiple TLRs rapidly appear upon activation of the cells. As in macrophages, TLRs are exclusively found within endosomal vesicles of microglia, illustrating their prime role in probing anything taken up by phagocytosis. While microglia express all TLRs at readily detectable levels (Bsibsi et al., 2002), TLRs 1–4 are the most dominant ones and, in particular, TLR2 is induced at somewhat higher levels than other family members; this appears to apply to microglia in rodents as well as in humans.

There is ample evidence to indicate that microglial TLRs are crucial as a first line of defense against bacterial or viral infection. In response to the appearance of multiple bacterial or viral TLR agonists, TLR-mediated signaling promotes production of a variety of inflammatory mediators (reviewed by Konat et al., 2006; Block et al., 2007). Also, phagocytosis is stimulated by TLR activation, which may be particularly relevant to clearance of bacteria, as well as of aggregated or abnormal proteins such as amyloid fibers from the CNS (Chen et al., 2006; Tahara et al., 2006). As in most cases of microglial activation, a balance exists between a mild TLR-mediated response by microglia, which is beneficial in combating infection, clearing debris, and promoting repair, and an exaggerated or persistent response that contributes to tissue damage and neurodegenerative processes. Thus, microglial activation via TLRs has been found to exert beneficial effects as well as deleterious effects depending on the context and strength of the challenge applied.

In studies of experimental cerebral ischemia, for example, such apparently contradictory effects have been clearly highlighted. Upon induction of experimental ischemia in mice, both TLR2 and 4 can markedly contribute to tissue damage as mice deficient for either of these TLRs develop much less brain damage and display increased neuronal survival (Lehnardt et al., 2007; Ziegler et al., 2007; Kilic et al., 2008; Caso et al., 2008). Yet, exposure of experimental animals to agonists for TLR2, 4, or 9 prior to induction of ischemia surprisingly increases their resistance to injury caused by cerebral ischemia and/or reperfusion (Hua et al., 2008; Stevens et al., 2008). Clearly, conditioning of the CNS by prior TLR activation limits the detrimental effects of subsequent inflammatory challenge. This effect is also well known in other situations where macrophage activation is key to disease development, and relies on TLR-induced downregulation of many different genes including those controlling TLR functions themselves (Mages et al., 2007). Thus, different timing of TLR activation relative to induction of ischemia can lead to radically different outcomes.

The TLR-mediated response after experimental nerve damage tends to work out more beneficially in most cases. Absence of either TLR2- or TLR4-mediated signaling after peripheral nerve injury results not only in delayed recruitment of macrophages and clearance of cellular debris but also in delayed axonal regeneration and locomotor recovery (Boivin et al., 2007). Conversely, activation of TLR signaling in this model promotes these regenerative effects. Also in the CNS these post-trauma effects of at least TLR2 and 4 are apparent. Their activation is beneficial after spinal cord injury (Kigerl et al., 2007), or surfactant-mediated experimental cytolysis (Glezer et al., 2006). Under such experimental conditions where pathogens are not in play, both TLRs play important roles in regulating gliosis and tissue repair processes and promote mobilization of oligodendrocyte progenitors. The full range of such microglia-mediated effects may still not be uncovered. Apart from controlling microbial invasion and neurodegeneration in response to various types of cellular injury including Wallerian degeneration (Lehnardt et al., 2008; Boivin et al., 2007), microglial TLR are also important for control of painful neuropathies (Tanga et al., 2005; Kim et al., 2007a, b), and for migration and differentiation of neural precursors (Aarum et al., 2003).

Intuitively, one might perhaps assume that the response by microglia is diverse and tailored to meet the unique defensive demands to counteract each structural challenge. Indeed, it has been suggested that viral agonists would trigger different types of microglial responses from bacterial stimuli and different responses would again be observed in responses to endogenous TLR agonists (Jack et al., 2005). Yet, other data indicate that differential responsiveness may very well not be a dominant trait of the microglial response. At least in culture, human microglia are not very sensitive at all to the type of stimulus they meet, and just respond to the mere fact that there is a stimulus. As illustrated in Fig. 1, activation of human microglia by agonists for either TLRs 3, 4, 8, or 9 produces hardly any differences in the patterns of cytokines and chemokines that are induced. In all of these four cases examined in detail, IL-13, TNF-$\alpha$, CCL5 (RANTES), CXCL10 (IP-10), IL-12p40, IL-15, IL-6 sR, TGF-$\beta$, and IL-10 are induced at rather similar levels with only few exceptions. In no case there is a unique mediator induced by a single TLR agonist, which is also not induced by other agonists.

Clearly, these first results should be extended to other response pathways as well. Yet, they do suggest that the prime role of microglia is not to combat a variety of microbial threats by sophisticated and differentiated response profiles tailored to fight off different challenges in different ways. Instead, microglia tend to "raise the red flag" and appear to simply mount a single common response to structurally different types of insults, for other mechanisms to take over the actual tailored host-defense role. Given the prominent presence of both TNF-$\alpha$ (as a prime activator of blood–brain barrier endothelial cells) and several chemokines that will help attract peripheral blood leukocytes (including CCL2, CCL5, and CXCL10) in the set of TLR-induced mediators, microglia appear to be particularly inclined to recruit the assistance by the peripheral immune repertoire to help resolve inflammatory insults or microbial invasion. On the other hand, many cells in the CNS including glial cells themselves also express receptors for the various chemokines secreted by activated microglia in response to TLR agonists. Therefore, TLR-mediated activation of microglia is also likely to trigger a wave of migratory activity by glial cells themselves to the site of insult or injury. Microglia do not seem to be particularly well equipped to solve any insult themselves, but rather call upon other cells to do this.

**Expression and function of TLRs in astrocytes**

Astrocytes from both human and rodent origin express TLRs1–4 in readily detectable amounts, but much lower levels of the other TLRs. Yet, the TLR profile of astrocytes is not a fixed quality, but a highly dynamic feature. In a healthy human CNS, TLR expression on astrocytes is very

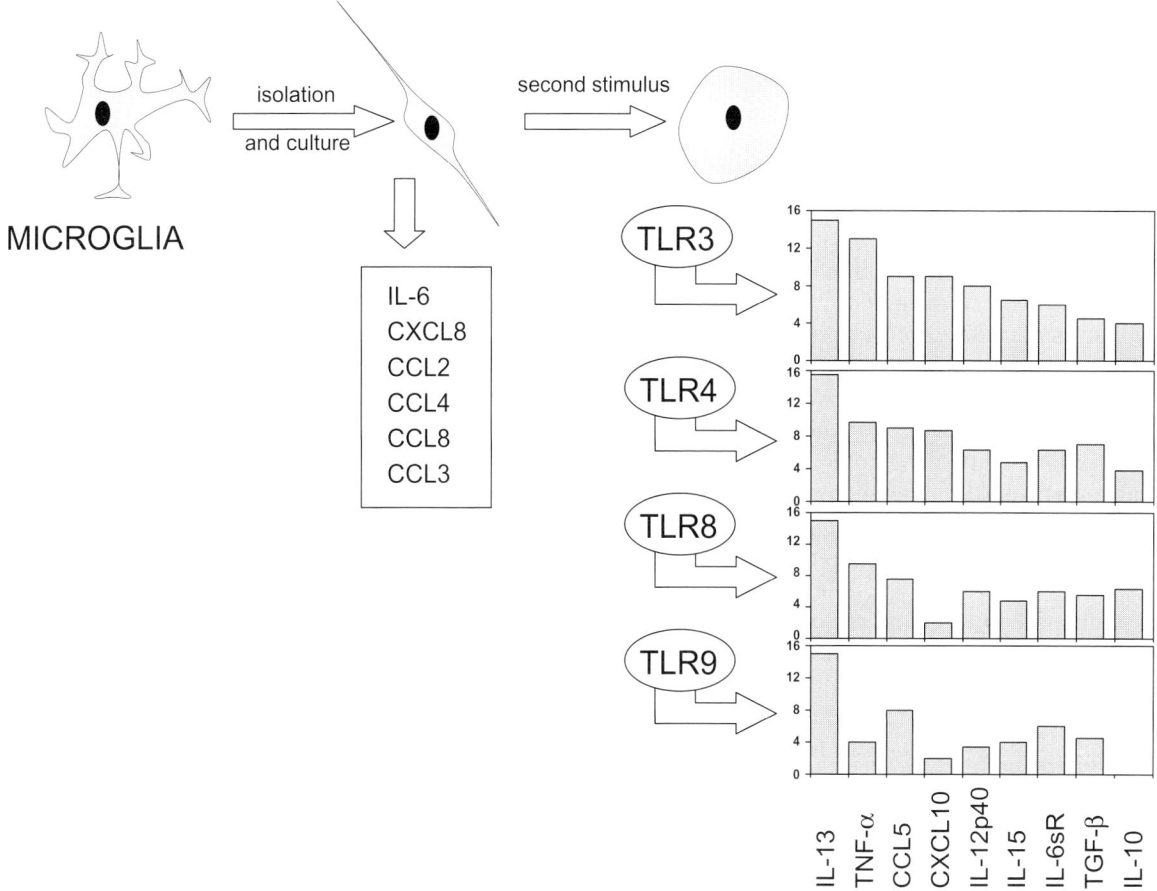

Fig. 1. Production of a standard set of cytokines and chemokines by cultured human microglia in response to different TLR agonists. When human microglia are isolated and cultured in the presence of GM-CSF, marked production of several chemokines occurs, as can be monitored by antibody arrays that allow simultaneous evaluation of 40 different soluble mediators. Following subsequent activation with selective TLR agonists, strikingly similar patterns of cytokines and chemokines are induced in all cases.

difficult to detect by immunohistochemistry, but when inflammation develops, TLRs readily emerge on the cell surface of astrocytes (Bsibsi et al., 2002). The data that have accumulated on the expression of TLRs on astrocytes of either rodent or human origin have consistently revealed a striking preference of astrocytes to express high levels of TLR3 (Jack et al., 2005; Farina et al., 2005; Bsibsi et al., 2006). While other TLR family members, notably including TLR2 and 4, are clearly detectable on astrocytes both in cell culture models and in vivo during trauma or inflammation, especially activated astrocytes produce much more TLR3 than any other TLR.

This intriguing preference of astrocytes to express up to 200-fold elevated levels of TLR3 upon activation is somewhat puzzling since the only currently known agonist for TLR3 is double-stranded RNA, which is believed to emerge as an intermediate during viral replication. Yet, double-stranded RNA is generally inside cells rather than secreted into the microenvironment, and its detectable presence in the microenvironment of CNS cells is rare. It is therefore difficult to envisage that the consistently dominant expression of TLR3 on the surface of astrocytes under a wide variety of conditions is designed solely to detect the very infrequent presence of extracellular double-stranded RNA. The quality of the astroglial response to TLR3-mediated signaling only adds to this apparent confusion since it is primarily neuroprotective rather than pro-inflammatory.

As dissected using microarray-based transcript analysis, the TLR3-mediated response in human astrocytes is far more comprehensive than the TLR4-mediated response. Also, it is striking that in response to TLR3-mediated activation astrocytes produce a variety of factors that are either well-known mediators of neuroprotection, or anti-inflammatory mediators, as is illustrated in Fig. 2. Examples of the first group include cilliary neurotrophic factor, neurotrophin-4, leukemia inhibitory factor, vascular endothelial growth factor, embryonic growth/differentiation factor, neuregulin, neurite growth-promoting factor (pleiotrophin), brain-derived neurotrophic factor, glial growth factors 1 and 2, and erythroid differentiation protein. An impressive number of studies have highlighted the neuroprotective qualities of all of these mediators. Several have been or still are under scrutiny as candidate therapeutic agents. Examples of the second group include tumor necrosis factor-inducible protein 6, TGF-β, IL-9, IL-10, and IL-11. Together, the mediators that are all selectively induced at least two-fold by TLR3 do not appear to represent a traditional pro-inflammatory host-defense response. Indeed, when poly I:C as an activator of such TLR3-mediated responses in astrocytes is added to organotypic human brain slice cultures, survival

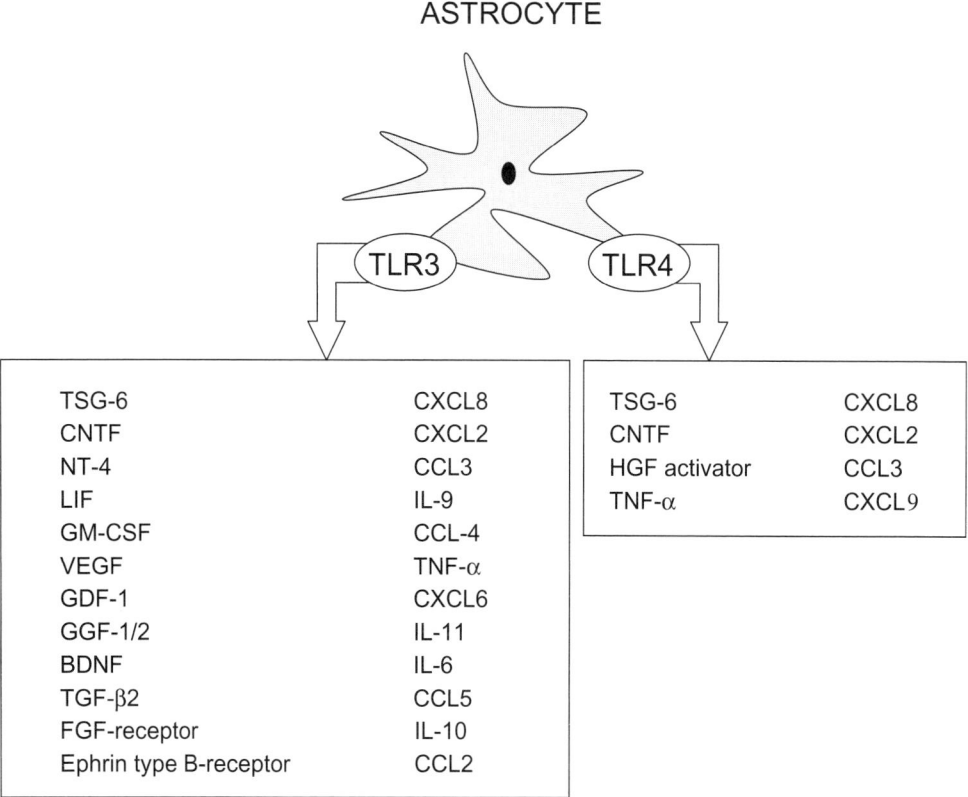

Fig. 2. Production of cytokines, chemokines, and growth factors by human astrocytes activated either via TLR3 or 4. Cultured human astrocytes were stimulated either with poly I:C or with ultrapure LPS as agonists for TLR3 and 4, respectively. After 1–2 days, gene transcript levels were evaluated by microarrays for several hundred genes encoding cytokines, chemokines, growth factors, and their receptors. While the TLR3-triggered response includes marked induction of a wide variety of anti-inflammatory or neuroprotective mediators, this is much less the case after TLR4-mediated activation (for details, see Bsibsi et al., 2006).

of neurons in such cultures significantly improves. When LPS is added, no such improvement is observed (Bsibsi et al., 2006).

These data strongly suggest that TLR3 on astrocytes may be an inducible regulator to activate tissue repair responses and limit inflammatory processes that develop in response to stress or trauma. As many different types of activators including pro-inflammatory cytokines and oxidative stress and TLR agonists themselves selectively induce TLR3 on astrocytes, the subsequent TLR3-mediated response is likely to represent a common secondary repair response designed to protect the CNS microenvironment after a first wave of host-defense activity. The dynamics of the glial response to stimuli, at least in cell culture models, is in line with such a view. Astrocytes take much longer than microglia to either upregulate TLRs or produce cytokines and growth factors in response to TLR activation. Also, TLR3-mediated activation of astrocytes leads to strong induction of indoleamine 2,3-dioxygenase (Suh et al., 2007). This enzyme converts extracellular tryptophan into kynurenine, thereby reducing its concentrations in the microenvironment, which in turn markedly enhances the sensitivity of any nearby T cell for Fas-ligand-induced apoptosis (Kwidzinksi et al., 2003). In this way, the TLR3-mediated induction of indoleamine 2,3-dioxygenase in astrocytes contributes to the elimination of any activated T cell or, in other words, acts as a local immune-suppressive factor. It appears unlikely that the potent anti-inflammatory and tissue repair responses that are apparently mediated by TLR3 would be uniquely dependent on an exogenous microbial signal like double-stranded viral RNA. Instead, a still unknown endogenous agonist for astroglial TLR3 to control this pathway in the absence of infection appears to be much more likely.

**Expression and function of TLRs in neurons**

After a first report in 2005 on the expression of TLR3 in cultured human neurons following viral infection (Prehaud et al., 2005), the expression of TLR3 on neurons in human brain tissue samples was confirmed in cases of rabies or herpes simplex virus infection (Jackson et al., 2006). Indeed, convincing evidence has now accumulated that neurons can express different functional TLRs, including TLR2, 3, 4, and 8 (Lafon et al., 2006; Ma et al., 2006; Tang et al., 2007; Cameron et al., 2007; Kim et al., 2007a, b; Acosta and Davies, 2008). As in other cells, levels of expression are dynamic, and influenced by soluble mediators including interferon-$\gamma$, or by energy deprivation.

A very interesting aspect of TLR-mediated signaling in neurons appears to be the role of a peculiar member of the MyD88 family, viz., MyD88-5 (previously known under the acronym SARM for the inconvenient "sterile alpha and HEAT/Armadillo motifs containing protein") which is preferentially expressed by neurons (Kim et al., 2007a, b). Most TLRs except TLR3 signal via the founding family member of the MyD88 family, which predominantly activates NF-$\kappa$B-mediated responses. The neuronal MyD88-5, on the other hand, associates with mitochondria, microtubules, and JNK3, and regulates neuronal death during deprivation of oxygen and glucose. Preferred expression of MyD88-5 in neurons confers a different quality of TLR responsiveness to these cells as compared to cells such as glial cells that do not express MyD88-5, but utilize other adaptors to relay TLR-mediated signaling. As a consequence, TLR3, which is concentrated in the growth cones of neurons, triggers growth cone collapse (Cameron et al., 2007), TLR2 and 4 induce apoptotic death (Tang et al., 2007), and TLR8 inhibits neurite outgrowth and triggers apoptosis (Ma et al., 2006). In all these cases, signaling pathways operate independently from NK-$\kappa$B. Clearly, by introducing MyD88-5 as the dominant adaptor for TLR-mediated intracellular signaling, neurons turn most TLR-mediated signals into negative signals for growth, development, and even survival. Yet, additional functions for neuronal TLR remain. For example, engagement of TLR4 on neurons induces the expression of nociceptin, an opioid-related neuropeptide (Acosta and Davies, 2008). Again, however, this response is somewhat different from TLR4-mediated responses in many other cells in that neurons apparently use the co-receptor MD-1 instead of the routinely used MD-2 along with CD14 as interaction partners for binding the TLR4 agonist LPS. This illustrates that neurons apparently modulate the TLR-signaling platform not only by introducing unusual intracellular adaptors to link TLRs to unique signaling pathways, but also by employing uncommon surface co-receptors which modulate the response.

**Concluding remarks and future perspectives**

As in other parts of the body TLRs are crucial to fight off microbial invasion also in the CNS. Yet, evidence is rapidly accumulating that TLRs fulfill functions well beyond orchestrating host-defense responses alone. TLRs are dynamically expressed on every type of neural cell, and influence inflammation and repair, also under sterile conditions (Bsibsi et al., 2002; Prinz et al., 2006; Tahara et al., 2006; Walter et al., 2007). In addition, they play a role in differentiation, migration, and development of neural cells. This suggests that a number of endogenous agonists must be available to allow for TLR-mediated processes to become activated also in the absence of microbial invaders. A few of such endogenous agonists are already known, including soluble CD14 that acts as an endogenous agonist for TLR2 (Bsibsi et al., 2007) and facilitates TLR4 activation by various ligands including brain gangliosides (Jou et al., 2006). Also heat shock proteins such as HSP60 that can be released by stressed or injured neural cells can promote TLR4-mediated responses in the absence of any microbial challenge (Lehnardt et al., 2008). Many other endogenous TLR agonists probably still await uncovering. When comparing expression and function of TLRs on different types of neural cells, striking differences emerge which emphasize that each type of cell has a different relationship with TLRs, and is likely to respond differently to any given TLR agonist. While microglia, for example, express many different TLRs and respond to their activation with a surprisingly standardized "danger" response, astrocytes clearly display differential responses to different TLR agonists. Neurons relay TLR-mediated signaling to completely different signaling pathways from other cells, by linking the TLR machinery to a unique intracellular adaptor. Along with effects of location, timing, and strength of TLR-mediated responses, and the fact that in most cases multiple TLRs will be activated at the same time, these aspects of TLR functioning in different cell types pose formidable challenges to the development of a coherent paradigm for TLR functions in the CNS as a whole. Both cell culture models and in vivo experiments will be required for a full clarification of these functions. A more profound understanding of TLR functions in the CNS, however, is very likely to make a crucial contribution to our understanding of the CNS itself, including issues of development and repair of brain functions, well beyond problems of infection and host-defense responses alone.

**Abbreviations**

| | |
|---|---|
| MyD88 | myeloid differentiation primary response gene 88 |
| TLR | toll-like receptor |
| TRIF | toll IL-1 receptor-domain-containing adapter-inducing interferon-β |

**Acknowledgements**

Our studies on TLRs in the CNS have been financially supported by the Foundation for the support of MS Research in The Netherlands. Human brain tissue samples used were kindly provided by the Netherlands Brain Bank, Amsterdam, The Netherlands.

**References**

Aarum, J., Sandberg, K., Budd Haeberlein, S. L., & Persson, M. A. A. (2003). Migration and differentiation of neuronal precursor cells can be directed by microglia. *Proceedings of the National Academy of Sciences of the United States of America, 100*, 15983–15988.

Acosta, C., & Davies, A. (2008). Bacterial lipopolysaccharide regulates nociceptin expression in sensory neurons. *Journal of Neuroscience Research, 86*, 1077–1086.

Block, M. I., Zecca, L., & Hong, J. S. (2007). Microglia-mediated neurotoxicity: uncovering the molecular mechanisms. *Nature Reviews Neuroscience*, 8, 57–69.

Boivin, A., Pineau, I., Barrette, B., Filali, M., Vallières, N., Rivest, S., et al. (2007). Toll-like receptor signaling is critical for Wallerian degeneration and functional recovery after peripheral nerve injury. *The Journal of Neuroscience*, 27, 12565–12576.

Bsibsi, M., Bajramovic, J. J., Van Duijvenvoorden, E., Persoon, C., Ravid, R., van Noort, J. M., et al. (2007). Identification of soluble CD14 as an endogenous agonist for Toll-like receptor 2 on human astrocytes by genome-scale functional screening of glial cell derived proteins. *Glia*, 55, 473–482.

Bsibsi, M., Persoon-Deen, C., Verwer, R. W. H., Meeuwsen, S., Ravid, R., & van Noort, J. M. (2006). Toll-like receptor 3 on adult human astrocytes triggers production of neuroprotective mediators. *Glia*, 53, 688–695.

Bsibsi, M., Ravid, R., Gveric, D., & van Noort, J. M. (2002). Broad expression of Toll-like receptors in the human central nervous system. *Journal of Neuropathology and Experimental Neurology*, 61, 1013–1021.

Cameron, J. S., Alexopoulou, L., Sloane, J. A., DiBernardo, A. B., Ma, Y., Kosaras, B., et al. (2007). Toll-like receptor 3 is a potent negative regulator of axonal growth in mammals. *The Journal of Neuroscience*, 27, 13033–13041.

Caso, J. R., Pradillo, J. M., Hurtado, O., Leza, J. C., Moro, M. A., & Lizasoain, I. (2008). Toll-like receptor 4 is involved in subacute stress-induced neuroinflammation and in the worsening of experimental stroke. *Stroke*, 39, 1314–1320.

Chen, K., Iribarren, P., Hu, J., Chen, J., Gong, W., Cho, E. H., et al. (2006). Activation of Toll-like receptor 2 on microglia promotes cell uptake of Alzheimer Disease-associated amyloid β peptide. *The Journal of Biological Chemistry*, 281, 3651–3659.

Farina, C., Krumbholz, M., Giese, T., Hartmann, G., Aloisi, F., & Meinl, E. (2005). Preferential expression and function of Toll-like receptor 3 in human astrocytes. *Journal of Neuroimmunology*, 159, 12–19.

Glezer, I., Lapointe, A., & Rivest, S. (2006). Innate immunity triggers oligodendrocyte progenitors reactivity and confines damages to brain injuries. *The FASEB Journal*, 20, 750–752.

Hua, F., Ma, J., Ha, T., Kelley, J., William, D. L., Kao, R. L., et al. (2008). Preconditioning with a TLR2-specific ligand increases resistance to cerebral ischemia/reperfusion injury. *Journal of Neuroimmunology*, 199, 75–82.

Jack, C. S., Arbour, N., Manusow, J., Montgrain, V., Blain, M., McCrea, E., et al. (2005). TLR signaling tailors innate immune responses in human microglia and astrocytes. *Journal of Immunology*, 175, 4320–4330.

Jackson, A. C., Rossiter, J. P., & Lafon, M. (2006). Expression of Toll-like receptor 3 in the human cerebellar cortex in rabies, herpes simplex encephalitis, and other neurological diseases. *Journal of Neurovirology*, 12, 229–234.

Jou, I., Lee, J. H., Park, S. Y., Yoon, H. J., Joe, E.-H., & Park, E. J. (2006). Gangliosides trigger inflammatory responses via TLR4 in brain glia. *The American Journal of Pathology*, 168, 1619–1630.

Kigerl, K. A., Lai, W., Rivest, S., Hart, R. P., Satoskar, A. R., & Popovich, P. G. (2007). Toll-like receptor (TLR)-2 and TLR4 regulate inflammation, gliosis, and myelin sparing after spinal cord injury. *Journal of Neurochemistry*, 102, 37–50.

Kilic, U., Kilic, E., Matter, C. M., Bassetii, C. L., & Hermann, D. M. (2008). TLR4-deficiency protects against focal cerebral ischemiia and axotomy-induced neurodegeneration. *Neurobiology of Disease*, 31, 33–40.

Kim, D., Kim, M. A., Cho, I.-H., Kim, M. S., Lee, S., Jo, E.-K., et al. (2007a). A critical role of Toll-like receptor 2 in nerve injury-induced spinal cord glial cell activation and pain hypersensitivity. *The Journal of Biological Chemistry*, 282, 14975–14983.

Kim, Y., Zhou, P., Qian, L., Chuang, J.-Z., Lee, J., Li, C., et al. (2007b). MyD88-5 links mitochondria, microtubules, and JNK3 in neurons and regulates neuronal survival. *The Journal of Experimental Medicine*, 204, 2063–2074.

Konat, G. W., Kielian, T., & Marriott, I. (2006). The role of Toll-like receptors in CNS responses to microbial challenge. *Journal of Neurochemistry*, 99, 1–12.

Kwidzinksi, E., Bunse, J., Kovac, A. D., Ulrich, O., Zipp, F., Nitsch, R., et al. (2003). IDO (indolamine 2,3-dioxygenase) expression and function in the CNS. *Advances in Experimental Medicine and Biology*, 527, 113–118.

Lafon, M., Megret, F., Lafage, M., & Prehaud, C. (2006). The innate immune facet of brain: human neurons express TLR-3 and sense viral dsRNA. *Journal of Molecular Neuroscience*, 29, 185–194.

Lehnardt, S., Lehmann, S., Kaul, D., Tschimmel, K., Hoffmann, O., Cho, S., et al. (2007). Toll-like receptor 2 mediates CNS injury in focal cerebral ischemia. *Journal of Neuroimmunology*, 190, 28–33.

Lehnardt, S., Schott, E., Trimbuch, T., Laubisch, D., Kreueger, C., Wulcyn, G., et al. (2008). A vicious cycle involving release of heat shock protein 60 from injured cells and activation of Toll-like receptor 4 mediates neurodegeneration in the CNS. *The Journal of Neuroscience*, 28, 2320–2331.

Ma, Y., Li, J., Chiu, I., Wang, Y., Sloane, J. A., Lü, J., et al. (2006). Toll-like receptor 8 functions as a negative regulator of neurite outgrowth and inducer of neuronal apoptosis. *The Journal of Cell Biology*, 175, 209–215.

Mages, J., Dietrich, H., & Lang, R. (2007). A genome-wide analysis of LPS tolerance in macrophages. *Immunobiology*, 212, 723–737.

Prehaud, C., Megret, F., Lafage, M., & Lafon, M. (2005). Virus infection switches TLR-3-positive human neurons to become strong producers of interferon beta. *Journal of Virology*, 79, 12893–12904.

Prinz, M., Garbe, F., Schmidt, H., Mildner, A., Gutcher, I., Wolter, K., et al. (2006). Innate immunity mediated by TLR9 modulates pathogenicity in an animal model of multiple sclerosis. *The Journal of Clinical Investigation*, 116, 456–464.

Stevens, S. L., Ciesielki, T. M., Marsh, B. J., Yang, T., Homen, D. S., Boule, J. L., et al. (2008). Toll-like receptor 9: a new target of ischemic preconditioning in the brain. *Journal of Cerebral Blood Flow and Metabolism, 28*, 1040–1047.

Suh, H.-S., Zhao, M.-L., Rivieccio, M., Choi, S., Connoly, E., Zhao, Y., et al. (2007). Astrocyte indoleamine 2,3-disoxygenase is induced by the TLR3 ligand poly (I:C): mechanism of induction and role in antiviral response. *Journal of Virology, 81*, 9838–9850.

Tahara, K., Kim, H. D., Jin, J. J., Maxwell, J. A., Li, L., & Fukuchi, K. (2006). Role of Toll-like receptor signaling in Abeta uptake and clearance. *Brain, 129*, 3006–3019.

Tang, S.-C., Arumugam, T. V., Xu, X., Cheng, A., Mughal, M. R., Jo, D. G., et al. (2007). Pivotal role for neuronal Toll-like receptors in ischemic brain injury and functional deficits. *Proceedings of the National Academy of Sciences of the United States of America, 104*, 13798–13803.

Tanga, F. Y., Nutile-McMenemy, N., & DeLeo, J. A. (2005). The CNS role of Toll-like receptor 4 in innate neuroimmunity and painful neuropathy. *Proceedings of the National Academy of Sciences of the United States of America, 102*, 5856–5861.

Walter, S., Letiembre, M., Liu, Y., Heine, H., Penke, B., Hao, W., et al. (2007). Role of Toll-like receptor 4 in neuroinflammation in Alzheimer's disease. *Cellular Physiology and Biochemistry, 20*, 947–956.

Ziegler, G., Harhausen, D., Schepers, C., Hoffmann, O., Röhr, C., Prinz, V., et al. (2007). TLR2 has a detrimental role in mouse transient focal cerebral ischemia. *Biochemical and Biophysical Research Communications, 359*, 574–579.

… SECTION III

# Gene Therapy

CHAPTER 10

# Gene therapy and transplantation in the retinofugal pathway

Alan R. Harvey[1],[*], Mats Hellström[1] and Jenny Rodger[2]

[1]*School of Anatomy and Human Biology, The University of Western Australia, Crawley, WA, Australia*
[2]*School of Animal Biology, The University of Western Australia, Crawley, WA, Australia*

**Abstract:** The mature CNS has limited intrinsic capacity for repair after injury; therefore, strategies are needed to enhance the viability and regrowth of damaged neurons. Here we review gene therapy studies in the eye, aimed at improving the survival and regeneration of injured retinal ganglion cells (RGCs). To target RGCs most current methods use recombinant adeno-associated viral vectors (AAV), usually serotype-2 (AAV2), that are injected into the vitreal chamber of the eye. This vector provides long-term transduction of adult RGCs. Strong, constitutive promoters such as CMV and/or β-actin are commonly used but cell-specific promoters have also been tested. Transgenes encoded by AAV have been selected to limit cell death, enhance growth factor expression, or promote growth cone responsiveness. We have assessed the effects of AAV vectors in adult rodent models (i) after optic nerve (ON) crush and (ii) after transplantation of peripheral nerve (PN) onto the cut ON, a procedure that induces injured RGCs to regenerate axons over longer distances. AAV–CNTF–GFP promotes RGC survival and axonal regrowth in mice after ON crush, and in rats after ON crush or PN transplantation. In rats, intravitreal injection of AAV–BDNF–GFP also increases RGC viability but does not promote regeneration. RGC viability and axonal regrowth is further enhanced when AAV–CNTF–GFP is injected into transgenic mice that over-express bcl-2. Reconstituted PN grafts containing Schwann cells that were transduced ex vivo with lentiviral (LV) vectors encoding a secretable form of CNTF support RGC axonal regrowth, however grafts containing Schwann cells transduced with LV–BDNF or LV–GDNF are less successful. We have also quantified the transduction efficiency and tropism of different AAV vectors injected intravitreally. AAV 2/2 and AAV 2/6 showed highest levels of transduction, AAV 2/8 the lowest, and each serotype displayed different transduction profiles for retinal cells. We are also studying the long-term impact of AAV2-mediated CNTF or BDNF expression on the dendritic morphology of RGCs in normal and PN grafted retinas. Analysis of regenerating RGCs intracellularly injected with lucifer yellow indicates gene-specific changes in dendritic structure that likely impact upon visual function.

**Keywords:** adeno-associated viral vectors; lentiviral vectors; retinal ganglion cells; optic nerve injury; peripheral nerve grafts; ciliary neurotrophic factor; brain-derived neurotrophic factor; AAV serotypes

[*]Corresponding author.
Tel.: 61 8 6488 3294; Fax: 61 8 6488 1051;
E-mail: alan.harvey@uwa.edu.au

## Introduction

The mature mammalian central nervous system (CNS) normally has only limited capacity for regenerative growth after traumatic injury and cannot adequately compensate for neuronal loss in neurodegenerative disease. The adult brain and spinal cord express a variety of inhibitory factors that together serve to limit plasticity (Yiu and He, 2006; Fitch and Silver, 2008), and there are maturational changes in neuronal responsiveness that may also limit regenerative growth potential (e.g. Shen et al., 1989; Goldberg et al., 2002). Therapeutic strategies are therefore needed to enhance the viability of compromised neurons and promote and maintain the regrowth of injured axons. The ultimate aim is to preserve or restore functional circuitries.

The retina and optic nerve (ON) are part of the CNS and the visual system is often used as a model in which to test experimental therapeutic approaches aimed at promoting neuronal survival and regeneration after CNS injury (Chierzi and Fawcett, 2001; Harvey et al., 2006; Benowitz and Yin, 2008). Retinal ganglion cells (RGCs) can be targeted by intravitreal injections and the ON is a discrete tract that is accessible within the orbit. Testing of therapies designed to improve RGC viability also has direct clinical relevance because of loss of these centrally projecting neurons in ophthalmic conditions such as glaucoma, retinal ischemia, optic neuropathy, and optic neuritis (reviewed in Harvey et al., 2006). In this report, we summarize gene therapy and transplantation studies aimed at enhancing adult RGC survival and axonal regeneration after ON injury in rodents. Benefits and some potential caveats to current gene therapy approaches in the retina will also briefly be considered.

## AAV-mediated transduction of retinal ganglion cells

Various pharmacotherapeutic strategies have been used in attempts to protect injured RGCs and, where necessary, stimulate axonal regrowth. Intravitreal injections of recombinant trophic factors such as brain-derived neurotrophic factor (BDNF), neurotrophin 4/5, and ciliary neurotrophic factor (CNTF) increase adult RGC survival after ON injury, but their protective effects are generally short-lived (Isenmann et al., 2004). The use of replication-deficient recombinant viral vectors to introduce neuroprotective genes into compromised neurons or perhaps into neighboring glial cells, resulting in a sustained supply of appropriate factors, may provide a better method for increasing the long-term viability of injured adult RGCs (Martin and Quigley, 2004; Harvey et al., 2006).

For gene therapy to be an effective tool in the CNS, vectors must be able to transduce postmitotic cells and they must induce no, or only minor, immune reactions. In the eye, a number of viral vectors have been tested for their ability to target RGCs. For some genes, direct infection of RGCs is essential; however, if a transgene encodes a secretable protein then transduction of other retinal cells may be sufficient, and such an approach may even offer some advantages (see later). Intraocular application of adenoviral (AdV) vectors encoding CNTF (Weise et al., 2000; van Adel et al., 2005), BDNF (Di Polo et al., 1998), or glial cell-line derived neurotrophic factor (GDNF) (Schmeer et al., 2002) have been shown to protect injured RGCs. In these studies, there was some RGC transduction, but Müller glia were also consistently transduced indicating indirect trophic effects on RGCs. Targeted transduction of Müller cells may now be possible using modified lentiviral (LV) vectors (Greenberg et al., 2007). Transduction of RGCs with AdV can also result from applying the vector directly to the ON stump (reviewed in Harvey et al., 2006). Note however that AdV vector injections into the eye can induce cell-mediated immune responses and inflammatory changes that may limit the duration of transgene expression (Isenmann et al., 2004). Modified AdV vectors are being designed in an attempt to ameliorate these immunogenic effects and provide more suitable vehicles for stable gene transfer (Lamartina et al., 2007).

Adeno-associated viruses (AAV) are non-enveloped single-stranded DNA viruses derived from parvoviruses and are generally regarded as being nonpathogenic and nontoxic. Recombinant, replication-deficient AAV vectors have been widely and successfully used in the eye, injected either subretinally or into the vitreal chamber (Dinculescu et al., 2005; Buch et al., 2008; Surace and Auricchio, 2008). Clinical trials involving subretinal injection of AAV to transduce cells in outer retina have begun (Bainbridge et al., 2008; Maguire et al., 2008).

There are many different serotypes of AAV (see below), but AAV2 displays excellent tropism for RGCs, especially when injected into the vitreous (Harvey et al., 2002). Strong, constitutive promoters such as CMV and/or β-actin are commonly used to drive transgene expression. In RGC studies, transgenes encoded by AAV have been selected to limit cell death, enhance growth factor expression, or promote growth cone responsiveness after various types of insult (e.g. Cheng et al., 2002; Martin et al., 2003; Sapieha et al., 2003; Fischer et al., 2004a, b; Schuettauf et al., 2004; Wu et al., 2004; Malik et al., 2005; Zhou et al., 2005). In our lab, we have examined the effects of intraocular injection of AAV2 vectors in adult rat and mouse models of ON injury, and also after transplantation of peripheral nerve (PN) onto the transected ON (Leaver et al., 2006a, b). Similar to most other groups we use bi-*cis* tronic AAV2 vectors that encode the gene of interest as well as the gene for green fluorescent protein (GFP). This allows identification of transduced cells in retinal wholemounts or sections, and also permits tracking of axons from transduced RGCs within the ON and into the brain (Harvey et al., 2002; Leaver et al., 2006a, b).

AAV2 encoding a modified secretable form of CNTF transduced a substantial number of RGCs and promoted RGC survival and axonal regrowth after ON crush in rats (Leaver et al., 2006b) and mice (Leaver et al., 2006a). These effects were studied 7 weeks after the initial ON injury. Intravitreal injections of AAV–BDNF–GFP also increased RGC viability in adult rats; however, while increased numbers of axons were seen proximal to the ON crush site, in these animals there was no regeneration into the nerve distal to the injury. No effect on RGC viability or axon growth was seen following (control) AAV–GFP injections or after transduction of retinal neurons with AAV encoding GAP43, a protein that is expressed at high levels in the growth cone during development and in mature neurons that are regenerating their axons (Schaden et al., 1994; see also Leaver et al., 2006b).

When using AAV vectors that encoded CNTF or BDNF there was an increase not only in GFP-labelled RGC numbers but also a substantial increase in the total number of surviving RGCs (Leaver et al., 2006b). Increased survival of nontransduced RGCs in these AAV-injected eyes is consistent with earlier AdV studies (see above), and supports the proposal that virally transduced cells can express and release trophic factors that protect neighboring nongenetically modified cells (Baumgartner and Shine, 1997).

In transgenic mice we have examined the effect of combining antiapoptotic and growth-promoting stimuli on adult RGC survival and axonal regeneration following intraorbital ON crush (Leaver et al., 2006a). AAV–CNTF–GFP was injected into eyes of wild-type mice or mice that had been engineered to over-express the antiapoptotic protein *bcl-2*. Five weeks after the ON crush, *bcl-2* overexpression by itself promoted the survival of axotomized RGCs but AAV-mediated expression of CNTF in adult retinas significantly increased the survival and axonal regeneration of injured RGCs. Note here that, while *bcl-2* and AAV–CNTF–GFP had a synergistic effect on RGC survival, the proportion of RGCs that regenerated an axon after ON crush was similar in wild-type and *bcl-2* transgenic mice (about 1 in 40 of surviving RGCs) (Leaver et al., 2006a). Thus, the greater number of regenerating axons in *bcl-2* mice was merely a consequence of the increased number of viable RGCs, and combined *bcl-2* and CNTF over-expression did not act synergistically to further enhance the regenerative ability of surviving RGCs. Why the vast majority of viable RGCs remained unable to regrow their axon is an important issue that requires further analysis. However, what these and other studies from our laboratory (Cui et al., 2003b; Park et al., 2004; Hu et al., 2007b) do emphasize is that therapies that promote neuronal survival do not necessarily also promote axonal regeneration, and vice-versa, indicating a clear dissociation between signaling pathways involved in cell survival and axogenesis (see also Goldberg and Barres, 2000).

After ON injury, RGC axon regeneration is increased by the attachment of a piece of autologous PN onto the cut ON (Bray and Aguayo, 1989; Dezawa and Adachi-Usami, 2000). These Schwann cell containing grafts alter the cellular responses of axotomized RGCs and provide a permissive environment that supports the long-distance regrowth of RGC axons. PN grafts have been used as bridges to reconnect the retina with central visual target areas, which can result in the restoration of some basic visual functions (Thanos, 1997; Coffey et al., 2000). Using the PN graft method we have tested the effects of AAV gene therapy on RGC axonal regeneration. We observed greater survival of adult RGCs and a substantial amount of axonal regrowth. After intraocular AAV–CNTF–GFP injections and PN-to-ON grafts about 25% of adult rat RGCs remained viable 7 weeks after surgery, and about half of these cells regenerated an axon into the PN graft (Leaver et al., 2006b). Similar to the paracrine effects of vector-delivered CNTF on RGC survival after ON crush, many nontransduced RGCs in AAV–CNTF–GFP-injected eyes regenerated an axon into the PN autografts.

Chlorphenylthio-cAMP (CPT-cAMP), a cell permeant cAMP analog, increases intraocular cAMP levels and potentiates RGC responsiveness to intravitreal injections of recombinant CNTF by increasing the proportion of viable RGCs that regenerate an axon into autologous PN grafts (Cui et al., 2003b; Park et al., 2004). Surprisingly, using the same PN autograft model, combining CPT-cAMP injections with AAV–CNTF–GFP transduction has no synergistic effect on axonal regrowth (Hellstrom and Harvey, in preparation). Possible reasons for this intriguing difference between vector-mediated CNTF release and intravitreal recombinant CNTF injections are currently under investigation, but may involve differences in the way negative regulators of cytokine signaling are activated under the two experimental conditions (Park et al., 2009).

## Use of LV to genetically modify Schwann cells in chimeric peripheral nerve grafts

Autologous PN transplants are a proven method for nerve repair and can be used as bridges to allow axons to by-pass areas of tissue damage in the CNS. Clinically however this approach may lead to complications due to problems associated with harvesting sufficient graft material and functional deficits that may arise from the additional surgery. We have therefore tested a new graft approach that involves the use of donor acellular PN sheaths (nerves are freeze-thawed to kill endogenous cells) that are repopulated ex vivo with purified adult Schwann cells before attachment to a transected ON (Cui et al., 2003a). The sheaths themselves do not appear to initiate major immune reactions and can be derived from rat strains that are different from the host. More recently, we extended these observations by genetically modifying the cultured Schwann cells prior to reconstitution of the PN grafts. Adult Schwann cells were exposed to LV vectors encoding different types of neurotrophic factor. This vector efficiently transduced a large proportion of the purified Schwann cell population. When grafted onto the cut ON, chimeric PN grafts containing Schwann cells transduced with LV-CNTF increased RGC viability and also enhanced the regrowth of RGC axons (Hu et al., 2005). The type of the cell used in the grafts is important because PN sheaths populated with CNTF-producing fibroblasts did not support RGC axonal regrowth (Hu et al., 2007a). The nature of the growth factor is also critical, because grafts containing Schwann cells transduced with LV–BDNF or LV–GDNF did not promote RGC axonal regeneration, although these grafts did support the ingrowth of large numbers of peripheral sensory axons (Hu et al., 2007a).

## Transduction and tropic properties of modified AAV vectors in the retina

We have seen that AAV2 delivered intravitreally successfully transduces RGCs and that this vector can be used to introduce a range of therapeutic genes into these CNS neurons. AAV2 transduction is, however, not totally confined to RGCs. After injection into adult eyes about one third of GFP⁺ cells remain in other retinal layers, and in the ganglion cell layer about 10–15% of transduced cells appear to be displaced amacrine cells (Harvey et al., 2006; Leaver et al., 2006b). Various cell types in the inner nuclear layer as well as occasional photoreceptors and sometimes Müller glia can be transfected (Harvey et al., 2002). This lack of specificity may complicate gene therapy treatment of ophthalmic pathologies. Greater selectivity may be obtained by using neuron-specific promoters (e.g. Kügler et al., 2003; Kerrison et al., 2005; Komáromy et al., 2008) although, in our knowledge, promoters that exclusively target RGCs have not been described and may require novel methods to ensure complete specificity (Luan and White, 2007).

Over 100 alternative isolates of AAV have been identified of which nine established serotypes have been tested in vivo (Rutledge et al., 1998; Broekman et al., 2006; Lebherz et al., 2008; Zaiss and Muruve, 2008; Zincarelli et al., 2008). The amino acid capsid structure varies slightly between the serotypes. This has an impact on viral tropism and affects expression kinetics in terms of the onset and expression levels of therapeutic genes (Rabinowitz et al., 2002; Surace et al., 2003; Gigout et al., 2005; Taymans et al., 2007; Lebherz et al., 2008; Pang et al., 2008; Zincarelli et al., 2008). Identification of different AAV serotypes may lead to the production of more cell-specific vectors for gene therapy, and while the retina has a degree of immune privilege, common neutralizing antibodies in human sera for several of the AAV serotype capsids also suggest the need to develop vector variants (e.g. Vandenberghe and Wilson, 2007; Zaiss and Muruve, 2008). The ability to incorporate larger genes by increasing AAV packaging capacity in different serotypes may also be advantageous (Allocca et al., 2008).

In the retina it is important to evaluate the tropism and transduction efficiency of AAV serotypes to discover if there are even more efficient and specific vectors for targeting RGCs. Improved targeting of photoreceptors after intravitreal injections might also be important. In photoreceptors, there is early onset and extensive transduction with many vector types after subretinal injections but, for the serotypes tested so far, photoreceptors are only infrequently transduced after intravitreal vector delivery (Auricchio et al., 2001; Harvey et al., 2002; Rabinowitz et al., 2002; Yang et al., 2002; Surace et al., 2003; Lebherz et al., 2008; Natkunarajah et al., 2008; Stieger et al., 2008). Intravitreal vector delivery is a less-invasive technique compared to subretinal delivery and more vector volume can be delivered. This approach also allows repeated injections and can be combined with delivery of other pharmacotherapeutic agents (Harvey et al., 2006). Thus being able to genetically modify photoreceptor cells in the outer retina via intravitreal vector injections may eventually be of clinical benefit.

In a study in collaboration with Joost Verhaagen and others, we have examined the tropism and transduction properties of titre matched AAV2/1, -2/2, -2/3, -2/4, -2/5, -2/6, and -2/8, after intravitreal injection into adult Wistar rat retinas (Hellström et al., 2009). All vectors possessed an AAV2 backbone and encoded GFP, cross packed into capsids from another serotype. Retinas were isolated 10 weeks post injection and were then cryosectioned. For each serotype, the number, morphology, and laminar distribution of transduced GFP positive cells was quantified using confocal and fluorescence microscopy (Fig. 1). The phenotype was also assessed by double-labeling with various immunomarkers for retinal cells. AAV2/2 and AAV2/6 possessed the highest transduction efficiency, followed by AAV2/3. AAV2/5, and AAV2/8. AAV2/6 had the most diverse tropism profile among the tested vectors, transfecting a range of neuronal types as well as Müller glia. The remaining six tested vectors transduced RGCs far more frequently than other cell types; about two thirds of AAV2/2 transduced cells were RGCs, and because of its high efficiency this vector transduced the highest absolute number of RGCs per retina. After AAV2/3 or AAV2/5 intravitreal injections, 20–25% of GFP⁺ cells were photoreceptors, but the total number of these cells was higher using the AAV2/3 serotype due to its

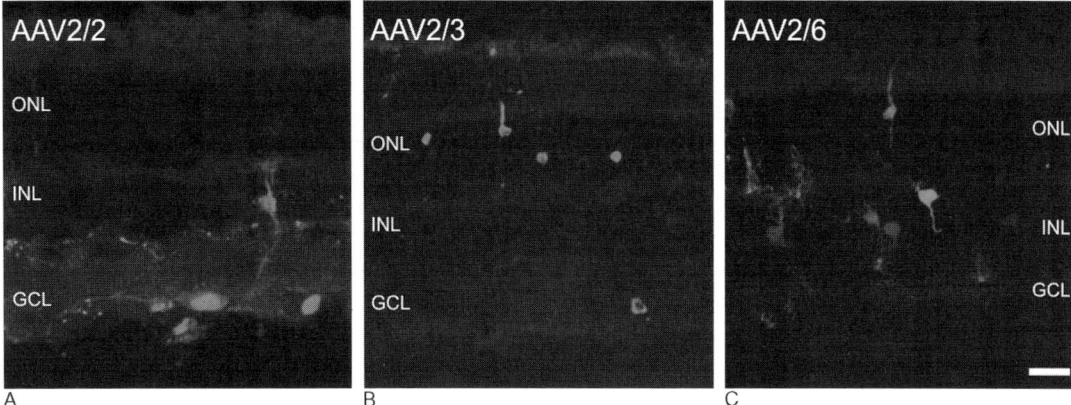

Fig. 1. Fluorescence photomicrographs of retinal sections from eyes that had been injected intravitreally with different AAV serotypes expressing GFP, 10 weeks post-injection. (A, B, and C) show GFP positive cells in retina from eyes injected with AAV 2/2, AAV 2/3, or AAV 2/6 respectively. Note differences in the distribution and morphology of transduced cells. GCL, retinal ganglion cell layer; INL, inner nuclear layer; ONL, outer nuclear layer. Scale bar for all figures = 20 μm. (See Color Plate 10.1 in color plate section.)

greater overall transduction efficiency (Hellström et al., 2009).

It is important to continuously evaluate novel recombinant AAV vectors for human gene therapy. New vector combinations, isolation of new serotypes, and sophisticated chemical modifications (Carlisle et al., 2008) may lead to novel vectors with greater cell specificity and minimal immunological impact. Our quantitative serotype data provide further information on tropism of different AAV vectors that may facilitate the development of improved gene therapy protocols and cell rescue for a variety of retinal disorders.

## Dendritic morphology of RGCs transduced with AAV encoding different genes

Despite many reports showing that appropriate AAV delivered genes can promote neuronal viability and axonal regeneration in the CNS, when using promoters such as CMV and/or β-actin it would seem prudent to determine whether sustained expression of virally introduced proteins affects the structure and function of transduced neurons. For example, all the products of the transgenes used in our AAV vectors (CNTF, BDNF, and GAP43) have been reported to affect dendritic growth and plasticity in cultured neurons (Guo et al., 1999; Gauthier-Campbell et al., 2004; Dijkuizen and Ghosh, 2005). Furthermore, if the transgene encodes a secretable factor then it may also be necessary to examine the impact of gene therapy on adjacent nontransduced cells or tissues (Buch et al., 2006; Leaver et al., 2006b). From a neurological perspective, while gene therapy will most likely benefit injured patients, it is important to know if there are any potential pitfalls in this promising technology, and whether it is necessary to use vector systems that allow the temporal regulation of transgene expression (e.g. Dejneka et al., 2001; McGee Sanftner et al., 2001; Kerrison et al., 2005; Stieger et al., 2006; Lamartina et al., 2007).

In a large-scale study we are in the process of examining if RGC dendritic morphology is altered after prolonged AAV2 vector transduction, and whether any such changes vary depending on (i) the type of gene that is introduced and (ii) whether or not the RGCs are injured. Adult rats received an intravitreal injection of saline, AAV–GFP, AAV–CNTF–GFP, AAV–BDNF–GFP, or AAV–GAP43–GFP. Some rats were not operated on further (normal retinas), while other AAV-injected animals from each vector group received, 2 weeks later, an autologous PN–ON graft. Animals were sacrificed 6–9 months later (Rodger et al., 2009).

In the PN grafted groups, fluorogold (FG) was applied to the distal end of each PN graft to allow identification of RGCs with regenerated axons. Retinas were wholemounted in oxygenated AMES medium and RGCs identified by FG label were injected by iontophoresis with lucifer yellow (2% LY, 20–50 cells injected per retina). Retinas were fixed and processed for anti-LY (Santa Cruz Biotech) immunohistochemistry with a Cy3 conjugated secondary antibody (Jackson). Transduced cells were identified by GFP expression (Fig. 2). The morphologies of transduced and untransduced (GFP negative) RGCs were documented by manual tracing at $400\times$ magnification using Neurolucida software. Regenerating, FG positive RGCs were classified into three major types according to morphological criteria used in previous studies of RGC morphology in the normal rat retina (Huxlin and Goodchild, 1997) and in adult rat retina following PN transplantation (Thanos and Mey, 1995). The classification was as follows: RI: large cell soma ($>400\,\mu m^2$), 5–6 thick dendrites; RII: small oval or round soma ($100–300\,\mu m^2$), 3–4 thin dendrites; RIII: medium polarized soma ($200–400\,\mu m^2$), 1–2 dendrites, large field size; unclassifiable cells possessed sparse or tangled dendrites. Parameters measured were: cell soma area, number and length of dendrites. Complexity was estimated by the number of nodes and tips, dendrite tortuosity, maximum branch order, and fractal count. We also used convex hull analysis to determine dendritic field area. Values were obtained separately for each RGC type in the four PN grafted groups that have been quantitatively analyzed so far (saline: 45 cells; AAV–GFP: 70 cells; AAV–CNTF–GFP: 147 cells; AAV–BDNF–GFP: 25 cells).

There were no significant differences between saline and GFP groups in the frequency distribution of the RGC types or in their morphologies. In contrast, AAV-mediated expression of CNTF or BDNF increased the proportion of unclassifiable RGCs to nearly 50%. BDNF and CNTF expression increased the proportion of RI and RIII relative to RII cells. Furthermore, in AAV–CNTF–GFP injected retinas, cells of all types had a significantly increased soma size and RII cells

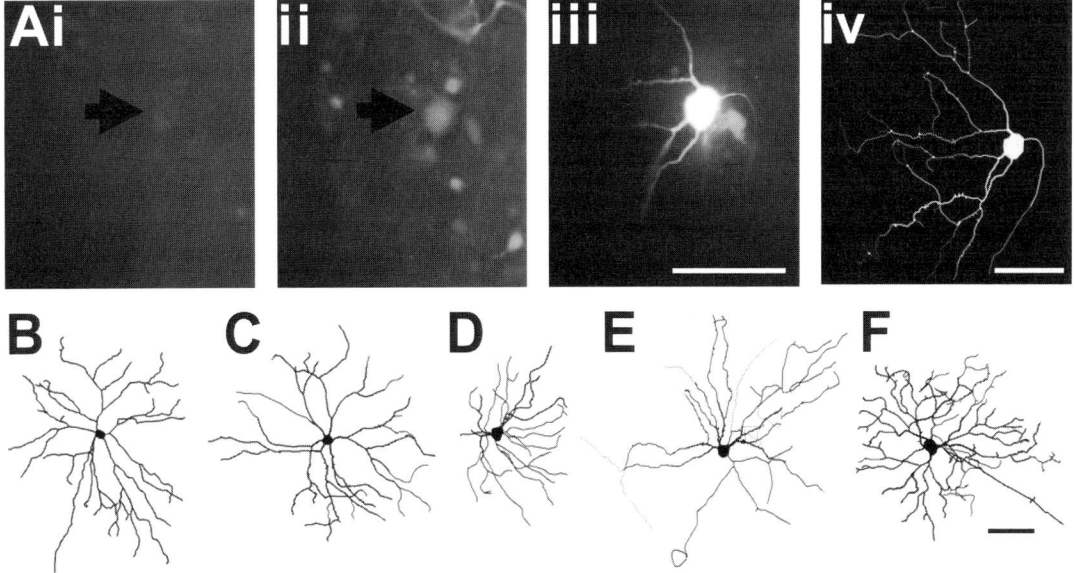

Fig. 2. (A(i)–A(iv)) Retinal ganglion cell from an AAV–CNTF–GFP-injected retina. The cell is identified by fluorogold labeling (A(i)), is transduced because it expresses GFP (A(ii)), and is injected with Lucifer yellow (A(iii)). The cell is traced using Neurolucida software for analysis. (B–F) traces of Type 1 RGCs from Saline (B), AAV–GFP (C), AAV–CNTF–GFP (D,E), and AAV–BDNF–GFP (F) injected eyes. Scale bar for all figures = 100 µm. (See Color Plate 10.2 in color plate section.)

showed increased branch complexity. In AAV–BDNF–GFP-injected retinas, all cells had an increased soma area and RII cells had increased tortuosity and larger dendritic fields. Separate analysis of transduced and nontransduced, FG positive RGCs is ongoing, but preliminary data suggest that there are differences between these two subgroups.

Alteration in specific cell morphologies in growth factor transduced retinas has implications for visual function, since each morphology is associated with a particular aspect of vision. RI cells, the most common in the normal intact rat retina (Huxlin and Goodchild, 1997) are thought to be homologous to alpha cells which provide sensitivity to moving stimuli. The smaller RII RGCs account for the majority of RGCs surviving following a PN graft, and their equivalents in the normal intact rat retina may be beta and gamma cells, which mediate visual acuity. RIII have thin axons, providing slow conduction and may be equivalent to W cells in the cat (Rowe and Palmer, 1995). Therefore, in addition to improved RGC survival and axonal regeneration, AAV-mediated CNTF over-expression could result in improved movement detection in PN grafted rats through preferential rescue of RI and RIII cells. Such an effect might however be offset by reported changes in photoreceptor function in CNTF treated retinas (Buch et al., 2006). In AAV–BDNF–GFP treated rats, increased RGC survival may not be functionally useful, not only because of a failure of axonal regeneration, but also due to increased dendritic field size and changes in dendritic architecture in RII cells that could compromise visual acuity.

## Concluding comments

Accumulated experimental evidence strongly suggests that combinations of gene- and cell-based therapies will eventually be of significant clinical benefit to injured patients. Phase 1 clinical trials using gene therapy in the treatment of neurological disease are planned or are underway (Tuszynski et al., 2005; Mandel et al., 2006; Eberling et al., 2008) and trials for AAV transduction of photoreceptors have also begun (Bainbridge et al., 2008; Maguire et al., 2008). Ex vivo genetic manipulation of cells prior to transplantation into the eye, brain, or spinal cord also holds much promise. For gene therapy, development of nonimmunogenic vector types continues to be important and there is likely to be a need for more refined targeting of specific cell populations. Animal studies should be followed over extended periods of time to determine if long-term transgene expression affects the morphology and function of transduced neurons and/or their neighbors. Ultimately, to achieve the best possible outcome it may be necessary to use regulatory viral vectors, allowing transgenes to be switched on or off when and where appropriate (e.g. Kerrison et al., 2005; Stieger et al., 2006; Chtarto et al., 2007; Liu et al., 2008).

## References

Allocca, M., Doria, M., Petrillo, M., Colella, P., Garcia-Hoyos, M., Gibbs, D., et al. (2008). Serotype-dependent packaging of large genes in adeno-associated viral vectors results in effective gene delivery in mice. *Journal of Clinical Investigation*, *118*, 1955–1964.

Auricchio, A., Kobinger, G., Anand, V., Hildinger, M., O'Connor, E., Maguire, A. M., et al. (2001). Exchange of surface proteins impacts on viral vector cellular specificity and transduction characteristics: the retina as a model. *Human Molecular Genetics*, *10*, 3075–3081.

Bainbridge, J. W., Smith, A. J., Barker, S. S., Robbie, S., Henderson, R., Balaggan, K., et al. (2008). Effect of gene therapy on visual function in Leber's congenital amaurosis. *New England Journal of Medicine*, *358*, 2231–2239.

Baumgartner, B. J., & Shine, H. D. (1997). Targeted transduction of CNS neurons with adenoviral vectors carrying neurotrophic factor genes confers neuroprotection that exceeds the transduced population. *Journal of Neuroscience*, *17*, 6504–6511.

Benowitz, L., & Yin, Y. (2008). Rewiring the injured CNS: lessons from the optic nerve. *Experimental Neurology*, *209*, 389–398.

Bray, G. M., & Aguayo, A. J. (1989). Exploring the capacity of CNS neurons to survive injury, regrow axons and form new synapses in adult mammals. In F. J. Seil (Ed.), *Neural regeneration and transplantation* (pp. 67–78). New York: Liss.

Broekman, M. L., Comer, L. A., Hyman, B. T., & Sena-Esteves, M. (2006). Adeno-associated virus vectors serotyped with AAV8 capsid are more efficient than AAV-1 or -2 serotypes for widespread gene delivery to the neonatal mouse brain. *Neuroscience*, *138*, 501–510.

Buch, P. K., Bainbridge, J. W., & Ali, R. R. (2008). AAV-mediated gene therapy for retinal disorders: from mouse to man. *Gene Therapy*, *15*, 849–857.

Buch, P. K., MacLaren, R. E., Durán, Y., Balaggan, K. S., MacNeil, A., Schlichtenbrede, F. C., et al. (2006). In contrast to AAV-mediated CNTF expression, AAV-mediated GDNF expression enhances gene replacement therapy in rodent models of retinal degeneration. *Molecular Therapy*, *14*, 700–709.

Carlisle, R. C., Benjamin, R., Briggs, S. S., Sumner-Jones, S., McIntosh, J., Gill, D., et al. (2008). Coating of adeno-associated virus with reactive polymers can ablate virus tropism, enable retargeting and provide resistance to neutralising antisera. *Journal of Gene Medicine*, *10*, 400–411.

Cheng, L., Sapieha, P., Kittlerova, P., Hauswirth, W. W., & Di Polo, A. (2002). TrkB gene transfer protects retinal ganglion cells from axotomy-induced death in vivo. *Journal of Neuroscience*, *22*, 3977–3986.

Chierzi, S., & Fawcett, J. W. (2001). Regeneration in the mammalian optic nerve. *Restorative Neurology and Neuroscience*, *19*, 109–118.

Chtarto, A., Yang, X., Bockstael, O., Melas, C., Blum, D., Lehtonen, E., et al. (2007). Controlled delivery of glial cell line-derived neurotrophic factor by a single tetracycline-inducible AAV vector. *Experimental Neurology*, *204*, 387–399.

Coffey, P. J., Whiteley, S. J., & Lund, R. D. (2000). Preservation and restoration of vision following transplantation. *Progress in Brain Research*, *127*, 489–499.

Cui, Q., Pollett, M. A., Symons, N. A., Plant, G. W., & Harvey, A. R. (2003a). A new approach to CNS repair using chimeric peripheral nerve grafts. *Journal of Neurotrauma*, *20*, 17–31.

Cui, Q., Yip, H. K., Zhao, R. C., So, K. F., & Harvey, A. R. (2003b). Intraocular elevation of cyclic AMP potentiates ciliary neurotrophic factor-induced regeneration of adult rat retinal ganglion cell axons. *Molecular and Cellular Neuroscience*, *22*, 49–61.

Dejneka, N. S., Auricchio, A., Maguire, A. M., Ye, X., Gao, G. P., Wilson, J. M., et al. (2001). Pharmacologically regulated gene expression in the retina following transduction with viral vectors. *Gene Therapy*, *8*, 442–446.

Dezawa, M., & Adachi-Usami, E. (2000). Role of Schwann cells in retinal ganglion cell axon regeneration. *Progress in Retinal and Eye Research*, *19*, 171–204.

Di Polo, A., Aigner, L. J., Dunn, R. J., Bray, G. M., & Aguayo, A. J. (1998). Prolonged delivery of brain-derived neurotrophic factor by adeno-virus-infected Müller cells temporarily rescues injured retinal ganglion cells. *Proceedings of the National Academy of Sciences (USA)*, *95*, 3978–3983.

Dijkuizen, P. A., & Ghosh, A. (2005). BDNF regulates primary dendrite formation in cortical neurons via the PI3-kinase and MAP kinase signaling pathways. *Journal of Neurobiology*, *62*, 278–288.

Dinculescu, A., Glushakova, L., Min, S. H., & Hauswirth, W. W. (2005). Adeno-associated virus-vectored gene therapy for retinal disease. *Human Gene Therapy*, *16*, 649–663.

Eberling, J. L., Jagust, W. J., Christine, C. W., Starr, P., Larson, P., Bankiewicz, K. S., et al. (2008). Results from a phase I safety trial of hAADC gene therapy for Parkinson disease. *Neurology*, *70*, 1980–1983.

Fischer, D., He, Z., & Benowitz, L. I. (2004a). Counteracting the Nogo receptor enhances optic nerve regeneration if retinal ganglion cells are in an active growth state. *Journal of Neuroscience*, *24*, 1646–1651.

Fischer, D., Petkova, V., Thanos, S., & Benowitz, L. I. (2004b). Switching mature retinal ganglion cells to a robust growth state in vivo: gene expression and synergy with RhoA inactivation. *Journal of Neuroscience*, *24*, 8726–8740.

Fitch, M. T., & Silver, J. (2008). CNS injury, glial scars, and inflammation: inhibitory extracellular matrices and regeneration failure. *Experimental Neurology*, *209*, 294–301.

Gauthier-Campbell, C., Bredt, D. S., Murphy, T. H., & El-Husseini, Ael-D. (2004). Regulation of dendritic branching and filopodia formation in hippocampal neurons by specific acylated protein motifs. *Molecular Biology of the Cell.*, *15*, 2205–2217.

Gigout, L., Rebollo, P., Clement, N., Warrington, K. H., Jr., Muzyczka, N., Linden, R. M., et al. (2005). Altering AAV tropism with mosaic viral capsids. *Molecular Therapy*, *11*, 856–865.

Goldberg, J. L., & Barres, B. A. (2000). The relationship between neuronal survival and regeneration. *Annual Review of Neuroscience*, *23*, 579–612.

Goldberg, J. L., Klassen, M. P., Hua, Y., & Barres, B. A. (2002). Amacrine-signaled loss of intrinsic axon growth ability by retinal ganglion cells. *Science*, *296*, 1860–1864.

Greenberg, K. P., Geller, S. F., Schaffer, D. V., & Flannery, J. G. (2007). Targeted transgene expression in Müller glia of normal and diseased retinas using lentiviral vectors. *Investigative Ophthalmology and Visual Science*, *48*, 1844–1852.

Guo, X., Chandrasekaran, V., Lein, P., Kaplan, P., & Higgins, D. (1999). Leukemia inhibitory factor and ciliary neurotrophic factor cause dendritic retraction in cultured rat sympathetic neurons. *Journal of Neuroscience*, *19*, 2113–2121.

Harvey, A. R., Hu, Y., Leaver, S. G., Mellough, C. B., Park, K., Verhaagen, J., et al. (2006). Gene therapy and transplantation in CNS repair: the visual system. *Progress in Retinal and Eye Research*, *25*, 449–489.

Harvey, A. R., Kamphuis, W., Eggers, R., Symons, N., Blits, B., Niclou, S., et al. (2002). Intravitreal injection of adeno-associated viral vectors results in the transduction of different types of retinal neurons in neonatal and adult rats: a comparison with lentiviral vectors. *Molecular and Cellular Neuroscience*, *21*, 141–157.

Hellström, M., Pollett, M. A., Ruitenberg, M. J., Ehlert, E. M. E., Twisk, J., Verhaagen, J., et al. (2009). Cellular tropism and transduction properties of 7 adeno-associated viral vector serotypes in adult retina after vitreal injection. *Gene Therapy*, *16*, 521–532.

Hu, Y., Arulpragasam, A., Plant, G. W., Hendriks, W. T. J., Cui, Q., & Harvey, A. R. (2007a). The importance of transgene and cell type on the regeneration of adult retinal

Hu, Y., Cui, Q., & Harvey, A. R. (2007b). Interactive effects of C3, cyclic AMP and ciliary neurotrophic factor on adult retinal ganglion cell survival and axonal regeneration. *Molecular and Cellular Neuroscience, 34*, 88–98.

Hu, Y., Leaver, S. G., Plant, G. W., Hendricks, W. T. J., Niclou, S. P., Verhaagen, J., et al. (2005). Lentiviral-mediated transfer of CNTF to Schwann cells within reconstructed peripheral nerve grafts enhances adult retinal ganglion cell survival and axonal regeneration. *Molecular Therapy, 11*, 906–915.

Huxlin, K. R., & Goodchild, A. K. (1997). Retinal ganglion cells in the albino rat: revised morphological classification. *Journal of Comparative Neurology, 385*, 309–323.

Isenmann, S., Schmeer, C., & Kretz, A. (2004). How to keep injured CNS neurons viable — strategies for neuroprotection and gene transfer to retinal ganglion cells. *Molecular and Cellular Neuroscience, 26*, 1–16.

Kerrison, J. B., Duh, E. J., Yu, Y., Otteson, D. C., & Zack, D. J. (2005). A system for inducible gene expression in retinal ganglion cells. *Investigative Ophthalmology and Visual Science, 46*, 2932–2939.

Komáromy, A. M., Alexander, J. J., Cooper, A. E., Chiodo, V. A., Acland, G. M., Hauswirth, W. W., et al. (2008). Targeting gene expression to cones with human cone opsin promoters in recombinant AAV. *Gene Therapy, 15*, 1049–1055.

Kügler, S., Lingor, P., Schöll, U., Zolotukhin, S., & Bähr, M. (2003). Differential transgene expression in brain cells in vivo and in vitro from AAV-2 vectors with small transcriptional control units. *Virology, 311*, 89–95.

Lamartina, S., Cimino, M., Roscilli, G., Dammassa, E., Lazzaro, D., Rota, R., et al. (2007). Helper-dependent adenovirus for the gene therapy of proliferative retinopathies: stable gene transfer, regulated gene expression and therapeutic efficacy. *Journal of Gene Medicine, 9*, 862–874.

Leaver, S. G., Cui, Q., Bernard, O., & Harvey, A. R. (2006a). Cooperative effects of Bcl-2 and AAV-mediated expression of CNTF on retinal ganglion cell survival and axonal regeneration in adult transgenic mice. *European Journal of Neuroscience, 24*, 3323–3332.

Leaver, S. G., Cui, Q., Plant, G. W., Verhaagen, J., & Harvey, A. R. (2006b). AAV-mediated expression of CNTF, but not BDNF or GAP-43, promotes long-term survival and regeneration of injured adult retinal ganglion cells. *Gene Therapy, 18*, 1328–1341.

Lebherz, C., Maguire, A., Tang, W., Bennett, J., & Wilson, J. M. (2008). Novel AAV serotypes for improved ocular gene transfer. *Journal of Gene Medicine, 10*, 375–382.

Liu, Y., Okada, T., Shimazaki, K., Sheykholeslami, K., Nomoto, T., Muramatsu, S., et al. (2008). Protection against aminoglycoside-induced ototoxicity by regulated AAV vector-mediated GDNF gene transfer into the cochlea. *Molecular Therapy, 16*, 474–480.

Luan, H., & White, B. H. (2007). Combinatorial methods for refined neuronal gene targeting. *Current Opinion in Neurobiology, 17*, 575–580.

Maguire, A. M., Simonelli, F., Pierce, E. A., Pugh, E. N., Jr., Mingozzi, F., Bennicelli, J., et al. (2008). Safety and efficacy of gene transfer for Leber's congenital amaurosis. *New England Journal of Medicine, 358*, 2240–2248.

Malik, J. M., Shevtsova, Z., Bähr, M., & Kügler, S. (2005). Long-term in vivo inhibition of CNS neurodegeneration by Bcl-XL gene transfer. *Molecular Therapy, 11*, 373–381.

Mandel, R. J., Manfredsson, F. P., Foust, K. D., Rising, A., Reimsnider, S., Nash, K., et al. (2006). Recombinant adeno-associated viral vectors as therapeutic agents to treat neurological disorders. *Molecular Therapy, 13*, 463–483.

Martin, K. R., & Quigley, H. A. (2004). Gene therapy for optic nerve disease. *Eye, 18*, 1049–1055.

Martin, K. R., Quigley, H. A., Zack, D. J., Levkovitch-Verbin, H., Kielczewski, J., Valenta, D., et al. (2003). Gene therapy with brain-derived neurotrophic factor as a protection: retinal ganglion cells in a rat glaucoma model. *Investigative Ophthalmology and Visual Science, 44*, 4357–4365.

McGee Sanftner, L. H., Rendahl, K. G., Quiroz, D., Coyne, M., Ladner, M., Manning, W. C., et al. (2001). Recombinant AAV-mediated delivery of a tet-inducible reporter gene to the rat retina. *Molecular Therapy, 3*, 688–696.

Natkunarajah, M., Trittibach, P., McIntosh, J., Duran, Y., Barker, S. E., Smith, A. J., et al. (2008). Assessment of ocular transduction using single-stranded and self-complementary recombinant adeno-associated virus serotype 2/8. *Gene Therapy, 15*, 463–467.

Pang, J. J., Lauramore, A., Deng, W. T., Li, Q., Doyle, T. J., Chiodo, V., et al. (2008). Comparative analysis of in vivo and in vitro AAV vector transduction in the neonatal mouse retina: effects of serotype and site of administration. *Vision Research, 48*, 377–385.

Park, K. K., Hu, Y., Muhling, J., Pollett, M. A., Dallimore, E. J., Turnley, A. M., et al. (2009). Cytokine-induced SOCS expression is inhibited by cAMP analogue: impact on regeneration of injured retina. *Molecular and Cellular Neuroscience, 41*, 313–324.

Park, K., Luo, J.-M., Hisheh, S., Harvey, A. R., & Cui, Q. (2004). Cellular mechanisms associated with spontaneous and ciliary neurotrophic factor/cAMP-induced survival and axonal regeneration of adult retinal ganglion cells. *Journal of Neuroscience, 24*, 10806–10815.

Rabinowitz, J. E., Rolling, F., Li, C., Conrath, H., Xiao, W., Xiao, X., et al. (2002). Cross-packaging of a single adeno-associated virus (AAV) type 2 vector genome into multiple AAV serotypes enables transduction with broad specificity. *Journal of Virology, 76*, 791–801.

Rodger, J., Hellström, M., Hu, Y., Heel, K. A., Robertson, D., & Harvey, A. R. (2009). Alteration in RGC dendritic morphology following AAV transduction with different growth factors. *Proceedings of the Australian Neuroscience Society, 19*, 53.

Rowe, M. H., & Palmer, L. A. (1995). Spatio-temporal receptive-field structure of phasic W cells in the cat retina. *Visual Neuroscience, 12*, 117–139.

Rutledge, E. A., Halbert, C. L., & Russell, D. W. (1998). Infectious clones and vectors derived from adeno-associated virus (AAV) serotypes other than AAV type 2. *Journal of Virology, 72*, 309–319.

Sapieha, P. S., Peltier, M., Rendahl, K. G., Manning, W. C., & Di Polo, A. (2003). Fibroblast growth factor-2 gene delivery stimulates axon growth by adult retinal ganglion cells after acute optic nerve injury. *Molecular and Cellular Neuroscience, 24*, 656–672.

Schaden, H., Stuermer, C. A. O., & Bähr, M. (1994). GAP-43 immunoreactivity and axon regeneration in retinal ganglion cells of the rat. *Journal of Neurobiology, 25*, 1570–1578.

Schmeer, C., Straten, G., Kügler, S., Gravel, C., Bähr, M., & Isenmann, S. (2002). Dose-dependent rescue of axotomized rat retinal ganglion cells by adenovirus-mediated expression of GDNF in vivo. *European Journal of Neuroscience, 15*, 637–643.

Schuettauf, F., Vorwerk, C., Naskar, R., Orlin, A., Quinto, K., Zurakowski, D., et al. (2004). Adeno-associated viruses containing bFGF or BDNF are neuroprotective against excitotoxicity. *Current Eye Research, 29*, 379–386.

Shen, S., Wiemelt, A. P., McMorris, F. A., & Barres, B. A. (1989). Retinal ganglion cells lose trophic responsiveness after axotomy. *Neuron, 23*, 285–295.

Stieger, K., Colle, M. A., Dubreil, L., Mendes-Madeira, A., Weber, M., Le Meur, G., et al. (2008). Subretinal delivery of recombinant AAV serotype 8 vector in dogs results in gene transfer to neurons in the brain. *Molecular Therapy, 16*, 916–923.

Stieger, K., Le Meur, G., Lasne, F., Weber, M., Deschamps, J. Y., Nivard, D., et al. (2006). Long-term doxycycline-regulated transgene expression in the retina of nonhuman primates following subretinal injection of recombinant AAV vectors. *Molecular Therapy, 13*, 967–975.

Surace, E. M., & Auricchio, A. (2008). Versatility of AAV vectors for retinal gene transfer. *Vision Research, 48*, 353–359.

Surace, E. M., Auricchio, A., Reich, S. J., Rex, T., Glover, E., Pineles, S., et al. (2003). Delivery of adeno-associated virus vectors to the fetal retina: impact of viral capsid proteins on retinal neuronal progenitor transduction. *Journal of Virology, 77*, 7957–7963.

Taymans, J. M., Vandenberghe, L. H., Haute, C. V., Thiry, I., Deroose, C. M., Mortelmans, L., et al. (2007). Comparative analysis of adeno-associated viral serotypes 1, 2, 3, 7 and 8 in mouse brain. *Human Gene Therapy, 18*, 195–206.

Thanos, S. (1997). Neurobiology of the regenerating retina and its functional reconnection with the brains by means of peripheral nerve transplants in adult rats. *Survey of Ophthalmology, 42*, S5–S26.

Thanos, S., & Mey, J. (1995). Type-specific stabilization and target-dependent survival of regenerating ganglion cells in the retina of adult rats. *Journal of Neuroscience, 15*, 1057–1079.

Tuszynski, M., Thal, L., Pay, M., Salmon, D. P., U, H. S., Bakay, R., et al. (2005). A phase 1 clinical trial of nerve growth factor gene therapy for Alzheimer disease. *Nature Medicine, 11*, 551–555.

van Adel., B. A., Arnold, J. M., Phipps, J., Doering, L. C., & Ball, A. K. (2005). Ciliary neurotrophic factor protects retinal ganglion cells from axotomy-induced apoptosis via modulation of retinal glia in vivo. *Journal of Neurobiology, 63*, 215–234.

Vandenberghe, L. H., & Wilson, J. M. (2007). AAV as an immunogen. *Current Gene Therapy, 7*, 325–333.

Weise, J., Isenmann, S., Klöcker, N., Kügler, S., Hirsch, S., Gravel, C., et al. (2000). Adenovirus-mediated expression of ciliary neurotrophic factor (CNTF) rescues axotomized rat retinal ganglion cells but does not support axonal regeneration in vivo. *Neurobiology of Disease, 7*, 212–223.

Wu, W. C., Lai, C. C., Chen, S. L., Sun, M. H., Xiao, X., Chen, T. L., et al. (2004). GDNF gene therapy attenuates retinal ischemic injuries in rats. *Molecular Vision, 10*, 93–102.

Yang, G. S., Schmidt, M., Yan, Z., Lindbloom, J. D., Harding, T. C., Donahue, B. A., et al. (2002). Virus-mediated transduction of murine retina with adeno-associated virus: effects of viral capsid and genome size. *Journal of Virology, 76*, 7651–7660.

Yiu, G., & He, Z. (2006). Glial inhibition of CNS axon regeneration. *Nature Reviews in Neuroscience, 7*, 617–627.

Zaiss, A. K., & Muruve, D. A. (2008). Immunity to adeno-associated virus vectors in animals and humans: a continued challenge. *Gene Therapy, 15*, 808–816.

Zhou, Y., Pernet, V., Hauswirth, W. W., & Di Polo, A. (2005). Activation of the extracellular signal-regulated kinase 1/2 pathway by AAV gene transfer protects retinal ganglion cells in glaucoma. *Molecular Therapy, 12*, 402–412.

Zincarelli, C., Soltys, S., Rengo, G., & Rabinowitz, J. E. (2008). Analysis of AAV serotypes 1–9 mediated gene expression and tropism in mice after systemic injection. *Molecular Therapy, 16*, 1073–1080.

# Controlled dissemination of AAV vectors in the primate brain

Vanja Varenika, Adrian P. Kells, Francisco Valles, Piotr Hadaczek,
John Forsayeth and Krystof S. Bankiewicz[*]

*Department of Neurological Surgery, University of California, San Francisco, CA, USA*

**Abstract:** Adeno-associated viral (AAV) vectors are currently the preeminent gene therapy vehicles for neurological application. However, issues regarding the trafficking of AAV vectors within the primate brain, and consequently control over the targeting of transgene expression, remain a matter of investigation. Studies in nonhuman primates have shown that distribution of AAV vectors is largely mediated by the flow of cerebrospinal fluid within perivascular space, trafficking of vector along axonal projections, and AAV receptor binding. Together these processes can result in transduction of cells in areas distant from the parenchymal site of infusion. Additionally, we have addressed the unique surgical issues concerning delivery of AAV vectors by convection-enhanced delivery and are working toward tailored delivery by means of real-time MRI.

**Keywords:** AAV; convection-enhanced deliver; MRI; perivasculature; cannula

## Introduction

The struggle to develop gene-based medicines has entered a new phase in recent years. In the face of many disappointments, either through lack of efficacy or adverse events, clinical development of viral vector-based therapeutic agents has begun to advance more certainly toward an eventual FDA-approved product. These advances have occurred chiefly in the field of neurology, particularly in Parkinson's disease (PD). An emerging body of safety data with adeno-associated virus serotype 2 (AAV2) in well over 100 subjects, at the time of writing, has established a vector system that can be produced in good manufacturing practice (GMP) facilities, has known biodistribution properties, and has negligible vector-dependent pathology. The preeminence of AAV2 vectors in neurological therapies has resolved a major issue in gene therapy for brain diseases. But there are still a number of other issues in neurological gene therapy that require investigation and development.

A consistently observed property of AAV2, apart from its in vivo specificity for neurons, is the extensive trafficking of vector from the site of infusion to distal anatomic locations. For example, infusion of AAV2 vector into putamen results in rapid transport to globus pallidus. The most remarkable example of AAV2 transport appears to be retrograde transport up motor neurons to the cell bodies in the spinal cord after injection of vector into rodent hind limb. In the design of

[*]Corresponding author.
Tel.: +1-415-502-3132;
E-mail: krystof.bankiewicz@ucsf.edu

nonclinical studies, consideration must be given to the issue of brain size and organization. The use of nonhuman primates (NHP) may be considered a prerequisite for clinical studies, since the ultimate anatomic location of AAV2 vectors is likely to be an important determinant of side effects and toxicity, and consequently of delivery parameters such as dose and volume.

A second important issue is one that our group has addressed for many years: gene transfer to central nervous system (CNS) requires surgery. In the case of neurological gene therapy of the CNS, new delivery hardware and procedures were essential to the development of a safe, widely reproducible procedure that could be practiced around the world beyond just a few premier research centers. Our early studies showed that convection-enhanced delivery (CED) was an effective means by which AAV2 could be distributed in brain parenchyma (Bankiewicz et al., 2000). Subsequently, we described an improved infusion cannula designed to prevent vector infused under pressure from refluxing up the outside of the cannula (Krauze et al., 2005b). More recently, we have investigated the use of magnetic resonance imaging (MRI) to guide infusions in real time in order to give the neurosurgeon rapid feedback on how a specific infusion is proceeding (Fiandaca et al., 2008a, b, c; Varenika et al., 2008). In our view, the importance of this more individualized approach to therapy is considerable. In our nonclinical experience of this technology, we have found that monitored infusions can alert us when infusate begins to leak into ventricles or when cannula reflux occurs (Varenika et al., 2008). Moreover, the precise volume of infusion (Vi) appropriate for each individual can be adjusted intraoperatively, and we hope that this approach will improve consistency and safety in the future.

Our rapidly advancing, detailed understanding of how AAV distributes within the brain, coupled with procedural improvements in delivery technology, will assist the field of neurological gene therapy to address diseases, such as Parkinson's (PD) and Lysosomal Storage Disorders (LSD), where broader gene delivery is required. In the following perspective, we describe some of the challenges and solutions addressed both by us and other investigators over the past few years in order to convert neurological gene therapy from a research project into a broadly applicable medicament.

**Trafficking of AAV in the brain**

The structure of the brain is complex and heterogeneous. Anatomical and physiological factors within the brain itself, such as myelination, cellular density, and fluid dynamics, all affect the ability of macromolecules and nanoparticles to distribute within the CNS. Although little is currently understood regarding the active transport of AAV vectors, it is becoming evident that, under the right conditions, gene delivery is not merely a localized delivery platform but can result in the expression and/or delivery of gene products to secondary areas of the CNS that are connected but spatially separated from the primary site of transduction. When combined with CED, it appears that widespread delivery of a gene product can be achieved after a limited number of infusions. Alternatively, it is likely that very focal AAV vector delivery to specific nuclei can result in widespread expression in areas that are directly connected by axonal projections, allowing for the targeting of gene products to specific neuronal populations while avoiding other areas of the CNS (Kells et al., 2009).

Infusion of a gene therapy vehicle such as the increasingly employed AAV vectors (primarily AAV2) requires a profound understanding of the mechanisms by which viral particles are disseminated from the infusion site often to remarkably distal locations in the primate brain. The pioneering work of Oldfield and his colleagues (Bobo et al., 1994) first advanced the concept of "convection." By pumping infusate into brain tissue slowly at a pressure that exceeds the hydrostatic pressure of the extracellular space it is possible to force infusate broadly into recipient tissue (Fig. 1). There appear to be at least three key factors at work in this mechanism of AAV vector distribution: (i) the vasculature; (ii) axonal projections; and (iii) AAV receptors.

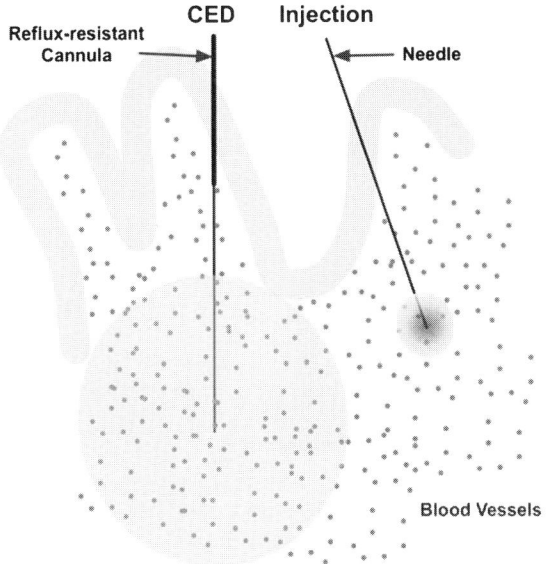

Fig. 1. Schematic comparison of CED and simple injection. Simple injection of AAV vector into parenchyma with a needle device results in poor distribution that depends on diffusion (right). In contrast, pressurized infusion of vector by the CED method pushes the AAV vector out into the interstitial space where viral particles can then engage the perivasculature that surrounds the dense network of blood vessels (red) pervading the brain. Pulsation of these vessels further enhances distribution (Hadaczek et al., 2006).

## *The perivascular pump*

Perivascular spaces are extensions of the subarachnoid space that accompany penetrating arteries into the brain all the way to the level of capillaries. The CSF flows freely into these spaces, allowing metabolites and small solutes to diffuse from the extracellular fluid. This circulation of CSF between subarachnoid space and perivascular space has been repeatedly documented (Rennels et al., 1985, 1990). The movement of fluid within the perivascular spaces seems to be driven by the pulsation of arteries in the brain. This rhythmic pulsation is described in our perivascular pump model of convective distribution (Hadaczek et al., 2006). It seems that fluid circulation through the perivascular space is the primary mechanism by which viral particles are distributed through the CNS during (and perhaps after) CED (Fig. 2). In one of our early experiments, we noticed the presence of viral capsids in the globus pallidus immediately after injection of AAV2 vector in the rat striatum. More recently, we documented strong hrGFP (green fluorescent protein) fluorescence in the external and internal globus pallidus, the subthalamic nucleus, medial forebrain bundle, the internal capsule, and anterior commissure after the infusion of AAV1-hrGFP into NHP striatum (Hadaczek et al., 2009). The perivascular pump is the most likely mechanism for the vector trafficking that we have seen in the basal ganglia. Manipulating cerebral vasomotor properties is a potential future direction for exerting additional control over therapeutic distribution (Varenika et al., 2008).

## *Axonal projections*

An often overlooked and poorly understood mechanism of gene delivery is axonal transport of vectors away from the site of infusion (Fig. 3). Recently, we have investigated gene expression in the frontal cortex of NHPs after thalamic AAV2 infusion (Kells et al., 2009). A single CED infusion of AAV2 vector encoding GFP or GDNF (glial cell-derived neurotrophic factor) that covered and transduced neurons across a large

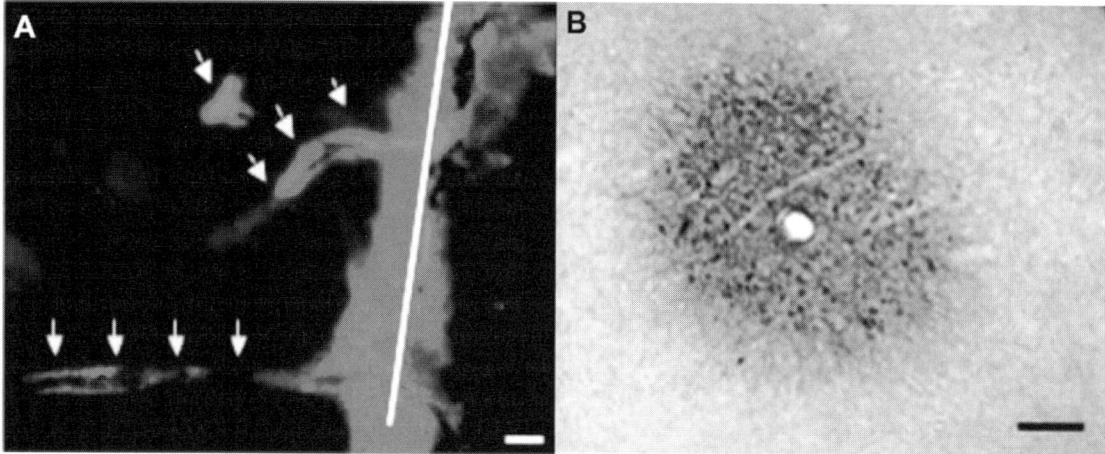

Fig. 2. The perivascular pump enhances the distribution of AAV2 vectors after CED infusion. The preferential movement of AAV vectors along perivascular space can be observed in the NHP brain by the presence of vector capsid or transgene expression in the immediate vicinity of blood vessels. (A) Directly after CED delivery of AAV2 vector to the putamen, vector capsids were found close to the catheter tract (white line) and surrounding nearby blood vessels (arrows). (B) Transgene expression in close proximity to blood vessels (white hole) in the thalamus after AAV2-GFP delivery to the putamen.

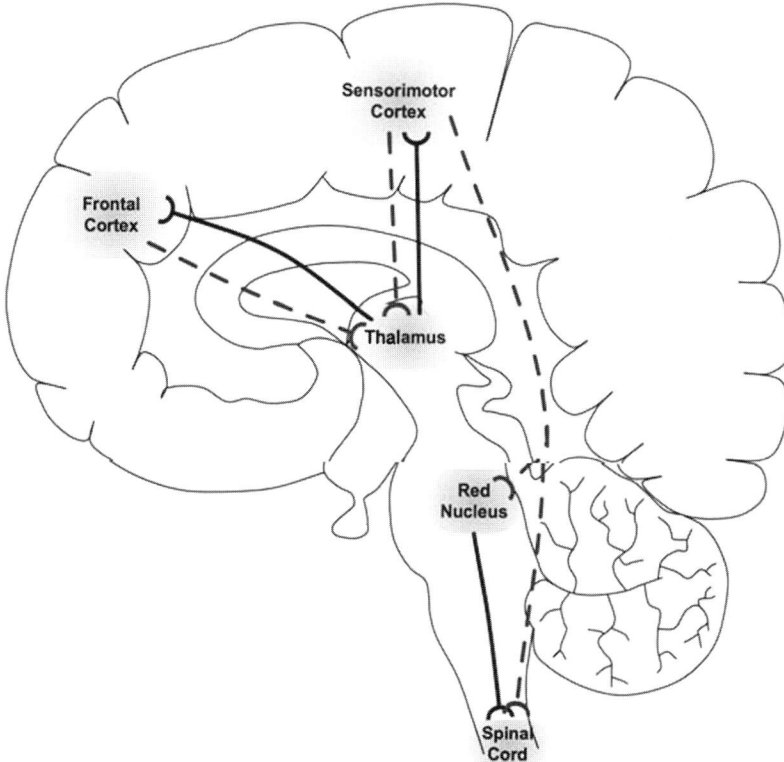

Fig. 3. Axonal trafficking of AAV vectors. Schematic diagram showing axonal pathways along which AAV vectors have been observed to be transported. Foust et al. (2008) showed anterograde transportation of AAV5 from the red nucleus to spinal cord. Transduction of cortical neurons after AAV2 vector infusion to the thalamus could be mediated via anterograde or retrograde transport mechanisms.

portion of the anterior and ventral thalamic nuclei also resulted in extensive transgenic protein expression in the cortex. Expression in the cortex was almost exclusively restricted to the gray matter layers, extending from the prefrontal cortex to the somatosensory and motor cortices. Identical patterns of expression were found with both secreted (GDNF) and nonsecreted (GFP) transgene products. Neuronal cell bodies and fibers in the cortex contained transgenic protein indicating both the AAV2 vector-mediated transduction of cortical neurons and anterograde transportation of transgenic proteins along thalamocortical projections. There are several means by which such trafficking might be mediated. Intraneuronal axonal transport is one possibility for which there is some evidence, either in an anterograde direction with subsequent release to transduce local neurons (Cearley and Wolfe, 2007), or in a retrograde direction to transduce the cell body as has been shown for motor neurons (Kaspar et al., 2002). Reciprocal projections between the thalamus and cortex are known to exist, and it will be important to determine whether anterograde or retrograde transportation is the predominant mechanism of AAV and transgene product transportation. Alternatively, it is possible that the periaxonal environment contributes to the migration of viral particles in a similar manner to the perivascular pump, and that neuronal tracts act as a conduit for the vectorial movement of macromolecules and nanoparticles.

A remarkable example of the clinical potential of axonal projections is that infusion of AAV5 vector into the red nucleus results in effective transduction of the rubro-spinal network (Foust et al., 2008). In primates, the rubro-spinal network seems somewhat vestigial. The attractiveness of this system is that each motor neuron in the ventral spinal cord is served by an afferent serotonergic neuron (Houk, 1991; Massion, 1967, 1988). Thus, as shown by Foust and colleagues, AAV5-GDNF delivery into the red nucleus resulted in secretion of GDNF into the extracellular space surrounding each motor unit, and sharply increased the concentration of GDNF in the CSF. This work has clear and important implications for treatment of ALS and other spinal disorders, including neuropathic pain. This study also effectively illustrates how targeting of small nuclei in the brain can result in effective targeting of large anatomical domains.

Another dramatic example can be seen in differential spread of CED infusions within gray and white matter regions of the brain. Our group has shown that the distribution of infusate increases more rapidly in white matter regions, e.g. corona radiata, than in the putamen that is predominantly gray matter (Krauze et al., 2005a). Additionally, gray matter infusions that gain access to white matter tracks will preferentially expand along the white matter, and concomitantly fail to increase in the targeted gray matter region (K. S. Bankiewicz, unpublished data), a phenomenon presumably due to the lower density of white matter.

*AAV receptors*

Cellular receptors also play a large role in determining the distribution and cellular specificity of viral vectors. Receptors like alpha-2,3-N-linked sialic acid (Wu et al., 2006) and heparan sulfate proteoglycan (Summerford and Samulski, 1998) facilitate efficient binding and transduction by AAV1 and AAV2, respectively. Although these receptors are absolutely necessary for transduction, they also have a tendency to retard the distribution of viral vector away from the infusion site. It has been shown that a more robust distribution of vector can be achieved by co-infusion of heparin and basic fibroblast growth factor because they temporarily block binding of AAV2 to its receptors, HSPG and bFGF receptors (Hadaczek et al., 2004; Mastakov et al., 2002; Nguyen et al., 2001).

**Real-time imaging**

Intraparenchymal infusion of AAV vector is fraught with the possibility of error either in cannula placement or aberrant distribution into the brain. We have shown that, even with flow rates that are less than 5 μl/min, reflux and leakage are detected in almost 20% of CED infusions in canines and NHP (Varenika et al., 2008). Approximately 10% of these infusions reach a large enough volume to cause detectable anatomical compression of the ventricles (Valles et al., 2009). Along with reflux, leakage, and anatomical compression, aberrant delivery is also a reality (Fiandaca et al., 2008a). All of these suboptimal CED infusion phenomena can only be observed, and potentially corrected if visualization of CED with real-time imaging is performed. In fact, variability of results, lack of efficacy, and serious complications have many potential ramifications within a clinical trial. These problems highlight the need to visualize infusions to ensure therapeutic safety and efficacy.

We have developed a real-time imaging technique that utilizes magnetic resonance (MR) imaging and co-infused MR contrast agents, such as free gadoteridol or gadolinium-loaded liposomes (GDL), to observe the progression of CED. This method has been developed extensively over the last 5 years, and has been used in the delivery of proteins (Hamilton et al., 2001) and liposomes (Dickinson et al., 2008). A combination of real-time imaging and reflux-resistant cannulae (Krauze et al., 2005b) has been used to increase the success rate of our infusions. Our group has actively promoted this technique and has shown that real-time CED is predictable and safe (Krauze et al., 2008).

Target structure coverage is another important feedback tool that real-time imaging provides during a CED infusion. Limiting CED to the target structure is necessary to ensure therapeutic dosage and to prevent undesired side effects, such as antibody development to therapeutic agent (Slevin et al., 2007). In our own AAV2-hAADC phase I clinical trial for PD, the treatment was safe and well tolerated (Eberling et al., 2008). As MRI of the trial was analyzed post-treatment, however, only partial coverage of the target structures was identified in some patients (Valles et al., 2009). Had real-time imaging been employed during these treatments, more extensive coverage might have been achieved. Furthermore, we have recently shown that GDL distribution seen with MRI corresponds well with the volume of AAV1 and AAV2 vector-mediated transduction, as detected by immunohistochemistry staining of histological sections (Fig. 4) (Fiandaca et al., 2008a). This suggests that GDL are not only good for visualizing the progress of infusions but also for marking future areas of gene transfer. These developments further emphasize the need for real-time imaging during CED.

Real-time MRI will allow neurosurgeons to have more confidence in their ability to cover the intended target by directly observing the progress of CED. The use of real-time MRI with a contrast agent will accurately define the volume of distribution (Vd) and provide feedback regarding reflux, leakage, or extension of the infusate beyond the target. Understanding therapeutic coverage allows the neurosurgeon to know when to stop the infusion to prevent leakage from occurring, or when maximal coverage has been reached. In the case of viral vector delivery, imaging markers can potentially predict areas of likely transduction in order to determine when the optimal therapeutic dose has been achieved. It is also important to note that mixed results of CED clinical trials have been reported in the absence of real-time visual guidance (Kordower et al., 1999; Kunwar, 2003; Levy et al., 2001). Visualizing the Vd in real time improves the ability of the surgeon to replicate this Vd in each patient, thereby, allowing comparisons of efficacy in similarly treated individuals.

## Clinical implications

The broad objective of our efforts over the past few years has been to assist in the maturation of neurological gene therapy from a clinical research tool into a widely available medical procedure. One of the key developments in the past 5–7 years in the field has been the gradual preeminence of a favored vector system, AAV2. One reason for this outcome has been merely historical. The more clinical experience we acquire for AAV2, the more compelling the rationale must be to deviate from this choice. The attractiveness of AAV2 vectors originally emerged because it is specific for neurons and, although this was not really appreciated at first, the fact that neurons are not

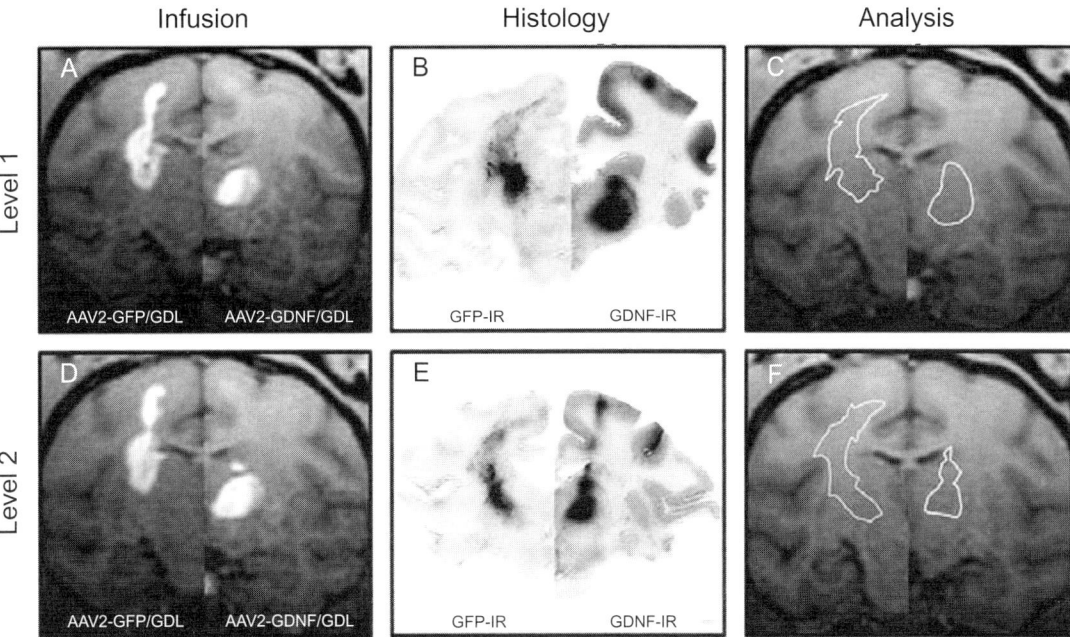

Fig. 4. Real-time MRI-guided infusion of AAV2 vector delivery. MRI correlation with histology in NHP with thalamic RCD, AAV2-GFP/GDL (left) and AAV2-GDNF/GDL (right). GFP and GDNF expression areas and GDL distribution at two levels are shown in panels (A) and (D). MR images and histology for each hemisphere were matched and composite panels were created. Two brain section levels, separated 1 mm from each other in thalamus, were examined. Area of GDL (A and D) as well as GFP and GDNF immunoreactivity (B and E) were examined. Areas of gene expression were drawn on MRI as shown in (C) and (F). Area of GDL distribution in the right thalamus measured $0.42\,cm^2$, whereas area of GDNF expression was $0.37\,cm^2$. Area of GDL distribution in the left thalamus was $0.67\,cm^2$ and area of GFP expression was $0.75\,cm^2$. While this type of analysis does not reveal total Vd of transgene expression and GDL, it strongly indicates very close codistribution between GDL and AAV2 vector-induced expression in the thalamus and supports use of image-guided CED of AAV2 vectors.

antigen-presenting cells (APC) may have made the use of AAV2 serendipitous (Forsayeth, 2006). This is not to say, however, that other serotypes should be avoided or are unlikely to be clinically useful, but merely that the use of AAV2 vectors in CNS applications has probably avoided problems that might have been encountered by serotypes with broader tropism. A further clinically relevant aspect of this investment in AAV2 as the clinical vector of choice has been that the path to the clinical trial for any AAV2-based vector has become easier. Regulatory agencies, such as the NIH Recombinant DNA Advisory Committee (RAC) and the United States Food and Drug Administration (FDA), have now seen a significant number of investigational new drugs (INDs) that propose to use AAV2 in clinical trials. The experience of clinical investigators in neurological Phase 1 safety trials so far has been excellent, although the same cannot be said with respect to peripheral use of AAV2 (Mingozzi and High, 2007). Manufacture of AAV2 vector has matured with the emergence of GMP facilities with both commercial and academic entities sponsoring clinical trials with AAV2 vector produced in-house. Our group has a long cumulative experience with the use of AAV2 in nonclinical studies, and our development of delivery technology is built around the use of this serotype. Extensive data on dosing, biodistribution, and immunology have served to increase our confidence that AAV2 is now becoming the default vector for neurological gene therapy (Cunningham et al., 2008; Forsayeth et al., 2006; Hackett et al., 2005; Herzog et al., 2008; Jacobson et al., 2006; Sanftner et al., 2004).

Whatever the application, it is of little consequence if the transgene cannot be delivered in a highly controlled manner to the target region(s) of the brain. Our recent efforts have been directed at accelerating the development of advanced cannula systems that will be qualified as medical devices and made available to the medical community through partnership with device manufacturers. We have substantially overcome the problem of reflux at the research level in a way that has enabled us to increase infusion rates fivefold from 1 ml/min to 5 ml/min (Krauze et al., 2005b). Clearly, a significant reduction in the time a patient spends on the operating table is a welcome improvement.

One of the most frustrating problems in intracranial vector delivery has been that, once an infusion starts, it is impossible to monitor what is happening. Has the cannula been placed accurately? Is vector leaking into ventricles, an event that causes CED into the parenchyma to cease (Varenika et al., 2008)? Has a sufficient volume of vector been delivered into the target region in order to achieve the desired volume of distribution? In order to develop a practically monitored delivery of AAV vector for clinical use, we have been combining MRI contrast reagents with AAV2 and have found that the two agents distribute very similarly in both gray and white matter. This raises the possibility that we will be able to validate in the coming years a system where infusions of AAV2 vector can be performed in an MRI suite by using the tracer distribution as a surrogate for AAV2. This approach is likely to have considerable potential in making neurological gene therapy a more predictable, therefore safer procedure. It also raises the prospect of personalizing neurological gene therapy to take account of patient-to-patient variability in delivery parameters.

## Conclusion

The development of AAV-based gene transfer vectors is providing an efficient route through which gene-based therapies can be targeted directly to the CNS. Intraparenchymal infusion of viral vectors through CED can provide long-lasting expression of therapeutic molecules, while avoiding issues related to systemic administration and blood–brain barrier transportation. The ability to target specific nuclei directly within the brain with stereotactic neurosurgery has lead to the investigation of several candidate genes for the treatment of PD (Fiandaca et al., 2008b). Development of delivery methods such as CED and the optimization of their parameters is the key to the future success of gene therapy in the CNS.

## References

Bankiewicz, K. S., Eberling, J. L., Kohutnicka, M., Jagust, W., Pivirotto, P., Bringas, J., et al. (2000). Convection-enhanced delivery of AAV vector in Parkinsonian monkeys; in vivo detection of gene expression and restoration of dopaminergic function using pro-drug approach. *Experimental Neurology, 164*, 2–14.

Bobo, R. H., Laske, D. W., Akbasak, A., Morrison, P. F., Dedrick, R. L., & Oldfield, E. H. (1994). Convection-enhanced delivery of macromolecules in the brain. *Proceedings of the National Academy of Sciences of the United States of America, 91*, 2076–2080.

Cearley, C. N., & Wolfe, J. H. (2007). A single injection of an adeno-associated virus vector into nuclei with divergent connections results in widespread vector distribution in the brain and global correction of a neurogenetic disease. *Journal of Neuroscience, 27*, 9928–9940.

Cunningham, J., Pivirotto, P., Bringas, J., Suzuki, B., Vijay, S., Sanftner, L., et al. (2008). Biodistribution of adeno-associated virus type-2 in nonhuman primates after convection-enhanced delivery to brain. *Molecular Therapy, 16*, 1267–1275.

Dickinson, P. J., LeCouteur, R. A., Higgins, R. J., Bringas, J. R., Roberts, B., Larson, R. F., et al. (2008). Canine model of convection-enhanced delivery of liposomes containing CPT-11 monitored with real-time magnetic resonance imaging: laboratory investigation. *Journal of Neurosurgery, 108*, 989–998.

Eberling, J. L., Jagust, W. J., Christine, C. W., Starr, P., Larson, P., Bankiewicz, K. S., et al. (2008). Results from a phase I safety trial of hAADC gene therapy for Parkinson disease. *Neurology, 70*, 1980–1983.

Fiandaca, M., Eberling, J., McKnight, T. R., Bringas, J., Pivirotto, P., Beyer, J., et al. (2008a). Real-time MR imaging of adeno-associated viral vector delivery to the primate brain. *NeuroImage* (in press).

Fiandaca, M., Forsayeth, J., & Bankiewicz, K. (2008b). Current status of gene therapy trials for Parkinson's disease. *Experimental Neurology, 209*, 51–57.

Fiandaca, M. S., Forsayeth, J. R., Dickinson, P. J., & Bankiewicz, K. S. (2008c). Image-guided convection-enhanced delivery platform in the treatment of neurological diseases. *Neurotherapeutics*, 5, 123–127.

Forsayeth, J. (2006). Influence of the immune system on central nervous system gene transfer. In M. G. Kaplitt & M. J. During (Eds.), *Gene therapy of the central nervous system: from bench to bedside* (pp. 47–51). Amsterdam: Academic Press.

Forsayeth, J. R., Eberling, J. L., Sanftner, L. M., Zhen, Z., Pivirotto, P., Bringas, J., et al. (2006). A dose-ranging study of AAV-hAADC therapy in Parkinsonian monkeys. *Molecular Therapy*, 14, 571–577.

Foust, K. D., Flotte, T. R., Reier, P. J., & Mandel, R. J. (2008). Recombinant adeno-associated virus-mediated global anterograde delivery of glial cell line-derived neurotrophic factor to the spinal cord: comparison of rubrospinal and corticospinal tracts in the rat. *Human Gene Therapy*, 19, 71–82.

Hackett, N. R., Redmond, D. E., Sondhi, D., Giannaris, E. L., Vassallo, E., Stratton, J., et al. (2005). Safety of direct administration of AAV2(CU)hCLN2, a candidate treatment for the central nervous system manifestations of late infantile neuronal ceroid lipofuscinosis, to the brain of rats and nonhuman primates. *Human Gene Therapy*, 16, 1484–1503.

Hadaczek, P., Forsayeth, J., Mirek, H., Munson, K., Bringas, J., Pivirotto, P., et al. (2009). Transduction of nonhuman primate brain with adeno-associated virus serotype 1: vector trafficking and immune response. *Human Gene Therapy*, 20, 225–237.

Hadaczek, P., Mirek, H., Bringas, J., Cunningham, J., & Bankiewicz, K. (2004). Basic fibroblast growth factor enhances transduction, distribution, and axonal transport of adeno-associated virus type 2 vector in rat brain. *Human Gene Therapy*, 15, 469–479.

Hadaczek, P., Yamashita, Y., Mirek, H., Tamas, L., Bohn, M. C., Noble, C., et al. (2006). The "perivascular pump" driven by arterial pulsation is a powerful mechanism for the distribution of therapeutic molecules within the brain. *Molecular Therapy*, 14, 69–78.

Hamilton, J. F., Morrison, P. F., Chen, M. Y., Harvey-White, J., Pernaute, R. S., Phillips, H., et al. (2001). Heparin coinfusion during convection-enhanced delivery (CED) increases the distribution of the glial-derived neurotrophic factor (GDNF) ligand family in rat striatum and enhances the pharmacological activity of neurturin. *Experimental Neurology*, 168, 155–161.

Herzog, C. D., Dass, B., Gasmi, M., Bakay, R., Stansell, J. E., Tuszynski, M., et al. (2008). Transgene expression, bioactivity, and safety of CERE-120 (AAV2-neurturin) following delivery to the monkey striatum. *Molecular Therapy*, 16, 1737–1744.

Houk, J. C. (1991). Red nucleus: role in motor control. *Current Opinion in Neurobiology*, 1, 610–615.

Jacobson, S. G., Boye, S. L., Aleman, T. S., Conlon, T. J., Zeiss, C. J., Roman, A. J., et al. (2006). Safety in nonhuman primates of ocular AAV2-RPE65, a candidate treatment for blindness in Leber congenital amaurosis. *Human Gene Therapy*, 17, 845–858.

Kaspar, B. K., Erickson, D., Schaffer, D., Hinh, L., Gage, F. H., & Peterson, D. A. (2002). Targeted retrograde gene delivery for neuronal protection. *Molecular Therapy*, 5, 50–56.

Kells, A. P., Hadaczek, P., Yin, D., Bringas, J., Varenika, V., Forsayeth, J., et al. (2009). Efficient gene therapy-based method for the delivery of therapeutics to primate cortex. *Proceedings of the National Academy of Sciences of the United States of America*, 106, 2407–2411.

Kordower, J. H., Palfi, S., Chen, E. Y., Ma, S. Y., Sendera, T., Cochran, E. J., et al. (1999). Clinicopathological findings following intraventricular glial-derived neurotrophic factor treatment in a patient with Parkinson's disease. *Annals of Neurology*, 46, 419–424.

Krauze, M. T., McKnight, T. R., Yamashita, Y., Bringas, J., Noble, C. O., Saito, R., et al. (2005a). Real-time visualization and characterization of liposomal delivery into the monkey brain by magnetic resonance imaging. *Brain Research. Brain Research Protocols*, 16, 20–26.

Krauze, M. T., Saito, R., Noble, C., Tamas, M., Bringas, J., Park, J. W., et al. (2005b). Reflux-free cannula for convection-enhanced high-speed delivery of therapeutic agents. *Journal of Neurosurgery*, 103, 923–929.

Krauze, M. T., Vandenberg, S. R., Yamashita, Y., Saito, R., Forsayeth, J., Noble, C., et al. (2008). Safety of real-time convection-enhanced delivery of liposomes to primate brain: a long-term retrospective. *Experimental Neurology*, 210, 638–644.

Kunwar, S. (2003). Convection enhanced delivery of IL13-PE38QQR for treatment of recurrent malignant glioma: presentation of interim findings from ongoing phase 1 studies. *Acta Neurochirurgica Supplement*, 88, 105–111.

Levy, R. M., Major, E., Ali, M. J., Cohen, B., & Groothius, D. (2001). Convection-enhanced intraparenchymal delivery (CEID) of cytosine arabinoside (AraC) for the treatment of HIV-related progressive multifocal leukoencephalopathy (PML). *Journal of Neurovirology*, 7, 382–385.

Massion, J. (1967). The mammalian red nucleus. *Physiological Reviews*, 47, 383–436.

Massion, J. (1988). Red nucleus: past and future. *Behavioural Brain Research*, 28, 1–8.

Mastakov, M. Y., Baer, K., Kotin, R. M., & During, M. J. (2002). Recombinant adeno-associated virus serotypes 2- and 5-mediated gene transfer in the mammalian brain: quantitative analysis of heparin co-infusion. *Molecular Therapy*, 5, 371–380.

Mingozzi, F., & High, K. A. (2007). Immune responses to AAV in clinical trials. *Current Gene Therapy*, 7, 316–324.

Nguyen, J. B., Sanchez-Pernaute, R., Cunningham, J., & Bankiewicz, K. S. (2001). Convection-enhanced delivery of AAV-2 combined with heparin increases TK gene transfer in the rat brain. *Neuroreport*, 12, 1961–1964.

Rennels, M. L., Blaumanis, O. R., & Grady, P. A. (1990). Rapid solute transport throughout the brain via paravascular fluid pathways. *Advances in Neurology*, 52, 431–439.

Rennels, M. L., Gregory, T. F., Blaumanis, O. R., Fujimoto, K., & Grady, P. A. (1985). Evidence for a 'paravascular' fluid circulation in the mammalian central nervous system, provided by the rapid distribution of tracer protein throughout the brain from the subarachnoid space. *Brain Research*, *326*, 47–63.

Sanftner, L. M., Suzuki, B. M., Doroudchi, M. M., Feng, L., McClelland, A., Forsayeth, J. R., et al. (2004). Striatal delivery of rAAV-hAADC to rats with preexisting immunity to AAV. *Molecular Therapy*, *9*, 403–409.

Slevin, J. T., Gash, D. M., Smith, C. D., Gerhardt, G. A., Kryscio, R., Chebrolu, H., et al. (2007). Unilateral intraputamenal glial cell line-derived neurotrophic factor in patients with Parkinson disease: response to 1 year of treatment and 1 year of withdrawal. *Journal of Neurosurgery*, *106*, 614–620.

Summerford, C., & Samulski, R. J. (1998). Membrane-associated heparan sulfate proteoglycan is a receptor for adeno-associated virus type 2 virions. *Journal of Virology*, *72*, 1438–1445.

Valles, F., Fiandaca, M. S., Bringas, J., Dickinson, P., LeCouteur, R., Higgins, R., et al. (2009). Anatomical compression due to high volume convection-enhanced delivery. *Neurosurgery*, (in press).

Varenika, V., Dickenson, P., Bringas, J., LeCouteur, R., Higgins, R., Park, J. W., et al. (2008). Real-time imaging of CED in the brain permits detection of infusate leakage. *Journal of Neurosurgery*, *109*, 874–880.

Wu, Z., Miller, E., Agbandje-McKenna, M., & Samulski, R. J. (2006). Alpha2,3 and alpha2,6 N-linked sialic acids facilitate efficient binding and transduction by adeno-associated virus types 1 and 6. *Journal of Virology*, *80*, 9093–9103.

CHAPTER 12

# From microsurgery to nanosurgery: how viral vectors may help repair the peripheral nerve

Martijn R. Tannemaat[1,2,]*, Gerard J. Boer[1], Ruben Eggers[1], Martijn J.A. Malessy[3] and Joost Verhaagen[1]

[1]*Laboratory for Neuroregeneration, Netherlands Institute for Neuroscience, an Institute of the Royal Academy of Arts and Sciences, Amsterdam, The Netherlands*
[2]*Department of Neurology, Leiden University Medical Center, Leiden, The Netherlands*
[3]*Department of Neurosurgery, Leiden University Medical Center, Leiden, The Netherlands*

**Abstract:** Reconstructive surgery of the peripheral nerve has undergone major technical improvements over the last decades, leading to a significant improvement in the clinical outcome of surgery. Nonetheless, functional recovery remains suboptimal in the majority of patients after nerve repair surgery. In this review, we first discuss the molecular mechanisms involved in peripheral nerve injury and regeneration, with a special emphasis on the role of neurotrophic factors. We then identify five major challenges that currently exist in the clinical practice of nerve repair and their molecular basis. The first challenge is the slow rate of axonal outgrowth after peripheral nerve repair. The second problem is that of scar formation at the site of nerve injury, which is detrimental to functional recovery. As a third issue, we discuss the difficulty in assessing the degree of injury in closed traction lesions without total loss of continuity of the involved nerve elements. The fourth challenge is the problem of misrouting of regenerating axons. As a fifth and final issue we discuss the potential drawbacks of using sensory nerve grafts to support the regeneration of motoneurons. For all these challenges, solutions are likely to emerge from (a) a better understanding of their molecular basis and (b) the ability to influence these processes at a molecular level, possibly with the aid of viral vectors. We discuss how lentiviral vectors have been applied in the peripheral nerve to express neurotrophic factors and summarize both the advantages and drawbacks of this approach. Finally, we discuss how lentiviral vectors can be used to provide new, molecular neurobiology-based, approaches to address the clinical challenges described above.

**Keywords:** peripheral nerve; regeneration; viral vectors; gene therapy; neurotrophic factors

## Introduction: peripheral nerve repair

The first series of successful surgical reconstructions of the peripheral nerve were described shortly after the Second World War (Sunderland, 1991). Since then, a better understanding of peripheral nerve anatomy, the evolution of surgical techniques, including epineurial suturing and the introduction of high power operation microscopes, have led to a significant improvement in the clinical outcome of surgery. Surgery is currently the preferred treatment for the transected nerve and consists of direct, tension-free,

*Corresponding author.
Tel.: +31 20 5665500; Fax: +31 20 5666121;
E-mail: m.r.tannemaat@lumc.nl

suture coaptation of nerve stumps. A detrimental factor in autonomous nerve regeneration is the occurrence of fibrosis in the gap between the proximal and distal nerve stump. This may lead to neuroma formation, thereby greatly reducing the chance of successful axonal regeneration. In such instances, the present gold standard of treatment is neuroma resection and the application of autologous nerve grafts as scaffolds to bridge the gap between the proximal and distal nerve stumps.

Recovery after peripheral nerve reconstruction is almost never complete, and a considerable degree of functional impairment usually remains. The clinical outcome of repair is especially limited (i) in adults over the age of around 50, (ii) in proximal nerve lesions such as brachial plexus lesions, and (iii) when long nerve transplants have to be used or (iv) when there is a long delay between trauma and surgery (Sunderland, 1991). It is generally held that microsurgery has now reached an optimal technical refinement and new concepts have to be developed to further promote recovery after surgical nerve repair (Lundborg, 2000). A new conceptual framework is currently emerging from a better understanding of the molecular basis of nerve regeneration and the application of novel intervention strategies, including viral vector-mediated gene transfer.

In this review, we first discuss the molecular mechanisms involved in peripheral nerve injury and regeneration, with a special emphasis on the role of neurotrophic factors. Subsequently, we discuss the challenges that currently exist in the clinical practice of nerve repair. We then discuss viral vector-mediated overexpression of neurotrophic factors and its effect on regeneration. Finally, we suggest new molecular neurobiology-based approaches to address the clinical challenges described above.

## Molecular mechanisms involved in peripheral nerve regeneration

Nerve regeneration starts within hours of peripheral axonal injury. Myelin is cleared by the Schwann cells distal to the lesion site as they change from their usual myelinating state to an activated, outgrowth-promoting state (Chen et al., 2007). Transected axons send out several sprouts with growth cones that contain receptors at their tips for basal lamina components and neurotrophic factors (Maggi et al., 2003). These regenerating axons are physically guided to the appropriate targets by the basal lamina tubes and attracted by neurotrophic factors secreted by Schwann cells. The chance for outgrowing axons to successfully re-establish contact with the appropriate end organs is greatly reduced when the continuity of basal lamina tubes is lost (Nguyen et al., 2002). However, successful regeneration requires more than just the outgrowth of axons into the distal stump. In proximal lesions, motoneurons show atrophy which impairs their regenerative capacity (Fu and Gordon, 1995b). Furthermore, a number of elongating axons are inevitably lost due to malcoaptation, suture line scarring, and discrepancies in the cross-sectional dimensions of the proximal and distal stump, respectively. In addition, misrouting of axons into functionally inappropriate endoneurial tubes results in regenerated, but useless axons. These aberrant patterns of innervation are responsible for phenomena such as cocontraction (Gramsbergen et al., 2000). Distal from the injured axon, Schwann cells express a wide range of neurotrophic proteins including nerve growth factor (NGF), neurotrophin-3 (NT-3), neurotrophin-4/5 (NT-4/5), and brain-derived neurotrophic factor (BDNF), glial cell line-derived neurotrophic factor (GDNF, a member of the transforming growth factor superfamily), and ciliary neurotrophic factor (CNTF, a neuroactive cytokine) (Boyd and Gordon, 2003b). Spinal motoneurons and different subtypes of sensory neurons in the dorsal root ganglion differ in their sensitivity for these neurotrophic factors. Motoneurons express the neurotrophin receptor p75, the receptor for BDNF, tropomyosin receptor kinase (trk) B, and the multicomponent receptor for GDNF composed of a common signal transduction subunit, ret, and the GDNF family receptor α (GFRα) (Boyd and Gordon, 2003b). Sensory neurons of the dorsal root ganglion exist at least as three subpopulations: (a) nociceptive, peptidergic neurons that express trkA, the receptor for NGF, (b) proprioceptive neurons that express trkC, the receptor for NT3, and (c) a population of neurons that are defined by their ability to bind to the *Griffiona Simplicifolia* isolectin B4, which express the receptors for GDNF, GFRα, and ret (Pezet and McMahon, 2006). The function of the latter population of neurons is not entirely clear, but they appear to play a role in nociception as well (Pezet and McMahon, 2006).

The initially elevated expression of neurotrophic factors by Schwann cells in the distal stump, and their corresponding capability to support regeneration decreases after a period of several weeks to months (Boyd and Gordon, 2003b). This contributes significantly to the lack of long-distance regeneration and poor functional recovery after proximal peripheral nerve lesions (Hoke, 2006). Interestingly, the expression of several of these factors is different in Schwann cells from sensory and motor nerves and this differential expression improves regeneration of motoneurons through motor nerve grafts compared to sensory grafts and vice versa (Hoke et al., 2006). Therefore, there are profound and differential effects of neurotrophic factors on the regeneration of subpopulations of neurons.

## Current challenges in peripheral nerve repair

In this section, we will briefly discuss the challenges that exist in the present clinical practice of peripheral nerve repair. This is by no means intended as an exhaustive review of all issues that determine the functional outcome of peripheral nerve surgery, but instead, we will focus on those issues (Fig. 1) that are relevant to new putative molecular neurobiology-based therapeutic strategies.

The most common clinical distinction is between axonotmetic nerve injuries, from which generally recovery is spontaneous and which are therefore not treated surgically, and neurotmetic injuries, in which surgical reconstruction is usually indicated (Sunderland, 1991). However, although regeneration takes place in axonotmetic injuries, this does not always result in full *functional* recovery. At a rate of regeneration of 1–5 mm/day (Sunderland, 1991), it may take several months for regenerating axons to reinnervate their distal targets and during this time these targets may have atrophied (Fu and Gordon, 1995a), resulting in a poor restoration of function (Hoke, 2006). This slow rate of regeneration and associated atrophy of denervated target organs is therefore a clinical problem and a potential target for therapeutic intervention. Therefore, a *first* challenge for improvement of functional recovery after severe peripheral nerve trauma is to increase the velocity of axonal outgrowth.

In cases of acute neurotmetic injuries (e.g., the complete transection of the median nerve in a stab wound), the clinical strategy to treat the injured nerve is relatively straightforward and involves immediate repair surgery, either through the direct tension-free coaptation of proximal and distal nerve stumps, or through the application of nerve grafts that act as scaffolds for regenerating axons. However, the treatment strategy is not always this clear-cut, especially in closed traction lesions without total loss of continuity of the involved nerve elements. These lesions are characterized by extensive intraneural fibrosis and the formation of a collagenous scar at the site of injury, the so-called neuroma-in-continuity. It is generally accepted that this process is detrimental to functional recovery, although the molecular basis of this phenomenon is not fully understood (MacKinnon and Dellon, 1988; Sunderland, 1991; Maggi et al., 2003). Myofibroblasts appear to proliferate at the site of nerve injury, contributing to scar formation by producing collagen (Badalamente et al., 1985). Fibroblasts also express the inhibitory proteoglycan NG2, which may play a role in blocking axon regeneration through scar tissue (Morgenstern et al., 2003). We have recently shown that the chemorepulsive protein Semaphorin 3A is also expressed by fibroblasts in this scar and surrounds the axons of regenerating neurons, again potentially contributing to the inhibitory properties of the scar (Tannemaat et al., 2007a). Prevention of scar formation at the site of nerve injury, and at the

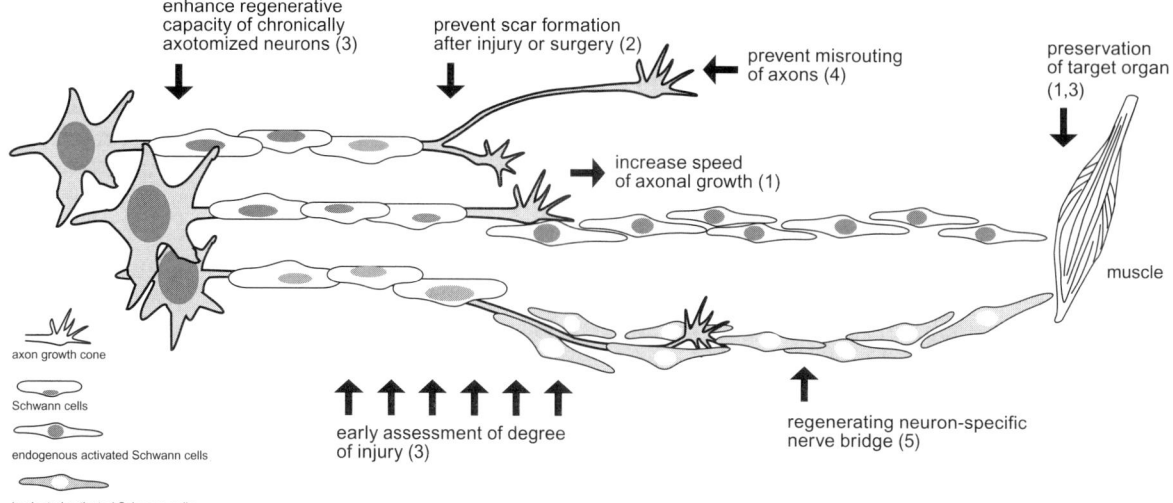

Fig. 1. Potential ways to improve the results of peripheral nerve surgery. In this paper, we identify five major challenges that currently exist in the clinical practice of nerve repair: (1) the slow rate of axonal outgrowth after peripheral nerve repair and associated target organ atrophy, (2) scar formation at the site of nerve injury, (3) the difficulty in assessing the degree of injury in closed traction lesions, (4) the problem of misrouting of regenerating axons, and (5) problems related to the use of sensory nerve grafts to support the regeneration of motoneurons. This schematic overview of injured and regenerating neurons shows the approaches that can be taken to overcome these challenges. The numbers in parentheses indicate which challenge is addressed by each particular approach.

coaptation site after nerve repair, is therefore a *second* challenge to be faced to increase the number of axons that can reach their target organ successfully.

In the neuroma-in-continuity, both axonotmetic and neurotmetic injury types can be present, and it is therefore difficult to predict the level of spontaneous regeneration and recovery of function (MacKinnon and Dellon, 1988). The absence of a sensitive method to determine the degree of nerve injury in closed lesions at an early time point implies that the indication for surgical exploration is based on the case history, age of the patient, involved nerve elements, localization of the lesion, repeated clinical examination in time, and electromyographic examinations (Malessy and Pondaag, 2009). When functional recovery does not take place after an empirically defined waiting period, surgery will be performed. However, the regenerative potential of, e.g., axotomized motoneurons diminishes over time (Fu and Gordon, 1995b; Furey et al., 2007) and therefore, this approach is in effect a trade-off between not waiting long enough (running the risk of performing surgery on injuries that would have displayed a certain degree of spontaneous recovery) and waiting too long (impairing the outcome of reconstructive surgery for severe lesions) (Gilbert and Tassin, 1984; Sunderland, 1991; Clarke and Curtis, 1995; Waters, 1999). The *third* clinical challenge, consists, therefore, of (i) improving the ability to assess the degree of injury immediately following the trauma, (ii) enhancing the regenerative capacity of chronically axotomized neurons (Fu and Gordon, 1995b), and/or (iii) preserving the condition of denervated target organs (Fu and Gordon, 1995a).

Although after nerve repair both motor and sensory neurons are able to regenerate and re-establish functional connections, there are differences in their regenerative response. Motoneurons preferentially reinnervate muscles, a phenomenon that is the result of both interaction between the motoneuron and the distal endoneurial tubes at the coaptation site (Redett et al., 2005) and of the "pruning" of motoneuron axons that have inaccurately entered endoneurial tubes that lead to sensory organs (Madison et al., 1999). However, the increased likelihood that motoneurons reinnervate muscles, called "preferential motor reinnervation" (Brushart, 1993), does not mean that these motoneurons are capable of selectively finding their original targets. Instead, the outgrowth of motor axons is a random process that frequently results in the reinnervation of inappropriate, antagonistic muscles (Gramsbergen et al., 2000) or even reinnervation of two muscles by branches from one motoneuron (de Ruiter et al., 2008a). This leads to the cocontraction of antagonizing muscles and subsequently a failure to recover functionally. The prevention of this "misrouting" phenomenon is the *fourth* challenge that provides an opportunity for future therapies.

Finally, the regeneration of motoneurons through grafts derived from purely motor nerves is better than through grafts from sensory nerves and vice versa (Hoke et al., 2006; Moradzadeh et al., 2008). Whereas some researchers have attributed this fact to the differential expression of several neurotrophic factors in motor and sensory nerve grafts (Hoke et al., 2006), others claim that physical differences play an important role. Axons of sensory neurons generally have a smaller diameter, and the endoneurial tubes in grafts derived from sensory nerves may not be ideally suited to accommodate the larger diameter axons of motoneurons (Moradzadeh et al., 2008). The most commonly used graft in human reconstructive surgery is derived from the sural nerve, because this sensory nerve can be harvested without causing major disturbing functional deficiencies, which is not the case for motor nerves. The putative drawback of using sensory nerve grafts to support the regeneration of motoneurons is the *fifth* and final problem in the clinical practice of nerve repair.

In light of the issues described above, we will now continue to discuss novel approaches to enhance recovery after peripheral nerve injury.

## Viral vectors: promising tools to express potentially therapeutic proteins in injured peripheral nerves

We have recently reviewed the properties of the various types of viral vectors and their suitability for peripheral nerve research (Tannemaat et al., 2008b). In summary, lentiviral (LV) vectors currently appear to be the vector of choice, as these vectors consistently and durably transduce significant numbers of Schwann cells in an injured rat peripheral nerve, without interfering with reconstructive surgery or impairing its functional outcome (Hendriks et al., 2007; Eggers et al., 2008). The relative ease with which new, potentially interesting genes can be cloned into these vectors means that the in vivo effect on regeneration of many proteins can now be studied in far more efficient ways than previously possible. This will have a remarkable impact on the future direction of peripheral nerve research, and in the long run perhaps even result in clinical application when recovery can be improved convincingly in animal studies.

Therapies aimed at influencing peripheral nerve regeneration are, inherently, only required temporarily. Therefore, it is not unreasonable to question whether it is clinically realistic to inject vectors that will permanently express a therapeutic gene in cells, when the need for a particular protein is only short-term. This issue could be resolved with viral vectors with regulatable expression, in which transgene expression could be regulated by the oral intake of doxycyclin (Gossen and Bujard, 1992; Blesch et al., 2005). However, there are issues related to the safety of the application viral vectors that need to be resolved: (i) the presently available tetracycline-controlled transactivator necessary for regulatable gene expression is of bacterial origin (Gossen and Bujard, 1992) and is therefore immunogenic (Lena et al., 2005), and (ii) LV vectors may cause unwanted mutational effects in transduced cells since these vectors insert the transgene into the host cells genome. Although this drawback of LV vectors is repeatedly mentioned in the literature (Beard et al., 2007; Cattoglio et al., 2007; Hargrove et al., 2008), no negative effects of

LV vector-induced insertional mutagenesis have been reported so far. Insertional mutagenesis is not an issue with the application of recently created nonintegrating LV vectors (Yanez-Munoz et al., 2006). However, one has to keep in mind that episomal vector genomes may be lost during cell division. Since Schwann cells in the distal nerve stump divide after an injury, insertion-deficient LV vector-mediated transgene expression may be rapidly lost as a result of Schwann cell proliferation.

Adeno-associated viral (AAV) vectors are probably the safest viral vector available to date (Kaplitt et al., 1994) and AAV vectors have been applied in a number of clinical trials for neurodegenerative disease (Kaplitt et al., 2007). To our knowledge, the currently available AAV vectors do not transduce Schwann cells, although it would be worthwhile to investigate whether some of the recently developed AAV serotypes have the capacity to transduce these cells. AAV vectors would however be the vector of choice for delivery of a therapeutic gene to motor or sensory neurons. Most AAV vectors preferentially transduce neurons and in the laboratory of one of the authors (Joost Verhaagen) a large effort is underway to determine the optimal AAV serotype for gene therapy for motor and sensory neurons. We also refer to the chapter by Alan Harvey and colleagues in this volume for more information on AAV serotypes and their capacity to transduce different cell types in the retina.

The most likely role for viral vectors will be that they can be used as powerful tools for fundamental research on peripheral nerve repair. We predict that it will take several years to definitively establish the beneficial effect of viral vector-mediated expression of a protein in clinically relevant animal models for peripheral nerve injury. During this time, the safety of viral vectors for human application may have been established in the clinical trials that are currently underway (Carter, 2005) and solutions will probably emerge for the issues described above. Depending on the progress that is made in the field of gene therapy, it may be possible to translate positive effects of viral vector-mediated gene transfer in experimental peripheral nerve injury models directly to a gene therapy approach in humans, but alternatively the effective therapeutic proteins may then also be delivered by a more conventional method such as biodegradable slow-release capsules (Fine et al., 2002). Either way, gene delivery by means of viral vectors will have a significant impact on the development of new therapeutic strategies to enhance the results of neurosurgical peripheral nerve repair.

## Lentiviral vector-mediated overexpression of neurotrophic factors to enhance nerve regeneration

We have recently performed a series of experiments in which we applied LV vectors to express neurotrophic factors in various rat models of peripheral nerve injury, in an attempt to enhance regeneration after surgical repair. However, first we investigated whether it was feasible to transduce human nerve grafts ex vivo. This should answer the question whether it will be possible to introduce LV vector-mediated gene transfer in the surgical arena. We were able to develop a protocol to genetically modify fibroblasts in cultured segments of human sural nerve using LV vectors expressing the marker gene for green fluorescent protein (GFP) (Tannemaat et al., 2007a). With the subsequent application of an LV vector encoding nerve growth factor (LV-NGF), long-term production of biologically active NGF could be directed to cultured human nerve segments. We then moved to rat studies of surgical repair of lesioned nerve, since the application of ex vivo LV vector-treated nerve implants in this species had already been shown to induce long-term transgene expression. Schwann cells within the peripheral nerve could easily be transduced. They express the transgene up to at least 16 weeks provided that the transgene encodes a nonimmunogenic protein (Hendriks et al., 2007). Following application of LV vector-mediated overexpression of BDNF or GDNF in the ventral root after avulsion and reimplantation in the cord, a complete reversal of spinal cord motoneuron atrophy and an increased number of regenerating axons entering the reimplanted roots

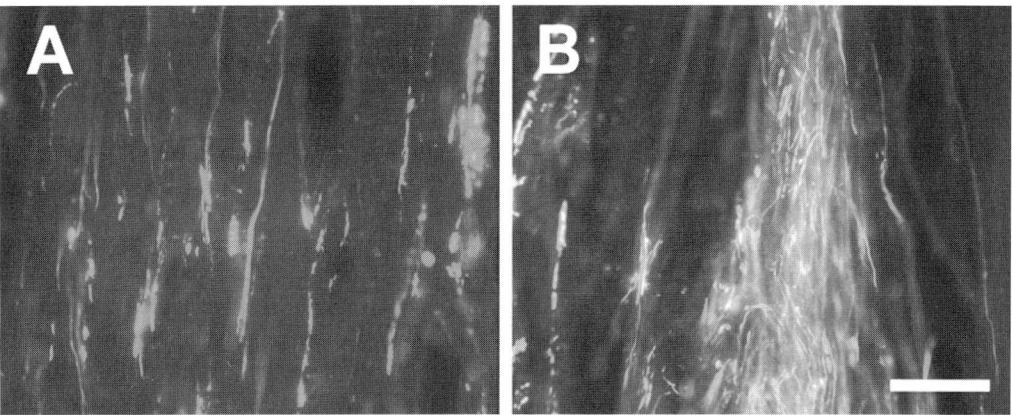

Fig. 2. The candy store effect. Locally increasing the concentration of neurotrophic factor (e.g., through viral vector-mediated overexpression) at the site of peripheral nerve injury does not improve regeneration. In this longitudinal section of a peripheral nerve after experimental lesioning and subsequent repair, high levels of transgenic GDNF expression at the site of repair (green: GDNF) induce a neuroma-like "candy store" of coiled motoneuron axons (red: choline acetyl transferase). Long-distance outgrowth of neurites towards their targets is thus impaired. (See Color Plate 12.2 in color plate section.)

was observed with LV-GDNF (Eggers et al., 2008). However, high levels of transgenic GDNF expression in the reimplanted ventral root induced a truly striking neuroma-like "oasis" of coiled axons (Fig. 2), and the number of regenerated fibers more distally was, in fact, lower in comparison with control treatments. Thus, long-term local expression of GDNF hampered the long-distance outgrowth of neurites towards their targets, previously also reported in the spinal cord as the "candy store" effect for regenerating fibers (Blits et al., 2004). Consequently, LV-GDNF treatment also failed to improve the recovery of hind limb function.

The effect of LV vector-mediated overexpression of NGF and GDNF was also investigated in a rat model for peripheral, i.e., sciatic nerve transection and coaptation repair (Tannemaat et al., 2008a). Being a mixed nerve (i.e., containing both sensory and motor axons), the sciatic nerve model allowed us to compare the regenerative response of sensory and motor neurons to increased levels of different neurotrophic factors. The regeneration of motoneurons was not affected by LV vector-mediated overexpression of NGF. However, again the LV vector-mediated expression of GDNF in the transected nerve impaired the long-distance regeneration of motoneurons: the number of regenerated motoneurons as assessed by retrograde tracing 1 cm distal from the transection/repair site was lower in comparison with controls. In the lesion-affected sensory neurons, LV vector-mediated overexpression of both NGF and GDNF causes profound phenotypic changes that are in line with the notion that these factors play an important role in the pathogenesis of pain (Tannemaat et al., 2008a). Although both the differential regenerative response of motor and sensory neurons upon application of these factors and the changes in nociceptive sensory neurons could form the starting point for new exciting investigations, these findings also highlight the unwanted side-effects associated with long-term local viral vector-mediated overexpression of neurotrophic factors.

**Exogenous neurotrophic factors: precise time- and location-specific application is necessary to promote outgrowth**

Since the discovery that NGF can promote neurite outgrowth by Rita Levi-Montalcini in 1952, neurotrophic factors have arguably been the most widely studied proteins in the field of neuroregeneration (Boyd and Gordon, 2003b). Interestingly, in the treatment of several neuro-*degenerative* diseases (e.g., Alzheimer's disease and Parkinson's disease), the application of neurotrophic factors has been very promising (Kordower et al., 2000; Tuszynski et al., 2005). Local expression of these factors prevents the degeneration of specific sets of neurons in animal models for these diseases. Clinical trials with neurotrophic factors for these diseases are currently underway and encouraging results have been reported (Tuszynski et al., 2005; Elsworth et al., 2008). In contrast, the application of neurotrophic factors to enhance axonal *regeneration* has proven to be far more difficult. In addition to the studies described above, neurotrophic factors have been applied by nonviral methods with only very modest effects on regeneration. The addition of recombinant NGF or GDNF to fibrin sealant applied at the site of nerve repair leads to minor improvements in functional recovery tests in rats, and only NGF increases the number of regenerated motoneurons (Jubran and Widenfalk, 2003). The application of NT4 in a similar model led to modest, but significant improvements in functional recovery, axon numbers, and myelination of axons (Yin et al., 2001). The application of NGF or NT3 through osmotic mini-pumps did not lead to long-lasting improvements of anatomical or functional recovery at 12 weeks post-lesion in a rat model for sciatic nerve transection (Young et al., 2001). A combination of GDNF and BDNF applied at the site of repair through osmotic mini-pumps increased the number of regenerated motoneurons in a model of chronic axotomy (Boyd and Gordon, 2003a), while the effect of BDNF alone appears to be either inhibitory or stimulatory, depending on the dose applied (Boyd and Gordon, 2002). In summary, these results of the experimental application of neurotrophic factors to enhance regeneration in rat nerve repair models have not been overwhelmingly positive.

One explanation for these findings can be that unlike promoting successful neuronal survival, successful regeneration depends on precise time- and location-dependent expression of these factors to create a gradually shifting gradient towards which axons continue to extend over periods of several months. The "candy store" effect initially described in the spinal cord following root avulsion (Blits et al., 2004) and recently also in reimplanted ventral roots (Eggers et al., 2008) and in the transected peripheral nerve (Tannemaat et al., 2008a), shows that higher levels of certain neurotrophic factors are not necessarily better when long-distance axon regeneration is required. Furthermore, we have recently obtained data on the temporal expression profile of NGF, NT-3, BDNF, and GDNF in avulsed nerve roots that clearly show the subtle temporal changes in the expression of these factors (unpublished observations). In other words, the spatial and temporal endogenous expression of these factors in the injured peripheral nerve may already be optimally regulated to support regeneration and one can envision that their exogenous application will have no other effect than disrupting this naturally occurring delicate balance of neurotrophic support.

An important caveat in this respect is that the robust regeneration in rat models for peripheral nerve injury does not necessarily translate to the human clinical situation, where the distance to be bridged from lesion site to target organ is usually much longer (Hoke, 2006). Nonetheless, the injured human peripheral nerve is also capable of regenerating over long distances (Sunderland, 1991; Maggi et al., 2003), suggesting a mechanism that shifts the elevated endogenous production of neurotrophic factors ahead of regenerating axons. Long-term exogenous and locally increased amounts of neurotrophic factors in the nerve therefore seems neither needed nor helpful. This is disappointing, because the transduction with LV vectors to overexpress a combination of motoneuron-specific neurotrophic factors (Hoke et al., 2006) could be an interesting strategy to render sensory sural nerve grafts more supportive of the regeneration of motoneurons. However, unless it becomes possible to create a time- and location-dependent gradient of these factors in the nerve (Blesch and Tuszynski, 2007), this approach is not likely to be beneficial for long-distance regeneration through nerve grafts.

## Enhancing motoneuron survival as an application of neurotrophic factors

The exogenous application of neurotrophic factors still faces significant challenges if the goal is to stimulate peripheral nerve *regeneration*. However, by exploiting the *survival*-enhancing properties of trophic factors, there may be several possibilities to enhance the functional outcome of nerve repair.

Firstly, the viral vector-mediated application of GDNF could be a promising approach to prevent motoneuron atrophy after nerve root avulsion, "keeping them in shape" prior to reconstructive surgery. This is of particular importance as the diminished regenerative capacity of chronically axotomized motoneurons (Furey et al., 2007) can be boosted by the temporary application of GDNF (Boyd and Gordon, 2003a). This could therefore be a way to compensate for the deleterious effect of the waiting period that is currently part of the surgical decision-making process.

Secondly, there may be a role for neurotrophic factors in the prevention of target muscle atrophy during the period of denervation. CNTF has a strong myotrophic effect on denervated skeletal muscle (Helgren et al., 1994; Huang et al., 2002), but systemic application of CNTF causes unwanted side-effects such as severe weight loss (Vergara and Ramirez, 2004). Local LV vector-mediated expression of CNTF in the muscle (Kaspar et al., 2003) may be a better method to prevent muscle atrophy without causing significant side-effects, thereby addressing the clinical issue of chronic denervation.

Apart from practical issues involved in applying viral vectors to, for instance, axotomized motoneurons in the spinal cord, the success of these strategies will depend on two factors: (i) the ability to tightly regulate the amount of neurotrophic factors produced and (ii) the ability to switch off transgene expression completely when it is no longer helpful. This is theoretically possible with viral vectors with regulatable gene expression. As discussed in the previous paragraphs, there are still unresolved issues regarding the safety and clinical applicability of vectors that direct regulatable transgene expression. It will therefore require several years of additional research before these strategies are ready to be tested in a clinical setting.

## Failed functional recovery: misrouting of regenerating axons is an important factor

As proposed in the paragraphs above, in most cases regeneration of the peripheral nerve may not need stimulation by the exogenous application of factors like NGF and GDNF as they are already tightly and autonomously regulated to optimize continuous outgrowth. Why then, is functional recovery of the peripheral nerve after reconstructive surgery often not complete? If this is not the result of an insufficient spontaneous regenerative response, there must be another reason why, for instance, patients with a brachial plexus lesion often fail to regain function of distal targets like the hand after reconstructive surgery (Pondaag and Malessy, 2006).

There is an increasing body of evidence suggesting that the misrouting problem is a strong contributing factor to the lack of functional recovery after nerve repair surgery (DeMatteo et al., 2006; van Dijk et al., 2007; de Ruiter et al., 2008). Younger patients may be able to partially compensate for misrouted axons due to the plasticity of their still maturing CNS (Malessy et al., 1998a, b), but this ability of the brain to adapt to the newly formed peripheral connections diminishes with aging. To eliminate the cocontraction of antagonizing muscles, Botulinum toxin type A has been injected in the triceps muscle of patients with brachial plexus injuries (DeMatteo et al., 2006). The purpose of this approach is to facilitate motor learning by temporarily inducing the relaxation of antagonist muscles and allowing increased activity in the reinnervated biceps muscle. Other than this, there are no pharmacological options to treat misrouting and the problem is usually addressed with intensive physical therapy (Ramos and Zell, 2000).

In research, the routing problem has mainly been addressed mechanistically, for instance by applying artificial scaffolds with a three-dimensional structure to enhance the physical guidance of regenerating axons (Stokols and Tuszynski, 2004; Stokols et al., 2006; Bozkurt et al., 2007; de Ruiter et al., 2008b). It is true that the physical properties of the distal endoneurial tubes determine the fate of regenerating axons and thus strongly influence the degree of functional recovery (Nguyen et al., 2002). However, surgical reconstruction of severe neurotmetic nerve injuries is inherently accompanied by a loss of continuity of nerve fascicles, impairing the ability of the nerve to physically guide regenerating axons. In addition, the sural nerve graft that is most commonly used for reconstructive surgery (see above) already contains thousands of aligned Schwann cells in longitudinally oriented endoneurial tubes, so it is difficult to imagine how this could be improved by artificial guides. Theoretically, one could envision a refinement of surgical techniques up to a level where each individual axon in the proximal stump is matched to one endoneurial tube in carefully prepared (artificial or sural) nerve grafts, but even then it is hard to imagine how the appropriate distal targets for these axons could be identified or how branching at the coaptation site could be prevented. Antibodies against neurotrophic factors have been applied in a rat model for peripheral nerve transection to reduce the branching of regenerating axons and thus improve the quality of regeneration (Streppel et al., 2002), but this approach carries the risk of interfering with the natural neurotrophic factor signaling that is needed for successful regeneration.

Another approach could be to allow regeneration to take place after repair as usual, but then enhance the ability to use the newly formed connections by increasing plasticity at the level of the spinal cord (Galtrey et al., 2007) or brain. Although this approach will not be able to compensate for synkinesias caused by the innervation of antagonizing muscles by branches from the same motoneuron (de Ruiter et al., 2008a), it shows the importance of including the CNS as a target for therapeutic interventions aimed at functional recovery of the injured peripheral nerve.

## Addressing the routing problem at a molecular level

We have recently discovered that the chemorepulsive protein semaphorin3A is present in the human neuroma and may contribute directly to the trapping and disorganization of axons in the human nerve scar (Tannemaat et al., 2007b). Furthermore, the expression of several class 3 semaphorins is upregulated in the injured rat peripheral nerve (Ara et al., 2004). These findings reinforce the emerging view that the routing of regenerating axons is not only influenced by (i) the physical alignment of regenerating axons in endoneurial tubes and (ii) the expression of neurotrophic factors with specificity for particular populations of peripheral neurons, but also (iii) by the lesion-induced expression of specific axon guidance genes in non-neuronal cells (Redett et al., 2005). Among the genes specifically expressed in the injured peripheral nerve are the carbohydrate epitope L2 (Martini et al., 1994), versican (Hattori et al., 2008), NG2 (Morgenstern et al., 2003), F-spondin (Burstyn-Cohen et al., 1998), and MMP-28 (which modulates the axonal–glial microenvironment) (Werner et al., 2007). These molecules directly interact with regenerating axons and may play a role in the fasciculation (the tendency of regenerating axons to grow in the same direction) and/or defasciculation of outgrowing axons (Huber et al., 2005). These phenomena are highly relevant for the misrouting problem, because by influencing fasciculation, an attempt can be made to diminish the likelihood of double innervation of antagonizing muscles and the subsequent development of unwanted co-contractions. Moreover, the expression of above factors in, e.g., a sural nerve graft may be manipulated with the use of an LV vector (Tannemaat et al., 2007a).

Furthermore, some of the axon guidance molecules described above may play a role in the phenomenon of "preferential motor reinnervation," which is in part caused by an interaction of axotomized motoneurons and endoneurial tubes at the site of axotomy (Redett et al., 2005). The carbohydrate epitope L2 may be of particular interest in this respect, as it is expressed specifically by Schwann cells originally associated with motoneurons (Martini et al., 1994) and L2 upregulation (probably mediated by BDNF/TrkB signaling) in regenerating motor nerves promotes functional recovery after peripheral nerve repair (Eberhardt et al., 2006). In contrast, F-Spondin is repulsive to embryonic motoneurons (Tzarfati-Majar et al., 2001) but promotes the outgrowth of sensory neurons (Burstyn-Cohen et al., 1998). Influencing the expression of L2, F-Spondin or similar genes could preferentially stimulate or inhibit the regeneration of motoneurons, stimulate or reduce motoneuron sprouting or perhaps even stimulate the pruning of redundant axons of regenerated motoneurons, once again increasing the likelihood of accurate reinnervation of target muscles.

Perhaps an even more promising approach to address the routing problem would be to investigate whether the injured peripheral nerve contains axon guidance cues that guide regenerating axons towards their original targets, similar to the patterning process that takes place during nervous system development. In the developing brachial plexus of the chick embryo, axonal outgrowth indeed leads to successful target finding due to an elaborate interplay between motoneurons and the differential expression of the repulsive guidance cues semaphorin 3A and semaphorin 3F in the ventral and dorsal parts of the developing limb (Huber et al., 2005). Provided that these motoneurons continue to express their respective receptors in the same pattern in the adult human nervous system, it may be possible to use semaphorin 3A and 3F to help guide them towards their original targets after a brachial plexus injury. It is currently not known whether such patterning cues are expressed during regeneration in the same way as during development. As this will be hard to study in human material, it will need to be investigated in animal models first (Reza et al., 1999). A study comparing gene expression patterns during development and regeneration of the rat sciatic nerve has already shown that approximately half of the regeneration-associated genes were also significantly regulated in development, suggesting that regeneration is indeed partly a recapitulation of development (Bosse et al., 2006).

## Future perspective: the continuing miniaturization of surgical and diagnostic tools

As described in the first paragraph, the field of peripheral nerve surgery has steadily evolved since World War II (Sunderland, 1991). The introduction of the operating microscope and improved microsuturing techniques has had a great impact on the outcome of surgery. Essentially, molecular-neurobiology-based approaches form a logical extension of this path of continuous miniaturization by identifying new molecular targets in the human peripheral nerve scar and by developing a viral vector-based strategy to influence the expression of these targets. In this final section, we will discuss other approaches that will likely have a great impact on the clinical practice of nerve repair.

A significant challenge is the absence of a reliable method to determine the degree of injury in neuroma in continuity lesions. The early diagnosis of the most severe lesions would enable rapid surgical intervention, probably enhancing functional recovery. Novel imaging techniques like diffusion-tensor imaging and diffusion-direction-dependent imaging are likely to make it possible in the near future to assess the integrity of individual nerve fascicles in the lesioned peripheral nerve (Skorpil et al., 2007). Another intriguing option stems from the recent development of radiolabeled probes that can be used to detect and quantify collagen in humans in vivo (Van den Borne et al., 2007). An early assessment of the degree of fibrosis in the peripheral nerve scar may constitute a highly reliable way to predict the chances of spontaneous recovery of nerve function. The latter putative diagnostic tool may therefore become one of the earliest novel applications based on insights obtained from the field of molecular neurobiology.

In our view, basic science will have an increasingly important impact on the treatment of peripheral nerve injuries. Despite the challenges described above, viral vectors offer exciting new opportunities to alter the molecular composition of the injured peripheral nerve, enabling the manipulation of regenerating axons on a much smaller scale than microsurgery alone. The application of viral vectors, in combination with microsurgical repair and novel in vivo imaging techniques will offer fascinating, nanoscale inroads into treatments for traumatically injured peripheral nerves. In the future, these developments will herald the development of the field of "molecular nerve repair."

## References

Ara, J., Bannerman, P., Hahn, A., Ramirez, S., & Pleasure, D. (2004). Modulation of sciatic nerve expression of class 3 semaphorins by nerve injury. *Neurochemical Research, 29*, 1153–1159.

Badalamente, M. A., Hurst, L. C., Ellstein, J., & McDevitt, C. A. (1985). The pathobiology of human neuromas: An electron microscopic and biochemical study. *Journal of Hand Surgery [British Volume], 10*, 49–53.

Beard, B. C., Dickerson, D., Beebe, K., Gooch, C., Fletcher, J., Okbinoglu, T., et al. (2007). Comparison of HIV-derived lentiviral and MLV-based gammaretroviral vector integration sites in primate repopulating cells. *Molecular Therapy, 15*, 1356–1365.

Blesch, A., Conner, J., Pfeifer, A., Gasmi, M., Ramirez, A., Britton, W., et al. (2005). Regulated lentiviral NGF gene transfer controls rescue of medial septal cholinergic neurons. *Molecular Therapy, 11*, 916–925.

Blesch, A., & Tuszynski, M. H. (2007). Transient growth factor delivery sustains regenerated axons after spinal cord injury. *Journal of Neuroscience, 27*, 10535–10545.

Blits, B., Carlstedt, T. P., Ruitenberg, M. J., de Winter, F., Hermens, W. T., Dijkhuizen, P. A., et al. (2004). Rescue and sprouting of motoneurons following ventral root avulsion and reimplantation combined with intraspinal adeno-associated viral vector-mediated expression of glial cell line-derived neurotrophic factor or brain-derived neurotrophic factor. *Experimental Neurology, 189*, 303–316.

Bosse, F., Hasenpusch-Theil, K., Kury, P., & Muller, H. W. (2006). Gene expression profiling reveals that peripheral nerve regeneration is a consequence of both novel injury-dependent and reactivated developmental processes. *Journal of Neurochemistry, 96*, 1441–1457.

Boyd, J. G., & Gordon, T. (2002). A dose-dependent facilitation and inhibition of peripheral nerve regeneration by brain-derived neurotrophic factor. *European Journal of Neuroscience, 15*, 613–626.

Boyd, J. G., & Gordon, T. (2003a). Glial cell line-derived neurotrophic factor and brain-derived neurotrophic factor sustain the axonal regeneration of chronically axotomized motoneurons in vivo. *Experimental Neurology, 183*, 610–619.

Boyd, J. G., & Gordon, T. (2003b). Neurotrophic factors and their receptors in axonal regeneration and functional recovery after peripheral nerve injury. *Molecular Neurobiology, 27*, 277–324.

Bozkurt, A., Brook, G. A., Moellers, S., Lassner, F., Sellhaus, B., Weis, J., et al. (2007). In vitro assessment of axonal growth using dorsal root ganglia explants in a novel three-dimensional collagen matrix. *Tissue Engineering, 13*, 2971–2979.

Brushart, T. M. (1993). Motor axons preferentially reinnervate motor pathways. *Journal of Neuroscience, 13*, 2730–2738.

Burstyn-Cohen, T., Frumkin, A., Xu, Y. T., Scherer, S. S., & Klar, A. (1998). Accumulation of F-spondin in injured peripheral nerve promotes the outgrowth of sensory axons. *Journal of Neuroscience, 18*, 8875–8885.

Carter, B. J. (2005). Adeno-associated virus vectors in clinical trials. *Human Gene Therapy, 16*, 541–550.

Cattoglio, C., Facchini, G., Sartori, D., Antonelli, A., Miccio, A., Cassani, B., et al. (2007). Hot spots of retroviral integration in human CD34+ hematopoietic cells. *Blood, 110*, 1770–1778.

Chen, Z. L., Yu, W. M., & Strickland, S. (2007). Peripheral regeneration. *Annual Review of Neuroscience, 30*, 209–233.

Clarke, H. M., & Curtis, C. G. (1995). An approach to obstetrical brachial plexus injuries. *Hand Clinics, 11*, 563–580. Discussion pp. 580–561.

DeMatteo, C., Bain, J. R., Galea, V., & Gjertsen, D. (2006). Botulinum toxin as an adjunct to motor learning therapy and surgery for obstetrical brachial plexus injury. *Developmental Medicine and Child Neurology, 48*, 245–252.

de Ruiter, G. C., Malessy, M. J., Alaid, A. O., Spinner, R. J., Engelstad, J. K., Sorenson, E. J., et al. (2008a). Misdirection of regenerating motor axons after nerve injury and repair in the rat sciatic nerve model. *Experimental Neurology, 211*, 339–350.

de Ruiter, G. C., Onyeneho, I. A., Liang, E. T., Moore, M. J., Knight, A. M., Malessy, M. J., et al. (2008b). Methods for in vitro characterization of multichannel nerve tubes. *Journal of Biomedical Materials Research A, 84*, 643–651.

Eberhardt, K. A., Irintchev, A., Al-Majed, A. A., Simova, O., Brushart, T. M., Gordon, T., et al. (2006). BDNF/TrkB signaling regulates HNK-1 carbohydrate expression in regenerating motor nerves and promotes functional recovery after peripheral nerve repair. *Experimental Neurology, 198*, 500–510.

Eggers, R., Hendriks, W. T., Tannemaat, M. R., van Heerikhuize, J. J., Pool, C. W., Carlstedt, T. P., et al. (2008). Neuroregenerative effects of lentiviral vector-mediated GDNF expression in reimplanted ventral roots. *Molecular and Cellular Neuroscience, 39*, 105–117.

Elsworth, J. D., Redmond, D. E., Jr., Leranth, C., Bjugstad, K. B., Sladek, J. R., Jr., Collier, T. J., et al. (2008). AAV2-mediated gene transfer of GDNF to the striatum of MPTP monkeys enhances the survival and outgrowth of co-implanted fetal dopamine neurons. *Experimental Neurology, 211*, 252–258.

Fine, E. G., Decosterd, I., Papaloizos, M., Zurn, A. D., & Aebischer, P. (2002). GDNF and NGF released by synthetic guidance channels support sciatic nerve regeneration across a long gap. *European Journal of Neuroscience, 15*, 589–601.

Fu, S. Y., & Gordon, T. (1995a). Contributing factors to poor functional recovery after delayed nerve repair: prolonged denervation. *Journal of Neuroscience, 15*, 3886–3895.

Fu, S. Y., & Gordon, T. (1995b). Contributing factors to poor functional recovery after delayed nerve repair: Prolonged axotomy. *Journal of Neuroscience, 15*, 3876–3885.

Furey, M. J., Midha, R., Xu, Q. G., Belkas, J., & Gordon, T. (2007). Prolonged target deprivation reduces the capacity of injured motoneurons to regenerate. *Neurosurgery, 60*, 723–732. Discussion pp. 732–723.

Galtrey, C. M., Asher, R. A., Nothias, F., & Fawcett, J. W. (2007). Promoting plasticity in the spinal cord with chondroitinase improves functional recovery after peripheral nerve repair. *Brain, 130*, 926–939.

Gilbert, A., & Tassin, J. L. (1984). Surgical repair of the brachial plexus in obstetric paralysis. *Chirurgie, 110*, 70–75.

Gossen, M., & Bujard, H. (1992). Tight control of gene expression in mammalian cells by tetracycline-responsive promoters. *Proceedings of the National Academy of Science USA, 89*, 5547–5551.

Gramsbergen, A., IJkema-Paassen, J., & Meek, M. F. (2000). Sciatic nerve transection in the adult rat: Abnormal EMG patterns during locomotion by aberrant innervation of hindleg muscles. *Experimental Neurology, 161*, 183–193.

Hargrove, P. W., Kepes, S., Hanawa, H., Obenauer, J. C., Pei, D., Cheng, C., et al. (2008). Globin lentiviral vector insertions can perturb the expression of endogenous genes in beta-thalassemic hematopoietic cells. *Molecular Therapy, 16*, 525–533.

Hattori, T., Matsuyama, Y., Sakai, Y., Ishiguro, N., Hirata, H., & Nakamura, R. (2008). Chondrotinase ABC enhances axonal regeneration across nerve gaps. *Journal of Clinical Neuroscience, 15*, 185–191.

Helgren, M. E., Squinto, S. P., Davis, H. L., Parry, D. J., Boulton, T. G., Heck, C. S., et al. (1994). Trophic effect of ciliary neurotrophic factor on denervated skeletal muscle. *Cell, 76*, 493–504.

Hendriks, W. T., Eggers, R., Carlstedt, T. P., Zaldumbide, A., Tannemaat, M. R., Fallaux, F. J., et al. (2007). Lentiviral vector-mediated reporter gene expression in avulsed spinal ventral root is short-term, but is prolonged using an immune "stealth" transgene. *Restorative Neurology and Neuroscience, 25*, 585–599.

Hoke, A. (2006). Mechanisms of disease: What factors limit the success of peripheral nerve regeneration in humans? *Natural Clinical Practices in Neurology, 2*, 448–454.

Hoke, A., Redett, R., Hameed, H., Jari, R., Zhou, C., Li, Z. B., et al. (2006). Schwann cells express motor and sensory phenotypes that regulate axon regeneration. *Journal of Neuroscience, 26*, 9646–9655.

Huang, S., Wang, F., Hong, G., Wan, S., & Kang, H. (2002). Protective effects of ciliary neurotrophic factor on denervated skeletal muscle. *Journal of Huazhong University of Science Technology and Medical Science, 22*, 148–151.

Huber, A. B., Kania, A., Tran, T. S., Gu, C., De Marco Garcia, N., Lieberam, I., et al. (2005). Distinct roles for secreted semaphorin signaling in spinal motor axon guidance. *Neuron, 48*, 949–964.

Jubran, M., & Widenfalk, J. (2003). Repair of peripheral nerve transections with fibrin sealant containing neurotrophic factors. *Experimental Neurology, 181*, 204–212.

Kaplitt, M. G., Feigin, A., Tang, C., Fitzsimons, H. L., Mattis, P., Lawlor, P. A., et al. (2007). Safety and tolerability of gene therapy with an adeno-associated virus (AAV) borne GAD gene for Parkinson's disease: An open label, phase I trial. *Lancet, 369*, 2097–2105.

Kaplitt, M. G., Leone, P., Samulski, R. J., Xiao, X., Pfaff, D. W., O'Malley, K. L., et al. (1994). Long-term gene expression and phenotypic correction using adeno-associated virus vectors in the mammalian brain. *Nature Genetics, 8*, 148–154.

Kaspar, B. K., Llado, J., Sherkat, N., Rothstein, J. D., & Gage, F. H. (2003). Retrograde viral delivery of IGF-1 prolongs survival in a mouse ALS model. *Science, 301*, 839–842.

Kordower, J. H., Emborg, M. E., Bloch, J., Ma, S. Y., Chu, Y., Leventhal, L., et al. (2000). Neurodegeneration prevented by lentiviral vector delivery of GDNF in primate models of Parkinson's disease. *Science, 290*, 767–773.

Lena, A. M., Giannetti, P., Sporeno, E., Ciliberto, G., & Savino, R. (2005). Immune responses against tetracycline-dependent transactivators affect long-term expression of mouse erythropoietin delivered by a helper-dependent adenoviral vector. *Journal of Gene Medicine, 7*, 1086–1096.

Lundborg, G. (2000). A 25-year perspective of peripheral nerve surgery: Evolving neuroscientific concepts and clinical significance. *Journal of Hand Surgery [American Version], 25*, 391–414.

MacKinnon, S. E., & Dellon, A. L. (1988). *Surgery of the peripheral nerve*. New York: Thieme Medical Publishers, Inc.

Madison, R. D., Archibald, S. J., Lacin, R., & Krarup, C. (1999). Factors contributing to preferential motor reinnervation in the primate peripheral nervous system. *Journal of Neuroscience, 19*, 11007–11016.

Maggi, S. P., Lowe, J. B., 3rd., & Mackinnon, S. E. (2003). Pathophysiology of nerve injury. *Clinical Plastic Surgery, 30*, 109–126.

Malessy, M. J., & Pondaag, W. (2009). Obstetric brachial plexus injuries. *Neurosurgery Clinics of North America, 20*, 1–14, v.

Malessy, M. J., Thomeer, R. T., & van Dijk, J. G. (1998a). Changing central nervous system control following intercostal nerve transfer. *Journal of Neurosurgery, 89*, 568–574.

Malessy, M. J., van der Kamp, W., Thomeer, R. T., & van Dijk, J. G. (1998b). Cortical excitability of the biceps muscle after intercostal-to-musculocutaneous nerve transfer. *Neurosurgery, 42*, 787–794. Discussion pp. 794–785.

Martini, R., Schachner, M., & Brushart, T. M. (1994). The L2/HNK-1 carbohydrate is preferentially expressed by previously motor axon-associated Schwann cells in reinnervated peripheral nerves. *Journal of Neuroscience, 14*, 7180–7191.

Moradzadeh, A., Borschel, G. H., Luciano, J. P., Whitlock, E. L., Hayashi, A., Hunter, D. A., et al. (2008). The impact of motor and sensory nerve architecture on nerve regeneration. *Experimental Neurology*, 212, 370–376.

Morgenstern, D. A., Asher, R. A., Naidu, M., Carlstedt, T., Levine, J. M., & Fawcett, J. W. (2003). Expression and glycanation of the NG2 proteoglycan in developing, adult, and damaged peripheral nerve. *Molecular and Cellular Neuroscience*, 24, 787–802.

Nguyen, Q. T., Sanes, J. R., & Lichtman, J. W. (2002). Preexisting pathways promote precise projection patterns. *Natural Neuroscience*, 5, 861–867.

Pezet, S., & McMahon, S. B. (2006). Neurotrophins: Mediators and modulators of pain. *Annual Review of Neuroscience*, 29, 507–538.

Pondaag, W., & Malessy, M. J. (2006). Recovery of hand function following nerve grafting and transfer in obstetric brachial plexus lesions. *Journal of Neurosurgery*, 105, 33–40.

Ramos, L. E., & Zell, J. P. (2000). Rehabilitation program for children with brachial plexus and peripheral nerve injury. *Seminars of Pediatric Neurology*, 7, 52–57.

Redett, R., Jari, R., Crawford, T., Chen, Y. G., Rohde, C., & Brushart, T. M. (2005). Peripheral pathways regulate motoneuron collateral dynamics. *Journal of Neuroscience*, 25, 9406–9412.

Reza, J. N., Gavazzi, I., & Cohen, J. (1999). Neuropilin-1 is expressed on adult mammalian dorsal root ganglion neurons and mediates semaphorin3a/collapsin-1-induced growth cone collapse by small diameter sensory afferents. *Molecular Cellular Neuroscience*, 14, 317–326.

Skorpil, M., Engstrom, M., & Nordell, A. (2007). Diffusion-direction-dependent imaging: A novel MRI approach for peripheral nerve imaging. *Magnetic Resonance Imaging*, 25, 406–411.

Stokols, S., Sakamoto, J., Breckon, C., Holt, T., Weiss, J., & Tuszynski, M. H. (2006). Templated agarose scaffolds support linear axonal regeneration. *Tissue Engineering*, 12, 2777–2787.

Stokols, S., & Tuszynski, M. H. (2004). The fabrication and characterization of linearly oriented nerve guidance scaffolds for spinal cord injury. *Biomaterials*, 25, 5839–5846.

Streppel, M., Azzolin, N., Dohm, S., Guntinas-Lichius, O., Haas, C., Grothe, C., et al. (2002). Focal application of neutralizing antibodies to soluble neurotrophic factors reduces collateral axonal branching after peripheral nerve lesion. *European Journal of Neuroscience*, 15, 1327–1342.

Sunderland, S. (1991). *Nerve injuries and their repair: A critical appraisal*. Melbourne: Chuchill Livingstone.

Tannemaat, M. R., Boer, G. J., Verhaagen, J., & Malessy, M. J. (2007a). Genetic modification of human sural nerve segments by a lentiviral vector encoding nerve growth factor. *Neurosurgery*, 61, 1286–1294. Discussion pp. 1294–1286.

Tannemaat, M. R., Eggers, R., Hendriks, W. T., de Ruiter, G. C., van Heerikhuize, J. J., Pool, C. W., et al. (2008a). Differential effects of lentiviral vector-mediated overexpression of nerve growth factor and glial cell line-derived neurotrophic factor on regenerating sensory and motor axons in the transected peripheral nerve. *European of Journal Neuroscience*, 28, 1467–1479.

Tannemaat, M. R., Korecka, J., Ehlert, E. M., Mason, M. R., van Duinen, S. G., Boer, G. J., et al. (2007b). Human neuroma contains increased levels of semaphorin 3A, which surrounds nerve fibers and reduces neurite extension in vitro. *Journal of Neuroscience*, 27, 14260–14264.

Tannemaat, M. R., Verhaagen, J., & Malessy, M. (2008b). The application of viral vectors to enhance regeneration after peripheral nerve repair. *Neurological Research*, 30, 1039–1046.

Tuszynski, M. H., Thal, L., Pay, M., Salmon, D. P., U, H. S., Bakay, R., et al. (2005). A phase 1 clinical trial of nerve growth factor gene therapy for Alzheimer disease. *Nature Medicine*, 11, 551–555.

Tzarfati-Majar, V., Burstyn-Cohen, T., & Klar, A. (2001). F-spondin is a contact-repellent molecule for embryonic motor neurons. *Proceedings of National Academy Science USA*, 98, 4722–4727.

Van den Borne, S. W., Isobe, S., Verjans, J., Petrov, A., Lovhaug, D., Li, P., et al. (2007). Molecular imaging of postinfarction cardiac remodeling and effects of anti-angiotensin in therapy. *Circulation*, 116, 289–290.

van Dijk, J. G., Pondaag, W., & Malessy, M. J. (2007). Botulinum toxin and the pathophysiology of obstetric brachial plexus lesions. *Developmental Medicine and Child Neurology*, 49, 318. Author reply pp. 318–319.

Vergara, C., & Ramirez, B. (2004). CNTF, a pleiotropic cytokine: Emphasis on its myotrophic role. *Brain Research and Brain Research Reviews*, 47, 161–173.

Waters, P. M. (1999). Comparison of the natural history, the outcome of microsurgical repair, and the outcome of operative reconstruction in brachial plexus birth palsy. *Journal of Bone Joint Surgery America*, 81, 649–659.

Werner, S. R., Mescher, A. L., Neff, A. W., King, M. W., Chaturvedi, S., Duffin, K. L., et al. (2007). Neural MMP-28 expression precedes myelination during development and peripheral nerve repair. *Developmental Dynamics*, 236, 2852–2864.

Yanez-Munoz, R. J., Balaggan, K. S., MacNeil, A., Howe, S. J., Schmidt, M., Smith, A. J., et al. (2006). Effective gene therapy with nonintegrating lentiviral vectors. *Nature Medicine*, 12, 348–353.

Yin, Q., Kemp, G. J., Yu, L. G., Wagstaff, S. C., & Frostick, S. P. (2001). Neurotrophin-4 delivered by fibrin glue promotes peripheral nerve regeneration. *Muscle Nerve*, 24, 345–351.

Young, C., Miller., E., Nicklous, D. M., & Hoffman, J. R. (2001). Nerve growth factor and neurotrophin-3 affect functional recovery following peripheral nerve injury differently. *Restorative Neurology and Neuroscience*, 18, 167–175.

CHAPTER 13

# Gene therapy for neurodegenerative diseases based on lentiviral vectors

## Aikaterini Nanou and Mimoun Azzouz*

*Academic Neurology Unit, University of Sheffield, Sheffield, UK*

**Abstract:** Gene therapy approaches to treat inherited and acquired disorders offer many unique advantages over conventional therapeutic approaches. For neurodegenerative diseases, gene therapy is particularly attractive due to the restricted bioavailability of conventional therapeutic substances to the affected structures of the brain and progressive nature of these diseases. With the development of lentiviral vector systems, many issues have been addressed and new delivery routes to the nervous system have been identified. Lentiviral vectors can efficiently deliver genes to postmitotic neuronal cell types offering long-term expression, can be generated in high titers, and do not give immunological complications. Various animal studies have demonstrated the effectiveness of these vectors to deliver therapeutic genes into the nervous system, as well as to model human diseases. This chapter will describe the basic features of lentiviral vectors, the progress, and their applications as a therapeutic strategy to treat diseases such as amyotrophic lateral sclerosis, spinal muscular atrophy, Parkinson's disease, and Huntington's disease.

**Keywords:** gene therapy; lentiviral vectors; neurodegeneration; amyotrophic lateral sclerosis/motor neuron disease; spinal muscular atrophy; Parkinson's disease; Huntington's disease

## Lentiviral vector-mediated gene therapy

### Principle of gene therapy and gene transfer vehicles

Gene therapy is not just limited to the replacement of a defective gene with a functional one, but it describes any nucleic acid transfer to treat or prevent disease and provides many advantages over conventional therapeutic strategies, such as the selective treatment of affected cells and long-term treatment after a single application. Different gene therapy systems have been generated, including viral vectors, nonviral synthetic vectors such as naked DNA and cationic liposomes (Li and Huang, 2000), and hybrid synthetic-viral systems (Kaneda, 1999), and delivery of cells manipulated ex vivo with therapeutic genes (Li and Huang, 2000). The rate-limiting step for the successful gene delivery using viral vectors is the ability to transfer efficiently the desired therapeutic gene to the target tissue, and different viral vectors have been developed with distinct advantages and disadvantages (Table 1).

The most common vectors used for gene delivery in clinical trials involve recombinant adenoviruses and retroviruses, followed closely by various forms of nonviral gene transfer (Table 2).

*Corresponding author.
Tel.: +44 (0) 1142 713204; Fax: +44 (0) 1142 261201;
E-mail: m.azzouz@sheffield.ac.uk

Table 1. Characteristics of the most common vectors used in gene therapy

| Vector | Insert size (kb) | Expression | Advantages | Disadvantages |
| --- | --- | --- | --- | --- |
| Adeno-associated virus | 4.5 (not all serotypes) | Stable | Little immunogenicity; integrates; transduces nondividing cells | Small cloning capacity |
| Adenovirus | 36 | Transient | High viral titer; wide host-cell range; transduces nondividing cells | Immunogenicity; does not integrate; short-term expression |
| Retrovirus | 8 | Stable | No immune response against the vector; integrates | Low viral titers; transduces only dividing cells; variable expression |
| Lentivirus | 8 | Stable | Transduces nondividing cells; high transduction efficiency | Potential safety risk |
| Herpes simplex virus | >30 | Stable | Large cloning capacity; transduces nondividing cells; wide host-cell range | Safety concerns; short-term expression; possibly immunogenic |
| Vaccinia virus | 186 | Transient | High expression, best for immunization strategies | Transient expression; transduces only dividing cells; safety concerns |

Table 2. Most common vectors and gene transfer systems used in clinical trials

| Vector system | Gene therapy clinical trials | |
| --- | --- | --- |
| | Number | Percentage |
| Viral delivery | | |
| Adenovirus | 337 | 25.0 |
| Retrovirus | 304 | 22.6 |
| Vaccinia virus | 65 | 4.9 |
| Poxvirus | 61 | 4.5 |
| Adeno-associated virus | 54 | 4.0 |
| Herpes simplex virus | 43 | 3.2 |
| Poxvirus+vaccinia virus | 27 | 2.0 |
| Lentivirus | 11 | 0.8 |
| Flavivirus | 8 | 0.6 |
| Adenovirus+retrovirus | 3 | 0.2 |
| Nonviral delivery | | |
| Naked/plasmid DNA | 244 | 18.1 |
| Lipofection | 102 | 7.6 |
| RNA transfer | 19 | 1.4 |
| Gene gun | 5 | 0.4 |

*Source:* Gene Therapy Clinical Trials Worldwide (2008).

## Lentiviral vector-mediated gene therapy in neurodegeneration

Lentiviral vectors can be derived from primates, such as HIV (human immunodeficiency virus) and SIV (simian immunodeficiency virus) vectors, and from nonprimates, such as EIAV (equine infectious anemia virus), FIV (feline immunodeficiency virus), and BIV (bovine immunodeficiency virus) vectors. Lentiviral vectors have the capacity to be pseudotyped with various envelope proteins, and thus acquire an altered tropism or mechanism of transport within the host cell. Pseudotyping HIV vectors with VSV-G (vesicular stomatitis virus) envelope (the first lentiviral system developed) enabled the generation of high-titer viral stocks and a broad tissue tropism (Naldini et al., 1996), whereas pseudotyping EIAV vectors with the rabies-G envelope enabled the retrograde transport of the vector allowing targeting of remote neuronal cells by delivering the vector to cell bodies at distal sites (Azzouz et al., 2004b; Mazarakis et al., 2001; Wong et al., 2004). In addition, by pseudotyping EIAV and HIV vectors with rabies-G glycoproteins, it is possible to mimic the route of entry of the rabies virus, and thus administer the virus intramuscularly in order to enter the central nervous system (CNS) by retrograde transport via the motor neuron axon (Mazarakis et al., 2001; Mentis et al., 2006). The latter system proved very efficient for gene transfer in both rodent and primate brains in order to target distant neuronal populations (Kato et al., 2007).

Two general strategies have been used to address neurodegeneration using lentiviruses:

either viral delivery of neuroprotective genes or viral delivery of transcription-based RNA silencing (small interfering RNA, siRNA) in order to limit the synthesis of mutant genes. These methods have been studied in both tissue culture and animal models of many neurodegenerative diseases. This chapter will review the progress and applications of lentiviral vector gene delivery for neurodegenerative diseases such as amyotrophic lateral sclerosis (ALS), spinal muscular atrophy (SMA), Parkinson's disease (PD), and Huntington's disease (HD).

**Amyotrophic lateral sclerosis**

*Clinical symptoms and current treatment*

Amyotrophic lateral sclerosis or motor neuron disease (ALS or MND, also known as Lou Gehrig's disease) is a progressive neurodegenerative disorder that is characterized by the loss of upper and lower motor neurons of the spinal cord, brain stem, and motor cortex (Mulder, 1982). The progressive manifestations of upper and lower motor neuron dysfunction include muscle weakness and wasting usually accompanied by pathologically brisk reflexes, eventually involving the limb and bulbar muscles. It is a late-onset disease (usually between age 45 and 60) with a typical disease course of 2–5 years, and generally, the fatal event is respiratory failure due to denervation of the respiratory muscles and diaphragm (Williams and Windebank, 1991). Approximately 5–10% of ALS cases are inherited in an autosomal dominant fashion, of which 15–20% of cases are caused by missense mutations in the gene encoding Cu/Zn superoxide dismutase 1 (SOD1) (Rosen, 1993). Since both sporadic and familial ALS affect the same neurons exhibiting similar pathological hallmarks, including progressive muscle weakness, atrophy, and spasticity, it is speculated that therapeutic interventions developed on mutant SOD1 models (well-characterized tissue culture and animal models) will translate to sporadic ALS cases as well. Various pathogenic hypotheses have been proposed for ALS over the years, including protein misfolding and aggregation, defective axonal transport, excitotoxicity, mitochondrial dysfunction, apoptosis, oxidative stress, and even toxicity caused by nonneuronal cells (Shaw, 2005). There is evidence supporting each hypothesis, and this probably suggests that there is not a unifying disease-causing mechanism, but ALS might be the common end-stage phenotype of diverse causes.

Partly due to the complex and possibly multifactorial etiology of ALS, and partly to inefficient delivery methods of therapeutic factors, there are no significant treatment options for ALS patients. The only approved drug for treatment of ALS is the antiglutamate riluzole, which prolongs survival in SOD1 transgenic mice (widely used ALS mouse model) by a modest 10% without affecting the disease onset (Gurney et al., 1996) and has a similarly modest effect in ALS patients (Miller et al., 2000). In addition, neurotrophic factors have been considered as a promising therapeutic approach for ALS since they could support neuronal maintenance and survival in stress conditions without targeting a specific pathogenic mechanism (Ekestern, 2004). However, little, if no, effect was seen either through conventional delivery routes or through intrathecal administration for a number of neurotrophic agents. The inefficiency of these treatments in addition to some undesirable side effects seen and the limited dosage options that could reduce bioavailability (BDNF Study Group, 1999; Borasio et al., 1998; Eriksdotter Jonhagen et al., 1998; Kordower et al., 1999b; Penn et al., 1997) lead to the development of more efficient viral-mediated gene delivery.

*Lentiviral vector-mediated gene therapy in amyotrophic lateral sclerosis*

The therapeutic potential of lentiviral delivery of neurotrophic factors in ALS animal models was demonstrated by the administration of VEGF (vascular endothelial growth factor). In direct protein delivery studies, recombinant VEGF administration through intraperitoneal injections in G93A SOD1 transgenic mice resulted in delayed disease progression and increased survival (Zheng et al., 2004), and a similar effect was shown in G93A SOD1 transgenic rats after intracerebroventricular delivery (Storkebaum et al., 2005). However, lentiviral delivery of VEGF into muscles of the same animal model achieved even higher therapeutic effects by preventing cell death and prolonging the survival of the transgenic mice by 30% (Azzouz et al., 2004b). Most importantly, this treatment was effective even when given at the onset of paralysis, which is relevant to the human clinical application since the majority of ALS cases appear sporadically and thus diagnosis and treatment are only possible after the onset of the clinical symptoms. Furthermore, the effect of GDNF (glial-derived neurotrophic factor) has also been studied in G93A SOD1 transgenic mice when delivered by direct facial nucleus or intraspinal injection of lentiviral vectors encoding for this particular neurotrophic factor (Guillot et al., 2004). Although a strong expression of GDNF was achieved at the lumbar level and diseased facial motor neurons were significantly rescued from cell death, no effect was seen on spinal motor neurons. Nonetheless, this study highlighted the ability of the lentiviral vectors to generate a strong and long-term expression of intracellular proteins.

In addition, RNA interference (RNAi) has been studied as a potential therapeutic strategy using lentiviruses. Lentiviral vectors expressing functional RNAi molecules directed against SOD1 expression were injected either intraspinally or intramuscularly in mutant SOD1 mice, and therapeutic efficacy was demonstrated (Ralph et al., 2005; Raoul et al., 2005). By specifically targeting the mutant human SOD1 expression while leaving the wild-type murine SOD1 expression intact, the life expectancy of the transgenic mice increased by an impressive 80%, highlighting the potential applications of this method in treating familial ALS (Ralph et al., 2005). Although these studies show very promising results, more research is needed to evaluate the use of lentiviruses in ALS treatment since currently the number of studies is limited.

## Spinal muscular atrophy

### Clinical manifestations and genetics of spinal muscular atrophy

Spinal muscular atrophy (SMA) is a common autosomal recessive neuromuscular disease characterized by the degeneration of the anterior horn motor neurons of the spinal cord and is associated with symmetrical limb and trunk paralysis with muscle atrophy. SMA disease severity is heterogeneous, and its clinical manifestations are divided into four types (Table 3) depending on age of onset and clinical course (Munsat and Davies, 1992; Russman, 2007).

All four types were found to be caused by mutations in chromosome 5q13, through linkage analysis (Brzustowicz et al., 1990; Melki et al., 1990), and the gene was later identified as survival of motor neuron (SMN) gene (Lefebvre et al., 1995). The human SMN gene exists in two copies, SMN1 and SMN2, which are localized within a duplicated and inverted chromosomal segment of around 500 kb, with one being a telomeric copy and the other being a centromeric copy, respectively. Although the two SMN isoforms are almost identical, their expression levels of full-length SMN protein are very different due to one functional difference (C→T transition within an exonic splicing region of SMN2) in the nucleotide composition (Lorson et al., 1999). SMN2 produces only approximately 10% of full-length transcripts, and the rest 90% of transcripts are alternatively

Table 3. Clinical manifestations of all types of spinal muscular atrophy

| SMA type | Age of onset | Age of death | Disease phenotype |
|---|---|---|---|
| Type I (Werdnig–Hoffmann disease) | 0–6 months | <2 years | Unable to sit unaided |
| Type II | 7–18 months | >2 years | Sit unaided, unable to stand or walk |
| Type III (Kugelberg–Welander disease) | >18 months | Normal life expectancy | Stand and walk unaided |
| Type IV | 20–30 years | Normal life expectancy | Stand and walk unaided |

spliced exons that lack exon 7, in contrast to SMN1 that produces full-length transcripts (Helmken et al., 2003; Lefebvre et al., 1995). The transcript lacking exon 7 encodes for a truncated and unstable form of the protein with impaired function (Lorson et al., 1998). Mutations or loss of SMN1 causes the disease, and although 5–10% of the normal population lack SMN2 copy, all SMA cases have retained the gene (Lefebvre et al., 1995). In addition, the SMN2 gene copy number varies in the population due to the unstable chromosomal region it resides in, and the SMN2 copy number directly correlates with the severity of the disease phenotype; most patients with SMA type I have one or two copies of SMN2, while patients with SMA type III have three or four copies (Feldkotter et al., 2002; Parsons et al., 1998). This clearly demonstrates that the disease is caused by the reduced levels of SMN protein, and that a low copy number of SMN2 cannot compensate for this loss. The SMN protein is ubiquitously expressed and is localized both in the cytoplasm and in structures in the nucleus termed gemini of coiled bodies (gems) in close proximity to or overlapping with Cajal bodies that contain a high level of factors involved in the transcription and processing of RNA (Liu and Dreyfuss, 1996; Young et al., 2000). SMN protein is involved in small nuclear ribonucleoprotein (snRNP) biogenesis and pre-mRNA splicing (Meister et al., 2000; Pellizzoni et al., 1998), although it has been suggested that it may have other cellular functions as well.

## *Lentiviral vector-mediated gene therapy in spinal muscular atrophy*

With the identification of the SMA gene, it was found that SMN is a single-copy gene in the mouse, and homozygous SMN disruption leads to massive cell death during embryonic development (Schrank et al., 1997). The fact that human SMN2, however, rescued embryonic lethality in SMN-deficient mice illustrated the importance of the protein for cell survival (Hsieh-Li et al., 2000; Monani et al., 2000). Thus, the replacement of SMN1 gene using a gene therapy approach to treat SMA patients appears to be an attractive therapeutic strategy. However, only a modest increase in life span was demonstrated when SMA mice were injected with lentiviral vectors expressing SMN gene (Azzouz et al., 2004a). Even though the mouse life span was only increased by 3–5 days, this study shows that this approach is attractive and a better therapeutic result might arise if the transduction is more widespread (restore SMN expression in other cell types as well, not only motor neurons) and/or the administration takes place at an earlier stage (in utero, for example) in order to allow the expression of the SMN protein expressed by the vector to compensate the loss of endogenous.

Only one attempt has been made using the neuroprotective approach of delivering neurotrophic factors. Thus, using an adenoviral vector, cardiotrophin-1, a member of the IL-6 cytokine family, was delivered in the SMA mouse model, and it managed to delay the disease progression and protect against cell loss (Lesbordes et al., 2003). More research is needed to assess the neuroprotective ability of neurotrophic factors in SMA, but it is possible that a combinatorial approach of gene therapy with the concomitant administration of a pharmacological agent (such as valproic acid that increases the cellular SMN protein levels of SMA patients) might be more effective (Sumner et al., 2003).

## Parkinson's disease

### *Clinical symptoms*

Parkinson's disease (PD) is the second most common progressive neurodegenerative disorder (after Alzheimer's disease), and it is primarily, but not exclusively, characterized by the selective loss of the dopaminergic neurons in the substantia nigra pars compacta (SNpc) and thus the depletion of the neurotransmitter dopamine (DA) in the striatum. The mean age at onset is 50–60 years, and the clinical symptoms include rigidity, resting tremor, bradykinesia, and postural instability (Lang and Lozano, 1998). Apart from the neuronal loss, a hallmark of the disease is the presence of intraneuronal cytoplasmic inclusions, termed Lewy bodies, whose primary components include alpha-synuclein and ubiquitin, as well as parkin and neurofilaments (Spillantini et al., 1997). However, the presence of Lewy bodies is not a feature only of PD, but is seen also in other neurological diseases such as Alzheimer's disease, as well as healthy individuals of advanced age (Gibb and Lees, 1988). Nonetheless, it is still not clear whether these inclusions are protective by sequestering harmful protein aggregates or toxic to neurons either by interfering with cellular processes or by sequestering essential proteins. The majority of PD cases are thought to be sporadic, although currently 13 PD loci have been identified in rare familial cases, including alpha-synuclein, parkin, leucine-rich repeat kinase 2 (LRRK2), PTEN-induced kinase 1 (PINK1), and DJ-1 (Belin and Westerlund, 2008).

### *Modeling Parkinson's disease*

Several hypotheses have been proposed concerning the pathogenesis of PD, including genetic factors, environmental factors, oxidative stress, and mitochondrial dysfunction, but still the exact mechanism of neurodegeneration is unknown (Dauer and Przedborski, 2003). In order to elucidate the pathogenic events and identify therapeutic compounds, an animal model of the disease is needed. The first animal model of PD that clearly showed a selectively dopaminergic neuronal death was introduced more than 30 years ago, and it was generated through the use of the neurotoxin 6-hydroxydopamine (6-OHDA) that mimicked the PD pathology, although not accurately (Ungerstedt, 1968). The 6-OHDA model is still used extensively, although many different neurotoxins have been identified since then, of which the MPTP (1-methyl-4-phenyl-1,2,3,6-tetrahydropyrindine) model is the most widely studied since it produces an irreversible PD-like pathology (Forno et al., 1993; Langston et al., 1983).

However, with the identification of the genes involved in the familial PD cases, the possibility of generating transgenic animal models arose. Thus, transgenic mouse and fly models of PD were developed by overexpressing the wild-type or mutant human alpha-synuclein (Feany and Bender, 2000; Kahle et al., 2000; van der Putten et al., 2000). These transgenic mouse models, although they manage to mimic many PD features such as decreased levels of DA, progressive motor neuron function, and presence of Lewy bodies, fail to accurately reproduce the selective pathophysiological features of substantia nigra. In order to address this issue, lentiviruses were used not as a therapeutic tool, but as a means of generating a better animal model of the disease. In particular, rats were injected with lentiviral vectors expressing either wild-type or mutant human alpha-synuclein, and in contrast to transgenic models, selective degeneration of the dopaminergic neurons was demonstrated (Lo Bianco et al., 2002). In addition, a mouse model of the disease was generated through the lentiviral delivery of human alpha-synuclein into alpha-synuclein knockout mice (Alerte et al., 2008). These reports reveal the usefulness of lentiviruses as a tool to create a model system in which further studies can elucidate the pathological mechanisms of the disease and identify therapeutic targets.

### *Lentiviral vector-mediated gene therapy in Parkinson's disease*

From human PD cases and animal models of the disease, it was evident that at the onset of the clinical manifestations of PD, the DA levels are already reduced by approximately 80% and more than 60% of the stratial dopaminergic neurons are lost (Dunnett and Bjorklund, 1999). This indicates that only a targeted therapeutic strategy will be able to affect the course of the disease, such as the use of lentiviral vectors. In nonhuman primates, the efficiency of lentiviral transduction was tested using a vector expressing LacZ, and it was shown that more than one million cells were transduced when injected into the striatum, and 190,000 cells when injected into the substantia nigra (Kordower et al., 1999a), highlighting the potential clinical application of this system. Currently, the most widespread treatment for PD is DA replacement through the oral administration of L-dopa, the precursor of DA. L-dopa increases the DA levels in the denervated striatum and is effective in the early stages of the disease, but it loses efficacy over time and severe side effects may develop (Brotchie et al., 2005). Similarly, one therapeutic approach that was investigated with the lentiviral system is the restoration of DA levels. Specifically, a tricistronic EIAV-based lentiviral vector expressing three essential enzymes (tyrosine hydroxylase, aromatic amino acid dopa decarboxylase, and GTP cyclohydrolase 1) for DA production was generated and delivered into the striatum of the 6-OHDA-lesioned rat model (Azzouz et al., 2002). Due to the sustained expression of each enzyme that was accomplished, as well as effective DA production and functional improvement of the rat model, this viral system entered clinical phase I at 2008, and the results are highly anticipated.

The delivery of neurotrophic factor approach has been investigated in PD, as in other neurodegenerative disorders. Many different groups have investigated the lentiviral delivery of GDNF due to the safety and the high efficiency this vector system offers. Lentiviral delivery of GDNF by intracerebral injection to the 6-OHDA-lesioned rat model (Azzouz et al., 2004c; Bensadoun et al., 2000; Dowd et al., 2005; Georgievska et al., 2002) and to the MPTP-treated monkey model (Kordower et al., 2000; Palfi et al., 2002) resulted in persistent neuroprotection and recovery of motor function, indicating the neuroprotective ability of GDNF as well as the applicability of the lentiviral vector system. Furthermore, another neurotrophic factor belonging to the GDNF family, neurturin, has shown persistent and efficient neuroprotection when delivered using adeno-associated viral (AAV) vectors (Gasmi et al., 2007; Kordower et al., 2006) and has moved to clinical trial phase II. The neuroprotection offered by neurturin was further demonstrated when inserted into a lentiviral vector (Fjord-Larsen et al., 2005); however, further studies are needed to draw any conclusions.

## Huntington's disease

### Huntington's disease: a polyglutamine disorder

Huntington's disease (HD) belongs to a heterogeneous class of at least nine genetically distinct disorders, termed polyglutamine disorders, which are caused by the expansion of trinucleotide repeats coding for polyglutamine tracts in respective proteins (Nakamura et al., 2001; Zoghbi and Orr, 2000). HD is an inherited progressive neurodegenerative disorder with the onset of disease being between the ages 35 and 50 years, and the clinical symptoms include involuntary (choreic) movements, psychiatric disturbances, and gradual mental decline with a typical disease course of 15–20 years (Vonsattel and DiFiglia, 1998). It is characterized by the selective gradual atrophy and cell loss of the striatum (caudate nucleus and putamen), particularly affecting the medium spiny GABAergic neurons, with possible later involvement of other nonstriatal brain structures such as the cerebral cortex, thalamus, subthalamic nucleus, substantia nigra, and cerebellum. The disease is caused by an expanded CAG (cytosine–adenine–guanine) repeat length in exon 1 of the gene huntingtin located on chromosome 4 (The Huntington's Disease Collaborative Research Group, 1993), and a neuropathological hallmark of HD is the presence of neuronal intranuclear inclusions (DiFiglia et al., 1997) and cytoplasmic protein aggregates in dystrophic neurites in striatal and cortical neurons (Sapp et al., 1999). Typically, there are approximately 35 or less CAG repeats encoding for glutamine in wild-type huntingtin. Incomplete penetrance is seen with 36–40 repeats, whereas when the repeats reach 41 or more, the disease is fully penetrant (Rubinsztein et al., 1996). The number of repeats is inversely correlated with the age of disease onset, and a high number of repeats (more than 60) lead to juvenile onset (Landles and Bates, 2004). Various hypotheses have been proposed concerning the pathogenesis of the disease, such as protein aggregation and misfolding, transcriptional dysregulation, and mitochondrial dysfunction (Gil and Rego, 2008). Due to the limited knowledge of the pathogenic mechanisms, the available treatments are only symptomatic.

### Modeling Huntington's disease

Due to the limited knowledge of the pathogenic events leading to neurodegeneration, cellular and animal models of the disease are needed. Various in vitro models have been generated over the years, such as those of neuronal origin (Lunkes and Mandel, 1998), transfected neuronal cell lines (Saudou et al., 1998), primary neurons derived from transgenic HD mouse models (Petersen et al., 2001), and striatal cells isolated from knock-in mice (Trettel et al., 2000). However, these in vitro models have limitations since some have not managed to model the cell death seen when expressing mutant huntingtin (Petersen et al., 2001), and others are generated in low yields, not allowing extensive experimentation (Saudou et al., 1998). As with PD, lentiviral vectors were used to create a more relevant cellular model to the human disease. In particular, rat primary striatal neurons were transduced with a lentiviral vector expressing a mutant huntingtin protein, and this model exhibited all the neuropathological hallmarks of HD (Zala et al., 2005).

As far as the animal models are concerned, the first animal models of the disease were generated through the use of toxins (McGeer and McGeer, 1976), and some of them are still widely used (Roberts et al., 1993). With the identification of the HD gene though, various genetic mouse models were developed, including transgenic, knock-in, knockout, and virally delivered mutant huntingtin models (Wang and Qin, 2006). All the different models mimic some of the symptoms and pathological events of HD, and even though none can replicate the massive loss of striatal neurons occurring in human HD cases, each model has unique advantages and limitations. The first animal HD model using lentiviral vectors was generated by injecting vectors coding for the first 171, 853, and 1520 amino acids of wild-type (19 GAG) or mutant (44, 66, and 82 GAG) huntingtin in rat striatum (de Almeida et al., 2002). In another study, the use of tetracycline-regulated lentiviral vectors to model HD in rats demonstrated that apart from the very high transgene levels this promoter generated, the system allows the conditional suppression of mutant huntingtin synthesis upon peripheral doxycycline administration, showing the constant expression of the mutant protein is needed for the progression of the disease (Regulier et al., 2003).

The advantages that these models offer include the precise localization of gene expression, giving rise to a more controlled environment for studying the effect of mutant huntingtin in specific brain areas, and the possibility of applying this technology to nonrodent species in a way that is not currently possible with the methods generating the transgene and knock-in models. The latter was demonstrated in a recent report aiming to develop a transgenic model of HD in a rhesus macaque that expresses mutant huntingtin (Yang et al., 2008). This approach emphasizes the usefulness of the lentiviral vectors as tools to develop valuable nonhuman primate models not only of HD, but also of other genetic disorders that will closely mimic the human disease.

## Lentiviral vector-mediated gene therapy in Huntington's disease

With the development of conditional models of HD, it became evident that the suppression of mutant huntingtin expression leads to a clearance of neuronal inclusions and amelioration of the behavioral phenotype (Regulier et al., 2003; Yamamoto et al., 2000); thus, silencing the mutant gene expression using RNAi has emerged as an attractive therapeutic approach for HD. Even though silencing mutant huntingtin has not yet being tested with the lentiviral vector system, studies using AAV vectors showed improvements in motor function and neuropathological abnormalities in HD mouse models (Franich et al., 2008; Harper et al., 2005; Wang et al., 2005).

A variety of neurotrophic factors have been virally delivered into HD animal models, of which the most studied is the ciliary neurotrophic factor (CNTF). Lentiviral delivery of CNTF in rat and mouse HD models showed significant neuroprotection of the striatum (de Almeida et al., 2001; Zala et al., 2004), and using the tetracycline-regulated lentiviral system, a dose-dependent effect of CNTF was demonstrated, highlighting the application prospects of a conditional gene transfer in the brain (Regulier et al., 2002). Furthermore, despite the neuroprotective effect seen in rodent HD models when injected with AAV vectors expressing GDNF (Kells et al.,

2004; McBride et al., 2003), the intrastriatal lentiviral vector transfer of GDNF in HD transgenic mice failed to replicate the findings (Popovic et al., 2005). Nonetheless, further extensive testing of lentiviral vector-mediated overexpression of neurotrophic factors is needed to assess the application of this therapeutic approach to the human disease.

Other gene therapy approaches have been investigated in HD, apart from silencing the mutant gene and delivering neurotrophic factors. Even though the exact pathogenic mechanism causing the disease phenotype is unknown, there is evidence that mutant huntingtin represses transcription (Dunah et al., 2002), alters the energy metabolism, and disrupts mitochondrial function (Browne and Beal, 2004). Thus, in an attempt to target the upstream events of neurodegeneration in HD, transgenic HD mice were injected with lentiviral vectors expressing PGC1alpha (peroxisome proliferator-activated receptor gamma coactivator-1 alpha), a transcriptional regulator regulating, among others, mitochondrial biogenesis and oxidative phosphorylation (Cui et al., 2006). The lentiviral vector-mediated delivery of PGC1alpha in the striatum provided neuroprotection in the mouse model, suggesting that mitochondrial dysfunction might be involved in the pathogenesis of the disease. More importantly, this study shows that all data emerging from HD animal models provide insights into the pathogenic mechanisms of the disease, and identification of these mechanisms will lead to more effective therapeutic targets.

## Clinical prospects and challenges of lentiviral vectors

Lentiviral vector-mediated gene therapy offers many advantages compared to other viral systems and, as a therapeutic approach, is very attractive particularly for neurodegenerative disorders since it can be delivered to the affected area, and one treatment offers long-term expression. Currently, only one clinical trial is ongoing with lentiviral vectors for PD patients, but with the growing knowledge of the pathogenic mechanisms of all neurodegenerative diseases, new candidates for gene therapy are emerging. A positive outcome from this first clinical trial would benefit future clinical applications of lentiviral vector systems in CNS disorders. For ALS, the identification of new disease-causing genes can lead to new animal models offering insight to the pathogenesis of the disease. In addition, heat-shock genes to target the protein aggregation, antioxidant genes, or antiapoptotic genes might prove to be promising candidates for gene therapy. In SMA, although it seems straightforward therapeutic strategy, the time window might not be sufficient to generate a therapeutic effect by replacing the SMN gene. Nonetheless, a combination of treatments or therapeutic targets delivered by lentiviral vectors might promote neuroprotection in a more efficient manner. For example, by the concomitant lentiviral vector expression of SMN and plastin 3, a therapeutic effect might be enhanced, since it has been reported that plastin 3 managed to rescue axonal defects associated with reduced SMN protein levels in motor neurons (Oprea et al., 2008). In HD, the development of a lentiviral vector suppressing mutant huntingtin is imminent, and all data to date show promising results. However, clinical application of this strategy might require siRNA targeting only mutant huntingtin allele.

## Abbreviations

| | |
|---|---|
| 6-OHDA | 6-hydroxydopamine |
| AAV | adeno-associated virus |
| ALS | amyotrophic lateral sclerosis |
| BIV | bovine immunodeficiency virus |
| CAG | cytosine–adenine–guanine |
| CNS | central nervous system |
| CNTF | ciliary neurotrophic factor |
| DA | dopamine |
| EIAV | equine infectious anemia virus |
| FIV | feline immunodeficiency virus |
| GABA | gamma aminobutyric acid |
| GDNF | glial-derived neurotrophic factor |
| HD | Huntington's disease |
| HIV | human immunodeficiency virus |
| LRRK2 | leucine-rich repeat kinase 2 |
| MND | motor neuron disease |

| | |
|---|---|
| MPTP | 1-methyl-4-phenyl-1,2,3,6-tetrahydropyrindine |
| PD | Parkinson's disease |
| PGC1alpha | peroxisome proliferator-activated receptor gamma coactivator-1 alpha |
| PINK1 | PTEN-induced kinase 1 |
| RNAi | RNA interference |
| siRNA | small interfering RNA |
| SIV | simian immunodeficiency virus |
| SMA | spinal muscular atrophy |
| SMN | survival motor neuron |
| SNpc | substantia nigra pars compacta |
| snRNP | small nuclear ribonucleoprotein |
| SOD1 | Cu/Zn superoxide dismutase 1 |
| VEGF | vascular endothelial growth factor |
| VSV-G | vesicular stomatitis virus |

## References

Alerte, T. N., Akinfolarin, A. A., Friedrich, E. E., Mader, S. A., Hong, C. S., & Perez, R. G. (2008). Alpha-synuclein aggregation alters tyrosine hydroxylase phosphorylation and immunoreactivity: lessons from viral transduction of knockout mice. *Neuroscience Letters, 435*, 24–29.

Azzouz, M., Le, T., Ralph, G. S., Walmsley, L., Monani, U. R., Lee, D. C., et al. (2004a). Lentivector-mediated SMN replacement in a mouse model of spinal muscular atrophy. *The Journal of Clinical Investigation, 114*, 1726–1731.

Azzouz, M., Martin-Rendon, E., Barber, R. D., Mitrophanous, K. A., Carter, E. E., Rohll, J. B., et al. (2002). Multicistronic lentiviral vector-mediated striatal gene transfer of aromatic L-amino acid decarboxylase, tyrosine hydroxylase, and GTP cyclohydrolase I induces sustained transgene expression, dopamine production, and functional improvement in a rat model of Parkinson's disease. *The Journal of Neuroscience, 22*, 10302–10312.

Azzouz, M., Ralph, G. S., Storkebaum, E., Walmsley, L. E., Mitrophanous, K. A., Kingsman, S. M., et al. (2004b). VEGF delivery with retrogradely transported lentivector prolongs survival in a mouse ALS model. *Nature, 429*, 413–417.

Azzouz, M., Ralph, S., Wong, L. F., Day, D., Askham, Z., Barber, R. D., et al. (2004c). Neuroprotection in a rat Parkinson model by GDNF gene therapy using EIAV vector. *Neuroreport, 15*, 985–990.

BDNF Study Group. (1999). A controlled trial of recombinant methionyl human BDNF in ALS: The BDNF Study Group (Phase III). *Neurology, 52*, 1427–1433.

Belin, A. C., & Westerlund, M. (2008). Parkinson's disease: a genetic perspective. *The FEBS Journal, 275*, 1377–1383.

Bensadoun, J. C., Deglon, N., Tseng, J. L., Ridet, J. L., Zurn, A. D., & Aebischer, P. (2000). Lentiviral vectors as a gene delivery system in the mouse midbrain: cellular and behavioral improvements in a 6-OHDA model of Parkinson's disease using GDNF. *Experimental Neurology, 164*, 15–24.

Borasio, G. D., Robberecht, W., Leigh, P. N., Emile, J., Guiloff, R. J., Jerusalem, F., et al. (1998). A placebo-controlled trial of insulin-like growth factor-I in amyotrophic lateral sclerosis. European ALS/IGF-I Study Group. *Neurology, 51*, 583–586.

Brotchie, J. M., Lee, J., & Venderova, K. (2005). Levodopa-induced dyskinesia in Parkinson's disease. *Journal of Neural Transmission, 112*, 359–391.

Browne, S. E., & Beal, M. F. (2004). The energetics of Huntington's disease. *Neurochemical Research, 29*, 531–546.

Brzustowicz, L. M., Lehner, T., Castilla, L. H., Penchaszadeh, G. K., Wilhelmsen, K. C., Daniels, R., et al. (1990). Genetic mapping of chronic childhood-onset spinal muscular atrophy to chromosome 5q11.2-13.3. *Nature, 344*, 540–541.

Cui, L., Jeong, H., Borovecki, F., Parkhurst, C. N., Tanese, N., & Krainc, D. (2006). Transcriptional repression of PGC-1alpha by mutant huntingtin leads to mitochondrial dysfunction and neurodegeneration. *Cell, 127*, 59–69.

Dauer, W., & Przedborski, S. (2003). Parkinson's disease: mechanisms and models. *Neuron, 39*, 889–909.

de Almeida, L. P., Ross, C. A., Zala, D., Aebischer, P., & Deglon, N. (2002). Lentiviral-mediated delivery of mutant huntingtin in the striatum of rats induces a selective neuropathology modulated by polyglutamine repeat size, huntingtin expression levels, and protein length. *The Journal of Neuroscience, 22*, 3473–3483.

de Almeida, L. P., Zala, D., Aebischer, P., & Deglon, N. (2001). Neuroprotective effect of a CNTF-expressing lentiviral vector in the quinolinic acid rat model of Huntington's disease. *Neurobiology of Disease, 8*, 433–446.

DiFiglia, M., Sapp, E., Chase, K. O., Davies, S. W., Bates, G. P., Vonsattel, J. P., et al. (1997). Aggregation of huntingtin in neuronal intranuclear inclusions and dystrophic neurites in brain. *Science, 277*, 1990–1993.

Dowd, E., Monville, C., Torres, E. M., Wong, L. F., Azzouz, M., Mazarakis, N. D., et al. (2005). Lentivector-mediated delivery of GDNF protects complex motor functions relevant to human Parkinsonism in a rat lesion model. *The European Journal of Neuroscience, 22*, 2587–2595.

Dunah, A. W., Jeong, H., Griffin, A., Kim, Y. M., Standaert, D. G., Hersch, S. M., et al. (2002). Sp1 and TAFII130 transcriptional activity disrupted in early Huntington's disease. *Science, 296*, 2238–2243.

Dunnett, S. B., & Bjorklund, A. (1999). Prospects for new restorative and neuroprotective treatments in Parkinson's disease. *Nature, 399*, A32–A39.

Ekestern, E. (2004). Neurotrophic factors and amyotrophic lateral sclerosis. *Neuro-degenerative Diseases, 1*, 88–100.

Eriksdotter Jonhagen, M., Nordberg, A., Amberla, K., Backman, L., Ebendal, T., Meyerson, B., et al. (1998). Intracerebroventricular infusion of nerve growth factor in three patients with Alzheimer's disease. *Dementia and Geriatric Cognitive Disorders, 9*, 246–257.

Feany, M. B., & Bender, W. W. (2000). A *Drosophila* model of Parkinson's disease. *Nature, 404*, 394–398.

Feldkotter, M., Schwarzer, V., Wirth, R., Wienker, T. F., & Wirth, B. (2002). Quantitative analyses of SMN1 and SMN2 based on real-time LightCycler PCR: fast and highly reliable carrier testing and prediction of severity of spinal muscular atrophy. *American Journal of Human Genetics, 70*, 358–368.

Fjord-Larsen, L., Johansen, J. L., Kusk, P., Tornoe, J., Gronborg, M., Rosenblad, C., et al. (2005). Efficient in vivo protection of nigral dopaminergic neurons by lentiviral gene transfer of a modified Neurturin construct. *Experimental Neurology, 195*, 49–60.

Forno, L. S., DeLanney, L. E., Irwin, I., & Langston, J. W. (1993). Similarities and differences between MPTP-induced parkinsonism and Parkinson's disease. Neuropathologic considerations. *Advances in Neurology, 60*, 600–608.

Franich, N. R., Fitzsimons, H. L., Fong, D. M., Klugmann, M., During, M. J., & Young, D. (2008). AAV vector-mediated RNAi of mutant huntingtin expression is neuroprotective in a novel genetic rat model of Huntington's disease. *Molecular Therapy, 16*, 947–956.

Gasmi, M., Herzog, C. D., Brandon, E. P., Cunningham, J. J., Ramirez, G. A., Ketchum, E. T., et al. (2007). Striatal delivery of neurturin by CERE-120, an AAV2 vector for the treatment of dopaminergic neuron degeneration in Parkinson's disease. *Molecular Therapy, 15*, 62–68.

Gene Therapy Clinical Trials Worldwide. (2008). *Journal of Gene Medicine*, Wiley, available at: http://www.wiley.co.uk/genmed/clinical/

Georgievska, B., Kirik, D., Rosenblad, C., Lundberg, C., & Bjorklund, A. (2002). Neuroprotection in the rat Parkinson model by intrastriatal GDNF gene transfer using a lentiviral vector. *Neuroreport, 13*, 75–82.

Gibb, W. R., & Lees, A. J. (1988). The relevance of the Lewy body to the pathogenesis of idiopathic Parkinson's disease. *Journal of Neurology, Neurosurgery, and Psychiatry, 51*, 745–752.

Gil, J. M., & Rego, A. C. (2008). Mechanisms of neurodegeneration in Huntington's disease. *The European Journal of Neuroscience, 27*, 2803–2820.

Guillot, S., Azzouz, M., Deglon, N., Zurn, A., & Aebischer, P. (2004). Local GDNF expression mediated by lentiviral vector protects facial nerve motoneurons but not spinal motoneurons in SOD1(G93A) transgenic mice. *Neurobiology of Disease, 16*, 139–149.

Gurney, M. E., Cutting, F. B., Zhai, P., Doble, A., Taylor, C. P., Andrus, P. K., et al. (1996). Benefit of vitamin E, riluzole, and gabapentin in a transgenic model of familial amyotrophic lateral sclerosis. *Annals of Neurology, 39*, 147–157.

Harper, S. Q., Staber, P. D., He, X., Eliason, S. L., Martins, I. H., Mao, Q., et al. (2005). RNA interference improves motor and neuropathological abnormalities in a Huntington's disease mouse model. *Proceedings of the National Academy of Sciences of the United States of America, 102*, 5820–5825.

Helmken, C., Hofmann, Y., Schoenen, F., Oprea, G., Raschke, H., Rudnik-Schoneborn, S., et al. (2003). Evidence for a modifying pathway in SMA discordant families: reduced SMN level decreases the amount of its interacting partners and Htra2-beta1. *Human Genetics, 114*, 11–21.

Hsieh-Li, H. M., Chang, J. G., Jong, Y. J., Wu, M. H., Wang, N. M., Tsai, C. H., et al. (2000). A mouse model for spinal muscular atrophy. *Nature Genetics, 24*, 66–70.

Kahle, P. J., Neumann, M., Ozmen, L., Muller, V., Jacobsen, H., Schindzielorz, A., et al. (2000). Subcellular localization of wild-type and Parkinson's disease-associated mutant alpha-synuclein in human and transgenic mouse brain. *The Journal of Neuroscience, 20*, 6365–6373.

Kaneda, Y. (1999). Development of a novel fusogenic viral liposome system (HVJ-liposomes) and its applications to the treatment of acquired diseases. *Molecular Membrane Biology, 16*, 119–122.

Kato, S., Inoue, K., Kobayashi, K., Yasoshima, Y., Miyachi, S., Inoue, S., et al. (2007). Efficient gene transfer via retrograde transport in rodent and primate brains using a human immunodeficiency virus type 1-based vector pseudotyped with rabies virus glycoprotein. *Human Gene Therapy, 18*, 1141–1151.

Kells, A. P., Fong, D. M., Dragunow, M., During, M. J., Young, D., & Connor, B. (2004). AAV-mediated gene delivery of BDNF or GDNF is neuroprotective in a model of Huntington disease. *Molecular Therapy, 9*, 682–688.

Kordower, J. H., Bloch, J., Ma, S. Y., Chu, Y., Palfi, S., Roitberg, B. Z., et al. (1999a). Lentiviral gene transfer to the nonhuman primate brain. *Experimental Neurology, 160*, 1–16.

Kordower, J. H., Emborg, M. E., Bloch, J., Ma, S. Y., Chu, Y., Leventhal, L., et al. (2000). Neurodegeneration prevented by lentiviral vector delivery of GDNF in primate models of Parkinson's disease. *Science, 290*, 767–773.

Kordower, J. H., Herzog, C. D., Dass, B., Bakay, R. A., Stansell, J., III, Gasmi, M., et al. (2006). Delivery of neurturin by AAV2 (CERE-120)-mediated gene transfer provides structural and functional neuroprotection and neurorestoration in MPTP-treated monkeys. *Annals of Neurology, 60*, 706–715.

Kordower, J. H., Palfi, S., Chen, E. Y., Ma, S. Y., Sendera, T., Cochran, E. J., et al. (1999b). Clinicopathological findings following intraventricular glial-derived neurotrophic factor treatment in a patient with Parkinson's disease. *Annals of Neurology, 46*, 419–424.

Landles, C., & Bates, G. P. (2004). Huntingtin and the molecular pathogenesis of Huntington's disease. Fourth in molecular medicine review series. *EMBO Reports, 5*, 958–963.

Lang, A. E., & Lozano, A. M. (1998). Parkinson's disease. First of two parts. *The New England Journal of Medicine, 339*, 1044–1053.

Langston, J. W., Ballard, P., Tetrud, J. W., & Irwin, I. (1983). Chronic Parkinsonism in humans due to a product of meperidine-analog synthesis. *Science, 219*, 979–980.

Lefebvre, S., Burglen, L., Reboullet, S., Clermont, O., Burlet, P., Viollet, L., et al. (1995). Identification and characterization of a spinal muscular atrophy-determining gene. *Cell, 80*, 155–165.

Lesbordes, J. C., Cifuentes-Diaz, C., Miroglio, A., Joshi, V., Bordet, T., Kahn, A., et al. (2003). Therapeutic benefits of cardiotrophin-1 gene transfer in a mouse model of spinal muscular atrophy. *Human Molecular Genetics, 12*, 1233–1239.

Li, S., & Huang, L. (2000). Nonviral gene therapy: promises and challenges. *Gene Therapy, 7*, 31–34.

Liu, Q., & Dreyfuss, G. (1996). A novel nuclear structure containing the survival of motor neurons protein. *EMBO Journal, 15*, 3555–3565.

Lo Bianco, C., Ridet, J. L., Schneider, B. L., Deglon, N., & Aebischer, P. (2002). Alpha-synucleinopathy and selective dopaminergic neuron loss in a rat lentiviral-based model of Parkinson's disease. *Proceedings of the National Academy of Sciences of the United States of America, 99*, 10813–10818.

Lorson, C. L., Hahnen, E., Androphy, E. J., & Wirth, B. (1999). A single nucleotide in the SMN gene regulates splicing and is responsible for spinal muscular atrophy. *Proceedings of the National Academy of Sciences of the United States of America, 96*, 6307–6311.

Lorson, C. L., Strasswimmer, J., Yao, J. M., Baleja, J. D., Hahnen, E., Wirth, B., et al. (1998). SMN oligomerization defect correlates with spinal muscular atrophy severity. *Nature Genetics, 19*, 63–66.

Lunkes, A., & Mandel, J. L. (1998). A cellular model that recapitulates major pathogenic steps of Huntington's disease. *Human Molecular Genetics, 7*, 1355–1361.

Mazarakis, N. D., Azzouz, M., Rohll, J. B., Ellard, F. M., Wilkes, F. J., Olsen, A. L., et al. (2001). Rabies virus glycoprotein pseudotyping of lentiviral vectors enables retrograde axonal transport and access to the nervous system after peripheral delivery. *Human Molecular Genetics, 10*, 2109–2121.

McBride, J. L., During, M. J., Wuu, J., Chen, E. Y., Leurgans, S. E., & Kordower, J. H. (2003). Structural and functional neuroprotection in a rat model of Huntington's disease by viral gene transfer of GDNF. *Experimental Neurology, 181*, 213–223.

McGeer, E. G., & McGeer, P. L. (1976). Duplication of biochemical changes of Huntington's chorea by intrastriatal injections of glutamic and kainic acids. *Nature, 263*, 517–519.

Meister, G., Buhler, D., Laggerbauer, B., Zobawa, M., Lottspeich, F., & Fischer, U. (2000). Characterization of a nuclear 20S complex containing the survival of motor neurons (SMN) protein and a specific subset of spliceosomal Sm proteins. *Human Molecular Genetics, 9*, 1977–1986.

Melki, J., Abdelhak, S., Sheth, P., Bachelot, M. F., Burlet, P., Marcadet, A., et al. (1990). Gene for chronic proximal spinal muscular atrophies maps to chromosome 5q. *Nature, 344*, 767–768.

Mentis, G. Z., Gravell, M., Hamilton, R., Shneider, N. A., O'Donovan, M. J., & Schubert, M. (2006). Transduction of motor neurons and muscle fibers by intramuscular injection of HIV-1-based vectors pseudotyped with select rabies virus glycoproteins. *Journal of Neuroscience Methods, 157*, 208–217.

Miller, R. G., Mitchell, J. D., & Moore, D. H. (2000). Riluzole for amyotrophic lateral sclerosis (ALS)/motor neuron disease (MND). *Cochrane Database of Systematic Reviews 2000, CD001447*.

Monani, U. R., Sendtner, M., Coovert, D. D., Parsons, D. W., Andreassi, C., Le, T. T., et al. (2000). The human centromeric survival motor neuron gene (SMN2) rescues embryonic lethality in Smn(−/−) mice and results in a mouse with spinal muscular atrophy. *Human Molecular Genetics, 9*, 333–339.

Mulder, D. W. (1982). Clinical limits of amyotrophic lateral sclerosis. *Advances in Neurology, 36*, 15–22.

Munsat, T. L., & Davies, K. E. (1992). International SMA consortium meeting. (26–28 June 1992, Bonn, Germany). *Neuromuscular Disorders, 2*, 423–428.

Nakamura, K., Jeong, S. Y., Uchihara, T., Anno, M., Nagashima, K., Nagashima, T., et al. (2001). SCA17, a novel autosomal dominant cerebellar ataxia caused by an expanded polyglutamine in TATA-binding protein. *Human Molecular Genetics, 10*, 1441–1448.

Naldini, L., Blomer, U., Gallay, P., Ory, D., Mulligan, R., Gage, F. H., et al. (1996). In vivo gene delivery and stable transduction of nondividing cells by a lentiviral vector. *Science, 272*, 263–267.

Oprea, G. E., Krober, S., McWhorter, M. L., Rossoll, W., Muller, S., Krawczak, M., et al. (2008). Plastin 3 is a protective modifier of autosomal recessive spinal muscular atrophy. *Science, 320*, 524–527.

Palfi, S., Leventhal, L., Chu, Y., Ma, S. Y., Emborg, M., Bakay, R., et al. (2002). Lentivirally delivered glial cell line-derived neurotrophic factor increases the number of striatal dopaminergic neurons in primate models of nigrostriatal degeneration. *The Journal of Neuroscience, 22*, 4942–4954.

Parsons, D. W., McAndrew, P. E., Iannaccone, S. T., Mendell, J. R., Burghes, A. H., & Prior, T. W. (1998). Intragenic telSMN mutations: frequency, distribution, evidence of a founder effect, and modification of the spinal muscular atrophy phenotype by cenSMN copy number. *American Journal of Human Genetics, 63*, 1712–1723.

Pellizzoni, L., Kataoka, N., Charroux, B., & Dreyfuss, G. (1998). A novel function for SMN, the spinal muscular atrophy disease gene product, in pre-mRNA splicing. *Cell, 95*, 615–624.

Penn, R. D., Kroin, J. S., York, M. M., & Cedarbaum, J. M. (1997). Intrathecal ciliary neurotrophic factor delivery for treatment of amyotrophic lateral sclerosis (phase I trial). *Neurosurgery, 40*, 94–99. discussion pp. 99–100.

Petersen, A., Larsen, K. E., Behr, G. G., Romero, N., Przedborski, S., Brundin, P., et al. (2001). Expanded CAG repeats in exon 1 of the Huntington's disease gene stimulate dopamine-mediated striatal neuron autophagy and degeneration. *Human Molecular Genetics, 10*, 1243–1254.

Popovic, N., Maingay, M., Kirik, D., & Brundin, P. (2005). Lentiviral gene delivery of GDNF into the striatum of R6/2 Huntington mice fails to attenuate behavioral and neuropathological changes. *Experimental Neurology, 193*, 65–74.

Ralph, G. S., Radcliffe, P. A., Day, D. M., Carthy, J. M., Leroux, M. A., Lee, D. C., et al. (2005). Silencing mutant SOD1 using RNAi protects against neurodegeneration and extends survival in an ALS model. *Nature Medicine, 11*, 429–433.

Raoul, C., Abbas-Terki, T., Bensadoun, J. C., Guillot, S., Haase, G., Szulc, J., et al. (2005). Lentiviral-mediated silencing of SOD1 through RNA interference retards disease onset and progression in a mouse model of ALS. *Nature Medicine, 11*, 423–428.

Regulier, E., Pereira de Almeida, L., Sommer, B., Aebischer, P., & Deglon, N. (2002). Dose-dependent neuroprotective effect of ciliary neurotrophic factor delivered via tetracycline-regulated lentiviral vectors in the quinolinic acid rat model of Huntington's disease. *Human Gene Therapy, 13*, 1981–1990.

Regulier, E., Trottier, Y., Perrin, V., Aebischer, P., & Deglon, N. (2003). Early and reversible neuropathology induced by tetracycline-regulated lentiviral overexpression of mutant huntingtin in rat striatum. *Human Molecular Genetics, 12*, 2827–2836.

Roberts, R. C., Ahn, A., Swartz, K. J., Beal, M. F., & DiFiglia, M. (1993). Intrastriatal injections of quinolinic acid or kainic acid: differential patterns of cell survival and the effects of data analysis on outcome. *Experimental Neurology, 124*, 274–282.

Rosen, D. R. (1993). Mutations in Cu/Zn superoxide dismutase gene are associated with familial amyotrophic lateral sclerosis. *Nature, 364*, 362.

Rubinsztein, D. C., Leggo, J., Coles, R., Almqvist, E., Biancalana, V., Cassiman, J. J., et al. (1996). Phenotypic characterization of individuals with 30–40 CAG repeats in the Huntington disease (HD) gene reveals HD cases with 36 repeats and apparently normal elderly individuals with 36–39 repeats. *American Journal of Human Genetics, 59*, 16–22.

Russman, B. S. (2007). Spinal muscular atrophy: clinical classification and disease heterogeneity. *Journal of Child Neurology, 22*, 946–951.

Sapp, E., Penney, J., Young, A., Aronin, N., Vonsattel, J. P., & DiFiglia, M. (1999). Axonal transport of N-terminal huntingtin suggests early pathology of corticostriatal projections in Huntington disease. *Journal of Neuropathology and Experimental Neurology, 58*, 165–173.

Saudou, F., Finkbeiner, S., Devys, D., & Greenberg, M. E. (1998). Huntingtin acts in the nucleus to induce apoptosis but death does not correlate with the formation of intranuclear inclusions. *Cell, 95*, 55–66.

Schrank, B., Gotz, R., Gunnersen, J. M., Ure, J. M., Toyka, K. V., Smith, A. G., et al. (1997). Inactivation of the survival motor neuron gene, a candidate gene for human spinal muscular atrophy, leads to massive cell death in early mouse embryos. *Proceedings of the National Academy of Sciences of the United States of America, 94*, 9920–9925.

Shaw, P. J. (2005). Molecular and cellular pathways of neurodegeneration in motor neurone disease. *Journal of Neurology, Neurosurgery, and Psychiatry, 76*, 1046–1057.

Spillantini, M. G., Schmidt, M. L., Lee, V. M., Trojanowski, J. Q., Jakes, R., & Goedert, M. (1997). Alpha-synuclein in Lewy bodies. *Nature, 388*, 839–840.

Storkebaum, E., Lambrechts, D., Dewerchin, M., Moreno-Murciano, M. P., Appelmans, S., Oh, H., et al. (2005). Treatment of motoneuron degeneration by intracerebroventricular delivery of VEGF in a rat model of ALS. *Nature Neuroscience, 8*, 85–92.

Sumner, C. J., Huynh, T. N., Markowitz, J. A., Perhac, J. S., Hill, B., Coovert, D. D., et al. (2003). Valproic acid increases SMN levels in spinal muscular atrophy patient cells. *Annals of Neurology, 54*, 647–654.

The Huntington's Disease Collaborative Research Group. (1993). A novel gene containing a trinucleotide repeat that is expanded and unstable on Huntington's disease chromosomes. The Huntington's Disease Collaborative Research Group. *Cell, 72*, 971–983.

Trettel, F., Rigamonti, D., Hilditch-Maguire, P., Wheeler, V. C., Sharp, A. H., Persichetti, F., et al. (2000). Dominant phenotypes produced by the HD mutation in STHdh(Q111) striatal cells. *Human Molecular Genetics, 9*, 2799–2809.

Ungerstedt, U. (1968). 6-Hydroxy-dopamine induced degeneration of central monoamine neurons. *European Journal of Pharmacology, 5*, 107–110.

van der Putten, H., Wiederhold, K. H., Probst, A., Barbieri, S., Mistl, C., Danner, S., et al. (2000). Neuropathology in mice expressing human alpha-synuclein. *The Journal of Neuroscience, 20*, 6021–6029.

Vonsattel, J. P., & DiFiglia, M. (1998). Huntington disease. *Journal of Neuropathology and Experimental Neurology, 57*, 369–384.

Wang, L. H., & Qin, Z. H. (2006). Animal models of Huntington's disease: implications in uncovering pathogenic mechanisms and developing therapies. *Acta Pharmacologica Sinica, 27*, 1287–1302.

Wang, Y. L., Liu, W., Wada, E., Murata, M., Wada, K., & Kanazawa, I. (2005). Clinico-pathological rescue of a model mouse of Huntington's disease by siRNA. *Neuroscience Research, 53*, 241–249.

Williams, D. B., & Windebank, A. J. (1991). Motor neuron disease (amyotrophic lateral sclerosis). *Mayo Clinic Proceedings, 66*, 54–82.

Wong, L. F., Azzouz, M., Walmsley, L. E., Askham, Z., Wilkes, F. J., Mitrophanous, K. A., et al. (2004). Transduction patterns of pseudotyped lentiviral vectors in the nervous system. *Molecular Therapy, 9*, 101–111.

Yamamoto, A., Lucas, J. J., & Hen, R. (2000). Reversal of neuropathology and motor dysfunction in a conditional model of Huntington's disease. *Cell, 101*, 57–66.

Yang, S. H., Cheng, P. H., Banta, H., Piotrowska-Nitsche, K., Yang, J. J., Cheng, E. C., et al. (2008). Towards a transgenic model of Huntington's disease in a non-human primate. *Nature, 453*, 921–924.

Young, P. J., Le, T. T., thi Man, N., Burghes, A. H., & Morris, G. E. (2000). The relationship between SMN, the spinal muscular atrophy protein, and nuclear coiled bodies in differentiated tissues and cultured cells. *Experimental Cell Research*, *256*, 365–374.

Zala, D., Benchoua, A., Brouillet, E., Perrin, V., Gaillard, M. C., Zurn, A. D., et al. (2005). Progressive and selective striatal degeneration in primary neuronal cultures using lentiviral vector coding for a mutant huntingtin fragment. *Neurobiology of Disease*, *20*, 785–798.

Zala, D., Bensadoun, J. C., Pereira de Almeida, L., Leavitt, B. R., Gutekunst, C. A., Aebischer, P., et al. (2004). Long-term lentiviral-mediated expression of ciliary neurotrophic factor in the striatum of Huntington's disease transgenic mice. *Experimental Neurology*, *185*, 26–35.

Zheng, C., Nennesmo, I., Fadeel, B., & Henter, J. I. (2004). Vascular endothelial growth factor prolongs survival in a transgenic mouse model of ALS. *Annals of Neurology*, *56*, 564–567.

Zoghbi, H. Y., & Orr, H. T. (2000). Glutamine repeats and neurodegeneration. *Annual Review of Neuroscience*, *23*, 217–247.

CHAPTER 14

# Trophic factors therapy in Parkinson's disease

Shilpa Ramaswamy, Katherine E. Soderstrom and Jeffrey H. Kordower[*]

*Department of Neurological Sciences, Rush University Medical Center, Chicago, IL, USA*

**Abstract:** Parkinson's disease (PD) is a progressive, neurodegenerative disorder for which there is currently no effective neuroprotective therapy. Patients are typically treated with a combination of drug therapies and/or receive deep brain stimulation to combat behavioral symptoms. The ideal candidate therapy would be the one which prevents neurodegeneration in the brain, thereby halting the progression of debilitating disease symptoms. Neurotrophic factors have been in the forefront of PD research, and clinical trials have been initiated using members of the GDNF family of ligands (GFLs). GFLs have been shown to be trophic to ventral mesencephalic cells, thereby making them good candidates for PD research. This paper examines the use of GDNF and neurturin, two members of the GFL, in both animal models of PD and clinical trials.

**Keywords:** neurotrophic factors; Parkinson's disease; glial cell line-derived neurotrophic factor family ligands; GDNF; neurturin; gene therapy; clinical trials

## Parkinson's disease and neurotrophic factors

Parkinson's disease (PD) is a progressive neurodegenerative disorder characterized by the cardinal motor symptoms of tremor, rigidity, bradykinesia, and postural instability (Parkinson's Disease Foundation, 2009). In PD, there is a loss of dopamine in the striatum and degeneration of dopaminergic neurons within the substantia nigra pars compacta (Hwang et al., 2003). This causes dysfunction in the basal ganglia, ultimately resulting in impoverished thalamocortical innervation and the manifestation of the cardinal motor symptoms of the disease. While the pathology of PD is not limited to the nigrostriatal circuitry, it has been, to date, the focus of most therapeutic interventions.

Due to the progressive nature of PD, many researchers have focused their efforts on the use of neuroprotective agents to rescue vulnerable nigral neurons before they are lost to disease. One particularly exciting novel therapy that has gained interest in recent decades has been the use of neurotrophic factors, molecules typically characterized for their role in neuronal development and maintenance. Neurotrophic factors have allowed researchers to expand their therapeutic reach, beyond merely augmenting dopaminergic function to replacing neurons lost to disease and rescuing intrinsic neuronal systems before they succumb.

The neurotrophic factors that have been commonly explored for use in the therapy of PD patients include the glial cell line-derived

[*]Corresponding author.
Tel.: +1 312 563 3570; Fax: +1 312 563 3571;
E-mail: jkordowe@rush.edu

neurotrophic factor (GDNF) family ligands (GFLs), neurotrophins, and cytokines. GDNF and neurturin (NTN) are the two main members of the GFLs that have been widely tested in animal models of PD and clinically tested in PD patients.

**Glial cell line-derived neurotrophic factor**

GDNF supports the survival of several different neuronal populations, in both the central and the peripheral nervous system. GDNF's potential therapeutic value for PD was first recognized in 1993 when it was purified and shown to promote the growth and survival of midbrain embryonic dopaminergic neurons (Lin et al., 1993). GDNF signaling is mediated via a multicomponent receptor complex consisting of a binding receptor (GDNF family receptor alpha, GFRα) that forms a ligand–receptor complex, which then is retrogradely transported from the target to the cell soma where it signals through a second receptor called Ret receptor tyrosine kinase (Sariola and Saarma, 2003). Serendipitously for PD therapy, all components of the GDNF signaling pathway are expressed at high levels in the striatum and substantia nigra pars compacta while the Ret receptor is in abundance only in the nigra. Therefore, GDNF can be injected into the striatum, and still provide trophic influence at the level of the midbrain.

GDNF is essential for the survival of dopaminergic neurons as shown in a conditional GDNF knockout mouse model. Down-regulation of GDNF, even by only 40% in adulthood, causes a marked reduction in dopaminergic neurons in the substantia nigra, the locus coeruleus, and ventral tegmental area (VTA). This neuronal loss is accompanied by a detectable hypokinetic movement disorder in mice (Pascual et al., 2008). These findings, illustrating that GDNF is protective for dopaminergic neurons, prompted several animal studies in PD models.

*Injections in animal PD models*

Initial studies using GDNF involved direct bolus injections of the trophic factor either into the striatum or lateral ventricle, or directly into the substantia nigra. Most of these studies initially used the 6-hydroxydopamine (6-OHDA) lesion model. 6-OHDA, when administered to the striatum, causes a progressive dying back of nigrostriatal fibers and eventually leads to cell loss in the substantia nigra (Rosenblad et al., 1999). In rats receiving lesions to either the striatum or the substantia nigra, GDNF administered to the nigra has differential effects on neuronal survival (Kearns and Gash, 1995). When injected directly into the striatum, GDNF preserves nigral neurons destined to die following the administration of 6-OHDA. In a second study, administration of GDNF to the region just above the substantia nigra 1-week post lesion results in a partial but significant protection of tyrosine hydroxylase (TH)-positive nigral neurons (Sauer et al., 1995). However, the neurons that do remain appear significantly atrophied, indicating that administering GDNF far from the site of lesion may cause a decrease in functionality of protected neurons. Furthermore, these studies did not examine the effects on TH-positive fibers in the nigrostriatal system, a critical component in preserving motor function.

Single bolus injections of GDNF have been used in other studies of rats receiving 6-OHDA lesions. As mentioned above, protection of nigral neurons is irrelevant if it is not accompanied by a preservation of function. A crucial study compared both the motor and the cellular benefits of administering GDNF to the striatum, nigra, or lateral ventricle prior to 6-OHDA delivery (Kirik et al., 2000a). When GDNF is administered to the striatum, both cell bodies in the nigra and TH-positive fibers in the striatum are significantly protected. More importantly, this neuroprotection is accompanied by a preservation of motor function, a far more relevant barometer of trophic factor efficacy. When administered directly to the nigra, GDNF protected nigral cell bodies and caused some local axonal sprouting, but did not promote the preservation of striatal TH levels. Furthermore, it did not significantly protect motor function in these rats. Finally, when GDNF is infused into the lateral ventricles, there is inefficient diffusion of the trophic factor from the cerebrospinal fluid. This causes GDNF-treated rats to be indistinguishable behaviorally from untreated rats receiving 6-OHDA.

## *Infusions in animal PD models*

While the therapeutic value of GDNF was evident in the above-mentioned studies, it was recognized that the previously used methods of administration were inefficient. Researchers began to explore more long-term, sustainable ways of getting GDNF into target regions. Catheters were inserted into the brain and GDNF was infused over prolonged periods of time using pumps. In one such study, a catheter was inserted into the putamen of aged rhesus monkeys (Ai et al., 2003). This catheter was connected to a pump that was implanted subcutaneously in the abdominal region. The pump was programmed to continuously infuse GDNF into the putamen over 8 weeks. This delivery method effectively distributed the trophic factor up to 11 mm away from the site of catheter insertion. GDNF diffused to the rostral putamen, internal capsule, external capsule, caudate nucleus, and globus pallidus. Additionally, retrograde transport of GDNF was seen in nigral cell bodies. This transport of GDNF into adjacent areas translated into a significant improvement in the overall motor performance of these aged monkeys in the last 3 weeks of the study compared to controls (Maswood et al., 2002). Additionally, there was a 50% increase in dopamine levels in the ipsilateral caudate nucleus and a 390% increase in dopamine in the ipsilateral globus pallidus. Similar encouraging results were seen in a PD model using 1-methyl-4-phenyl-1,2,3,6-tetrahydropyridine (MPTP)-lesioned rhesus monkeys that received continuous infusion of GDNF to either the lateral ventricle or the putamen (Grondin et al., 2002). GDNF promoted a significant anti-parkinsonian effect in monkeys that received the trophic factor to both the ventricle and the striatum. This positive motor effect was brought on by a very modest increase in overall TH-positive fiber density throughout the striatum. However, TH-positive fiber density was increased five-fold only in the immediate area surrounding the lateral ventricles, indicating that small areas that are efficiently delivered GDNF can experience robust trophic effects.

## *Experimental administration in PD human subjects*

Based on the seemingly encouraging results from animal studies, the first clinical trial using GDNF in PD patients was initiated in 1996. This was a randomized, double-blinded study administering recombinant GDNF protein into the lateral ventricle using mechanical pumps (Nutt et al., 2003). Fifty patients, between the ages of 35 and 75 years, with moderate or advanced idiopathic PD were chosen for this study. Patients received either placebo or doses of GDNF varying between 25 and 4000 μg into the ventricles once a month over 8 months. Sixteen of these patients then received 4000 μg of GDNF for an additional 20 months in an open-labeled trial. When the study was unblinded after the first 8 months, results were disappointing. Not only did patients not improve but they also experienced several adverse events including nausea, vomiting, and anorexia for several days after GDNF administration. Patients who received higher doses of GDNF also experienced weight loss and depression symptoms. Even patients who received 4000 μg of GDNF in the 20-month open-label continuation of the trial did not show any improvements in either the "on" or "off" Unified Parkinson's Disease Rating Scale (UPDRS) scores. The trial initiated by AMGEN was halted in September 2004 (Slevin et al., 2007). The lack of symptomatic relief seen in these patients may have been attributed to the inadequate penetration of GDNF from the cerebrospinal fluid into the adjacent striatum. Postmortem analysis in one patient from this study demonstrated that GDNF did not efficiently diffuse out of the lateral ventricles and thus was unable to elicit any effect in the striatum or the nigra (Kordower et al., 1999). It was evident that administering direct bolus injections of GDNF into the lateral ventricles was an inadequate method of trophic factor delivery. Studies have shown that intraputamental injection of GDNF in rhesus monkeys causes a variable distribution of the trophic factor with only 2–9% of the area receiving GDNF infusion (Salvatore et al., 2006). Furthermore, it has been shown that even convection-enhanced delivery of GDNF to slowly diffuse the factor into the target region is not desirable. There is a great deal of variability in GDNF diffusion using this method (Gash et al., 2005). There is little consistency in the distribution of GDNF in MPTP-treated rhesus monkeys even using the convection-enhanced delivery method. Researchers saw a volume of GDNF ranging anywhere from 59 to 325 mm$^3$ in the putamen. This may be because GDNF easily binds to receptor sites in the extracellular matrix, impeding its even distribution (Hamilton et al., 2001). Despite these varying results in animal models using intraputamenal infusion, Amgen optimistically conducted an initial Phase I open-labeled trial using this method of delivery in five patients (Gill et al., 2003). All but one of the patients received bilateral infusion of GDNF into the posterior putamen for 43 months. The dose of GDNF was increased from 14.4 to 28.8 μg/putamen/day because of a decrease in benefit. This increased dose resulted in a sustained and progressive improvement (Patel et al., 2005). No adverse side effects were reported after 1 year of treatment, and in fact significant decreases were reported in both "on" and "off" UPDRS scores. Additionally, there was a 39% decrease in the off-medication motor score, a 61% improvement in the activities of daily living sub-score, a 20% decrease in severe immobility, a decrease in medication-induced dyskinesias, and a 28% increase in fluoro-dopa ($^{18}$F-dopa) uptake in the posterior putamen. One of the five patients, who started receiving unilateral (right putamen) infusions of GDNF at the age of 62, died from a myocardial infarct 3 months after drug withdrawal (Love et al., 2005). Postmortem analyses of brain tissue indicated that there was a more than five-fold increase in tyrosine hydroxylase immunoreactivity in the right versus left putamen. However, due to the asymmetrical pathology seen in most PD patients, there was a higher level of TH immunoreactivity and higher numbers of neurons were seen in the left versus the right substantia nigra. Interestingly, there was an increase in growth-associated protein 43 (GAP43) staining in the right putamen, indicating that GDNF induces sprouting in substantia nigra fibers. The researchers note that the increase in TH staining in the putamen may either be a result of sprouting of fibers or an upregulation of the enzyme in spared fibers. These exciting results prompted a double-blinded, placebo-controlled study using 34 subjects, half of whom received placebo and the other half received 15 μg/putamen/day of GDNF. Unfortunately, bilateral infusion of GDNF into the putamen in this study did not significantly reduce UPDRS scores even after 6 months of treatment. Surprisingly, there was a 23% increase in $^{18}$F-dopa uptake in the posterior putamen. The discrepancy between the increase in $^{18}$F-dopa uptake and a lack of clinical benefit might indicate that although there is an increase in dopamine in the putamen as a result of GDNF treatment, it is not being efficiently released. The researchers state that they used a different-sized catheter to administer GDNF in this study compared to the initial open-labeled trial. Results from both the ventricular and the putamenal infusion studies indicate that direct administration of GDNF to the brain is not an efficient method of treatment. Therefore, a novel vehicle is needed to aid in the administration of GDNF to the striatal parenchyma.

## Gene therapy in animal PD models

As trophic factor therapy in PD models was evolving, gene therapy approaches were evolving in parallel. Gene therapy employs viral vectors, which provide a safe and robust way to deliver trophic factors such as GDNF uniformly over very long periods of time.

### Adenoviral vector-mediated gene therapy

One of the first vehicles to be used for the administration of GDNF was the adenoviral (Ad) vector. Marty Bohn and colleagues showed that a single injection of Ad-GDNF near the rat substantia nigra could sustain trophic factor expression for at least 7 weeks (Choi-Lundberg et al., 1997). Additionally, Ad-GDNF significantly protected TH-positive neurons in the substantia nigra from 6-OHDA-induced toxicity. It did not however, alter the expression of TH-positive fibers in the striatum, indicating that there may not have been any therapeutic consequences from solely treating the nigra. They repeated this study but this time administered the Ad-GDNF to the striatum, the site of dopamine fiber loss (Choi-Lundberg et al., 1998). Again they protected 40% of cells in the substantia nigra but did not maintain levels of TH in fibers of the striatum. Interestingly, Ad-GDNF administration to the striatum improved motor performance in treated rats, indicating that either a preservation of striatal TH-positive fibers was not necessary for behavioral improvement in this model or TH levels were increased to a level undetected by the methods used. A second group conducted a similar study and found that both nigral cells and striatal fibers were protected (Bilang-Bleuel et al., 1997). They also showed an attenuation of behavioral deficits as seen using the amphetamine-induced rotational paradigm. Adenoviral delivery of GDNF was the first to be used in animal studies of PD to successfully transfect cells with GDNF and protect both nigral neurons and fibers. However, the original version of this vector caused severe immune responses in the injected region (Bilang-Bleuel et al., 1997). Thus alternative, less immunogenic vectors had to be developed and tested. The two vectors that emerged were the lentiviral (LV) and the adeno-associated viral (AAV) vectors.

*Lentiviral vector-mediated gene therapy*

Subsequent to the adenoviral era, recombinant lentiviral (rLV) vectors were used to express GDNF in both the striatum and the substantia nigra in a 6-OHDA model of PD (Georgievska et al., 2002b). When expressed in the striatum, GDNF was successfully transported to nigral neurons where it protected 65–77% of these cells. This protection was dose-dependent and rats receiving a higher dose of GDNF showed a greater magnitude of cellular protection. However, fibers in the striatum were not significantly protected. Encouragingly though, fibers along the striato-nigral pathway were conserved as seen by intact fibers in the globus pallidus. Additionally, sprouting was seen in areas where GDNF was expressed at very high levels like in the globus pallidus and the immediately surrounding striatum. Irrespective of a lack of striatal fiber preservation, deficits in amphetamine-induced rotational behavior were prevented in rLV-GDNF rats compared to lesioned controls, indicating an increase in dopamine function on the GDNF-treated side. The lack of striatal preservation may have been due to the short time course of GDNF treatment. To confirm this theory, this group repeated this study and expressed GDNF in the striatum for 9 months using the rLV vector (Georgievska et al., 2002a). Unfortunately, they saw similar results, namely, neuroprotection in the nigra, striato-nigral fiber protection through the globus pallidus, but no fiber protection in the striatum. They reported a lack of functional recovery, which was attributed to a lack of dopamine in the striatum.

While the use of rodent models in the testing of therapies for PD is essential, ultimately any potential therapy likely must be tested in nonhuman primates before it can be used in the clinic. Therefore, the efficacy of LV-GDNF was tested both in aged (Fig. 1) rhesus monkeys and in monkeys lesioned 1 week before using the toxin MPTP (Kordower et al., 2000). In aged monkeys, LV-GDNF enhanced dopaminergic function. Aged monkeys receiving LV-GDNF treatment to the striatum showed an enhanced $^{18}$F-dopa uptake ipsilaterally. These monkeys had an increase in TH immunoreactivity in the striatum, an 85% increase in the number of TH-immunoreactive neurons within the substantia nigra, and a 35% increase in the volume of these neurons (Fig. 2). In MPTP-treated monkeys, LV-GDNF reversed functional deficits and completely prevented nigrostriatal degeneration. Monkeys receiving striatal LV-GDNF showed significant improvements in clinical rating scale scores during the 3-month period after GDNF treatment. Additionally, LV-GDNF treatment reversed motor deficits in an operant hand-reach task. LV-GDNF-treated monkeys also showed robust

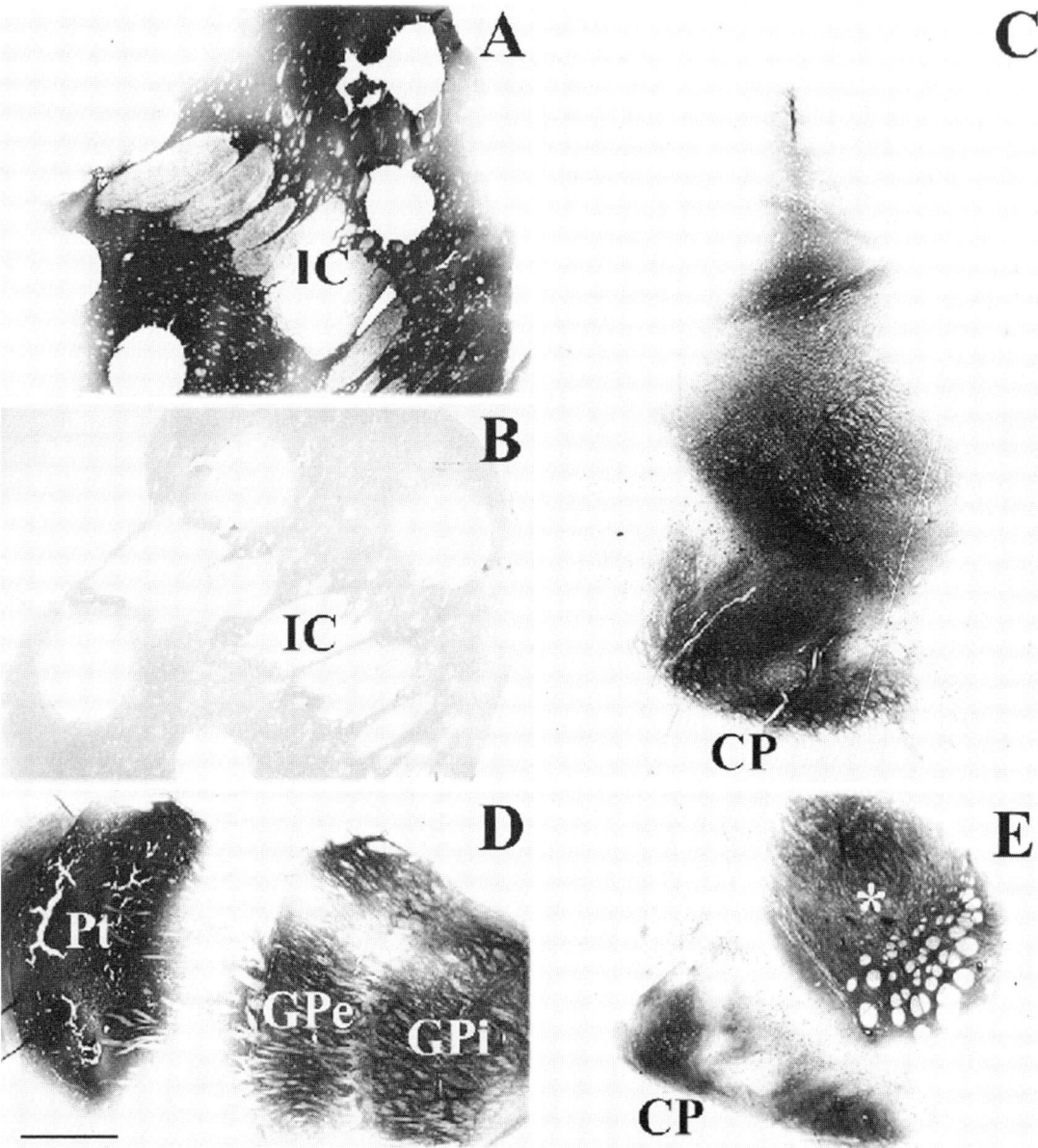

Fig. 1. GDNF immunohistochemistry in aged monkeys receiving LV-GDNF or LV-βGal (control) in the striatum and substantia nigra. (A) Robust GDNF immunoreactivity is seen within the caudate and putamen in a LV-GDNF-treated aged monkey. (B) In contrast, no GDNF immunoreactivity is observed in a control LV-βGal-treated animal (IC, internal capsule). (C) Robust GDNF immunoreactivity is also observed within the midbrain of a LV-GDNF-treated monkey. (D) GDNF immunoreactivity within the forebrain of a LV-GDNF-treated monkey. Staining is seen within the injection site in the putamen (Pt) and within both segments of the globus pallidus (GPe and GPi) from anterograde transport. (E) GDNF immunohistochemistry is also seen in the substantia nigra pars reticulata from anterograde transport. Holes in the tissue are from postmortem for HPLC analysis. Asterisk in (E) represents a LV-GDNF injection site (CP, cerebral peduncle). Scale bar in (D) represents 1600 μm for panels A, B, and D; 1150 μm for panel C; and 800 μm for panel E.

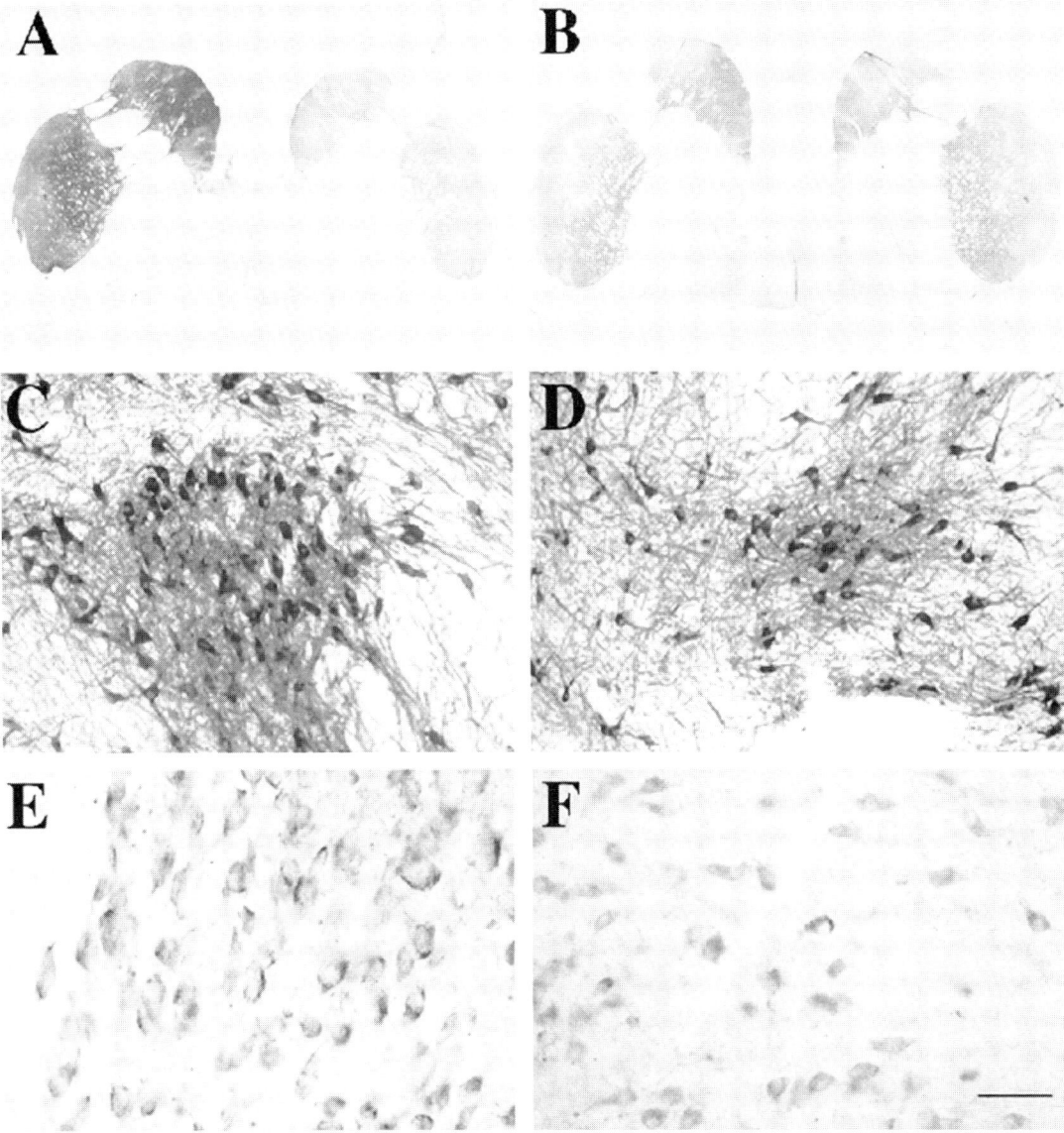

Fig. 2. Tyrosine hydroxylase staining in aged monkeys receiving LV-GDNF to the right striatum. (A) LV-GDNF administration to the right striatum increases TH immunoreactivity within the right caudate and putamen in aged monkeys. (B) In monkeys receiving control LV-βGal injections to the right striatum, there is symmetrical and less intense staining for TH. (C) There are greater numbers and larger TH-immunoreactive neurons within the substantia nigra (SN) of LV-GDNF-treated animals relative to (D) a LV-βGal-treated monkeys. (E) LV-GDNF-treated aged monkeys display increased TH mRNA relative to (F) LV-βGal-treated monkeys in the SN. Scale bar in (F) represents 4500 μm for panels A and B; 250 μm for panels C and D; and 100 μm for panels E and F.

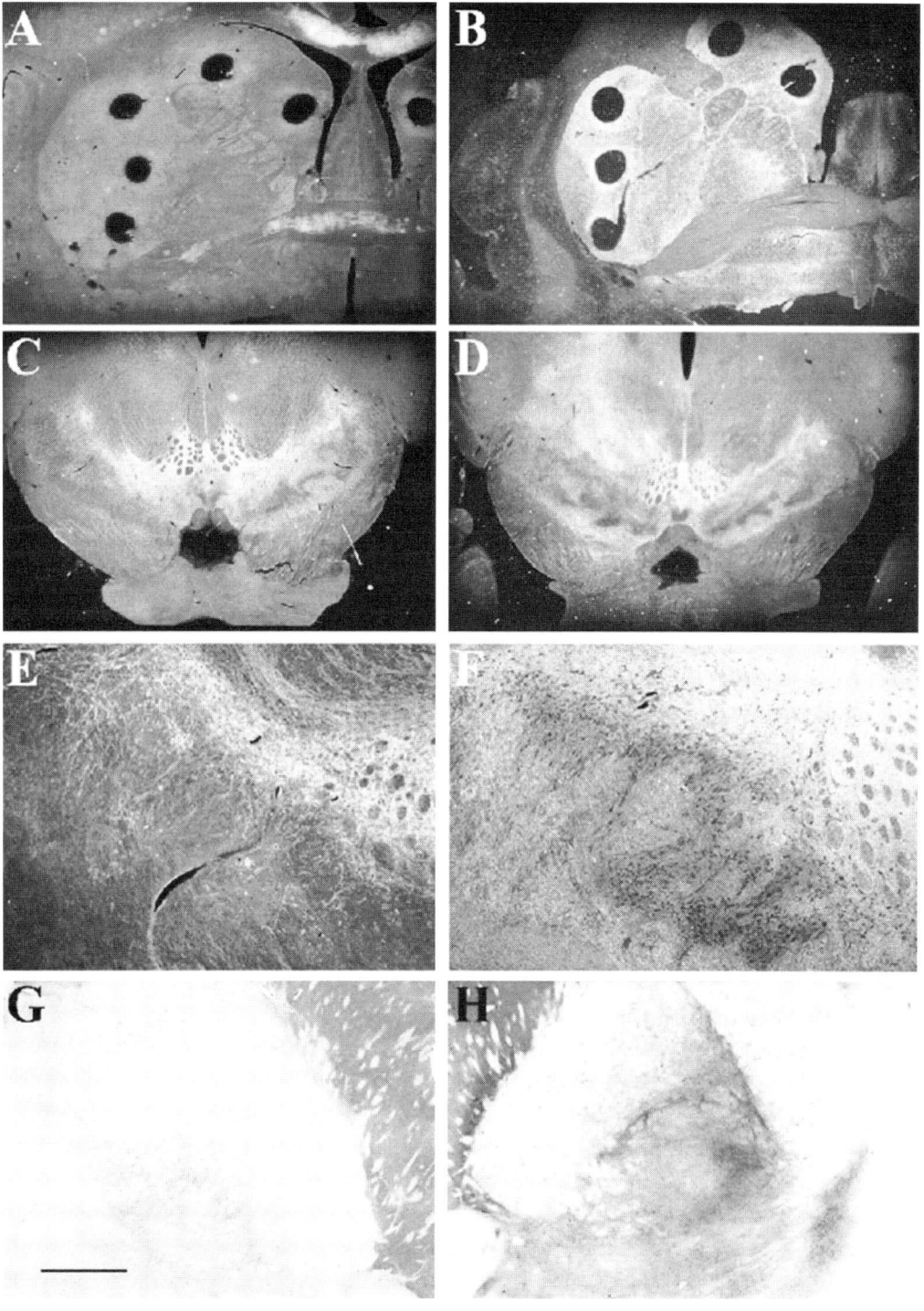

increases in $^{18}$F-dopa uptake on the impaired side compared to untreated controls. All LV-GDNF-treated monkeys displayed enhanced striatal TH levels and 32% more TH-positive nigral neurons compared to the intact side (Fig. 3).

While gene delivery of GDNF has been shown to potently protect against toxin-induced models of PD, it has been ineffective in protecting neurons in the novel, disease-relevant transgenic model. A relatively new genetic model of PD attempts to mimic one of the major pathological hallmarks of the disease — α-synuclein-positive Lewy bodies. Aebischer and colleagues have created a rat model using a LV vector to administer mutated α-synuclein to the substantia nigra, thereby inducing dopaminergic neuronal loss (Lo et al., 2002). This group administered the LV-α-synuclein vector to the nigra 2 weeks after LV-GDNF treatment (Lo et al., 2004). They found robust expression of GDNF but no beneficial neuroprotective effects on nigral neurons. These findings may present a major challenge to the use of GDNF for the treatment of PD, as it has been proposed that nigral degeneration may result from α-synuclein toxicity. If true, then the use of GDNF as a means for nigral neuroprotection may need to be reconsidered. However, it is also important to consider the shortcomings of the LV-α-synuclein model. The levels of α-synuclein achieved from LV vector administration are supraphysiological. Therefore, although GDNF may not be able to overcome the toxic effects induced by the high level of α-synuclein expressed in the LV-α-synuclein rat PD model, it may not need to render such vigorous trophism in the clinic.

Overall, the use of LV vectors to administer GDNF in animal models of PD has been quite successful. While LV vectors are safe, they are derived from the human immunodeficiency virus (HIV). The virus is altered in the laboratory prior to use, rendering it virtually impossible to transmit HIV to the host following its use. Still the association to HIV has slowed the clinical development of this vector system and led to the search for equally effective vectors unrelated to HIV. In this regard, Dowd et al. (2005) administered GDNF using an equine infectious anaemia virus (EIAV) vector and saw preservation of nondrug-induced complex behaviors like performances on an operant, corridor, staircase, stepping, and cylinder tasks.

*Adeno-associated viral vector-mediated gene therapy*

Another viral vector system that is under widespread use is the AAV vector. AAV is known to be safe to humans, many of whom already harbor the virus. When AAV-GDNF is administered to the rat nigra 3 weeks before a partial 6-OHDA lesion, there is a 94% protection of nigral neurons compared to a 50% loss in controls (Mandel et al., 1997). This group also showed stable expression of GDNF using AAV vector for up to 10 weeks. They did not however show any functional recovery using AAV-GDNF. Kirik et al. (2000b) conducted a long-term study showing that site of

Fig. 3. TH immunoreactivity in unilaterally MPTP-lesioned young monkeys. (A and B) Low-power dark-field photomicrographs through the right striatum of TH-immunostained sections of MPTP-treated monkeys treated with (A) LV-βGal or (B) LV-GDNF. (A) There is a comprehensive loss of TH immunoreactivity in the caudate and putamen of LV-βGal-treated animal. In contrast, near normal level of TH immunoreactivity is seen in LV-GDNF-treated animals. Low-power (C and D) and intermediate-power (E and F) photomicrographs of TH-immunostained section through the substantia nigra of animals treated with LV-βGal (C and E) and LV-GDNF (D and F). There is a loss of TH-immunoreactive neurons in the LV-βGal-treated animals on the side of the MPTP injection. TH-immunoreactive sprouting fibers as well as a supranormal number of TH-immunoreactive nigral perikarya are seen in LV-GDNF-treated animals on the side of the MPTP injection. (G and H) Bright-field low-power photomicrographs of a TH-immunostained section from a LV-GDNF-treated monkey. (G) Note the normal TH-immunoreactive fiber density through the globus pallidus on the intact side, which was not treated with LV-GDNF. (H) In contrast, an enhanced network of TH-immunoreactive fibers is seen on the side treated with both MPTP and LV-GDNF. Scale bar in (G) represents the following magnifications: panels A–D at 3500 μm and panels E–H at 1150 μm.

GDNF delivery is crucial to functional recovery. They showed that stable GDNF expression can be achieved for up to 6 months with a single injection of AAV-GDNF in rats. AAV-GDNF administered to both the substantia nigra and the striatum is capable of providing complete neuroprotection of nigral neurons from 6-OHDA-induced toxicity. However, only AAV-GDNF administered to the striatum can provide functional recovery mediated by a regeneration of striatal TH-positive fibers. This striatal fiber regeneration was gradual and occurred over 4–5 months.

AAV-GDNF has also been studied in various nonhuman primate models of PD. In a 6-OHDA-induced marmoset monkey model, AAV-GDNF was injected into the striatum and the substantia nigra 4 weeks prior to a unilateral lesion (Eslamboli et al., 2003). GDNF treatment protected 40% of the TH-positive cells in the lesioned substantia nigra compared to 21% of cells that remained in untreated monkeys. Upon close observation, at 5 weeks after the 6-OHDA lesion, dopamine-immunoreactive fibers were seen in the treated striatum of some monkeys. This suggests that AAV-GDNF may have partially prevented the loss of striato-nigral innervation of the striatum. This partial dopamine innervation may have been responsible for the amelioration of behavioral deficits in amphetamine- and apomorphine-induced rotations, and deficits in a PD rating scale.

Most of the aforementioned studies have administered GDNF in order to protect cells prior to performing a lesion. While these studies are essential for testing the efficacy of GDNF, a study in which the trophic factor is administered after a lesion has greater clinical relevance. One such study tested AAV-GDNF when administered 4 weeks after a 6-OHDA lesion (Wang et al., 2002). AAV-GDNF-treated rats showed a higher density of TH-positive fibers in the striatum and a greater number of TH-positive neurons in the substantia nigra compared to untreated controls. Additionally, levels of dopamine and its metabolites were higher in the treated compared with the control striata. These neuroprotective effects correlated with a significant behavioral recovery that started at 4 weeks after AAV-GDNF treatment. These data indicate that AAV-GDNF treatment is efficacious in this model even when it is administered after the neurodegenerative process is initiated.

## Clinical studies with glial cell-derived neurotrophic factor upheld

The above studies using GDNF gene therapy in animal models of PD have yielded encouraging results for clinic trials. However, due to intellectual property issues, the use of GDNF has been limited by AMGEN who owns the patent for the GDNF gene. As mentioned earlier, clinical trials using GDNF have been undertaken. Studies have been conducted by infusing GDNF either into the lateral ventricles (Nutt et al., 2003) or into the putamen (Gill et al., 2003; Lang et al., 2006; Love et al., 2005; Patel et al., 2005). Initial results were encouraging from both a safety and efficacy standpoint, and AMGEN performed double-blinded GDNF trials. The failure of the Phase II trials and an expression of antibodies against GDNF in 10% of the patients, likely occurring due to the leakage of GDNF from the pump upon refilling (Slevin et al., 2007), prompted AMGEN to abandoned GDNF as a therapeutic interest. The failure of these trials is likely attributed to inefficient methods of gene delivery. It remains to be seen whether future studies led by others will refuel the interest in GDNF for PD therapy. Until then, new trophic factors must be explored.

## Neurturin

Fortunately, GDNF is not the only trophic factor that supports the viability of nigral neurons. NTN is also a member of the GDNF family of ligands and shares a 40% homology with GDNF. Endogenously, NTN binds to the GFRα-2 receptor, which is robustly expressed in the adult substantia nigra but not in the adult striatum (Burazin and Gundlach, 1999). However, at the levels trophic factors are expressed following gene delivery, NTN is promiscuous and binds to GFRα-1 receptors, which are abundant in the striatum (Burazin and Gundlach, 1999). Thus although GDNF is not available for clinical use at this time,

NTN can be used in its place, since both function using the same receptors and thereby the same signaling pathways.

## Administration in animal PD models

The NTN mRNA is expressed in the ventral midbrain and striatum during development (Horger et al., 1998). In vitro, NTN promotes the survival of developing dopaminergic neurons and, in vivo, NTN can protect nigral neurons from 6-OHDA-induced toxicity. When tested in cell culture, both NTN and GDNF exert comparable neuroprotective effects on ventral mesencephalic cells (Akerud et al., 1999). However, only GDNF induces sprouting of dopaminergic neurons. Also, when fibroblasts genetically engineered to deliver either GDNF or NTN are grafted supranigrally in a 6-OHDA rat model, both trophic factors are just as effective at preventing the death of nigral dopaminergic neurons. Again, only GDNF was able to induce the sprouting of nigral TH-positive neurons. In a separate study, the route of NTN administration and its effects on cell survival were also shown to be crucial. Either GDNF or NTN was injected every third day for 3 weeks starting on the day after the 6-OHDA lesion (Rosenblad et al., 1999). In GDNF-treated rats, there was a 90–92% protection of nigral neurons after both striatal and ventricular administration. In NTN-treated animals, in contrast, there was only a 72% protection after striatal administration and no visible protection after ventricular administration. Neither treatment was able to rescue the reduced levels of TH in nigral neurons or the extent of dopamine denervation in the striatum. The importance of time at which NTN is administered has also been studied by examining its neuroprotective and neuroregenerative effects when administered to the striatum (Oiwa et al., 2002). NTN given 3 days before a 6-OHDA lesion was able to protect both neurons in the nigra and TH-positive fibers in the striatum. NTN given 12 weeks after a 6-OHDA lesion can promote the survival and trophism of TH-positive fibers in the striatum and increase striatal dopamine levels, although it is unable to protect neurons in the nigra.

NTN has also been administered in the 6-OHDA model using bone marrow stromal cells transfected with the NTN gene (Ye et al., 2007). One month after transplant, the number of amphetamine-induced rotations was lower in the NTN-transplant group compared to the untreated group, which may have been a result of an increase in dopamine content in the striatum of NTN-treated animals. Similar results were obtained when neural stem cells engineered to express NTN were administered (Liu et al., 2007). Striatal transplantation of these cells resulted in neuroprotection in the substantia nigra and improvements in motor tasks that lasted for 4 months after treatment. Another early study using NTN gene delivery administered the trophic factor via polymer-encapsulated cells implanted near the substantia nigra 1 week before a unilateral medial forebrain bundle axotomy (Tseng et al., 1998). A week after axotomy, animals were sacrificed and postmortem analyses revealed a significant increase in TH levels in the nigra. This protection was not accompanied by behavioral improvements.

## Gene therapy in animal PD models

The delivery system used to administer NTN may have been responsible for the mediocre results obtained in previous studies. Ceregene Inc. has developed an AAV2-NTN gene delivery system (commercially known as CERE-120). When AAV2-NTN is administered to the brain, expression of NTN has been shown to be rapid, increasing significantly up to 4 weeks and remaining stable for at least 1 year (Gasmi et al., 2007b). AAV2-NTN delivery to the striatum provides neuroprotection of TH-positive nigral neurons in the rat 6-OHDA lesion model in a dose-dependent manner (Gasmi et al., 2007a). This method of gene delivery also expresses NTN in the striatum for up to 12 months with no visible toxicity. AAV2-NTN was also tested in aged rhesus monkeys that received unilateral injections of AAV2-NTN into the striatum (Herzog et al., 2007). Robust expression of NTN within the nigrostriatal system was observed for up to 8 months. PET scanning revealed increased $^{18}$F-dopa uptake in the treated striatum. Additionally, in

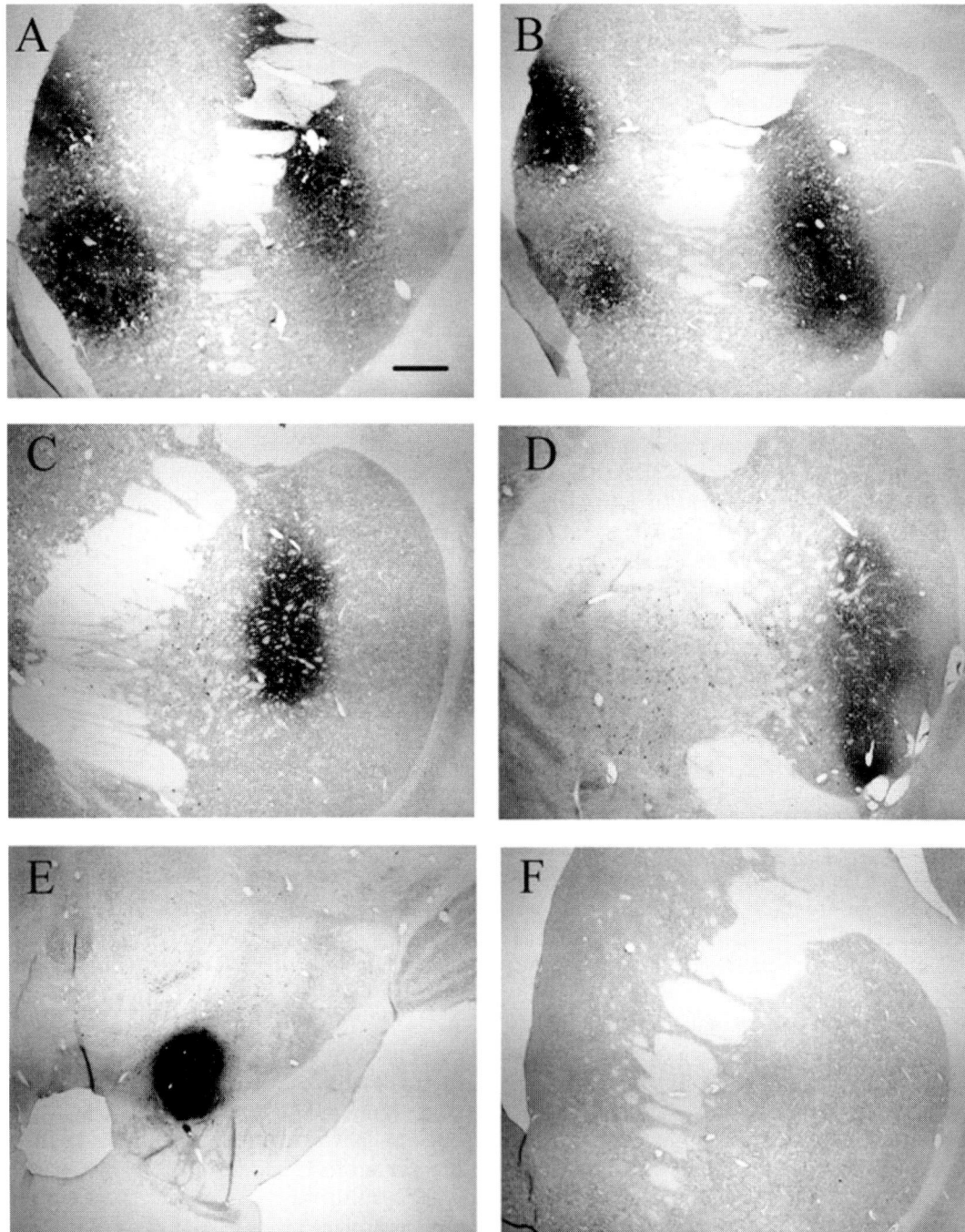

Fig. 4. Neurturin (NTN) immunohistochemistry illustrating the appropriate targeting and spread of NTN protein for each of the five injection sites. (A) Head of caudate and rostral putamen. (B) Commissural putamen. (C and D) Postcommissural putamen. (E) Substantia nigra. (F) Control-treated monkey stained for NTN, illustrating the lack of endogenous upregulation of NTN after control injection. Scale bar = 1 mm.

AAV2-NTN-treated animals, 8 months post-administration, there was a significant increase in TH-positive fibers in the striatum and an increase in the number of TH-positive cells in the ipsilateral nigra. This study also provided insight into the mechanism of NTN action, as it activated pERK, a marker for the downstream signaling mechanism for both NTN and GDNF.

AAV2-NTN has also been tested in hemiparkinsonian monkeys lesioned with MPTP. In one such study, AAV2-NTN was administered 4 days after MPTP lesion to five different sites in the caudate and rostral putamen, the commissural putamen, the postcommissural putamen, and the substantia nigra (Kordower et al., 2006) (Fig. 4).

The AAV2-NTN-treated monkeys showed an 80–90% improvement in MPTP-induced motor deficits starting 4 months after treatment and lasting until the end of the study at 10 months. Also AAV2-NTN-treated monkeys had significantly higher TH levels in the striatum compared to untreated controls (Fig. 5) and a 62% nigral cell protection versus 14% protection in untreated controls (Fig. 6). This study also showed an upregulation of pERK in the substantia nigra.

## Experimental gene therapy in human PD subjects

Based upon these and other such studies, Ceregene Inc. initiated a Phase I clinical trial using 12

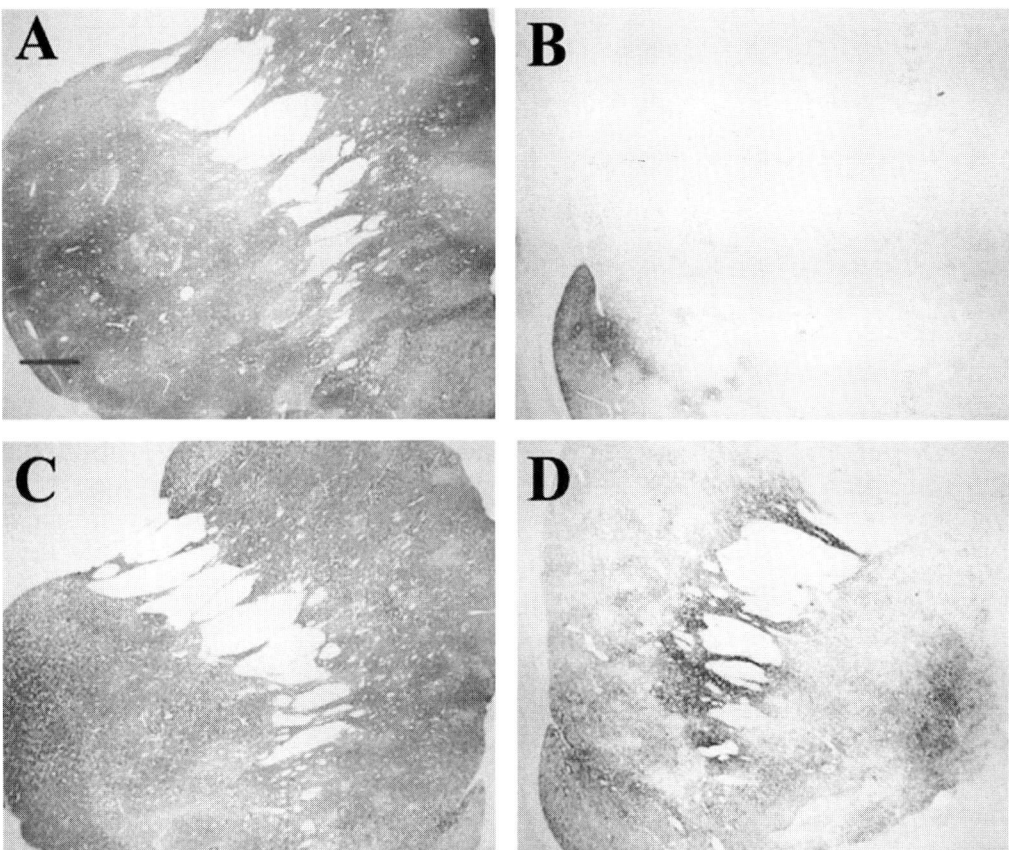

Fig. 5. Preservation of striatal tyrosine hydroxylase (TH) after AAV2-NTN delivery. (A and C) Intact sides. (B) There is a comprehensive loss of striatal TH immunoreactivity on the MPTP side with control treatment. (D) In contrast, there is a preservation of striatal TH on the MPTP side with AAV2-NTN treatment. Scale bar = 1 mm.

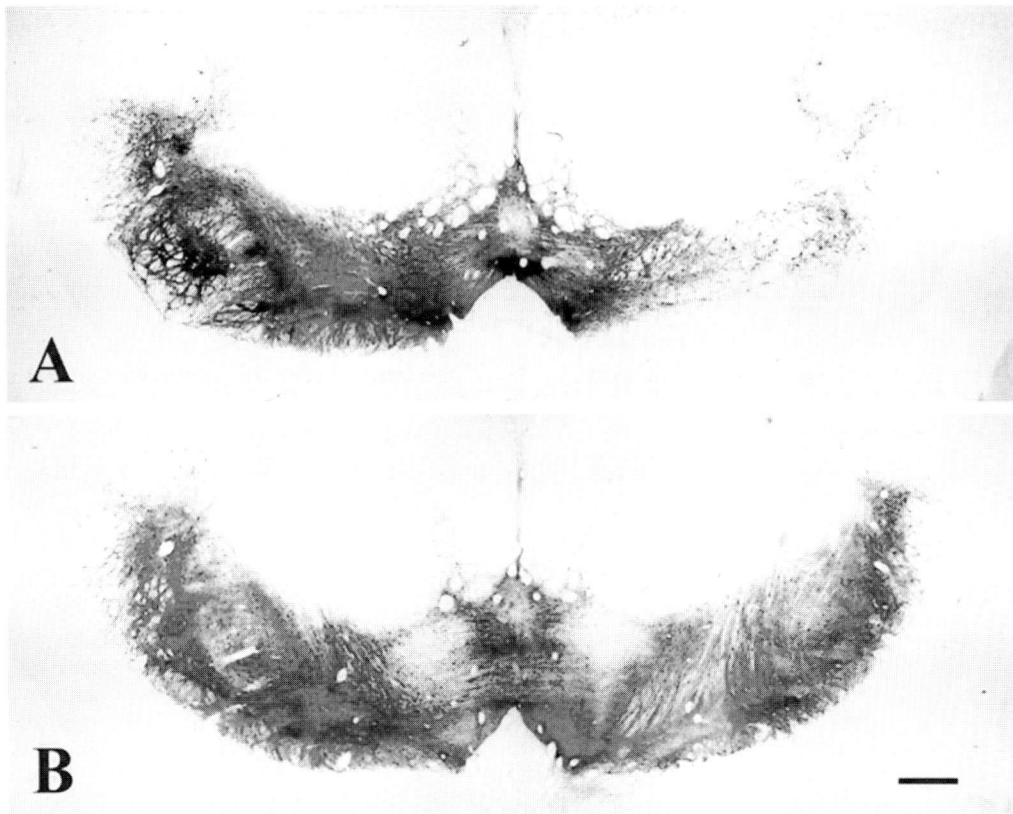

Fig. 6. Preservation of nigral TH after AAV2-NTN. (A) In control-treated monkeys, there is a comprehensive loss of TH-immunoreactive neurons on the lesioned (right) side relative to the intact (left) side. (B) In contrast, there is a prevention of loss of TH-immunoreactive nigral neurons on the side of MPTP treatment in AAV2-NTN monkeys. Scale bar = 1 mm.

advanced PD patients. All patients received bilateral injections of AAV2-NTN into the putamen using either a low dose ($n = 6$) or a high dose ($n = 6$) paradigm (Marks et al., 2008). Patients were tracked over 12 months and no serious adverse events were observed. There was a statistically significant improvement in the UPDRS score in the "off" state and no differences between the two treatment doses. Based on the safety of AAV2-NTN, a Phase II double-blinded trial was initiated including 58 patients, two-thirds of whom received CERE-120 and one-thirds who received a placebo. This study failed. However, technical issues concerning the delivery of CERE-120 to the human striatum likely underlie this failure and future clinical trials that will better test the hypothesis are presently being planned.

## Conclusion

In a field like PD where there is no effective therapy to combat the progression of the disease, neurotrophic factors have provided some hope. While it is essential to assist patients in reducing disease symptoms for immediate relief, these approaches are often fraught with side effects over time. The "holy grail" of disease-modifying therapies needs further investigation. Neurons in the brains of PD patients are destined to die. Once cell numbers are reduced below a critical level, drug therapies may no longer be effective. Therefore, it is important to stop cell death in its track, and neurotrophic factors, especially those in the GDNF family of ligands, serve to do just that. For the benefit of PD patients who are still awaiting the arrival of a disease-modifying therapy, it is crucial that efforts using these trophic factors not be abandoned.

## References

Ai, Y., Markesbery, W., Zhang, Z., Grondin, R., Elseberry, D., Gerhardt, G. A., et al. (2003). Intraputamenal infusion of GDNF in aged rhesus monkeys: distribution and dopaminergic effects. *The Journal of Comparative Neurology, 461*, 250–261.

Akerud, P., Alberch, J., Eketjall, S., Wagner, J., & Arenas, E. (1999). Differential effects of glial cell line-derived neurotrophic factor and neurturin on developing and adult substantia nigra dopaminergic neurons. *Journal of Neurochemistry, 73*, 70–78.

Bilang-Bleuel, A., Revah, F., Colin, P., Locquet, I., Robert, J. J., Mallet, J., et al. (1997). Intrastriatal injection of an adenoviral vector expressing glial-cell-line-derived neurotrophic factor prevents dopaminergic neuron degeneration and behavioral impairment in a rat model of Parkinson disease. *Proceedings of the National Academy of Sciences of the United States of America, 94*, 8818–8823.

Burazin, T. C., & Gundlach, A. L. (1999). Localization of GDNF/neurturin receptor (c-ret, GFRalpha-1 and alpha-2) mRNAs in postnatal rat brain: differential regional and temporal expression in hippocampus, cortex and cerebellum. *Brain Research. Molecular Brain Research, 73*, 151–171.

Choi-Lundberg, D. L., Lin, Q., Chang, Y. N., Chiang, Y. L., Hay, C. M., Mohajeri, H., et al. (1997). Dopaminergic neurons protected from degeneration by GDNF gene therapy. *Science, 275*, 838–841.

Choi-Lundberg, D. L., Lin, Q., Schallert, T., Crippens, D., Davidson, B. L., Chang, Y. N., et al. (1998). Behavioral and cellular protection of rat dopaminergic neurons by an adenoviral vector encoding glial cell line-derived neurotrophic factor. *Experimental Neurology, 154*, 261–275.

Dowd, E., Monville, C., Torres, E. M., Wong, L. F., Azzouz, M., Mazarakis, N. D., et al. (2005). Lentivector-mediated delivery of GDNF protects complex motor functions relevant to human Parkinsonism in a rat lesion model. *The European Journal of Neuroscience, 22*, 2587–2595.

Eslamboli, A., Cummings, R. M., Ridley, R. M., Baker, H. F., Muzyczka, N., Burger, C., et al. (2003). Recombinant adeno-associated viral vector (rAAV) delivery of GDNF provides protection against 6-OHDA lesion in the common marmoset monkey (*Callithrix jacchus*). *Experimental Neurology, 184*, 536–548.

Gash, D. M., Zhang, Z., Ai, Y., Grondin, R., Coffey, R., & Gerhardt, G. A. (2005). Trophic factor distribution predicts functional recovery in parkinsonian monkeys. *Annals of Neurology, 58*, 224–233.

Gasmi, M., Brandon, E. P., Herzog, C. D., Wilson, A., Bishop, K. M., Hofer, E. K., et al. (2007a). AAV2-mediated delivery of human neurturin to the rat nigrostriatal system: long-term efficacy and tolerability of CERE-120 for Parkinson's disease. *Neurobiology of Disease, 27*, 67–76.

Gasmi, M., Herzog, C. D., Brandon, E. P., Cunningham, J. J., Ramirez, G. A., Ketchum, E. T., et al. (2007b). Striatal delivery of neurturin by CERE-120, an AAV2 vector for the treatment of dopaminergic neuron degeneration in Parkinson's disease. *Molecular Therapy, 15*, 62–68.

Georgievska, B., Kirik, D., & Bjorklund, A. (2002a). Aberrant sprouting and downregulation of tyrosine hydroxylase in lesioned nigrostriatal dopamine neurons induced by long-lasting overexpression of glial cell line derived neurotrophic factor in the striatum by lentiviral gene transfer. *Experimental Neurology, 177*, 461–474.

Georgievska, B., Kirik, D., Rosenblad, C., Lundberg, C., & Bjorklund, A. (2002b). Neuroprotection in the rat Parkinson model by intrastriatal GDNF gene transfer using a lentiviral vector. *Neuroreport, 13*, 75–82.

Gill, S. S., Patel, N. K., Hotton, G. R., O'Sullivan, K., McCarter, R., Bunnage, M., et al. (2003). Direct brain infusion of glial cell line-derived neurotrophic factor in Parkinson disease. *Nature Medicine, 9*, 589–595.

Grondin, R., Zhang, Z., Yi, A., Cass, W. A., Maswood, N., Andersen, A. H., et al. (2002). Chronic, controlled GDNF infusion promotes structural and functional recovery in advanced parkinsonian monkeys. *Brain, 125*, 2191–2201.

Hamilton, J. F., Morrison, P. F., Chen, M. Y., Harvey-White, J., Pernaute, R. S., Phillips, H., et al. (2001). Heparin coinfusion during convection-enhanced delivery (CED) increases the distribution of the glial-derived neurotrophic factor (GDNF) ligand family in rat striatum and enhances the pharmacological activity of neurturin. *Experimental Neurology, 168*, 155–161.

Herzog, C. D., Dass, B., Holden, J. E., Stansell, J., III, Gasmi, M., Tuszynski, M. H., et al. (2007). Striatal delivery of CERE-120, an AAV2 vector encoding human neurturin, enhances activity of the dopaminergic nigrostriatal system in aged monkeys. *Movement Disorders, 22*, 1124–1132.

Horger, B. A., Nishimura, M. C., Armanini, M. P., Wang, L. C., Poulsen, K. T., Rosenblad, C., et al. (1998). Neurturin exerts potent actions on survival and function of midbrain dopaminergic neurons. *The Journal of Neuroscience, 18*, 4929–4937.

Hwang, D. Y., Ardayfio, P., Kang, U. J., Semina, E. V., & Kim, K. S. (2003). Selective loss of dopaminergic neurons in the substantia nigra of Pitx3-deficient aphakia mice. *Brain Research. Molecular Brain Research, 114*, 123–131.

Kearns, C. M., & Gash, D. M. (1995). GDNF protects nigral dopamine neurons against 6-hydroxydopamine in vivo. *Brain Research, 672*, 104–111.

Kirik, D., Rosenblad, C., & Bjorklund, A. (2000a). Preservation of a functional nigrostriatal dopamine pathway by GDNF in the intrastriatal 6-OHDA lesion model depends on the site of administration of the trophic factor. *The European Journal of Neuroscience, 12*, 3871–3882.

Kirik, D., Rosenblad, C., Bjorklund, A., & Mandel, R. J. (2000b). Long-term rAAV-mediated gene transfer of GDNF in the rat Parkinson's model: intrastriatal but not intranigral transduction promotes functional regeneration in the lesioned nigrostriatal system. *The Journal of Neuroscience, 20*, 4686–4700.

Kordower, J. H., Emborg, M. E., Bloch, J., Ma, S. Y., Chu, Y., Leventhal, L., et al. (2000). Neurodegeneration prevented by lentiviral vector delivery of GDNF in primate models of Parkinson's disease. *Science, 290,* 767–773.

Kordower, J. H., Herzog, C. D., Dass, B., Bakay, R. A., Stansell, J., III, Gasmi, M., et al. (2006). Delivery of neurturin by AAV2 (CERE-120)-mediated gene transfer provides structural and functional neuroprotection and neurorestoration in MPTP-treated monkeys. *Annals of Neurology, 60,* 706–715.

Kordower, J. H., Palfi, S., Chen, E. Y., Ma, S. Y., Sendera, T., Cochran, E. J., et al. (1999). Clinicopathological findings following intraventricular glial-derived neurotrophic factor treatment in a patient with Parkinson's disease. *Annals of Neurology, 46,* 419–424.

Lang, A. E., Gill, S., Patel, N. K., Lozano, A., Nutt, J. G., Penn, R., et al. (2006). Randomized controlled trial of intraputamenal glial cell line-derived neurotrophic factor infusion in Parkinson disease. *Annals of Neurology, 59,* 459–466.

Lin, L. F., Doherty, D. H., Lile, J. D., Bektesh, S., & Collins, F. (1993). GDNF: a glial cell line-derived neurotrophic factor for midbrain dopaminergic neurons. *Science, 260,* 1130–1132.

Liu, W. G., Lu, G. Q., Li, B., & Chen, S. D. (2007). Dopaminergic neuroprotection by neurturin-expressing c17.2 neural stem cells in a rat model of Parkinson's disease. *Parkinsonism & Related Disorders, 13,* 77–88.

Lo, B. C., Deglon, N., Pralong, W., & Aebischer, P. (2004). Lentiviral nigral delivery of GDNF does not prevent neurodegeneration in a genetic rat model of Parkinson's disease. *Neurobiology of Disease, 17,* 283–289.

Lo, B. C., Ridet, J. L., Schneider, B. L., Deglon, N., & Aebischer, P. (2002). α-Synucleinopathy and selective dopaminergic neuron loss in a rat lentiviral-based model of Parkinson's disease. *Proceedings of the National Academy of Sciences of the United States of America, 99,* 10813–10818.

Love, S., Plaha, P., Patel, N. K., Hotton, G. R., Brooks, D. J., & Gill, S. S. (2005). Glial cell line-derived neurotrophic factor induces neuronal sprouting in human brain. *Nature Medicine, 11,* 703–704.

Mandel, R. J., Spratt, S. K., Snyder, R. O., & Leff, S. E. (1997). Midbrain injection of recombinant adeno-associated virus encoding rat glial cell line-derived neurotrophic factor protects nigral neurons in a progressive 6-hydroxydopamine-induced degeneration model of Parkinson's disease in rats. *Proceedings of the National Academy of Sciences of the United States of America, 94,* 14083–14088.

Marks, W. J., Jr., Ostrem, J. L., Verhagen, L., Starr, P. A., Larson, P. S., Bakay, R. A., et al. (2008). Safety and tolerability of intraputaminal delivery of CERE-120 (adeno-associated virus serotype 2-neurturin) to patients with idiopathic Parkinson's disease: an open-label, phase I trial. *Lancet Neurology, 7,* 400–408.

Maswood, N., Grondin, R., Zhang, Z., Stanford, J. A., Surgener, S. P., Gash, D. M., et al. (2002). Effects of chronic intraputamenal infusion of glial cell line-derived neurotrophic factor (GDNF) in aged Rhesus monkeys. *Neurobiology of Aging, 23,* 881–889.

Nutt, J. G., Burchiel, K. J., Comella, C. L., Jankovic, J., Lang, A. E., Laws, E. R., Jr., et al. (2003). Randomized, double-blind trial of glial cell line-derived neurotrophic factor (GDNF) in PD. *Neurology, 60,* 69–73.

Oiwa, Y., Yoshimura, R., Nakai, K., & Itakura, T. (2002). Dopaminergic neuroprotection and regeneration by neurturin assessed by using behavioral, biochemical and histochemical measurements in a model of progressive Parkinson's disease. *Brain Research, 947,* 271–283.

Parkinson's Disease Foundation. (2009). Available at www.pdf.org

Pascual, A., Hidalgo-Figueroa, M., Piruat, J. I., Pintado, C. O., Gomez-Diaz, R., & Lopez-Barneo, J. (2008). Absolute requirement of GDNF for adult catecholaminergic neuron survival. *Nature Neuroscience, 11,* 755–761.

Patel, N. K., Bunnage, M., Plaha, P., Svendsen, C. N., Heywood, P., & Gill, S. S. (2005). Intraputamenal infusion of glial cell line-derived neurotrophic factor in PD: a two-year outcome study. *Annals of Neurology, 57,* 298–302.

Rosenblad, C., Kirik, D., Devaux, B., Moffat, B., Phillips, H. S., & Bjorklund, A. (1999). Protection and regeneration of nigral dopaminergic neurons by neurturin or GDNF in a partial lesion model of Parkinson's disease after administration into the striatum or the lateral ventricle. *The European Journal of Neuroscience, 11,* 1554–1566.

Salvatore, M. F., Ai, Y., Fischer, B., Zhang, A. M., Grondin, R. C., Zhang, Z., et al. (2006). Point source concentration of GDNF may explain failure of phase II clinical trial. *Experimental Neurology, 202,* 497–505.

Sariola, H., & Saarma, M. (2003). Novel functions and signalling pathways for GDNF. *Journal of Cell Science, 116,* 3855–3862.

Sauer, H., Rosenblad, C., & Bjorklund, A. (1995). Glial cell line-derived neurotrophic factor but not transforming growth factor beta 3 prevents delayed degeneration of nigral dopaminergic neurons following striatal 6-hydroxydopamine lesion. *Proceedings of the National Academy of Sciences of the United States of America, 92,* 8935–8939.

Slevin, J. T., Gash, D. M., Smith, C. D., Gerhardt, G. A., Kryscio, R., Chebrolu, H., et al. (2007). Unilateral intraputamenal glial cell line-derived neurotrophic factor in patients with Parkinson disease: response to 1 year of treatment and 1 year of withdrawal. *Journal of Neurosurgery, 106,* 614–620.

Tseng, J. L., Bruhn, S. L., Zurn, A. D., & Aebischer, P. (1998). Neurturin protects dopaminergic neurons following medial forebrain bundle axotomy. *Neuroreport, 9,* 1817–1822.

Wang, L., Muramatsu, S., Lu, Y., Ikeguchi, K., Fujimoto, K., Okada, T., et al. (2002). Delayed delivery of AAV-GDNF prevents nigral neurodegeneration and promotes functional recovery in a rat model of Parkinson's disease. *Gene Therapy, 9,* 381–389.

Ye, M., Wang, X. J., Zhang, Y. H., Lu, G. Q., Liang, L., Xu, J. Y., et al. (2007). Transplantation of bone marrow stromal cells containing the neurturin gene in rat model of Parkinson's disease. *Brain Research, 1142,* 206–216.

SECTION IV

# Pharmacotherapy

CHAPTER 15

# Progesterone as a neuroprotective factor in traumatic and ischemic brain injury

Iqbal Sayeed[1,*] and Donald G. Stein[1]

[1]*Brain Research Laboratory, Department of Emergency Medicine, Emory University, Atlanta, GA, USA*

**Abstract:** The search for a "magic bullet" drug targeting a single receptor for the treatment of stroke or traumatic brain injury (TBI) has failed thus far for a variety of reasons. The pathophysiology of ischemic brain injury and TBI involves a number of mechanisms leading to neuronal injury, including excitotoxicity, free radical damage, inflammation, necrosis, and apoptosis. Brain injury also triggers auto-protective mechanisms, including the up-regulation of anti-inflammatory cytokines and endogenous antioxidants. In these conditions an agent with pleiotropic consequences is more likely to provide effective neuroprotection and repair than one operating primarily on a single, or a small number of, injury mechanisms. There is growing evidence, including recently published clinical trials, that progesterone and perhaps its metabolite allopregnanolone exert neuroprotective effects on the injured central nervous system (CNS). Laboratories around the world have shown that progesterone and allopregnanolone act through numerous metabolic and physiological pathways that can affect the injury response in many different tissues and organ systems. Furthermore, progesterone is a natural hormone, synthesized in both males and females, that can act as a pro-drug for other metabolites with their own distinct mode of action in CNS repair. These properties make progesterone a unique and compelling natural agent to consider for testing in clinical trial for CNS injuries including TBI and stroke.

**Keywords:** ischemia; progesterone; neuroprotection; neurosteroids; stroke; traumatic brain injury

## Introduction

Traumatic and ischemic brain injuries are a global health problem. Much progress has been made in developing novel therapeutic therapies to treat these injuries, including glutamate receptor antagonists, calcium channel blockers, radical scavengers, and anti-inflammatory and anti-apoptotic agents. Yet despite progress in identifying many of the critical injury mechanisms and processes, the effort to develop novel pharmacological agents leading to clinical trials in stroke and traumatic brain injury (TBI) has met with little success. Over the last several years, preclinical studies around the world have reported that progesterone (PROG), given in the acute stage of injury, limits tissue damage and improves functional outcome after blunt TBI, stroke, spinal cord injury, diabetic neuropathies, and other types of acute neuroinjury in several species (Thomas et al., 1999; Stein, 2001; Labombarda et al., 2006a, b;

*Corresponding author.
Tel.: 404 727 3639; Fax: 404 727 2388;
E-mail: isayeed@emory.edu

Sayeed et al., 2006, 2007; Garay et al., 2007; Leonelli et al., 2007). Progesterone appears to protect or rebuild the blood–brain barrier (BBB), improve vascular tone, reduce cerebral edema, down-regulate the inflammatory cascade, up-regulate gamma-amino butyric acid (GABA), reduce excitotoxicity and seizure activity, stimulate myelination in damaged axons, modulate hemostatic proteins, and decrease apoptosis (Wright et al., 2001; He et al., 2004a; O'Connor et al., 2005; Pettus et al., 2005; Guo et al., 2006; Leonelli et al., 2007; Roof et al., 1997; Vanlandingham et al., 2008; Stein and Hurn, 2009). Among the advantages of progesterone treatments are: a good safety profile with few adverse events; no need for long-term dosing; a large window of opportunity for treatment without increased risk; effectiveness in the treatment of both males and females; ready availability; and ease of administration in emergency situations (Stein, 2005).

Since much of the published work has been devoted to the treatment of TBI, this review will focus on what is known about progesterone as a potential treatment for ischemic stroke. A more detailed and recent review of the role of progesterone in TBI can be found in Stein and Hurn (2009).

## Neuroprotective effects of progesterone in TBI

TBI is a complex series of events that unfolds over time and involves a relatively large number of biochemical pathways that can take months to years before complete stabilization (Fig. 1).

No new pharmacotherapy for TBI has entered clinical practice in over 30 years (Roberts et al., 1998). Glucocorticoids, once a mainstay of TBI treatment, are now known to be harmful (Vink and Van Den Heuvel, 2004; Watson et al., 2004; Gomes et al., 2005). The Corticosteroids After Significant Head Injury (CRASH) trial, with more than 10,000 subjects enrolled in 40 countries, was halted because of a higher mortality rate in the treatment group (Roberts et al., 2004; Edwards et al., 2005). For similar reasons a magnesium sulfate trial was terminated (see Temkin et al., 2007, for details). Hypothermia is useful for treating global ischemia, and has produced favorable effects in some brain-injured subjects, but it may be harmful to patients over 45 (Clifton et al., 2001a,b;

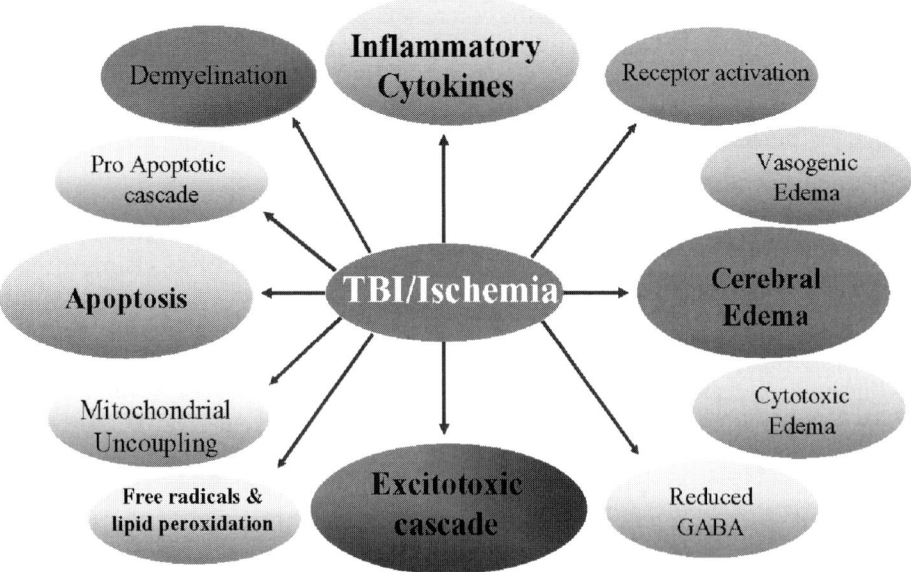

Fig. 1. Flow chart summarizing pathophysiological events triggered following traumatic and ischemic brain injury.

Anderson et al., 2004). Another recent study with 225 children with TBI found no evidence of benefit from hypothermia (Hutchison et al., 2008), suggesting that it may be too problematic for widespread clinical use.

## *Progesterone as a therapeutic candidate for TBI*

The conventional view that progesterone is just a female sex hormone affecting the reproductive and endocrine systems has changed greatly over the past few years. We now understand that progesterone exerts a wide range of physiological actions, depending on the target tissue. Progesterone and its metabolites are classed as neurosteroids because there is substantial evidence that the biosynthetic enzymes required for their synthesis are present in brain tissue of both men and women at similar levels (Baulieu and Robel, 1990; Guerra-Araiza et al., 2002; Meffre et al., 2005; Reddy et al., 2005). The progesterone receptor is abundant and widely distributed throughout the forebrain, limbic system, and hypothalamic regions very early in development (Quadros et al., 2007). Progesterone is synthesized from its precursor, pregnenolone, a derivative of cholesterol (Mellon et al., 2001; Schumacher et al., 2007). Cholesterol is converted first through the enzymatic action of cytochrome P450 to pregnenolone, which is then converted to progesterone by 3-beta-hydroxysteroid dehydrogenase. In this paper we will examine whether this natural neurosteroid has the potential for therapeutic use in the treatment of stroke and other types of brain injuries.

Sex differences in cerebral recovery in response to experimentally induced TBI led to the discovery of progesterone's neuroprotective properties. In the late 1980s Stein and colleagues conducted an experiment to test the hypothesis that sex-related differences in brain injury have a hormonal basis and that the magnitude of recovery depends, at least in part, on where a female rat is in her estrous cycle at injury. It was observed that female rats in the proestrus stage (when endogenous levels of progesterone are high) at the time of injury developed significantly less brain edema than male rats (Roof et al., 1993a). Also, pseudopregnant females (with high circulating levels of progesterone and comparatively lower levels of estrogen) developed almost no post-injury cerebral edema. Subsequent experiments tested whether exogenous progesterone was beneficial (Roof et al., 1993a,b, 1994). This research led to a series of studies demonstrating progesterone's effectiveness in the treatment of TBI and then elucidating many of the molecular and physiological mechanisms underlying the neuroprotection conferred by this hormone and its metabolites. These preclinical findings have been extensively reviewed (see, e.g., Stein, 2008; Stein et al., 2008). As a result of the laboratory data, phase II, single-site clinical studies were done to evaluate the effects of progesterone in patients with moderate-to-severe TBI.

## *Human trials evaluating progesterone as a treatment for TBI*

To date, two human studies have been performed. The first was Progesterone for Traumatic Brain Injury — Experimental Clinical Treatment (ProTECT), a single-site, double-blinded trial that randomly allocated 100 adult patients with moderate-to-severe acute TBI who received infusion of intravenous progesterone for 3 days. The results showed that intravenous administration achieves predictable steady-state serum concentrations that do not appreciably differ by sex or injury severity (Wright et al., 2005). Furthermore, the clinical study also produced compelling preliminary evidence that the drug is safe, even in the setting of polytrauma (Wright et al., 2007). Although the Wright et al. trial lacked sufficient power to assess definitively the efficacy of progesterone treatment, promising signs of activity were observed: a statistically significant decrease in 30-day mortality among progesterone-treated patients who sustained a severe TBI compared to placebo (13.4% vs. 33.6%) and improved 30-day functional outcome (Disability Rating Scale) scores among progesterone-treated patients with a moderate TBI compared to the placebo group. These findings represent the first success in treating acute TBI. Recently, the Wright et al. results were replicated and extended in a trial with 159 severely brain-injured patients which reported a 45% reduction in mortality, lower intracranial pressure (72 h and 7 days), and improved functional outcomes at 30 days and 6 months post-injury (Xiao et al., 2008). The promising results of these two single-site trials need to be confirmed in a much larger, multicenter trial now being planned.

## Stroke and progesterone

### Scope of stroke problem

Stroke is another major cause of injury and disability worldwide (Durukan and Tatlisumak, 2007) and the third most frequent cause of death in adults in the United States (American Heart Association, 2005). Stroke is most commonly the result of an obstruction of blood flow in a major cerebral vessel (e.g., the middle cerebral artery [MCA]), which, if not quickly resolved, will lead to an infarcted area of tissue that cannot be therapeutically salvaged (Murray and Lopez, 1997; Hankey, 1999; Muntner et al., 2006). An effective therapeutic strategy for stroke has been a priority of neuroscientists for decades. To date, the several hundred industry-sponsored clinical trials for the treatment of stroke by medication after it has occurred have been disappointing, with genetically engineered tissue plasminogen activators (tPAs) still the only agents approved by the Food and Drug Administration (FDA). tPA has limited applicability and is currently used in fewer than 5% of stroke victims (Grotta et al., 2001), although there is now some evidence that the window of opportunity for use of tPA may be extended to 4 h after the event (Lyden, 2008).

Another candidate for stroke therapy has been thrombolysis (Furlan, 2006). Recently Dubinsky and Lai (2006) used the Nationwide Inpatient Sample (1999–2002) for acute ischemic stroke admissions to examine the rate of mortality in community hospital patients given thrombolysis therapy. In the treatment cohort, hospital mortality was significantly greater (10.1% vs. 5.8%) and secondary intracranial hemorrhage much higher (4.2% vs. 0.4%) compared to controls. These findings will be debated, but this type of therapy will not see widespread use until the issues of safety, mortality, and functional efficacy can be resolved.

Thus, the problem of what, if any, neuroprotective treatment to give immediately after a stroke remains an increasingly serious concern. At a recent International Stroke Conference (2007), Mark Goldberg concluded that "there remains an urgent need for additional therapeutic approaches to acute ischemic stroke" (Goldberg, 2007). It was suggested that, in addition to failures in large-scale trials, part of the problem is the lack of appropriate preclinical models to test such agents, the 1999 Stroke Therapy Academic Industry Roundtable (STAIR) recommendations not having been met. This issue of appropriate animal models was revisited by Savitz et al. (2005) in discussing the recent failure of NXY-059, a free radical spin trap drug developed by AstraZeneca as a neuroprotective agent for acute ischemic stroke: "Before NXY-059 every neuroprotective agent brought forward to a clinical trial had failed to improve outcome from stroke on the pre-specified endpoint. In retrospect all prior pre-clinical drug development programs of neuroprotective agents had been inadequate" (p. 21). Savitz asserted that the trial of NXY-059, too, failed because AstraZeneca moved too quickly to clinical evaluation and did not conduct the preclinical animal studies according to STAIR guidelines.

### Preclinical stroke studies with progesterone

Almost 20 years ago, Betz and Coester (1990) published the first study examining progesterone as a neuroprotective agent in a focal permanent occlusion model. Sixty-one male Sprague–Dawley rats were pretreated 1 h before MCA occlusion (MCAO) with dexamethasone, progesterone (2 mg/kg intraperitoneally, IP), or control. The rats underwent 4 h of MCAO by cauterization; the brains were then immediately removed. The results demonstrated that treatment with either steroid decreased brain edema, but found no significant reduction of the BBB in the ischemic brain. Jiang, Chopp, Stein, and Feit (1996) studied progesterone in a rat focal temporary ischemia model. Forty-eight male Wistar rats underwent temporary MCAO (tMCAO; suture method) for 2 h. Progesterone (IP, 4 mg/kg) was administered to rats in four groups: progesterone in saline with pretreatment dose ($t$: −30 min, +6 h, +24 h), progesterone in dimethylsulfoxide (DMSO) with pretreatment dose ($t$: −30 min, +6 h, +24 h), progesterone in DMSO without pretreatment ($t$: +2, +6 h, +24 h), and control ($t$: −30 min, +6 h, +24 h). Outcomes included neurologic deficit measured at 24 and 48 h by forelimb flexion (Zea Longa) and infarct size measured at 48 h. The researchers found that percent infarct was reduced in both pre- and delayed-treated groups (by 39 and 34%, respectively) compared to controls, and neurologic score was improved in pre- and delayed-treatment groups. No significant differences in saline versus DMSO vehicle for administration of progesterone were found.

Chen et al. (1999) also used a 2-h temporary occlusion model to study the neuroprotective effects of progesterone. Twenty-eight male Wistar rats received water-soluble progesterone IV in doses of 4, 8, and 32 mg/kg at $t$+2, 6, and 24 h. Outcomes included neurologic deficit measured daily for 7 days with three tests: Zea Longa, rotarod (motor), and an adhesive-backed paper test (sensory). Infarct size was also measured at 7 days. The investigators found that lesion volume was significantly reduced in the progesterone 8 mg/kg group (19% lesion volume vs. 34% in controls) and neurologic outcomes were significantly improved in the 8 mg/kg group, with most benefit demonstrated between 4 and 7 days. This study was the first to demonstrate a dose–response. Kumon et al. (2000) also investigated the neuroprotective effects of progesterone using a temporary 2-h occlusion rat model. Forty-eight male spontaneously hypertensive rats (SHRs) were given progesterone (4 or 8 mg/kg, IP) in DMSO with one dose 10 min prior to reperfusion and doses after reperfusion at 2, 6, 24, 48, 72, 96, 120, and 144 h. Outcome measures included neurologic deficits measured daily for 2 or 7 days with a Zea Longa test and infarct size measured at 2 and 7 days, which was separated into cortical, striatal, and total percent volume. The investigators found no statistically significant difference in lesion size at either progesterone dose at 2 days, but did observe decreased lesion size (9.9% vs. 21%) at 7 days in the 8 mg/kg group with the result isolated to the cortex (striatal regions were not different between groups). Additionally, significantly better neurologic scores in the 8 mg/kg group were demonstrated. An important aspect of this project was the examination of the locus of the neuroprotective effect.

In the same year, Alkayed et al. (2000) also examined progesterone using a temporary occlusion model. Here, the subjects were 48 male (25%) and female (75%) Wistar rats (16 months old) given 2 h of MCAO (suture method). The rats were separated into male and female controls and females treated with estrogen (25 μg subcutaneously [SC]) or progesterone (10 mg SC) for 7 days prior to ischemia. Serum drug levels were monitored. Outcomes were infarct size measured at 22 h and separated into cortical, striatal, and total percent volume; and regional cerebral blood flow in MCA territories. The investigators observed a reduction of cortical infarct from 31 to 16% in progesterone pretreated females. However, striatal infarct was unaffected by progesterone pretreatment and regional blood flow was also not different between groups. Murphy et al. (2000) used a temporary occlusion model in 56 sexually mature, ovariectomized female Wistar rats. These subjects received 30 or 60 mg/kg of progesterone IP either 30 min prior to MCAO or 30 mg/kg daily for 7–10 days before MCAO. The only outcome measure was infarct size at 22 h after reperfusion. They found that progesterone treatment exacerbated the size of the ischemic injury in progesterone-deficient ovariectomized female rats. Here, the use of progesterone pretreatment does not provide a realistic approach to the acute post-injury treatment of human stroke. These seemingly discrepant findings may be explained by the possible confounding effect of hormone withdrawal, by the use of ovariectomized rats, and by the relatively high doses of progesterone administered compared to previous studies.

Sayeed et al. (2006) recently determined that both progesterone and its metabolite allopregnanolone (AP) were beneficial treatments for stroke caused by tMCAO. Two injections of systemic treatment with either agent resulted in significantly smaller infarcts compared to vehicle-treated controls. Although progesterone produced favorable effects, AP was more effective for reducing stroke infarct compared to progesterone (Fig. 2). These data can be interpreted to suggest that progesterone's effects in ischemic brain tissue may be expressed through its metabolite AP. However, dose–response studies are needed to determine whether AP is in fact more effective than progesterone as a treatment for ischemic stroke and also to determine whether the mechanisms of action may be different for the hormone and its metabolite. Sayeed et al. (2007) also assessed whether progesterone would show neuroprotective efficacy on cerebral infarction and functional deficits induced by permanent stroke (pMCAO) in the rat. Focal brain ischemia was induced in Sprague–Dawley rats by pMCAO with a silicone-coated monofilament inserted through the right internal carotid artery. The authors then assessed the effect of progesterone on behavioral dysfunction caused by the pMCAO. They showed that repeated progesterone treatment significantly reduced the size of the cerebral infarct following 72 h pMCAO (Fig. 3) and reduced functional deficits as assessed by accelerating rotarod testing. The observation that progesterone is neuroprotective in permanent focal ischemia suggests that reperfusion is not essential. This is particularly relevant clinically since the majority of human stroke victims suffer from permanent occlusion (Hacke et al., 1996; Pantano et al., 1999; Kassem-Moussa and Graffagnino, 2002).

Using a mouse model of tMCAO ischemic injury (60 min occlusion), Gibson and Murphy (2004) found that post-injury treatment with progesterone IP at 1, 6, and 24 h significantly reduced lesion volume and enhanced motor recovery in males. A subsequent report (Gibson et al., 2005) using both transient and permanent occlusion models in mice followed by magnetic resonance imaging (MRI) evaluation found a dramatic reduction in the extent of cerebral edema in the progesterone-treated animals. Polymerase chain reaction (PCR) evaluation showed that the treatment suppressed interleukin-1β (IL-1β), transforming growth factors (TGF-βs), and nitrous oxide (NOS) — all markers of injury-induced inflammation.

After causing global ischemia in rats, Morali et al. (2005) found that post-injury IV administration of progesterone at 20 min and 2, 6, and 24 h after the injury was able to reduce neuronal cell loss in the hippocampus and reduce cerebral ventricle dilation and cortical shrinkage significantly.

Examination of progesterone as a neuroprotective agent in higher animals has been limited to the work of Gonzalez-Vidal et al. (1998). In this experiment, 18 adult ovariectomized female cats underwent temporary global ischemia for 15 min during cardiorespiratory arrest. The treatment group received progesterone (10 mg/kg SC) for 7 days both before and after the ischemia. Outcome measures included neurologic deficit measured daily for survival (with parameters such as level of consciousness, cranial nerves, behavioral reactions, etc.) and the number of surviving neurons measured at day 14. In this global ischemia model, neurological function was significantly better in the progesterone group and there was an increased number of surviving neurons as a result of the treatment (21–49% loss vs. 54–85% loss in control).

Most recently, Aggarwal et al. (2008) reported the possible neuroprotective effect of progesterone on acute phase changes in a mouse model of cerebral ischemia induced by bilateral common carotid artery occlusion (BCAO). The BCAO model was used to induce partial global cerebral ischemia. Progesterone (15 mg/kg IP) administration significantly reduced the cerebral infarct size compared to the controls.

In a different model of injury, it has been suggested that microvascular myocardial ischemia could be a consequence of reduction of progesterone levels (Miyagawa et al., 1997; Minshall et al., 2001; Hermsmeyer et al., 2004). Hermsmeyer et al. (2004) found that coronary hyperreactivity was prevented by subphysiological levels of progesterone in preatherosclerotic primates. Also, it has been shown that menopausal rhesus monkeys were protected (>90%) against excessive and prolonged coronary vasoconstriction when progesterone was continuously delivered at a rate that maintained defined blood levels.

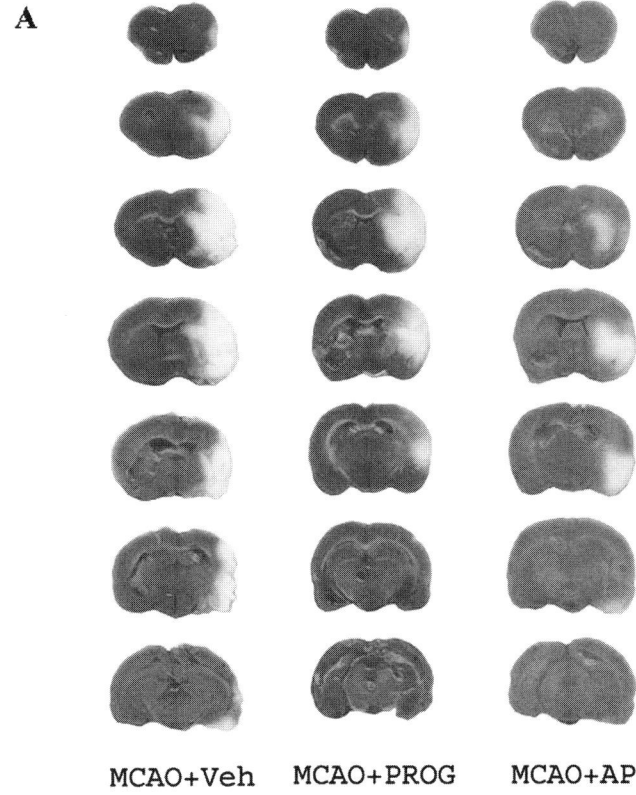

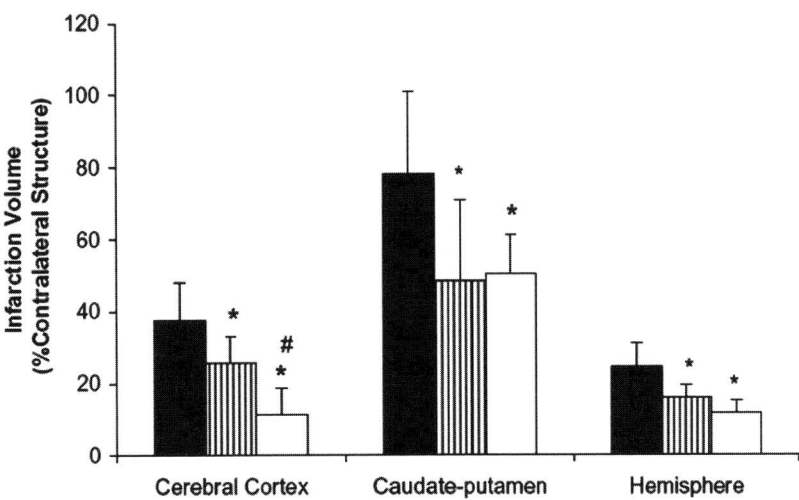

Fig. 2. Progesterone (PROG) and allopregnanolone (AP) reduce infarct volume following transient MCAO. (A) TTC stained coronal sections from representative animals given either vehicle (veh), or PROG or AP, brains harvested at 72 h post occlusion. Infarcts are shown as pale (unstained) regions involving striatum and overlying cortex. The infarct area in AP-treated animals is substantially reduced. (B) Infarct volumes after 2 h occlusion followed by 72 h reperfusion. Compared to vehicle alone, either progesterone or AP significantly reduced cortical, caudate-putamen, and hemispheric infarct volumes (percentage of contralateral structure). The data are represented as mean ± SD; *, significant difference compared to MCAO+Vehicle; #, significant difference compared to MCAO+progesterone (Sayeed et al., 2006).

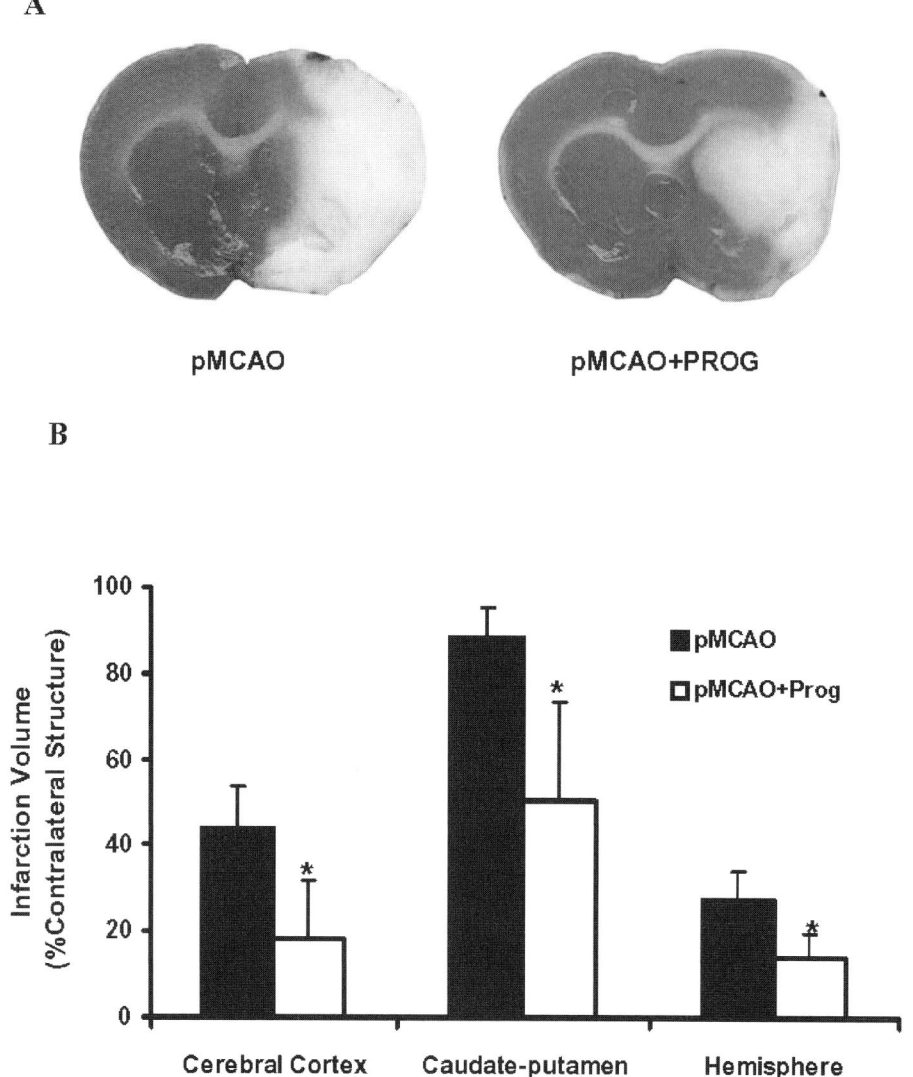

Fig. 3. Progesterone reduces infarct volume following permanent MCAO. (A) TTC-stained coronal section from representative animals given either vehicle or progesterone, brains harvested at 72 h post occlusion. Infarcts are shown as pale (unstained) regions involving striatum and overlying cortex. The infarct area in progesterone-treated animals is substantially reduced. (B) Infarct volumes after 72 h pMCAO. Compared to vehicle alone, progesterone significantly reduced cortical, caudate-putamen, and hemispheric infarct volumes (percentage of contralateral structure). The data are represented as mean±SD; *($p<0.001$), significant difference compared to pMCAO+Vehicle (Sayeed et al., 2007).

Progesterone has accumulated substantial reputation as a neuroprotectant in animal stroke models. Nevertheless, there are gaps between experimental data and application to the clinic. For example, most experimental stroke studies have been performed almost exclusively in rodents, and tests in other species are needed. Moreover, outcomes have focused on short-term histology, and studies of long-term neurological/behavioral outcomes are needed. Further bench studies of progesterone in stroke models is required to generate sufficient preclinical data to support the safety and efficacy of progesterone in stroke patients.

## *Is progesterone different from other recently tried drugs for stroke treatment?*

Protecting the brain from ischemic damage remains a major research priority, and many neuroprotective agents have been studied, but none of them have proven safe and effective for stroke. These failures have been attributed to flawed trial design, insufficient preclinical studies, inappropriate animal modeling, small sample size, and other problems; however, it is also possible that the trials failed because their targets, and their hypotheses about mechanisms of action, were too narrowly focused. For example, NXY-059, a spin trap agent designed to scavenge free radicals, could not cross the BBB (Dehouck et al., 2002) and, although it attenuated oxidative stress-mediated cell death, it did little to ameliorate the inflammatory cascade or exert anti-apoptotic effects. Dizocilpine (MK-801), a noncompetitive antagonist of the *N*-methyl-D-aspartic acid (NMDA) receptor, also failed in patients because it produced too many detrimental side effects related to psychoses. Magnesium acts as a noncompetitive NMDA receptor blocker and blocks voltage-gated calcium channels, but in a recent clinical TBI trial its administration caused a 25% increase in mortality compared to patients given placebo (Temkin et al., 2007).

The pathophysiology of stroke involves a number of mechanisms leading to neuronal injury, implying that in these conditions an agent with a multitude of beneficial actions is more likely to provide effective neuroprotection. We now recognize that progesterone and its metabolites exert their beneficial effects through a number of metabolic and physiological pathways that can affect the injury response in many different tissues and organ systems. Moreover, progesterone is not a drug but a natural hormone, synthesized in both males and females, which can act as a pro-drug for other metabolites with their own distinct mode of action in central nervous system (CNS) repair (Djebaili et al., 2004; He et al., 2004a,b; Roof and Fritts, 1997; Vanlandingham et al., 2008). These properties, now combined with results from two independent clinical trials showing marked benefit in reducing mortality and improving functional outcomes after TBI (Wright et al., 2007; Xiao et al., 2008), make this neurosteroid hormone a unique and compelling natural agent to consider for testing in a clinical trial for stroke. It now remains to resolve any preclinical questions about best dose, duration, and therapeutic window of treatment for ischemic stroke, and any potential differences in the acute and chronic response to the treatment in males and females.

## *Improving animal study design for human translation*

Many studies on the efficacy of neuroprotective agents for the treatment of stroke in humans have failed despite being successful in animal models. The only approved therapy for acute ischemic stroke remains intravenous recombinant tPA initiated within 3–4 h of stroke onset, following a computerized tomography (CT) scan to exclude intracerebral hemorrhage. Many other therapies have been evaluated in phase III clinical trials, including more than 50 neuroprotective agents, but the results have been either inconclusive or negative. Are these drugs truly ineffective in humans or are animal models not accurately simulating the human stroke? These trials have provided valuable lessons for the design of future studies in acute ischemic stroke, including the importance of adequate testing in preclinical studies, window of treatment, dose, patient selection, sample size, and the outcome measures used. In an effort to develop better animal studies, it is important to examine retrospectively previous neuroprotective studies using animals to see what went wrong. Keeping this in view, the STAIR Committee issued several recommendations for optimal preclinical development of neuroprotective agents based on previous failures to translate successful animal studies into successful clinical trials. The STAIR Committee's recommendations (1999) addressed all aspects of study design from subject selection to outcome measures. Applying these STAIR recommendations to the previously published studies on progesterone in ischemic stroke may lead to better design of animal studies and successful translation of progesterone from the laboratory bench to clinical use.

NXY-059, a novel free radical-trapping neuroprotectant, was one of the first compounds to meet all preclinical STAIR criteria and progress to phase III stroke clinical trials but then failed. The failed Stroke Acute Ischemic NXY-059 Treatment Trial (SAINT III) led to the realization that stroke preclinical study designs should go beyond the STAIR recommendations. Given the recently reported increased risk of stroke and heart attacks with the popular COX-2 inhibitors Vioxx and Celebrex, and the failure of SAINT III (Papadakis et al., 2008), it is likely that much more scrutiny and preclinical testing will be required by the FDA before any new treatments for stroke will be approved for clinical testing.

There are calls to perform retrospective reviews on animal experiments aimed at modeling stroke and TBI (Sandercock and Roberts, 2002; Dirnagl, 2006). Recently, Gibson et al. (2008) conducted a systematic review and performed meta analyses with the purpose of identifying key factors, such as timing of treatment, therapeutic dose, and effectiveness according to sex and age, behavioral measures, and other parameters used to evaluate the neuroprotective properties of progesterone treatment on lesion volume after experimental cerebral injury. They identified and analyzed studies using Cochrane Review Manager software. This approach provided some supporting evidence for a neuroprotective role of progesterone following either cerebral ischemia or TBI but, importantly, it highlighted areas which need further preclinical investigation. The authors concluded that the methodological quality of many of the animal studies was poor and did not include appropriate randomization procedures.

A safe and effective post-injury treatment for ischemic stroke should meet certain criteria: it should quickly reduce or prevent secondary neuronal loss; reduce or prevent mortality and morbidity and improve functional recovery; have few or no long-term side effects; be effective in both males and females; be effective across the developmental spectrum; act rapidly without the need for long-term dosing; have a large window of opportunity for treatment; be easy to administer in an emergency situation; and if at all possible, be readily available and inexpensive. Although progesterone has been shown in principle to be effective in stroke and other CNS injuries, more preclinical studies following guidelines going beyond STAIR recommendations are needed before it can be tested in clinical trials for the different kinds of stroke. The following features of study design may provide the best hope for translating successful animal studies with progesterone into successful human clinical trials.

*Systematic preclinical dose–response studies*

Progesterone has been shown to be efficacious following stroke but systematic preclinical dose–response studies in different stroke animal models are lacking. In order to establish a safety profile further studies are needed to establish dose–response relationships in both global transient and permanent stroke models.

*Physiological monitoring*

A lack of careful monitoring of variables such as cerebral blood flow, blood pressure, heart rate, blood gases, hemoglobin levels, blood cell numbers, glucose concentration, and other relevant biochemical factors during the surgeries has been recognized as a major flaw in preclinical pharmacological studies, and such gaps may have contributed significantly to the failure of compounds in clinical trials. For example, hypothermia or hyperthermia can reduce or augment, respectively, ischemic brain damage (Coimbra et al., 1996). To reduce such potentially confounding factors it is important that all physiological variables be monitored and controlled by using blood gas analyzers, infrared heating lamps, and homeothermic blanket systems.

*Window of treatment*

Animal studies have shown convincingly that neuroprotective drugs given after stroke onset can reduce infarct size and improve functional outcome measures. Unfortunately, modeling has not been ideal. Initiation of therapy has, in most experiments, not been sufficiently delayed to mimic what takes place in the clinical setting. Experimental drug administration has often followed early effective reperfusion, which does not represent a typical clinical situation. STAIR guidelines emphasize the importance of developing agents with a clinically relevant window of opportunity. Study is needed to look at the effects of progesterone on morphological and functional outcomes when progesterone treatment is delayed after ischemic stroke. Although in two recent clinical trials in TBI treatment was not provided for 6–8 h, both studies resulted in significant reductions in mortality and better functional outcomes. While TBI is by no means the same as stroke, there is reason for optimism that progesterone may have salutary effects if given well beyond the "golden hour" required for other stroke treatments such as tPA. Another possible advantage of progesterone is that it can be administered quickly without involving skilled personnel, whereas tPA must be given with great care in a hospital setting, thus decreasing the window of treatment opportunity. Progesterone could be given at home or in an EMS vehicle, increasing the chances of getting it to larger numbers of patients sooner after stroke onset.

*Long-term functional recovery*

None of the studies done so far have assessed long-term functional recovery following progesterone treatment for stroke. Most animal studies complete any complex functional and behavioral assays within only a few weeks after injury — with little attention to whether the treatment effects endure over time, as expected in human outcome studies. As in clinical practice (Rothrock et al., 1995), most animals present with a certain degree of spontaneous recovery after experimental cerebral ischemia (Hunter et al., 1998; Roof and Hall, 2000; Zausinger et al., 2000; Zhang et al., 2000). For new drug evaluation, both acute (spanning a few days) and long-term (spanning several weeks or months) outcomes should be evaluated to measure a stable neuroprotection. Moreover, whereas the correlation between acute histological lesion and early behavioral impairment is generally accepted (Rogers et al., 1997), less is known about the long-term evolution of this relationship. To model the clinical setting more appropriately it is critical to examine the effects of progesterone treatment on a panel of behavioral tests known to be sensitive to ischemic insult performed several months after the injury.

*Study population (age, gender, comorbid disease condition)*

Given the higher incidence of stroke in older people it is essential to evaluate the effectiveness of progesterone (or any other putative) treatment in older male and female animals. Strokes in women have important differences from those in males, in risk factors and etiology, and in preventive and therapeutic treatment at each stage of the hormonal cycle. Recently several highly publicized initiatives have called for the systematic study of gender differences in disease conditions, and several edited volumes on sex differences and brain functions (McIntosh, 1996; Kimura, 1999; Legato, 2004; Morrison et al., 2004) have been published. Nonetheless, the amount of empirical research reversing the male-only approach in animal-based studies has been very limited. Young premenopausal women typically experience lower rates of vascular disease and atherosclerosis-related ischemic stroke than their male counterparts (Sudlow and Warlow, 1997) and recanalize more frequently after stroke than men after they have been treated with tPA (Savitz et al., 2005). However, this epidemiological advantage is lost by the time women reach the perimenopausal years, indicating that female reproductive hormones play some role in this sex difference. Thus, both age and gender may affect the timing and extent of recovery and few preclinical studies take this into consideration in their experimental design.

Aging women sustain a large burden for stroke, a fact frequently overlooked in the public's view of breast cancer as the main killer of women. Compared to ischemic males, females with ischemic stroke may require different dosing, duration of treatment, or window of opportunity for treatment. The overall neuroprotective effect of gender on ischemic injury is evident even in the presence of specific stroke risk factors such as diabetes (Vannucci et al., 2001; Toung et al., 2004) and hypertension (Alkayed et al., 1998; Lindner et al., 1998; Carswell et al., 2000) but most of these variables are neglected in preclinical designs. Based on evaluations of other outcomes such as length of survival or incidence/mortality rates in spontaneous or induced stroke (Yamori et al., 1976a, b), researchers have generally reported that female rodents have greater survival times and lower rates of stroke incidence/mortality than males. Thus, it is very important to conduct more preclinical studies with progesterone in female rats as well, although given the existing human and animal data from TBI research there is good reason to believe that it is very likely to have similar beneficial effects in both sexes.

Subjects with prior conditions placing them at more risk for stroke than normotensive individuals may not be as susceptible to the benefits of progesterone treatment, but they may represent a better model of the clinical population at risk for stroke. Several risk factors predisposing to cerebral infarction have been reported, including chronic hypertension, diabetes, smoking, and hypercholesteremia (Wolf et al., 1991). Among these, hypertension is a risk factor for atherothrombotic cerebral infarction. Chronic arterial hypertension greatly reduces the capacity of the cerebral circulation to dilate in the face of stimuli such as hemorrhagic or drug-induced hypotension. Experimentally, infarct size is greater in SHRs subjected to focal cerebral ischemia compared to that in normotensive rat strains (Duverger and MacKenzie, 1988; Dogan et al., 1998). Some drugs show efficacy in normotensive rats but not in their hypertensive counterparts. For example, dizocilpine demonstrates weak tissue-sparing effects in SHRs compared to non-hypertensive strains (Roussel et al., 1992). Yet dizocilpine is as neuroprotective in conditions of exaggerated infarction volume (following hyperglycemia and subacute diabetes) as in normal animals, suggesting that the condition must be contrasted with chronic arterial hypertension (e.g., SHR), in which effective neuroprotection was difficult to show (Bomont and MacKenzie, 1995). It would be interesting to look at the efficacy of progesterone in these comorbid disease models. The information gained will help determine whether progesterone treatment is equally effective in comorbid conditions, or if additional combined pharmacological or behavioral therapies are needed to go forward with any future clinical trial including these subsets of the population at stroke risk. Replicating the experiments in higher order species such as rabbits, marmosets, primates, etc. would add considerable strength to the case for testing progesterone in human clinical trials.

*Combination therapy*

Given the complex pathophysiology of stroke, it may be unrealistic to hope for a single "magic bullet" that will result in neuroprotection and rescue of damaged but not-yet-destroyed neurons. A major reason cited for the inability to demonstrate clinically effective pharmaceutical compounds is that the large number of mechanisms associated with different types of stroke are not being addressed with a single drug agent. Multidrug combinations that target at least some defined and distinct and/or repair processes following stroke may offer a solution. To date, no single agent has been shown to improve stroke outcome in humans, and the current standard of care remains primarily supportive. It is important to select drugs judiciously to carry out combination treatment along with progesterone. The selection of those drugs should be based on identifying missing mechanisms of action of progesterone and selecting drugs with those actions. Although progesterone is a pleiotropic hormone, which in itself meets the spirit of a combination therapy, there are several pathways of the injury process where it has limited activity. For example, progesterone does block excitotoxicity by reducing GABA, but a compound with stronger NMDA and glutamate actions given in combination with progesterone may produce better synergistic effects. Combination treatment with progesterone and a clot-busting drug such as tPA might be a better strategy. Progesterone modulates calcium homeostasis and mitochondrial disturbances but there may be therapies with even stronger mitochondrial modulatory effects, for example, transcranial photonic laser therapy, which robustly increases mitochondrial viability and activity. It may be advantageous to combine progesterone with agents that enhance free radical scavenging and cerebral blood flow. Thus, using combinatorial pharmacotherapies to improve neuroprotection and behavioral outcomes following experimental stroke may be a route to future successful stroke therapy. It should not be automatically assumed that such combinations have to be given at the same time. For example, progesterone could be used to reduce excitotoxicity, edema, and inflammation in the acute stage of the injury and followed days or weeks later with a neurotrophic agent to stimulate cellular and axonal repair and regeneration.

## Progesterone's neuroprotective mechanisms

Progesterone and its metabolites exert their beneficial effects through a number of metabolic and physiological pathways that can affect the injury response in many different tissues and organ systems (Fig. 4).

Based on numerous reports, we now know that progesterone given to both males and females can: (1) cross the BBB (Schumacher et al., 2003) and reduce edema levels after TBI (Roof and Stein, 1992; Roof et al., 1996a, b); (2) reduce lipid peroxidation and isoprostanes, which in turn contribute to post-injury ischemic conditions (Roof et al., 1997); generate metabolites which (3) reduce pro-apoptotic and increase anti-apoptotic enzymes (Djebaili et al., 2004); (4) affect the expression of proinflammatory genes and their protein products (Pettus et al., 2005; Vanlandingham et al., 2007); (5) influence the expression of aquaporins implicated in the resolution of edema (Guo et al., 2006); (7) protect neurons distal to the injury which would normally die (Roof et al., 1994); (8) enhance oligodendrocyte-induced remyelination in young and aged rats with demyelinating disorders (Ghoumari et al., 2003; Ibanez et al.,

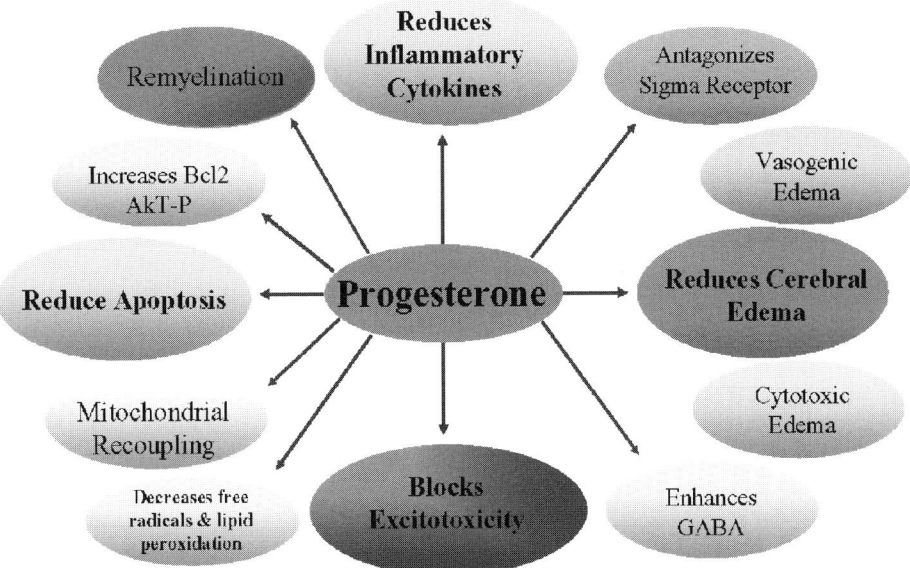

Fig. 4. Progesterone's neuroprotective mechanisms: Progesterone exerts its beneficial effects through a number of metabolic and physiological pathways.

2003; Labombarda et al., 2006a); (9) produce significant sparing of cognitive, sensory, and spatial learning performance in laboratory rats after bilateral injury to the medial frontal cortex (Roof et al., 1994); and (10) produce effects repeatable across species (mice, rats, cats, and humans) with comparable effective doses (Lowery et al., 2002; Goss et al., 2003). Progesterone has been shown to lead to improvements via a variety of molecular mechanisms (Djebaili et al., 2004; Pettus et al., 2005; Vanlandingham et al., 2006, 2007, 2008), making it likely that interacting pleiotropic actions are responsible for its observed benefits.

The multiple mechanisms through which progesterone exerts its beneficial effects have been extensively reviewed (Stein, 2008; Stein and Hurn, 2009). Recently further evidence of the effects of progesterone and AP on hemostatic proteins has been reported. Vanlandingham et al. (2008) examined TBI-induced coagulopathy, an important contributor to the secondary destruction of nervous tissue, to see whether progesterone and AP administration would have a beneficial effect on expression of coagulatory proteins after TBI. Adult male Sprague–Dawley rats were given bilateral contusions of the medial frontal cortex followed by treatments with progesterone (16 mg/kg), AP (8 mg/kg), or vehicle (22.5% hydroxypropyl-β-cyclodextrin). Controls received no injury or injections. Progesterone and AP were observed to have different effects on the injury cascade. Progesterone maintained a procoagulant state by up-regulating thrombin, fibrinogen, and coagulation factor XIII, while AP acted as an anticoagulant by increasing tPA expression. Based on data from this study the authors suggested that it may be preferable to treat TBI patients with progesterone, where blood loss may exacerbate injury, whereas it might be more appropriate to use AP as a treatment for thrombotic stroke, where a reduction in coagulation would be more beneficial. These new insights into the actions of neurosteroids on the coagulation cascade and clotting time may impact the design of safe and effective clinical trials for the treatment of different types of brain injury.

**Conclusion**

The processes of stroke and TBI share similar pathogenic mechanisms which include, among others, excitotoxicity, free radical damage, inflammation, and apoptosis. Cerebral ischemia and trauma also trigger a similar auto-protective responses, including anti-inflammatory cytokines and endogenous antioxidants (Leker and Shohami, 2002; Bramlett and Dietrich, 2004). The fact that there are more similarities than differences in the pathology of cerebral ischemia and TBI may imply that neuroprotective compounds found to be active against one condition may also be protective in the other. However, there are still several important etiological differences between stroke and TBI. Tissue responses to different injury severities and types (e.g., ischemic vs. traumatic) may differ, complicating treatment strategies. Given progesterone's demonstrated potential to affect many aspects of neural repair in TBI, the success of the TBI clinical trials (Wright et al., 2007; Xiao et al., 2008), the planning for a multicenter trial of progesterone in TBI, and in light of the important similarities between TBI and stroke, the logical next step will be to determine whether the hormone and its metabolites warrant clinical testing for different kinds of stroke.

Multiple molecular mechanisms have been identified but underlying mechanisms explaining progesterone's actions in vivo are not yet completely understood, especially for the various kinds of stroke. Progesterone has manifold actions but its most direct beneficial effect appears to be in reducing cerebral edema and inflammation, thereby decreasing intracranial pressure that leads to the secondary loss of nerve cells. There is also more and more evidence that progesterone exerts anti-inflammatory, anti-apoptotic, and perhaps antioxidant effects. These actions may work synergistically to prevent the death of neurons and glia, leading to improved functional outcomes.

Besides progesterone's neuroprotective effects, it also shows beneficial effects in other organs following trauma-induced damage. Progesterone administration following trauma-hemorrhage has been reported to ameliorate the proinflammatory response and, subsequently, hepatocellular injury via direct action on immunocompetent cells (Kuebler et al., 2003). It is evident that progesterone can regulate the systemic changes that cause post-injury "illness" and frailty by reducing cytokines in the gut and liver (Kuebler et al., 2003). Thus progesterone, an agent with pleiotropic consequences, is more likely to provide effective neuroprotection and repair for a range of CNS disorders including TBI and stroke than one operating primarily on a single, or a small number of, injury mechanisms.

In conclusion, among the many neuroprotective agents that have been tested in preclinical and clinical settings, progesterone may prove to be the most successful in meeting the criteria for application that we outlined above. Thus far, its safety profile in clinical trial has been excellent and proof-of-principle for efficacy has been demonstrated. It now seems likely that a major, multicenter national trial for progesterone in TBI will begin in 2009 which, if successful, could stimulate further interest in testing the hormone in patients with ischemic stroke. In the meantime, it will be important to develop and test appropriate stroke models in animals to assure that most, if not all, of the important parameters needed to assess efficacy can be appropriately translated to use in human subjects in order to avoid yet another disappointing and costly failure.

## Abbreviations

| | |
|---|---|
| AP | allopregnanolone |
| BBB | blood–brain barrier |
| BCAO | bilateral common carotid artery occlusion |
| CNS | central nervous system |
| CT | computerized tomography |
| DMSO | dimethylsulfoxide |
| FDA | Food and Drug Administration |
| GABA | gamma-amino butyric acid |
| IL-1β | interleukin 1β |
| IP | intraperitoneally |
| MCA | middle cerebral artery |
| MCAO | MCA occlusion |
| MRI | magnetic resonance imaging |
| NMDA | $N$-methyl-D-aspartic acid |
| NOS | nitrous oxide |
| PCR | polymerase chain reaction |
| pMCAO | permanent MCAO |
| SC | subcutaneously |
| STAIR | Stroke Therapy Academic Industry Roundtable |
| TBI | traumatic brain injury |
| TGF-βs | transforming growth factors |
| tMCAO | temporary MCAO |
| tPA | tissue plasminogen activator |

## Acknowledgment

The authors would like to thank Leslie McCann for editorial assistance.

## References

Aggarwal, R., Medhi, B., Pathak, A., Dhawan, V., & Chakrabarti, A. (2008). Neuroprotective effect of progesterone on acute phase changes induced by partial global cerebral ischaemia in mice. *Journal of Pharmacy and Pharmacology*, 60, 731–737.

Alkayed, N. J., Harukuni, I., Kimes, A. S., London, E. D., Traystman, R. J., & Hurn, P. D. (1998). Gender-linked brain injury in experimental stroke. *Stroke*, 29, 159–165. discussion p. 166

Alkayed, N. J., Murphy, S. J., Traystman, R. J., Hurn, P. D., & Miller, V. M. (2000). Neuroprotective effects of female gonadal steroids in reproductively senescent female rats. *Stroke*, 31, 161–168.

American Heart Association. (2005). *Stroke and heart statistical update*. Dallas, TX: American Heart Association.

Anderson, G. L., Limacher, M., Assaf, A. R., Bassford, T., Beresford, S. A., Black, H., et al. (2004). Effects of conjugated equine estrogen in postmenopausal women with hysterectomy: the Women's Health Initiative randomized controlled trial. *JAMA*, 291, 1701–1712.

Baulieu, E. E., & Robel, P. (1990). Neurosteroids: a new brain function? *Journal of Steroid Biochemistry and Molecular Biology*, 37, 395–403.

Betz, A. L., & Coester, H. C. (1990). Effect of steroids on edema and sodium uptake of the brain during focal ischemia in rats. *Stroke*, 21, 1199–1204.

Bomont, L., & MacKenzie, E. T. (1995). Neuroprotection after focal cerebral ischaemia in hyperglycaemic and diabetic rats. *Neuroscience Letters*, 197, 53–56.

Bramlett, H. M., & Dietrich, W. D. (2004). Pathophysiology of cerebral ischemia and brain trauma: similarities and differences. *Journal of Cerebral Blood Flow and Metabolism, 24*, 133–150.

Carswell, H. V., Anderson, N. H., Morton, J. J., McCulloch, J., Dominiczak, A. F., & Macrae, I. M. (2000). Investigation of estrogen status and increased stroke sensitivity on cerebral blood flow after a focal ischemic insult. *Journal of Cerebral Blood Flow and Metabolism, 20*, 931–936.

Chen, J., Chopp, M., & Li, Y. (1999). Neuroprotective effects of progesterone after transient middle cerebral artery occlusion in rat. *Journal of Neurological Sciences, 171*, 24–30.

Clifton, G. L., Choi, S. C., Miller, E. R., Levin, H. S., Smith, K. R., Jr., Muizelaar, J. P., et al. (2001a). Intercenter variance in clinical trials of head trauma—experience of the National Acute Brain Injury Study: Hypothermia. *Journal of Neurosurgery, 95*, 751–755.

Clifton, G. L., Miller, E. R., Choi, S. C., Levin, H. S., McCauley, S., Smith, K. R., Jr., et al. (2001b). Lack of effect of induction of hypothermia after acute brain injury. *The New England Journal of Medicine, 344*, 556–563.

Coimbra, C., Drake, M., Boris-Moller, F., & Wieloch, T. (1996). Long-lasting neuroprotective effect of postischemic hypothermia and treatment with an anti-inflammatory/antipyretic drug. Evidence for chronic encephalopathic processes following ischemia. *Stroke, 27*, 1578–1585.

Dehouck, M. P., Cecchelli, R., Richard Green, A., Renftel, M., & Lundquist, S. (2002). In vitro blood-brain barrier permeability and cerebral endothelial cell uptake of the neuroprotective nitrone compound NXY-059 in normoxic, hypoxic and ischemic conditions. *Brain Research, 955*, 229–235.

Dirnagl, U. (2006). Bench to bedside: the quest for quality in experimental stroke research. *Journal of Cerebral Blood Flow and Metabolism, 26*, 1465–1478.

Djebaili, M., Hoffman, S. W., & Stein, D. G. (2004). Allopregnanolone and progesterone decrease cell death and cognitive deficits after a contusion of the rat pre-frontal cortex. *Journal of Neuroscience, 123*, 349–359.

Dogan, A., Baskaya, M. K., Rao, V. L., Rao, A. M., & Dempsey, R. J. (1998). Intraluminal suture occlusion of the middle cerebral artery in spontaneously hypertensive rats. *Neurological Research, 20*, 265–270.

Dubinsky, R., & Lai, S. M. (2006). Mortality of stroke patients treated with thrombolysis: analysis of nationwide inpatient sample. *Neurology, 66*, 1742–1744.

Durukan, A., & Tatlisumak, T. (2007). Acute ischemic stroke: overview of major experimental rodent models, pathophysiology, and therapy of focal cerebral ischemia. *Pharmacology, Biochemistry, and Behavior, 87*, 179–197.

Duverger, D., & MacKenzie, E. T. (1988). The quantification of cerebral infarction following focal ischemia in the rat: influence of strain, arterial pressure, blood glucose concentration, and age. *Journal of Cerebral Blood Flow and Metabolism, 8*, 449–461.

Edwards, P., Arango, M., Balica, L., Cottingham, R., El-Sayed, H., Farrell, B., et al. (2005). Final results of MRC CRASH, a randomised placebo-controlled trial of intravenous corticosteroid in adults with head injury—outcomes at 6 months. *Lancet, 365*, 1957–1959.

Furlan, A. J. (2006). IV tissue plasminogen activator for stroke in the community: what we know and don't know 10 years after FDA approval. *Stroke, 37*, 281.

Garay, L., Deniselle, M. C., Lima, A., Roig, P., & De Nicola, A. F. (2007). Effects of progesterone in the spinal cord of a mouse model of multiple sclerosis. *Journal of Steroid Biochemistry and Molecular Biology, 107*, 228–237.

Ghoumari, A. M., Ibanez, C., El-Etr, M., Leclerc, P., Eychenne, B., O'Malley, B. W., et al. (2003). Progesterone and its metabolites increase myelin basic protein expression in organotypic slice cultures of rat cerebellum. *Journal of Neurochemistry, 86*, 848–859.

Gibson, C. L., Constantin, D., Prior, M. J., Bath, P. M., & Murphy, S. P. (2005). Progesterone suppresses the inflammatory response and nitric oxide synthase-2 expression following cerebral ischemia. *Experimental Neurology, 193*, 522–530.

Gibson, C. L., Gray, L. J., Bath, P. M., & Murphy, S. P. (2008). Progesterone for the treatment of experimental brain injury; a systematic review. *Brain, 131*, 318–328.

Gibson, C. L., & Murphy, S. P. (2004). Progesterone enhances functional recovery after middle cerebral artery occlusion in male mice. *Journal of Cerebral Blood Flow and Metabolism, 24*, 805–813.

Goldberg, M. P. (2007). New approaches to clinical trials in neuroprotection: introduction. *Stroke, 38*, 789–790.

Gomes, J. A., Stevens, R. D., Lewin, J. J., III, Mirski, M. A., & Bhardwaj, A. (2005). Glucocorticoid therapy in neurologic critical care. *Critical Care Medicine, 33*, 1214–1224.

Gonzalez-Vidal, M. D., Cervera-Gaviria, M., Ruelas, R., Escobar, A., Morali, G., & Cervantes, M. (1998). Progesterone: protective effects on the cat hippocampal neuronal damage due to acute global cerebral ischemia. *Archives of Medical Research, 29*, 117–124.

Goss, C. W., Hoffman, S. W., & Stein, D. G. (2003). Behavioral effects and anatomic correlates after brain injury: a progesterone dose-response study. *Pharmacology, Biochemistry, and Behavior, 76*, 231–242.

Grotta, J. C., Burgin, W. S., El-Mitwalli, A., Long, M., Campbell, M., Morgenstern, L. B., et al. (2001). Intravenous tissue-type plasminogen activator therapy for ischemic stroke: Houston experience 1996 to 2000. *Archives of Neurology, 58*, 2009–2013.

Guerra-Araiza, C., Coyoy-Salgado, A., & Camacho-Arroyo, I. (2002). Sex differences in the regulation of progesterone receptor isoforms expression in the rat brain. *Brain Research Bulletin, 59*, 105–109.

Guo, Q., Sayeed, I., Baronne, L. M., Hoffman, S. W., Guennoun, R., & Stein, D. G. (2006). Progesterone administration modulates AQP4 expression and edema after traumatic brain injury in male rats. *Experimental Neurology, 198*, 469–478.

Hacke, W., Schwab, S., Horn, M., Spranger, M., De Georgia, M., & von Kummer, R. (1996). 'Malignant' middle cerebral artery territory infarction: clinical course and prognostic signs. *Archives of Neurology, 53*, 309–315.

Hankey, G. J. (1999). Stroke: how large a public health problem, and how can the neurologist help? *Archives of Neurology, 56*, 748–754.

He, J., Evans, C. O., Hoffman, S. W., Oyesiku, N. M., & Stein, D. G. (2004a). Progesterone and allopregnanolone reduce inflammatory cytokines after traumatic brain injury. *Experimental Neurology, 189*, 404–412.

He, J., Hoffman, S. W., & Stein, D. G. (2004b). Allopregnanolone, a progesterone metabolite, enhances behavioral recovery and decreases neuronal loss after traumatic brain injury. *Restorative Neurology and Neuroscience, 22*, 19–31.

Hermsmeyer, R. K., Mishra, R. G., Pavcnik, D., Uchida, B., Axthelm, M. K., Stanczyk, F. Z., et al. (2004). Prevention of coronary hyperreactivity in preatherogenic menopausal rhesus monkeys by transdermal progesterone. *Arteriosclerosis, Thrombosis, and Vascular Biology, 24*, 955–961.

Hunter, A. J., Mackay, K. B., & Rogers, D. C. (1998). To what extent have functional studies of ischaemia in animals been useful in the assessment of potential neuroprotective agents? *Trends in Pharmacological Sciences, 19*, 59–66.

Hutchison, J. S., Ward, R. E., Lacroix, J., Hebert, P. C., Barnes, M. A., Bohn, D. J., et al. (2008). Hypothermia therapy after traumatic brain injury in children. *The New England Journal of Medicine, 358*, 2447–2456.

Ibanez, C., Shields, S. A., El-Etr, M., Leonelli, E., Magnaghi, V., Li, W. W., et al. (2003). Steroids and the reversal of age-associated changes in myelination and remyelination. *Progress in Neurobiology, 71*, 49–56.

Jiang, N., Chopp, M., Stein, D., & Feit, H. (1996). Progesterone is neuroprotective after transient middle cerebral artery occlusion in male rats. *Brain Research, 735*, 101–107.

Kassem-Moussa, H., & Graffagnino, C. (2002). Nonocclusion and spontaneous recanalization rates in acute ischemic stroke: a review of cerebral angiography studies. *Archives of Neurology, 59*, 1870–1873.

Kimura, D. (1999). *Sex and cognition*. Cambridge, MA: MIT University Press.

Kuebler, J. F., Yokoyama, Y., Jarrar, D., Toth, B., Rue, L. W., III, Bland, K. I., et al. (2003). Administration of progesterone after trauma and hemorrhagic shock prevents hepatocellular injury. *Archives of Surgery, 138*, 727–734.

Kumon, Y., Kim, S. C., Tompkins, P., Stevens, A., Sakaki, S., & Loftus, C. M. (2000). Neuroprotective effect of postischemic administration of progesterone in spontaneously hypertensive rats with focal cerebral ischemia. *Journal of Neurosurgery, 92*, 848–852.

Labombarda, F., Gonzalez, S., Gonzalez Deniselle, M. C., Garay, L., Guennoun, R., Schumacher, M., et al. (2006a). Progesterone increases the expression of myelin basic protein and the number of cells showing NG2 immunostaining in the lesioned spinal cord. *Journal of Neurotrauma, 23*, 181–192.

Labombarda, F., Pianos, A., Liere, P., Eychenne, B., Gonzalez, S., Cambourg, A., et al. (2006b). Injury elicited increase in spinal cord neurosteroid content analyzed by gas chromatography mass spectrometry. *Endocrinology, 147*, 1847–1859.

Legato, M. (Ed.) (2004). *Principles of gender-specific medicine*. New York, NY: Academic Press.

Leker, R. R., & Shohami, E. (2002). Cerebral ischemia and trauma—different etiologies yet similar mechanisms: neuroprotective opportunities. *Brain Research, 39*, 55–73.

Leonelli, E., Bianchi, R., Cavaletti, G., Caruso, D., Crippa, D., Garcia-Segura, L. M., et al. (2007). Progesterone and its derivatives are neuroprotective agents in experimental diabetic neuropathy: a multimodal analysis. *Journal of Neuroscience, 144*, 1293–1304.

Lindner, M. D., Plone, M. A., Cain, C. K., Frydel, B., Francis, J. M., Emerich, D. F., et al. (1998). Dissociable long-term cognitive deficits after frontal versus sensorimotor cortical contusions. *Journal of Neurotrauma, 15*, 199–216.

Lowery, D. W., Logan, J. E., Shear, D. A., Hoffman, S. W., & Stein, D. G. (2002). Progesterone improves behavioral and morphological outcomes after traumatic brain injury in male C57BL6 mice. *Journal of Neurotrauma, 19*, 1286.

Lyden, P. (2008). Thrombolytic therapy for acute stroke—not a moment to lose. *The New England Journal of Medicine, 359*, 1393–1395.

McIntosh, T. K. (1996). Neuropathological sequelae of traumatic brain injury: relationship to neurochemical and biomechanical mechanisms. *Laboratory Investigation, 74*, 315–342.

Meffre, D., Delespierre, B., Gouezou, M., Leclerc, P., Vinson, G. P., Schumacher, M., et al. (2005). The membrane-associated progesterone-binding protein 25-Dx is expressed
in brain regions involved in water homeostasis and is up-regulated after traumatic brain injury. *Journal of Neurochemistry, 93*, 1314–1326.

Mellon, S. H., Griffin, L. D., & Compagnone, N. A. (2001). Biosynthesis and action of neurosteroids. *Brain Research, 37*, 3–12.

Minshall, R. D., Pavcnik, D., Halushka, P. V., & Hermsmeyer, K. (2001). Progesterone regulation of vascular thromboxane A(2) receptors in rhesus monkeys. *American Journal of Physiology. Heart and Circulatory Physiology, 281*, H1498–H1507.

Miyagawa, K., Rosch, J., Stanczyk, F., & Hermsmeyer, K. (1997). Medroxyprogesterone interferes with ovarian steroid protection against coronary vasospasm. *Nature Medicine, 3*, 324–327.

Morali, G., Letechipia-Vallejo, G., Lopez-Loeza, E., Montes, P., Hernandez-Morales, L., & Cervantes, M. (2005). Postischemic administration of progesterone in rats exerts neuroprotective effects on the hippocampus. *Neuroscience Letters, 382*, 286–290.

Morrison, W. E., Arbelaez, J. J., Fackler, J. C., De Maio, A., & Paidas, C. N. (2004). Gender and age effects on outcome after pediatric traumatic brain injury. *Pediatric Critical Care Medicine, 5*, 145–151.

Muntner, P., DeSalvo, K. B., Wildman, R. P., Raggi, P., He, J., & Whelton, P. K. (2006). Trends in the prevalence, awareness, treatment, and control of cardiovascular disease risk factors among noninstitutionalized patients with a history of myocardial infarction and stroke. *American Journal of Epidemiology, 163*, 913–920.

Murphy, S. J., Traystman, R. J., Hurn, P. D., & Duckles, S. P. (2000). Progesterone exacerbates striatal stroke injury in progesterone-deficient female animals. *Stroke, 31*, 1173–1178.

Murray, C. J., & Lopez, A. D. (1997). Mortality by cause for eight regions of the world: Global Burden of Disease Study. *Lancet, 349*, 1269–1276.

O'Connor, C. A., Cernak, I., & Vink, R. (2005). Both estrogen and progesterone attenuate edema formation following diffuse traumatic brain injury in rats. *Brain Research, 1062*, 171–174.

Pantano, P., Caramia, F., Bozzao, L., Dieler, C., & von Kummer, R. (1999). Delayed increase in infarct volume after cerebral ischemia: correlations with thrombolytic treatment and clinical outcome. *Stroke, 30*, 502–507.

Papadakis, M., Nagel, S., & Buchan, A. M. (2008). Development and efficacy of NXY-059 for the treatment of acute ischemic stroke. *Future Neurology, 3*, 229–240.

Pettus, E. H., Wright, D. W., Stein, D. G., & Hoffman, S. W. (2005). Progesterone treatment inhibits the inflammatory agents that accompany traumatic brain injury. *Brain Research, 1049*, 112–119.

Quadros, P. S., Pfau, J. L., & Wagner, C. K. (2007). Distribution of progesterone receptor immunoreactivity in the fetal and neonatal rat forebrain. *Journal of Comparative Neurology, 504*, 42–56.

Reddy, D. S., O'Malley, B. W., & Rogawski, M. A. (2005). Anxiolytic activity of progesterone in progesterone receptor knockout mice. *Neuropharmacology, 48*, 14–24.

Roberts, I., Schierhout, G., & Alderson, P. (1998). Absence of evidence for the effectiveness of five interventions routinely used in the intensive care management of severe head injury: a systematic review. *Journal of Neurology, Neurosurgery, and Psychiatry, 65*, 729–733.

Roberts, I., Yates, D., Sandercock, P., Farrell, B., Wasserberg, J., Lomas, G., et al. (2004). Effect of intravenous corticosteroids on death within 14 days in 10008 adults with clinically significant head injury (MRC CRASH trial): randomised placebo-controlled trial. *Lancet, 364*, 1321–1328.

Rogers, D. C., Campbell, C. A., Stretton, J. L., & Mackay, K. B. (1997). Correlation between motor impairment and infarct volume after permanent and transient middle cerebral artery occlusion in the rat. *Stroke, 28*, 2060–2065. discussion p. 2066

Roof, R. L., Duvdevani, R., Braswell, L., & Stein, D. G. (1994). Progesterone facilitates cognitive recovery and reduces secondary neuronal loss caused by cortical contusion injury in male rats. *Experimental Neurology, 129*, 64–69.

Roof, R. L., Duvdevani, R., Heyburn, J. W., & Stein, D. G. (1996a). Progesterone rapidly decreases brain edema: treatment delayed up to 24 hours is still effective. *Experimental Neurology, 138*, 246–251.

Roof, R. L., Duvdevani, R., & Stein, D. G. (1993a). Gender influences outcome of brain injury: progesterone plays a protective role. *Brain Research, 607*, 333–336.

Roof, R. L., & Fritts, M. E. (1997). Progesterone metabolites may mediate its neuroprotective effects after traumatic brain injury. *Journal of Neurotrauma, 14*, 760.

Roof, R. L., Fritts, M. E., Castro, E. A., Powell, R. A., & Stein, D. G. (1996b). Progesterone is more effective than methylprednisolone at reducing edema after cortical contusion in male rats. *Society for Neurscience Abstracts*, p. 1186, Washington, DC.

Roof, R. L., & Hall, E. D. (2000). Gender differences in acute CNS trauma and stroke: neuroprotective effects of estrogen and progesterone. *Journal of Neurotrauma, 17*, 367–388.

Roof, R. L., Hoffman, S. W., & Stein, D. G. (1997). Progesterone protects against lipid peroxidation following traumatic brain injury in rats. *Molecular and Chemical Neuropathology, 31*, 1–11.

Roof, R. L., & Stein, D. G. (1992). Progesterone treatment attenuates brain edema following contusion injury in male and female rats. *Restorative Neurology and Neuroscience, 4*, 425–427.

Roof, R. L., Zhang, Q., Glasier, M. M., & Stein, D. G. (1993b). Gender-specific impairment on Morris water maze task after entorhinal cortex lesion. *Behavioural Brain Research, 57*, 47–51.

Rothrock, J. F., Clark, W. M., & Lyden, P. D. (1995). Spontaneous early improvement following ischemic stroke. *Stroke, 26*, 1358–1360.

Roussel, S., Pinard, E., & Seylaz, J. (1992). Effect of MK-801 on focal brain infarction in normotensive and hypertensive rats. *Hypertension, 19*, 40–46.

Sandercock, P., & Roberts, I. (2002). Systematic reviews of animal experiments. *Lancet, 360*, 586.

Savitz, S. I., Schlaug, G., Caplan, L., & Selim, M. (2005). Arterial occlusive lesions recanalize more frequently in women than in men after intravenous tissue plasminogen activator administration for acute stroke. *Stroke, 36*, 1447–1451.

Sayeed, I., Guo, Q., Hoffman, S. W., & Stein, D. G. (2006). Allopregnanolone, a progesterone metabolite, is more effective than progesterone in reducing cortical infarct volume after transient middle cerebral artery occlusion. *Annals of Emergency Medicine, 47*, 381–389.

Sayeed, I., Wali, B., & Stein, D. G. (2007). Progesterone inhibits ischemic brain injury in a rat model of permanent middle cerebral artery occlusion. *Restorative Neurology and Neuroscience, 25*, 151–159.

Schumacher, M., Guennoun, R., Ghoumari, A., Massaad, C., Robert, F., El-Etr, M., et al. (2007). Novel perspectives for progesterone in hormone replacement therapy, with special reference to the nervous system. *Endocrine Reviews, 28*, 387–439.

Schumacher, M., Weill-Engerer, S., Liere, P., Robert, F., Franklin, R. J., Garcia-Segura, L. M., et al. (2003). Steroid hormones and neurosteroids in normal and pathological aging of the nervous system. *Progress in Neurobiology, 71*, 3–29.

Stein, D. G. (2001). Brain damage, sex hormones and recovery: a new role for progesterone and estrogen? *Trends in Neuroscience, 24*, 386–391.

Stein, D. G. (2005). The case for progesterone. *Annals of the New York Academy of Sciences, 1052*, 152–169.

Stein, D. G. (2008). Progesterone exerts neuroprotective effects after brain injury. *Brain Research Reviews, 57*, 386–397.

Stein, D. G., & Hurn, P. D. (2009). Effects of sex steroids on damaged neural systems. In D. W. Pfaff, A. P. Arnold, A. M. Etgen, S. E. Fahrbach, & R. T. Rubin (Eds.), *Hormones, brain and behavior* (Vol. 4, 2nd ed., pp. 2223–2258). Oxford: Elsevier.

Stein, D. G., Wright, D. W., & Kellermann, A. L. (2008). Does progesterone have neuroprotective properties? *Annals of Emergency Medicine, 51*, 164–172.

Stroke Therapy Academic Industry Roundtable (STAIR). (1999). Recommendations for standards regarding preclinical neuroprotective and restorative drug development. *Stroke, 30*, 2752–2758.

Sudlow, C. L., & Warlow, C. P. (1997). Comparable studies of the incidence of stroke and its pathological types: results from an international collaboration. International stroke incidence collaboration. *Stroke, 28*, 491–499.

Temkin, N. R., Anderson, G. D., Winn, H. R., Ellenbogen, R. G., Britz, G. W., Schuster, J., et al. (2007). Magnesium sulfate for neuroprotection after traumatic brain injury: a randomised controlled trial. *Lancet Neurology, 6*, 29–38.

Thomas, A. J., Nockels, R. P., Pan, H. Q., Shaffrey, C. I., & Chopp, M. (1999). Progesterone is neuroprotective after acute experimental spinal cord trauma in rats. *Spine, 24*, 2134–2138.

Toung, T. J., Chen, T. Y., Littleton-Kearney, M. T., Hurn, P. D., & Murphy, S. J. (2004). Effects of combined estrogen and progesterone on brain infarction in reproductively senescent female rats. *Journal of Cerebral Blood Flow and Metabolism, 24*, 1160–1166.

Vanlandingham, J. W., Cekic, M., Cutler, S., Hoffman, S. W., & Stein, D. G. (2007). Neurosteroids reduce inflammation after TBI through CD55 induction. *Neuroscience Letters, 425*, 94–98.

Vanlandingham, J. W., Cekic, M., Cutler, S. M., Hoffman, S. W., Washington, E. R., Johnson, S. J., et al. (2008). Progesterone and its metabolite allopregnanolone differentially regulate hemostatic proteins after traumatic brain injury. *Journal of Cerebral Blood Flow and Metabolism, 28*(11), 1786–1794.

Vanlandingham, J. W., Cutler, S. M., Virmani, S., Hoffman, S. W., Covey, D. F., Krishnan, K., et al. (2006). The enantiomer of progesterone acts as a molecular neuroprotectant after traumatic brain injury. *Neuropharmacology, 51*, 1078–1085.

Vannucci, S. J., Willing, L. B., Goto, S., Alkayed, N. J., Brucklacher, R. M., Wood, T. L., et al. (2001). Experimental stroke in the female diabetic, db/db, mouse. *Journal of Cerebral Blood Flow and Metabolism, 21*, 52–60.

Vink, R., & Van Den Heuvel, C. (2004). Recent advances in the development of multifactorial therapies for the treatment of traumatic brain injury. *Expert Opinion on Investigational Drugs, 13*, 1263–1274.

Watson, N. F., Barber, J. K., Doherty, M. J., Miller, J. W., & Temkin, N. R. (2004). Does glucocorticoid administration prevent late seizures after head injury? *Epilepsia, 45*, 690–694.

Wolf, P. A., D'Agostino, R. B., Belanger, A. J., Kannel, W. B., & Wolf, P. A. (1991). Probability of stroke: a risk profile from the Framingham Study. *Stroke, 22*, 312–318.

Wright, D. W., Bauer, M. E., Hoffman, S. W., & Stein, D. G. (2001). Serum progesterone levels correlate with decreased cerebral edema after traumatic brain injury in male rats. *Journal of Neurotrauma, 18*, 901–909.

Wright, D. W., Kellermann, A. L., Hertzberg, V. S., Clark, P. L., Frankel, M., Goldstein, F. C., et al. (2007). ProTECT: a randomized clinical trial of progesterone for acute traumatic brain injury. *Annals of Emergency Medicine, 49*, 391–402.

Wright, D. W., Ritchie, J. C., Mullins, R. E., Kellermann, A. L., & Denson, D. D. (2005). Steady-state serum concentrations of progesterone following continuous intravenous infusion in patients with acute moderate to severe traumatic brain injury. *Journal of Clinical Pharmacology, 45*, 640–648.

Xiao, G., Wei, J., Yan, W., Wang, W., & Lu, Z. (2008). Improved outcomes from the administration of progesterone for patients with acute severe traumatic brain injury: a randomized controlled trial. *Critical Care, 12*, R61.

Yamori, Y., Horie, R., Sato, M., & Fukase, M. (1976a). Hypertension as an important factor for cerebrovascular atherogenesis in rats. *Stroke, 7*, 120–125.

Yamori, Y., Horie, R., Sato, M., & Fukase, M. (1976b). Studies of etiology and prophylaxis of stroke-prone spontaneously hypertensive rats. *Nippon Rinsho, 34*, 25–34.

Yamori, Y., Horie, R., Sato, M., & Ohta, K. (1976c). Proceedings: prophylactic trials for stroke in stroke-prone SHR: effect of sex hormones. *Japanese Heart Journal, 17*, 404–406.

Zausinger, S., Hungerhuber, E., Baethmann, A., Reulen, H., & Schmid-Elsaesser, R. (2000). Neurological impairment in rats after transient middle cerebral artery occlusion: a comparative study under various treatment paradigms. *Brain Research, 863*, 94–105.

Zhang, L., Chen, J., Li, Y., Zhang, Z. G., & Chopp, M. (2000). Quantitative measurement of motor and somatosensory impairments after mild (30 min) and severe (2 h) transient middle cerebral artery occlusion in rats. *Journal of Neurological Sciences, 174*, 141–146.

CHAPTER 16

# Estrogen and testosterone therapies in multiple sclerosis

Stefan M. Gold[1] and Rhonda R. Voskuhl[2],*

[1]Multiple Sclerosis Program, Department of Neurology, and Cousins Center for Psychoneuroimmunology, Geffen School of Medicine, University of California Los Angeles, Los Angeles, CA, USA
[2]Multiple Sclerosis Program, Department of Neurology, Geffen School of Medicine, University of California Los Angeles, Los Angeles, CA, USA

**Abstract:** It has been known for decades that females are more susceptible than men to inflammatory autoimmune diseases, including multiple sclerosis (MS), rheumatoid arthritis, and psoriasis. In addition, female patients with these diseases experience clinical improvements during pregnancy with a temporary "rebound" exacerbation postpartum. These clinical observations indicate an effect of sex hormones on disease and suggest the potential use of the male hormone testosterone and the pregnancy hormone estriol, respectively, for the treatment of MS. A growing number of studies using the MS animal model experimental autoimmune encephalomyelitis (EAE) support a therapeutic effect of these hormones. Both testosterone and estriol have been found to induce anti-inflammatory as well as neuroprotective effects. Findings from two recent pilot studies of transdermal testosterone in male MS patients and oral estriol in female MS patients are encouraging. In this paper, we review the preclinical and clinical evidence for sex hormone treatments in MS and discuss potential mechanisms of action.

**Keywords:** multiple sclerosis; inflammation; atrophy; immunomodulation; neuroprotection; gonadal steroids; clinical trials

## Introduction

### Inflammation versus neurodegeneration in multiple sclerosis

Multiple sclerosis (MS) is a heterogeneous inflammatory, demyelinating, and degenerative disease of a presumed Th1-autoimmune origin that occurs in genetically susceptible individuals (Hemmer et al., 2002). The exact pathogenetic mechanisms are unknown, but peripheral activation of autoreactive CD4+ T cells targeting proteins of the myelin sheath of neurons has been hypothesized as a key process in the development of the disease (McFarland and Martin, 2007). Upon activation, these cells cross the blood-brain barrier to enter the central nervous system (CNS), recognize myelin antigens, and initiate a chronic inflammatory cascade that results in demyelination of axons, mainly by macrophages (Sospedra and Martin, 2005). Involvement of humoral (antibodies and complement) and cellular mechanisms, as well as primary oligodendroglial degeneration and apoptosis, has also been proposed (Lassmann et al., 2001). The pathological hallmark is the

*Corresponding author.
Tel.: +1 310-206-7313; Fax: +1 310-206-7282;
E-mail: rvoskuhl@ucla.edu

DOI: 10.1016/S0079-6123(09)17516-7

demyelinated plaque, which consists of well-demarcated areas characterized by the loss of myelin and formation of astrocytic scars. However, it is becoming increasingly clear that axonal loss may be the major determinant for long-term, permanent disability. It is unclear whether all neurodegeneration is directly related to acute inflammation, since diffuse axonal damage may occur separately from pathological lesions (Evangelou et al., 2000), and even robust and effective immunosuppression with chemotherapeutic agents is not sufficient to stop accumulation of disability, in particular during later disease stages (Coles et al., 2006). Thus, it appears that MS has both an inflammatory and a neurodegenerative component in its pathogenesis. Over the last decade, abundant neuroimaging and neuropathological studies have indicated a significant neurodegenerative process in MS. Neuroimaging has demonstrated atrophy (Brex et al., 2000; Filippi et al., 2003; Ge et al., 2000; Losseff et al., 1996; Rudick et al., 1999; Stevenson et al., 1998), particularly in gray matter (Bakshi et al., 2001; Catalaa et al., 1999; Rudick et al., 1999). This gray matter atrophy has been shown to correlate better with permanent disability than does the white matter inflammatory marker of gadolinium-enhancing lesions (Ge et al., 2000; Rudick et al., 1999; Stevenson et al., 1998). Also, abnormalities beyond classic white matter T2 hyperintensities, within "normal-appearing white matter" (NAWM), have been shown using magnetization transfer, spectroscopy, and diffusion-weighted imaging (Catalaa et al., 2000; De Stefano et al., 1999; Filippi et al., 2000a, b; Gasperini et al., 1996; Narayanan et al., 1997; Santos et al., 2002; Tortorella et al., 2000). Furthermore, the degree of change in the NAWM may be a predictor of future clinical progression (Santos et al., 2002). Pathological findings in MS have described cortical lesions that were characterized by transected neurites (both axons and dendrites) and apoptosis with very little T and B cell infiltration (Bo et al., 2003; Peterson et al., 2001). Axonal transection has also been described within white matter lesions, raising the possibility of Wallerian degeneration in white matter tracts.

In light of these observations, there is now a consensus by MS investigators that there is a need to discover novel treatment options that combine neuroprotective properties with anti-inflammatory effects. In this paper, we will outline the scientific basis for sex hormones as putative treatment options in MS as well as other CNS diseases with both an inflammatory and a neurodegenerative component and review potential mechanisms of action.

## Rationale for sex hormones as treatment options for MS

The concept that sex hormones may play a role in MS pathogenesis and disease activity and could, therefore, potentially be used for therapeutic interventions is based on two well-established clinical observations: a higher prevalence of MS in females compared to males and a decrease in disease activity during pregnancy, in particular in the third trimester. In the following text, we will briefly outline the evidence for these two phenomena and their relevance for sex hormone treatments in MS. For a comprehensive overview of this area, we refer the reader to a recent review published elsewhere (Voskuhl, 2009).

### Gender gap

Many autoimmune diseases are more prevalent in women than in men. In MS, there is a female-to-male preponderance approaching 2:1 to 3:1 (Duquette et al., 1992), and recent evidence seems to suggest that the gender gap is widening (Orton et al., 2006). The causes for the gender bias in MS and other autoimmune diseases may include sex-linked genetic factors, sex differences in immune responsiveness, and/or sex steroid effects (Whitacre et al., 1999). Interestingly, a later onset of disease in male patients compared to female patients (Weinshenker, 1994) coincides with a decline in bioavailable testosterone in men (Swerdloff and Wang, 2004). Although only a minority of male patients with MS have demonstrated testosterone levels significantly below the normal range (Foster et al., 2003; Wei and Lightman, 1997), these findings suggest that testosterone may be protective in young men genetically susceptible to MS. There is an ongoing controversy whether established MS progresses at different speeds in men and women. A detailed review of the empirical evidence in this area can be found elsewhere (Voskuhl, 2009). Taken together, the data suggest that men are less likely to develop clinical relapses and enhancing lesions on magnetic resonance imaging (MRI), but it remains unclear if there is a gender difference regarding progression of clinical disease or neurodegeneration on MRI. Generally, this is in line with a beneficial, anti-inflammatory effect of endogenous testosterone in MS.

*The protective effects of pregnancy*

It has been appreciated for decades that symptoms of patients with autoimmune diseases are affected by pregnancy and the postpartum period. MS patients as well as individuals with other inflammatory autoimmune diseases such as rheumatoid arthritis (RA) and psoriasis experience clinical improvement during pregnancy, with a temporary "rebound" exacerbation postpartum (Abramsky, 1994; Birk et al., 1990; Confavreux et al., 1998; Da Silva and Spector, 1992; Damek and Shuster, 1997; Nelson et al., 1992; Runmarker and Andersen, 1995). The most definitive study of the effect of pregnancy on MS came in 1998 by the Pregnancy in Multiple Sclerosis (PRIMS) Group (Confavreux et al., 1998). This study followed 254 women with MS for up to 1 year postdelivery and showed that relapse rates were significantly reduced from 0.7 per woman per year in the year before pregnancy to 0.2 during the third trimester. Rates then increased to 1.2 during the first 3 months postpartum before returning to pre-pregnancy rates. Together these data clearly demonstrated that late pregnancy is associated with a significant reduction in relapses, while there is a rebound increase in relapses postpartum. It is, however, unclear if this effect on relapse rate translates into a beneficial effect on long-term disability. One short-term 2-year follow-up study indicated that there is no "net" effect of a single pregnancy on disability (Vukusic et al., 2004). However, a long-term study in 200 women showed that patients who had at least one pregnancy after onset were wheelchair dependent after 18.6 years, versus 12.5 years for the other women (Verdru et al., 1994), indicating a protective effect of pregnancy on long-term disability accumulation. Thus, there is clear evidence that pregnancy has a potent short-term effect on inflammation and relapse rate, but data regarding long-term effects on disability are inconclusive.

Pregnancy is characterized by an array of biological changes that could mediate both immunomodulatory and neuroprotective effects. First, a pronounced systemic shift from Th1-type cellular immunity toward Th2-type humoral immunity can be observed during pregnancy (Whitacre et al., 1999). This immune shift, rather than a general immune suppression, is beneficial during pregnancy for two reasons: The fetus represents an "allograft" in immunological terms, since it harbors antigens inherited from the father, and the natural immunomodulation is thus important to prevent fetal rejection. However, the developing fetus depends on the mother for the passive transport of antibodies in light of its immature immune system, and this antibody production is supported by a shift toward Th2-type humoral immunity. Second, pregnancy is characterized by the presence of potentially neuroprotective hormones, including estrogens, progesterone, and prolactin. The secretion of these factors are thought to play a crucial role for the CNS neuronal and oligodendroglial cell lineages during development (Craig et al., 2003).

From an evolutionary standpoint, biological changes during pregnancy are generally aimed at protecting the fetus and promoting its development. However, the same mechanisms, that is, suppression of cellular immunity and promotion of neuroprotection, may coincidentally also be highly beneficial for a mother with an autoimmune inflammatory CNS disease. One could therefore consider the advantageous effects in MS a side effect of pregnancy. Importantly, this "side effect" can provide valuable insight into MS pathology as well as highlight new therapeutic avenues.

Numerous factors that have been identified in blood during pregnancy have been shown to be immunomodulatory, including estrogens, cortisol, progesterone, vitamin D, early pregnancy factor (EPF), α-fetoprotein, and others, some of which also have neuroprotective properties. Estriol is one of the major candidates as a therapeutic agent in MS since it has potent effects on both the immune system as well as the CNS and peaks during the last trimester, that is, when the most pronounced decrease in relapse rate occurs.

## Potential mechanisms of sex hormones

### Immunomodulatory properties of sex hormones

*Testosterone*

The protective role of testosterone in autoimmunity in vivo has been demonstrated by the deleterious effect of castration of male animals on disease susceptibility and severity in numerous models of autoimmune diseases, including experimental autoimmune encephalomyelitis (EAE), diabetes in nonobese mice, thyroiditis, and adjuvant arthritis (Ahmed and Penhale, 1982; Bebo et al., 1998; Fitzpatrick et al., 1991; Fox, 1992; Harbuz et al., 1995; Smith et al., 1999). Conversely, testosterone treatment of females can ameliorate a variety of autoimmune disease models (Dalal et al., 1997; Fox, 1992; Sato et al., 1992).

In vitro, naive T cells stimulated with CNS autoantigens in the presence of testosterone produce higher levels of interleukin 5 (IL-5) and IL-10 but decreased levels of interferon γ (IFNγ) (Bebo et al., 1999), indicating a Th2-like shift. Similar changes were seen after in vivo treatment of EAE mice with testosterone (Dalal et al., 1997). Studies have also shown that testosterone can reduce the in vitro production of inflammatory cytokines such as tumor necrosis factor α (TNFα) and IL-1β by human macrophages (D'Agostino et al., 1999) and monocytes (Li et al., 1993; Liva and Voskuhl, 2001). These studies further support the hypothesis that testosterone treatment may induce an immune shift in vivo and exert beneficial effects in Th1-mediated autoimmune diseases.

*Estrogen*

It has been previously shown by numerous laboratories that the clinical severity of both active and adoptive EAE is reduced by estrogen (estriol or 17β-estradiol) treatment in several strains of mice (SJL, C57BL/6, B10.PL, B10.RIII) (Bebo et al., 2001; Ito et al., 2001; Jansson et al., 1994; Kim et al., 1999; Liu et al., 2002, 2003; Matejuk et al., 2001; Polanczyk et al., 2003; Subramanian et al., 2003). Estriol treatment has also been shown to be effective in EAE when administered after disease onset (Kim et al., 1999).

Protective mechanisms of estrogen treatment (both estriol and estradiol) in EAE clearly involve anti-inflammatory processes. Estrogen treatment has been shown to affect cytokines, chemokines, matrix metalloproteinase-9 (MMP-9), antigen presentation, and dendritic cell function (Bebo et al., 2001; Ito et al., 2001; Liu et al., 2003; Matejuk et al., 2001; Palaszynski et al., 2004; Subramanian et al., 2003). Estrogen treatment has also recently been shown to induce CD4+CD25+ regulatory T cells in EAE (Matejuk et al., 2004; Polanczyk et al., 2004). The anti-inflammatory effects of estrogen treatment on pathological mechanisms in EAE and MS are illustrated in Fig. 1.

Estrogens regulate gene transcription by nuclear estrogen receptors (ERs), and the two ERs, ERα and ERβ, exhibit distinct transcriptional properties. Plasma membrane-associated and cytoplasmic ERs as well as the recently identified G-protein coupled high affinity ER GPR30 can mediate rapid, nongenomic signaling. Although both ERα and ERβ are expressed in the immune system and the CNS, studies using ERα signaling-deficient mouse strains have shown that clinical protection from EAE by estradiol (Polanczyk et al., 2003) and estriol (Liu et al., 2003) depends on signaling through ERα. Correspondingly, anti-inflammatory mechanisms of estrogens have been found to be mediated by ERα: selective ERα ligand treatment was sufficient to ameliorate EAE, induced

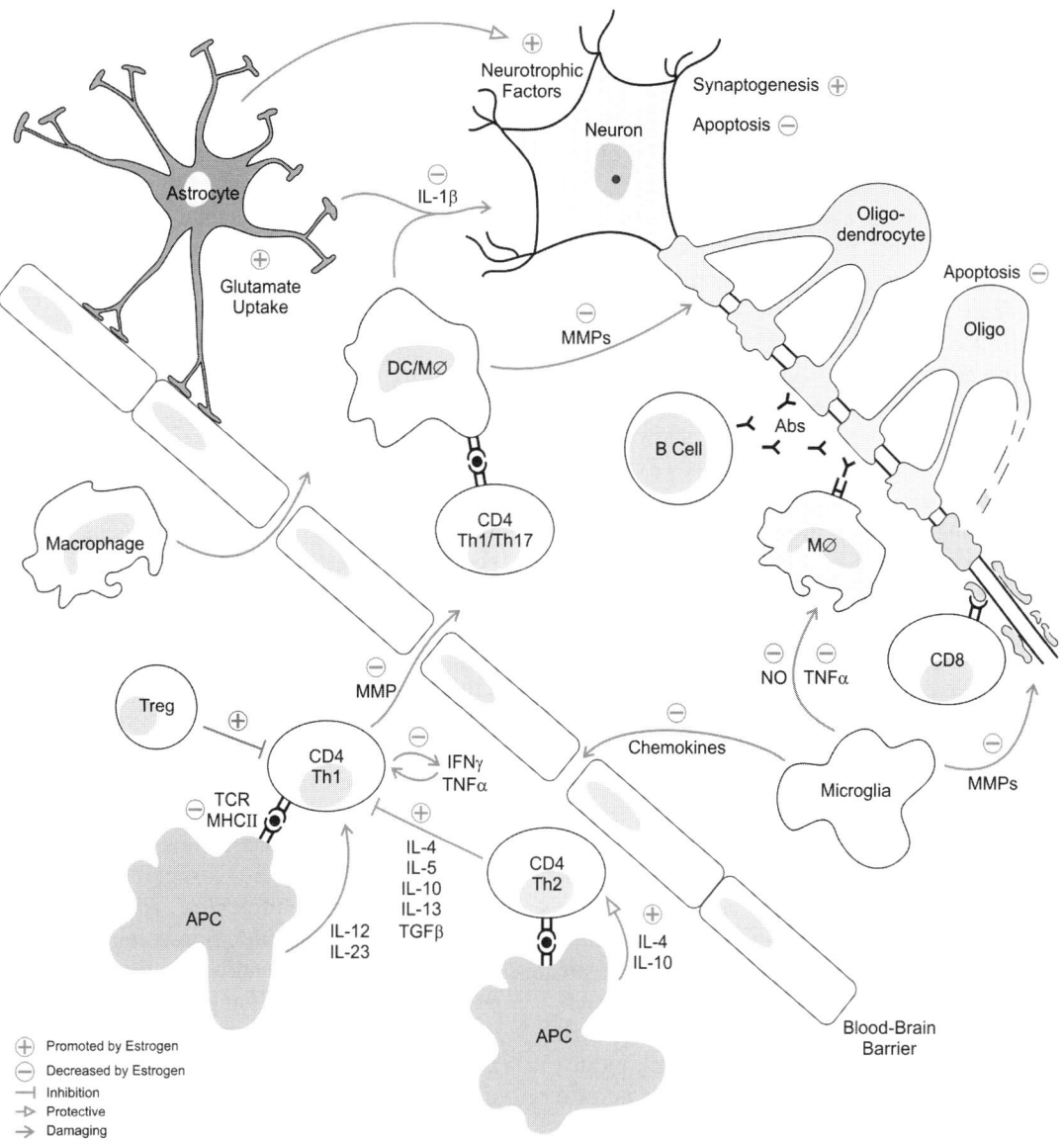

Fig. 1. Mechanisms of estrogen treatment in multiple sclerosis and experimental autoimmune encephalomyelitis. In the peripheral immune system, estrogens exert immunomodulatory effects by inhibiting antigen presentation by APCs and shifting immune responses from a Th1 toward a Th2 pattern. This process is further promoted by enhancing regulatory T cell function. The migratory capacity of peripheral immune cells is inhibited by downregulating MMP production, in particular MMP-9, and reducing chemokines. In addition, estrogens limit immune-mediated damage by reducing cytokines, MMP, and NO production in the CNS. The neuroprotective effects of estrogens include protection of neurons and oligodendrocytes from apoptosis, which may be linked to the ability of estrogens to decrease excitotoxicity by increasing glutamate uptake of astrocytes. Estrogens also induce neurotrophic factors and promote synaptogenesis. APC: antigen-presenting cell; TCR: T cell receptor; MHC: major histocompatibility complex; NO: nitric oxide; MMP: matrix metalloproteinase; MØ: macrophage; DC: dendritic cell.

favorable changes in autoantigen-specific cytokine production in the peripheral immune system (decreased TNFα, IFNγ, and IL-6, with increased IL-5), and decreased CNS white matter inflammation and demyelination in EAE (Morales et al., 2006). Selective ERα ligand treatment also decreased CNS infiltration in EAE, whereas a selective ERβ ligand had no effect on peripheral cytokine production or CNS infiltration (Tiwari-Woodruff et al., 2007). Estriol treatment effects on MMP-9 bioactivity and CNS infiltration by T cells and monocytes in EAE were also mediated via ERα (Gold et al., 2008b). In addition to these peripheral effects, it has been shown that ERα-mediated regulation of resident CNS cells, including microglia, is important for amelioration of EAE using a bone marrow chimera model (Garidou et al., 2004). Overall, these results suggest that the anti-inflammatory effect of estrogens is mediated by ERα.

## Neuroprotective properties of sex hormones

### Testosterone

Recently, studies on a possible neuroprotective effect of testosterone have begun to accumulate. Testosterone in its free form can cross the blood-brain barrier (Iqbal et al., 1983) and thus directly influence neuronal cells. It has been shown to protect spinal cord neurons in culture from glutamate toxicity (Ogata et al., 1993). Testosterone as well as dehydrotestosterone (DHT), which cannot be converted to estrogen, can induce neuronal differentiation and increases in neurite outgrowth in cultured neuronal cells (Lustig, 1994). In addition, testosterone has been shown to protect from oxidative stress in neuronal cell lines (Chisu et al., 2006a, b). Also, both testosterone and DHT protected cultured neurons against β-amyloid toxicity-induced cell death, and this protective effect of testosterone was not blocked by droloxifene, an ER antagonist (Pike, 2001). This indicates that at least some neuroprotective effects of testosterone are not dependent upon conversion to estrogen. While numerous mechanisms of testosterone-mediated neuroprotection may exist, it is possible that some are mediated through an increase in the expression of neurotrophic factors such as brain-derived neurotrophic factor (BDNF). Increased survival of neurons during testosterone treatment in the adult avian brain was shown to be abrogated when BDNF was blocked (Rasika et al., 1999). A recent article reviews the neuroprotective effects of testosterone, in vitro as well as in vivo, in animal models (Bialek et al., 2004).

### Estrogen

Numerous reviews have described estrogen's neuroprotective effects, both in vitro and in vivo (Garcia-Segura et al., 2001; Sribnick et al., 2003; Wise et al., 2001). In vitro, estrogens have been shown to protect neurons in a variety of models of neurodegeneration, including those induced by excitotoxicity and oxidative stress (Behl et al., 1995, 1997; Goodman et al., 1996; Harms et al., 2001). Treatment with estrogen decreased glutamate-induced apoptosis and preserved electrophysiological function in neurons (Sribnick et al., 2004; Zhao et al., 2004). Estrogens may also protect neurons from excitotoxicity by increasing glutamate uptake by astrocytes (Liang et al., 2002; Pawlak et al., 2005). Estrogen treatment also protected oligodendrocytes from cytotoxicity (Cantarella et al., 2004; Sur et al., 2003; Takao et al., 2004) as well as accelerated oligodendrocyte process formation (Zhang et al., 2004). In vivo studies have shown that estrogen treatment can be neuroprotective in animal models of Parkinson's disease, cerebellar ataxia, late-onset leukodystrophy, stroke, and spinal cord injury, often by reducing apoptosis (Dubal et al., 2001; Jover et al., 2002; Leranth et al., 2000; Matsuda et al., 2001; Rau et al., 2003; Sierra et al., 2003; Yune et al., 2004). Estrogens have also been shown in vitro and in vivo to increase dendritic spine formation and synapses on CA1 pyramidal cells of the hippocampus in healthy rats, resulting in improved spatial working memory (Murphy et al., 1998; Rudick and Woolley, 2001; Sandstrom and Williams, 2001; Yankova et al., 2001). The neuroprotective effects of estrogens relevant to EAE and MS are illustrated in Fig. 1.

As described previously, the anti-inflammatory effects of estrogens in EAE are mediated via ERα. Since anti-inflammatory and neuroprotective effects are not mutually exclusive, it remains possible that some neuroprotective mechanisms may also be mediated through ERα. However, it is difficult to prove direct neuroprotection by ERα ligand treatment in EAE in a setting of such profound anti-inflammatory effects. In contrast, recent data suggest that the ERβ pathway mediates neuroprotective effects in EAE in the absence of an anti-inflammatory effect (Tiwari-Woodruff et al., 2007). In our study, ERα ligand treatment abrogated EAE at the onset and throughout the disease course. In contrast, ERβ ligand treatment had no effect at disease onset but promoted recovery during the chronic phase of the disease and was not anti-inflammatory in the peripheral immune system. Also, ERα ligand treatment reduced CNS inflammation, whereas ERβ ligand treatment did not. Interestingly, treatment with either the ERα or the ERβ ligand was neuroprotective, as evidenced by reduced demyelination and preservation of axon numbers in white matter, as well as decreased neuronal abnormalities in gray matter. This is in line with other recent studies using transgenic mice (Rissman et al., 2002) and selective ERβ agonists (Rhodes and Frye, 2006) that indicate that the beneficial effects of estrogen on cognitive function are dependent on the ERβ pathway. Selective ERβ agonist effects on cognition have been linked to increased dendritic branching and upregulation of key synaptic proteins, including PSD-95, synaptophysin, and AMPA-receptor subunit GluR1, in the hippocampus (Liu et al., 2008).

**Sex hormone treatments in MS**

*Testosterone*

In a pilot clinical trial, 10 male MS patients were treated with 10 g of gel containing 100 mg of testosterone in a crossover design (6-month observation period followed by 12 months of treatment) (Sicotte et al., 2007). Clinical measures of disability and cognition (the Multiple Sclerosis Functional Composite and the 7/24 Spatial Recall Test) were obtained every 3 months. In addition, monthly MRI measures of enhancing lesion activity and whole brain volumes were acquired. In addition, blood was drawn every 3 months during the entire study period for immunological analysis.

Treatment with testosterone gel was well tolerated and associated with improvement in cognitive performance as measured by the paced auditory serial-addition task (PASAT), a test of processing speed and attention, widely used in MS. In addition, treatment was associated with a slowing down of brain atrophy as measured by MRI. There was no significant effect of testosterone treatment on gadolinium-enhancing lesions (Sicotte et al., 2007). Testosterone treatment also significantly reduced delayed-type hypersensitivity (DTH) skin recall responses, a functional in vivo measure of inflammatory immune responses, and induced a shift in peripheral lymphocyte composition by decreasing CD4+ T cells and increasing NK cells (Gold et al., 2008a). In addition, peripheral blood mononuclear cell (PBMC) production of IL-2 was significantly decreased while transforming growth factor β1 (TGFβ1) production was increased. Furthermore, PBMCs obtained during the treatment period produced significantly more BDNF and platelet-derived growth factor (PDGF-BB). The concentrations of BDNF and PDGF-BB in PBMC cultures were in the biologically active range, as shown by their ability to reduce glutamate-induced neuronal cell death in vitro. These results are consistent with an immunomodulatory as well as a potentially neuroprotective effect of testosterone treatment in MS.

*Estriol*

Estriol was administered in a pilot clinical trial to women with MS in an attempt to recapitulate the protective effect of pregnancy on disease (Sicotte et al., 2002). A crossover study was used whereby patients were followed for 6 months pretreatment to establish baseline disease activity, which included cerebral MRI every month and neurological examination every 3 months. The patients were then treated with oral estriol (8 mg/day) for 6 months, and observed for 6 more months in the posttreatment period followed by another 4-month re-treatment period. Six relapsing-remitting MS (RRMS) patients and four secondary-progressive MS (SPMS) patients finished the entire 22-month study period.

As compared with pretreatment baseline, relapsing-remitting patients treated with oral estriol (8 mg/day) demonstrated significant decreases in DTH responses. Treatment also decreased gadolinium-enhancing lesion numbers and volumes on MRI. When estriol treatment was stopped, enhancing lesions increased to pretreatment levels. When estriol treatment was reinstituted, enhancing lesions again were significantly decreased. This improvement in the group as a whole was driven by the beneficial effect of estriol treatment in the RRMS, not the SPMS, group. Interestingly, estriol treatment also significantly increased cognitive function as measured by the PASAT in the RRMS group but not in the SPMS group.

Immunological studies (Soldan et al., 2003) revealed that oral estriol treatment was associated with significant decreases in CD4+ and CD8+ T cells and an increase in CD19+ B cells, with no changes in CD64+ monocytes/macrophages. Significant decreases in CD4+CD45Ro+ (memory T cells) and increases in CD4+CD45Ra+ (naive T cells) were also observed. Significantly increased levels of IL-5 and IL-10 and decreased TNFα were observed in stimulated PBMCs isolated during estriol treatment. These changes in cytokines correlated with reductions of enhancing lesions on MRI in RRMS. Further studies were conducted in a subgroup of three of the RRMS patients in this study. Here, supernatants from stimulated PBMCs obtained during treatment showed decreased levels and bioactivity of MMP-9 (Gold et al., 2008b).

**Conclusions and future directions**

A large body of evidence supports the therapeutic potential of testosterone and estrogens in animal models of MS. Mechanisms of action include both immunomodulatory and neuroprotective pathways, thus suggesting that sex hormones represent novel treatment options that could beneficially affect the inflammatory as well as the neurodegenerative component of the disease. We now also have first clinical evidence for the effectiveness of testosterone and estriol in MS from two completed pilot studies. As a result, a phase II trial is underway for oral estriol treatment in female patients with RRMS. Both testosterone and estriol have a favorable safety profile in men and women, respectively. Both hormones also have an advantageous route of administration compared to available treatments in MS since testosterone can be applied transdermally and estriol may be taken orally. Thus, these treatments, tailored to each gender, represent an attractive alternative to currently approved therapeutic agents such as IFNβ and glatiramer acetate, which are each taken by injection only.

More research is needed to understand the pathways and mechanisms underlying the beneficial effects of sex hormones on MS pathology. For estrogens, there is accumulating evidence that the anti-inflammatory and neuroprotective effects are selectively mediated via ERα and ERβ pathways. One must consider the risk–benefit ratio of any estrogen treatment when considering its use in MS. The goal is to optimize efficacy and minimize toxicity. Hence, determining which ER mediates the neuroprotective effect of estrogen treatment is of central importance. The reviewed data demonstrating that treatment with an ERβ ligand is neuroprotective are of clinical relevance, because both breast and uterine endometrial cancers are mediated through ERα, not ERβ. Thus, treatment could be tailored to minimize the risk–benefit ratio for individual patients. If certain conditions such as a known risk for breast or uterine cancer prohibit the use of estriol, the patient may benefit from a standard anti-inflammatory treatment in combination with ERβ ligand treatment. This way, the neuroprotective properties of estrogen treatment could be maintained while avoiding the increased risk of cancer in the breast and the uterus.

Comparatively little is known about the anti-inflammatory and neuroprotective mechanisms of testosterone. Testosterone is converted to estrogen in the brain by aromatase, and the neuroprotective properties of testosterone treatment in vivo may be due at least in part to this conversion. However, some studies using the nonconvertible DHT have also shown testosterone can be directly beneficial.

Testosterone therapy has potentially harmful side effects, as it may worsen preexisting prostate cancer in some men. Testing of prostate-specific antigen levels is recommended before and during testosterone therapy. However, testosterone replacement is widely used in aging and hypogonadal men, and there is no clear evidence that higher levels of circulating testosterone, within the physiological range, are linked to an increased risk of prostate cancer.

In this review, we have focused on hormonal influences on MS. The gender gap in MS, however, may be due to the effects of sex hormones, genetic differences, or a combination of the two. A nonmutually exclusive, alternative hypothesis includes a direct genetic effect on the immune system and/or the CNS. That is, specific gene products, which are not induced by gonadal hormones, yet are expressed in a sexually dimorphic manner, could induce gender differences in MS pathogenesis and progression. In human studies, these factors cannot be dissected since men and women differ with regard to both sex chromosomes as well as sex hormones. However, there are now sophisticated transgenic mouse models available that allow the examination of the effects of sex hormones versus sex chromosomes independently. Recently, our laboratory has employed this model to examine the contribution of gonadal gene complement on immune responses (Palaszynski et al., 2005) as well as susceptibility to autoimmune disease (Smith-Bouvier et al., 2008). Findings suggest that the XX sex chromosome complement, as compared to XY complement, can indeed promote autoimmunity. Taken together, one must consider the contribution of both sex hormones and sex chromosomes in complex autoimmune diseases such as MS.

## Abbreviations

| | |
|---|---|
| BDNF | brain-derived neurotrophic factor |
| CNS | central nervous system |
| DHT | dehydrotestosterone |
| DTH | delayed-type hypersensitivity |
| EAE | experimental autoimmune encephalomyelitis |
| ER | estrogen receptor |
| IL | interleukin |
| IFN | interferon |
| MMP | matrix metalloproteinase |
| MRI | magnetic resonance imaging |
| MS | multiple sclerosis |
| NAWM | normal-appearing white matter |
| PASAT | paced auditory serial-addition task |
| PBMC | peripheral blood mononuclear cell |
| PDGF | platelet-derived growth factor |
| RA | rheumatoid arthritis |
| RRMS | relapsing-remitting MS |
| SPMS | secondary-progressive MS |
| TGF | transforming growth factor |
| TNF | tumor necrosis factor |

## Acknowledgments

The authors would like to thank Donna Crandall for producing the artwork for Fig. 1. We would also like to thank all members of the UCLA MS program for their conceptual contributions to the model in Fig. 1.

## References

Abramsky, O. (1994). Pregnancy and multiple sclerosis. *Annals of Neurology*, *36*(Suppl.), S38–S41.

Ahmed, S. A., & Penhale, W. J. (1982). The influence of testosterone on the development of autoimmune thyroiditis in thymectomized and irradiated rats. *Clinical and Experimental Immunology*, *48*, 367–374.

Bakshi, R., Benedict, R. H., Bermel, R. A., & Jacobs, L. (2001). Regional brain atrophy is associated with physical disability in multiple sclerosis: semiquantitative magnetic resonance imaging and relationship to clinical findings. *Journal of Neuroimaging*, *11*, 129–136.

Bebo, B. F., Jr., Fyfe-Johnson, A., Adlard, K., Beam, A. G., Vandenbark, A. A., & Offner, H. (2001). Low-dose estrogen therapy ameliorates experimental autoimmune encephalomyelitis in two different inbred mouse strains. *Journal of Immunology, 166*, 2080–2089.

Bebo, B. F., Jr., Schuster, J. C., Vandenbark, A. A., & Offner, H. (1999). Androgens alter the cytokine profile and reduce encephalitogenicity of myelin-reactive T cells. *Journal of Immunology, 162*, 35–40.

Bebo, B. F., Jr., Zelinka-Vincent, E., Adamus, G., Amundson, D., Vandenbark, A. A., & Offner, H. (1998). Gonadal hormones influence the immune response to PLP 139-151 and the clinical course of relapsing experimental autoimmune encephalomyelitis. *Journal of Neuroimmunology, 84*, 122–130.

Behl, C., Skutella, T., Lezoualc'h, F., Post, A., Widmann, M., Newton, C. J., et al. (1997). Neuroprotection against oxidative stress by estrogens: structure-activity relationship. *Molecular Pharmacology, 51*, 535–541.

Behl, C., Widmann, M., Trapp, T., & Holsboer, F. (1995). 17-beta estradiol protects neurons from oxidative stress-induced cell death in vitro. *Biochemical and Biophysical Research Communications, 216*, 473–482.

Bialek, M., Zaremba, P., Borowicz, K. K., & Czuczwar, S. J. (2004). Neuroprotective role of testosterone in the nervous system. *Polish Journal of Pharmacology, 56*, 509–518.

Birk, K., Ford, C., Smeltzer, S., Ryan, D., Miller, R., & Rudick, R. A. (1990). The clinical course of multiple sclerosis during pregnancy and the puerperium. *Archives of Neurology, 47*, 738–742.

Bo, L., Vedeler, C. A., Nyland, H., Trapp, B. D., & Mork, S. J. (2003). Intracortical multiple sclerosis lesions are not associated with increased lymphocyte infiltration. *Multiple Sclerosis, 9*, 323–331.

Brex, P. A., Jenkins, R., Fox, N. C., Crum, W. R., O'Riordan, J. I., Plant, G. T., et al. (2000). Detection of ventricular enlargement in patients at the earliest clinical stage of MS. *Neurology, 54*, 1689–1691.

Cantarella, G., Risuglia, N., Lombardo, G., Lempereur, L., Nicoletti, F., Memo, M., et al. (2004). Protective effects of estradiol on TRAIL-induced apoptosis in a human oligodendrocytic cell line: evidence for multiple sites of interactions. *Cell Death and Differentiation, 11*, 503–511.

Catalaa, I., Fulton, J. C., Zhang, X., Udupa, J. K., Kolson, D., Grossman, M., Wei, L., et al. (1999). MR imaging quantitation of gray matter involvement in multiple sclerosis and its correlation with disability measures and neurocognitive testing. *American Journal of Neuroradiology, 20*, 1613–1618.

Catalaa, I., Grossman, R. I., Kolson, D. L., Udupa, J. K., Nyul, L. G., Wei, L., et al. (2000). Multiple sclerosis: magnetization transfer histogram analysis of segmented normal-appearing white matter. *Radiology, 216*, 351–355.

Chisu, V., Manca, P., Lepore, G., Gadau, S., Zedda, M., & Farina, V. (2006a). Testosterone induces neuroprotection from oxidative stress. Effects on catalase activity and 3-nitro-L-tyrosine incorporation into alpha-tubulin in a mouse neuroblastoma cell line. *Archives Italiennes de Biologie, 144*, 63–73.

Chisu, V., Manca, P., Zedda, M., Lepore, G., Gadau, S., & Farina, V. (2006b). Effects of testosterone on differentiation and oxidative stress resistance in C1300 neuroblastoma cells. *Neuro Endocrinology Letters, 27*, 807–812.

Coles, A. J., Cox, A., Le Page, E., Jones, J., Trip, S. A., Deans, J., et al. (2006). The window of therapeutic opportunity in multiple sclerosis: evidence from monoclonal antibody therapy. *The Journal of Neurology, 253*, 98–108.

Confavreux, C., Hutchinson, M., Hours, M. M., Cortinovis-Tourniaire, P., & Moreau, T. (1998). Rate of pregnancy-related relapse in multiple sclerosis. Pregnancy in Multiple Sclerosis Group. *The New England Journal of Medicine, 339*, 285–291.

Craig, A., Ling Luo, N., Beardsley, D. J., Wingate-Pearse, N., Walker, D. W., Hohimer, A. R., et al. (2003). Quantitative analysis of perinatal rodent oligodendrocyte lineage progression and its correlation with human. *Experimental Neurology, 181*, 231–240.

D'Agostino, P., Milano, S., Barbera, C., Di Bella, G., La Rosa, M., Ferlazzo, V., et al. (1999). Sex hormones modulate inflammatory mediators produced by macrophages. *Annals of the New York Academy of Sciences, 876*, 426–429.

Da Silva, J. A., & Spector, T. D. (1992). The role of pregnancy in the course and aetiology of rheumatoid arthritis. *Clinical Rheumatology, 11*, 189–194.

Dalal, M., Kim, S., & Voskuhl, R. R. (1997). Testosterone therapy ameliorates experimental autoimmune encephalomyelitis and induces a T helper 2 bias in the autoantigen-specific
T lymphocyte response. *Journal of Immunology, 159*, 3–6.

Damek, D. M., & Shuster, E. A. (1997). Pregnancy and multiple sclerosis. *Mayo Clinic Proceedings, 72*, 977–989.

De Stefano, N., Narayanan, S., Matthews, P. M., Francis, G. S., Antel, J. P., & Arnold, D. L. (1999). In vivo evidence for axonal dysfunction remote from focal cerebral demyelination of the type seen in multiple sclerosis. *Brain, 122*, 1933–1939.

Dubal, D. B., Zhu, H., Yu, J., Rau, S. W., Shughrue, P. J., Merchenthaler, I., et al. (2001). Estrogen receptor alpha, not beta, is a critical link in estradiol-mediated protection against brain injury. *Proceedings of the National Academy of Sciences of the United States of America, 98*, 1952–1957.

Duquette, P., Pleines, J., Girard, M., Charest, L., Senecal-Quevillon, M., & Masse, C. (1992). The increased susceptibility of women to multiple sclerosis. *The Canadian Journal of Neurological Sciences, 19*, 466–471.

Evangelou, N., Esiri, M. M., Smith, S., Palace, J., & Matthews, P. M. (2000). Quantitative pathological evidence for axonal loss in normal appearing white matter in multiple sclerosis. *Annals of Neurology, 47*, 391–395.

Filippi, M., Bozzali, M., Rovaris, M., Gonen, O., Kesavadas, C., Ghezzi, A., et al. (2003). Evidence for widespread axonal damage at the earliest clinical stage of multiple sclerosis. *Brain, 126*, 433–437.

Filippi, M., Iannucci, G., Cercignani, M., Assunta, R. M., Pratesi, A., & Comi, G. (2000a). A quantitative study of water diffusion in multiple sclerosis lesions and normal-appearing white matter using echo-planar imaging. *Archives of Neurology, 57*, 1017–1021.

Filippi, M., Inglese, M., Rovaris, M., Sormani, M. P., Horsfield, P., Iannucci, P. G., et al. (2000b). Magnetization transfer imaging to monitor the evolution of MS: a 1-year follow-up study [In Process Citation]. *Neurology, 55,* 940–946.

Fitzpatrick, F., Lepault, F., Homo-Delarche, F., Bach, J. F., & Dardenne, M. (1991). Influence of castration, alone or combined with thymectomy, on the development of diabetes in the nonobese diabetic mouse. *Endocrinology, 129,* 1382–1390.

Foster, S. C., Daniels, C., Bourdette, D. N., & Bebo, B. F., Jr. (2003). Dysregulation of the hypothalamic-pituitary-gonadal axis in experimental autoimmune encephalomyelitis and multiple sclerosis. *Journal of Neuroimmunology, 140,* 78–87.

Fox, H. S. (1992). Androgen treatment prevents diabetes in nonobese diabetic mice. *The Journal of Experimental Medicine, 175,* 1409–1412.

Garcia-Segura, L. M., Azcoitia, I., & DonCarlos, L. L. (2001). Neuroprotection by estradiol. *Progress in Neurobiology, 63,* 29–60.

Garidou, L., Laffont, S., Douin-Echinard, V., Coureau, C., Krust, A., Chambon, P., et al. (2004). Estrogen receptor alpha signaling in inflammatory leukocytes is dispensable for 17beta-estradiol-mediated inhibition of experimental autoimmune encephalomyelitis. *Journal of Immunology, 173,* 2435–2442.

Gasperini, C., Horsfield, M. A., Thorpe, J. W., Kidd, D., Barker, G. J., Tofts, P. S., et al. (1996). Macroscopic and microscopic assessments of disease burden by MRI in multiple sclerosis: relationship to clinical parameters. *Journal of Magnetic Resonance Imaging, 6,* 580–584.

Ge, Y., Grossman, R. I., Udupa, J. K., Wei, L., Mannon, L. J., Polansky, M., et al. (2000). Brain atrophy in relapsing-remitting multiple sclerosis and secondary progressive multiple sclerosis: longitudinal quantitative analysis. *Radiology, 214,* 665–670.

Gold, S. M., Chalifoux, S., Giesser, B. S., & Voskuhl, R. R. (2008a). Immune modulation and increased neurotrophic factor production in multiple sclerosis patients treated with testosterone. *Journal of Neuroinflammation, 5,* 32.

Gold, S. M., Manda, S. V., Morales, L. B., Sicotte, N. L., & Voskuhl, R. R. (2008b). Estriol treatment reduces matrix metalloprotease-9 activity in multiple sclerosis and experimental autoimmune encephalomyelitis. *Multiple Sclerosis, 14,* S29.

Goodman, Y., Bruce, A. J., Cheng, B., & Mattson, M. P. (1996). Estrogens attenuate and corticosterone exacerbates excitotoxicity, oxidative injury, and amyloid beta-peptide toxicity in hippocampal neurons. *Journal of Neurochemistry, 66,* 1836–1844.

Harbuz, M. S., Perveen-Gill, Z., Lightman, S. L., & Jessop, D. S. (1995). A protective role for testosterone in adjuvant-induced arthritis. *British Journal of Rheumatology, 34,* 1117–1122.

Harms, C., Lautenschlager, M., Bergk, A., Katchanov, J., Freyer, D., Kapinya, K., et al. (2001). Differential mechanisms of neuroprotection by 17 beta-estradiol in apoptotic versus necrotic neurodegeneration. *The Journal of Neuroscience, 21,* 2600–2609.

Hemmer, B., Archelos, J. J., & Hartung, H. P. (2002). New concepts in the immunopathogenesis of multiple sclerosis. *Nature Reviews. Neuroscience, 3,* 291–301.

Iqbal, M. J., Dalton, M., & Sawers, R. S. (1983). Binding of testosterone and oestradiol to sex hormone binding globulin, human serum albumin and other plasma proteins: evidence for non-specific binding of oestradiol to sex hormone binding globulin. *Clinical Science (London), 64,* 307–314.

Ito, A., Bebo, B. F., Jr., Matejuk, A., Zamora, A., Silverman, M., Fyfe-Johnson, A., et al. (2001). Estrogen treatment down-regulates TNF-alpha production and reduces the severity of experimental autoimmune encephalomyelitis in cytokine knockout mice. *Journal of Immunology, 167,* 542–552.

Jansson, L., Olsson, T., & Holmdahl, R. (1994). Estrogen induces a potent suppression of experimental autoimmune encephalomyelitis and collagen-induced arthritis in mice. *Journal of Neuroimmunology, 53,* 203–207.

Jover, T., Tanaka, H., Calderone, A., Oguro, K., Bennett, M. V., Etgen, A. M., et al. (2002). Estrogen protects against global ischemia-induced neuronal death and prevents activation of apoptotic signaling cascades in the hippocampal CA1. *The Journal of Neuroscience, 22,* 2115–2124.

Kim, S., Liva, S. M., Dalal, M. A., Verity, M. A., & Voskuhl, R. R. (1999). Estriol ameliorates autoimmune demyelinating disease: implications for multiple sclerosis. *Neurology, 52,* 1230–1238.

Lassmann, H., Bruck, W., & Lucchinetti, C. (2001). Heterogeneity of multiple sclerosis pathogenesis: implications for diagnosis and therapy. *Trends in Molecular Medicine, 7,* 115–121.

Leranth, C., Roth, R. H., Elsworth, J. D., Naftolin, F., Horvath, T. L., & Redmond, D. E., Jr. (2000). Estrogen is essential for maintaining nigrostriatal dopamine neurons in primates: implications for Parkinson's disease and memory. *The Journal of Neuroscience, 20,* 8604–8609.

Li, Z. G., Danis, V. A., & Brooks, P. M. (1993). Effect of gonadal steroids on the production of IL-1 and IL-6 by blood mononuclear cells in vitro. *Clinical and Experimental Rheumatology, 11,* 157–162.

Liang, Z., Valla, J., Sefidvash-Hockley, S., Rogers, J., & Li, R. (2002). Effects of estrogen treatment on glutamate uptake in cultured human astrocytes derived from cortex of Alzheimer's disease patients. *Journal of Neurochemistry, 80,* 807–814.

Liu, F., Day, M., Muniz, L. C., Bitran, D., Arias, R., Revilla-Sanchez, R., et al. (2008). Activation of estrogen receptor-beta regulates hippocampal synaptic plasticity and improves memory. *Nature Neuroscience, 11,* 334–343.

Liu, H. B., Loo, K. K., Palaszynski, K., Ashouri, J., Lubahn, D. B., & Voskuhl, R. R. (2003). Estrogen receptor alpha mediates estrogen's immune protection in autoimmune disease. *Journal of Immunology, 171,* 6936–6940.

Liu, H. Y., Buenafe, A. C., Matejuk, A., Ito, A., Zamora, A., Dwyer, J., et al. (2002). Estrogen inhibition of EAE involves effects on dendritic cell function. *The Journal of Neuroscience Research, 70,* 238–248.

Liva, S. M., & Voskuhl, R. R. (2001). Testosterone acts directly on CD4+ T lymphocytes to increase IL-10 production. *Journal of Immunology, 167*, 2060–2067.

Losseff, N. A., Wang, L., Lai, H. M., Yoo, D. S., Gawne, C. M., McDonald, W. I., et al. (1996). Progressive cerebral atrophy in multiple sclerosis. A serial MRI study. *Brain, 119*, 2009–2019.

Lustig, R. H. (1994). Sex hormone modulation of neural development in vitro. *Hormones and Behavior, 28*, 383–395.

Matejuk, A., Adlard, K., Zamora, A., Silverman, M., Vandenbark, A. A., & Offner, H. (2001). 17beta-estradiol inhibits cytokine, chemokine, and chemokine receptor mRNA expression in the central nervous system of female mice with experimental autoimmune encephalomyelitis. *The Journal of Neuroscience Research, 65*, 529–542.

Matejuk, A., Bakke, A. C., Hopke, C., Dwyer, J., Vandenbark, A. A., & Offner, H. (2004). Estrogen treatment induces a novel population of regulatory cells, which suppresses experimental autoimmune encephalomyelitis. *The Journal of Neuroscience Research, 77*, 119–126.

Matsuda, J., Vanier, M. T., Saito, Y., & Suzuki, K. (2001). Dramatic phenotypic improvement during pregnancy in a genetic leukodystrophy: estrogen appears to be a critical factor. *Human Molecular Genetics, 10*, 2709–2715.

McFarland, H. F., & Martin, R. (2007). Multiple sclerosis: a complicated picture of autoimmunity. *Nature Immunology, 8*, 913–919.

Morales, L. B., Loo, K. K., Liu, H. B., Peterson, C., Tiwari-Woodruff, S., & Voskuhl, R. R. (2006). Treatment with an estrogen receptor alpha ligand is neuroprotective in experimental autoimmune encephalomyelitis. *The Journal of Neuroscience, 26*, 6823–6833.

Murphy, D. D., Cole, N. B., Greenberger, V., & Segal, M. (1998). Estradiol increases dendritic spine density by reducing GABA neurotransmission in hippocampal neurons. *The Journal of Neuroscience, 18*, 2550–2559.

Narayanan, S., Fu, L., Pioro, E., De Stefano, N., Collins, D. L., Francis, G. S., et al. (1997). Imaging of axonal damage in multiple sclerosis: spatial distribution of magnetic resonance imaging lesions. *Annals of Neurology, 41*, 385–391.

Nelson, J. L., Hughes, K. A., Smith, A. G., Nisperos, B. B., Branchaud, A. M., & Hansen, J. A. (1992). Remission of rheumatoid arthritis during pregnancy and maternal-fetal class II alloantigen disparity. *American Journal of Reproductive Immunology, 28*, 226–227.

Ogata, T., Nakamura, Y., Tsuji, K., Shibata, T., & Kataoka, K. (1993). Steroid hormones protect spinal cord neurons from glutamate toxicity. *Neuroscience, 55*, 445–449.

Orton, S. M., Herrera, B. M., Yee, I. M., Valdar, W., Ramagopalan, S. V., Sadovnick, A. D., et al. (2006). Sex ratio of multiple sclerosis in Canada: a longitudinal study. *Lancet Neurology, 5*, 932–936.

Palaszynski, K. M., Liu, H., Loo, K. K., & Voskuhl, R. R. (2004). Estriol treatment ameliorates disease in males with experimental autoimmune encephalomyelitis: implications for multiple sclerosis. *Journal of Neuroimmunology, 149*, 84–89.

Palaszynski, K. M., Smith, D. L., Kamrava, S., Burgoyne, P. S., Arnold, A. P., & Voskuhl, R. R. (2005). A Yin-Yang effect between sex chromosome complement and sex hormones on the immune response. *Endocrinology, 146*, 3277–3279.

Pawlak, J., Brito, V., Kuppers, E., & Beyer, C. (2005). Regulation of glutamate transporter GLAST and GLT-1 expression in astrocytes by estrogen. *Brain Research. Molecular Brain Research, 138*, 1–7.

Peterson, J. W., Bo, L., Mork, S., Chang, A., & Trapp, B. D. (2001). Transected neurites, apoptotic neurons, and reduced inflammation in cortical multiple sclerosis lesions. *Annals of Neurology, 50*, 389–400.

Pike, C. J. (2001). Testosterone attenuates beta-amyloid toxicity in cultured hippocampal neurons. *Brain Research, 919*, 160–165.

Polanczyk, M., Zamora, A., Subramanian, S., Matejuk, A., Hess, D. L., Blankenhorn, E. P., et al. (2003). The protective effect of 17beta-estradiol on experimental autoimmune encephalomyelitis is mediated through estrogen receptor-alpha. *The American Journal of Pathology, 163*, 1599–1605.

Polanczyk, M. J., Carson, B. D., Subramanian, S., Afentoulis, M., Vandenbark, A. A., Ziegler, S. F., et al. (2004). Cutting edge: estrogen drives expansion of the CD4+CD25+ regulatory T cell compartment. *Journal of Immunology, 173*, 2227–2230.

Rasika, S., Alvarez-Buylla, A., & Nottebohm, F. (1999). BDNF mediates the effects of testosterone on the survival of new neurons in an adult brain. *Neuron, 22*, 53–62.

Rau, S. W., Dubal, D. B., Bottner, M., Gerhold, L. M., & Wise, P. M. (2003). Estradiol attenuates programmed cell death after stroke-like injury. *The Journal of Neuroscience, 23*, 11420–11426.

Rhodes, M. E., & Frye, C. A. (2006). ERbeta-selective SERMs produce mnemonic-enhancing effects in the inhibitory avoidance and water maze tasks. *Neurobiology of Learning and Memory, 85*, 183–191.

Rissman, E. F., Heck, A. L., Leonard, J. E., Shupnik, M. A., & Gustafsson, J. A. (2002). Disruption of estrogen receptor beta gene impairs spatial learning in female mice. *Proceedings of the National Academy of Sciences of the United States of America, 99*, 3996–4001.

Rudick, C. N., & Woolley, C. S. (2001). Estrogen regulates functional inhibition of hippocampal CA1 pyramidal cells in the adult female rat. *The Journal of Neuroscience, 21*, 6532–6543.

Rudick, R. A., Fisher, E., Lee, J. C., Simon, J., & Jacobs, L. (1999). Use of the brain parenchymal fraction to measure whole brain atrophy in relapsing-remitting MS. Multiple Sclerosis Collaborative Research Group. *Neurology, 53*, 1698–1704.

Runmarker, B., & Andersen, O. (1995). Pregnancy is associated with a lower risk of onset and a better prognosis in multiple sclerosis. *Brain, 118*, 253–261.

Sandstrom, N. J., & Williams, C. L. (2001). Memory retention is modulated by acute estradiol and progesterone replacement. *Behavioral Neuroscience, 115*, 384–393.

Santos, A. C., Narayanan, S., de Stefano, N., Tartaglia, M. C., Francis, S. J., Arnaoutelis, R., et al. (2002). Magnetization transfer can predict clinical evolution in patients with multiple sclerosis. *The Journal of Neurology, 249*, 662–668.

Sato, E. H., Ariga, H., & Sullivan, D. A. (1992). Impact of androgen therapy in Sjogren's syndrome: hormonal influence on lymphocyte populations and Ia expression in lacrimal glands of MRL/Mp-lpr/lpr mice. *Investigative Ophthalmology & Visual Science, 33*, 2537–2545.

Sicotte, N. L., Giesser, B. S., Tandon, V., Klutch, R., Steiner, B., Drain, A. E., et al. (2007). Testosterone treatment in multiple sclerosis: a pilot study. *Archives of Neurology, 64*, 683–688.

Sicotte, N. L., Liva, S. M., Klutch, R., Pfeiffer, P., Bouvier, S., Odesa, S., et al. (2002). Treatment of multiple sclerosis with the pregnancy hormone estriol. *Annals of Neurology, 52*, 421–428.

Sierra, A., Azcoitia, I., & Garcia-Segura, L. (2003). Endogenous estrogen formation is neuroprotective in model of cerebellar ataxia. *Endocrine, 21*, 43–51.

Smith, M. E., Eller, N. L., McFarland, H. F., Racke, M. K., & Raine, C. S. (1999). Age dependence of clinical and pathological manifestations of autoimmune demyelination. Implications for multiple sclerosis. *The American Journal of Pathology, 155*, 1147–1161.

Smith-Bouvier, D. L., Divekar, A. A., Sasidhar, M., Du, S., Tiwari-Woodruff, S. K., King, J. K., et al. (2008). A role for sex chromosome complement in the female bias in autoimmune disease. *The Journal of Experimental Medicine, 205*, 1099–1108.

Soldan, S. S., Alvarez Retuerto, A. I., Sicotte, N. L., & Voskuhl, R. R. (2003). Immune modulation in multiple sclerosis patients treated with the pregnancy hormone estriol. *Journal of Immunology, 171*, 6267–6274.

Sospedra, M., & Martin, R. (2005). Immunology of multiple sclerosis. *Annual Review of Immunology, 23*, 683–747.

Sribnick, E. A., Ray, S. K., Nowak, M. W., Li, L., & Banik, N. L. (2004). 17beta-estradiol attenuates glutamate-induced apoptosis and preserves electrophysiologic function in primary cortical neurons. *The Journal of Neuroscience Research, 76*, 688–696.

Sribnick, E. A., Wingrave, J. M., Matzelle, D. D., Ray, S. K., & Banik, N. L. (2003). Estrogen as a neuroprotective agent in the treatment of spinal cord injury. *Annals of the New York Academy of Sciences, 993*, 125–133. discussion pp.159–160

Stevenson, V. L., Leary, S. M., Losseff, N. A., Parker, G. J., Barker, G. J., Husmani, Y., et al. (1998). Spinal cord atrophy and disability in MS: a longitudinal study. *Neurology, 51*, 234–238.

Subramanian, S., Matejuk, A., Zamora, A., Vandenbark, A. A., & Offner, H. (2003). Oral feeding with ethinyl estradiol suppresses and treats experimental autoimmune encephalomyelitis in SJL mice and inhibits the recruitment of inflammatory cells into the central nervous system. *Journal of Immunology, 170*, 1548–1555.

Sur, P., Sribnick, E. A., Wingrave, J. M., Nowak, M. W., Ray, S. K., & Banik, N. L. (2003). Estrogen attenuates oxidative stress-induced apoptosis in C6 glial cells. *Brain Research, 971*, 178–188.

Swerdloff, R. S., & Wang, C. (2004). Androgens and the ageing male. *Best Practice & Research. Clinical Endocrinology & Metabolism, 18*, 349–362.

Takao, T., Flint, N., Lee, L., Ying, X., Merrill, J., & Chandross, K. J. (2004). 17beta-estradiol protects oligodendrocytes from cytotoxicity induced cell death. *Journal of Neurochemistry, 89*, 660–673.

Tiwari-Woodruff, S., Morales, L. B., Lee, R., & Voskuhl, R. R. (2007). Differential neuroprotective and antiinflammatory effects of estrogen receptor (ER)alpha and ERbeta ligand treatment. *Proceedings of the National Academy of Sciences of the United States of America, 104*, 14813–14818.

Tortorella, C., Viti, B., Bozzali, M., Sormani, M. P., Rizzo, G., Gilardi, M. F., et al. (2000). A magnetization transfer histogram study of normal-appearing brain tissue in MS. *Neurology, 54*, 186–193.

Verdru, P., Theys, P., D'Hooghe, M. B., & Carton, H. (1994). Pregnancy and multiple sclerosis: the influence on long term disability. *Clinical Neurology and Neurosurgery, 96*, 38–41.

Voskuhl, R. R. (2009). Sex differences in autoimmune diseases. In A. Arnold et al. (Eds.), *Hormones, brain and behaviour* (2nd ed.). Academic Press.

Vukusic, S., Hutchinson, M., Hours, M., Moreau, T., Cortinovis-Tourniaire, P., Adeleine, P., et al. (2004). Pregnancy and multiple sclerosis (the PRIMS study): clinical predictors of post-partum relapse. *Brain, 127*, 1353–1360.

Wei, T., & Lightman, S. L. (1997). The neuroendocrine axis in patients with multiple sclerosis. *Brain, 120*(6), 1067–1076.

Weinshenker, B. G. (1994). Natural history of multiple sclerosis. *Annals of Neurology, 36*(Suppl.), S6–S11.

Whitacre, C. C., Reingold, S. C., & O'Looney, P. A. (1999). A gender gap in autoimmunity. *Science, 283*, 1277–1278.

Wise, P. M., Dubal, D. B., Wilson, M. E., Rau, S. W., & Bottner, M. (2001). Minireview: neuroprotective effects of estrogen—new insights into mechanisms of action. *Endocrinology, 142*, 969–973.

Yankova, M., Hart, S. A., & Woolley, C. S. (2001). Estrogen increases synaptic connectivity between single presynaptic inputs and multiple postsynaptic CA1 pyramidal cells: a serial electron-microscopic study. *Proceedings of the National Academy of Sciences of the United States of America, 98*, 3525–3530.

Yune, T. Y., Kim, S. J., Lee, S. M., Lee, Y. K., Oh, Y. J., Kim, Y. C., et al. (2004). Systemic administration of 17beta-estradiol reduces apoptotic cell death and improves functional recovery following traumatic spinal cord injury in rats. *Journal of Neurotrauma, 21*, 293–306.

Zhang, Z., Cerghet, M., Mullins, C., Williamson, M., Bessert, D., & Skoff, R. (2004). Comparison of in vivo and in vitro subcellular localization of estrogen receptors alpha and beta in oligodendrocytes. *Journal of Neurochemistry, 89*, 674–684.

Zhao, L., Wu, T. W., & Brinton, R. D. (2004). Estrogen receptor subtypes alpha and beta contribute to neuroprotection and increased Bcl-2 expression in primary hippocampal neurons. *Brain Research, 1010*, 22–34.

CHAPTER 17

# Treatment of retinal diseases with VEGF antagonists

R.O. Schlingemann* and A.N. Witmer

*Medical Retina Unit and Ocular Angiogenesis Group, Department of Ophthalmology, University of Amsterdam, Academic Medical Centre, Amsterdam, The Netherlands*

**Abstract:** Diabetic retinopathy (DR) and age-related macular degeneration (AMD) are the most prevalent causes of blindness in the Western world. The pathogenesis of neovascularization and vascular leakage, both hallmarks of these diseases, appears to have one common denominator: vascular endothelial growth factor (VEGF). Since the recent introduction of anti-VEGF therapy, intravitreal injections with these agents have become standard care in neovascular AMD, and have been found to be a valuable additional treatment strategy in several other vascular retinal diseases. This review provides an overview of the history of anti-VEGF treatment in the eye, its rationale, its efficacy, and its potential drawbacks.

**Keywords:** vascular endothelial growth factor; angiogenesis; neovascularization; diabetic retinopathy; age-related macular degeneration; therapy

## Introduction

Angiogenesis and vascular leakage in the eye are major pathogenic causes of visual loss and blindness in the Western world, and deeply affect the lives of millions of people. Antagonists of vascular endothelial growth factor-A (VEGF) have created a landslide in the treatment and clinical outcome of ocular conditions associated with these processes.

Here we will review the history of anti-VEGF treatment in the eye, its rationale, its efficacy, and its potential drawbacks.

## Ocular angiogenesis

Angiogenesis (also termed neovascularization) can occur as a final common pathway in the course of a variety of pathological conditions in the eye (D'Amore, 1994; Witmer et al., 2003). Ocular angiogenesis is usually the visible sign of wound-healing-like responses that will ultimately cause fibrosis and scarring. Its detrimental effect on vision has two causes. First, new vessels or hemorrhages can obscure the visual axis in normally avascular transparent ocular tissues such as the cornea and vitreous. Later, the final fibrovascular scar formation causes degeneration of the retina and other ocular tissues.

The two most important forms of ocular angiogenesis are preretinal angiogenesis, originating from the retinal vasculature, and subretinal (or choroidal) neovascularization. These two forms of

*Corresponding author.
Tel.: +31205663682; Fax: +31205669048;
E-mail: r.schlingemann@amc.uva.nl

neovascularization mirror the vascular anatomy of the normal retina, which has a dual vascular supply. A delicate network of retinal vessels feeds the inner layers of the retina. Their endothelial cells have tight junctions and form the inner blood–retinal barrier, which means that transport from the blood to the interstitium is highly restricted and regulated by specific transport mechanisms. The outer retina, in which photoreceptors are located, is avascular and depends on the extensively fenestrated rich capillary plexus of the adjacent choroid (choriocapillaris). The choroidal vasculature does not have a barrier function. The retinal pigment epithelial (RPE) layer separates the retina from the choroidal nonbarrier endothelium and forms the outer blood–retinal barrier (Campochiaro, 2000; Raviola, 1977).

*Preretinal angiogenesis* occurs as a final common pathway in several diseases associated with capillary nonperfusion and ischemia of the neuroretina, so-called proliferative retinopathies. The ischemic retinal areas incite growth of new vessels along the interface of the vitreous and the optic disc and retina, finally forming large contractile fibrovascular membranes within the vitreous cavity. These membranes and the associated hemorrhages cause blindness by obscuration of the visual axis and retinal detachment. When the retinal ischemia is widespread, angiogenesis and scarring can also occur on the iris and cause an untreatable form of glaucoma. The best example of these conditions is diabetic retinopathy (DR), but similar vascular proliferation on the retina can cause blindness in venous retinal vascular occlusions, sickle cell retinopathy, systemic lupus erythematosus, hypertensive retinopathy, and many others. Destruction of the ischemic retinal areas with laser, so-called panretinal laser treatment, can be effective in inducing regression and fibrosis of the newly formed vessels. Despite these treatment options, and because of the fact that often the treatment cannot be carried out due to obscuration of the retina by hemorrhages and fibrovascular tissue, many patients with proliferative retinopathies still lose vision.

*Subretinal (or choroidal) neovascularization (CNV)* also occurs as a final common pathway, but in conditions affecting the layers of the eye beneath the neuroretina: the RPE, Bruch's membrane, and the choroid. Typically, CNV is a wound-healing response that occurs only when an anatomical discontinuation (break) of Bruch's membrane is present, in combination with a driving force such as inflammation, hypoxia, and oxidative stress (reviewed in Penn et al., 2008; Witmer et al., 2003). For most conditions it is unknown, however, to what extent these three mechanisms contribute to the initiation of a CNV response. CNV is seen in the course of age-related macular degeneration (AMD), located either between the RPE and Bruch's membrane (occult CNV), or between the RPE and the neuroretina (classic CNV), but also, as classic CNV, in the course of high myopia, pseudoxanthoma elasticum, inflammatory conditions such as punctate inner choroidopathy, and many others.

Later in CNV development, the new vessels can regress, leaving an atrophic retinal area, seen often in the course of occult CNV in AMD, or an angio-fibrotic switch can occur leading to formation of a fibrotic scar (Kuiper et al., 2008). In both cases, the overlying neuroretina will slowly degenerate leading to loss of sharp sight, contrast sensitivity, and color vision.

In eyes with otherwise healthy RPE, such as in young patients with CNV associated with chorioretinal scars or moderate myopia, CNV membranes are usually encapsulated by migrating and proliferating RPE, which leads to arrest of growth of the membrane preventing further damage (Witmer et al., 2003). Typically, this phenomenon does not occur in AMD, explaining why in this condition patients are often left with large macular scars and very poor vision.

## Mechanisms of angiogenesis and wound healing

Angiogenesis occurs by formation of new vessels from preexistent vasculature (Folkman, 1997; Hanahan and Folkman, 1996). It is a multistep process in which the normally highly specialized endothelial cells and pericytes of small vessels acquire a new set of functions enabling:

- the enzymatic breakdown of the vascular basal lamina and tissue matrix,

- formation of filopodia and migration toward the angiogenic stimulus and toward other sprouts,
- merging of sprouts,
- proliferation of endothelial cells, and
- maturation of the newly formed vessel.

Angiogenesis is an essential and early process in most wound-healing processes, allowing further influx of leukocytes and deposition of a fibrin matrix, and is followed by myofibroblast formation, fibrosis, and scarring.

A large number of paracrine and autocrine growth factors (GF) have been implied in the different steps of angiogenesis, among which are members of the VEGF family, insulin-like GF, and the angiopoietins (Carmeliet, 2000; Shibuya, 2008). In the fibrotic phase following ocular angiogenesis, other factors such as connective tissue growth factor (CTGF) plays an important role (Kuiper et al., 2008).

## *VEGF family*

The VGEF family includes placenta growth factor (PlGF), VEGF-A, VEGF-B, VEGF-C, VEGF-D, and the viral VEGF homologue VEGF-E (reviewed in Witmer et al., 2003).

VEGF-A, which is the main focus of this review, is a dimeric 36–46 kd glycosylated protein with a *N*-terminal signal sequence and a heparin-binding domain. In the human, nine relatively abundant VEGF-A isoforms have been identified with varying numbers of amino acids: $VEGF_{121}$, $VEGF_{145}$, $VEGF_{148}$, $VEGF_{162}$, $VEGF_{165b5}$, $VEGF_{165}$, $VEGF_{183}$, $VEGF_{189}$, and $VEGF_{206}$ (versus $VEGF_{120}$, $VEGF_{164}$, and $VEGF_{188}$ in the mouse). They arise by alternative splicing of mRNA. The longer forms are matrix-bound and the shorter forms are freely diffusible. $VEGF_{165}$ is the predominant isoform (reviewed in Nagy et al., 2007). The differential role of the various VEGF isoforms has been investigated in mice genetically engineered to express only one isoform. In these studies, the various VEGF isoforms appeared redundant for initial vessel assembly and growth. However, they differ greatly in providing critical spatial guidance cues for vascular remodeling (Stalmans et al., 2002; Stalmans, 2005). In addition, one experimental study provided evidence that isoform $VEGF_{164/165}$ plays a predominant role in pathological, but not in physiological angiogenesis (Ishida et al., 2003).

## *VEGF receptors*

VEGFs exert their functions by binding to three VEGF receptors (VEGFR) (reviewed in Shibuya, 2008; Witmer et al., 2003). Two high-affinity tyrosine kinase receptors have been identified for VEGF-A, namely VEGFR-1 (*fms*-like tyrosine kinase-1 or Flt-1) and VEGFR-2 (kinase insert domain-containing receptor or KDR). The third high-affinity receptor VEGFR-3 (*fms*-like tyrosine kinase-4 or Flt-4) is a receptor for VEGF-C and -D. The VEGFR-1 gene encodes two polypeptides: a full-length membrane protein (receptor form of VEGFR-1) and a short-length soluble VEGF-binding protein (soluble form of VEGFR-1). In addition, neuropilin-1 (NP-1) was identified as a coreceptor for $VEGF_{165}$ enhancing VEGF binding to VEGFR-2.

VEGFR-2 is the receptor that mediates the mitogenic, angiogenic, and permeability-enhancing signaling effects of VEGF-A in endothelial cells.

It remains controversial which function VEGFR-1 has on endothelial cells and whether VEGFR-1 is able to transmit a meaningful signal in endothelial cells when it binds VEGF-A. There is evidence suggesting that VEGFR-1 functions only as a "decoy" receptor, regulating the activity of VEGF-A by rendering it less available for VEGFR-2. In contrast, VEGFR-1 expressed on pericytes (Grosskreutz et al., 1999; Witmer et al., 2004) may act as a functional receptor with intracellular signal transduction. PlGF binding to VEGFR-1 on endothelial cells and/or pericytes plays an important enhancing role in angiogenesis under pathological conditions (Carmeliet et al., 2001; Hiratsuka et al., 1998).

## *VEGF in pathology outside the eye*

VEGF-A is strongly upregulated by hypoxia (Detmar et al., 1997). It plays a predominant role in pathological angiogenesis in cancer and in ischaemic and inflammatory diseases (reviewed in Kerbel, 2008).

## *Role of VEGF in (ocular) physiology*

VEGFs probably have physiological functions in and outside the eye. First, VEGF is indispensable in embryonic development as VEGF gene knockout in mice is lethal early in embryonic life. Second, in the adult, VEGF appears to be a survival factor for quiescent blood vessels, and a stimulatory factor for physiologic angiogenesis, such as in the female reproductive system. That VEGF may act as a vascular survival factor is suggested by several observations: VEGF-A and its receptors are expressed in various normal tissues in adult humans and monkeys, such as the epithelium of choroid plexus in the brain, glomerulus epithelium in the kidney and gastrointestinal mucosa (reviewed in Witmer et al., 2003). VEGF-A is produced by epithelial cells in these tissues in close vicinity to endothelial cells which preferentially express VEGFRs (Breier et al., 1992; Carmeliet and Collen, 1998; Esser et al., 1998) suggesting that VEGF-A maintains the integrity of adjacent endothelial cells (Witmer et al., 2002a, b).

In the eye, it has long been known that loss of RPE cells causes loss of fenestrations and later atrophy of the choriocapillaris (Korte et al., 1984). VEGF-A is secreted basally by RPE cells toward the fenestrated choriocapillaris, which expresses VEGFRs (Blaauwgeers et al., 1999). Therefore, RPE-derived VEGF seems to act as a permeability/survival factor in quiescent choriocapillaris endothelium. Recent studies in mice and monkeys support this role: extensive systemic VEGF inhibition in mice led to regression of fenestrated capillaries in several tissues (Kamba et al., 2006), and intra-ocular injection of an anti-VEGF agent in monkeys caused a temporary decrease in the number of fenestrations of the choriocapillaris (Peters et al., 2007). This paracrine relation between epithelia and endothelium may include a hypoxia-driven feedback mechanism promoting vessel formation to compensate for insufficient tissue oxygenation (Blaauwgeers et al., 1999; Dor and Keshet, 1997). It has been suggested that in AMD derailment of such a feedback mechanism in the choroid and subsequent excessive VEGF production may be the cause of subretinal neovascularization (Blaauwgeers et al., 1999).

## *Neuroprotective role of VEGF*

In addition to its function as a survival factor for quiescent endothelium, VEGF may also have a role as a neuroprotective factor (Oosthuyse et al., 2001). This was shown in transgenic mice lacking the hypoxia-responsive gene of VEGF-A which suffer from a neurodegenerative disorder, suggesting that a neurotrophic function of hypoxia-induced VEGF-A is essential under physiological conditions. Some evidence also points at a neurotrophic function of VEGF in the retina, which has relevance for the use of VEGF antagonists in the eye: in the retina, all three VEGFRs are expressed by neural elements (Witmer et al., 2002a), and experimental work in rats has shown that VEGF-A is necessary for neural survival in ischemia–reperfusion model (Nishijima et al., 2007).

## *VEGF antagonists as therapeutic agents*

Judah Folkman's proposed in 1971 that an antibody to a putative "tumor angiogenesis factor" could be a therapeutic agent in oncology. In line with this suggestion, Bevacizumab (Avastin®), a humanized mouse IgG effective in metastatic colorectal cancer (Hurwitz et al., 2004), was the first VEGF antagonist used in patients. Several other methods have been developed to antagonize VEGF-A: inhibition of VEGF-A production, specific inhibition of the VEGF receptors, and inhibition of the signaling cascade activated by the VEGF–VEGFR complex in target cells (Table 1; Fig. 1). In addition, other members of the VEGF family such as PlGF and VEGFR-3 are explored as targets for therapeutic anti-angiogenesis.

In ophthalmology, a large number of VEGF antagonists is in clinical development, most of which act by binding and thereby inactivation of free VEGF-A. Others employ alternative

Table 1. Therapeutic strategies of anti-VEGF treatment

| Strategy | Target | Generic name |
|---|---|---|
| Anti-VEGF production | Anti-VEGF SiRNA | Bevasiranib |
| Anti-VEGF | Receptor decoy | VEGF Trap |
|  | IgG | Bevacizumab |
|  | IgG fragment | Ranibizumab |
|  | Aptamer | Pegaptanib |
| VEGF receptors | Anti-VEGF receptor SiRNA | AGN211745 |
| VEGF signal transduction pathways | Tyrosine kinase cascade | Vatalanib |
|  |  | TG100801 |
|  |  | Pazopanib |
|  |  | AG013958 |
|  |  | AL39324 |

*Source*: Chappelow and Kaiser (2008).

approaches such as inhibition of VEGF production by small interfering RNA (siRNA) or VEGF receptor inhibition (Table 1).

Two drugs are registered for ophthalmic use. Pegaptanib sodium (Macugen®), approved for use in AMD in 2004, is a 28-base RNA aptamer which specifically binds $VEGF_{165}$ (Gragoudas et al., 2004). Ranibizumab (Lucentis®) is a Fab fragment of bevacizumab, and was developed specifically
for use in the eye. Its first registration was for AMD in 2006.

**Role of VEGF in ocular pathology and clinical use of VEGF antagonists**

*VEGF and eye disease*

As early as in 1948, it was suggested by Michaelson that the developing retina induces vascular ingrowth by the release of a diffusible "metabolic" factor, and that this factor could also play a role in retinal disease associated with vascular insufficiency such as DR. A few years later Ashton hypothesized that this factor "X" is induced by hypoxia.

VEGF-A was discovered as a protein in 1983 by Senger et al. (1983), who named it Vascular Permeability Factor (VPF), because of its marked ability to increase vascular permeability, but its angiogenic properties went initially unnoticed. Six years later, the protein was rediscovered under the name of VEGF, cloned, and its major role in angiogenesis elucidated by Ferrara et al. (1992) and others, followed by the discovery by Shweiki et al. (1992) that VEGF is induced by hypoxia.

As VEGF-A had all the properties of the factor "X" proposed in the early 1950s, it was not surprising that from 1994 onward convincing evidence became available that VEGF is the major causal agent in vascular leakage and angiogenesis in the eye. This was first shown by the clear association of VEGF levels in ocular fluids with angiogenic activity in patients with proliferative retinopathies (Aiello et al., 1994). Further support came from a series of experiments in monkey eyes that demonstrated that VEGF is both necessary and sufficient for the typical neovascularization of the retina and iris associated with retinal ischemia (Adamis et al., 1996; Tolentino et al., 1996a, b, 2002).

However, the best evidence for involvement of VEGF-A in ocular disease is the clinical effects of VEGF antagonists, both from state-of-the-art clinical trials and from observations in case series.

*VEGF-induced retinopathy in experimental animals*

Interestingly, repeated injections of VEGF-A in monkey eyes cause widespread capillary nonperfusion similar to ischemic retinopathies such as DR (Tolentino et al., 1996b). This showed that VEGF is not only the result of capillary nonperfusion in DR but also a possible cause. Several hypotheses have been proposed to explain this observation. Plugging of retinal capillaries by adherence of VEGF-activated leukocytes to retinal endothelial cells is considered an important mechanism (McLeod et al., 1995; Miyamoto et al., 1997, 2000; Miyamoto and Ogura, 1999; Paques et al., 2000; Schroder et al., 1991). A second possible cause is VEGF-A-induced hypertrophy of endothelial cells of retinal

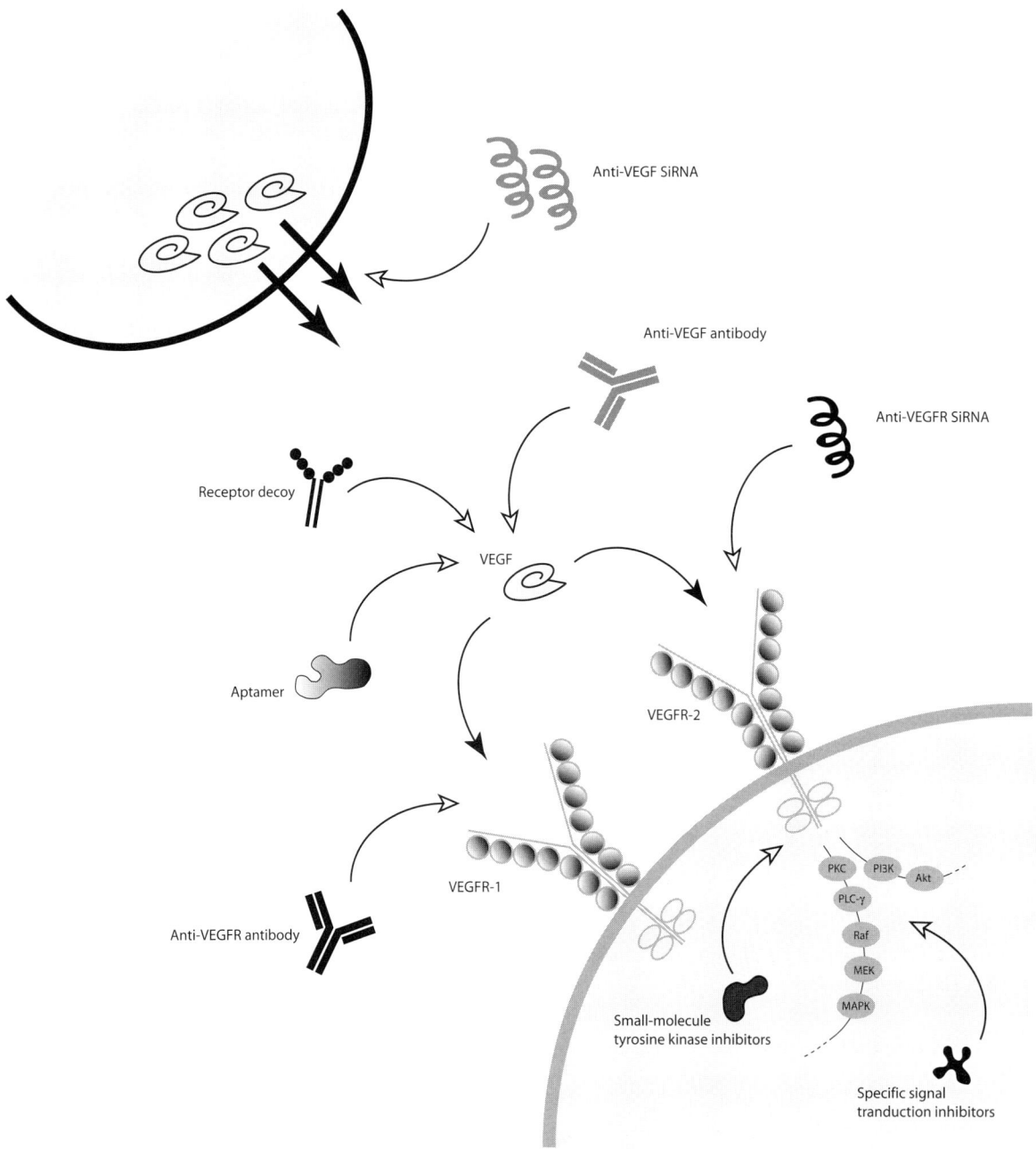

Fig. 1. Therapeutic strategies in VEGF treatment. Open arrowhead: inhibiting strategy.

capillaries, which leads to narrowing of capillary lumina (Hofman et al., 2001b; Tolentino et al., 2002) and capillary nonperfusion.

## VEGF antagonists in conditions with preretinal neovascularization

### Diabetic retinopathy

The most common proliferative retinopathy occurs in an advanced form of DR. Proliferative diabetic retinopathy (PDR) is preceded by preclinical and nonproliferative DR (Fig. 2). A variety of metabolic imbalances and vascular changes, such as thickening of the basement membrane, apoptosis of pericytes and endothelial cells, and diffusely increased vascular permeability, occur in the retina in diabetes, long before clinical disease is recognized. Hyperglycemia is the major risk factor for this early development and progression of DR (Klein et al., 1998). The high intracellular glucose levels result in an increased flux of glucose via the sorbitol pathway (Crabbe and Goode, 1998; Ola et al., 2006), protein kinase C (PKC-β) activation (Koya and King, 1998), formation of free radicals (oxidative stress), NADH depletion causing a state of intracellular

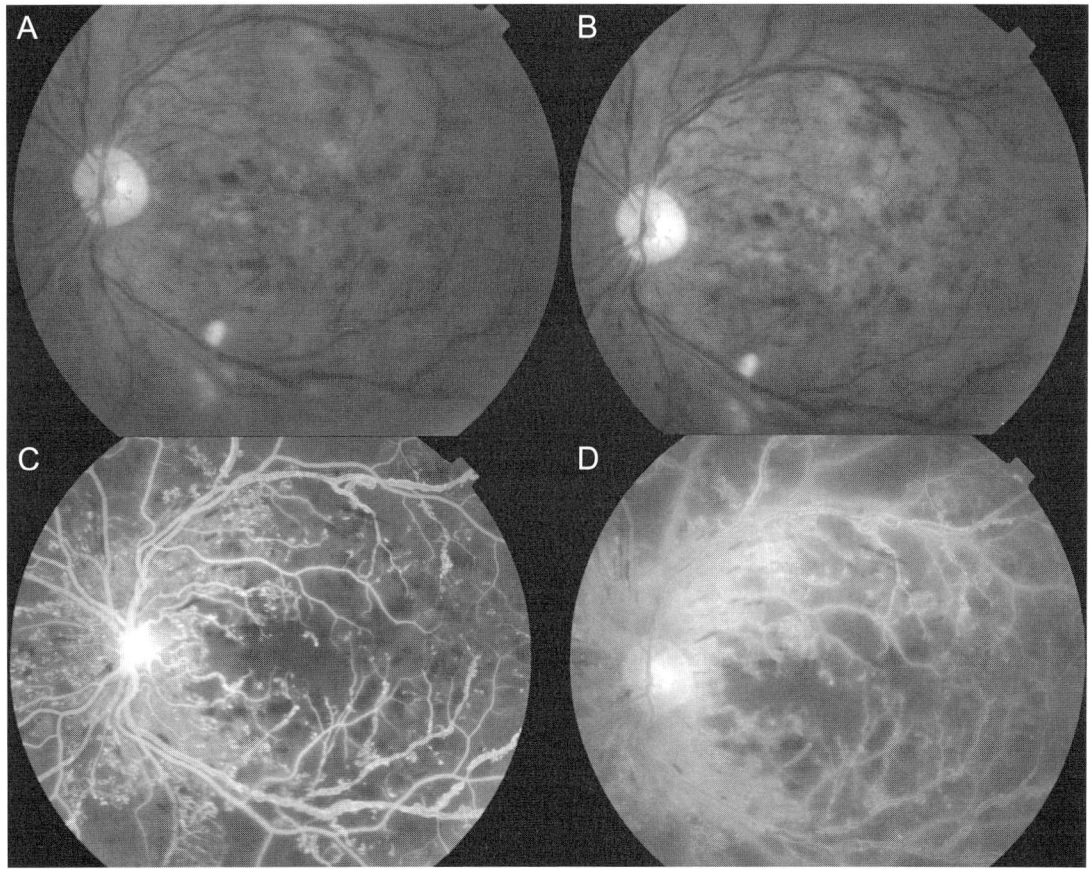

Fig. 2. Fundus photograph (A), red-free photograph (B), early-phase (C), and late-phase (D) fluorescein angiographic imaging of a patient with severe proliferative diabetic retinopathy (PDR). Note the extensive intraretinal hemorrhages (A, B); a hyperfluorescent lesion on the optic nerve, which is suggestive of a neovascularization (C); extensive venous changes (C, D) and vascular leakage (D). The dark aspect of the retina in the early phase (C) represents widespread capillary nonperfusion and, thus, ischemia. (See Color Plate 17.2 in color plate section.)

"pseudohypoxia" (Williamson et al., 1993), and the formation of advanced glycation end products (AGEs) from the nonenzymatic glycation of proteins (Stitt et al., 1997). All these effects of hyperglycaemia together lead to a gradual loss of vascular and neural retinal cells, which after years of diabetes mellitus will cause acellular capillaries and ischaemic areas of capillary nonperfusion.

VEGF-A expression is already increased in the retina in preclinical DR, where it enhances permeability and may act as a vascular survival factor (reviewed in Witmer et al., 2003). In later stages, high VEGF-A production in expanding areas of ischaemia, where VEGFR-2 is also upregulated, then leads to the vision-threatening signs of DR, i.e. vascular leakage and neovascularization (Smith et al., 1999; Witmer et al., 2002a). As VEGF-induced vascular closure may enhance ischemia and could cause further VEGF upregulation, a vicious circle of spreading capillary nonperfusion in DR may develop (Hofman et al., 2001b). This is supported by the clinical observation that focal laser treatment of ischemic areas in diabetic maculopathy can arrest further progression of DR. In PDR, VEGF-A levels are elevated in the vitreous and correlate with angiogenesis activity (Adamis et al., 1994; Aiello et al., 1994; Ishida et al., 2000; Pe'er et al., 1996; Shinoda et al., 1999; Wells et al., 1996).

At this moment, panretinal photocoagulation is the standard of care for PDR. It often prevents further retinal neovascularization (Early Treatment Diabetic Retinopathy Study Research Group, 1991). The treatment is effective, but painful and may induce night blindness and a decreased visual acuity. Often, laser cannot be applied due to vitreous hemorrhages and in other cases, fibrovascular proliferation has already progressed to a stage where laser cannot prevent blindness. Laser treatment probably works by destruction of photoreceptors, which decreases the oxygen demand of the retina and/or increases oxygenation of the inner retina from the choroid (Landers et al., 1982).

VEGF antagonists have a striking short-term effect on preretinal neovascularization in PDR (Avery et al., 2006a; Mason et al., 2006; Spaide and Fisher, 2006). After only one intravitreal injection with bevacizumab or ranibizumab, the new vessels regress or transform to fibrotic membranes (Kuiper et al., 2008). This effect has been widely observed and is used in clinical practice by most retinal specialists, in particular in patients with severe PDR that is complicated by hemorrhages or as an adjunct treatment before surgical delamination of fibrovascular membranes. However, the data reported in the literature are mostly case series with short-term follow-up, and no high-quality randomized controlled trials (RCTs) are available showing the true benefit of intravitreal anti-VEGF adjunct over standard care.

### Diabetic macular edema (DME)

Vascular leakage from retinal vessels in the macula causes DME, which is the major cause of vision loss in diabetic patients. Based on RCTs carried out more than 10 years ago, focal and grid laser treatment is the standard care in this condition. Experimental evidence suggests that VEGF is involved in DME, where it is probably produced in small areas with focal ischemia. This has led to the initiation of large-scale studies exploring the clinical use of ranibizumab and other VEGF antagonists in DME, with or without laser, several of which are still going on. The published results so far show improved visual acuity, a decrease or resolution of macula edema, and a decrease in the need for further photocoagulation (Campochiaro et al., 2006; Chun et al., 2006; Haritoglou et al., 2006; Lee et al., 2008; Shimura et al., 2008). However, although these studies have confirmed that VEGF is an important causal factor in DME development, the final role of this short-term treatment in the daily clinical management of chronic DME remains to be established.

### Other conditions with preretinal neovascularization or macular edema

Branch and central retinal vein occlusions, sickle cell disease, and retinopathy of prematurity are other conditions where blindness is caused by macular edema or preretinal neovascularization (Fig. 3), respectively. In each of these conditions,

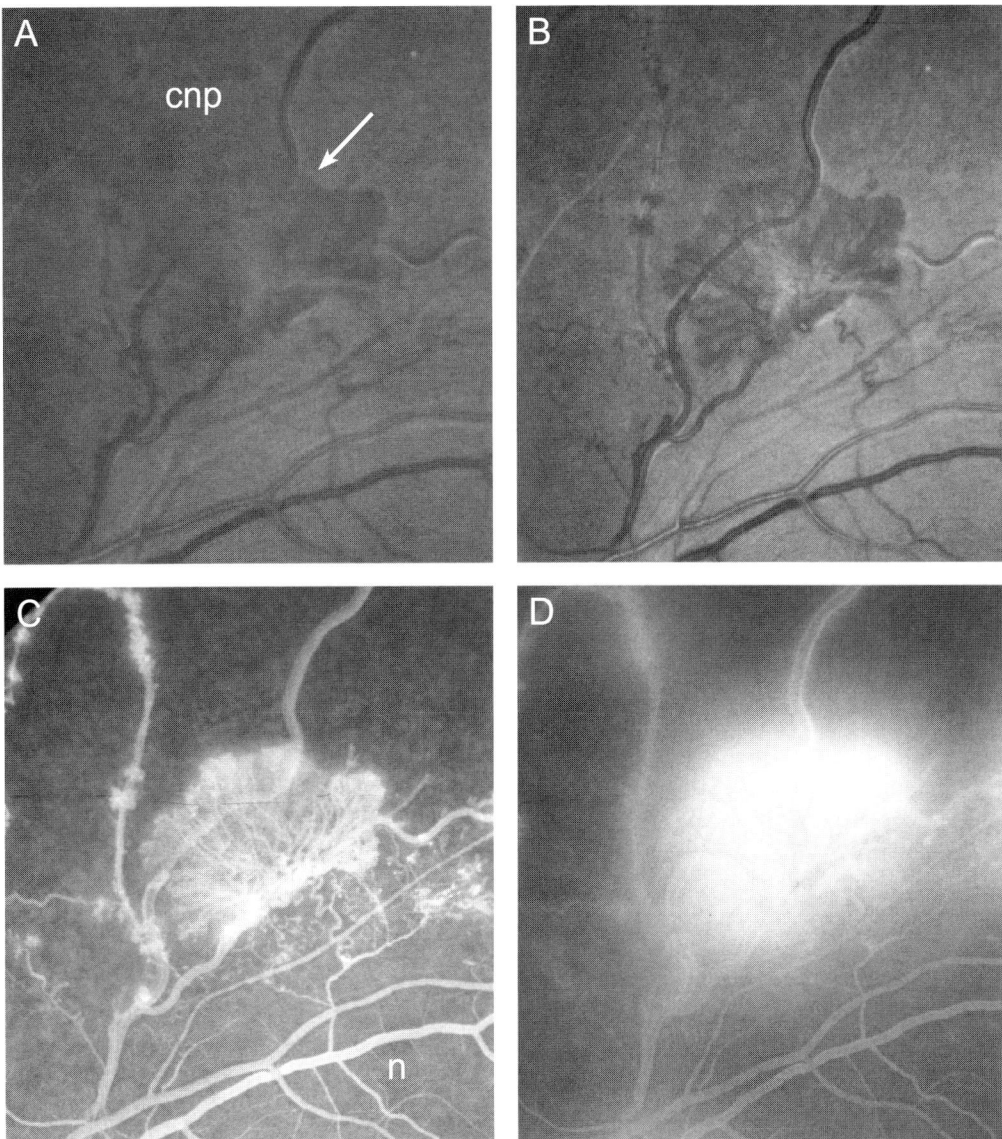

Fig. 3. Fundus photograph (A), red-free photograph (B), early-phase (C), and late-phase (D) fluorescein angiographic imaging of a patient with preretinal neovascularization after a branch retinal vein occlusion. Note the large neovascularization (white arrow) adjacent to an extensive area of capillary nonperfusion (cnp). (See Color Plate 17.3 in color plate section.)

the value of VEGF antagonists is being evaluated in clinical studies. Early case series with short-term follow-up are promising, showing resolution of the associated neovascularization or macula edema as well as functional improvement (Pieramici et al., 2008; Spaide et al., 2009), but state-of-the-art RCTs are still in progress. Nevertheless, VEGF antagonists are widely used off-label in several of these conditions.

*Iris neovascularization and neovascular glaucoma* occur typically acutely in the course of ischemic central retinal vein occlusions, or subacutely in ocular ischemic syndrome and very severe PDR. The risk of iris neovascularization is increased in aphakic or pseudophakic eyes due to the facilitated diffusion of retina-derived angiogenic factors to the anterior parts of the eye. Iris neovascularization can lead to rapid closure of the chamber angle by scar tissue causing intractable glaucoma. That VEGF-A is the main factor in this process was shown in experimental studies in monkeys (Hofman et al., 2001a; Tolentino et al., 1996a), and confirmed by the astounding effect of VEGF antagonists on iris neovascularization in patients. When VEGF-A itself is injected intravitreally in Cynomolgus monkeys, iris neovascularization develops after 4–9 days, whereas in the retina initially only massive leakage is observed without any sign of neovascularization (Tolentino et al., 1996a; Witmer et al., 2002a). In patients, VEGF antagonists such as bevacizumab can lead to complete regression of iris neovascularization in early cases (Avery et al., 2006a; Avery, 2006; Batioglu et al., 2008; Jiang et al., 2008; Oshima et al., 2006; Wakabayashi et al., 2008). In patients with early-stage neovascular glaucoma without angle closure, bevacizumab is able to control the intraocular pressure as well. In advanced neovascular glaucoma, intraocular pressure could not be controlled, but bevacizumab may be used adjunctively to improve results of surgical procedures (Wakabayashi et al., 2008). Further evaluation in controlled randomized studies is warranted.

## VEGF antagonists in AMD and other conditions with subretinal neovascularization

As described in the Introduction, subretinal or choroidal neovascularization (CNV) is the growth of new blood vessels from the choroid beneath the retina. It is classified according to its appearance on fluorescein angiography and its presumed localization. Classic CNV is located between the RPE and the neuroretina, occult CNV between Bruch's membrane and the RPE (occult), and mixed type consists of both types of CNV. Exudative AMD and high myopia are the most common causes of CNV.

### Pegaptanib sodium (Macugen®)

The effect of pegaptanib sodium in exudative AMD was tested in two RCTs. Three doses of pegaptanib sodium were investigated (0.3, 1.0, and 3.0 mg) over a period of 54 weeks in 1186 patients with subfoveal neovascularization of any lesion type, with intravitreal injections given every 6 weeks for a total of 9 treatments. The results were slightly disappointing: Only 206 out of 296 (70%) patients receiving 0.3 mg pegaptanib sodium ($p < 0.001$) lost fewer than 15 letters of visual acuity between baseline and week 54 (the prespecified primary endpoint for efficacy), as compared to 164 of 296 (55%) patients receiving sham injection (Gragoudas et al., 2004). The average loss of letters after 1 year was 10 letters (2 lines) in the treated groups as compared to 17 letters in the sham injection group. Some patients gained vision, but most patients deteriorated despite treatment.

### Ranibizumab (Lucentis®)

In contrast, the clinical results with Ranibizumab, tested in several large trials, represent a breakthrough in the treatment of exudative AMD (Brown et al., 2006; Rosenfeld et al., 2006). The MARINA study compared ranibizumab (either 0.3 or 0.5 mg) with placebo in minimally classic and occult CNV due to AMD, with monthly injections over a period of 24 months. The primary endpoint of this study was again defined as the percentage of patients with a loss of vision of less than 15 letters on the ETDRS chart at month 12. The ANCHOR study compared the safety and efficacy of ranibizumab with photodynamic therapy (PDT) in patients with predominantly classic subfoveal CNV. In both trials, ranibizumab stabilized vision in 92–96% of patients during the 2-year treatment period, compared to stable vision in only 52–62% of patients treated with sham injections or PDT (Brown et al., 2006; Heier et al., 2006; Rosenfeld et al., 2006). The maximum effect of ranibizumab was observed at 3 months, after which the effect stabilized. Parallel to its effect on visual acuity, ranibizumab markedly decreased retinal thickness as measured by optical coherence tomography (OCT). Patients treated with ranibizumab gained on average 5–10 letters on the ETDRS chart, compared to a loss of 10–15 letters loss in patients on placebo or PDT. More than 40% of the treated patients had a visual acuity of 0.5 or better after 2 years, a level that would allow reading and driving, compared to less than 6% of patients on the comparative treatments.

The US Food and Drug Administration gave approval for the treatment of exudative AMD with monthly injections of ranibizumab 0.5 mg on June 30, 2006. European approval has also been granted since January 2007. It is registered in Europe to be used as three monthly injections with additional injections when needed, although the available evidence suggests that treatment regimens based on disease recurrence are less effective than monthly dosing.

### Bevacizumab (Avastin®)

This drug was first used off-label in neovascular AMD in a study in nine patients treated with intravenous bevacizumab at a similar dose (Michels et al., 2005) as used in metastatic colorectal cancer patients (Kabbinavar et al., 2003). After two or three treatments, a significant increase in visual acuity (median +8 letters, $p = 0.011$), a significant decrease in central retinal thickness on OCT (median $-157\,\mu m$, $p = 0.008$), and a decrease in retinal leakage on fluorescein angiography was noted (Michels et al., 2005). In patients with colorectal cancer, arterial thromboembolic events are a major concern in the treatment with systemic bevacizumab. Therefore, Rosenfeld et al. (2005b) first explored the possibility to treat patients with intravitreal bevacizumab to minimize the possibility of systemic adverse effects.

Since then, bevacizumab has been used off-label on a wide scale by ophthalmologists in the treatment of exudative AMD (Avery et al., 2006b; Lazic and Gabric, 2007; Rosenfeld et al., 2005a). No randomized controlled studies, but numerous case series have been published, describing a total of more than 1000 patients treated with bevacizumab. The reported results suggest that bevacizumab is as effective and safe as ranibizumab (Kiss et al., 2006; Manzano et al., 2006; Maturi et al., 2006; Shahar et al., 2006; Spandau and Jonas, 2006). In the largest case series, bevacizumab was found to improve visual acuity, as defined by a halving of the visual angle, in 38.3% of 144 patients at 3 months follow-up. If confirmed in a state-of-the-art trial, this would be comparable to the effect of ranibizumab. In addition, bevacizumab markedly reduced retinal thickness on OCT, to a similar extent as reported for ranibizumab (Lazic and Gabric, 2007). The other published case series support these findings.

Approximately 40,000 patients have been treated worldwide with bevacizumab. However, conclusive evidence from RCTs directly comparing bevacizumab and ranibizumab is lacking. Such RCTs are not likely to be funded by the pharmaceutical industry, but have been started recently with public money in the UK, USA, and in the Netherlands (www.clinicaltrials.gov).

The costs of bevacizumab are 20- to 50-fold lower than those of ranibizumab, and it has been estimated that in the Netherlands alone, the costs of ranibizumab treatment are around 30–50 million Euros higher than treatment with bevacizumab.

### Side effects of ocular use of VEGF antagonists

The main safety concerns regarding ocular use of VEGF antagonists are associated with the injection procedure, which causes bacterial infection in the eye (endophthalmitis) or retinal detachment in approximately 1 in 1500 injections.

However, as systemic bevacizumab, when given to patients with metastatic colorectal cancer, has a significantly increased risk of systemic hypertension and arterial thromboembolic events including stroke and myocardial infarction, there is a chance that the 1000-fold lower ocular dose could have similar side effects in some patients. This has been studied to some extent in an international survey, which collected safety data on 7113 injections in 5228 patients (Rich et al., 2006). In this open label, retrospective and uncontrolled study, only 0.21% of patients experienced mild increases in systemic blood pressure, and cerebrovascular events were noted in 0.07% (3 patients). Due to the design of the study, these surprisingly low figures, which are lower than the background risk in this age group, may not represent the true incidence of systemic side effects.

More convincingly, in the MARINA and ANCHOR trials, and in the postmarketing SAILOR study, arterial thromboembolic events were not significantly higher in patients receiving ranibizumab compared to placebo, but most comparisons did show a trend toward a higher risk for the higher ranibizumab dose, particularly in patients with a history of stroke or of cardiac arrhythmias (Brown et al., 2006; Rosenfeld et al., 2006).

## Conclusion

VEGF antagonists are gradually finding their place in the clinical management of a large variety of ocular conditions involving neovascularization and vascular leakage. Potential ocular and systemic side effects, as well as the emerging need for chronic treatment are important concerns, but the introduction of these drugs into clinical practice has meant hope for many patients for whom previously no therapeutic options were available.

## References

Adamis, A. P., Miller, J. W., Bernal, M. T., D'Amico, D. J., Folkman, J., Yeo, T. K., et al. (1994). Increased vascular endothelial growth factor levels in the vitreous of eyes with proliferative diabetic retinopathy. *American Journal of Ophthalmology*, *118*, 445–450.

Adamis, A. P., Shima, D. T., Tolentino, M. J., Gragoudas, E. S., Ferrara, N., Folkman, J., et al. (1996). Inhibition of vascular endothelial growth factor prevents retinal ischemia-associated iris neovascularization in a nonhuman primate. *Archives of Ophthalmaology*, *114*, 66–71.

Aiello, L. P., Avery, R. L., Arrigg, P. G., Keyt, B. A., Jampel, H. D., Shah, S. T., et al. (1994). Vascular endothelial growth factor in ocular fluid of patients with diabetic retinopathy and other retinal disorders. *New England Journal of Medicine*, *331*, 1480–1487.

Avery, R. L. (2006). Regression of retinal and iris neovascularization after intravitreal bevacizumab (Avastin) treatment. *Retina*, *26*, 352–354.

Avery, R. L., Pearlman, J., Pieramici, D. J., Rabena, M. D., Castellarin, A. A., Nasir, M. A., et al. (2006a). Intravitreal bevacizumab (Avastin) in the treatment of proliferative diabetic retinopathy. *Ophthalmology*, *113*, 1695. e1–15

Avery, R. L., Pieramici, D. J., Rabena, M. D., Castellarin, A. A., Nasir, M. A., & Giust, M. J. (2006b). Intravitreal bevacizumab (Avastin) for neovascular age-related macular degeneration. *Ophthalmology*, *113*, 363–372.

Batioglu, F., Astam, N., & Ozmert, E. (2008). Rapid improvement of retinal and iris neovascularization after a single intravitreal bevacizumab injection in a patient with central retinal vein occlusion and neovascular glaucoma. *International Ophthalmology*, *28*, 59–61.

Blaauwgeers, H. G., Holtkamp, G. M., Rutten, H., Witmer, A. N., Koolwijk, P., Partanen, T. A., et al. (1999). Polarized vascular endothelial growth factor secretion by human retinal pigment epithelium and localization of vascular endothelial growth factor receptors on the inner choriocapillaris. Evidence for a trophic paracrine relation. *American Journal of Pathology*, *155*, 421–428.

Breier, G., Albrecht, U., Sterrer, S., & Risau, W. (1992). Expression of vascular endothelial growth factor during embryonic angiogenesis and endothelial cell differentiation. *Development*, *114*, 521–532.

Brown, D. M., Kaiser, P. K., Michels, M., Soubrane, G., Heier, J. S., Kim, R. Y., et al. (2006). Ranibizumab versus verteporfin for neovascular age-related macular degeneration. *New England Journal of Medicine*, *355*, 1432–1444.

Campochiaro, P. A. (2000). Retinal and choroidal neovascularization. *Journal of Cell Physiology*, *184*, 301–310.

Campochiaro, P. A., Nguyen, Q. D., Tatlipinar, S., Shah, S. M., Haller, J. A., Singletary, P., et al. (2006). Results of an open label phase 1/2 study assessing the effects of multiple intravitreous injections of Ranibizumab in patients with diabetic macular edema. *Investigative Ophthalmology & Visual Science*, *47*, 5443.

Carmeliet, P. (2000). Mechanisms of angiogenesis and arteriogenesis. *Nature Medicine*, *6*, 389–395.

Carmeliet, P., & Collen, D. (1998). Vascular development and disorders: molecular analysis and pathogenic insights. *Kidney International*, *53*, 1519–1549.

Carmeliet, P., Moons, L., Luttun, A., Vincenti, V., Compernolle, V., De Mol, M., et al. (2001). Synergism between vascular endothelial growth factor and placental growth factor contributes to angiogenesis and plasma extravasation in pathological conditions. *Nature Medicine*, *7*, 575–583.

Chappelow, A. V., & Kaiser, P. K. (2008). Neovascular age-related macular degeneration: Potential therapies. *Drugs*, *68*(8), 1029–1036.

Chun, D. W., Heier, J. S., Topping, T. M., Duker, J. S., & Bankert, J. M. (2006). A pilot study of multiple intravitreal injections of ranibizumab in patients with center-involving clinically significant diabetic macular edema. *Ophthalmology*, *113*, 1706–1712.

Crabbe, M. J., & Goode, D. (1998). Aldose reductase: a window to the treatment of diabetic complications? *Progress in Retinal and Eye Research*, *17*, 313–383.

D'Amore, P. A. (1994). Mechanisms of retinal and choroidal neovascularization. *Investigative Ophthalmology & Visual Science, 35*, 3974–3979.

Detmar, M., Brown, L. F., Berse, B., Jackman, R. W., Elicker, B. M., Dvorak, H. F., et al. (1997). Hypoxia regulates the expression of vascular permeability factor/vascular endothelial growth factor (VPF/VEGF) and its receptors in human skin. *Journal of Investigative Dermatology, 108*, 263–268.

Dor, Y., & Keshet, E. (1997). Ischemia-driven angiogenesis. *Trends in Cardiovascular Medicine, 7*, 289–294.

Early Treatment Diabetic Retinopathy Study Research Group. (1991). Early photocoagulation for diabetic retinopathy. ETDRS Report Number 9. *Ophthalmology, 98*, 766–785.

Esser, S., Wolburg, K., Wolburg, H., Breier, G., Kurzchalia, T., & Risau, W. (1998). Vascular endothelial growth factor induces endothelial fenestrations in vitro. *Journal of Cell Biology, 140*, 947–959.

Ferrara, N., Houck, K., Jakeman, L., & Leung, D. W. (1992). Molecular and biological properties of the vascular endothelial growth factor family of proteins. *Endocrine Reviews, 13*, 18–32.

Folkman, J. (1997). Angiogenesis and angiogenesis inhibition: an overview. *EXS, 79*, 1–8.

Gragoudas, E. S., Adamis, A. P., Cunningham, E. T., Jr., Feinsod, M., & Guyer, D. R. (2004). VEGF inhibition study in ocular neovascularization clinical trial group. Pegaptanib for neovascular age-related macular degeneration. *New England Journal of Medicine, 351*, 2805–2816.

Grosskreutz, C. L., Anand-Apte, B., Dupláa, C., Quinn, T. P., Terman, B. I., Zetter, B., et al. (1999). Vascular endothelial growth factor-induced migration of vascular smooth muscle cells in vitro. *Microvascular Research, 58*, 128–136.

Hanahan, D., & Folkman, J. (1996). Patterns and emerging mechanisms of the angiogenic switch during tumorigenesis. *Cell, 86*, 353–364.

Haritoglou, C., Kook, D., Neubauer, A., Wolf, A., Priglinger, S., Strauss, R., et al. (2006). Intravitreal bevacizumab (Avastin) therapy for persistent diffuse diabetic macular edema. *Retina, 26*, 999–1005.

Heier, J. S., Boyer, D. S., Ciulla, T. A., Ferrone, P. J., Jumper, J. M., Gentile, R. C., et al. (2006). Ranibizumab combined with verteporfin photodynamic therapy in neovascular age-related macular degeneration: year 1 results of the FOCUS Study. *Archives of Ophthalmology, 124*, 1532–1542.

Hiratsuka, S., Minowa, O., Kuno, J., Noda, T., & Shibuya, M. (1998). Flt-1 lacking the tyrosine kinase domain is sufficient for normal development and angiogenesis in mice. *Proceedings of the National Academy of Sciences of the United States of America, 95*, 9349–9354.

Hofman, P., Blaauwgeers, H. G., Vrensen, G. F., & Schlingemann, R. O. (2001a). Role of VEGF-A in endothelial phenotypic shift in human diabetic retinopathy and in VEGF-A induced retinopathy in monkeys. *Ophthalmic Research, 33*, 156–162.

Hofman, P., van Blijswijk, B. C., Gaillard, P. J., Vrensen, G. F., & Schlingemann, R. O. (2001b). Endothelial cell hypertrophy induced by vascular endothelial growth factor in the retina: new insights into the pathogenesis of capillary nonperfusion. *Archives of Ophthalmology, 119*, 861–866.

Hurwitz, H., Fehrenbacher, L., Novotny, W., Cartwright, T., Hainsworth, J., Heim, W., et al. (2004). Bevacizumab plus irinotecan, fluorouracil, and leucovorin for metastatic colorectal cancer. *New England Journal of Medicine, 350*, 2335–2342.

Ishida, S., Shinoda, K., Kawashima, S., Oguchi, Y., Okada, Y., & Ikeda, E. (2000). Coexpression of VEGF receptors VEGF-R2 and neuropilin-1 in proliferative diabetic retinopathy. *Investigative Ophthalmology & Visual Science, 41*, 1649–1656.

Ishida, S., Usui, T., Yamashiro, K., Kaji, Y., Amano, S., Ogura, Y., et al. (2003). VEGF164-mediated inflammation is required for pathological, but not physiological, ischemia-induced retinal neovascularization. *Journal of Experimental Medicine, 198*, 483–489.

Jiang, Y., Liang, X., Li, X., Tao, Y., & Wang, K. (2008). Analysis of the clinical efficacy of intravitreal bevacizumab in the treatment of iris neovascularization caused by proliferative diabetic retinopathy. *Acta Ophthalmology* (in press).

Kabbinavar, F., Hurwitz, H. I., Fehrenbacher, L., Meropol, N. J., Novotny, W. F., Lieberman, G., et al. (2003). Phase II, randomized trial comparing bevacizumab plus fluorouracil (FU)/leucovorin (LV) with FU/LV alone in patients with metastatic colorectal cancer. *Journal of Clinical Oncology, 21*, 60–65.

Kamba, T., Tam, B. Y., Hashizume, H., Haskell, A., Sennino, B., Mancuso, M. R., et al. (2006). VEGF-dependent plasticity of fenestrated capillaries in the normal adult microvasculature. *American Journal of Physiology. Heart and Circulatory Physiology, 290*, H560–H576.

Kerbel, R. S. (2008). Tumor Angiogenesis. *New England Journal of Medicine, 358*, 2039–2049.

Kiss, C., Michels, S., Prager, F., Weigert, G., Geitzenauer, W., & Schmidt-Erfurth, U. (2006). Evaluation of anterior chamber inflammatory activity in eyes treated with intravitreal bevacizumab. *Retina, 26*, 877–881.

Klein, R., Klein, B. E., Moss, S. E., & Cruickshanks, K. J. (1998). The Wisconsin Epidemiologic Study of Diabetic Retinopathy: XVII. The 14-year incidence and progression of diabetic retinopathy and associated risk factors in type 1 diabetes. *Ophthalmology, 105*, 1801–1815.

Korte, G. E., Reppucci, V., & Henkind, P. (1984). RPE destruction causes choriocapillary atrophy. *Investigative Ophthalmology & Visual Science, 25*, 1135–1145.

Koya, D., & King, G. L. (1998). Protein kinase C activation and the development of diabetic complications. *Diabetes, 47*, 859–866.

Kuiper, E. J., Van Nieuwenhoven, F. A., de Smet, M. D., van Meurs, J. C., Tanck, M. W., Oliver, N., et al. (2008). The angio-fibrotic switch of VEGF and CTGF in proliferative diabetic retinopathy. *PLoS ONE, 3*, e2675.

Landers, M. B., III., Stefansson, E., & Wolbarsht, M. L. (1982). Panretinal photocoagulation and retinal oxygenation. *Retina*, *2*, 167–175.

Lazic, R., & Gabric, N. (2007). Intravitreally administered bevacizumab (Avastin) in minimally classic and occult choroidal neovascularization secondary to age-related macular degeneration. *Graefe's Archive for Clinical and Experimental Ophthalmology*, *245*, 68–73.

Lee, E., Hubschman, J.-P., Gonzales, C., & Schwartz, S. (2008). Combination therapy: targeted retinal photocoagulation (TRP) and anti-VEGF treatment for retinal vascular macular edema. *Investigative Ophthalmology & Visual Science*, *49*, 3494.

Manzano, R. P., Peyman, G. A., Khan, P., & Kivilcim, M. (2006). Testing intravitreal toxicity of bevacizumab (Avastin). *Retina*, *26*, 257–261.

Mason, J. O., III., Nixon, P. A., & White, M. F. (2006). Intravitreal injection of bevacizumab (Avastin) as adjunctive treatment of proliferative diabetic retinopathy. *American Journal of Ophthalmology*, *142*, 685–688.

Maturi, R. K., Bleau, L. A., & Wilson, D. L. (2006). Electrophysiologic findings after intravitreal bevacizumab (Avastin) treatment. *Retina*, *26*, 270–274.

McLeod, D. S., Lefer, D. J., Merges, C., & Lutty, G. A. (1995). Enhanced expression of intracellular adhesion molecule-1 and P-selectin in the diabetic human retina and choroid. *American Journal of Pathology*, *147*, 642–653.

Michels, S., Rosenfeld, P. J., Puliafito, C. A., Marcus, E. N., & Venkatraman, A. S. (2005). Systemic bevacizumab (Avastin) therapy for neovascular age-related macular degeneration: twelve-week results of an uncontrolled open-label clinical study. *Ophthalmology*, *112*, 1035–1047.

Miyamoto, K., Khosrof, S., Bursell, S. E., Moromizato, Y., Aiello, L. P., Ogura, Y., et al. (2000). Vascular endothelial growth factor (VEGF)-induced retinal vascular permeability is mediated by intercellular adhesion molecule-1 (ICAM-1). *American Journal of Pathology*, *156*, 1733–1739.

Miyamoto, K., & Ogura, Y. (1999). Pathogenetic potential of leukocytes in diabetic retinopathy. *Seminars in Ophthalmology*, *14*, 233–239.

Miyamoto, K., Ogura, Y., Kenmochi, S., & Honda, Y. (1997). Role of leukocytes in diabetic microcirculatory disturbances. *Microvascular Research*, *54*, 43–48.

Nagy, J. A., Dvorak, A. M., & Dvorak, H. F. (2007). VEGF-A and the induction of pathological angiogenesis. *Annual Review of Pathology: Mechanisms of Disease*, *2*, 251–275.

Nishijima, K., Ng, Y. S., Zhong, L., Bradley, J., Schubert, W., Jo, N., et al. (2007). Vascular endothelial growth factor-A is a survival factor for retinal neurons and a critical neuroprotectant during the adaptive response to ischemic injury. *American Journal of Pathology*, *171*, 53–67.

Ola, M. S., Berkich, D. A., Xu, Y., King, M. T., Gardner, T. W., Simpson, I., et al. (2006). Analysis of glucose metabolism in diabetic rat retinas. *American Jouranl of Physiology. Endocrinology and Metabolism*, *290*, E1057–E1067.

Oosthuyse, B., Moons, L., Storkebaum, E., Beck, H., Nuyens, D., Brusselmans, K., et al. (2001). Deletion of the hypoxia-response element in the vascular endothelial growth factor promoter causes motor neuron degeneration. *Nature Genetics*, *28*, 131–138.

Oshima, Y., Sakaguchi, H., Gomi, F., & Tano, Y. (2006). Regression of iris neovascularization after intravitreal injection of bevacizumab in patients with proliferative diabetic retinopathy. *American Journal of Ophthalmology*, *142*, 155–158.

Paques, M., Boval, B., Richard, S., Tadayoni, R., Massin, P., Mundler, O., et al. (2000). Evaluation of fluorescein-labeled autologous leukocytes for examination of retinal circulation in humans. *Current Eye Research.*, *21*, 560–565.

Pe'er, J., Folberg, R., Itin, A., Gnessin, H., Hemo, I., & Keshet, E. (1996). Upregulated expression of vascular endothelial growth factor in proliferative diabetic retinopathy. *British Journal of Ophthalmology*, *80*, 241–245.

Penn, J. S., Madan, A., Caldwell, R. B., Bartoli, M., Caldwell, R. W., & Hartnett, M. E. (2008). Vascular endothelial growth factor in eye disease. *Progress in Retinal and Eye Research*, *27*, 331–371.

Peters, S., Heiduschka, P., Julien, S., Ziemssen, F., Fietz, H., Bartz-Schmidt, K. U., et al. (2007). Ultrastructural findings in the primate eye after intravitreal injection of bevacizumab. *American Journal of Ophthalmology*, *143*, 995–1002.

Pieramici, D. J., Rabena, M., Castellarin, A. A., Nasir, M., See, R., Norton, T., et al. (2008). Ranibizumab for the treatment of macular edema associated with perfused central retinal vein occlusions. *Ophthalmology*, *115*, e47–e54.

Raviola, G. (1977). The structural basis of the blood-ocular barriers. *Experimental Eye Research*, *25*(Suppl.), 27–63.

Rich, R. M., Rosenfeld, P. J., Puliafito, C. A., Dubovy, S. R., Davis, J. L., Flynn, H. W., Jr., et al. (2006). Short-term safety and efficacy of intravitreal bevacizumab (Avastin) for neovascular age-related macular degeneration. *Retina*, *26*, 495–511.

Rosenfeld, P. J., Brown, D. M., Heier, J. S., Boyer, D. S., Kaiser, P. K., Chung, C. Y., et al. (2006). Ranibizumab for neovascular age-related macular degeneration. *New England Journal of Medicine*, *355*, 1419–1431.

Rosenfeld, P. J., Moshfeghi, A. A., & Puliafito, C. A. (2005a). Optical coherence tomography findings after an intravitreal injection of bevacizumab (Avastin) for neovascular age-related macular degeneration. *Ophthalmic Surgery, Lasers & Imaging*, *36*, 331–335.

Rosenfeld, P. J., Schwartz, S. D., Blumenkranz, M. S., Miller, J. W., Haller, J. A., Reimann, J. D., et al. (2005b). Maximum tolerated dose of a humanized anti-vascular endothelial growth factor antibody fragment for treating neovascular age-related macular degeneration. *Ophthalmology*, *112*, 1048–1053.

Schroder, S., Palinski, W., & Schmid-Schonbein, G. W. (1991). Activated monocytes and granulocytes, capillary nonperfusion, and neovascularization in diabetic retinopathy. *American Journal of Pathology*, *139*, 81–100.

Senger, D. R., Galli, S. J., Dvorak, A. M., Perruzzi, C. A., Harvey, V. S., & Dvorak, H. F. (1983). Tumor cells secrete a vascular permeability factor that promotes accumulation of ascites fluid. *Science, 219,* 983–985.

Shahar, J., Avery, R. L., Heilweil, G., Barak, A., Zemel, E., Lewis, G. P., et al. (2006). Electrophysiologic and retinal penetration studies following intravitreal injection of bevacizumab (Avastin). *Retina, 26,* 262–269.

Shibuya, M. (2008). Vascular endothelial growth factor-dependent and –independent regulation of angiogenesis. *BMB Reports, 41,* 278–286.

Shimura, M., Nakazawa, T., Yasuda, K., Shiono, T., Iida, T., Sakamoto, T., et al. (2008). Comparative therapy evaluation of intravitreal bevacizumab and triamcinolone acetonide on persistent diffuse diabetic macular edema. *American Jouranl of Ophthalmology, 145,* 854–861.

Shinoda, K., Ishida, S., Kawashima, S., Wakabayashi, T., Matsuzaki, T., Takayama, M., et al. (1999). Comparison of the levels of hepatocyte growth factor and vascular endothelial growth factor in aqueous fluid and serum with grades of retinopathy in patients with diabetes mellitus. *British Journal of Ophthalmology, 83,* 834–837.

Shweiki, D., Itin, A., Soffer, D., & Keshet, E. (1992). Vascular endothelial growth factor induced by hypoxia may mediate hypoxia-initiated angiogenesis. *Nature, 359,* 843–845.

Smith, L. E., Shen, W., Perruzzi, C., Soker, S., Kinose, F., Xu, X., et al. (1999). Regulation of vascular endothelial growth factor-dependent retinal neovascularization by insulin-like growth factor-1 receptor. *Nature Medicine, 5,* 1390–1395.

Spaide, R. F., Chang, L. K., Klancnik, J. M., Yannuzzi, L. A., Sorenson, J., Slakter, J. S., et al. (2009). Prospective study of intravitreal ranibizumab as a treatment for decreased visual acuity secondary to central retinal vein occlusion. *American Journal of Ophthalmology, 147*(2), 298–306.

Spaide, R. F., & Fisher, Y. L. (2006). Intravitreal bevacizumab (Avastin) treatment of proliferative diabetic retinopathy complicated by vitreous hemorrhage. *Retina, 26,* 275–278.

Spandau, U. H., & Jonas, J. B. (2006). Retinal pigment epithelium tear after intravitreal bevacizumab for exudative age-related macular degeneration. *American Journal of Ophthalmology, 142,* 1068–1070.

Stalmans, I. (2005). Role of the vascular endothelial growth factor isoforms in retinal angiogenesis and DiGeorge syndrome. *Verhandelingen-Koninklijke Academie voor Geneeskunde van Belgie, 67,* 229–276.

Stalmans, I., Ng, Y. S., Rohan, R., Fruttiger, M., Bouché, A., Yuce, A., et al. (2002). Arteriolar and venular patterning in retinas of mice selectively expressing VEGF isoforms. *Journal of Clinical Invest., 109,* 327–336.

Stitt, A. W., Li, Y. M., Gardiner, T. A., Bucala, R., Archer, D. B., & Vlassara, H. (1997). Advanced glycation end products (AGEs) co-localize with AGE receptors in the retinal vasculature of diabetic and of AGE-infused rats. *American Journal of Pathology, 150,* 523–531.

Tolentino, M. J., McLeod, D. S., Taomoto, M., Otsuji, T., Adamis, A. P., & Lutty, G. A. (2002). Pathologic features of vascular endothelial growth factor-induced retinopathy in the nonhuman primate. *American Journal of Ophthalmology, 133,* 373–385.

Tolentino, M. J., Miller, J. W., Gragoudas, E. S., Chatzistefanou, K., Ferrara, N., & Adamis, A. P. (1996a). Vascular endothelial growth factor is sufficient to produce iris neovascularization and neovascular glaucoma in a nonhuman primate. *Archives of Ophthalmology, 114,* 964–970.

Tolentino, M. J., Miller, J. W., Gragoudas, E. S., Jakobiec, F. A., Flynn, E., Chatzistefanou, K., et al. (1996b). Intravitreous injections of vascular endothelial growth factor produce retinal ischemia and microangiopathy in an adult primate. *Ophthalmology, 103,* 1820–1828.

Wakabayashi, T., Oshima, Y., Sakaguchi, H., Ikuno, Y., Miki, A., Gomi, F., et al. (2008). Intravitreal bevacizumab to treat iris neovascularization and neovascular glaucoma secondary to ischemic retinal diseases in 41 consecutive cases. *Ophthalmology, 115,* 1571–1580.

Wells, J. A., Murthy, R., Chibber, R., Nunn, A., Molinatti, P. A., Kohner, E. M., et al. (1996). Levels of vascular endothelial growth factor are elevated in the vitreous of patients with subretinal neovascularisation. *British Journal of Ophthalmology, 80,* 363–366.

Williamson, J. R., Chang, K., Frangos, M., Hasan, K. S., Ido, Y., Kawamura, T., et al. (1993). Hyperglycemic pseudohypoxia and diabetic complications. *Diabetes, 42,* 801–813.

Witmer, A. N., Blaauwgeers, H. G., Weich, H. A., Alitalo, K., Vrensen, G. F., & Schlingemann, R. O. (2002a). Altered expression patterns of VEGF receptors in human diabetic retina and in experimental VEGF-induced retinopathy in monkey. *Investigative Ophthalmology & Visual Science, 43,* 849–857.

Witmer, A. N., Dai, J., Weich, H. A., Vrensen, G. F., & Schlingemann, R. O. (2002b). Expression of vascular endothelial growth factor receptors 1, 2, and 3 in quiescent endothelia. *Journal of Histochemistry and Cytochemistry, 50,* 767–778.

Witmer, A. N., van Blijswijk, B. C., van Noorden, C. J., Vrensen, G. F., & Schlingemann, R. O. (2004). In vivo angiogenic phenotype of endothelial cells and pericytes induced by vascular endothelial growth factor-A. *Journal of Histochemical of Cytochemistry, 52,* 39–52.

Witmer, A. N., Vrensen, G. F., Van Noorden, C. J., & Schlingemann, R. O. (2003). Vascular endothelial growth factors and angiogenesis in eye disease. *Progress in Retinal and Eye Research, 22,* 1–29.

CHAPTER 18

# Pharmacological modification of the extracellular matrix to promote regeneration of the injured brain and spinal cord

Nicole Brazda and Hans Werner Müller[*]

*Molecular Neurobiology Laboratory, Department of Neurology, Heinrich Heine University of Düsseldorf, Düsseldorf, Germany*

**Abstract:** This chapter focuses on the role of the fibrous lesion scar as a major impediment for axonal regeneration in the injured central nervous system (CNS). We describe the appearance and complementary distribution of the glial and fibrous scar components in spinal cord lesions focusing on the morphology as well as on axon growth inhibitory molecular components accumulating in the collagenous and basement membrane-rich fibrous scar. We further report on the differential responses to fibrous scar of distinct fiber tracts in the injured spinal cord including the rubrospinal and corticospinal tracts as well as serotonergic, dopaminergic, and calcitonin gene-related peptide (CGRP) systems. Finally, we discuss therapeutic strategies to suppress fibrous scarring in traumatic CNS injury with particular emphasis on a unique pharmacological treatment using iron chelators and cyclic adenosine monophosphate (cAMP) to inhibit collagen biosynthesis. The latter treatment has been shown to promote long-distance axon growth, retrograde protection of injured neurons, and significant functional improvement.

**Keywords:** axon regeneration; basement membrane; collagen IV; fibrous scar; functional recovery; glial scar; iron chelator; spinal cord repair

## Introduction

Traumatic brain and spinal cord injury (SCI) have devastating consequences, because, in contrast to the peripheral nervous system (PNS), interrupted neuronal interconnections do not restore in the adult mammal. Although lesioned axons of central nervous system (CNS) neurons are capable to regrow in a favorable environment such as PNS tissue implants, they do not spontaneously regenerate within the CNS. At least they do not extend far enough to reestablish lost functional connections with their original targets. The reason for this inherent difference of CNS versus PNS neurons might be an interplay between the following factors: (a) deficits in intrinsic neuronal capacities to start a regeneration program after lesion, (b) lack of growth support, for example, of neurotrophic factors, (c) suppression of axonal outgrowth by inhibitory molecules present in the environment. Moreover, some neuronal subpopulations of the CNS might respond more rigorously to distinct inhibitory molecules than others.

[*]Corresponding author.
Tel.: +49 211 81 18410; Fax: +49 211 81 18411;
E-mail: hanswerner.mueller@uni-duesseldorf.de

DOI: 10.1016/S0079-6123(09)17518-0

This review focuses on a specific molecular barrier that appears at the lesion site after brain and SCI, the fibrous lesion scar. In contrast to the glial scar, which is located at the penumbra of a lesion and is characterized by reactive astrogliosis, the fibrous scar resides in the center of the lesion and is virtually free of reactive astrocytes, as revealed by immunohistological staining using antibodies directed to the astroglial marker protein GFAP (glial fibrillary acidic protein). The impact of the fibrous scar on regeneration failure of different axonal populations in brain and spinal cord lesions will be reviewed. Furthermore, a new and effective pharmacological approach to suppress fibrous scar formation will be described, and consequences with respect to long-distance axon growth, neuronal protection and functional recovery will be discussed.

## Lesion scarring after CNS injury

It is commonly accepted that lesion scarring after CNS injury is a major impediment for axonal regeneration (Fawcett, 2006; Grimpe and Silver, 2002; Shearer and Fawcett, 2001; Silver and Miller, 2004; Stichel and Müller, 1998). The lesion scar is composed of the fibrous scar consisting of a dense extracellular matrix (ECM) network which is deposited in the lesion center (Fig. 1). The latter is surrounded by the glial scar, which is *par definitionem* the area of astrogliosis and characterized by high immunoreactivity to GFAP. The backbone of the fibrous scar tissue is a collagen type IV (Coll IV) network, which forms a sheet-like cicatrix (Timpl, 1989; Yurchenco and Schittny, 1990) and renders the fibrous scar detectable by Coll IV immunoreactivity (Hermanns and Müller, 2001; Kawano et al., 2005; Stichel et al., 1999a, b). Like in other tissues, the CNS wound healing scar delimits healthy tissue from the non-CNS environment (Shearer and Fawcett, 2001). The predominant supramolecular structure of the fibrous scar is the basement membrane in which Coll IV is a prominent key component forming a polygonal network. Coll IV and the basement membrane are not inhibitory for axonal outgrowth per se as revealed by tissue culture studies, where Coll IV supported outgrowth of explanted mouse dorsal root ganglia (DRG) without addition of growth factors (Tonge et al., 1997). Instead, the growth barrier function of the fibrous lesion scar in CNS is most likely due to axon growth inhibitory molecules which bind to

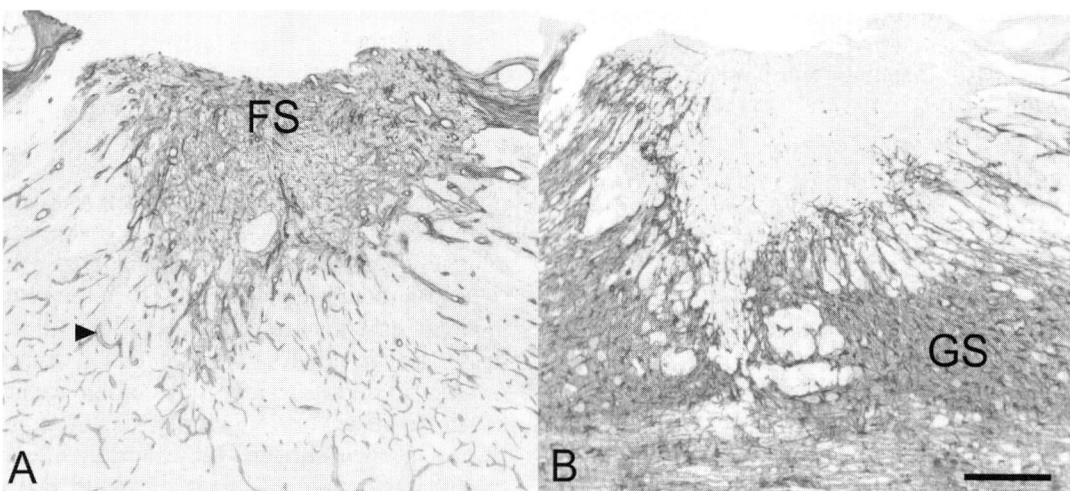

Fig. 1. Immunohistological staining of the collagenous scar with antibodies against Coll IV (A) and the glial scar with antibodies against GFAP (B) in adjacent sections at 7 days pL in a transection lesion at thoracic level of rat spinal cord. The two different scar types are located complementary to each other: FS, fibrous scar; GS, glial scar. Arrowhead: Coll IV-positive blood vessel in uninjured tissue. Magnification bar: 500 μm.

the sticky basement membrane in the scar and thus accumulate at the lesion site to high local concentrations.

## Morphology of the fibrous scar

The fibrous scar in transection injuries of tracts in the brain of adult rats or young mice appears as a thin line of appoximately 50 μm thickness (Hermanns et al., 2001a; Kawano et al., 2005). In contrast, after transection injury in spinal cord, the fibrous scar exhibits a more expansive morphology with a rostro–caudal dimension of 2–3 mm (Hermanns et al., 2001a). Besides transection injuries, other types of brain trauma also yield fibrous scar formation, as was shown for ischemia in rat brain (Ellison et al., 1999) and in mice suffering from experimental autoimmune encephalomyelitis (Brazda and Müller, unpublished observation).

The morphological appearance of the fibrous scar in different rat SCI models is rather variable. While transection lesions reproducibly result in evenly distributed ECM tissue in the lesion site in most experiments, contusion or crush injuries often exhibit cystic fluid-filled cavity formation (Steward et al., 1999). On the one hand, injuries with closed dura might increase the risk of enhanced intraspinal pressure due to accumulation of tissue liquids leaking out of the destroyed central canal or damaged blood vessels which, in turn, could cause the formation of fibrous scarring around the fluid-filled cyst. On the other hand, tissue-processing methods have great impact on the preservation of the fragile fibrous scar particularly at early timepoints after CNS injury (Hermanns and Müller, 2001).

The density of fibrous scarring depends on the lesion model. Seitz et al. (2002) showed that disruption of the dura and dislocation of the spinal cord stumps in mice enhanced fibrous scarring, while scarring was less intense in animals with intact dura. Functional recovery in these animals could be directly correlated with the amount of fibrous scarring.

The morphology of the fibrous scar also depends on the animal lesion model, since there seems to be major differences in scarring after crush injury between rats and mice, the latter exhibiting very few cystic cavities after crush injury (Inman and Steward, 2003b). Nevertheless, other features of the fibrous scar, like the molecular composition and complementary spatial distribution to the glial scar after transection is very similar in mice (Camand et al., 2004), compared to rat and even in cat (Risling et al., 1993).

A remaining but very important question is whether fibrous scarring in human patients is comparable to that in animal models of SCI. There are still only very limited data on collagenous lesion scarring in human SCI. A recent paper by Buss et al. (2007) investigated postmortem spinal cord tissue from human patients who died between 2 days and 30 years after SCI. The presented data reveal that a fibrous scar is formed after human SCI. The authors describe staining patterns for fibronectin as loose extracellular network and for Coll IV as sheet-like lamellae, structures which appear within 24 days after injury in the lesion epicenter and where Schwann cell infiltration is observed. These findings are important with regard to the transfer of potential therapies directed to alteration of fibrous scarring in SCI toward clinical application.

## Molecular composition of the fibrous scar

As mentioned above, after CNS lesion, a plethora of secreted molecules bind to the Coll IV network which is the backbone of the fibrous scar. Therefore, the fibrous scar offers a mixture of growth promoting and inhibiting molecules (as reviewed in Condic and Lemons, 2002; Klapka and Müller, 2006; Niclou et al., 2006). Besides structural molecules like collagens IV, I, and III (Maxwell et al., 1984), laminin, fibronectin, and nidogen/entactin, putative growth-inhibiting molecules like NG2, phosphacan, versican, syndecan-2, heparan sulfate proteoglycan (HSPG), keratan sulfate proteoglycan (KSPG), chondroitin sulfate proteoglycans (CSPG; Snow et al., 1990; Morgenstern et al., 2002; Jones et al., 2002; Tang et al., 2003; Davies et al., 2004), ephrins (Bundesen et al., 2003), tenascin-C (TN-C; Tang et al., 2003; Camand et al., 2004), TN-R (Deckner et al., 2000), Sema 3A (De Winter et al., 2002; Niclou et al., 2003), Sema 3B, Sema 3C, Sema 3E, Sema 3F (Niclou et al., 2006), and repulsive guidance molecule (RGM; Schwab et al., 2005) are found in the fibrous scar after CNS injury.

As observed in sophisticated long-time experiments, these associated molecules are not implicitly stable in their concentration inside scar tissue, which is a highly dynamic structure. For the following molecules, a change in concentration in the fibrous scar could be detected:

- TN-C peaked at 8 days pL, went down at later timepoints in mouse (Camand et al., 2004) and rat (Tang et al., 2003)
- CS-56 peaked at 8 days pL, went down to control levels at later stages in mouse (Camand et al., 2004)
- NG2 peaked at 8 days pL, decreased over time (up to 6 months pL) in rat (Tang et al., 2003)
- Phosphacan peaked at 1 month pL in rat (Tang et al., 2003)

The fibrous scar is associated with nonneural cells, for example, meningeal fibroblasts, inflammatory cells (Camand et al., 2004), and endothelial cells, which secrete extracellular matrix molecules (Berry et al., 1983; Schwab et al., 2001) and NG2-positive oligodendrocyte precursor cells (Fawcett and Asher, 1999).

## Differences in axonal growth responses to scar formation

Ramon y Cajal (1928) was the first to describe the CNS lesion scar in the year 1928 and postulated a major role as barrier for axonal regeneration, because he noticed aberrant turning back of axons encountering the scar. Since this time, numerous studies strengthened this hypothesis (for review see Stichel and Müller, 1998; Shearer and Fawcett, 2001; Grimpe and Silver, 2002).

In a very elegant study, Davies et al. (1997) demonstrated that dissociated P8 and adult DRG neurons were able to grow in brain's white matter (corpus callosum and fimbria) when no fibrous scarring was present. If a fibrous scar existed at the microtransplantation site, DRG axons stopped or actively turned away from the scar.

In a consecutive experiment (Davies et al., 1999), they implanted these neurons into degenerating white matter of the spinal cord acutely, 2 weeks or 3 months after a dorsal column lesion at several millimeters distance. Dissociated DRG axons grew out in degenerating white matter, but reaching the lesion site, they stopped at the fibrous scar, where their growth cones became dystrophic.

Furthermore, labeled corticospinal tract (CST) axons grow out after SCI, but abruptly stop at the fibrous scar (as shown, e.g., by Jones et al., 2002; Klapka et al., 2005).

Tracing studies of motor fiber tracts confirm the repulsive effect of the fibrous scar, but general neuron markers yield inconsistent results, since neurofilament immunoreactivity is outspread in the fibrous scar area after CNS injury. To avoid false interpretation of immunohistological experimental data by extrapolating such observations to draw general conclusions about the repulsive effect of the scar on axons, one must take into account that brain and spinal cord axons are not homogenous and uniform, but split into subpopulations according to their function, development, region of origin, neurotransmitters, receptors, etc. A definite review of the literature exposes profound differences in axonal responses to scar contact as shown in Table 1.

The literature overview reveals that the spinal subpopulation of ascending sensory fibers, including calcitonin gene-related peptide (CGRP)-positive neurons, glycinergic neurons, and neurons immunopositive for Substance P, grow to some extent in the fibrous scar after SCI, while descending serotonergic axons, the CST, rubrospinal tract (RST), and reticulospinal tract, are very rarely detected in the fibrous scar. Recent investigations in our laboratory elicited the inherent differences in growth arrest on fibrous scar encounter of five axon populations, that is, serotonergic, dopaminergic, ascending sensory, CST, and RST axons, in a quantitative direct comparative study (Schiwy et al., 2008). Quantification of axon numbers in the fibrous scar after SCI revealed major differences between these subpopulations when no treatment was applied.

Table 1. Spontaneous differential responses of neuronal subpopulations to fibrous lesion scar in rodent CNS

| Neuron type | Lesion type | Growth in fibrous scar | Reference |
| --- | --- | --- | --- |
| 5-HT (serotonergic) | Transection SCI | − | Camand et al. (2004) |
| | | +/− | Risling et al. (1993) |
| | | + | de Castro et al. (2005) |
| | Crush/compression SCI | +/− | Brook et al. (1998), Inman and Steward (2003a) |
| Ascending sensory BDA-labeled fibers | Transection SCI | + | Frisen et al. (1993) |
| | Crush/compression SCI | ++ | Inman and Steward (2003a) |
| CGRP (sensory) | Transection SCI | ++ | de Castro et al. (2005) |
| | Crush/compression SCI | − | Inman and Steward (2003a) |
| | | + | Brook et al. (1998) |
| CST | Transection SCI | − | Jones et al. (2002) |
| | | − | Hermanns et al. (2001a), Klapka et al. (2005) |
| | Crush/compression SCI | − | Inman and Steward (2003a) |
| RST | Transection SCI | − | Houle and Jin (2001) |
| | Crush/compression SCI | − | Inman and Steward (2003a) |
| Reticulospinal tract | Crush/compression SCI | − | Inman and Steward (2003a) |
| Glycinergic neurons | Crush/compression SCI | + | Brook et al. (1998) |
| Substance P | Crush/compression SCI | + | Brook et al. (1998) |
| Postcomissural fornix | Brain transection | − | Stichel et al. (1999a) |
| Nigrostriatal dopaminergic | Brain transection | − | Kawano et al. (2005) |

*Abbreviations*: 5-HT, serotonin; BDA, biotinylated dextran amine (tracer substance); CGRP, calcitonin gene-related peptide; RST, rubrospinal tract; "−," no fibres; "+/−," very few fibres; "+," many fibres; "++," high number of fibers.

Suppression of fibrous scar formation by a pharmacological treatment developed in our laboratory significantly enhanced axonal growth in the fibrous scar and back into distal spinal cord of all investigated spinal tracts, those which did not enter scar tissue in control lesioned animals and even those which inherently grew into the scar without treatment. These results will be shown in detail later on.

The reason for the inherent differences of neuronal subpopulations to the fibrous scar is not known, but it is interesting in this context that sensory neurons can upregulate integrin receptors enabling them to efficiently grow on laminin in vitro (Condic and Letourneau, 1997). Integrins interact with CSPG and semaphorins, and their expression and stability can be influenced by extracellular molecules as reviewed by Lemons and Condic (2008). Manipulation of integrin expression enhances axonal regeneration (Condic, 2001), and myelin inhibitors might also act via integrin (Hu and Strittmatter, 2008). Moreover [as reviewed in Condic and Lemons (2002); Klapka and Müller (2006)] CSPG and tenscain molecules exhibit different effects on neuronal subpopulations, depending on neuron type and molecular environment, for example, cerebellar neurons are inhibited in vitro by NG2, while DRG neurons are not under the same conditions, that is, grown on L1 (Ng-CAM). CNS neurons are inhibited by phosphacan when grown on L1, but not on fibronectin or poly-L-lysin.

The interplay between nonrepulsive ECM molecules and known inhibitors, such as Sema 5A, which is attractive or repulsive depending on the presence of HSPG or CSPG (Kantor et al., 2004), might also add to the complexity of the mechanism of fibrous scar induced growth arrest.

**Therapies to prevent fibrous scar formation**

To overcome inhibition by the fibrous scar, most experimental approaches target single inhibitors of the scar, for example, by using specific antibodies or enzymes such as neutralization of the CSPG

NG2, by antibody application (Tan et al., 2006), or by enzymatic degradation of CSPG, or by chondroitinase ABC (Bradbury et al., 2002). A more general therapeutic intervention is the suppression of collagenous basement membrane formation. As discussed before (see above), Coll IV forms the backbone of the fibrous scar, which offers binding sites for a plethora of growth-inhibiting molecules, which can be stained immunohistologically in the fibrous scar. Up to date, the impact of single scar-associated inhibitors, like NG2, Ehrins, or Sema 3A, on functional outcome after SCI cannot be assessed directly. Blocking one single inhibitor may yield regeneration and functional benefit up to a certain extent, as was shown for Sema 3A (Kaneko et al., 2006), but regarding the multiplicity of inhibitors, it seems unlikely that a single key component solves the problem of axonal growth failure for all neuronal populations affected. On the other hand, application of a cocktail of blockers to diverse inhibitory molecules of the fibrous scar was not tried yet, probably due to low clinical feasibility. Therefore, suppression of collagenous scar formation might not influence secretion of inhibitory molecules by astrocytes; nevertheless, it deprives inhibitors of their binding sites and prevents accumulation at the lesion center. Related therapeutic approaches aim at the prevention and transient suppression of fibrous scar formation and do not destroy existing extracellular matrix. Early experiments directed toward the enzymatic digestion of the fibrous scar by protease treatment also affected the extracellular matrix of intact blood vessels, causing massive bleeding and thus counteracted the potential beneficial effect of the strategy (Matinian and Andreasian, 1973; Guth et al., 1980).

Recent strategies to suppress fibrous scar formation involve (for references see table 2):

(a) inhibition of Coll IV network formation by either Coll IV antibody infusion or iron chelator application,
(b) molecular inhibition of key molecules of ECM production, that is, transforming growth factor β (TGFβ) by antibodies or decorin infusion,
(c) surgical methods, that is, sealing off the dura for inhibition of meningeal fibroblast invasion and thus ECM deposition after SCI,
(d) immunosuppression for reduction of ECM secreting cells,
(e) direct inhibition of cell proliferation after SCI by X-irradiation, affecting also ECM secreting cells.

Those studies that achieved suppression of ECM accumulation at the lesion site and described axonal behavior reported axonal regeneration in the case of brain tracts even into former target areas (Stichel et al., 1999a; Kawano et al., 2005). These data strongly support the reasoning that suppression of the underlying ECM network of the fibrous scar leads to a beneficial environment for axonal regeneration. There is compelling evidence that fibrous scar suppression not only allows axonal regeneration, but also yields functional recovery after SCI, as will be shown in detail below.

Very interestingly, two recent studies examined the "side"-effect of different important therapeutic strategies on fibrous scar formation. Li et al. (2007) and Teng et al. (2008) tested the application of chondroitinase ABC on the one hand and implantation of olfactory ensheathing glia cells on the other, on their influence on fibrous scar formation after brain lesion. The major outcome of these experiments was that both interventions prevent collagenous scar formation at the lesion site, which was considered to be important for the beneficial effects of the treatments on axon regeneration.

## Pharmacological inhibition of scarring with iron chelators

### Principle of scar suppression by iron chelators

One of the key enzymes of Coll IV biosynthesis is prolyl 4-hydroxylase (P4H), which catalyzes hydroxylation of procollagen chains in fibroblasts and other Coll IV-producing cells (Yurchenco and O'Rear, 1994; Myllyharju, 2003). Underhydroxylated procollagen chains are proteolytically

Table 2. Therapies aiming at the prevention of fibrous scar formation after CNS lesion

| Putative mechanism of scar reduction | Therapy | Lesion model | Coll IV-IR in FS | FS components | Axonal regeneration | Functional benefit | Reference |
|---|---|---|---|---|---|---|---|
| Inhibition of Coll IV matrix formation | Iron chelator injection | TX brain | ↓ | Laminin ↓ | Postcommissural fornix[a] | NA | Stichel et al. (1999a) |
| | | | ↓ | NG2 ↓ | Nigrostriatal dopaminergic neurons[a] | NA | Kawano et al. (2005) |
| | | TX SC | ↓ | NG2 ↓ | CST (>10 mm into distal cord) | BBB, gridwalk, regular walking pattern (catwalk) | Klapka et al. (2005) |
| | Coll IV antibody injection | TX brain | ↓ | Laminin ↓ | Postcommissural fornix[a] | Electrophysiological conductance restored | Stichel et al. (1999a) |
| Inhibition of TGFβ and/or EGFR signaling to suppress ECM production | Decorin infusion | TX SC | NA | CSPG ↓, NG2 ↓; phosphacan ↓[b]; brevican ↓[b]; neurocan ↓[b] | Transplanted adult DRG neurons grow across lesion site, enter distal cord | NA | Davies et al. (2004) |
| | Antibody against TGFβ1 and TGFβ2 | TX brain | NA | No change in CS-56 | Injured nigrostriatal tract did not regenerate | NA | Moon and Fawcett (2001) |
| Reduction of connective tissue producing cells | Dural repair | TX SC | ↓ | Laminin ↓ | NA | NA | Iannotti et al. (2006) |
| | Immunosuppression | TX SC | NA | Basal lamina ↓ | NA | NA | Feringa et al. (1985) |
| | X-irradiation | Contusion SCI | NA | CSPG ↓ | NA | No recovery of BBB[c] | Zhang et al. (2005) |

All studies were performed in the rat, except for Kawano et al. (2005), who used mice.

*Abbreviations*: BBB, Basso Beattie Bresnahan locomotor score; CSPG, chondroitin sulfate proteoglycans; CS-56, antibody to CSPG; EGFR, epidermal growth factor receptor; FS, fibrous scar; NA, not analyzed; SC, spinal cord; TX, transection lesion.

[a] Regeneration into former target area.
[b] In the lesion penumbra.
[c] X-irradiation after SCI resulted in significant recovery of function (BBB) at 6 weeks post lesion in another study (Zeman et al. 2001).

degraded for the most part before they could be release into the extracellular space (Eleftheriades et al., 1995). A decrease of hydroxyproline content lowers the denaturation temperature of collagens (Burjanadze, 2000), which is normally above body temperature (Kühn, 1995). P4H is dependent on its cofactors iron, ascorbate, and 2-oxoglutarat. Iron chelators are therefore known and established inhibitors of P4H (Hales and Beattie, 1993; Wang et al., 2002).

In a rat brain lesion model, the transection of postcommissural fornix fibers, fibrous scarring could be suppressed up to 2 weeks post lesion (pL) by a single injection of the iron chelator 2,2′-dipyridyl (Stichel et al., 1999a, b), a result which could successfully be reproduced for nigrostriatal neurons by an indepenent laboratory (Kawano et al., 2005). The lack of fibrous scarring in the first 2 weeks after lesion facilitated axonal regeneration for both brain tracts into their former target area (Stichel et al., 1999a; Kawano et al., 2005). Transferring the scar suppressing treatment into the clinically relevant SCI model, the protocol used in deep brain lesions had to be adapted to the condition of a far more profound fibrous scarring reaction in the spinal cord, as described earlier. Therefore, a more potent inhibitor of the P4H, the iron chelator 2,2′-bipyridine-5,5′-dicarboxylic acid (BPY-DCA; Hales and Beattie, 1993) was selected (Hermanns et al., 2001b). Injection of BPY-DCA alone was not sufficient to reduce fibrous scarring, most probably due to the massive infiltration of the lesion area by ECM producing fibroblasts of the meninges (Carbonell and Boya, 1988). Since proliferation of fibroblasts is a TGFβ-induced autocrine process mediated by CTGF (connective tissue growth factor), a molecule expressed in fibroblasts and astrocytes after SCI (Conrad et al., 2005), cyclic adenosine monophosphate (cAMP) was additionally applied to the lesion area, which inhibits this proliferation process (Grotendorst, 1997; Duncan et al., 1999). Thus, a combination of direct BPY-DCA injection and cAMP into the midthoracic dorsal column transection site in the adult rat lead to fibrous scar suppression of up to 12–14 days (Hermanns et al., 2001b; Klapka et al., 2005). Concomitantly, the growth inhibitor NG2, which was tested as ECM-associated growth-inhibiting molecule, was found to be reduced in the lesion site by immunohistological staining methods (Klapka et al., 2005).

## Axonal regeneration following scar suppression with iron chelators

The CST retracts after SCI for 1–2 mm, then spontaneously forms growth cones and regenerates along the old path up to the fibrous scar where it abruptly stops (Schnell and Schwab, 1993; Jones et al., 2002; Klapka et al., 2005). Following scar suppressing treatment after dorsal column transection, anterogradely labeled CST axons were found to regenerate through and beyond the lesion site, which was deprived of ECM-associated inhibitors (Klapka et al., 2005). CST axons regenerated long distances (<1 cm) into the distal spinal cord at 3–4 months postinjury and treatment. Arborizations were found at the ends of these axons, decorated with varicosities, resembling presynaptic boutons (Fig. 2C).

To investigate the response of other spinal cord fiber tracts on scar suppressing treatment, elaborate tracing and neurotransmitter staining studies were performed recently in our laboratory (Schiwy et al., 2008). Quantification of axon numbers of five different neuronal subpopulations was performed at 5 and 12 weeks pL. The experiments revealed significantly enhanced ingrowth of serotonergic, CGRP-positive (sensory), tyrosin hydroxylase-positive, anterogradely labeled CST, and RST fibers into the lesion center of treated animals in comparison to control lesioned (Fig. 2A, B, D, and Schiwy et al., 2008). These results show for the first time a direct comparison of the growth repulsive effect of the fibrous scar on diverse neuronal subpopulations and a beneficial effect of scar suppression on all investigated fiber tracts.

## Neuroprotective effect of scar suppressing treatment

As was shown before (Hains et al., 2003), 35–42% of pyramidal neurons in the motor cortex of the adult rat projecting into the CST undergo

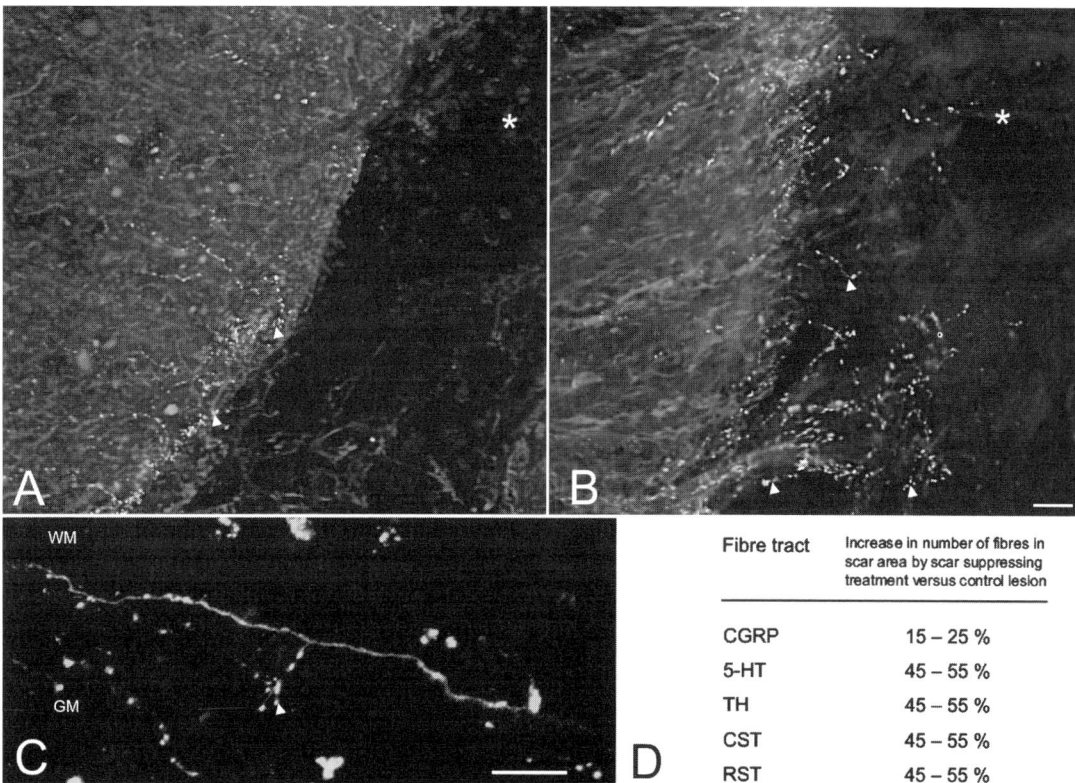

Fig. 2. Axonal regeneration of different fiber tracts after dorsal hemisection of the rat spinal cord at thoracic level 8 and scar suppressing treatment. Immunohistological double staining of the lesion site at 5 weeks postinjury with antibodies against serotonin (green) and GFAP (red) of a control lesioned rat (A) and a rat receiving the scar suppressing treatment. (B) In contrast to the control, numerous labeled axons enter the GFAP-negative collagenous scar after treatment in B. (C) Anterogradely BDA-labeled CST axon regenerating beyond the lesion site in the distal spinal cord at 4 months after injury and scar suppressing treatment (modified from Klapka et al., 2005). (D) Quantification of axon fragments of different fiber populations detected in the lesion center. The increase in fiber number within the collagenous scar area following the scar suppressing treatment versus treatment with buffer (control) is shown (see also Schiwy et al., 2008). Arrowheads in A and B: serotonergic axons. Arrowhead in C: varicosities on regenerated fiber. Asterisks in A and B: collagenous scar area. GM, gray matter; WM, white matter. Magnification bars: 50 μm. (See Color Plate 18.2 in color plate section.)

apoptosis after midthoracic SCI, a result we could confirm in our former study (Klapka et al., 2005). Application of the scar suppressing treatment rescued this proportion of neurons, which normally die after lesion as shown by quantification of retrogradely traced cortex layer V pyramidal neurons. The mechanism underlying this phenomenon is unclear, but could be explained by an influence of the applied iron chelator on oxidative stress reactions. Iron plays a key role in the formation of hydroxyl radicals and concomitantly other reactive oxidative species by the Fenton's reaction (Liu et al., 2003, 2004). Recent biochemical analysis of lesioned spinal cord tissue with and without treatment showed early differences (4 h pL) in the concentration of peroxidized lipids and oxidized proteins (own observation). Thus, further studies on oxidative stress reactions after treatment could possibly enlighten the neuroprotective effect of the scar suppressing treatment.

## Functional recovery following scar suppressing treatment

The dorsal CST of the rat is involved in fine motor control and placing response tasks (Metz and Whishaw, 2002; Whishaw and Metz, 2002).

Different functional locomotor tests were used to rate the deficits of rats receiving a dorsal column transection lesion at thoracic level 8 and following scar suppressing treatment (Klapka et al., 2005). The overall performance of lesioned rats was assessed using the Basso Beattie Bresnahan (BBB) scale in the open field; fine motor control and placing response was reflected by errors on the horizontal ladder (gridwalk test); and the regular walking pattern, revealing interlimb coordination, was quantified by the catwalk device, where rats run on a glass runway are video recorded from below. These tests were performed weekly and blinded to the experimentator over 4 months. At the end of the experiment, correct lesion depth of all animals was confirmed by immunohistological staining of spinal cord tissue slices. The scar suppressing treatment led to functional recovery in all locomotor tasks. Recovery of the BBB started between 7 and 10 weeks and in contrast to control animals with BBB levels of 13, treated animals reached sham-operated niveau with a BBB of 21. At the endpoint of the study, regular walking patterns of treated animals were also restored to preoperation baseline. The gridwalk test, which is the most suited for CST lesions, showed enduring deficits in the performance of spinal cord lesioned and treated animals. Nevertheless, beginning from week 1, treated animals performed significantly better in this task compared to control lesioned, reflecting better fine motor control and placing response. The early onset of these differences cannot be explained by axonal regeneration, but might originate from the above-discussed neuroprotective effects of the treatment.

## Conclusion

The fibrous scar is a major barrier to axonal regeneration in traumatic CNS injury, eliciting its effect by axon growth inhibitors which accumulate in the Coll IV-rich supramolecular network of the basement membrane. The response of lesioned axons which encounter the fibrous scar is mainly characterized by abrupt growth arrest, but shows specific differences depending on the fiber systems affected. Different fiber tracts vary in their scar penetrating abilities. For various experimental approaches aiming at suppression of fibrous scarring after SCI and brain lesions, positive effects on axonal regeneration have been described. A unique pharmacological scar suppressing treatment has been developed using an iron chelator and cAMP to inhibit collagen biosynthesis, which leads to neuroprotection and functional recovery of treated animals. Taken together, suppression of fibrous scarring is a promising therapeutic strategy for SCI that could be combined with most, if not all, complementary treatments known so far.

## Abbreviations

| | |
|---|---|
| 5-HT | serotonin |
| BBB | Basso Beattie Bresnahan locomotor score |
| BDA | biotinylated dextran amine |
| BPY-DCA | 2,2'-bipyridine-5,5'-dicarboxylic acid |
| cAMP | cyclic adenosine monophosphate |
| CGRP | calcitonin gene-related peptide |
| Coll IV | collagen type IV |
| CSPG | chondroitin sulfate proteoglycan |
| CST | corticospinal tract |
| CTGF | connective tissue growth factor |
| DRG | dorsal root ganglion |
| ECM | extracellular matrix |
| EGFR | epidermal growth factor receptor |
| FS | fibrous scar |
| GFAP | glial fibrillary acidic protein |
| GM | gray matter |
| GS | glial scar |
| HSPG | heparan sulfate proteoglycan |
| KSPG | keratan sulfate proteoglycan |
| P4H | prolyl 4-hydroxylase |
| PNS | peripheral nervous system |
| RGM | repulsive guidance molecule |
| RST | rubrospinal tract |
| SCI | spinal cord injury |
| Sema | semaphorin |
| TGFβ | transforming growth factor β |
| TH | tyrosin hydroxylase |
| TN | tenascin |

| | |
|---|---|
| TX | transection lesion |
| WM | white matter |
| wpL | weeks post lesion |

## Acknowledgments

Work in the authors' laboratory is supported by the Deutsche Forschungsgemeinschaft (DFG), Federal Research Ministry (BMBF), German Paraplegia Foundation (DSQ), and International Institute for Research in Paraplegia (IRP, Switzerland)

## References

Berry, M., Maxwell, W. L., Logan, A., Mathewson, A., McConnell, P., Ashhurst, D. E., et al. (1983). Deposition of scar tissue in the central nervous system. *Acta Neurochirurgica Supplement (Wien)*, 32, 31–53.

Bradbury, E. J., Moon, L. D., Popat, R. J., King, V. R., Bennett, G. S., Patel, P. N., et al. (2002). Chondroitinase ABC promotes functional recovery after spinal cord injury. *Nature*, 416, 636–640.

Brook, G. A., Plate, D., Franzen, R., Martin, D., Moonen, G., Schoenen, J., et al. (1998). Spontaneous longitudinally orientated axonal regeneration is associated with the Schwann cell framework within the lesion site following spinal cord compression injury of the rat. *Journal of Neuroscience Research*, 53, 51–65.

Bundesen, L. Q., Scheel, T. A., Bregman, B. S., & Kromer, L. F. (2003). Ephrin-B2 and EphB2 regulation of astrocyte-meningeal fibroblast interactions in response to spinal cord lesions in adult rats. *Journal of Neuroscience*, 23, 7789–7800.

Burjanadze, T. V. (2000). New analysis of the phylogenetic change of collagen thermostability. *Biopolymers*, 53, 523–528.

Buss, A., Pech, K., Kakulas, B. A., Martin, D., Schoenen, J., Noth, J., et al. (2007). Growth-modulating molecules are associated with invading Schwann cells and not astrocytes in human traumatic spinal cord injury. *Brain*, 130, 940–953.

Camand, E., Morel, M. P., Faissner, A., Sotelo, C., & Dusart, I. (2004). Long-term changes in the molecular composition of the glial scar and progressive increase of serotoninergic fibre sprouting after hemisection of the mouse spinal cord. *European Journal of Neuroscience*, 20, 1161–1176.

Carbonell, A. L., & Boya, J. (1988). Ultrastructural study on meningeal regeneration and meningo-glial relationships after cerebral stab wound in the adult rat. *Brain Research*, 439, 337–344.

Condic, M. L. (2001). Adult neuronal regeneration induced by transgenic integrin expression. *Journal of Neuroscience*, 21, 4782–4788.

Condic, M. L., & Lemons, M. L. (2002). Extracellular matrix in spinal cord regeneration: getting beyond attraction and inhibition. *Neuroreport*, 13, A37–A48.

Condic, M. L., & Letourneau, P. C. (1997). Ligand-induced changes in integrin expression regulate neuronal adhesion and neurite outgrowth. *Nature*, 389, 852–856.

Conrad, S., Schluesener, H. J., Adibzahdeh, M., & Schwab, J. M. (2005). Spinal cord injury induction of lesional expression of profibrotic and angiogenic connective tissue growth factor confined to reactive astrocytes, invading fibroblasts and endothelial cells. *Journal of Neurosurgery Spine*, 2, 319–326.

Davies, J. E., Tang, X., Denning, J. W., Archibald, S. J., & Davies, S. J. (2004). Decorin suppresses neurocan, brevican, phosphacan and NG2 expression and promotes axon growth across adult rat spinal cord injuries. *European Journal of Neuroscience*, 19, 1226–1242.

Davies, S. J., Goucher, D. R., Doller, C., & Silver, J. (1999). Robust regeneration of adult sensory axons in degenerating white matter of the adult rat spinal cord. *Journal of Neuroscience*, 19, 5810–5822.

Davies, S. J. A., Fitch, M. T., Memberg, S. P., Hall, A. K., Raisman, G., & Silver, J. (1997). Regeneration of adult axons in white matter tracts of the central nervous system. *Nature*, 390, 680–683.

de Castro, R., Jr., Tajrishi, R., Claros, J., & Stallcup, W. B. (2005). Differential responses of spinal axons to transection: influence of the NG2 proteoglycan. *Experimental Neurology*, 192, 299–309.

De Winter, F., Oudega, M., Lankhorst, A. J., Hamers, F. P., Blits, B., Ruitenberg, M. J., et al. (2002). Injury-induced class 3 semaphorin expression in the rat spinal cord. *Experimental Neurology*, 175, 61–75.

Deckner, M., Lindholm, T., Cullheim, S., & Risling, M. (2000). Differential expression of tenascin-C, tenascin-R, tenascin/J1, and tenascin-X in spinal cord scar tissue and in the olfactory system. *Experimental Neurology*, 166, 350–362.

Duncan, M. R., Frazier, K. S., Abramson, S., Williams, S., Klapper, H., Huang, X., et al. (1999). Connective tissue growth factor mediates transforming growth factor beta-induced collagen synthesis: down-regulation by cAMP. *FASEB Journal*, 13, 1774–1786.

Eleftheriades, E. G., Ferguson, A. G., Spragia, M. L., & Samarel, A. M. (1995). Prolyl hydroxylation regulates intracellular procollagen degradation in cultured rat cardiac fibroblasts. *Journal of Molecular and Cellular Cardiology*, 27, 1459–1473.

Ellison, J. A., Barone, F. C., & Feuerstein, G. Z. (1999). Matrix remodeling after stroke. De novo expression of matrix proteins and integrin receptors. *Annals of the New York Academy of Sciences*, 890, 204–222.

Fawcett, J. W. (2006). Overcoming inhibition in the damaged spinal cord. *Journal of Neurotrauma*, 23, 371–383.

Fawcett, J. W., & Asher, R. A. (1999). The glial scar and central nervous system repair. *Brain Research Bulletin*, 49, 377–391.

Feringa, E. R., Kowalski, T. F., & Vahlsing, H. L. (1985). Basal lamina at the site of spinal cord injury in normal, immunotolerant and immunosuppressed rats. *Neuroscience Letters, 54,* 225–230.

Frisen, J., Fried, K., Sjogren, A. M., & Risling, M. (1993). Growth of ascending spinal axons in CNS scar tissue. *International Journal of Developmental Neuroscience, 11,* 461–475.

Grimpe, B., & Silver, J. (2002). The extracellular matrix in axon regeneration. *Progress in Brain Research, 137,* 333–349.

Grotendorst, G. R. (1997). Connective tissue growth factor: a mediator of TGF-beta action on fibroblasts. *Cytokine and Growth Factor Reviews, 8,* 171–179.

Guth, L., Albuquerque, E. X., Deshpande, S. S., Barrett, C. P., Donati, E. J., et al. (1980). Ineffectiveness of enzyme therapy on regeneration in the transected spinal cord of the rat. *Journal of Neurosurgery, 52,* 73–86.

Hains, B. C., Black, J. A., & Waxman, S. G. (2003). Primary cortical motor neurons undergo apoptosis after axotomizing spinal cord injury. *Journal of Comparative Neurology, 462,* 328–341.

Hales, N. J., & Beattie, J. F. (1993). Novel inhibitors of prolyl 4-hydroxylase. 5. The intriguing structure–activity relationships seen with 2,2′-bipyridine and its 5,5′- dicarboxylic acid derivatives. *Journal of Medicinal Chemistry, 36,* 3853–3858.

Hermanns, S., Klapka, N., & Müller, H. W. (2001a). The collagenous lesion scar—an obstacle for axonal regeneration in brain and spinal cord injury. *Restorative Neurology and Neuroscience, 19,* 139–148.

Hermanns, S., & Müller, H. W. (2001). Preservation and detection of lesion-induced collagenous scar in the CNS depend on the method of tissue processing. *Brain Research. Brain Research Protocols, 7,* 162–167.

Hermanns, S., Reiprich, P., & Müller, H. W. (2001b). A reliable method to reduce collagen scar formation in the lesioned rat spinal cord. *Journal of Neuroscience Methods, 110,* 141–146.

Houle, J. D., & Jin, Y. (2001). Chronically injured supraspinal neurons exhibit only modest axonal dieback in response to a cervical hemisection lesion. *Experimental Neurology, 169,* 208–217.

Hu, F., & Strittmatter, S. M. (2008). The N-terminal domain of Nogo-A inhibits cell adhesion and axonal outgrowth by an integrin-specific mechanism. *Journal of Neuroscience, 28,* 1262–1269.

Iannotti, C., Zhang, Y. P., Shields, L. B., Han, Y., Burke, D. A., Xu, X. M., et al. (2006). Dural repair reduces connective tissue scar invasion and cystic cavity formation after acute spinal cord laceration injury in adult rats. *Journal of Neurotrauma, 23,* 853–865.

Inman, D. M., & Steward, O. (2003a). Ascending sensory, but not other long-tract axons, regenerate into the connective tissue matrix that forms at the site of a spinal cord injury in mice. *Journal of Comparative Neurology, 462,* 431–449.

Inman, D. M., & Steward, O. (2003b). Physical size does not determine the unique histopathological response seen in the injured mouse spinal cord. *Journal of Neurotrauma, 20,* 33–42.

Jones, L. L., Yamaguchi, Y., Stallcup, W. B., & Tuszynski, M. H. (2002). NG2 is a major chondroitin sulfate proteoglycan produced after spinal cord injury and is expressed by macrophages and oligodendrocyte progenitors. *Journal of Neuroscience, 22,* 2792–2803.

Kaneko, S., Iwanami, A., Nakamura, M., Kishino, A., Kikuchi, K., Shibata, S., et al. (2006). A selective Sema3A inhibitor enhances regenerative responses and functional recovery of the injured spinal cord. *Nature Medicine, 12,* 1380–1389.

Kantor, D. B., Chivatakarn, O., Peer, K. L., Oster, S. F., Inatani, M., Hansen, M. J., et al. (2004). Semaphorin 5A is a bifunctional axon guidance cue regulated by heparan and chondroitin sulfate proteoglycans. *Neuron, 44,* 961–975.

Kawano, H., Li, H. P., Sango, K., Kawamura, K., & Raisman, G. (2005). Inhibition of collagen synthesis overrides the age-related failure of regeneration of nigrostriatal dopaminergic axons. *Journal of Neuroscience Research, 80,* 191–202.

Klapka, N., Hermanns, S., Straten, G., Masanneck, C., Duis, D., Hamers, F. P., et al. (2005). Suppression of fibrous scarring in spinal cord injury of rat promotes long-distance regeneration of corticospinal tract axons, rescue of primary motoneurons in somatosensory cortex and significant functional recovery. *European Journal of Neuroscience, 22,* 3047–3058.

Klapka, N., & Müller, H. W. (2006). Collagen matrix in spinal cord injury. *Journal of Neurotrauma, 23,* 422–436.

Kühn, K. (1995). Basement membrane (type IV) collagen. *Matrix Biology, 14,* 439–445.

Lemons, M. L., & Condic, M. L. (2008). Integrin signaling is integral to regeneration. *Experimental Neurology, 209,* 343–352.

Li, H. P., Homma, A., Sango, K., Kawamura, K., Raisman, G., & Kawano, H. (2007). Regeneration of nigrostriatal dopaminergic axons by degradation of chondroitin sulfate is accompanied by elimination of the fibrotic scar and glia limitans in the lesion site. *Journal of Neuroscience Research, 85,* 536–547.

Liu, D., Liu, J., Sun, D., Alcock, N. W., & Wen, J. (2003). Spinal cord injury increases iron levels: catalytic production of hydroxyl radicals. *Free Radical Biology and Medicine, 34,* 64–71.

Liu, D., Liu, J., Sun, D., & Wen, J. (2004). The time course of hydroxyl radical formation following spinal cord injury: the possible role of the iron-catalyzed Haber–Weiss reaction. *Journal of Neurotrauma, 21,* 805–816.

Matinian, L. A., & Andreasian, A. S. (1973). *Akademia Nauk Armenian SSR.* [Enzyme therapy in organic lesions of the spinal cord in English.] Los Angeles: Brain Information Service, University of California (1976, p. 156).

Maxwell, W. L., Duance, V. C., Lehto, M., Ashurst, D. E., & Berry, M. (1984). The distribution of types I, III, IV and V collagens in penetrant lesions of the central nervous system of the rat. *Histochemical Journal, 16,* 1215–1229.

Metz, G. A., & Whishaw, I. Q. (2002). Cortical and subcortical lesions impair skilled walking in the ladder rung walking test: a new task to evaluate fore- and hindlimb stepping, placing, and co-ordination. *Journal of Neuroscience Methods, 115,* 169–179.

Moon, L. D., & Fawcett, J. W. (2001). Reduction in CNS scar formation without concomitant increase in axon regeneration following treatment of adult rat brain with a combination of antibodies to TGFbeta1 and beta2. *European Journal of Neuroscience, 14*, 1667–1677.

Morgenstern, D. A., Asher, R. A., & Fawcett, J. W. (2002). Chondroitin sulphate proteoglycans in the CNS injury response. *Progress in Brain Research, 137*, 313–332.

Myllyharju, J. (2003). Prolyl 4-hydroxylases, the key enzymes of collagen biosynthesis. *Matrix Biology, 22*, 15–24.

Niclou, S. P., Ehlert, E. M. E., & Verhaagen, J. (2006). Chemorepellent axon guidance molecules in spinal cord injury. *Journal of Neurotrauma, 23*, 409–421.

Niclou, S. P., Franssen, E. H., Ehlert, E. M., Taniguchi, M., & Verhaagen, J. (2003). Meningeal cell-derived semaphorin 3A inhibits neurite outgrowth. *Molecular and Cellular Neuroscience, 24*, 902–912.

Ramon y Cajal, S. (1928). *Degeneration and regeneration of the nervous system*. New York: Hafner.

Risling, M., Fried, K., Linda, H., Carlstedt, T., & Cullheim, S. (1993). Regrowth of motor axons following spinal cord lesions: distribution of laminin and collagen in the CNS scar tissue. *Brain Research Bulletin, 30*, 405–414.

Schiwy, N., Brazda, N., Estrada, V., & Müller, H. W. (2008). Differential axon growth capacities of various spinal cord fiber tracts following suppression of fibrous scarring. *Society for Neuroscience Abstract*, 74.2.

Schnell, L., & Schwab, M. E. (1993). Sprouting and regeneration of lesioned corticospinal tract fibres in the adult rat spinal cord. *European Journal of Neuroscience, 5*, 1156–1171.

Schwab, J. M., Beschorner, R., Nguyen, T. D., Meyermann, R., & Schluesener, H. J. (2001). Differential cellular accumulation of connective tissue growth factor defines a subset of reactive astrocytes, invading fibroblasts, and endothelial cells following central nervous system injury in rats and humans. *Journal of Neurotrauma, 18*, 377–388.

Schwab, J. M., Conrad, S., Monnier, P. P., Julien, S., Mueller, B. K., & Schluesener, H. J. (2005). Spinal cord injury-induced lesional expression of the repulsive guidance molecule (RGM). *European Journal of Neuroscience, 21*, 1569–1576.

Seitz, A., Aglow, E., & Heber-Katz, E. (2002). Recovery from spinal cord injury: a new transection model in the C57Bl/6 mouse. *Journal of Neuroscience Research, 67*, 337–345.

Shearer, M. C., & Fawcett, J. W. (2001). The astrocyte/meningeal cell interface—a barrier to successful nerve regeneration?. *Cell and Tissue Research, 305*, 267–273.

Silver, J., & Miller, J. H. (2004). Regeneration beyond the glial scar. *Nature Reviews Neuroscience, 5*, 146–156.

Snow, D. M., Lemmon, V., Carrino, D. A., Caplan, A. I., & Silver, J. (1990). Sulfated proteoglycans in astroglial barriers inhibit neurite outgrowth in vitro. *Experimental Neurology, 109*, 111–130.

Steward, O., Schauwecker, P. E., Guth, L., Zhang, Z., Fujiki, M., Inman, D., et al. (1999). Genetic approaches to neurotrauma research: opportunities and potential pitfalls of murine models. *Experimental Neurology, 157*, 19–42.

Stichel, C. C., Hermanns, S., Luhmann, H. J., Lausberg, F., Niermann, H., D'Urso, D., et al. (1999a). Inhibition of collagen IV deposition promotes regeneration of injured CNS axons. *European Journal of Neuroscience, 11*, 632–646.

Stichel, C. C., & Müller, H. W. (1998). The CNS lesion scar: new vistas on an old regeneration barrier. *Cell and Tissue Research, 294*, 1–9.

Stichel, C. C., Niermann, H., D'Urso, D., Lausberg, F., Hermanns, S., & Müller, H. W. (1999b). Basal membrane-depleted scar in lesioned CNS: characteristics and relationships with regenerating axons. *Neuroscience, 93*, 321–333.

Tan, A. M., Colletti, M., Rorai, A. T., Skene, J. H., & Levine, J. M. (2006). Antibodies against the NG2 proteoglycan promote the regeneration of sensory axons within the dorsal columns of the spinal cord. *Journal of Neuroscience, 26*, 4729–4739.

Tang, X., Davies, J. E., & Davies, S. J. (2003). Changes in distribution, cell associations, and protein expression levels of NG2, neurocan, phosphacan, brevican, versican V2, and tenascin-C during acute to chronic maturation of spinal cord scar tissue. *Journal of Neuroscience Research, 71*, 427–444.

Teng, X., Nagata, I., Li, H. P., Kimura-Kuroda, J., Sango, K., Kawamura, K., et al. (2008). Regeneration of nigrostriatal dopaminergic axons after transplantation of olfactory ensheathing cells and fibroblasts prevents fibrotic scar formation at the lesion site. *Journal of Neuroscience Research, 86*, 3140–3150.

Timpl, R. (1989). Structure and biological activity of basement membrane proteins. *European Journal of Biochemistry, 180*, 487–502.

Tonge, D. A., Golding, J. P., Edbladh, M., Kroon, M., Ekstrom, P. E., & Edstrom, A. (1997). Effects of extracellular matrix components on axonal outgrowth from peripheral nerves of adult animals in vitro. *Experimental Neurology, 146*, 81–90.

Wang, J., Buss, J. L., Chen, G., Ponka, P., & Pantopoulos, K. (2002). The prolyl 4-hydroxylase inhibitor ethyl-3,4-dihydroxybenzoate generates effective iron deficiency in cultured cells. *FEBS Letters, 529*, 309–312.

Whishaw, I. Q., & Metz, G. A. (2002). Absence of impairments or recovery mediated by the uncrossed pyramidal tract in the rat versus enduring deficits produced by the crossed pyramidal tract. *Behavioural Brain Research, 134*, 323–336.

Yurchenco, P. D., & O'Rear, J. J. (1994). Basal lamina assembly. *Current Opinion in Cell Biology, 6*, 674–681.

Yurchenco, P. D., & Schittny, J. C. (1990). Molecular architecture of basement membranes. *FASEB Journal, 4*, 1577–1590.

Zeman, R. J., Feng, Y., Peng, H., Visintainer, P. F., Moorthy, C. R., Couldwell, W. T., et al. (2001). X-irradiation of the contusion site improves locomotor and histological outcomes in spinal cord-injured rats. *Experimental Neurology, 172*, 228–234.

Zhang, S. X., Geddes, J. W., Owens, J. L., & Holmberg, E. G. (2005). X-irradiation reduces lesion scarring at the contusion site of adult rat spinal cord. *Histology and histopathology, 20*, 519–530.

CHAPTER 19

# The placebo response: neurobiological and clinical issues of neurological relevance

Antonella Pollo[1,*] and Fabrizio Benedetti[2]

[1]Department of Neuroscience, Faculty of Pharmacy, University of Turin; National Institute of Neuroscience, Turin, Italy
[2]Department of Neuroscience, University of Turin Medical School; National Institute of Neuroscience, Turin, Italy

**Abstract:** The recent upsurge in placebo research has demonstrated the sound neurobiological substrate of a phenomenon once believed to be only patient mystification, or at best a variable to control in clinical trials, bringing about a new awareness of its potential exploitation to the patient's benefit and framing it as a positive context effect, with the power to influence the therapy outcome.

Placebo effects have been described both in the experimental setting and in different clinical conditions, many of which are of neurological interest. Multiple mechanisms have been described, namely conditioning and cognitive factors like expectation, desire, and reward. A body of evidence from neurochemical, pharmacological, and neuroimaging studies points to the involvement of neural pathways specific to single conditions, such as the activation of the endogenous antinociceptive system during placebo analgesia or the release of dopamine in the striatum of parkinsonian patients experiencing placebo reduction of motor impairment.

The possible clinical applications of placebo studies range from the design of clinical trials incorporating specific recommendations and minimizing the use of placebo arms to the optimization of the context surrounding the patient, in order to maximize the placebo component present in any treatment.

**Keywords:** placebo; expectation; conditioning; pain; neurotherapy

## Introduction

The negative connotation carried by the word placebo is the heritage of decades of clinical trials in which the evaluation of the specific effect of a new treatment can be achieved only by eliminating the interfering nonspecific effects, grouped together in the placebo arm.

In recent times, however, interest in the neurobiology of the placebo effect has shed light on the positive aspect of the placebo, bringing about a new awareness of its potential exploitation to the patient's advantage and framing it as a positive context effect, with the power to influence the therapy outcome.

Placebo effects have been described in many different clinical and experimental settings, many of which are of neurological interest, ranging from

*Corresponding author.
Tel.: +39 011 6708491; Fax: +39 011 6708174;
E-mail: antonella.pollo@unito.it

motor disorders, like Parkinson's disease, and neuropsychiatric conditions, like depression and anxiety, to endocrine and immune systems changes (Benedetti, 2008a, b). Major effects have been observed in pain, and indeed placebo analgesia, that is, the lessening of pain experienced in response to a therapeutic act devoid of intrinsic analgesic activity, is widely used as a model to investigate the nature of the placebo response.

This chapter provides a short overview of neurochemical, pharmacological, and neuroimaging studies in the current lines of placebo research. Emphasis will be placed on mechanisms of action, enlightening the role of expectation and conditioning in activating neural pathways leading to specific clinical outcomes during both pharmacological and procedural placebo administration. Also, stress will be put on how this knowledge can translate into better clinical practice, optimizing the psychosocial context surrounding the patient, enhancing the environmental factors eliciting expectation of improvement, and designing new types of clinical trials.

In depth discussion of these topics can be found in a number of recent specific reviews (Benedetti, 2007, 2008a, b; Benedetti et al., 2005; Colloca and Benedetti, 2005; Finniss and Benedetti, 2005; Price et al., 2008; Enck et al., 2008).

## Two different meanings of the term "placebo effect"

The treatment nonspecific effects in the placebo arm of a clinical trial can be due to a number of different factors, which can be present in variable proportion and give a more or less important contribution to the total effect. The placebo biological phenomenon is one of them. Other factors are natural history (the time course of the symptom or disease, in the absence of any external intervention), regression to the mean (the tendency of a second assessment to give a value closer to the distribution mean), biases (e.g., the desire to please the clinician with the expected answers), and judgement errors. When the aim is the evaluation of a new treatment, as in clinical trials, it is not important to measure these factors separately. It is sufficient to subtract their sum from the overall effect. The term "placebo effect" is then used to refer to this sum of factors. On the other hand, when the aim is to study the biological phenomenon underlying the placebo effect, care must be taken to dissect it from confounding factors, by including a natural history group as a control for the placebo condition. The term "placebo effect" is in this case more restrictive, and in the single individual, it is more precisely called "placebo response." Confusion can arise if one attempts to compare results in placebo arms of clinical trials without a natural history group with neurobiological studies specifically targeted on placebo (Hrobjartsson and Gotzsche, 2001; Vase et al., 2002; Thorn, 2007).

## Evolution of the sugar pill

Traditionally, a placebo was a carbohydrate tablet given with the intent of soothing an otherwise incurable patient or detecting a mystifying one through the success of the sham therapy. By definition, however, the tablet content is absolutely irrelevant: the placebo effect is triggered not by the sugar but by the symbolic significance that the patient attaches to it (Brody, 2000). In fact, virtually anything can work as a placebo, by inducing expectations of improvement which in turn trigger internal changes resulting in specific experiences (e.g., analgesia or motor improvement) (Kirsch, 1999). Thus, all aspects of a therapy (the physician's words, the sight and smell of the environment, the memory of past experiences in similar situations, etc.) can carry healing meaning, rendering the placebo effect a context effect (Di Blasi et al., 2001; Benedetti, 2002). At the limit, research protocols can be applied where no placebo is actually given, but the context effect is elicited only by verbal suggestions or other environmental clues inducing expectation of benefit. In this way, the simulation of a therapeutic situation is an effective substitute for the sugar pill, as shown by sham acupuncture studies where consistent placebo effects could be achieved when the complex healing environment was reproduced in all details except needle insertion (Bausell et al., 2005). To say that anything can work as a placebo does not mean that all placebos are equal. In fact, differences in the magnitude of the response have been reported comparing routes of administration (De Craen et al., 2000) or inert pills versus sham devices (Kaptchuk et al., 2006). In all probability, these differences are attributable to the variable potency in raising expectations. Depriving the patient of the contextual clues about a therapy can also represent a means of evaluating the placebo effect. In this case, a comparison is made between open and hidden administration of a drug, in full view of the patient or by a computer-controlled infusion pump, respectively. When the patient does not expect a treatment, the placebo pathway is shut down, and the clinical outcome is reduced compared to the open administration. The specific action of the drug can then be calculated as the difference between the effects in the two conditions, without the need of a placebo intervention (Amanzio et al., 2001; Colloca et al., 2004). Reconceptualizing the placebo effect as a context effect should help us shift the focus from the sugar pill to the patient, whose brain is the primary mediator of the specific physiological changes provoking the response.

**The placebo response: reflex or cognitive?**

The early finding that a placebo effect could be reproduced in animals (Herrnstein, 1962; Ader and Cohen, 1982) suggested its interpretation as an acquired reflex, similar to the digestive reflexes of Pavlov's dogs. Thus, the repeated co-occurrence in the patient medical history of aspects of the clinical setting associated with drug assumption (such as taste, color, shape of a tablet, as well as white coats, or the peculiar hospital smell) and therapy outcome (e.g., analgesia) can provoke what is called a *conditioned response*, that is, the therapy outcome induced by the clinical setting alone, in the absence of the pharmacologically active principle (Wikramasekera, 1985; Siegel, 2002; Ader, 1997). Seminal experiments were carried out in humans by Voudouris et al. (1989, 1990),

who devised a protocol whereby conditioning was achieved by pairing a placebo analgesic cream with a painful stimulation, which was surreptitiously reduced with respect to a baseline condition to make the subjects believe that the cream was effective. In this way, a direct comparison could be made between a conditioned and an unconditioned group, with the former invariably showing a larger pain reduction. This kind of protocol is still widely used today, to boost placebo effects through conditioning in the experimental setting. Although the conditioning model is too simplistic to thoroughly explain placebo effects in all situations, it has found support for physiological functions outside the conscious control, like those involving the immune system or neuroendocrine secretions (Giang et al., 1996; Goebel et al., 2002; Benedetti et al., 2003). The physiological basis for these responses can conceivably be provided by autonomic nervous system activity and the release of neuroendocrine substances from the pituitary gland, induced by the psychosocial context through neural circuits including limbic and hypothalamic relays (Ader, 2003; Pacheco-López et al., 2006; Riether et al., 2008).

When the response falls in the conscious domain, conditioning is still possible, but a dominant role is played by cognitive aspects, such as expectations, motivations, and emotions (Kirsch, 1999). Grading expectancies resulted in graded placebo effects both in experimental (Price et al., 1999) and clinical conditions (Pollo et al., 2001). The desire to achieve a goal (e.g., pain reduction) and the emotional states associated with it also contribute to determine the magnitude of placebo effect (Vase et al., 2003, 2005). Actually, the two mechanisms of conditioning and expectation are not mutually exclusive, and can act simultaneously with additive effects (Amanzio and Benedetti, 1999). It has been argued that during the conditioning process, the subject learns what to expect (Reiss, 1980; Rescorla, 1988; Montgomery and Kirsch, 1997), and in keeping with this, memory of prior experience (i.e., learning) is crucial, as demonstrated in healthy volunteers undergoing electrical painful stimulation with different conditioning protocols, who exhibited small, medium, or large effects, depending on previous experience and time lag between conditioning and response assessment (Colloca and Benedetti, 2006; Colloca et al., 2008a). Learning (by exposure to prior positive experience) potentiated not only behavioral (subject pain report) but also neurophysiological placebo analgesic responses, as measured by laser-evoked potentials (LEPs) after verbal suggestion of analgesia (Colloca et al., 2008b).

A different cognitive mechanism, which has recently been proposed as a contributor to the genesis of placebo effects, is the recruitment of the reward circuitry. It has been argued that placebos have reward properties. Rewards are usually directed to increase survival, like food and sex, and so are the placebos with the beneficial outcome they provide. In fact, until recently, only few active treatments existed and the history of medicine largely matched that of placebo effects (de la Fuente-Fernández et al., 2004). According to this view, the expected clinical benefit is a form of reward, which triggers the placebo response (de la Fuente-Fernández and Stoessl, 2002; Lidstone and Stoessl, 2007).

## Neurological disorders showing prominent placebo effects

Most relevant to the neurologist are the aspects of placebo research concerning the correct evaluation of placebo effects in clinical trials and the exploitation of the surrounding clinical context to the patient's benefit. In fact, in the search for specific brain mechanisms generating the placebo response, scientists have focused on a number of neurological disorders which represent interesting models, thanks to the known molecular dysfunctions underlying them. In particular, pain, Parkinson's disease, and depression have been the target of recent experimental placebo research (de la Fuente-Fernández et al., 2002; Cavanna et al., 2007).

### *Pain*

It was in the field of placebo analgesia that neuropharmacological evidence of a chemical substrate for the placebo phenomenon was first obtained (Levine et al., 1978). Much subsequent work has corroborated the model whereby the secretion of endogenous opioids in the brain is the central event of the pain modulation by a placebo, with the activation of the descending antinociceptive pathway as its anatomical substrate (Fields and Levine, 1984; Lipman et al., 1990; Benedetti et al., 1999a, b; Pollo et al., 2003). In fact, in many of these studies, placebo analgesia was reversed by naloxone, although the presence of some naloxone-insensitive effects points to the involvement of other antinociceptive mechanisms as well, our understanding of which is still scarce (Gracely et al., 1983; Amanzio and Benedetti, 1999; Vase et al., 2005). Enhancing effects on placebo analgesia have been obtained with proglumide, a cholecystokinin (CCK) antagonist (Benedetti et al., 1995; Benedetti, 1996). It would seem that placebo analgesia is under the opposing actions of promoting endogenous opioids and inhibiting endogenous CCK, two systems which show overlapping distribution of brain receptors (Noble and Roques, 2003) and the opposing role of which has also been suggested for the emotional modulation of other incoming signals, like visual input (Gospic et al., 2008).

Recently, brain imaging and mapping techniques, such as positron emission tomography (PET), functional magnetic resonance imaging (fMRI), magneto-electroencephalography (MEG), and electroencephalography (EEG), have brought important contributions to the understanding of where and when is placebo analgesia generated in the central nervous system (Rainville and Duncan, 2006; Kong et al., 2007; Colloca et al., 2008). Initially, the focus was maintained on the top-down pain regulatory system already implicated by neuropharmacological studies. This endogenous opioid system has adaptive value, being called into action by fear or threat, as in stress-induced analgesia, depressing the incoming nociceptive signals (Millan, 2002; Fields, 2004). A PET study first showed that brain areas activated during opioid- or placebo-induced analgesia largely overlapped, involving part of the anatomical substrate of this system, that is, the rostral anterior cingulate cortex (rACC), the orbitofrontal cortex (OrbC), and the periaqueductal gray (PAG) (Petrovic et al., 2002). Subsequently, direct evidence for endogenous opioid release during a placebo intervention was provided by another PET study, measuring μ-opioid receptor availability (Zubieta et al., 2005). In this study, [$^{11}$C] carfentanil was displaced by the activation of opioid neurotransmission, showing significant binding decrease after placebo in pregenual rACC, insula, nucleus accumbens, and dorsolateral prefrontal cortex (DLPFC); in all areas except DLPFC, this decrease was correlated with placebo reduction of pain intensity reports. A number of other studies brought more contributions: by reporting dampened activation during placebo analgesia in brain areas of the so-called "pain matrix," like thalamus, anterior insula, and caudal rACC (Wager et al., 2004; but see also Kong et al. 2006, for a contrasting report); by attempting to correlate the activation of rACC with that of the antinociceptive system, suggesting a crucial cognitive control role for rACC (Bingel et al., 2006); or by advocating a modulation by placebo of spinal activity (Matre et al., 2006; Goffaux et al., 2007). Scalp LEPs amplitude was also found to be reduced during the placebo analgesic response, namely in the N2–P2 components, thought to be originated in the bilateral insula and cingulate gyrus (Wager et al., 2006; Watson et al., 2007).

As mentioned before, placebo effects can be induced by expectation of benefit even without the physical administration of a placebo. Thus, along a different line of research, knowledge of placebo analgesia can also be gained by focusing on changes in brain activity which take place during pain anticipation. In the anticipatory phase of the placebo analgesic response, increased activity in DLPFC and other frontal regions was positively correlated with increase in a midbrain region containing the PAG, and negatively correlated with the signal reduction in pain regions and with reported pain intensity. The interpretation could be that just before the onset of placebo analgesia, prefrontal cortical evaluation could drive the activation of the descending antinociceptive system (Wager et al., 2004). Similarly, a comparison of high and low expectations before a painful stimulus showed changes in activity in many areas of the descending inhibitory pathway (Keltner et al., 2006). In order to discriminate whether expectancy exerts its psychophysical effect through changes of the perceptual sensitivity of early cortical processes [i.e., in the primary (SI) and secondary (SII) somatosensory areas] or on later evaluative elaborations, such as stimulus identification and response selection (represented in ACC), Lorenz et al. (2005) used a combined application of the high temporal resolution techniques of EEG and MEG. They found that the amplitude of the laser-evoked MEG fields in SII was highly correlated to the expected stimulus intensity as signaled by an auditory cue, while the ensuing evoked responses with source in the caudal ACC varied with stimulus intensity (requiring a varying level of task engagement) but failed to show any cue validity effects.

Pain is associated with many neurological disorders, and it is hoped that a better understanding of the mechanisms underlying placebo analgesia can be reflected in the clinical practice, with clinicians sharing with experimenters the awareness of the usefulness of the placebo tool at their disposal.

### Parkinson's disease and motor performance

The placebo effect in Parkinson's disease is usually obtained through the administration of an inert substance, which the patient believes to be an effective antiparkinsonian drug. The assessment of the ensuing motor performance improvement is somewhat more objective than the self-reported variation of pain, as it can be evaluated by a blinded examiner with the Unified Parkinson's Disease Rating Scale (UPDRS). However, recent experimental work has also exploited the technique of subthalamic nucleus-deep brain stimulation (STN-DBS), manipulating the electrodes activity to configure different expectation and conditioning protocols. In an early such study, patients with the stimulator turned off showed faster hand movements when they mistakenly believed it to be on than when they were correctly informed (Pollo et al., 2002). An influence of expectation on UPDRS scores was also found by Mercado et al. (2006) comparing aware and unaware conditions of the stimulator status, both for the on and off situations. Thus, expectation plays an important role not only for placebo effects affecting sensory input but also for motor output. Subsequently, intraoperative recording of single neuron activity in the subthalamic nucleus in patients conditioned with apomorphine showed that placebo responders exhibited a significant decrease of neuronal firing rate associated to a shift from bursting to a more physiological pattern of discharge. These changes were coupled to rigidity reduction and subjective reports of well-being (Benedetti et al., 2004; Fig. 1). This study demonstrated for the first time a link between a placebo intervention and

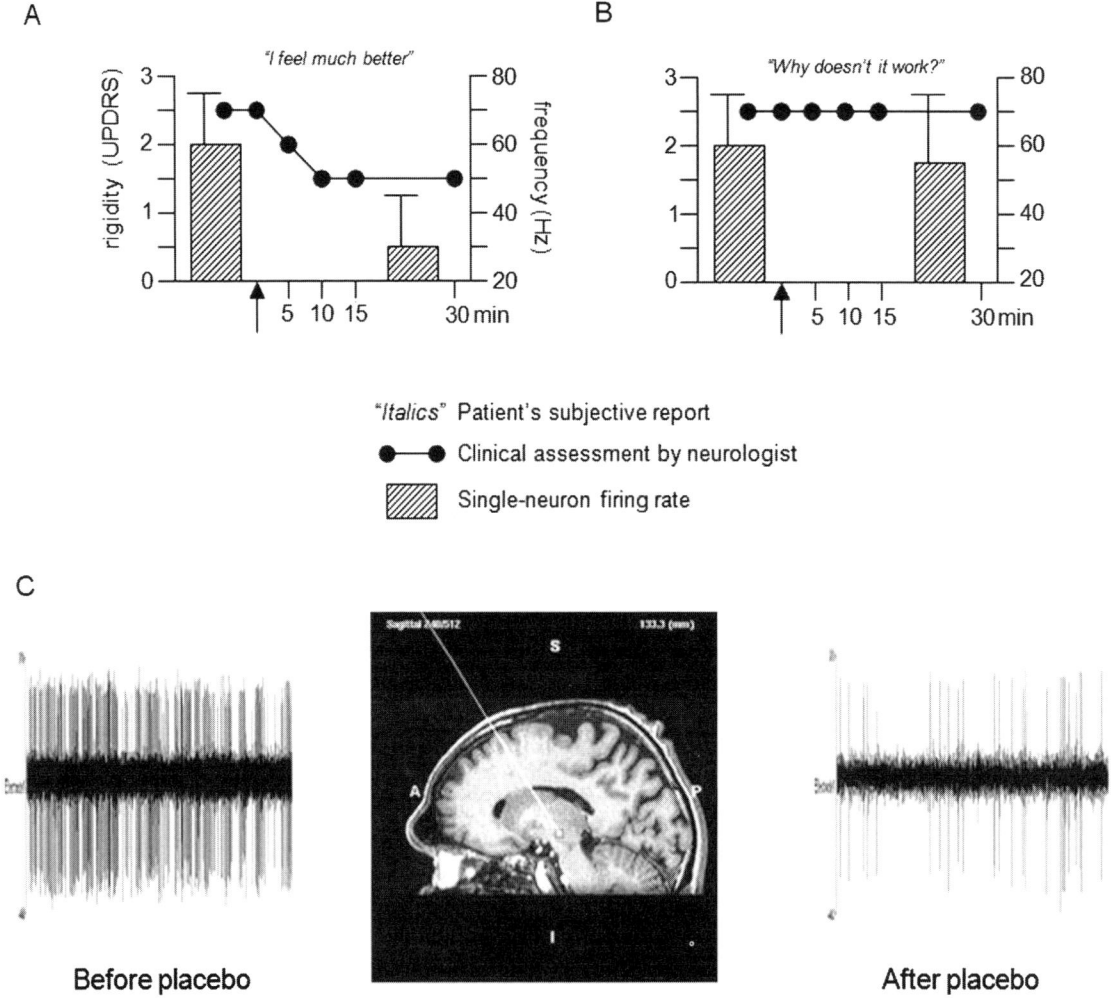

Fig. 1. The placebo effect in Parkinson's disease patients during subthalamic nucleus (STN) electrode implantation. Correlation among arm rigidity (black circles), STN neuronal frequency discharge (shaded columns) and subjective report (italics) in the case of a placebo responder (A) and nonresponder (B). The black arrow on the abscissa indicates placebo administration. Bars represent standard deviations. The placebo responder shows a decrease in neuronal activity and arm rigidity, together with subjective improvement, all of which are absent in the placebo nonresponder. (C) Single-neuron electrical activity in the STN, before and after placebo, in a placebo responder (left and right side) and magnetic resonance image showing the STN implanted electrode (middle). Modified from Benedetti et al. (2004).

a single cell activity, proving the influence of an expectation-inducing procedure on a specific neuronal population. In spite of the similarities with placebo analgesia, however, the neuropharmacological substrate is, in this case, quite different. In a PET study employing the $D_2$–$D_3$ dopamine receptor agonist [$^{11}$C]raclopride as a radiotracer, de la Fuente-Fernández et al. (2001) obtained the first evidence that endogenous dopamine is released in the striatum after pharmacological placebo administration. Their finding was later corroborated by similar results obtained with the use of sham transcranial magnetic stimulation as a placebo (Strafella et al., 2006).

Dopamine has been proposed as a possible mediator of placebo effects not strictly pertaining to the motor context. Its release has, in fact, been observed not only in the dorsal but also in the ventral striatum (e.g., in the nucleus accumbens), an area known to be involved in the reward circuitry (de la Fuente-Fernández et al., 2002). According to the authors, while dorsal striatum release is directly linked to performance improvement, ventral release could rather be connected to expectation of reward, that is, of clinical benefit (de la Fuente-Fernández et al., 2004). As such, it could well be implicated also in other types of placebo effect, including placebo analgesia. In fact, some of the cortical and subcortical areas activated during sustained pain are known to receive dopaminergic projections (Zubieta et al., 2001). In support of the role of reward mechanisms in generating placebo responses, Scott et al. (2007) found, in a combined PET and fMRI study, a correlation between individual responsiveness to placebo analgesia and monetary reward. The stronger the nucleus accumbens activation (fMRI) during the placebo response, the larger the dopamine release in the same nucleus during the monetary task (PET with [$^{11}$C]raclopride). These results also prompt the reward system as a possible neurological substrate in the search for "placebo responders," whose traits still elude research efforts (Kaptchuk et al., 2008; Oken, 2008). Finally, in a within-subject design PET study using both [$^{11}$C]carfentanil and [$^{11}$C]raclopride, both opioid and dopamine neurotransmission were found coupled with the placebo response, with changes of activity induced in several brain regions associated with the opioid and dopamine networks (Scott et al., 2008).

The relevance of motor placebo responses is not confined to damaged systems as in Parkinson's disease, but it can be extended to intact motor systems. In a recent study testing ergogenic placebos, muscle performance of healthy subjects was improved and their subjective rate of perceived exertion lessened, following a conditioning and expectation-raising procedure (Pollo et al., 2008). As the subjects were required to perform leg extensions until complete exhaustion, it is tempting to speculate that the placebo effect could be exerted on a putative central governor of fatigue, which integrates peripheral and central signals to determine maximal exercise (St Clair Gibson et al., 2006). The development of rehabilitation protocols which take into account the additional instrument of placebos to stimulate ailing patients to set higher goals in their physical treatments could be an interesting development of this field.

## Depression

Clinical trials for antidepressants show very high rates of placebo responses, with an increasing trend over time (Walsch et al., 2002). This posits a challenge for the development of new drugs, with the added complication of the subjectivity of the primary outcome measure, that is, mood rating. Imaging and brain mapping studies of placebo responses in depression have thus been stimulated by the need of physiologic indicators of treatment effectiveness. In an attempt to differentiate the therapeutic from the placebo response in patients with major depressive disorder, Leuchter et al. (2002) have used quantitative EEG to show changes in brain function of placebo responders (increase in frontal cordance) that are distinct from those associated with antidepressant medication (decrease), and from those observed in nonresponders to either placebo or medication (no change). In a PET study on the serotonin reuptake inhibitor fluoxetine, Mayberg et al. (2002) reported a pattern of activity changes (including increases in prefrontal, anterior cingulate, and other cortical and subcortical regions) which is similar for placebo and drug responders, with the latter exhibiting more pronounced changes. However, drug responders also showed additional subcortical and limbic variations in glucose metabolism which were not seen after placebo. Whether the serotonin system is directly affected by placebo is still a matter of speculation, as no displacement studies have yet been conducted, and long-term effects of antidepressant drugs are more difficult to study than motor or pain improvements.

## From research to clinical practice

The importance of placebo studies for clinical practice is manifold. An initial repercussion is on clinical trial design: here, the first recommendation is to control for patient expectations as an important variable influencing trial outcome. In acupuncture studies on osteoarthritis and other acute or chronic pain conditions, group assignment (to real or sham treatment) was unrelated to effects on pain, but regrouping the subjects according to perceived group assignment resulted in significant less pain for the subjects believing to be in the real treatment group (Bausell et al., 2005; Linde et al., 2007). Similar results were obtained in a long-term study on parkinsonian patients transplanted with human embryonic dopamine neurons, in which better scores on quality of life assessment were reported by patients believing to have received the real transplant (McRae et al., 2004). A second recommendation concerns the design of clinical trials, where the employment of the open/hidden design offers the opportunity to avoid the inclusion of a placebo group, thus circumventing potential ethical limitations. Also, possible direct interactions between the drug and the patient's expectations (i.e., the potential activation of nonspecific brain pathways by the mere act of drug administration) can be ruled out (Colloca and Benedetti, 2005).

In everyday clinical practice, placebos are still widely exploited (Sherman and Hickner, 2007; Nitzan and Lichtenberg, 2008), spurring a lively ethical debate on the opportunity of their use (Lichtenberg et al., 2004). While deception should generally be avoided, clinicians must be aware of the potential for cure of the psychosocial context. In fact, every real treatment administered has two distinct components: the active constituent and the placebo factor. Every effort should be made to enhance the latter in order to maximize the benefit of the therapeutic act. This represents a perfectly acceptable behavior, which does not challenge ethical imperatives. The patient/provider relationship is central, with both correct attitudes, skills of empathy and appropriate words on one side, and nonverbal clues intentionally or unintentionally conveyed on the other. A demonstration of the additive effect of the two components of treatment is offered by the lower analgesia obtained with hidden administration of analgesics (i.e., in the absence of psychosocial context) compared to open (Amanzio et al., 2001). Interestingly, this difference disappears in cognitively impaired patients, unable to communicate and purposefully interact with the caregiver. This loss might be taken into account when devising therapy plans for Alzheimer or other demented patients (Benedetti et al., 2006). The context influence has prompted Barrett et al. (2006) to propose a list of eight specific clinical actions: speak positively about treatments, provide encouragement, develop trust, provide reassurance, support relationships, respect uniqueness, explore values, and create ceremony. The "contextual healing" has been off the radar screen of scientific medicine, which has focused on therapeutic benefit produced by medical technology. We should instead redirect our attention to the wholeness of the act of administering treatment, especially for those conditions in which existing treatments are only partially effective in relieving symptoms (Miller and Kaptchuk, 2008).

## Conclusions

The roots of placebo effect extend into brain circuitry and biochemistry. Its mechanisms of actions are providing us with a holistic vision of mind–body interactions, enabling us to believe that we can influence our well-being, if not by sheer willpower, at least by employing means known to act on our central nervous system. Awareness of and feelings about treatments are able to influence patient responses both psychologically (in the mind) and physiologically (in the body), two realms no longer segregated by Cartesian dualism.

Application of knowledge gained in placebo research also stretches to medical conditions outside the neurological domain, like asthma (Kemeny et al., 2007) or rheumatic diseases (Pollo and Benedetti, 2008). Results so far obtained point to the activation of specific mechanisms, with release of opioids, dopamine, hormones, immune mediators, and possibly serotonin to meet the organism need in each case. We can thus speak not of a single but many placebo effects. Whether they are triggered independently or by a common pathway is still a matter of debate.

## Acknowledgment

This work was supported by grants from Istituto San Paolo di Torino and Regione Piemonte.

## References

Ader, R. (1997). The role of conditioning in pharmacotherapy. In A. Harrington (Ed.), *The placebo effect: an interdisciplinary exploration* (pp. 138–165). Cambridge, MA: Harvard University Press.

Ader, R. (2003). Conditioned immunomodulation: research needs and directions. *Brain, Behavior, and Immunity, 17*, S51–S57.

Ader, R., & Cohen, N. (1982). Behaviorally conditioned immunosuppression and murine systemic lupus erythematosus. *Science, 215*, 1534–1536.

Amanzio, M., & Benedetti, F. (1999). Neuropharmacological dissection of placebo analgesia: expectation-activated opioid systems versus conditioning-activated specific sub-systems. *Journal of Neuroscience, 19*, 484–494.

Amanzio, M., Pollo, A., Maggi, G., & Benedetti, F. (2001). Response variability and to analgesics: a role for non-specific activation of endogenous opioids. *Pain, 90*, 205–215.

Barrett, B., Muller, D., Rakel, D., Rabago, D., Marchand, L., & Scheder, J. (2006). Placebo, meaning and health. *Perspectives in Biology and Medicine, 49*, 178–198.

Bausell, R. B., Lao, L., Bergman, S., Lee, W. L., & Berman, B. M. (2005). Is acupuncture analgesia an expectancy effect? *Evaluation and the Health Professions, 28*, 9–26.

Benedetti, F. (1996). The opposite effects of the opiate antagonist naloxone and the cholecystokinin antagonist proglumide on placebo analgesia. *Pain, 64*, 535–543.

Benedetti, F. (2002). How the doctor's words affect the patient's brain. *Evaluation and the Health Professions, 25*, 369–386.

Benedetti, F. (2007). Placebo and endogenous mechanisms of analgcsia. *Handbook of Experimental Pharmacology, 177*, 393–413.

Benedetti, F. (2008a). Mechanisms of placebo and placebo-related effects across diseases and treatments. *Annual Review of Pharmacology and Toxicology, 48*, 33–60.

Benedetti, F. (2008b). *Placebo effects: understanding the mechanisms in health and disease*. Oxford University Press.

Benedetti, F., Amanzio, M., Baldi, S., Casadio, C., & Maggi, G. (1999a). Inducing placebo respiratory depressant responses in humans via opioid receptors. *European Journal of Neuroscience, 11*, 625–631.

Benedetti, F., Amanzio, M., & Maggi, G. (1995). Potentiation of placebo analgesia by proglumide. *Lancet, 346*, 1231.

Benedetti, F., Arduino, C., & Amanzio, M. (1999b). Somatotopic activation of opioid systems by target-directed expectations of analgesia. *Journal of Neuroscience, 19*, 3639–3648.

Benedetti, F., Arduino, C., Costa, S., Vighetti, S., Tarenzi, L., Rainero, I., et al. (2006). Loss of expectation-related mechanisms in Alzheimer's disease makes analgesic therapies less effective. *Pain, 121*, 133–144.

Benedetti, F., Colloca, L., Torre, E., Lanotte, M., Melcarne, A., Pesare, M., et al. (2004). Placebo-responsive Parkinson patients show decreased activity in single neurons of subthalamic nucleus. *Nature Neuroscience, 7*, 587–588.

Benedetti, F., Mayberg, H. S., Wager, T. D., Stolher, C. S., & Zubieta, J. K. (2005). Neurobiological mechanisms of the placebo effect. *Journal of Neuroscience, 25*, 10390–10402.

Benedetti, F., Pollo, A., Lopiano, L., Lanotte, M., Vighetti, S., & Rainero, I. (2003). Conscious expectation and unconscious conditioning in analgesic, motor and hormonal placebo/nocebo responses. *Journal of Neuroscience, 23*, 4315–4323.

Bingel, U., Lorenz, J., Schoell, E., Weiller, C., & Büchel, C. (2006). Mechanisms of placebo analgesia: rACC recruitment of a subcortical antinociceptive network. *Pain, 120*, 8–15.

Brody, H. (2000). *The placebo response*. New York: Harper Collins.

Cavanna, A. E., Strigaro, G., & Monaco, F. (2007). Brain mechanisms underlying the placebo effect in neurological disorders. *Functional Neurology, 22*, 89–94.

Colloca, L., & Benedetti, F. (2005). Placebos and painkillers: is mind as real as matter? *Nature Reviews. Neuroscience, 6*, 545–552.

Colloca, L., & Benedetti, F. (2006). How prior experience shapes placebo analgesia. *Pain, 124*, 126–133.

Colloca, L., Benedetti, F., & Porro, C. A. (2008). Experimental designs and brain mapping approaches for studying the placebo analgesic effect. *European Journal of Applied Physiology, 102*, 371–380.

Colloca, L., Lopiano, L., Lanotte, M., & Benedetti, F. (2004). Overt versus covert treatment for pain, anxiety, and Parkinson's disease. *Lancet Neurology, 3*, 679–684.

Colloca, L., Sigaudo, M., & Benedetti, F. (2008a). The role of learning in nocebo and placebo effects. *Pain, 136*, 211–218.

Colloca, L., Tinazzi, M., Recchia, S., Le Pera, D., Fiaschi, A., Benedetti, F., et al. (2008b). Learning potentiates neurophysiological and behavioral placebo analgesic responses. *Pain, 139*, 306–314.

De Craen, A., Tijssen, J. G. P., de Gans, J., & Kleijnen, J. (2000). Placebo effect in the acute treatment of migraine: subcutaneous placebos are better than oral placebos. *Journal of Neurology, 247*, 183–188.

de la Fuente-Fernández, R., Phillips, A. G., Zamburlini, M., Sossi, V., Calne, D. B., & Stoessl, A. J. (2002). Dopamine release in human ventral striatum and expectation of reward. *Behavioural Brain Research, 136*, 359–363.

de la Fuente-Fernández, R., Ruth, T. J., Sossi, V., Schulzer, M., Calne, D. B., & Stoessl, A. J. (2001). Expectation and dopamine release: mechanism of the placebo effect in Parkinson's disease. *Science, 293*, 1164–1166.

de la Fuente-Fernández, R., Schulzer, M., & Stoessl, A. J. (2002). The placebo effect in neurological disorders. *Lancet Neurology, 1*, 85–91.

de la Fuente-Fernández, R., Schulzer, M., & Stoessl, A. J. (2004). Placebo mechanisms and reward circuitry: clues from Parkinson's disease. *Biological Psychiatry, 56*, 67–71.

de la Fuente-Fernández, R., & Stoessl, A. J. (2002). The placebo effect in Parkinson's disease. *Trends in Neurosciences, 6*, 302–306.

Di Blasi, Z., Harkness, E., Ernst, E., Georgiou, A., & Kleijnen, J. (2001). Influence of context effect on health outcomes: a systematic review. *Lancet, 357*, 757–762.

Enck, P., Benedetti, F., & Schedlowski, M. (2008). New insights into the placebo and nocebo responses. *Neuron, 59*, 195–206.

Fields, H. (2004). State-dependent opioid control of pain. *Nature Reviews. Neuroscience, 5*, 565–575.

Fields, H. L., & Levine, J. D. (1984). Placebo analgesia—a role for endorphins?. *Trends in Neurosciences, 7*, 271–273.

Finniss, D. G., & Benedetti, F. (2005). Mechanisms of the placebo response and their impact on clinical trials and clinical practice. *Pain, 114*, 3–6.

Giang, D. W., Goodman, A. D., Schiffer, R. B., Mattson, D. H., Petrie, M., Cohen, N., et al. (1996). Conditioning of cyclophosphamide-induced leukopenia in humans. *Journal of Neuropsychiatry and Clinical Neurosciences, 8*, 194–201.

Goebel, M. U., Trebst, A. E., Steiner, J., Xie, Y. F., Exton, M. S., Frede, S., et al. (2002). Behavioral conditioning of immunosuppression is possible in humans. *FASEB Journal, 16*, 1869–1873.

Goffaux, P., Redmond, W. J., Rainville, P., & Marchand, S. (2007). Descending analgesia—when the spine echoes what the brain expects. *Pain, 130*, 137–143.

Gospic, K., Gunnarsson, T., Fransson, P., Ingvar, M., Lindefors, N., & Petrovic, P. (2008). Emotional perception modulated by an opioid and a cholecystokinin agonist. *Psychopharmacology, 197*, 295–307.

Gracely, R. H., Dubner, R., Wolskee, P. J., & Deeter, W. R. (1983). Placebo and naloxone can alter post-surgical pain by separate mechanisms. *Nature, 306*, 264–265.

Herrnstein, R. J. (1962). Placebo effect in the rat. *Science, 138*, 677–678.

Hrobjartsson, A., & Gotzsche, P. C. (2001). Is the placebo powerless? An analysis of clinical trials comparing placebo with no treatment. *The New England Journal of Medicine, 344*, 1594–1602.

Kaptchuk, T. J., Kelley, J. M., Deykin, A., Wayne, P. M., Lasagna, L. C., Epstein, I. O., et al. (2008). Do "placebo responders" exist?. *Contemporary Clinical Trials, 29*, 587–595.

Kaptchuk, T. J., Stason, W. B., Davis, R. B., Legedza, A. T. R., Schnyer, R. N., Kerr, C. E., et al. (2006). Sham device v. inert pill: randomized controlled trial of two placebo treatments. *BMJ, 332*, 391–397.

Keltner, J. R., Furst, A., Fan, C., Redfern, R., Inglis, B., & Fields, H. L. (2006). Isolating the modulatory effect of expectation on pain transmission: a functional magnetic imaging study. *Journal of Neuroscience, 26*, 4437–4443.

Kemeny, M. E., Rosenwasser, L. J., Panettieri, R. A., Rose, R. M., Berg-Smith, S. M., & Kline, J. N. (2007). Placebo response in asthma: a robust and objective phenomenon. *Journal of Allergy and Clinical Immunology, 119*, 1375–1381.

Kirsch, I. (1999). *How expectancies shape experience*. Washington, DC: American Psychological Association.

Kong, J., Gollub, R. L., Rosman, I. S., Webb, J. M., Vangel, M. G., Kirsch, I., et al. (2006). Brain activity associated with expectancy-enhanced placebo analgesia as measured by functional magnetic resonance imaging. *Journal of Neuroscience, 26*, 381–388.

Kong, J., Kaptchuk, T. J., Polich, G., Kirsch, I., & Gollub, R. L. (2007). Placebo analgesia: findings from brain imaging studies and emerging hypothesis. *Review of Neuroscience, 18*, 173–190.

Leuchter, A. F., Cook, I. A., Witte, E. A., Morgan, M., & Abrams, M. (2002). Changes in brain function of depressed subjects during treatment with placebo. *American Journal of Psychiatry, 159*, 122–129.

Levine, J. D., Gordon, N. C., & Fields, H. L. (1978). The mechanisms of placebo analgesia. *Lancet, 2*, 654–657.

Lichtenberg, P., Heresco-Levy, U., & Nitzan, U. (2004). The ethics of the placebo in clinical practice. *Journal of Medical Ethics, 30*, 551–554.

Lidstone, S. C., & Stoessl, A. J. (2007). Understanding the placebo effect: contributions from neuroimaging. *Molecular Imaging and Biology, 9*, 176–185.

Linde, K., Witt, C. M., Streng, A., Weidenhammer, W., Wagenpfeil, S., Brinkhaus, B., et al. (2007). The impact of patient expectations on outcomes in four randomized controlled trials of acupuncture in patients with chronic pain. *Pain, 128*, 264–271.

Lipman, J. J., Miller, B. E., Mays, K. S., Miller, M. N., North, W. C., & Byrne, W. L. (1990). Peak B endorphin concentration in cerebrospinal fluid: reduced in chronic pain patients and increased during the placebo response. *Psychopharmacology, 102*, 112–116.

Lorenz, J., Hauch, M., Paur, R. C., Nakamura, Y., Zimmermann, R., Bromm, B., et al. (2005). Cortical correlates of false expectations during pain intensity judgments—a possible manifestation of placebo/nocebo cognitions. *Brain, Behavior, and Immunity, 19*, 283–295.

Matre, D., Casey, K. L., & Knardahl, S. (2006). Placebo-induced changes in spinal cord pain processing. *Journal of Neuroscience, 26*, 559–563.

Mayberg, H. S., Silva, J. A., Brannan, S. K., Tekell, J. L., Mahurin, R. K., McGinnis, S., et al. (2002). The functional neuroanatomy of the placebo effect. *American Journal of Psychiatry, 159*, 728–737.

McRae, C., Cherin, E., Yamazaki, G. T., Diem, G., Vo, A. H., Russel, D., et al. (2004). Effects of perceived treatment on quality of life and medical outcomes in a double-blind placebo surgery trial. *Archives of General Psychiatry, 61*, 412–420.

Mercado, R., Constantoyannis, C., Mandat, T., Kumar, A., Schulzer, M., Stoessl, A. J., et al. (2006). Expectation and the placebo effect in Parkinson's disease patients with subthalamic nucleus deep brain stimulation. *Movement Disorders, 21*, 1457–1461.

Millan, M. J. (2002). Descending control of pain. *Progress in Neurobiology, 66*, 355–474.

Miller, F. G., & Kaptchuk, T. J. (2008). The power of context: reconceptualizing the placebo effect. *Journal of the Royal Society of Medicine, 101*, 222–225.

Montgomery, G. H., & Kirsch, I. (1997). Classical conditioning and the placebo effect. *Pain, 72*, 107–113.

Nitzan, U., & Lichtenberg, P. (2008). Questionnaire survey on use of placebo. *BMJ, 329*, 944–946.

Noble, F., & Roques, B. P. (2003). The role of CCK2 receptors in the homeostasis of the opioid system. *Drugs of Today, 39*, 897–908.

Oken, B. S. (2008). Placebo effects: clinical aspects and neurobiology. *Brain, 131*, 2812–2823.

Pacheco-López, G., Engler, H., Niemi, M. B., & Schedlowski, M. (2006). Expectations and associations that heal: immunomodulatory placebo effects and its neurobiology. *Brain, Behavior, and Immunity, 20*, 430–446.

Petrovic, P., Kalso, E., Petersson, K. M., & Ingvar, M. (2002). Placebo and opioid analgesia—imaging a shared neuronal network. *Science, 295*, 1737–1740.

Pollo, A., Amanzio, M., Arslanian, A., Casadio, C., Maggi, G., & Benedetti, F. (2001). Response expectancies in placebo analgesia and their clinical relevance. *Pain, 93*, 77–83.

Pollo, A., & Benedetti, F. (2008). Placebo response: relevance to the rheumatic diseases. *Rheumatic Disease Clinics of North America, 34*, 331–349.

Pollo, A., Carlino, E., & Benedetti, F. (2008). The top-down influence of ergogenic placebos on muscle work and fatigue. *European Journal of Neuroscience, 28*, 379–388.

Pollo, A., Torre, E., Lopiano, L., Rizzone, M., Lanotte, M., Cavanna, A., et al. (2002). Expectation modulates the response to subthalamic nucleus stimulation in Parkinsonian patients. *Neuroreport, 13*, 1383–1386.

Pollo, A., Vighetti, S., Rainero, I., & Benedetti, F. (2003). Placebo analgesia and the heart. *Pain, 102*, 125–133.

Price, D. D., Finniss, D. G., & Benedetti, F. (2008). A comprehensive review of the placebo effects: recent advances and current thought. *Annual Review of Psychology, 59*, 565–590.

Price, D. D., Milling, L. S., Kirsch, I., Duff, A., Montgomery, G. H., & Nicholls, S. S. (1999). An analysis of factors that contribute to the magnitude of placebo analgesia in an experimental paradigm. *Pain, 83*, 147–156.

Rainville, P., & Duncan, G. H. (2006). Functional brain imaging of placebo analgesia: methodological challenges and recommendations. *Pain, 121*, 177–180.

Reiss, S. (1980). Pavlovian conditioning and human fear: an expectancy model. *Behavior Therapy, 11*, 380–396.

Rescorla, R. A. (1988). Pavlovian conditioning: it is not what you think it is. *American Psychologist, 43*, 151–160.

Riether, C., Doenlen, R., Pacheco-López, G., Niemi, M. B., Engler, A., Engler, H., et al. (2008). Behavioural conditioning of immune functions: how the CNS controls peripheral immune responses by evoking associative learning processes. *Review of Neuroscience, 19*, 1–17.

Scott, D. J., Stoher, C. S., Egnatuk, C. M., Wang, H., Koeppe, R. A., & Zubieta, J. K. (2007). Individual differences in reward responding explains placebo-induced expectations and effects. *Neuron, 55*, 325–336.

Scott, D. J., Stoher, C. S., Egnatuk, C. M., Wang, H., Koeppe, R. A., & Zubieta, J. K. (2008). Placebo and nocebo effects are defined by opposite opioid and dopaminergic responses. *Archives of General Psychiatry, 65*, 220–231.

Sherman, R., & Hickner, J. (2007). Academic physicians use placebos in clinical practice and believe in the mind–body connection. *Journal of General Internal Medicine, 23*, 7–10.

Siegel, S. (2002). Explanatory mechanisms for placebo effects: Pavlovian conditioning. In H. A. Guess, A. Kleinman, J. W. Kusek, & L. W. Engel (Eds.), *The science of the placebo: toward an interdisciplinary research agenda* (pp. 133–157). London: BMJ Books.

St Clair Gibson, A., Lambert, E. V., Rauch, L. H., Tucker, R., Baden, D. A., Foster, C., et al. (2006). The role of information processing between the brain and peripheral physiological systems in pacing and perception of effort. *Sports Medicine, 36*, 705–722.

Strafella, A. P., Ko, J. H., & Monchi, O. (2006). Therapeutic application of transcranial magnetic stimulation in Parkinson's disease: the contribution of expectation. *Neuroimage, 31*, 1666–1672.

Thorn, B. (2007). Commentaries on the placebo concept in psychotherapy. *Journal of Clinical Psychology*, *63*, 371–372.

Vase, L., Riley, J. L., 3rd, & Price, D. D. (2002). A comparison of placebo effects in clinical analgesic trials versus studies of placebo analgesia. *Pain*, *99*, 443–452.

Vase, L., Robinson, M. E., Verne, G. N., & Price, D. D. (2003). The contribution of suggestion, desire, and expectation to placebo effects in irritable bowel syndrome patients. An empirical investigation. *Pain*, *105*, 17–25.

Vase, L., Robinson, M. E., Verne, G. N., & Price, D. D. (2005). Increased placebo analgesia over timein irritable bowel syndrome (IBS) patients is associated with desire and expectation but not endogenous opioid mechanisms. *Pain*, *115*, 338–347.

Voudouris, N. J., Peck, C. L., & Coleman, G. (1989). Conditioned response models of placebo phenomena: further support. *Pain*, *38*, 109–116.

Voudouris, N. J., Peck, C. L., & Coleman, G. (1990). The role of conditioning and verbal expectancy in the placebo response. *Pain*, *43*, 121–128.

Wager, T. D., Matre, D., & Casey, K. L. (2006). Placebo effects in laser-evoked pain potentials. *Brain, Behavior, and Immunity*, *20*, 219–230.

Wager, T. D., Rilling, J. K., Smith, E. E., Sokolik, A., Casey, K. L., Davidson, R. J., et al. (2004). Placebo-induced changes in fMRI in the anticipation and experience of pain. *Science*, *303*, 1162–1166.

Walsch, B. T., Seidman, S. N., Sysko, R., & Gould, M. (2002). Placebo response in studies of major depression: variable, substantial, and growing. *JAMA*, *287*, 1840–1847.

Watson, A., El-Dereby, W., Vogt, B. A., & Jones, A. K. (2007). Placebo analgesia is not due to compliance or habituation: EEG and behavioural evidence. *Neuroreport*, *18*, 771–775.

Wikramasekera, I. (1985). A conditioned response model of the placebo effect: predictions of the model. In L. White, B. Tursky, & G. E. Schwartz (Eds.), *Placebo: theory, research and mechanisms*. New York: Guilford Press.

Zubieta, J. K., Bueller, J. A., Jackson, L. R., Scott, D. J., Xu, Y., Koeppe, R. A., et al. (2005). Placebo effects mediated by endogenous opioid activity on μ-opioid receptors. *Journal of Neuroscience*, *25*, 7754–7762.

Zubieta, J. K., Smith, Y. R., Bueller, J. A., Xu, Y., Kilbourn, M. R., Jewett, D. M., et al. (2001). Regional mu opioid receptor regulation of sensory and affective dimensions of pain. *Science*, *293*, 311–315.

SECTION V

# Neuroprostheses

*J. Verhaagen et al. (Eds.)*
*Progress in Brain Research*, Vol. 175
ISSN 0079-6123
Copyright © 2009 Elsevier B.V. All rights reserved

CHAPTER 20

# Brain–computer interfaces: an overview of the hardware to record neural signals from the cortex

Thomas Stieglitz[1,*], Birthe Rubehn[1], Christian Henle[1], Sebastian Kisban[2], Stanislav Herwik[2], Patrick Ruther[2] and Martin Schuettler[1]

[1]*Laboratory for Biomedical Microtechnology, Department of Microsystems Engineering – IMTEK, University of Freiburg, Freiburg, Germany*
[2]*Microsystems Materials Laboratory, Department of Microsystems Engineering – IMTEK, University of Freiburg, Freiburg, Germany*

**Abstract:** Brain–computer interfaces (BCIs) record neural signals from cortical origin with the objective to control a user interface for communication purposes, a robotic artifact or artificial limb as actuator. One of the key components of such a neuroprosthetic system is the neuro–technical interface itself, the electrode array. In this chapter, different designs and manufacturing techniques will be compared and assessed with respect to scaling and assembling limitations. The overview includes electroencephalogram (EEG) electrodes and epicortical brain–machine interfaces to record local field potentials (LFPs) from the surface of the cortex as well as intracortical needle electrodes that are intended to record single-unit activity. Two exemplary complementary technologies for micromachining of polyimide-based arrays and laser manufacturing of silicone rubber are presented and discussed with respect to spatial resolution, scaling limitations, and system properties. Advanced silicon micromachining technologies have led to highly sophisticated intracortical electrode arrays for fundamental neuroscientific applications. In this chapter, major approaches from the USA and Europe will be introduced and compared concerning complexity, modularity, and reliability. An assessment of the different technological solutions comparable to a strength weaknesses opportunities, and threats (SWOT) analysis might serve as guidance to select the adequate electrode array configuration for each control paradigm and strategy to realize robust, fast, and reliable BCIs.

**Keywords:** brain–computer interface (BCI); neuroprostheses; epicortical; intracortical; electrode; MEMS

## Introduction to neuroprostheses

Brain–computer interfaces (BCIs) belong to the field of neuroprostheses, a research direction that started in the 1960s with the objective to restore deficits from neural disorders by means of technology. Neuroprostheses aim to establish an interface between a technical device and the neural information backbone of the body, at the level of either the peripheral nervous system (PNS) or the central nervous system (CNS). This neuro–technical interface will enable a mono- or bidirectional information transfer by recording

*Corresponding author.
Tel.: +49 761 203 7471; Fax: +49 761 203 7472;
E-mail: stieglitz@imtek.uni-freiburg.de

DOI: 10.1016/S0079-6123(09)17521-0

the electrical signals of neural structures or by electrical stimulation of nerve cells. As medical devices, these technical systems have to be safe for the patient, reliable over the envisioned lifetime of the system, and efficient in terms of their intended use.

With respect to clinical terminology, neuroprostheses can be assigned to therapeutical applications if functions are modulated due to the electrical stimulation and to rehabilitation if a (partial) restoration of a lost function can be achieved (Fig. 1). The most known electrical stimulation device that replaces lost neural activity is the cardiac pacemaker. Invented in the late 1950s, it is implanted over 350,000 times per year nowadays. The success stories of neuroprostheses are less known, but increasing numbers (Rijkhoff, 2004), mainly of implantable devices, show their potential in therapy and rehabilitation (Table 1). Cochlear implants helped to restore hearing and understand free speech in adults and children with inner ear diseases for more than 100,000 patients; auditory brainstem implants help patients who have lost their auditory nerve in sound perception and lip reading in more than 300 cases. Spinal cord stimulation alleviates chronic pain and treats urge incontinence in more than 130,000 persons. Deep brain stimulation in patients suffering from Parkinson's disease or obsessive–compulsive disorders reduces the symptoms of the diseases with good success and fewer side effects than pharmacotherapy. Vagal nerve stimulation interfaces with the parasympathetic nervous system and helps to treat medically refractory epilepsy and severe depression. Other fields, especially motor neuroprostheses, are still in the clinical trial phase or have less patient numbers.

The most successful product on the market, the Freehand System, which allowed spinal cord-injured subjects with a C5/C6 lesion to hand grasp, is no longer available. The company NeuroControl was shut down due to financial reasons after successful US Federal Drug Administration (FDA) approval and European Conformity (CE) marking of the system and nearly 300 implantations worldwide with extremely satisfied users. The sacral anterior root stimulator, developed in the early 1970s by Brindley (1994), has been commercialized by Finetech Medical and helps more than 2500 patients with paraplegia to stay continent and manage micturition. One of the most challenging approaches today is the realization of implantable vision prostheses. The race is open and several companies worldwide are in the first phase of clinical trials to be the first on the market with a commercially available system to restore vision in blind persons to navigate in unknown environment. More and more application

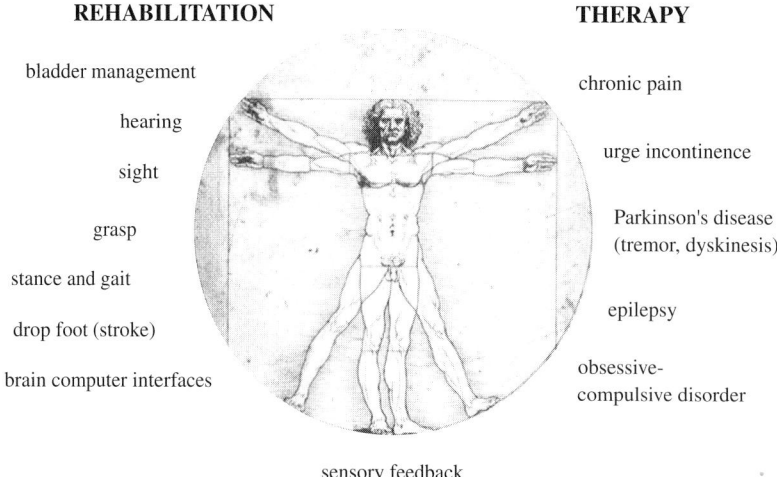

Fig. 1. Neuroprostheses in research and clinical applications.

Table 1. Fully implantable neuroprostheses with more than 1000 systems

| Application | Number of implants | Manufacturers |
| --- | --- | --- |
| Spinal cord stimulator to treat intractable pain and motor disorders | >130,000 | Advanced Neuromodulation Systems, USA<br>Medtronic, USA |
| Auditory nerve stimulator to restore hearing (cochlear implant) | >100,000[a] | Advanced Bionics, USA<br>AllHear, USA<br>Cochlear, Australia<br>MED-EL, Austria<br>MXM lab, France |
| Deep brain stimulator for tremor, Parkinson's disease, and pain | >20,000 | Medtronic, USA |
| Vagal nerve stimulator to treat intractable epilepsy | >17,000 | Cyberonics, USA |
| Sacral nerve stimulator for urinary urge incontinence, urinary retention, pelvic pain and fecal incontinence | >10,000 | Medtronic, USA |
| Sacral nerve stimulator for bladder emptying | >2,500 | Finetech Medical, UK |
| Phrenic nerve stimulator for respiration | >1,600 | Avery Laboratories, USA<br>Atrotech, Finland<br>MedImplant, Austria |

[a]Following Rijkhoff (2004), updated.

scenarios have arisen in the last years even though underlying physiological working mechanisms are still unclear and timescales are between 5 and 10 years at minimum between the first experiments and an approved product on the market.

Fundamental requirements for a medical device to establish a reliable interface between a technical system and the nervous system are challenging, mainly by means of implantable electrodes as the material interface. Material sciences and chemistry, mechanical and electrical engineering, biology, and medicine contribute to solutions that lead to properties that are termed "biocompatible." The biocompatibility of an implantable device addresses a variety of aspects. It is not a general property of a material or device but strongly depends on the medical application. In the case of neural interfaces, the fundamental requirements can be described as follows. The interface between the recording or stimulating electrode and the neural target tissue should be as stable as possible and the nerve activity should not be altered due to the implantation procedure. Inflammation must be as low and short as possible. A good surface biocompatibility is mandatory to limit foreign body response and encapsulation of the implant. Hence, materials and manufactured devices have to be evaluated with respect to material toxicity, residues, eludates, and systemic toxicity according to international standards (e.g., ISO 10993 — Biological Evaluation of Medical Devices). In addition, the devices must show a good structural biocompatibility, that is, they must match mechanically the surrounding environment as good as possible. Thus, design engineers prevent sharp edges and corners in the medical devices. Mechanical mismatch of nerve interfaces, especially in CNS applications, is still not completely resolved. However, coatings of soft materials and flexible substrate materials are often sufficient for a medical device that works reliably over decades.

BCIs are a relatively new area in the field of neuroprostheses. Their background lies often either in neurology and the use of electroencephalogram (EEG) in diagnosis or in basic neurosciences where results from the pursuit of understanding the brain are eventually transferred into applications. In this chapter, we would like to focus on the physical part of the BCIs for recording neural signals, that is, the electrode array at the material–tissue interface and the technical transmission system and not on different coding and signal processing paradigms to extract information out of the recorded data. However, the information extraction paradigm in recording

BCI as well as the application site in recording and stimulation BCI will strongly influence the physical design of the system. In order to understand the interdependency of both sides some questions might help to follow the different approaches in BCI development:

- What are the intention and the indication to use a BCI?
- Where is the origin of the signals of interest?
- How do I record these signals best?
- How do I process these signals and extract the information?

The engineering side in an interdisciplinary team might add some more:

- How many channels and signals do I really need?
- What performance speed is necessary?
- What bandwidth do these signals have?
- How might the related interface look like?
- Is there any difference in the design for fundamental animal experiments and devices for human clinical applications?

The clinical application fields of BCI comply with common objectives in medicine. They are used in diagnosis to detect neurological pathophysiologies of the brain during EEG recordings and presurgical epilepsy diagnosis. Therapeutical use includes deep brain stimulation to treat the symptoms of Parkinson's disease and severe psychiatric disorders (e.g., compulsive–obsessive-disorder). The application of BCI in rehabilitation is still in an early stage with some human clinical trials to establish command interfaces for communication and motor control in paralyzed persons with high spinal cord lesion level or patients suffering from amyotrophic lateral sclerosis (ALS, Lou Gehrig's disease). Here, the major focus is set on BCI in rehabilitation to establish an interface between the intact brain and the technical system, either for motor control or for communication, for example, as a "brain switch." Independently from the site of recording the neural signals from the brain and the particular application, any BCI must transfer the electrical signals from the brain into command signals with the objective to control in real time, that is, only a minor time delay is allowed between the recording of the signal and the action calculated on the computer (Fig. 2). Depending on the tasks to be performed, the realization of such a BCI as command interface will be different. A large variety of paradigms have to be selected carefully: recording modality (skull surface, epicortical, intracortical), feedback strategy (abstract, realistic, continuous, discontinuous), operation mode (synchronous, asynchronous), mental strategy (motor imagery, focused attention), and brain signal selection (real-time nerve signals, oscillations[1], cognitive evoked potentials like P300, slow cortical potentials) (Birbaumer, 2006a, b). The requirements for the application with respect to spatial and temporal resolution finally narrow the decision to a few alternative opportunities. For relatively coarse spatial resolution (mm to cm range), noninvasive and nonelectrical imaging techniques become an alternative to electrical recordings. However, magnetencephalography (MEG) and functional magnetic resonance imaging (fMRI) still need large apparatus and do not allow an application under circumstances of activities of daily living. Therefore, only BCI based on electrical data acquisition will be considered in this chapter.

**Classification of BCI**

One possible classification of BCIs is done with respect to the degree of invasiveness of the neuro–technical interfaces and the number of neurons that contribute to the recorded signal. Three different classes (I, II, III) can be differentiated. Class I interfaces record neural mass activity with noninvasive techniques like EEG or MEG. Class II and III are invasive interfaces that record signals from populations of neurons from local field potentials (LFPs) or electrocorticogram (ECoG) and single-unit activity from single neurons, respectively. All three classes differ significantly in the realization of the neuro–technical interface of the BCI, the

---
[1] ERS (event-related synchronization) or ERD (event-related desynchrnonization).

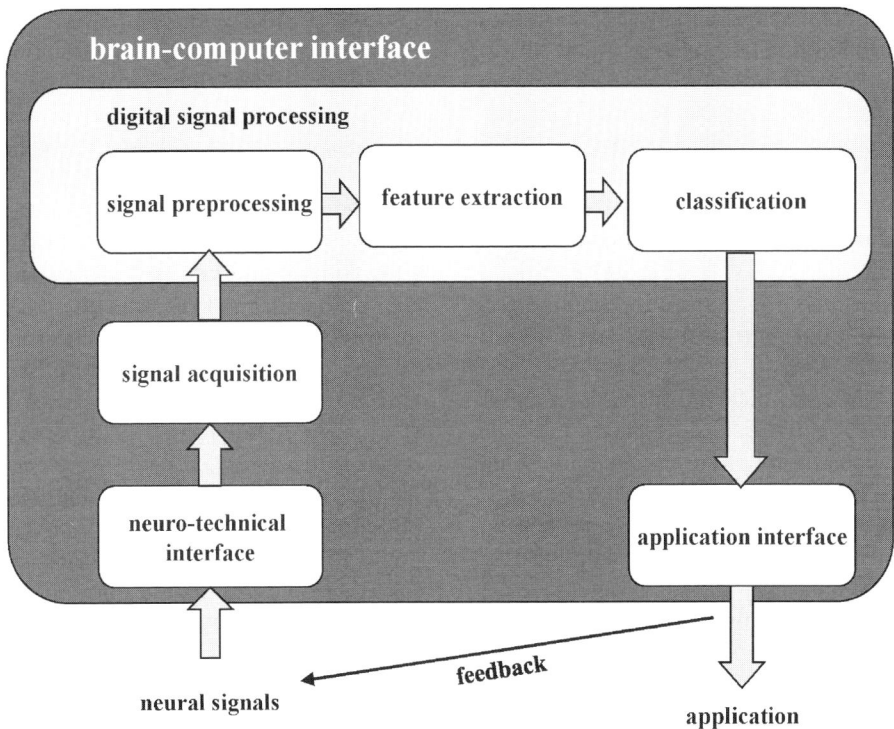

Fig. 2. Principal components in a brain–computer interface.

bandwidth of the recorded signals and finally the mathematical algorithms to extract the information and assign it to an adequate output signal. The electrode arrays act as direct interface between the neural signal and the control system. The number of electrodes, their size and spatial arrangement, and the signal transfer function should be tailored according to the envisioned application. The complete system requirements depend on the location and implantation site, the implantation time and the neural structure that shall be interfaced, respectively. These technical aspects that definitely influence the performance of a BCI will be discussed with various examples of the three BCI classes.

**Extracorporeal electrodes for class I BCI**

The noninvasive approach of BCI uses the components and techniques developed for EEGs. Surface electrodes are placed on skin of the skull according to the 10/20 scheme for EEG. These silver/silver chloride electrodes need a gel for an electrical contact to deliver extremely stable signals. However, skin, fatty tissue, and skull as well as the dura mater attenuate the electrical signals and act as an electrical low-pass filter that cuts out high-frequency components. Therefore, signals are limited to about 100 Hz in frequency. The tissues act as a spatial filter due to their electrical volume conductivity and thus also limit spatial selectivity. BCI with EEG electrodes use different brain signals and mental strategies (Wolpaw et al., 2002). In many approaches (e.g., Wolpaw et al., 2000, 2006), features were extracted from the signals to act as a kind of brain switch (Pfurtscheller et al., 2000) to select letters on a computer screen, to move a cursor on a computer screen (Birbaumer, 2006b), or to switch sequentially between the different states of a command control, for example, to control a functional electrical stimulation system for grasp function (Pfurtscheller et al., 2003). The learning procedure is only working in few persons immediately. Most patients need up to a month to be able to control a user interface in an adequate manner. Using the BCI as typewriter, only few letters per minute are feasible so far. The advantage of these systems, for example, with motor imagery is that the number of electrodes might be reduced, in some cases down to two (Pfurtscheller et al., 2003). A different approach works with a large set of electrodes, the so-called Berlin BCI. Without long training, subjects are able to obtain more than 15 bits per minute data rate (Blankertz et al., 2006). The assessment of class I BCI is different depending on the point of view: from the clinical side, the approach is advantageous because it can be started at any point of time and does not afford surgical intervention. The spatiotemporal resolution is low but sufficient to establish a stable user interface for patients who are not able to communicate with their environment in any way, for example, patients with ALS with locked-in syndrome. For this reason, the limited data rate of the interface is accepted by the patient. However, if the approach is combined with neuroprostheses, it shows some significant drawbacks. Paralyzed persons would use this approach to become more independent in their activities of daily living but need a caretaking person to apply the electrodes every morning. Apart from the societal accepted wheelchair, visible cabling from the head to a control unit has to be done that might be a cause of failure in daily life and the appearance might influence the motivation of the patient to use the system in the public.

**Epicortical electrodes for class II BCI**

The application of epicortical electrode arrays as medical device in clinical practice is limited to presurgical monitoring of seizures in patients with temporal lobe epilepsy to identify the focal area that should be removed. In fundamental research, epicortical arrays have been also used to record evoked potentials in different spatial resolutions using electrodes made by means of various technologies. Depending on the implantation site, that is, the "modality" of the cortex with its underlying anatomical organization and the research objective, number and spatial resolution of the electrode arrays vary over a wide geometrical range. In many cases, a large number of electrodes are combined with low electrode-to-electrode distances, resulting in an array that covers a relatively small area (Tsytsarev et al., 2006; Malkin and Pendley, 2000; Takahashi et al., 2003). In other studies, an intermediate number of electrodes are combined with quite large electrode sizes and electrode-to-electrode distances (Molina-Luna et al., 2007; Kitzmiller et al., 2006; Hollenberg et al., 2006; Ball et al., 2008) (Table 2). However, only few groups investigated the use of epicortical recordings to control a neural prosthesis.

One of the earliest experiments on movement control based on motor cortex recordings was carried out in monkeys by Michael Craggs, as described in his PhD thesis in 1974 (Craggs, 1974). He implanted 60-channel electrode arrays fabricated by platinum wires whose tips were melted into a ball-shaped electrode contact, sitting in a spherical rubber carrier. The electrodes could be linked to the lab equipment by a socket connector that was placed on the skull. Craggs proposed design rules based on the anatomical findings of the motor cortex, estimating the best electrode size to be smaller than 1 mm in diameter and the center-to-center spacing of larger than 1 mm. In the last years, the idea of prediction of movement with the help of direction-sensitive neurons (Taylor et al., 2002) based on single spike recordings from intracortical electrodes have been transferred to epicortical recordings (Rickert et al., 2005). The detection and prediction of the movement via LFPs has been subject of investigation in premotor and motor cortex of monkeys (Mehring et al., 2004) and humans (Pistohl et al., 2007), respectively.

The electrode sizes and electrode-to-electrode distance, however, are optimized with respect to the clinical application of presurgical epilepsy monitoring. For their use in motor BCI, they should cover at least the motor and premotor cortex. For fundamental neuroscientific investigations, even larger brain arrays should be under

Table 2. Comparison of epicortical electrode array properties for BCI

| Manufacturer/Reference | Electrode diameter (μm) | Electrode pitch (μm) | Conductor path pitch (μm) | Number of electrodes | Area of the array (mm$^2$) |
| --- | --- | --- | --- | --- | --- |
| Ad-Tech | 2300 | 5000 | Wire diameter: 70, isolation layer: 20 | 64 | 80 × 80 |
| Craggs (1974) | 500 | 2000 | Polyimide-insulated platinum wire; diameter: 76.2 | 60 | 200 (diameter: 16 mm) |
| Tsytsarev et al. (2006) | 50 | 100 | 100 | 64 | 0.8 × 0.8 |
| Malkin and Pendley (2000) | 50 | 100 | 100 | 400 | 10 × 10 |
| Takahashi et al. (2003) | 80 square | 225 | 50 | 69 | 2 × 2 |
| Molina-Luna et al. (2007) | 100 | 640/750 | 35 | 72 | 6.1 × 4.6 |
| Kitzmiller et al. (2006) | 200 square | 400 | Bonding wires | 16 | 1.4 × 1.4 |
| Hollenberg et al. (2006) | 150 | 900 | 100 | 64 | 6.5 × 6.5 |
| Rubehn et al. (2008) | 1000 | 2000/3000 | 30 | 252 | 35 × 60 |
| Schuettler et al. (2008b) | 600 | 1200 | 100 | 29 | 8.3 × 7.0 |

*Source*: Modified after Rubehn et al. (2008).

continuous monitoring, if eye-arm-hand coordination (Orban et al., 1995), or coherence of nerve activity as measure for long-distance communication in the brain (Engel et al., 2001), is subject of research. These arrays should smoothly adapt to the cortical surface with only low mechanical interaction and should follow the natural brain movement. For chronic implantations in animals and (so far) subchronic implantations in humans, that is, up to 30 days, the cables and/or plugs to recording systems have to be robust and mechanically flexible to obtain a stable percutaneous connection.

We have established two complementary technologies that match the requirements for experimental research and envisioned human preclinical studies: micromachining of polyimide substrate with embedded thin-film metallization and laser structuring of silicone rubber with sandwiched metal sheets. Micromachining and microsystems engineering have the advantages that high integration densities of structures with smallest feature sizes can be obtained. Batch processing using photolithography to transfer patterns, physical and chemical vapor deposition techniques to deposit metals and insulation layers, and wet and dry etching techniques allows the manufacturing of many devices on one substrate in parallel. However, all fabrication steps have to be performed with relatively expensive machinery in a clean room environment. All in all, design of photolithography masks and the various process steps are time-consuming. This technology is only advantageous if the designed devices are of high complexity or smallest structure sizes are mandatory.

We manufactured thin sheets of polyimide (10 μm) with integrated metal layers (300 nm) that form electrodes, interconnect lines, and connection pads. The technology (Stieglitz et al., 2000, 2005) has been proven to be biocompatible and stable and effective for neural microimplants in the PNS (Rodriguez et al., 2000; Klinge et al., 2001; Lago et al., 2005) and for retinal vision implants (Walter et al., 2005; Schanze et al., 2007). An epicortical electrode array was designed and developed with the target objective to record LFPs from a complete hemisphere of a monkey with equidistant electrode spacing. Since the adhesion between the metal layer and the polyimide is crucial for implant lifetime and electrode impedance for functionality, we used platinum for both interconnection lines and electrodes. Adhesion was sufficient and we did not have to consider the biocompatibility of additional adhesion promoters that are necessary when using gold tracks (Yeager et al., 2008). The designed epicortical array (Rubehn et al., 2008) covers an area of 35 mm × 60 mm (Fig. 3). Two hundred and fifty-two electrode sites with a diameter of 1 mm and an electrode-to-electrode distance (pitch) of 2 mm are located on 14 finger structures. This finger design allows good adaptation to the shape of the brain's surface. Monolithically, integrated interconnects with a width of 15 μm are arranged on a 15 mm wide percutaneous ribbon cable that ends in eight 32-channel plugs (Omnetics, NPD series, Omnetics Connector Corp., Minneapolis, MN) that can be connected to commercially available recording systems, for example, from Plexon. The electrode impedance (1.5–5 kΩ at 1 kHz, cutoff frequency of 300 Hz) is lower than those of other epicortical electrode arrays (Rubehn et al., 2008) and predicts good recording properties from the technical point of view. The device has been chronically implanted recently and LFPs have been recorded from all 252 electrodes (Rubehn et al., 2009). Since polyimide is not approved as material for chronic implantation, that is, no USP (U.S. Pharmacopeia) class VI approval, the procedure to transfer such a device into human clinical trials will take a long time and large monetary efforts.

If a new diagnosis method, therapy concept, or rehabilitation procedure has to be evaluated in human clinical trials, often the proof of concept only needs implants with few electrodes and medium integration densities. Other materials than polyimide and other manufacturing techniques might be more advantageous to decrease the development time from the first idea to the first human implantation.

For this purpose, a technology for microfabricating electrode arrays by laser processing has been developed (Schuettler et al., 2005), in which we restricted ourselves to the exclusive use of medical-grade materials such as silicone rubber and high-purity platinum foil, for which long-term stability in the body is proven. This technology allows quick and flexible fabrication of electrode arrays in many shapes and sizes. Designs can be easily transferred from a computer-aided design (CAD) file into prototypes without the need of masking technologies (Fig. 4). The electrodes and interconnects are structured by a laser out of a metal sheet that is sandwiched between layers of silicone rubber. Minimum feature sizes and pitches of 100 μm allow medium-scale integration of devices. A design example for an epicortical BCI has been developed (Fig. 5): 162 electrode sites with a diameter of 350 μm and an electrode-to-electrode distance of 2 mm have been manufactured. The device is flexible and easy to handle. However, the assembling of cables for percutaneous cables or a wireless telemetry unit for the long-term run is still under development. A much simpler device with only eight electrodes of similar size (320 μm diameter) and pitch (1 mm) has been implanted to investigate the recording capability after implantation. Signal amplitudes were relatively stable over four-week implantation time with an amplitude of about 90–100 μV. The electrode-tissue impedance between two electrodes with a distance of 3 mm increased mainly within the first week and stayed relatively stable in the range of 40–70 kΩ at 1 kHz (Cordeiro et al., 2008). First results are promising but the major challenge is still the assembling of cables. Several joining techniques are currently under investigation with respect to reliability, strength, reproducibility, and biocompatibility (Schuettler et al., 2008a).

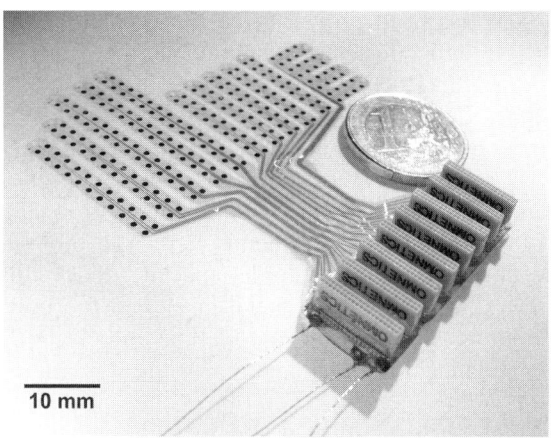

Fig. 3. Micromachined epicortical electrode array of 252 sites to record from a complete brain hemisphere.

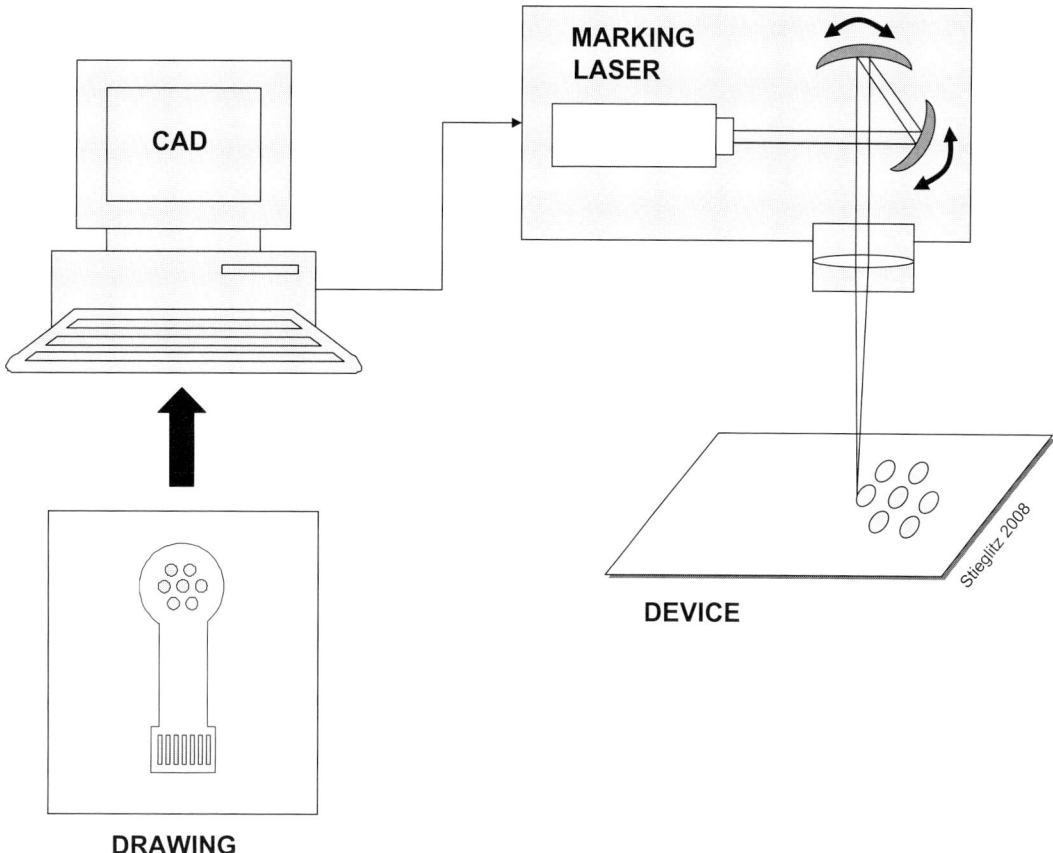

Fig. 4. Drawing is directly transferred to a computer (computer-aided design: CAD) that controls mirrors of a marking laser to transfer patterns into metal foils for electrodes and interconnection tracks as well as into silicon insulation.

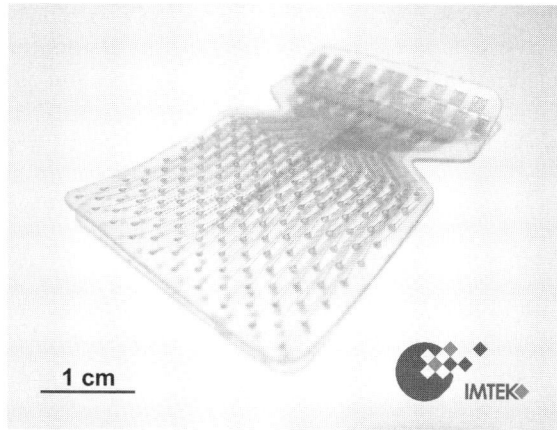

Fig. 5. Laser-structured epicortical electrode array with 162 sites before cable and connector assembly.

Both technologies are currently used in first implantation studies to evaluate the stability of the device as well as signal amplitude and the signal-to-noise ratio over the implantation time. Before first human clinical trials can be scheduled, legal requirements regarding the FDA approval or the CE mark have to be fulfilled. These work packages include risk assessment (ISO 14971 – Risk Assessment of Medical Devices) of the devices for the intended use and quality management during device manufacturing (ISO 13485) to reduce the hazard for the patient in novel applications of implantable medical devices. The application and the complexity of the device will eventually have a strong impact on the technology to be selected.

## Intracortical electrodes for class III BCI

Intracortical electrodes for class III BCI are intended to record single-unit activity in fundamental neuroscientific investigations. In the last years, the knowledge about movement encoding could be used to control motor BCI in animal experiments (Schwartz et al., 2006) and also in an investigator study in a single paralyzed person (Hochberg et al., 2006). We do not wish to discuss the different paradigms of signal processing and feature extraction that have been reviewed in detail elsewhere (Wolpaw et al., 2002; McFarland et al., 2006) but would like to present the hardware that is necessary to interface with the cortical neurons. In a cubicle of $1 \times 1 \times 2 \, mm^3$ of cortical tissue, one can count about 60,000 neurons and more than 1.4 billion synapses. A small number of electrodes record the signals to be used for a BCI application, for example, for motor control. Sometimes less than 10 single-unit signals are sufficient. However, it is not clear a priori where to find these signals. Therefore, more electrodes should be implanted to get the "right" signals stable over a long time. A second challenge is the mismatch of the mechanical properties of the material–tissue interface. The Young's modulus that describes the flexibility of a material is quite low for the brain ($\sim 100 \, kPa$) while it is orders of magnitudes higher for technical materials, for example, 1.2 GPa for silicon. So far, systems have been developed that are stiff enough to be implanted into the brain, sometimes without removing the dura mater. Three basic approaches of neuro–technical interfaces have been established: wire electrodes, micromachined shanks with multiple electrodes per shank, and a micromachined "nail bed" with one electrode at the tip of each shank (Table 3). Wire electrodes are the oldest approach that is widespread and established among many research groups. Electrodes can be purchased from commercial suppliers or simply manufactured by soldering an insulated wire on a connector that fits to the recording equipment of the lab. Self-made screw drives might allow adjustment in acute and chronic implantations, which have been made with up to 740 wires in a single cortex (Nicolelis et al., 2003). The temporal stability of the recordings from these wire electrodes is high. Single-unit recordings over 18 months have been obtained (Nicolelis et al., 2003). The tip of each wire serves as the electrode. The ends of the wires are directly connected to the pins of connectors and can be stacked to three-dimensional arrays. The wires can be arranged as stereotrodes (McNaughton et al., 1983) or tetrodes (Gray et al., 1995) to be able to discriminate the signals from more than one neuron, from one electrode site. However, recordings from several sites along one wire are not possible.

The use of micromachining techniques to manufacture microelectrodes for extracellular recordings has been proposed 40 years ago (Wise et al., 1969). Nowadays, two different concepts have been established. One is the so-called Utah array (Nordhausen et al., 1996), recently commercialized as BrainGate™ system from Cyberkinetics. Needles were manufactured in an out-of-plane approach from a block of silicon. The monolithic system consists of 100 needles with one electrode on each tip (Campbell et al., 1991; for details see Table 3). Recordings have been done in CNS (Nordhausen et al., 1996) and PNS (Warwick et al., 2003). The Utah system has also been proposed as hardware within a cortical vision prosthesis (Normann et al., 1999). It is the only system worldwide that proved its performance in a spinal cord-injured human subject (lesion level: C3) for 18 months (Hochberg et al., 2006), as hardware within a motor BCI to control a virtual keyboard. The 100 cables for the electrodes have been routed to a percutaneous connector that has to be connected via a cable-bound preamplifier unit to a computer-controlled signal acquisition system. Recent developments at the University of Utah focused on a wireless system with inductive energy supply and integrated circuitry to record and transmit the data from the needle array (Harrison et al., 2007). The system components have been developed, and the assembled and encapsulated system is ready for use now (Kim et al., 2007; Hsu et al., 2007a, b).

A competitive approach has been developed at the University of Michigan since 1969 (Wise et al., 1969). A large variety of probes has

Table 3. Comparison of main concepts for intracortical electrode arrays

| Manufacturer/Type | Wire | Utah/Cyberkinetics | Michigan/Neuronexus | Neuroprobes |
|---|---|---|---|---|
| Substrate material | Tungsten, stainless steel | Silicon | Silicon | Silicon |
| Possibility of monolithic integration of electronic circuitry | No | No | Yes | Yes |
| Number of electrodes | 1 per wire, 8/16 per device, per stack: 128 | 1 per shank, up to 100 per device | Up to 64 per probe, up to 1024 per device | 8 of 188 per shank, 4 shanks per comb, number per device to be determined |
| Electrode material | Tungsten, stainless steel | Iridium | Iridium oxide | Platinum, iridium oxide |
| Electrode diameter ($\mu m$) | 50, tip of wire | Tip of shank, 50 $\mu m$ from tip deinsulated | 15–40 | 20 |
| Electrode site spacing ($\mu m$) | Identical to shank spacing | Identical to shank spacing | 50–500 | Minimum: 40; typical: 187/500/625 for 2/4/8 mm shanks |
| Shank spacing ($\mu m$) | Variable, typical: 635 | 400 | 125–500 | |
| Shank thickness/width or diameter ($\mu m$) | 50 | 80 $\mu m$ at bottom, shank tapers to tip | ~15 × ~150 | 100 × 160 |
| Shank length (mm) | Variable, typical length: 4–10 | 1.0–1.5 | Up to 10 | 2, 4, 8 |
| Array size | Variable, stacks of 2-D needle arrays, upto 672 wires | 4 × 4 | Mainly 2-D shank arrays, 16 × 16 shank 3-D array about 6 × 6 | 2-D shank arrays to be mounted on platform, 4 × 4 in the current design |
| Variability of electrode design/arrangement | Medium | Low | High | Medium |
| Electrode development | Nicolelis Lab, Duke University (USA) | University of Utah (USA) | University of Michigan (USA) | IMTEK, University of Freiburg (Germany) |
| Commercially available | NB Labs | Cyberkinetics | Neuronexus | – |
| Website | http://nblabs.net | http://www.cyberkineticsinc.com | http://www.neuronexustech.com | http://www.neuroprobes.org |
| References | Nicolelis et al. (2003) | Nordhausen et al. (1996) | Wise et al. (2004) | Kisban et al. (2007), Seidl et al. (2009) |

been developed and tested over the years. These probes are now commercially available from NeuroNexus Technologies after the NIH funding of the Center for Neural Communication Technologies has run out some years ago. The Michigan probes also use silicon as substrate, but they are fabricated in a so-called in-plane approach, that is, the electrode sites are arranged along one surface of the shank. Dimensions of electrode size and arrangement as well as shank distance cover a wide range (Table 3). This technological approach allows the monolithic integration of electronic circuitry, for example, for multiplexing and amplification in the probes (Najafi and Wise, 1986). Planar comb-like structures (Kim and Wise, 1996) can be assembled to three-dimensional arrays (Hoogerwerf and Wise, 1994; Bai et al., 2000) with up to 1024 electrode sites (Wise et al., 2004). Electronic circuitry for spike detection of single-unit activity reduces the amount of data to be transferred. In addition to the electrodes for recording of bioelectrical signals, microfluidic channels can be integrated to deliver pharmaceutical agents (Papageorgiou et al., 2006).

Recently, the European NeuroProbes consortium (www.neuroprobes.org) developed yet another competitive approach for silicon-based intracortical microelectrodes (Neves et al., 2007). The probes also have an in-plane electrode arrangement on the shanks but with variable pitch and shank length (Table 3). The combs of shanks are either directly connected to a flexible polyimide cable or are assembled on a platform to a three-dimensional array (Fig. 6). The platform approach allows variable assembly of silicon probes with electrode sites for recording of single-unit activity from cortical neurons, probes with biosensors to monitor glutamate and dopamine, and probes with integrated microfluidic channels for sampling or drug delivery. The basic components of this system have been developed, including recording chambers for chronic implantations (Fig. 7). Single-unit activity was recorded after implantation (Kisban et al., 2007, 2008), but the assessment of long-term stability of the recordings is still under investigation. Researchers often report on the loss of single-unit recordings over time (Rousche and Normann, 1998) and propose adaptive systems that allow readjustment of the probes after implantation to be able to get another cortical cell to record from. Instead of an electrode drive with a micromachined actuator (Muthuswamy et al., 2005), we have developed an electronic depth control (Seidl et al., 2009) that electronically selects the best electrodes over the entire shank length and might adapt to a changing biological environment electronically.

All intracortical electrode approaches work well in acute recording conditions but often fail functionally with respect to reliability in chronic implantation. The materials have been proven to be noncytotoxic (Stensaas, 1978), but the brain tissue reaction after implantation against the probes is the major problem (Polikov et al., 2005). Manufacturing technology and mechanical properties limit minimal shaft size of the devices due to fragility and electrical resistance. Electrodes can only be inserted into the tissue in parallel to the shaft axis. Therefore, a minimum critical stiffness is necessary. These electrodes are not flexible and cannot adapt to curvature of anatomical structures and pulsatile changes of position due to heartbeat and respiration movements. Research focuses on different strategies to improve the tissue–material interface: geometrical variations like lattices structures (Seymour and Kipke, 2007), tubular substrates with resorbable cores (Takeuchi et al., 2005), surface modifications (Zhong et al., 2001), and hydrogel coatings (Winter et al., 2007; Kim et al., 2004) on stiff substrates. However, with these state-of-the-art developments, the basic problems in penetrating microelectrodes are not solved. Mechanical mismatch of the brain–material interface as described above in combination with the micromotions induces, with time, extensive scar formation around the electrodes, resulting in low nerve signal recording amplitudes due to electrical insulation properties of the scar tissue (Subbaroyan et al., 2005; Seymour and Kipke, 2007). We suggested the use of more flexible materials for neural implants (Stieglitz, 2004; Stieglitz et al., 2005; Navarro et al., 2005) and addressed the challenge to design polymer-based shaft electrodes (Fig. 8) and insert flexible materials into the

Fig. 6. Micromachined intracortical electrode array. Shafts arranged as combs (a: left) can be scaled up and mounted onto three-dimensional assemblies on a platform (b: right). Omnetics connector length is about 14 mm.

brain. While flat shaft electrodes were not able to penetrate the dura or even the pia mater (Haj Hosseini et al., 2007), pre-shaped shaft electrodes could be inserted into brain tissue (Stieglitz et al., 2008) due to an increased moment of inertia parallel to the insertion axis. However, the selection and tailoring of adequate materials that do not induce strong brain tissue reactions is still subject of intensive research in micromachining and material sciences.

## Comparison of BCI approaches

Multiple factors influence a BCI system: the intended use of the system, the decision to choose an application in fundamental neuroscientific investigations or in clinical practice, the selection of detection paradigms with underlying algorithms, and many more detailed decisions. So far, reviews have compared the performance of the algorithms (Bashashati et al., 2007) and project the course of BCI developments in the future (Lebedev and Nicolelis, 2006). SWOT analysis has been carried out for the three BCI classes that have been presented in the preceding text. The noninvasive EEG, the ECoG, and the intracortical approach have their specific strength but also carry some inherent risks for certain applications (Table 4). A careful comparison of their properties with respect to the underlying hardware realization of the neuro–technical interface (Fig. 9)

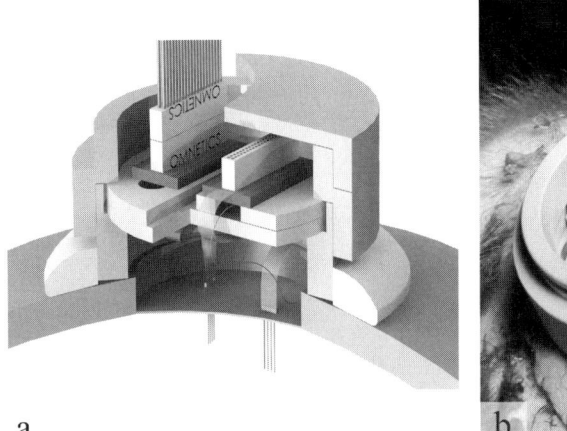 

Fig. 7. Assembly of electrodes into a chronic recording chamber for fundamental neuroscientific investigations. Schematic drawing (a: left) and realization (b: right).

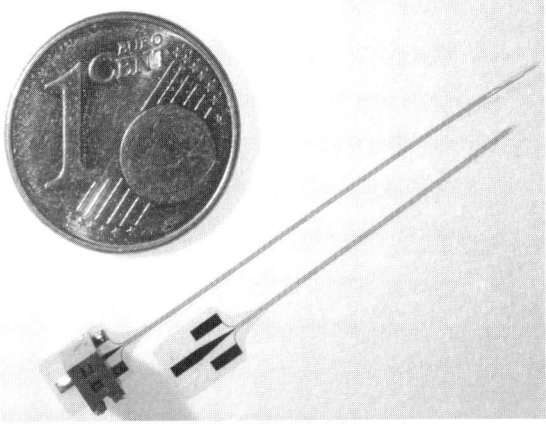

Fig. 8. Flexible, micromachined shaft electrode. Material: polyimide of 10 μm thickness.

might serve as an additional tool in the decision process to find the best BCI approach for the envisioned application.

## Concluding remarks

BCIs for recording have been investigated in many technical variations in fundamental studies in animals and in carefully selected patients in clinical trials. It has been proven that they definitely can serve as a command interface for motor control and communication. Paralyzed subjects with an intact brain or healthy animals used the BCI without limitation in performance. In noninvasive BCI, most data seem to come even from able body subjects, that is, with non-severed brains. Nearly, all trials and investigations have been performed in a controlled environment in the hospital or laboratory but not under conditions during activities of daily living.

First trials in completely locked-in patients (Kuebler and Birbaumer, 2008) showed no or only very limited performance. Studies with higher patient numbers have to show how the severely damaged brain might interact with BCI and how much cognitive capability and plasticity of the brain is needed for useful communication interfaces.

In the long-term run, implantable systems will be only transferred into clinical practice if they go wireless. Energy supply and data transmission must become reliable without handling infection risks. As the cardiac pacemaker, the cochlea implant and the deep brain stimulator show that patients will undergo a surgical intervention if the risk is well outweighed by the benefits of the implant.

In clinical practice, electrical stimulation of the CNS, which can be also assigned as "stimulation

Table 4. SWOT analysis for BCI: class I (EEG), class II (ECoG), class III (intracortical)

| | Strengths | | | Weaknesses | | | Opportunities | | | Threats/Risks | |
|---|---|---|---|---|---|---|---|---|---|---|---|
| EEG | ECoG | Intracortical | EEG | ECoG | Intracortical | EEG | ECoG | Intracortical | EEG | ECoG | Intracortical |
| • Noninvasive<br>• Surface electrodes<br>• Relatively low cost<br>• Established instrumentation | • Resolution of cortical columns possible<br>• No injury of brain tissue<br>• Relatively large frequency band | • Recording of single-unit activity<br>• Calculation of movement trajectories from single neuron recording<br>• Systems with high integration density | • Low spatial resolution<br>• Performance depends on electrode–skin contact<br>• Low speed with few electrodes<br>• Need of skilled person for electrode placement | • Wireless systems not available<br>• No clinical approved electrode array available that matches requirements for motor controls, i.e., electrode pitch too large, electrodes too large | • Chronic stability of recordings for human applications not yet proven<br>• Wireless system with CE mark/FDA approval not yet available | • Widespread use in simple rehabilitation systems possible<br>• Evaluation system to identify "responders" before implantation of other systems | • Higher resolution for applications in diagnosis, e.g., presurgical epilepsy monitoring<br>• Interface approach in which large arrays of brain surface might be covered | • Real-time control of complex motor systems without the need of nonadequate control signals<br>• Miniaturized wireless implants (electrodes and electronics in one piece) | • Reliability of function, if used in systems for motor control | • Benefit/risk ratio in patients with degenerating diseases too low<br>• Risk of infection in chronic implantation due to percutaneous cables | • Risk of nerve tissue damage in the brain<br>• Loss of single-unit signals after some months<br>• Long-term stability of microimplants |

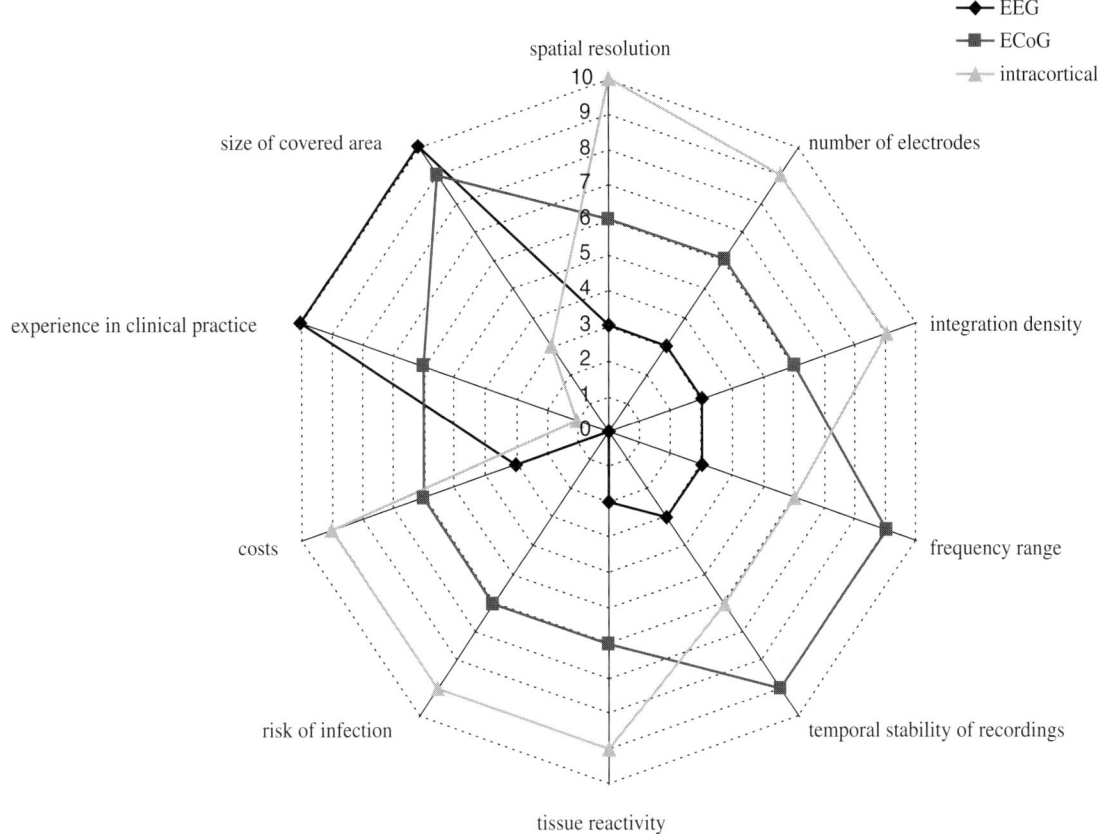

Fig. 9. Comparison of different brain–computer interface (BCI) approaches. Scoring: 0 = low; 10 = high.

BCI," has been successful for many applications, especially for therapeutical use. The success stories like treatment of dyskinesis and tremor in Parkinson's disease work with extremely simple systems and without the knowledge of the underlying pathophysiological principles. Electrical stimulation treatment of severe psychiatric disorders, especially mood disorders and obsessive–compulsive disorders, can be somehow linked to the limbic system and the subthalamic nuclei. However, stimulus–reaction mechanisms still have to be found. The first steps for sensory CNS BCI or sensory neuroprostheses, respectively, have been taken with respect to hearing in auditory brainstem and midbrain implants (Schwartz et al., 2008), and vision prosthesis (Normann et al., 1999). The journey from fundamental investigations and first clinical trials toward products that are successfully used in standard clinical treatments is long and expensive. Scientific excellence, engineering skills, entrepreneurship, and a portion of good luck are necessary to create success stories and help patients with BCI to live their life as independently as possible.

**Acknowledgments**

Part of the work has been funded by the German Federal Ministry of Education and Research (BMBF) in the "Exist Go Bio" Project (grant 01GQ0420), by a grant of the European Union in the sixth framework program within the "Neuroprobes" Project (IST-027017), and a collaboration with P. Fries from F.C. Donders Centre for Neurosciences (Nijmegen, the Netherlands).

# References

Bai, Q., Wise, K. D., & Anderson, D. J. (2000). A high-yield microassembly structure for three-dimensional microelectrode arrays. *IEEE Transactions on Bio-medical Engineering*, 47, 281–289.

Ball, T., Demandt, E., Mutschler, I., Neitzel, E., Mehring, C., Vogt, K., et al. (2008). Movement related activity in the high gamma range of the human EEG. *NeuroImage*, 41, 302–310.

Bashashati, A., Fatourechi, M., Ward, R. K., & Birch, G. E. (2007). A survey of signal processing algorithms in brain-computer interfaces based on electrical brain signals. *Journal of Neural Engineering*, 4, R32–R57.

Birbaumer, N. (2006a). Brain–computer interface research: coming of age. *Clinical Neurophysiology*, 117, 479–483.

Birbaumer, N. (2006b). Breaking the silence: brain–computer interfaces (BCI) for communication and motor control. *Psychophysiology*, 43, 517–532.

Blankertz, B., Dornhege, G., Krauledat, M., Muller, K.-R., Kunzmann, V., Losch, F., et al. (2006). The Berlin brain–computer interface: EEG-based communication without subject training. *IEEE Transactions on Neural Systems and Rehabilitation Engineering*, 14, 147–152.

Brindley, G. S. (1994). The first 500 patients with sacral anterior root stimulator implants: general description. *Paraplegia*, 32, 795–805.

Campbell, P. K., Jones, K. E., Huber, R. J., Horch, K. W., & Normann, R. A. (1991). A silicon-based, three-dimensional neural interface: manufacturing processes for an intracortical electrode array. *IEEE Transactions on Bio-medical Engineering*, 38, 758–768.

Cordeiro, J. G., Henle, C., Raab, M., Meier, W., Stieglitz, T., Schulze-Bonhage, A., et al. (2008). Micromachined electrodes for cortical field potential recordings: In vivo study. In: J. van der Sloten, P. Verdonck, M. Nyssen, & L. Haueisen (Eds.), *4th European Conference of the International Federation for Medical and Biological Engineering (ECIFMBE 2008, IFMBE Proceedings 22)*, November 23–27, 2008, Antwerp, Belgium (pp. 2375–2378).

Craggs, M. D. (1974). *The cortical control of limb prostheses*. PhD thesis, University of London, pp. 21–27.

Engel, A. K., Fries, P., & Singer, W. (2001). Dynamic predictions: oscillations and synchrony in top-down processing. *Nature Reviews Neuroscience*, 2, 704–716.

Gray, C. M., Maldonado, P. F., Wilson, M., & McNaughton, B. L. (1995). Tetrodes markedly improve the reliability and yield of multiple single-unit isolation from multi-unit recordings in cat striate cortex. *Journal of Neuroscience Methods*, 63, 43–54.

Haj Hosseini, N., Hoffmann, R., Kisban, S., Stieglitz, T., Paul, O., & Ruther, P. (2007) Comparative study on the insertion behavior of cerebral microprobes. The 29th Annual International Conference of the IEEE EMBS, Aug. 23–26, 2007, Lyon, France, 2007, *Proceedings of the 29th Annual International Conference of the IEEE EMBS* (pp. 4711–4714).

Harrison, R. R., Watkins, P. T., Kier, R. J., Lovejoy, R. O., Black, D. J., Greger, B., et al. (2007). A low-power integrated circuit for a wireless 100-electrode neural recording system. *IEEE Journal of Solid-State Circuits*, 42, 123–133.

Hochberg, L. R., Serruya, M. D., Friehs, G. M., Mukand, J. A., Saleh, M., Caplan, A. H., et al. (2006). Neuronal ensemble control of prosthetic devices by a human with tetraplegia. *Nature*, 442, 164–171.

Hollenberg, B. A., Richards, C. D., Richards, R., Bahr, D. F., & Rector, D. M. (2006). A MEMS fabricated flexible electrode array for recording surface field potentials. *Journal of Neuroscience Methods*, 153, 147–153.

Hoogerwerf, A. C., & Wise, K. D. (1994). A three-dimensional microelectrode array for chronic neural recording. *IEEE Transactions on Biomedical Engineering*, 41, 1136–1146.

Hsu, J.-M., Kammer, S., Jung, E., Rieth, L., Normann, R. A., & Solzbacher, F. (2007a). Characterization of parylene-C film as an encapsulation material for neural interface devices. In S. Dimov, W. Menz, and Y. Toshev (Eds.), Conference on Multi-Material Micro Manufacture, Oct. 3–5, 2007, Borovets, Bulgaria, *Proceedings of the 3rd international conference on multi-material micro manufacture*.

Hsu, J.-M., Tathireddy, P., Rieth, L., Normann, R. A., & Solzbacher, F. (2007b). Characterization of a-SiC$_x$: H thin films as an encapsulation material for integrated silicon based neural interface devices. *Thin Solid Films*, 516, 34–41.

Kim, C., & Wise, K. D. (1996). A 64-site multishank CMOS low-profile neural stimulating probe. *IEEE Journal of Solid-State Circuits*, 31, 1230–1238.

Kim, D. H., Abidian, M. R., & Martin, D. C. (2004). Conducting polymers grown in hydrogel scaffolds coated on neural prosthetic devices. *Journal of Biomedical Material Research-A*, 71, 577–585.

Kim, S., Zoschke, K., Klein, M., Black, D., Buschick, K., Toepper, M., et al. (2007). Switchable polymer-based thin film coils as a power module for wireless neural interfaces. *Sensor and Actuators A-Physical*, 136, 467–474.

Kisban, S., Herwik, S., Seidl, K., Rubehn, B., Jezzini, A., Umilta, M. A., et al. (2007). Microprobe array with low impedance electrodes and highly flexible polyimide cables for acute neural recording. 29th Annual International Conference of the IEEE EMBS, Aug. 23–26, Lyon, France, *Proceedings of the 29th Annual International Conference of the IEEE EMBS* (pp. 175–178).

Kisban, S., Moser, D., Rubehn, B., Stieglitz, T., Paul, O., & Ruther, P. (2008). Fatigue testing of polyimide-based micro implants. In: J. van der Sloten, P. Verdonck, M. Nyssen, & L. Haueisen (Eds.), *4th European Conference of the International Federation for Medical and Biological Engineering (ECIFMBE 2008, IFMBE Proceedings 22)*, November 23–27, 2008, Antwerp, Belgium (pp. 1594–1597).

Kitzmiller, J., Beversdorf, D., & Hansford, D. (2006). Fabrication and testing of microelectrodes for small-field cortical surface recordings. *Biomedical Microdevices*, 8, 81–85.

Klinge, P. M., Vafa, M. A., Brinker, T., Brandis, T., Walter, G. F., Stieglitz, T., et al. (2001). Immunohistochemical characterization of axonal sprouting and reactive tissue changes after long-term implantation of a polyimide sieve electrode to the transected adult rat sciatic nerve. *Biomaterials*, 22, 2333–2343.

Kuebler, A., & Birbaumer, N. (2008). Brain–computer interfaces and communication in paralysis: extinction of goal directed thinking in completely paralysed patients? *Clinical Neurophysiology*, 119, 2658–2666.

Lago, N., Ceballos, D., Rodriguez, F. J., Stieglitz, T., & Navarro, X. (2005). Long term assessment of axonal regeneration through polyimide regenerative electrodes to interface the peripheral nerve. *Biomaterials*, 26, 2021–2031.

Lebedev, M. A., & Nicolelis, M. A. L. (2006). Brain–machine interfaces: past, present and future. *Trends in Neurosciences*, 29, 536–546.

Malkin, R. A., & Pendley, B. D. (2000). Construction of a very high-density extracellular electrode array. *American Journal of Physiology-Heart and Circulatory physiology*, 279, H437–H442.

McFarland, D. J., Anderson, C. W., Muller, K.-R., Schlogl, A., & Krusienski, D. J. (2006). BCI meeting 2005-workshop on BCI signal processing: feature extraction and translation. *IEEE Transactions on Neural Sysetms and Rehabilitation Engineering*, 14, 135–138.

McNaughton, B. L., O'Keefe, J., & Barnes, C. A. (1983). The stereotrode: a new technique for simultaneous isolation of several single units in the central nervous system from multiple unit records. *Journal of Neuroscience Methods*, 8, 391–397.

Mehring, C., Nawrot, M., de Olivera, S. C., Vaadia, E., Schulze-Bonhage, A., Aertsen, A., et al. (2004). Comparing information about arm movement direction in single channels of local and epicortical field potentials from monkey and human motor cortex. *Journal of Physiology and Pharmacology*, 98, 498–506.

Molina-Luna, K., Buitrago, M. M., Hertler, B., Schubring, M., Haiss, F., Nisch, W., et al. (2007). Cortical stimulation mapping using epidurally implanted thin-film microelectrode arrays. *Journal of Neuroscience Methods*, 161, 118–125.

Muthuswamy, J., Okandan, M., Jain, T., & Gilletti, A. (2005). Electrostatic microactuators for precise positioning of neural microelectrodes. *IEEE Transactions on Biomedical Engineering*, 52, 1748–1755.

Najafi, K., & Wise, K. D. (1986). An implantable multielectrode array with on-chip signal processing. *IEEE Journal of Solid-State Circuits*, 21, 1035–1044.

Navarro, X., Krueger, T. B., Lago, N., Micera, S., Stieglitz, T., & Dario, P. (2005). A critical review of interfaces with the peripheral nervous system for the control of neuroprotheses and hybrid bionic systems. *Journal of the Peripheral Nervous System*, 10, 229–258.

Neves, H. P., Orban, G. A., Koudelka-Hep, M., Stieglitz, T., & Ruther, P. (2007). Development of modular multifunctional probe arrays for cerebral applications. The 3rd International IEEE EMBS Conference on Neural Engineering, Kohala Coast, Hawaii, USA, May 2–5, 2007, *Proceedings on 3rd International IEEE EMBS Conference on Neural Engineering* (pp. 104–109).

Nicolelis, M. A. L., Dimitrov, D., Carmena, J. M., Crist, R., Lehew, G., Kralik, J. D., et al. (2003). Chronic, multisite, multielectrode recordings in macaque monkeys. *Proceedings of the National Academy of Sciences USA*, 100, 11041–11046.

Nordhausen, C. T., Maynard, E. M., & Normann, R. A. (1996). Single unit recording capabilities of a 100 microclectrode array. *Brain Research*, 726, 129–140.

Normann, R. A., Maynard, E. M., Rousche, P. J., & Warren, D. J. (1999). A neural interface for a cortical vision prosthesis. *Vision Research*, 39, 2577–2587.

Orban, G. A., Dupont, P., De Bruyn, B., Vogels, R., & Vandenberghe, R. (1995). A motion area in human visual cortex. *Neurobiology*, 92, 993–997.

Papageorgiou, D. P., Shore, S. E., Bledsoe, S. C., Jr., & Wise, K. D. (2006). A shuttered neural probe with on-chip flowmeters for chronic in vivo drug delivery. *Journal of Microelectromechanical Systems*, 15, 1025–1033.

Pfurtscheller, G., Mueller, G. R., Pfurtscheller, J., Gerner, H. J., & Rupp, R. (2003). 'Thought'-control of functional electrical stimulation to restore hand grasp in a patient with tetraplegia. *Neuroscience Letters*, 351, 33–36.

Pfurtscheller, G., Neuper, C., Guger, C., Harkam, W., Ramoser, H., Schlogl, A., et al. (2000). Current trends in Graz brain–computer interface (BCI) research. *IEEE Transactions on Rehabilitation Engineering*, 8, 216–219.

Pistohl, T., Ball, T., Schulze-Bonhage, A., Aertsen, A., & Mehring, C. (2007). Prediction of arm movement trajectories from ECoG-recordings in humans. *Journal of Neuroscience Methods*, 167, 105–114.

Polikov, V. S., Tresco, P. A., & Reichert, W. M. (2005). Response of brain tissue to chronically implanted neural electrodes. *Journal of Neuroscience Methods*, 148, 1–18.

Rickert, J., Cardoso de Oliveira, S., Vaadia, E., Aertsen, A., Rotter, S., & Mehring, C. (2005). Encoding of movement direction in different frequency ranges of motor cortical local field potentials. *Journal of Neuroscience*, 25, 8815–8824.

Rijkhoff, N. J. M. (2004). Neuroprostheses to treat neurogenic bladder dysfunction: current status and future perspectives. *Child's Nervous System*, 20, 75–86.

Rodriguez, F. J., Ceballos, D., Schuettler, M., Valero-Cabre, A., Valderrama, E., Stieglitz, T., et al. (2000). Polyimide cuff electrodes for peripheral nerve stimulation. *Journal of Neuroscience Methods*, 98, 105–118.

Rousche, P. J., & Normann, R. A. (1998). Chronic recording capability of the Utah intracortical electrode array in cat sensory cortex. *Journal of Neuroscience Methods*, 82, 1–15.

Rubehn, B., Bosmann, C., Oostenfeld, R., Fries, P., & Stieglitz, T. (2009). A MEMS-based flexible multichannel ECoG-electrode array. *Journal of Neural Engineering*, 6(3), 1–10 (epub ahead of print, doi:10.1088/1741-2560/6/3/036003).

Rubehn, B., Fries, P., & Stieglitz, T. (2008). MEMS-technology for large-scale, multichannel ECoG-electrode array manufacturing. In: J. van der Sloten, P. Verdonck, M. Nyssen, & L. Haueisen (Eds.), *4th European Conference of the International Federation for Medical and Biological Engineering (ECIFMBE 2008, IFMBE Proceedings 22)*, November 23–27, 2008, Antwerp, Belgium (pp. 2413–2416).

Schanze, T., Hesse, L., Lau, C., Greve, N., Haberer, W., Kammer, S., et al. (2007). An optically powered single-channel stimulation implant as test system for chronic biocompatibility and biostability of miniaturized retinal vision prostheses. *IEEE Transactions on Biomedical Engineering, 54*, 983–992.

Schuettler, M., Henle, C., Ordonez, J. S., Meier, W., Guenther, T., & Stieglitz, T. (2008a). Interconnection technologies for laser-patterned electrode arrays. *Proceeding of the 30th Annual International Conference of IEEE Engineering Medicine Biological Society*, August 20–24, 2008, Vancouver, Canada (pp. 3212–3215).

Schuettler, M., Henle, C., Ordonez, J. S., Oh, D., Gilat, O., & Holder, S.A. (2008b). A Flexible 29 Channel Epicortical Electrode Array, *Biomedical Technological Journal, 53*, 232–234.

Schuettler, M., Stiess, S., & Suaning, G. J. (2005). Fabrication of implantable microelectrode arrays by laser cutting of silicone rubber and platinum foil. *Journal of Neural Engineering, 2*, S121–S128.

Schwartz, A. B., Cui, X. T., Weber, D. J., & Moran, D. W. (2006). Brain-controlled interfaces: movement restoration with neural prosthetics. *Neuron, 52*, 205–220.

Schwartz, M. S., Otto, S. R., Shannon, R. V., Hitselberger, W. E., & Brackmann, D. E. (2008). Auditory brainstem implants. *Neurotherapeutics, 5*, 128–136.

Seidl, K., Herwik, S., Nurcahyo, Y., Torfs, T., Keller, M., Schuttler, M., et al. (2009). CMOS-based high-density silicon microprobe array for electronic depth control in neural recording. *Micro Electro Mechanical Systems, 2009 (MEMS 2009). IEEE 22nd International Conference*, January 25–29, 2009, Sorrento, Italy (pp. 232–235) (Digital Object Identifier 10.1109/MEMSYS.2009.4805361).

Seymour, J. P., & Kipke, D. R. (2007). Neural probe design for reduced tissue encapsulation in CNS. *Biomaterials, 28*, 3594–3607.

Stensaas, S. S., & Stensaas, L. J. (1978). Histopathological evaluation of materials implanted in the cerebral cortex. *Acta Neuropathologica, 41*, 145–155.

Stieglitz, T. (2004). Considerations on surface and structural biocompatibility as prerequisite for long-term stability of neural prostheses. *Journal of Nanoscience and Nanotechnology, 4*, 496–503.

Stieglitz, T., Beutel, H., Schuettler, M., & Meyer, J.-U. (2000). Micromachined, polyimide-based devices for flexible neural interfaces. *Biomedical Microdevices, 2*, 283–294.

Stieglitz, T., Hoffmann, R., & Kaminsky, J. (2008) Investigations on mechanical properties of polyimide-based shaft electrodes for intracortical neural interfaces, in prep.

Stieglitz, T., Schuettler, M., & Koch, K. P. (2005). Implantable biomedical microsystems for neural prostheses. Flexible, polyimide-based, and modular. *IEEE Engineering in Medicine and Biology Magazine, 24*, 58–65.

Subbaroyan, J., Martin, D. C., & Kipke, D. R. (2005). A finite-element model of the mechanical effects of implantable microelectrodes in the cerebral cortex. *Journal of Neural Engineering, 2*, 103–113.

Takahashi, H., Ejiri, T., Nakao, M., Nakamura, N., Kaga, K., & Herve, T. (2003). Microelectrode array on folding polyimide ribbon for epidural mapping of functional evoked potentials. *IEEE Transactions on Biomedical Engineering, 50*, 510–516.

Takeuchi, S., Ziegler, D., Yoshida, Y., Mabuchi, K., & Suzuki, T. (2005). Parylene flexible neural probes integrated with microfluidic channels. *Lab on a Chip, 5*, 519–523.

Taylor, D. M., Tillery, S. I. H., & Schwartz, A. B. (2002). Direct cortical control of 3D neuroprosthetic devices. *Science, 296*, 1829–1832.

Tsytsarev, V., Taketani, M., Schottler, F., Tanaka, S., & Hara, M. (2006). A new planar multielectrode array: recording from a rat auditory cortex. *Journal of Neural Engineering, 3*, 293–298.

Walter, P., Kisvarday, Z. F., Goertz, M., Alteheld, N., Rossler, G., Stieglitz, T., et al. (2005). Cortical activation via an implanted wireless retinal prosthesis. *Investigative Ophthalmology and Visual Science, 46*, 1780–1785.

Warwick, K., Gasson, M., Hutt, B., Goodhew, I., Kyberd, P., Andrews, B. J., et al. (2003). The application of implant technology for cybernetic systems. *Archives of Neurology, 60*, 1369–1373.

Winter, J. O., Cogan, S. F., & Rizzo, J. F. (2007). Neurotrophin-eluting hydrogel coatings for neural stimulating electrodes. *Journal of Biomedical Materials Research-B, 81*, 551–563.

Wise, K. D., Anderson, D. J., Hetke, J. F., Kipke, D. R., & Najafi, K. (2004). Wireless implantable microsystems: high-density electronic interfaces to the nervous system. *Proceedings of the IEEE, 92*, 76–96.

Wise, K. D., Angell, J. B., & Starr, A. (1969) *An integrated circuit approach to extracellular microelectrodes*. The 8th ICMBE, Palmer House, Chicago, IL, July 20, 1969. Digest of the 8th ICMBE 1, 14.

Wolpaw, J. R., Birbaumer, N., Heetderks, W. J., McFarland, D. J., Peckham, P. H., Schalk, G., et al. (2000). Brain–computer interface technology: a review of the first international meeting. *IEEE Transactions on Rehabilitation Engineering, 8*, 164–173.

Wolpaw, J. R., Birbaumer, N., McFarland, D. J., Pfurtscheller, G., & Vaughan, T. M. (2002). Brain–computer interfaces for communication and control. *Clinical Neurophysiology, 113*, 767–791.

Wolpaw, J. R., Loeb, G. E., Allison, B. Z., Donchin, E., do Nascimento, O. F., Heetderks, W. J., et al. (2006). BCI meeting 2005—workshop on signals and recording methods. *IEEE Transactions on Neural Systems and Rehabilitation Engineering, 14*, 138–141.

Yeager, J. D., Phillips, D. J., Rector, D. M., & Bahr, D. F. (2008). Characterization of flexible ECoG electrode arrays for chronic recording in awake rats. *Journal of Neuroscience Methods, 173*, 279–285.

Zhong, Y., Yu, X., Gilbert, R., & Bellamkonda, R. V. (2001). Stabilizing electrode–host interfaces: a tissue engineering approach. *Journal of Rehabilitation Research and Development, 38*, 627–632.

CHAPTER 21

# Artificial vision: needs, functioning, and testing of a retinal electronic prosthesis

Gerald J. Chader[*], James Weiland and Mark S. Humayun

*Doheny Retina Institute, USC School of Medicine, Los Angeles, CA, USA*

**Abstract:** Hundreds of thousands around the world have poor vision or no vision at all due to inherited retinal degenerations (RDs) like retinitis pigmentosa (RP). Similarly, millions suffer from vision loss due to age-related macular degeneration (AMD). In both of these allied diseases, the primary target for pathology is the retinal photoreceptor cells that dysfunction and die. Secondary neurons though are relatively spared. To replace photoreceptor cell function, an electronic prosthetic device can be used such that retinal secondary neurons receive a signal that simulates an external visual image. The composite device has a miniature video camera mounted on the patient's eyeglasses, which captures images and passes them to a microprocessor that converts the data to an electronic signal. This signal, in turn, is transmitted to an array of electrodes placed on the retinal surface, which transmits the patterned signal to the remaining viable secondary neurons. These neurons (ganglion, bipolar cells, etc.) begin processing the signal and pass it down the optic nerve to the brain for final integration into a visual image. Many groups in different countries have different versions of the device, including brain implants and retinal implants, the latter having epiretinal or subretinal placement. The device furthest along in development is an epiretinal implant sponsored by Second Sight Medical Products (SSMP). Their first-generation device had 16 electrodes with human testing in a Phase 1 clinical trial beginning in 2002. The second-generation device has 60+ electrodes and is currently in Phase 2/3 clinical trial. Increased numbers of electrodes are planned for future versions of the device. Testing of the device's efficacy is a challenge since patients admitted into the trial have little or no vision. Thus, methods must be developed that accurately and reproducibly record small improvements in visual function after implantation. Standard tests such as visual acuity, visual field, electroretinography, or even contrast sensitivity may not adequately capture some aspects of improvement that relate to a better quality of life (QOL). Because of this, some tests are now relying more on "real-world functional capacity" that better assesses possible improvement in aspects of everyday living. Thus, a new battery of tests have been suggested that include (1) standard psychophysical testing, (2) performance in tasks that are used in real-life situations such as object discrimination, mobility, etc., and (3) well-crafted questionnaires that assess the patient's own feelings as to the usefulness of the device. In the Phase 1 trial of the SSMP 16-electrode device, six subjects with severe RP were implanted with ongoing, continuing testing since then. First, it was evident that even limited sight restoration is a slow, learning process that takes months for improvement to become evident.

---

[*]Corresponding author.
Tel.: 323-442-6767; Fax: 323-442-6755;
E-mail: gchader@doheny.org

DOI: 10.1016/S0079-6123(09)17522-2

However, light perception was restored in all six patients. Moreover, all subjects ultimately saw discrete phosphenes and could perform simple visual spatial and motion tasks. As mentioned above, a Phase 2/3 trial is now ongoing with a 60+ device. A 250+ device is on the drawing board, and one with over 1000 electrodes is being planned. Each has the possibility of significantly improving a patient's vision and QOL, being smaller and safer in design and lasting for the lifetime of the patient. From theoretical modeling, it is estimated that a device with approximately 1000 electrodes could give good functional vision, i.e., face recognition and reading ability. This could be a reality within 5–10 years from now. In summary, no treatments are currently available for severely affected patients with RP and dry AMD. An electrical prosthetic device appears to offer hope in replacing the function of degenerating or dead photoreceptor neurons. Devices with new, sophisticated designs and increasing numbers of electrodes could allow for long-term restoration of functional sight in patients with improvement in object recognition, mobility, independent living, and general QOL.

**Keywords:** neural retina; brain; retinal degeneration; electronic prosthetic devices; artificial vision; retinitis pigmentosa; age-related macular degeneration; visual performance; low vision; photoreceptor replacement

## Retinal degenerative diseases: an overview

One of the most feared disabilities or diseases around the world is blindness, ranking close to cancer. Among sight-robbing conditions, some like cataract can be usually satisfactorily addressed through interventions like surgery. On the other hand, most of the intractable blinding conditions are of retinal origin, the most common type being the inherited retinal degenerations (RDs). These conditions form a broad, heterogeneous family of diseases that primarily affects retinal photoreceptor cells and thus might even better be called photoreceptor degenerations. These are all inherited diseases or at least have a strong genetic component. They broadly fall into two categories: (1) degenerations like retinitis pigmentosa (RP) that begin by primarily affecting rod photoreceptor cells; and (2) macular degenerations that mainly affect cone photoreceptors. An example of the latter is age-related macular degeneration (AMD), although retinal pigment epithelial (RPE) cells are also affected early in this disease process. Along with these specific disease entities, there are many more variations, usually spoken of as the rare RDs. These can be relatively cone specific (at least early on), as is Stargardt's disease, or rod specific, as are diseases such as Leber congenital amaurosis (LCA), Batten disease, or Usher syndrome.

The prevalence of the RP-like degenerations is estimated to be about 1:3500 around the world (Haim, 2002). This estimate is based on data obtained in a single country, Denmark, and awaits more global confirmation. Most evidence indicates that ethnic origin and geographic locale seem to play little or no role in the prevalence of RP. AMD, on the other hand, is much more prevalent than RP but has a more specific pattern of occurrence. Most AMD is seen in Europe, and in countries like the USA, Canada, and Australia that have mainly European-based populations. In the USA, for example, it is estimated that about 2 million Americans above the age of 55 have AMD, with another 7 million being "presymptomatic," i.e., having no significant vision loss but exhibiting clinical signs of the disease such as the presence of drusen upon careful fundus examination. In RP and allied diseases, although the number of affected individuals around the world is relatively small, the disease usually is apparent at birth, in early childhood, or at least in the second or third decade of life. Thus, otherwise healthy individuals are severely affected socially and economically as well as often in need of substantial specialized care from governmental agencies for most of their lives. In contrast, AMD mainly affects those over 55 years of age but the large number affected and the effects on issues such as mobility, independent living, and injuries such as falls related to poor vision make this a costly disease in terms of loss of quality of life (QOL) and economic costs to both the individual and the government.

**Prospects for therapies available to RD patients**

To date, patients with an inherited RD have had few possibilities for therapy. This is especially true for patients with RP and allied diseases where use of the nutritional supplement vitamin A has been the only possibility of treatment (Berson et al., 1993). However, the vitamin A regimen only helps a subset of RP, and even in these, it only slows the course of the disease. Dry AMD patients also have the possibility of nutritional therapy with the AREDS clinical trial, demonstrating that a combination of antioxidants can slow the disease process (AREDS Study Research Group, 2001). Here again though, the antioxidants are only recommended for a specific stage of the disease process and only slow the disease course. Wet AMD, the neovascular form of advanced AMD, does have some antineovascular drug options but it constitutes only about 10% of all AMD patients.

New treatments for the RDs, though, are on the horizon including the use of electronic devices that take the place of degenerating, dysfunctional, or dead photoreceptor cells. In fact, these electronic devices might be the best opportunity for therapy in most cases in comparison to other possibilities such as gene therapy, pharmaceutical therapy, nutritional therapy, and stem cell transplantation. If one considers the bulk of RP and dry AMD patients in relation to these five therapeutic options, they will fall into one of two categories: those patients with some viable photoreceptor cells remaining in their retinas and those in which most or all of the photoreceptor cells have died. For the former category, gene therapy is an appealing possibility for treatment since replacement of the defective gene theoretically can slow or even reverse the course of the disease (Hauswirth et al., 2004). In specific cases, long-term experiments on animal models of RP have been very successful (Acland et al., 2001). With only about 50% of the RD gene mutations known, however, a large number of patients would be excluded from treatment. In a similar vein, some genes are difficult to work with in vector applications due to large size, etc., limiting the patient pool even further. Moreover, safety issues yet remain with the general technique of gene therapy.

Pharmaceutical therapy is applicable only when a sufficient number of photoreceptors remain. It can be defined as the use of any agent (natural or synthetic) that prolongs the life of a photoreceptor cell — maybe even enhancing function and performance. A large number of such agents have now been identified (La Vail et al., 1992), including ciliary neurotrophic factor (CNTF), and successfully tested in animal models of RD. The half-life of such agents, though, is relatively short, necessitating frequent replenishment over a lifetime. Some can have serious side effects due to the multiplicity of their actions. The use of nutrients in slowing the disease process in inherited RDs has received widespread attention over the last few years, not only the AREDS work demonstrating some efficacy in slowing AMD but, more recently, the work showing positive results in slowing RP. Specifically, van Veen and Campochiaro with their respective groups have demonstrated that antioxidants slow photoreceptor cell death in a number of animal models of RP (Komeima et al., 2007; Sanz et al., 2007). As with the use of pharmaceutical agents though, more work needs to be done on both safety and efficacy issues as well as the issue of sufficient numbers of remaining photoreceptors to warrant using such treatment as anything other than a "holding action" while waiting for more effective treatments and cures.

When all photoreceptors are dead or otherwise not functioning (e.g., in advanced disease), two major tactics can be taken to replace the photoreceptors or at least their function. A direct route would be the use of photoreceptor cell transplantation and stem cell therapy. Despite substantial time and effort though, only modest results have been seen with photoreceptor cell transplantation from donor eyes into animal models of RP (Sagdullaev et al., 2003). Transplantation of stem cells to replace the dead photoreceptors is an attractive alternative to transplantation of retinal sheets or dispersed photoreceptor cells from donor eyes. Stem cells have been found in adult mammalian tissues, e.g., retinal stem cells in the ciliary margin (Tropepe et al., 2000). Also, progress is being made in defining conditions in which stem cells assume a more adult photoreceptor morphology and function. Encouraging results from MacLaren et al. (2006) demonstrate some functional photoreceptor replacement in an animal model of RP but significant safety questions as well as questions of function need to be addressed before the technique can be deemed available for general RD therapy. The alternative to stem cell transplantation is the use of electronic prosthetic devices that are able to translate a photic image into an electrical response with the ultimate perception of a visual image. This could be by direct stimulation of remaining retinal cells (inner retinal layer) or by bypassing the eye completely and directly stimulating the brain. With the former, this affords artificial vision with an electronic implant functionally taking the place of the photoreceptor cells.

## Electron prosthetic devices: general considerations

The most advanced prosthesis project is led by Dr. Mark Humayun at the Doheny Eye Institute, USC Medical School in conjunction with Second Sight Medical Products (SSMP). This is an effort initiated originally by Dr. Humayun with Dr. Eugene de Juan Jr. about two decades ago and is now in Phase 2 clinical trial. Early work in this area is summarized in Humayun (2001). Simply put, a retinal electronic prosthetic device takes the place of dead or nonfunctional photoreceptor cells. It translates outside photic images into electrical signals in the retina that can ultimately be perceived by the brain as visual images. Technically, a small camera is mounted behind the patient's eyeglasses, which captures an image, and in some models of the device, wirelessly sends the image to a microprocessor for conversion to an electronic signal (Fig. 1). This signal moves to a specialized receiver and then to the prosthetic microelectrode implant on the retina. The implant transmits the signal to the underlying retinal cells, which, after some preliminary processing, send the signal down the optic nerve for final processing in the brain and synthesis of a visual image.

Compelling arguments can be made for use of electronic prosthetic devices — both now and in the future. One reason for hope that a retinal prosthesis could be successful comes from the dramatic success of the cochlear implant (Jones et al., 2008). Even though fairly simple, its success in restoring hearing demonstrated that at least some secondary neurons were viable and functional, i.e., survivors of trans-synaptic degeneration, and are able to pass on sensory input from the implanted device. Of course, success of the ocular implant depends on the viability and functionality of secondary neurons of the retina. One way of noninvasively assessing at least the presence of viable inner retinal neurons if not their functionality is by optical coherence tomography (OCT). Matsuo and Morimoto (2007), for example, have examined the retinas of a number of subjects with RP using OCT and correlated visual acuity with retinal thickness. They conclude that OCT "may be used as a clinical test to assess the feasibility of retinal prostheses in the future."

There are currently many groups around the world that are working on visual prosthetic devices and each has its own approaches and technologies. Some of these approaches bypass the eye completely and, after electronic processing of a video camera image, send the visual signal directly to the brain. One of the first to test for direct electrical stimulation of the brain in blind patients was Brindley and Lewin (1968). After stimulation of the occipital pole of the right cerebral hemisphere by "an array of radio receivers," the patient was "caused to experience sensations of light (phosphenes) in the left half of the visual field." Dobelle (2000) has also been a pioneer in this effort, trying to connect a television camera to the visual cortex. More basic work on a cortical visual prosthesis has demonstrated that sensory percepts can indeed be elicited by "modest levels of electrical currents passed into the cortex" by the Utah electrode array (Normann et al., 1999). Similarly, work on a sophisticated cortical implant by Troyk et al. (2005) has progressed to the point where there is hope for a future clinical trial.

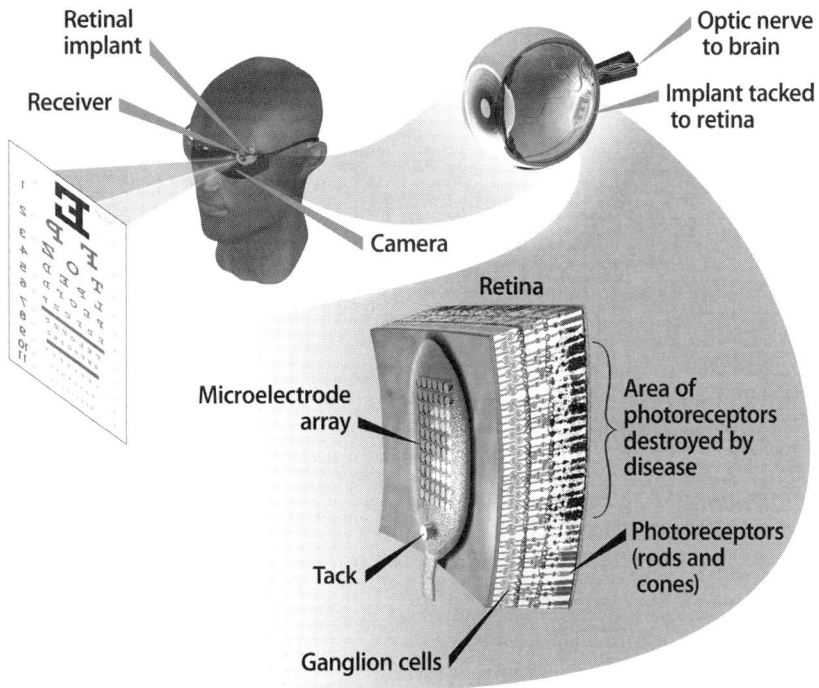

Adapted, with permission, from *IEEE Engineering in Medicine and Biology* **24**:15 (2005).

Fig. 1. Components of the retinal electronic prosthesis. (Left top) External images such as from an eye chart are captured by a miniature camera mounted behind the eyeglasses of the patient. (Right top) These signals are sent to a microprocessor that converts the data into an electronic signal, then to a receiver in the eye, and finally to a microelectrode implant tacked to the retina. The array stimulates underlying retinal cells and this biological signal is sent through the optic nerve to the brain for the creation of a visual image. (Bottom) The enlarged area of the retina shows a theoretical microelectrode array tacked to the front, vitreal (ganglion cell) side of the retina. (Adapted with permission from the Department of Energy newsletter, 5 January 2008.)

Most groups, though, focus on the eye and employ an intraocular implant. Some designs for the implant are quite new, for example, employing an optoelectronic system that theoretically could give a stimulating pixel density of up to 2500 pix/mm$^2$ (Palanker et al., 2005). Others have taken novel surgical approaches. In this regard, Tokuda et al. (2007) have implanted a multichip stimulator into a scleral pocket to achieve suprachoroidal transretinal stimulation. They were able to elicit electrically evoked potentials (EEPs) in the visual cortex using this device. The use of localized chemical release (e.g., neurotransmitters) from the implanted device has been proposed (Peterman et al., 2004) as well as a neural interface that utilizes vertically aligned multiwalled carbon nanotube pillars as microelectrodes (Wang et al., 2006).

Many other approaches from groups in different countries are under investigation but a summary of these activities is outside the scope of the present article. Most groups of researchers, however, have taken a more direct approach with the implantation of more conventional electrodes directly on the retina using a subretinal or epiretinal approach. In the subretinal approach, the electronic implant is placed in the subretinal space between the pigment epithelial cells and the dead/dying photoreceptors. In the epiretinal approach, the implant is placed on the front surface of the retina, i.e., the ganglion cell layer. Applicability of such approaches has been reviewed in several publications including Zrenner (2002); Weiland et al. (2005), Winter et al. (2007) and the recent book by Humayun et al. (2007).

Optobionics was the first company to attempt a government-approved clinical trial in the USA by using a subretinal implantation approach with a semiconductor-based microphotodiode array (Peachey and Chow, 1999). The subretinal space is perhaps theoretically the most logical place for implanting the array since it places the prosthetic device juxtaposed to the dying or dead photoreceptor cells. However, this is a surgically more challenging route compared to others like the sclera pocket route or the epiretinal route (described below), leading to the possibility of more severe surgical complications. The Optobionics device also apparently suffered from the fact that the current it generated was only from light energy, i.e., it is passive with no external power supply. In spite of these problems, initial reports were that this Artificial Silicone Retina (ASR) was both safe and efficacious (Chow et al., 2004). In fact, all implanted RP patients demonstrated unexpected improvements in visual function. Interestingly, this improvement included areas relatively far from the implants, suggesting a "possible generalized neurotrophic-like rescue effect on the damaged retina caused by the presence of the ASR" (Chow et al., 2004). The fact that a number of neuron-survival agents, normally found in the retina, can prolong photoreceptor cell life and function is well known (La Vail et al., 1992). Subsequent studies on the ASR in the RCS rat model of RD confirmed this hypothesis in that implantation of either active or inactive ASR chips resulted in photoreceptor rescue (Pardue et al., 2005). Since Optobionics did not meet the endpoints in the human trial, the company is now defunct.

A more successful version of a subretinal prosthetic device is seen in the implant developed by Retina Implant GmbH in Germany, which also contains light-sensing microelectrodes. This device has received extensive animal testing (Gekeler et al., 2007) and has been implanted in human subjects. Dr. E. Zrenner, University of Tuebingen Eye Clinic, heads the research effort and clinical testing, which, to date, has included eight patients for a 30-day period of time. This implant contains an array of 1500 light-sensitive microphotodiodes in the subretinal space, but unlike the Optobionics device, it has an external power source. The microphotodiodes serve to adjust the strength of the stimulus pulse based on the intensity of light incident on the photodiode. This experimental device has a percutaneous cable through which stimulus pulses and control signals are applied. Because of packaging though, the device cannot be used for long periods each day but only turned on for a few hours. The group hopes to have a permanent implant in the near future. A long-time leader in implant science has also been the Boston Retinal Implant Project led by Drs. Joseph Rizzo and John Wyatt Jr. They have developed novel strategies in engineering, surgical approaches, functional neuroimaging, and human testing, for example, studying the perceptual efficacy of array stimulation in short-term surgical trials in humans (Rizzo et al., 2003).

The alternative to a subretinal implantation approach is an epiretinal approach where the implant is placed on the vitreal (front) side of the retina. Here, there is close juxtaposition to the retinal ganglion cells, although the specific cell/cells stimulated by this approach is/are not known. Attention, however, is being given to possibly localizing the electrically stimulated cells through the use of techniques such as mathematical modeling (Ziv et al., 2005). Specifically for epiretinal implants though, three major efforts have progressed to the point of clinical testing. One is with IIP-Technologies GmbH, which has an implant called the Learning Retina Implant designed such that the patients can optimize their visual perceptions in a computer-mediated dialog. Human work has started in this arena with implantation studies on legally blind patients (Feucht et al., 2005). There has also been a demonstration in cats that activation of the cortex is achieved by epiretinal stimulation (Walter et al., 2005). Another effort has been called the EPI-RET project. This implant has a "learning neural computer" called a Retina Encoder, which works interactively with the user to achieve the best image possible. After implantation in two rabbits, Gerding et al. (2007) stated that "Retinal implant areas in contact to implanted devices presented a severe structural damage and disorganization." A report from the University Eye Hospital in

Aachen, Germany, states that six subjects have been implanted with the 25-electrode device that is relatively large and includes a part that replaces the ocular lens as well. The third effort, the USC-SSMP consortium is, as mentioned above, probably the most advanced with a government-approved clinical trial starting in 2002. Results from this work are given in more detail in the section A clinical trial and testing of an epiretinal prosthetic device.

## Morphological and neuronal bases for implantation of a retinal prosthesis

The morphological basis that demonstrates the feasibility of a retinal implant was established in publications designed to determine if suitable numbers of inner retinal neurons remained in RP and AMD patients after photoreceptor degeneration and death to act as a "platform" for the prosthetic implant. In RP, it was indeed found that there was significant preservation of the inner retinal layers well after onset of the disease process. For example, Stone et al. (1992) reported that, in the macular regions of donor eyes from patients with different types of RP, there was indeed ganglion cell loss but that a significant number of cells remained at increasing eccentricities. Santos et al. (1997) found that, in the macular region of the retina, 30% of the ganglion cells were "histologically intact" along with "78% and 88% of the inner nuclear layer cells ….in groups of patients with severe and moderate RP respectively." Subsequent morphometric analyses of the extramacular retina demonstrated less preservation in the inner nuclear layer and ganglion cell layer (Humayun, 1999). Parallel work on the macular region in patients with AMD demonstrated that the ganglion cell layer and the inner nuclear layers of patients with geographic atrophy (GA) (Kim et al., 2002a) and disciform AMD (Kim et al., 2002b) are relatively well preserved compared to the outer nuclear layer.

Even though morphometry shows significant numbers of cells remaining in the inner retina, marked abnormalities can be found in most of the remaining cell types of the retina in more advanced cases of RD. Studying retinas of donor patients with RP, Fariss et al. (2000), for example, found that remaining rod, amacrine, and horizontal cells demonstrate neurite sprouting. Many of these abnormal neurites were seen to contact the surfaces of GFAP-positive Muller glia. Other aspects of "neural remodeling" have now been reported in animal models of RD (Marc et al., 2003). Along with neurite sprouting, the formation of cryptic connections, and self-signaling, there is movement of amacrine and horizontal cells into the ganglion cell layer. Significantly, Muller cells also increase intermediate filament synthesis with the formation of a dense fibrotic layer in the subretinal space, effectively sealing the retina from the choroid. All of these pathological changes have ramifications for the ultimate success of retinal prosthetic implants. It would seem, however, from studies described below on CNS connections, that enough inner retinal neurons (e.g., ganglion cells) do remain such that they can pass a visual signal down the optic nerve to the brain. Similarly, neurite sprouting and inappropriate connections between neurons and neurons, and neurons and glia, can diminish the passage of proper signals, but again, human studies in patients with advanced RP, which have been cited below, demonstrate that at least some signals get through. Also, there is the possibility that, with neuronal plasticity, the imposition of a visual signal from the implant might lead to realignment of the connections in retina and/or brain to a more normal configuration. Finally, the formation of a gliotic seal at the level of the edge of the degenerating photoreceptor cells could be a significant impediment to restoration of function by prostheses placed in the subretinal space but would probably have little impact on epiretinal placement of the prosthetic device.

## Central connections in retinal degeneration and prosthesis implantation

Given that enough inner retinal cells are present in cases of RD on which the microelectrode device can be implanted, a key question is whether the brain, in fact, can "see" an appropriate visual image. Specifically, can the brain receive the visual signal from the remaining retinal inner layers and interpret it as a fair representation of the image input from the video camera? It is possible that central connections are damaged in the degenerative process or degenerate in response to disuse (i.e., lack of visual input) over years of effective blindness. To at least partially answer this question, Humayun et al. (1996) evaluated the direct stimulation of the retinal surface of RP subjects who had little or no light perception. Focal electrical stimulation was effected using monopolar and bipolar conductors in a controlled manner. The results demonstrated that the electrical stimulation did elicit visual perception, viewed by the subjects as discrete spots of light (phosphenes). Some subjects could track movement of the stimulating electrode and perceive two phosphenes in response to the stimulation of two independent electrodes. Importantly, the phosphenes were perceived in the appropriate stimulated area of the inner retina. In follow-up studies, Humayun et al. (1999) again examined pattern stimulation of the human retina in subjects with end-stage RP or AMD. These studies confirmed that the subjects could indeed see phosphenes in response to the electrodes and yielded valuable information as to the amplitude of current needed to elicit a percept, the need for close proximity of the electrode to the retina, and threshold differences between macular and extramacular areas of the retina. Taken together, all these studies show that the brain can respond to retinal stimulation, even after long years of little or no formed sight or even light perception. The effects of visual deprivation and the relative plasticity of the visual system have been recently reviewed (Fine, 2007).

In a parallel set of experiments, Weiland et al. (1999) examined the possible retinal site(s) of electrically elicited visual perception. In this case, eyes from two normal subjects were subjected to laser damage, thus creating an area of retinal "degeneration" surrounded by normal retina. These areas were then tested with a hand-held stimulating device placed within the eye over the damaged or normal portions of the retina. The tested eyes had been scheduled for exenteration due to cancer near the eye. The laser procedure was performed a few days before the exenteration surgery, and the stimulation procedure was performed prior to surgery. It was found that a variety of percepts could be seen by the subjects treated with krypton red laser, which ablated the outer retina but left the inner retinal cells relatively intact. No percepts were perceived, though, in areas treated with the argon green laser, which damaged both the outer and inner nuclear layers. This suggests that electrical stimulation can be effective in the damaged retina, that the site of such stimulation is the inner retinal neurons, and that the signal can be transmitted to the brain. More recently, Schiefer and Grill (2006) studied the retinal sites of excitation after epiretinal electrical stimulation. They found that stimulation was highly dependent on the physical geometry between the electrode and the underlying ganglion cells. Thresholds were lowest when the electrode was placed close to the characteristic 90° bend in the ganglion cell axon, perhaps explaining why epiretinal stimulation "results in the production of punctuate rather than diffuse or streaky phosphenes."

**Visual perception: measurements in low vision and brain processing after therapeutic intervention**

Along with establishing the morphological basis for prosthesis implantation in the retina as well as whether functional central connections yet remain, the problem of reliable and reproducible testing for small improvements in vision in subjects with advanced RD must be overcome. As defined by Dagnelie (2008) in a recent review, vision loss to RP patients is not a "simple, discrete variable" with "normal vision, low vision and blindness." Rather, it is a "near endless gradation of ever decreasing vision levels." This continuum then determines to a great extent the level of activity, performance, independence, QOL, etc. of the affected individual. Thus, sight restoration through an electronic prosthetic device must be considered in this continuum for each individual — starting with the most severely impaired (totally blind) patients but hopefully applicable to the partially sighted — both RP and AMD patients. Testing of patients with retinal implants is similarly complex in that it must also be individualized depending on the initial level of impairment and the level of sight restoration. However, this is not an insoluble problem. In his review, Dagnelie defines three basic approaches to measuring visual function in very low vision patients. These are simple light detection, light localization, and the perception of movement, usually of a specific light source. There are also features of vision, even very low vision, that are measurable. In Table 1, Dagnelie provides us with a hierarchy of functions, from light perception to stereoacuity. Each of these has a definable performance from simple orientation to relatively complex threading. Thus, simple tasks can be designed such that each of the eight separate levels of function can be assessed and therefore provide a good measure of visual function.

Simulations of prosthetic vision have also been made in subjects with full or partial vision to estimate the performance level that might be expected with actual patients using the prosthetic device (Walter et al., 2007). In this way, information can be gained not only on performance and the effects of learning on the measurement of low vision but also on the usefulness of simple experimental paradigms such as the use of checkerboard square patterns in testing implanted patients.

Table 1. Examples of measurable aspects of vision

| Measurable aspects of vision | |
|---|---|
| Function | Performance |
| Light | Orienting |
| Projection | Pointing |
| Movement | Following |
| Color | Selecting |
| Shape/pattern | Classifying |
| 3-D structure | Navigating/manipulating |
| Hyperacuity | Aligning |
| Stereoacuity | Threading |

*Note:* The examples are ordered from simplest to most complex. Each line in the left column ("Function") lists a specific visual function. To the right of it is a corresponding visual task ("Performance") for which the function is a prerequisite. (Reprinted, with permission, from the *Annual Review of Biomedical Engineering*, Volume 10 © 2008 by Annual Reviews www.annualreviews.org).

Wilke et al. (2007) point out that the use of artificial vision devices (AVDs) could yield results quite different from normal vision and that "novel test strategies" might be needed to adequately assess visual performance with these devices. Certainly, the most widely used tests for visual function are those for visual acuity and visual field but these are insufficient measures when confronted with very low vision as in bare light perception and with the small changes that might be encountered with at least the first generation of visual prosthetic devices. Similarly, electro-retinography (ERG) and pattern ERG are often used to assess outer and inner retinal function, respectively, but again, these may not be of sufficient resolution to detect small improvements in vision with the help of the implanted devices. To minimize these problems, Wilke and colleagues propose a battery of tests based on a three-pillar approach such that "sufficient information" is gained about the "efficacy of an AVD in severely visually impaired patients as well as information needed for further development." The first pillar consists of standardized psychophysical tests that might be applicable to the visual condition but, as pointed out by the authors, these tests are probably the least relevant to the real-life conditions faced by the patient in their daily living. The second pillar consists of tests related to the day-to-day activities of the patient, such as mobility and navigational skills. These may be the most relevant in assessing the helpfulness of the retinal prosthesis but can be less objective than the standard testing unless carefully monitored and controlled. The third pillar is the most subjective in that it solicits the patient's own evaluation and impressions as to the usefulness of the AVD in the format of a carefully crafted questionnaire. Taken together though, these three, very different testing methods could give the best evaluation possible of even a marginal or incremental improvement in vision with use of the device. Overall, what is needed is an accurate and reproducible method to link visual testing with real-world functional capacity in individual patients with very low vision.

In spite of these barriers, a number of psychophysical tests have been devised in animals that are important in preclinical testing of the prosthetic device and other treatment strategies and also might be applicable to human subjects. Optokinetic testing of visual acuity, for example, which, by inference, gives information about brain function (Thomas et al., 2004b). Also, direct responses can be measured from the superior colliculus after a therapeutic intervention in rats with RD (Thomas et al., 2004a). Smirnakis et al. (2005) used fMRI to see if there were changes in area V1 of the monkey brain after binocular retinal lesioning. They showed that the cortical topography was unchanged. Eckhorn et al. (2006) have measured cortical responses to prosthetic stimulation of the cat retina and determined that the resultant temporal and spatial resolutions are "sufficient for useful object recognition and visuomotor behavior." Indeed, the ultimate testing in vision must be done in the brain rather than in the retina as even robust retinal signals do not guarantee a coherent cortical visual image. In fact, much evidence points to loss of neural function over long-term blindness in a "use or lose" scenario. Also troublesome is the possibility that those affected very early in development with severe photoreceptor dysfunction or dysplasia, i.e., the loss of afferent input, might not even form the initial proper visual pathways between retina and brain since functional signals are never perceived by the retina. Although the latter situation has yet to be directly assessed in situations such as the early blindness seen in human LCA, it is clear from animal experiments that some measure of vision is possible. For example, sight restoration (both rod and cone function) through gene replacement therapy has been observed in a dog LCA model of retinal dysfunction. These are RPE65 mutants, in which photoreceptors remain relatively intact morphologically but lose function (Acland et al., 2005). This restoration of vision implies that there is rescue of already formed connections or that there is the formation of new, functional synaptic connections once the perception of the signal (i.e., light) is restored after successful gene transfer. Aguirre et al. (2007) explored visual processing in these mutant dogs using functional MRI (fMRI). They found that, before therapy, minimal cortical response could be detected in the primary visual areas of the lateral gyrus. Following therapy, though, cortical responses were markedly improved.

In parallel, human subjects with LCA were studied with structural MRI by Aguirre and collaborators and were found to have "preserved visual pathway anatomy and detectable cortical activation despite limited visual experience." Similarly, central visual pathways were found to be intact in a second model of LCA, the CEP290 mouse mutant (Cideciyan et al., 2007). Schoth et al. (2006) have used diffusion tensor imaging (DTI) to assess the level of organization of the optic radiation in subjects with acquired blindness compared with normally sighted subjects. DTI evaluates the integrity of large fiber tracts such as the optic radiation. The investigators found that both the visual fiber and pyramidal tracts appeared to be normal in the blind subjects with no axonal degeneration of the optic radiation observed. Taken together, all these data indicate that imposition of a visual signal even on the long-term blind patient with an electronic prosthetic device could result not only in restoration of retinal function but also in brain recognition and processing of the signal, yielding a visual image.

## A clinical trial and testing of an epiretinal prosthetic device

As mentioned above, the prosthetic device furthest along in development at this time is that engineered and in current clinical testing by SSMP as originally conceived by Dr. Mark Humayun with Dr. Eugene de Juan, Jr. at Duke University. This work was continued by Dr. Humayun with Dr. James Weiland, Dr. Robert Greenberg (now with SSMP), and others at Johns Hopkins University and currently at the Doheny Eye Institute, USC School of Medicine. The first-generation design of the prosthetic implant was simple; a silicone–platinum array with 16 electrodes ($4 \times 4$ array) touching or at least close to the retina. A schematic view is shown in Fig. 1. A small tack secures the array to the retina. As outlined above, an external camera and an image-processing chip are mounted on the eyeglasses of the patient. These capture the visual image, pixelize it, and send the signal through a telemetry link to the electronic retinal implant. The implant produces a pattern of small electrical currents that approximate the initial visual image. Appropriate underlying retinal neurons are activated, resulting in a dot pattern at each point of stimulation. Taken together, they theoretically yield an image akin to that formed by a dot-matrix printer. Although it is not yet known which cells of the retina (e.g., ganglion, bipolar, both, etc.) are active in accepting the signal, it is presumed that these remaining cells process the signal in as normal a manner as possible (based on the individual's severity of disease) and pass it down the optic nerve for final brain processing and the putative perception of a specific image.

In 2002, Phase 1 of an FDA-approved clinical trial began with the 16-electrode device (Argus-1, A-16). Ultimately, six patients with advanced RP who had little or no remaining light perception received the implant. Although this safety phase of the trial has been successfully completed, testing of these patients has continued as much as possible to the present time. Importantly, safety has been seen with all the implanted devices with no major sequelae, although one device was removed because of unrelated health problems. Surprisingly, some efficacy was also observed in the implanted patients. All patients had restoration of light perception and all saw discrete phosphenes. After a period of time, they also could perform visual spatial and motion tasks. The remaining patients are currently using their devices at home with continuing success. The first publication on an implanted patient reported on the initial 10 weeks of testing (Humayun et al., 2003). The patient had X-linked RP with no light perception in his implanted, right eye (50 years) and bare light perception in his control, left eye. Subsequent to a training period, this patient could describe the relative location of phosphenes generated by the activation of selected individual electrodes under laboratory, double-masked test conditions. For example, in a two-alternative forced-choice test, the patient was highly successful in distinguishing between pairs of vertically or horizontally aligned activated electrodes. Cortical-evoked potentials were also evoked by retinal stimulation through the device. The relative locations of the percepts were found to correspond to the position of the particular electrode(s) activated. Also, there was a good correlation found between percept brightness and stimulation level, demonstrating that, along with simple light perception, the ability to discern between different light levels can also be restored. Finally, testing of the use of the camera unit in conjunction with the retinal implant was also successful along with the tests described above when the electrodes were directly activated. With the camera unit activated, the presence or absence of ambient light could be ascertained as well as the direction of motion of test objects. One of the important lessons from this initial testing was that sight restoration is a learning process that takes time. Specifically, the ability to locate phosphenes in the correct visual field was markedly enhanced with use. Similarly, the learning effect was observed with increased use of the camera unit.

Mahadevappa et al. (2005) expanded the initial report on a single patient to longer-term results on three implanted patients. Specifically, threshold and impedance values were investigated in an attempt to gain critical information as to the charge needed to induce a percept without causing damage to the delicate underlying retinal tissue. Previous short-term studies on human subjects (Humayun et al., 1999) had indicated that a relatively high level of current was needed to induce a phosphene, one that, with the implantation of a permanent device, might cause retinal damage over a period of time. For these studies, the retinal array was connected to a stimulator that gave precise control of each individual electrode in the array. Concurrently, OCT measurements were made to determine the distance between the array and the retina as this geometry certainly could play a role in the current requirements. It was found that thresholds varied greatly between the three subjects (24–702 $\mu$A with a 1 ms pulse). However, these values were lower than those seen in the original, short-term studies, i.e., those values needed to elicit a percept. Thresholds were found to increase with time. This could be due to a number of factors, most likely to the lifting off of the array from the retinal surface. Variability was also seen in the impedance values. As with the thresholds, this is probably due to differences in the distance between the array and the retina in the different subjects and to changes with time. The studies underscore the importance of controlling the distance between the implant and the underlying tissue. If the array is within 0.5 mm of the retina, no correlation is seen between this gap distance and either threshold or impedance. As one might expect, at greater distances, higher thresholds and lower impedance values were observed. The underlying message from these studies, however, is that the relatively low threshold values observed permit the continuation of testing in a safe manner.

As outlined in section Visual perception: measurements in low vision and brain processing after therapeutic intervention, the ultimate test of efficacy of the retinal device is improved performance. Assuming that the implanted patients start with very low vision or none at all, this can be best assessed through the scoring of simple tasks that are relevant to everyday living. Yani et al. (2007) have done this with three subjects with RP who had the Argus 1 implant, i.e., the 16-electrode array. Because of the design of the device, the implant could be controlled either by the head-worn video camera or by an independent computer. Threshold and impedance values were measured to be able to correlate these values with the level of task performance. To assess the operation of the device, preliminary tests were performed with the electrode implant controlled by a computer operated by an investigator. In this way, several areas of function could be assessed: discrimination between individual electrodes, sequential activation of paired electrodes, and subject discrimination of activated rows versus columns of four electrodes. In all these activities, the patients scored significantly better than chance. Subsequent to this, a series of simple tasks were designed to be performed by the patient using the video camera. These were similar to those theoretically outlined by Dagnelie (2008) in Table 1. Table 2 summarizes these tasks. Operationally, white bars or other objects (with a

Table 2. Examples of measurable specific tasks

| Measurable task | |
|---|---|
| Task | Example |
| Motion discrimination | White bar movement |
| Spatial detection | Placement of white bar |
| Object counting | Detection of 0–3 objects |
| Form discrimination | Angle discrimination of white bars |
| Object identification | Identification of common objects |

*Note:* Tasks are designed to reliably measure specific functions of everyday living and are described more fully in the text. (Modified from Yani et al., 2007.)

black background) were passed in front of the video camera under conditions of ambient room lighting. On the whole, the patients scored well in multiple iterations of most tasks. In the motion discrimination task though, the patients were asked to keep their head motionless such that the perception of motion was not confounded. Under these conditions, they did not do as well. This was probably due to the small field of vision afforded by the camera (15° of visual angle). In cases where head scanning was permitted (other tasks listed in Table 2), all patients scored above the level of chance. Particularly compelling were results from the object discrimination task in which patients were asked to discriminate between a plate, knife, or cup placed before them on a dark background. Repeated testing gave scores well above chance — 67%, 73%, and 63% with $P<0.001$. Thus, on the whole, patients performed reliably better with the implant, although differences were seen between the patients. Such variation might be expected though due to many reasons such as differences in age, disease type/stage/severity, implant placement, patient attention, etc. That any positive results were obtained with the three end-stage RP patients and the initial low-resolution device, though, is heartening and bodes well for possible success in further studies with improved models of the device. As the authors conclude — "the results do suggest that a low-resolution epiretinal prosthesis can provide visual information that can be used to accomplish simple visual tasks that are impossible with the subject's natural light perception vision."

## Future studies

Since publication of these encouraging results, progress has been made in improving the USC-SSMP retinal prosthetic device and in clinical testing of other prosthetic devices by many of the other excellent groups working in this field. For example, basic studies on the functioning of the implant continue in patients receiving the SSMP Argus 1 (16 electrode) device. de Balthasar et al. (2008) investigated the relationship between perceptual thresholds, electrical impedance, electrode size, and distance between the device and the retina in six RP patients who had received the implant. Distance between the retina and the implant was measured by OCT. Interestingly, the investigators found a strong correlation between stimulation thresholds and the distance between the retina and the electronic implant but not with the other parameters. These data reinforce the importance of "maintaining close proximity between the electrode array and the retinal surface..." In a related study, Wang et al. (2008) investigated the effects of implantation of the device in different areas of the retina on pursuit eye movements in normally sighted subjects with normal vision and in those using a simulated prosthetic device. As expected, pursuit movements using the device were slower and less smooth than in normal vision but yet "functional, even if the prosthesis is implanted in the peripheral retina." In a rat model of RD, Kent et al. (2008) investigated the possible protective effects of neurotrophic agents that could be used in conjunction with the implantation of a retinal prosthesis. They found that retinal sensitivity was higher in eyes treated

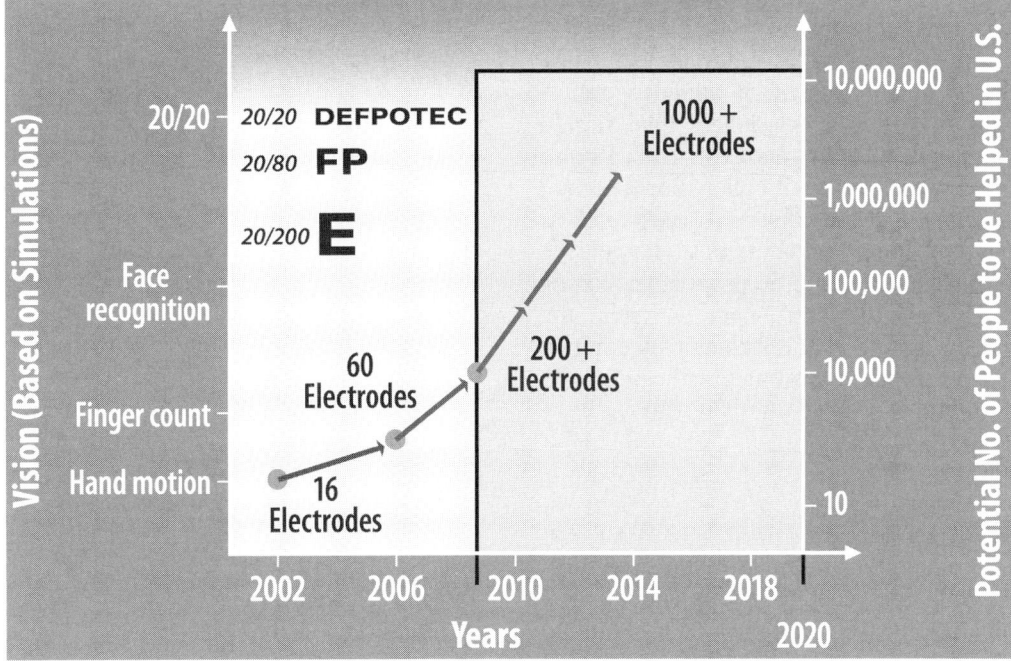

Fig. 2. Timeline for progress of the artificial retina. Progress started with the installation of the first 16-electrode device in 2002, which restored light perception and the ability to perform simple visual spatial and motion tasks (hand motion level). Theoretically, there is improvement with 60 and 200+ electrode implants (finger count level) with the possibility of face recognition and reading ability with a 1000+ electrode device. (Left scale) Progression in vision from bare hand motion to face recognition. At 20/200 visual acuity, large letters can be recognized. At 20/20, small letters can be seen with good reading ability. (Right scale) Number of patients potentially helped with the different generations of prosthetic devices. (Adapted with permission from the Department of Energy newsletter, 5 January 2008.)

with CNTF. Thus, agents such as CNTF might not only protect against apoptosis but help to maintain thresholds at lower levels in implant patients. It is interesting to point out that Caffe et al. (2001) have found that the combination of two neurotrophic agents (CNTF+BDNF) is even better than just one in rescuing photoreceptor cells in an animal model of RD. One can thus expect that improvements in implant design (e.g., numbers of electrodes), implant testing (e.g., better low vision assessment), as well as innovative ways of improving function of the implant (e.g., use of multiple neurotrophic agents) will continue to increase the overall performance and usefulness of the prosthetic device.

Figure 2 outlines the progress of the Artificial Retina Program. In 2002, the first of six patients received the 16-electrode implant (Argus I), which, besides safety, demonstrated the surprising results of restoring light perception and rudimentary vision. Currently, a 60-electrode device (Argus II) is being tested in Phase 2 of a clinical trial sponsored by SSMP. A device with 200+ electrodes is being planned. If all goes well, a 1000+ device can be envisioned that should restore a patient's ability to recognize faces and read large letters. This would allow fairly normal functioning of the patients in society with a marked improvement in their QOL. To achieve this though, there needs to be proper functioning of the two parts of the central nervous system, the brain and the neural retina, brought together in this case by a retinal electronic prosthesis.

## References

Acland, G., Aguirre, G., Bennett, J., Aleman, T., Cideciyan, A., Bennicelli, J., et al. (2005). Long-term restoration of rod and cone vision by single dose rAAV-mediated gene transfer to the retina in a canine model of childhood blindness. *Molecular Therapy*, *12*, 1072–1082.

Acland, G., Aguirre, G., Ray, J., Zhang, Q., Aleman, T., Cideciyan, A., et al. (2001). Gene therapy restores vision in a canine model of childhood blindness. *Nature Genetics*, *28*, 92–95.

Aguirre, G., Komaromy, A., Cideciyan, A., Brainard, D., Aleman, T., Roman, A., et al. (2007). Canine and human visual cortex intact and responsive despite early retinal blindness from RPE65 mutation. *PLoS Medicine*, *4*, e230.

AREDS Study Research Group. (2001). A randomized, placebo-controlled clinical trial of high-dose supplementation with vitamins C and E, beta-carotene and zinc for age-related macular degeneration and vision loss. *Archives of Ophthalmology*, *119*, 1417–1436.

Berson, E., Rosner, B., Sandberg, M., Hayes, K., Nicholson, B., Weigel-DiFranco, C., et al. (1993). A randomized trial of vitamin A and vitamin E supplementation for retinitis pigmentosa. *Archives of Ophthalmology*, *111*, 761–772.

Brindley, G., & Lewin, W. (1968). The sensation produced by electrical stimulation of the visual cortex. *Journal of Physiology*, *196*, 479–493.

Caffe, A., Soderpalm, A., Holmqvist, I., & van Veen, T. (2001). A combination of CNTF and BDNF rescues *rd* photoreceptors but changes rod differentiation in the presence of RPE in retinal explants. *Investigative Ophthalmology and Visual Science*, *42*, 275–282.

Chow, A., Chow, V., Packo, K., Pollack, J., Peyman, G., & Schuchard, R. (2004). The artificial silicone retina microchip for the treatment of vision loss from retinitis pigmentosa. *Archives of Ophthalmology*, *122*, 1156–1157.

Cideciyan, A., Aleman, T., Jacobson, S., Khanna, H., Sumaroka, A., Maguirre, G., et al. (2007). Centrosomal-cilliary gene CEP290/NPHP6 mutation result in blindness with unexpected sparing of photoreceptors and visual brain: Implications for therapy of Leber Congenital Amaurosis. *Human Mutation*, *28*, 1074–1083.

Dagnelie, G. (2008). Psychophysical evaluation for visual prosthesis. *Annual Review of Biomedical Engineering*, *10*, 15.1–15.30.

de Balthasar, C., Patel, S., Roy, A., Freda, R., Greenwald, S., Horsager, A., et al. (2008). Factors affecting perceptual thresholds in epiretinal prostheses. *Investigative Ophthalmology and Visual Science*, *49*, 2303–2314.

Dobelle, W. (2000). Artificial vision for the blind by connecting a television camera to the visual cortex. *ASAIO Journal*, *46*, 3–9.

Eckhorn, R., Wilms, M., Schanze, T., Eger, M., Hesse, L., Eysel, U., et al. (2006). Visual resolution with retinal implants estimated from recordings in cat visual cortex. *Vision Research*, *46*, 2675–2690.

Fariss, R., Li, Z.-Y., & Milam, A. (2000). Abnormalities in rod photoreceptors, amacrine cells and horizontal cells in human retinas with retinitis pigmentosa. *American Journal of Ophthalmology*, *129*, 215–223.

Feucht, M., Laube, T., Bornfeld, N., Walter, P., Velikay-Parel, M., Hornig, R., et al. (2005). Development of an epiretinal prosthesis for stimulation of the human retina. *Ophthalmologe*, *102*, 688–691.

Fine, I. (2007). The effects of visual deprivation: Implications for sensory prostheses. In M. Humayun, J. Weiland, G. Chader, & E. Greenbaum (Eds.), *Artificial sight: Basic research, biomedical engineering and clinical advances* (pp. 47–70). New York, NY: Springer.

Gekeler, F., Szurman, P., Grisanti, S., Weiler, U., Claus, R., Greiner, T., et al. (2007). Compound subretinal prosthesis with extra-ocular parts designed for human trials: Successful long-term implantation in pigs. *Graefes Archives of Clinical Experimental Ophthalmology, 24*, 230–241.

Gerding, H., Benner, F., & Taneri, S. (2007). Experimental implantation of epiretinal retina implants (EPI-RET) with an IOL-type receiver unit. *Journal of Neural Engineering, 4*, S38–S49.

Haim, M. (2002). Epidemiology of retinitis pigmentosa in Denmark. *Acta Ophthalmologica Scandinavica, 233*(Suppl), 1–34.

Hauswirth, W., Li, Q., Raisler, B., Timmers, A., Berns, K., Flannery, J., et al. (2004). Range of retinal diseases potentially treatable by AAV-vectored gene therapy. *Novartis Foundation Symposium, 255*, 179–188.

Humayun, M. (1999). Morphometric analysis of the extramacular retina from postmortem eyes with retinitis pigmentosa. *Investigative Ophthalmology and Visual Science, 40*, 143–148.

Humayun, M. (2001). Intraocular retinal prosthesis. *Transactions of the American Ophthalmologic Society, 99*, 271–300.

Humayun, M., de Juan, E., Jr., Dagnelie, G., Greenberg., R., Probst, R., & Phillips, H. (1996). Visual perception elicited by electrical stimulation of retina in blind humans. *Archives of Ophthalmology, 114*, 40–46.

Humayun, M., de Juan, E., Jr., Weiland, J., Dagnelie, G., Katona, S., Greenberg, R., et al. (1999). Pattern electrical stimulation of the human retina. *Visual Research, 39*, 2569–2576.

Humayun, M., Weiland, J., Chader, G., & Greenbaum, E. (Eds.). (2007). *Artificial vision: Basic research, biomedical engineering and clinical advances*. New York, NY: Springer.

Humayun, M., Weiland, J., Fujii, G., Greenberg, R., Williamson, R., Little, J., et al. (2003). Visual perception in a blind subject with a chronic microelectronic retinal prosthesis. *Vision Research, 43*, 2573–2581.

Jones, S., Harris, D., Estill, A., & Mikulec, A. (2008). Implantable hearing devices. *Molecular Medicine, 105*, 235–239.

Kent, T., Glybina, I., Abrams, G., & Iezzi, R. (2008). Chronic intravitreous infusion of ciliary neurptrophic factor modulates electrical retinal stimulation thresholds in the RCS rat. *Investigative Ophthalmology and Visual Science, 49*, 372–379.

Kim, S., Sadda, S., Humayun, M., de Juan, E., Jr., Melia, B., & Green, W. (2002a). Morphometric analysis of the macula in eyes with geographic atrophy due to age-related macular degeneration. *Retina, 46*, 4–10.

Kim, S., Sadda, S., Pearlman, J., Humayun, M., de Juan, E., Jr., Melia, B., et al. (2002b). Morphometric analysis of the macula in eyes with disciform age-related macular degeneration. *Retina, 47*, 1–7.

Komeima, K., Rogers, B., & Campochiaro, P. (2007). Antioxidants slow photoreceptor cell death in mouse models of retinitis pigmentosa. *Journal of Cellular Physiology, 213*, 809–815.

La Vail, M., Unoki, K., Yasumura, D., Matthes, M., Pancopoulos, G., & Steinberg, R. (1992). Multiple growth factors, cytokines and neurotrophins rescue photoreceptors from damaging effects of constant light. *Proceedings of National Academy of Science USA, 89*, 11249–11253.

MacLaren, R., Pearson, R., MacNeil, A., Douglas, R., Salt, T., Akimoto, M., et al. (2006). Retinal repair by transplantation of photoreceptor precursors. *Nature, 444*, 156–157.

Mahadevappa, M., Weiland, J., Yani, D., Fine, I., Greenberg, R., & Humayun, M. (2005). Perceptual thresholds and electrical impedance in 3 retinal prosthesis subjects. *IEEE Transactions on Neural System Rehabilitation Engineering, 13*, 201–206.

Marc, R., Jones, B., Watt, C., & Strettoi, E. (2003). Neural remodeling in retinal degeneration. *Progress in Retinal Eye Research, 22*, 607–655.

Matsuo, T., & Morimoto, N. (2007). Visual acuity and perimacular retinal layers detected by optical coherence tomography in patients with retinitis pigmentosa. *Investigative Ophthalmology and Visual Science, 91*, 888–890.

Normann, R., Maynard, E., Rousche, P., & Warren, D. (1999). A neural interface for a cortical vision prosthesis. *Visual Research, 39*, 2577–2587.

Palanker, D., Vankov, A., Huie, P., & Baccus, S. (2005). Design of a high resolution optoelectronic retinal prosthesis. *Journal of Neural Engineering, 2*, . S1-5-120

Pardue, M., Phillips, M., Yin, H., Fernandes, A., Cheng, Y., Chow, A., et al. (2005). Possible sources of neuroprotection following subretinal silicon chip implantation in RCS rats. *Journal of Neural Engineering, 2*, S39–S47.

Peachey, N., & Chow, A. (1999). Subretinal implantation of semiconductor-based photodiodes: Progress and challenges. *Journal of Rehabilitation and Development, 36*, 371–376.

Peterman, M., Noolandi, J., Blumenkranz, M., & Fishman, H. (2004). Localized chemical release from an artificial synapse chip. *Proceedings National Academy of Science USA, 101*, 9951–9954.

Rizzo, J., III, Wyatt, J., Jr., Lowenstein, J., Kelly, S., & Shire, D. (2003). Perceptual efficacy of electrical stimulation of human retina with microelectrode array during short-term surgical trials. *Investigative Ophthalmology and Visual Science, 44*, 5362–5369.

Sanz, M., Johnson, L., Ahuja, S., Ekstrom, P., Romero, J., & van Veen, T. (2007). Significant photoreceptor rescue by treatment with a combination of antioxidants in an animal model for retinitis pigmentosa. *Neuroscience, 145*, 1120–1129.

Sagdullaev, B., Aramat, R., Seiler, M., Woch, G., & McCall, M. (2003). Retinal transplantation-induced recovery of retinotectal visual function in a rodent model of retinitis pigmentosa. *Investigative Ophthalmology and Visual Science, 44*, 1686–1695.

Santos, A., Humayun, M., de Juan, E., Jr., Greenberg, R., Marsh, M., Klock, I., et al. (1997). Preservation of the inner retina in retinitis pigmentosa. A morphometric analysis. *Archives of Ophthalmology, 115*, 511–515.

Schiefer, M., & Grill, W. (2006). Sites of neuronal excitation by epiretinal electrical stimulation. *IEEE Transactions on Neural System Rehabilitation Engineering, 14*, 5–13.

Schoth, F., Burgel, U., Dorsch, R., Reinges, M., & Krings, T. (2006). Diffusion tensor imaging in acquired human blindness. *Neuroscience Letters, 398*, 178–182.

Smirnakis, S., Brewer, A., Schmid, M., Tollias, A., Schuz, A., Augath, M., et al. (2005). Lack of long-term cortical reorganization after macaque retinal lesions. *Nature, 435*, 300–307.

Stone, J., Barlow, W., Humayun, M., de Juan, E., Jr., & Milam, A. (1992). Morphometric analysis of macular photoreceptors and ganglion cells in retinas with retinitis pigmentosa. *Archives of Ophthalmology, 110*, 1634–1639.

Thomas, B., Seiler, M., Sadda, S., & Aramant, R. (2004a). Superior colliculus responses to light — preserved by transplantation in a slow degenerating rat model. *Experimental Eye Research, 79*, 29–39.

Thomas, B., Seiler, M., Sadda, S., Coffey, P., & Aramant, R. (2004b). Optokinetic test to evaluate visual acuity of each eye independently. *Journal of Neuroscience Methods, 138*, 7–13.

Tokuda, T., Asano, R., Sugitani, S., Terasawa, Y., Nunishita, M., Nakauchi, K., et al. (2007). In vivo stimulation on rabbit retina using CMOS LS1-based multi-chip flexible stimulator for retinal prosthesis. *Conference Proceedings of IEEE Engineering Medical Biological Society*, 5791–5794.

Tropepe, V., Coles, B., Chiasson, B., Horsford, D., Elia, A., McInnes, R., et al. (2000). Retinal stem cells in the adult mammalian eye. *Science, 287*, 2032–2036.

Troyk, P., Bradley, D., Bak, M., Cogan, S., Erickson, R., Hu, A., et al. (2005). Intracortical visual prosthesis research — approach and progress. *Conference Proceedings of IEEE Medical Biological Society, 7*, 7376–7379.

Wang, K., Fishman, H., Dai, H., & Harris, J. (2006). Neural stimulation with a carbon nanotube microelectrode array. *Nano Letters, 6*, 2043–2048.

Wang, L., Yang, L., & Dagnelie, G. (2008). Initiation and stabilization of pursuit eye movements in simulated retinal prosthesis at different implant locations. *Investigative Ophthalmology and Visual Science, 28*, 33–39.

Walter, M., Yang, L., & Dagnelie, G. (2007). Prosthetic vision simulation in fully and partially sighted individuals. In M. Humayun, J. Weiland, G. Chader, & E. Greenbaum (Eds.), *Artificial vision: Basic research, biomedical engineering and clinical advances* (pp. 71–90). New York, NY: Springer.

Walter, P., Kisvardy, Z., Gortz, M., Altheld, N., Rossler, G., Stiglitz, T., et al. (2005). Cortical activation via an implanted wireless retinal prosthesis. *Investigative Ophthalmology and Visual Science, 46*, 1780–1785.

Weiland, J., Humayun, M., Dagnelie, G., de Juan, E., Jr., Greenberg, R., & Iliff, N. (1999). Understanding the origin of visual percepts elicited by electrical stimulation of the human retina. *Graefes Archive of Experimental Ophthalmology, 237*, 1007–1013.

Weiland, J., Liu, T., & Humayun, M. (2005). Retinal prosthesis. *Annual Review of Biomedical Engineering, 7*, 361–401.

Wilke, R., Bach, M., Wilhelm, B., Durst, W., Trauzettel-Klosinski, S., & Zrenner, E. (2007). Testing visual functions in patients with visual prostheses. In M. Humayun, J. Weiland, G. Chader, & E. Greenbaum (Eds.), *Artificial sight: Basic research, biomedical engineering and clinical advances* (pp. 91–110). New York, NY: Springer.

Winter, J., Cogan, S., & Rizzo, J., 3rd (2007). Retinal prostheses: Current challenges and future outlook. *Journal of Biomaterial Science Polymer Edition, 18*, 1031–1055.

Yani, D., Weiland, J., Mahadevappa, M., Greenberg, R., Fine, I., & Humayun, M. (2007). Visual performance using a retinal prosthesis in three subjects with retinitis pigmentosa. *American Journal of Ophthalmology, 143*, 820–827.

Ziv, O., Rizzo, J., & Jensen, R. (2005). In vitro activation of retinal cells: Estimating location of stimulated cell by using a mathematical model. *Journal of Neural Engineering, 2*, S5–S15.

Zrenner, E. (2002). Will retinal implants restore vision? *Science, 295*, 1022–1025.

CHAPTER 22

# Progress in restoration of hearing with the auditory brainstem implant

Vittorio Colletti[1,*], Robert V. Shannon[2], Marco Carner[1], Sheila Veronese[1] and Liliana Colletti[1]

[1]*ENT Department, University of Verona, Verona, Italy*
[2]*House Ear Institute, Los Angeles, CA, USA*

**Abstract:** Fifty years ago auditory scientists were very skeptical about the potential of new prosthetic approaches that electrically stimulated the auditory nerve, the cochlear nuclei (CN), and the inferior colliculus (IC). In those decades, the basilar membrane was considered to play a fundamental and irreplaceable role as a fine spectrum analyzer in hearing physiology, and therefore it was thought that electrical stimulation of the auditory system would have never produced functionally useful hearing.

Over the last 30 years, cochlear implants (CIs) have improved steadily to the point where the average sentence recognition with modern multichannel devices is better than 90% correct. More recently, similar performance has been observed with electric stimulation of the brainstem with auditory brainstem implants (ABIs). However, it is clear that to fully understand hearing and to design the next generation of prosthetic devices we must better understand the ear–brain relationship. Indeed some aspects of hearing do not require the intricate complexities of cochlear physiological responses, while other auditory tasks rely critically on specialized details of cochlear processing. The progress in electrical stimulation of the central auditory system requires us to reconsider the patient selection criteria for different implant devices, in particular to evaluate the possibility of ABIs for etiologies with poor outcomes with CIs.

In the present review, the latest outcomes in restoration of hearing with ABI are presented. New guidelines are proposed for device selection for different etiologies and future research is suggested to further refine the process of matching an individual patient to the most appropriate implant device.

**Keywords:** cochlear implant; auditory brainstem implant; children and adults; tumor; nontumor

## Introduction

For many years, auditory scientists have been very skeptical about the potential that new prosthetic approaches bypassing the cochlea could provide functional hearing to profoundly deaf patients. It was assumed that the normal auditory system makes use of both (power) spectral and temporal information for frequency discrimination and the cochlea was regarded to be fundamental for the discrimination of sounds, including speech sounds. Spectral information is represented by the place on the basilar membrane that generates the

*Corresponding author.
Tel.: +39 045 8124275; Fax: +39 045 8027491;
E-mail: vittoriocolletti@yahoo.com

largest amplitude of vibration on the basilar membrane. The neural representation of cochlear place, as evidenced in tuning curves, shows exquisite selectivity as a function of frequency. However, evidence has been presented that the temporal representation of frequency is more robust than the place representation and is thus more important for speech discrimination. In the 1960s, electrical stimulation of the auditory system was considered a too crude a modality to reconstruct the complex spectral-temporal patterns of nerve activity produced by normal hearing. In addition, the role of the brain in processing complex patterns of information from the cochlea was greatly underestimated. At that time, most auditory researchers were fixated on the complexities of cochlear processing, and it was thought that the highly unnatural patterns of neural activation provided by electrical stimulation would only allow rudimentary auditory sensations.

## Cochlear implants

Over the last 30 years, cochlear implants (CIs) have improved steadily to the point where the average sentence recognition with modern multichannel devices is better than 90% correct and patients are able to understand speech well enough that most can communicate easily by telephone. Many patients are able to recognize 100% of simple sentences presented in a quiet environment and must now be tested in conditions of added noise to evaluate the limits of their performance. It is now clear that some aspects of hearing *do not* require the intricate complexities of cochlear physiological responses. Other auditory capabilities, such as complex pitch and binaural localization, do rely critically on specialized details of cochlear processing and so are not well represented by CIs (Litovsky et al., 2006). The improved outcomes of CIs have motivated an extension of implant candidacy to additional patient populations with considerable residual hearing. Combined auditory and electric stimulation (AES) devices deliver electric stimulation to the basal region of the cochlea in patients with residual acoustic low-frequency hearing (Gantz et al., 2006). Recent results show improvements in performance when combining acoustic and electric stimulation. Bilateral implants attempt to restore binaural cues to allow implant users to localize sounds in space and improve speech recognition in noise (Litovsky et al., 2004, 2006). Infants fitted with CIs before age 1 develop normal speech and language and report musical abilities that are beyond those available to adult implant patients (Svirsky et al., 2000; Robbins et al., 2004; Colletti and Zoccante, 2008). It is now clear that congenitally deaf children who receive the implant as adolescents derive limited benefit from the implant even after many years of daily use.

All these data clearly indicate that central auditory processing allows for high levels of speech pattern recognition, even though the peripheral pattern of activation is spectrally impoverished and highly unnatural in terms of the fine time structure, and even in cases where the pattern of peripheral information is distorted (Shannon et al., 1995, 1998, 2004). In this context, auditory implants can provide a tool for research on critical periods for complex pattern recognition, plasticity of early development, and the central plasticity in adults.

## Brainstem implants

For some profoundly deaf individuals with absent or destroyed cochlea or auditory nerve, CIs are not an option. Electrical stimulation must bypass the damaged cochlea and the auditory nerve and directly stimulate the auditory processing centers of the brainstem.

Auditory brainstem implants (ABIs) have been under development since the late 1970s, pioneered by physicians and researchers at the House Ear Institute in Los Angeles (Edgerton et al., 1982; Brackmann et al., 1993; Shannon et al., 1993; Otto et al., 2002) to provide auditory sensations to patients who could not be fitted with a CI owing to destruction of the auditory nerve as a sequellae of neurofibromatosis type 2 (NF2).

NF2 is a genetic disease that occurs in about 1 in 40,000 births and produces Schwann cell tumors along afferent nerve tracts as they enter the brainstem and spinal cord (Evans et al., 1992; Baser et al., 2003). One of the defining symptoms of NF2 is the growth of bilateral tumors along the VIII cranial nerve (composed of the auditory and vestibular nerves). Removal of the tumors almost always necessitates a transection of the auditory nerve and thus results in total deafness. If the tumors are not removed, they produce compression of the brainstem that is ultimately fatal. Faced with the tragic choice between total deafness and premature death, most patients, usually young people in the prime of life, opt for surgery, knowing very well that they will wake up without hearing. Their only hope is an auditory prosthesis such as an ABI that stimulates the brainstem directly to provide at least some limited hearing capability.

In 1979, Hitselberger et al. first implanted a pair of ball electrodes in the cochlear nucleus (CN) (Hitselberger et al., 1984). In 1981, a two-plate electrode device was used by the same surgeons and the system was upgraded to an 8-electrode system manufactured by Cochlear Ltd. in 1993. In 1992, a pilot study was initiated by Laszig et al. (1997) that used a multichannel ABI design with 20 electrodes, based on the Nucleus CI22 M CI (Cochlear Ltd., Lane Cove, Australia). The final design was a 21-electrode device (Nucleus 22 ABI) that was used until the introduction of the "Nucleus 24 ABI" that has been used since 1999. The Nucleus 24 ABI differs from the Nucleus 22 ABI in the stimulation strategies that can be used, the possibility of performing intraoperative electrical monitoring of the neural interface (neural response telemetry, NRT), and the possibility of removing the magnet. The Nucleus 22 ABI device uses only the spectral peak coding (SPEAK) strategy; the Nucleus 24 ABI has new speech processing strategies such as continuous interleaved sampling (CIS) and advanced combination encoder (ACE).

Until recently, ABIs have been fitted exclusively in patients with NF2. These patients receive sound awareness, some environmental sound discrimination and identification, and a significant improvement in face-to-face communication when the ABI is combined with lip reading (Lenarz et al., 2001; Otto et al., 2002; Nevison et al., 2002). However, out of more than 600 ABIs worldwide, no ABI patients with NF2 have been able to achieve the high level of speech recognition that is common with CIs. The overall performance is comparable to that achieved by single-channel CIs. Multichannel CIs, in contrast, typically restore speech understanding in postlingually deafened adults to a level where most can converse on the telephone.

The cause of the large difference in performance of the CI and ABI has not been clear because both use similar signal processing and a similar number of stimulating electrodes. Both devices consist of an external microphone, which collects sound waves, and a speech processor, which converts the sound waves into electrical impulses that are then transmitted to a receiver implanted under the skin. The receiver sends the electrical impulses to a microelectrode array implanted within the cochlea in the CI and on the CN in the ABI.

In CI devices, multiple stimulating electrodes are inserted into the cochlea, so that the electrodes are situated at different locations along the cochlea's basilar membrane. The mechanics of the cochlea are such that the basilar membrane is tonotopically organized; that is, high-frequency sounds activate hair cells at the base of the basilar membrane, whereas low-frequency sounds activate hair cells near the apex of the basilar membrane. The sound, received by an external microphone, is first analyzed by a microprocessor and then an electrical signal representing the information from each frequency region is transmitted to the appropriate electrode at each tonotopic location. Thus, CIs can access most of the frequency range of the cochlea and directly stimulate the correct populations of auditory nerve fibers so that electrical signals are propagated to the appropriate (tonotopic) regions of the CN of the brainstem and then on to higher auditory processing centers.

With the ABI, the microelectrode array is inserted through an opening in the retro-mastoid area, advanced into the lateral recess of the fourth ventricle of the brain, and placed over the surface of the CN. The CN are part of the mainstream (lateral lemniscal) auditory system that transmits sound frequency information to higher auditory centers (inferior colliculus, medial geniculate nucleus, and auditory cortex). The CN are tonotopically organized with a map oriented at a shallow angle to the surface. Due to this anatomical structure, it is difficult or impossible to achieve tonotopically selective activation with different electrodes on a surface-electrode array. Stimulation of surface ABI electrodes produces weak pitch sensations with no clear tonotopic ordering, and most ABI patients never reach open-set understanding of speech even after months and years of practice (Otto et al., 2002).

Another factor that might contribute to the lack of tonotopic resolution in surface-electrode ABIs is the high current needed for stimulation: Current fields spread broadly from each electrode and may not stimulate distinct neural populations.

One possibility for improving ABI performance is to use an array of microelectrodes penetrating into the ventral CN to achieve better tonotopic selectivity in the stimulation. This approach has been developed and tested in an animal model and recently initiated clinical trials. The presumed theoretical advantages of these electrodes should include the following: (1) direct access to the three-dimensional tonotopic gradients of the CN; (2) lower threshold for electrical stimulation of the auditory system and reduction of side effects, by reducing the spread of the current along the pial surface layers and by closer contact with a larger population of secondary auditory neurons; and (3) wider dynamic ranges and lower operating currents for electrical stimulation. McCreery et al. (1998) successfully implanted penetrating electrode arrays in cats and demonstrated the ability of the electrode arrays to evoke tonotopically localized neural activation in the next auditory relay station of the brainstem, the inferior colliculus.

The hope that the improved tonotopic selectivity provided by penetrating microelectrodes could result in improved speech understanding motivated a clinical trial of the penetrating ABI in NF2 patients. Recently, an ABI with penetrating microelectrodes (PABI) has been developed to achieve selective stimulation across the tonotopic axis of the human CN. Results with PABI fitted in NF2 patients showed that the penetrating microelectrodes were correctly placed in the posterior ventral cochlear nuclei (PVCN), as indicated by auditory percepts and low thresholds (below 1 nC-nanocoulomb). No interference could be measured between penetrating electrodes as determined by forward masking, indicating that the penetrating electrodes also achieve good selectivity. However, none of the patients with multiple penetrating electrodes have achieved significant open-set speech recognition, even after a year of implant experience (Otto et al., 2008). Both ABI and PABI provide basic sound perceptions, but the PABI provides excellent tonotopic selectivity. In spite of this, no NF2 patient has been able to achieve high levels of open-set speech recognition.

The limited success of the current ABIs has also been attributed to the possibility that stimulation of the CN with surface electrodes bypasses the processing of specialized neural circuitry that extracts information on sound modulation/periodicity and information on sound onset/offset. However, even PABIs with penetrating electrodes have not resulted in a significant improvement in speech recognition, which implies that even highly selective activation of small regions of specialized cells in the CN is not adequate for speech. Such a failure to produce speech recognition at the level of the CN would not bode well for a future auditory prosthesis that directly stimulates higher levels of the auditory pathway: the inferior colliculus or auditory cortex. Both structures, particularly the inferior colliculus, would be more surgically accessible than the CN, which is located deep in the brainstem. Regarding direct stimulation of the auditory cortex, attempts to directly stimulate the visual cortex have been disappointing, with patients seeing only phosphenes (spots of light) or other disconnected and meaningless sensations. However, with the gains made by research on both the visual and auditory cortex, it may be possible to design new electrode arrays that stimulate the auditory cortex appropriately.

## ABIs in nontumor patients

A basic prerequisite for the success of a CI in providing open-set speech understanding is the presence of an auditory nerve to connect the cochlea to the brainstem. The average sentence recognition with modern multichannel devices is better than 90% correct (Spahr and Dorman, 2006) and many patients are able to recognize 100% of simple sentences presented in quiet, but the variability in performance across listeners is still quite large. The variability in CI patient outcomes has been assumed to be due to nonoptimized speech processor settings, differences in electrode array positioning, or differences in patients' neural survival. However, studies in which speech processor parameters were optimized, performance improved for all patients but did not significantly reduce the variability in patient outcomes. These results suggest that some physiological factor may be responsible for the wide variation in CI outcomes.

Although auditory performance of NF2 ABI patients was poor compared to the better outcomes with a CI, some patients receive little or no benefit from a CI. If the low performance with a CI is due to damage to the VIII nerve, then an ABI might provide a better auditory outcome in such patients. The limited outcomes observed in NF2 ABI patients suggested that open-set speech recognition was not possible with an ABI, but even the limited performance of an ABI might be useful to those who failed to benefit from a CI due to the absence of a cochlea and CN. Most auditory researchers and implant clinicians did not think it was possible to restore open-set speech recognition with an ABI. It was thought that the ABI stimulating the CN bypasses too many functions: the spectrum analyzer function of the basilar membrane, the neural transduction in the hair cells, the auditory nerve, and probably some neural processing that normally occurs in the CN. In addition, the pattern of activation that an ABI produces in the CN is highly unnatural and may not even preserve the tonotopic ordering of stimulation across electrodes. Most auditory researchers were so focused on the cochlea–cochlear nerve complex and its complexity that they probably overlooked a highly important part of hearing — the ear–brain assembly.

In 1997, several patients with failed CIs were implanted with ABIs as a clinical trial. These patients had cochlear abnormalities, received a CI, and initially experienced good-to-excellent open-set speech recognition. Over time, presumably as the cochlear disease process progressed and damaged the VIII nerve (for example, by cochlear ossification), they exhibited a progressive deterioration in speech understanding to the point that patients discontinued use of the CI. Following contralateral ABI implantation, a good proportion of these patients regained excellent open-set speech understanding, including fluent conversational use of the telephone. This unexpected finding demonstrated for the first time that effective prosthetic electrical stimulation of the CN is indeed possible despite the missing cochlea, auditory nerve, and the highly unnatural activation patterns delivered by the ABI to the CN.

Following this preliminary experience, other patients with profound hearing loss who were not candidates for a CI were fitted with ABIs. Etiology of hearing loss included temporal bone fracture, prolific ossification following meningitis, severe ossification not necessarily associated with meningitis, and congenital malformations or absence of cochleas and VIII nerves. At the most recent follow-up, nontumor (NT) adults scored from 10% to 100% in open-set speech perception tests (average 59%), and tumor (NF2) patients scored from 5% to 31% (average 10%). The differences between these results are statistically significant ($p = .0007$). The best performance was observed in patients who lost their nerve VIII from head trauma or severe ossification. Lowest performance (although still highly beneficial to the patient) was observed in patients with neurological disorders, neuropathy, and cochlear malformations associated with cochlear nerve hypoplasia (Tables 1 and 2) (Colletti, 2006).

This finding was quite surprising because all the teams fitting ABIs in NF2 patients have reported that very few NF2 ABI listeners were able to

Table 1. Results of open-set sentence recognition at the last follow-up (1–10 years) postoperatively in adult patients

| Group | Number of subjects | Range (%) | X | Median | SD |
|---|---|---|---|---|---|
| T (NF2) | 32 | 5–31 | 10 | 16 | 21.34 |
| NT | 48 | 10–100 | 59 | 53 | 15.21 |
| T (NF2) vs. NT (t-test) | $p < .0007$ | | | | |

Table 2. Results of open-set sentence recognition at the last follow-up (1–10 years) postoperatively in the different subgroups of NT patients

| Etiology | Number of subjects | Range (%) | X | Median | SD | T (NF2) vs. NT subgroup (t-test) |
|---|---|---|---|---|---|---|
| Head trauma | 7 | 32–80 | 62 | 57 | 23.41 | $p = .005$ |
| Auditory neuropathy | 4 | 12–18 | 15 | 16 | 2.52 | $p = .07$ |
| Cochlear malformations | 6 | 37–61 | 44 | 61 | 11.12 | $p = .006$ |
| Altered cochlear patency | 31 | 34–100 | 60 | 64 | 19.81 | $p = .0048$ |

recognize a significant amount of speech when tested in the open-set recognition task. Only a few NF2 ABI patients can consistently recognize 20% of the words in sentences with only the sound from their ABI. Most NF2 patients recognize less than 5% of the words in sentences, even after many years of ABI experience.

These findings challenge one common interpretation of the low levels of speech recognition in NF2 patients fitted with ABIs. These results demonstrate that high levels of speech recognition can be achieved with an ABI, in spite of bypassing the cochlea and auditory nerve and in spite of the highly unnatural tonotopic pattern of activation of the ABI electrode array. With a modest amount of experience NT ABI patients were able to understand speech well enough to communicate by telephone. This result demonstrates the power of the brain's pattern recognition ability. However, the contrast in results between NT and NF2 patients poses an interesting puzzle. NT and NF2 patients use the same electrode array and the same signal processing, yet the outcomes are quite different. It appears that something related to NF2 interferes with the ability to recognize speech with an ABI.

To determine the cause of the large difference in open-set speech perception with ABI between NF2 and NT, a number of subjects from each group fitted with ABI were submitted to a series of psychophysical tests [1 – electrical stimulation thresholds, as an index of electrode proximity to stimulate neurons, 2 – electrode selectivity evaluated with forward masking to quantify interaction between electrodes, 3 – amplitude modulation detection, as an indication of temporal resolution, 4 – speech understanding of single phonemes (vowels) and simple sentences (HINT)]. For a detailed description of the single test procedures see Colletti and Shannon (2005).

The results of the study demonstrated that both etiology groups had a wide range of pitch perception across electrodes and a full range of loudness percepts, suggesting that both groups had sufficient surviving neurons in the CN to support pitch and loudness. Both groups also showed excellent electrode placement and little interference across electrodes. However, NF2 ABIs had significantly poorer modulation detection and speech understanding than NT patients. The difference in modulation detection between the two etiology groups suggests that there may be a difference in the survival of a specific neuronal pathway that is critical for modulation detection. The difference in speech recognition between NF2 and NT groups suggests that this putative pathway is also linked to speech recognition. Thus, NT ABI patients, using the same device implanted at the same CN location, using the same speech processing strategy as NF2 ABI patients, and with similar psychophysical capabilities, are capable of high levels of open-set speech recognition. The dichotomy in performance between NF2 and NT ABI patients suggests that simple perceptual features and more complex speech patterns may be processed differently, possibly even by different processing streams.

This pattern of results suggests that the CN may be damaged by the tumor (or its removal), causing NF2 ABI speech performance to be much poorer than NT ABI performance. The damage is probably limited to the CN because NF2 has no known central pathology. Some portion of the CN must survive the tumor removal surgery intact because the basic perceptual capabilities (loudness, pitch, temporal acuity, tonotopic selectivity) are still intact in NF2 ABIs. One possibility is that the tumor removal process may damage specialized neurons in the CN that are critical for speech pattern processing. Damage may not necessarily be to a "system" or distinct "pathway," but rather to a specialized subpopulation of cells that provide critical input for later processing stages. While the exact cause and location of damage are as yet uncertain, the large difference in speech performance between these two groups of ABI users provides a unique opportunity to investigate the relation between speech perception and basic perceptual capabilities. In addition, the difference in performance between NF2 and NT ABI patients has a profound impact on patient selection criteria.

If damage to the CN is limiting speech recognition performance, then bypassing this region of damage might provide improved performance for NF2 patients. Theoretically, prosthetic stimulation at a site that is more central than the damage might produce good speech recognition, while stimulation near the site of damage might produce relatively normal basic percepts but poor speech recognition. For example, if poor performance in CI patients is due to damage to the VIII nerve, a CI will provide no benefit because the necessary signals are not transmitted to the central auditory system. In such patients, an ABI may provide some benefit by bypassing the point of damage. The good speech recognition performance by NT ABI patients (who have no functioning auditory nerve) suggests that the ABI may provide a new option for patients who receive little benefit from a CI. Indeed, greatly improved speech recognition was observed in several NT ABI patients who previously received little benefit from their CI (Colletti and Shannon, 2005; Spahr and Dorman, 2006). However, if poor performance in CI patients is due to more central damage, the ABI may not provide any more benefit than the CI. Similarly, if poor ABI performance is due to damage at the level of the CN, then stimulation at a more central auditory nucleus, such as the IC, might provide improved performance by bypassing the site of damage.

The high levels of open-set speech understanding that have been observed in CIs and in NT ABIs show that good speech recognition is possible with electrical stimulation, even with stimulation at the level of the brainstem. More work is clearly needed to explore the limits of electrical stimulation. We should first of all be able to determine the site of lesion in an individual patient so that the appropriate stimulation site could be selected. This could help define whether good speech recognition can be achieved with implants at more central structures in the auditory system.

### *Analysis of possible sites of physiological damage and suggestions for the appropriate implant for each condition*

Most types of deafness result from damage to hair cells in the cochlea and so are suitable for a CI. However, etiologies that cause damage to the auditory nerve may produce poor performance with a CI and may achieve better outcomes with an ABI. Consider ossification from otosclerosis or as sequelae to meningitis. Cochlear ossification may progress in severity from mild ossification in the cochlea to severe ossification within the cochlea, to ossification of the modiolus, and finally to ossification of the internal auditory meatus (IAM). In some patients ossification will progress across all four levels, while in others the ossification may terminate at the more peripheral level. Mild or severe ossification in the cochlea may allow cochlear implantation with a drill-out and might provide a good outcome initially, but performance may decline over time if the ossification progresses and ultimately damages the nerve. Implantation will not halt the progression of the biological process of ossification. In some cases of ossification the nerve may survive within the internal meatus but the ossification may attenuate or block the implant signal from the scala tympani. In such cases, an ABI may provide better access to the remaining auditory nerve at its terminus in the CN. If the ossification fills the modiolus and/or IAM, then the nerve may be damaged or destroyed. High-resolution CT or MRI can resolve the entire length of the IAM and so determine if it contains an auditory nerve or not (Casselman, 2002). When cochlear ossification is extensive the IAM should be examined for the presence of an auditory nerve to help determine whether to select a CI or ABI (Govaerts et al., 2003). Even if the nerve is intact and the modiolus is free of ossification at the time of CI implantation, the progression of the disease process may eventually damage the nerve. Such cases should be carefully monitored over time for progression.

Auditory neuropathy (AN)/auditory dissynchrony disorder is a condition or collection of conditions in which the cochlea appears to function (as indicated by normal otoacoustic emissions) but no evoked neural response can be measured. In addition, speech recognition is poorer than would be expected from the audiogram. It is likely that the term auditory neuropathy is applied to a variety of physiological conditions that represent damage to different points in the auditory transmission pathway. In some cases a CI can provide improved speech recognition (Zeng and Liu, 2006), while in other cases a CI is of little benefit (Gibson and Sanli, 2007). Conditions that disrupt hair cells and hair cell–neuron synaptic transmission may respond well to the application of a CI because the nerve is still intact. In contrast, conditions that reflect damage to the auditory nerve may show little benefit from a CI, but may show benefit from an ABI. Physiological measures may be able to discriminate patients with different types of damage and so guide the selection of the appropriate implant (Gibson and Sanli, 2007). Further work is needed to differentiate the locus of damage in cases diagnosed as AN.

NF2 tumors and their removal do not always disrupt the CN. In some cases, a small vestibular schwannoma (VS) can be dissected from the auditory portion of the VIII nerve and the nerve function can be preserved even if hearing is lost. In such cases, a positive promontory response can indicate the survival of viable nerve and indicate the possibility of a CI (Neff et al., 2007). Additional screening tests are needed to reliably indicate when such patients might benefit from a CI and which should receive an ABI. In addition, studies are needed to determine if any etiological factors predict which patients can successfully use a CI in such cases. NF2 patients whose VIII nerve is damaged or lost during VS removal have the option of choosing an ABI. At the present time, the ABI is the standard of care for NF2 patients having the second VS removed. In NF2 patients an ABI can provide sound awareness, environmental sound discrimination, and an enhancement to lipreading (Otto et al., 2002; Hitselberger et al., 1984; Nevison et al., 2002; Lenarz et al., 2002). However, few NF2 ABI patients are able to recognize words and sentences without the aid of lipreading.

An ABI is also possible at the time of the first VS removal at selected clinical centers. Although these patients may not require the ABI initially, they can adjust to the sound quality of the ABI as their hearing declines on the second tumor side. In addition, if the initial ABI does not provide useful sound sensations the patient will have another chance to obtain a functioning ABI at the time of the second VS surgery.

A recent indication for an ABI is for patients in whom a CI has failed (Colletti et al., 2004a, b). In some cases, a CI was implanted following meningitis. Progressive ossification obliterated the cochlear and eventually filled the modiolus, destroying the nerve. In other cases, the ossification is familial/genetic and progressive. In some cases, a CI can be effective for several years followed by declining performance as the ossification progresses. In several such patients, an ABI was placed successfully and the performance level was similar to the previous excellent performance with the CI (Colletti et al., 2004a, b). Further clinical studies are needed of larger populations to determine which etiologies are most favorable for good ABI outcomes and what diagnostic tests might predict good performance.

Other NT conditions that damage the VIII nerve may also be candidates for an ABI, e.g., bilateral skull fracture severing the VIII nerve bilaterally. One patient received an ABI at age 34 after being deafened by bilateral skull fracture in a motorcycle accident at age 17 (Colletti et al., 2004a). He was able to achieve excellent open-set speech recognition and routine conversational use of the telephone within 3 months of initial activation.

However, if there is reason to suspect that damage has occurred to the CN then an ABI may not produce a satisfactory outcome. Most NF2 ABI recipients receive sound awareness, some sound discrimination, and enhancement of lip-reading (Otto et al., 2002). However, there are only a few reports of open-set speech recognition in NF2 ABIs out of more than 600 implants worldwide. In contrast with the ABI results in NT patients, this suggests that the NF2 tumor removal damages not only the VIII nerve, but possibly some portion of the CN, limiting the perceptual outcome. Although an ABI is presently the standard of care for such patients, it may be desirable to target a higher auditory center for implant, bypassing the putative damage to the CN. Two devices are presently in limited clinical trials to stimulate the IC in the auditory midbrain: the ICI using surface electrodes (Colletti et al., 2007) and the auditory midbrain implant (AMI) using a penetrating array of electrodes (Lenarz et al., 2006a, b; Lim and Anderson, 2006). It is not clear at the present time if electrical stimulation of the IC will produce a better outcome in NF2 patients than an ABI.

## *ABI in children*

Based on the excellent speech recognition results in adults, clinical trials are now underway to evaluate the efficacy of ABIs in children (Colletti et al., 2001, 2005a, b; Colletti, 2007; Eisenberg et al., 2008). Based on the experience with CIs in children, ABIs in children could be successful as well. Children with cochlear nerve aplasia or severe ossification following meningitis may have no auditory nerve available for activation by a CI. In some cases, children with these conditions have received CIs and have received no benefit (Colletti et al., 2001, 2005). The early plasticity of the nervous system could allow young infants to integrate the auditory pattern of neural activity from an ABI for speech recognition. However, it is probably important that children have a normally developed CN to process the ABI activation. Conditions that produce abnormal development of the CN might not be suitable for an ABI. Imaging and electrophysiological assessment tools are necessary to determine if an individual infant has a normally developed CN or not. Of course, the same critical periods for development apply equally to the CI and ABI. Studies have shown that CI performance is better when the device is implanted before 2 years of age (Svirsky et al., 2000; Manrique et al., 2004; Colletti et al., 2005). Congenitally deaf children who receive the ABI prior to age 2 also would probably be more likely to learn to use the stimulation of the ABI for speech understanding. Children who were hearing at birth and then lost their auditory nerves due to ossification or trauma would also be more likely to utilize the ABI for speech understanding. As in CIs, congenitally deaf children who receive an ABI after the age of 5 would be less likely to develop speech understanding with the ABI alone.

The existing data show that ABIs in children produce mixed results, with some children showing higher levels of auditory performance and learning than others (Colletti et al., 2001, 2005). However, ABI children of all etiologies do show improved auditory performance and improved cognitive development (Colletti, 2007). Eisenberg et al. (2008) evaluated the performance of an ABI in a child with congenital cochlear malformation and cochlear nerve aplasia (Goldenhar syndrome). The child was implanted at age 3.5 years and evaluated after 6 and 12 months of ABI experience. The child's ABI performance was comparable to that of a large cohort of children implanted at a similar age with CIs. The CI children used for comparison were also congenitally deaf and implanted between 3 and 4 years of age. This result shows that the ABI can provide CI-like auditory development even in children with congenital absence of the auditory nerve. Additional measures have shown CI-like auditory development in several other congenitally deaf children with ABIs (Colletti and Zoccante, 2008).

*Role of learning*

One of the important factors in assessing implant outcomes is the role of learning. When the brain is presented with a new pattern of neural activity representing sound, it may not immediately be able to correctly interpret the pattern. If the pattern of neural activation is similar to the acoustic pattern with which the brain is familiar, then recognition of sounds may take place soon after initial implant stimulation. There are reports of CI recipients who are able to communicate by telephone within minutes of their initial stimulation. Presumably, this indicates that the prosthetic pattern of stimulation is recreating the normal acoustic pattern of activation. However, many CI recipients improve in performance over several months, with most of the improvement coming in the first 3 months (Tyler et al., 1997). However, in some cases performance may only increase in small amount and reach a plateau at a performance level that is far below the average implant performance. In such cases it may not be clear whether the limitation is due to a damaged peripheral system, or to a misadjustment in the speech processor. If repeated adjustments of a CI speech processor continue to result in low performance levels, this may indicate damage to the auditory nerve and an ABI should be considered (Colletti et al., 2004a, b).

Auditory training is an under-utilized technique that can result in dramatically improved performance (Evans et al., 1992). In many cases, recognition of words and sentences can be improved by 20 percentage points with a short period of training, even in patients with long-term experience (Fu et al., 2005a, b). Auditory training programs should be implemented for all implant patients to take advantage of these performance gains. While CI performance typically reaches an asymptote after 3–6 months of experience, ABI performance in NF2 patients continues to improve over many years (Lenarz et al., 2001). ABI performance in NF2 patients starts at a low level and slowly improves, but never reaches a level comparable to that of CIs, even after many years. This limitation in NF2 ABIs may indicate damage to the CN region so that the central auditory pattern recognition system never receives a sufficient pattern of peripheral information. Furthermore, the pattern of peripheral information is presumably so degraded or distorted that high levels of speech recognition is not achieved even after many years of learning.

In contrast, NT ABI users have shown dramatic improvements in performance over the initial 6 months of activation (Colletti et al., 2005), much like CI patients. Some NT ABI users have achieved CI-like levels of performance, including open-set speech recognition and conversational use of the telephone (Colletti and Shannon, 2005). The difference between NF2 and non-NF2 ABI outcomes suggests that NF2 tumor removal disrupts a portion of the peripheral auditory system that is essential for speech recognition. Even after as much as 20 years of ABI use, NF2 ABI users cannot achieve high levels of open-set speech recognition. For NF2 patients, it may be necessary to bypass the CN and stimulate the IC to obtain better speech recognition. Training and learning may not be able to overcome a damaged peripheral system. If CIs do not provide satisfactory outcomes after a sufficient period of learning and training, an ABI should be considered.

**Conclusions**

Some etiological factor appears to be limiting NF2 ABI patients' ability to synthesize speech from the stimulation patterns provided by the ABI. Since there is no known central manifestation of NF2, the problem is most likely localized to the CN; this suggests that some physiological mechanism/structure/pathway in the CN may be damaged by the tumor or during the removal of the tumor.

Excellent hearing results may be obtained in NT patients fitted with ABI and this demonstrates that effective prosthetic stimulation is possible at the CN, despite the missing cochlea and auditory nerve processing and the highly unnatural activation patterns in the CN. At almost every stage of prosthesis development we have underestimated the potential benefit of the devices. For many years, it was thought that CIs would never be able to allow normal conversation by telephone, but that is now a normal outcome. It was thought that ABIs would never allow the same level of speech recognition as CIs, but many NT ABIs show comparable performance to CIs.

ABIs may be successful because much of the natural speech signal is redundant and implants only need to transmit a small fraction of the information that is contained in speech sounds to achieve good speech intelligently (Shannon et al., 1995). Furthermore, many of the processing capabilities of the ear and the auditory nervous system are redundant and individuals with normal hearing can understand speech with highly degraded signals. In addition, the CNS has an enormous ability to adapt (re-wire) to changing demands through expression of neural plasticity.

The cause of the large difference in performance between NF2 and NT ABI patients, and among NT ABI patients, needs further investigation. We must continue to push forward the boundaries of knowledge in prosthetic stimulation of the auditory system. Will the ICI or AMIs allow open-set speech recognition for NF2 patients or others with damage to the CN? Will children with cochlear and cochlear nerve disorders, not suitable for CI, be able to understand speech using the stimulation of an ABI, with outcomes comparable to those of children implanted with CI? There are many questions still to be solved in the application of electrical stimulation to restore hearing. Properly controlled scientific and clinical studies are needed to "push the envelope" of outcomes. Implant technology has been a great success story in otolaryngology and the ultimate limits of the technology are still not known.

# References

Baser, M. E., Evans, D. G., & Gutmann, D. H. (2003). Neurofibromatosis 2. *Current Opinions in Neurology*, *16*(1), 27–33.

Brackmann, D. E., Hitselberger, W. E., Nelson, R. A., Moore, J. K., Waring, M., Portillo, F., et al. (1993). Auditory brainstem implant. I: Issues in surgical implantation. *Otolaryngology and Head and Neck Surgery*, *108*, 624–634.

Casselman, J. W. (2002). Diagnostic imaging in clinical neuro-otology. *Current Opinions in Neurology*, *15*(1), 23–30.

Colletti, L. (2007). Beneficial auditory and cognitive effects of auditory brainstem implantation in children. *Acta Oto-Laryngologica*, *9*, 1–4.

Colletti, L., & Zoccante, L. (2008). Nonverbal cognitive abilities and auditory performance in children fitted with auditory brainstem implants: preliminary report. *Laryngoscope*, *118*(2), 1443–1448.

Colletti, V. (2006). Auditory outcomes in tumor vs. non-tumor patients fitted with auditory brainstem implants. *Advanced Otorhinolaryngology*, *64*, 167–185.

Colletti, V., Carner, M., Morelli, V., Guida, M., Colletti, L., & Fiorino, F. G. (2004a). Auditory brainstem implant in post-traumatic cochlear nerve avulsion. *Audiology Neuro-otology*, *9*, 247–255.

Colletti, V., Carner, M., Miorelli, V., Guida, M., Colletti, L., & Fiorino, F. (2005a). Auditory brainstem implant (ABI): new frontiers in adults and children. *Otolaryngology and Head and Neck Surgery*, *133*(1), 126–138.

Colletti, V., Carner, M., Miorelli, V., Guida, M., Colletti, L., & Fiorino, F. G. (2005b). Cochlear implantation at under 12 months: report on 10 patients. *Laryngoscope*, *115*, 445–449.

Colletti, V., Fiorino, F., Sacchetto, L., Miorelli, V., & Carner, M. (2001). Hearing habilitation with auditory brainstem implantation in two children with cochlear nerve aplasia. *International Journal of Pediatric Otorhinolaryngology*, *60*, 99–111.

Colletti, V., Fiorino, F. G., Carner, M., Miorelli, V., Guida, M., & Colletti, L. (2004b). Auditory brainstem implant as a salvage treatment after unsuccessful cochlear implantation. *Otology and Neurotology*, *25*(4), 485–496.

Colletti, V., & Shannon, R. (2005). Open set speech perception with auditory brainstem implant? *Laryngoscope*, *115*, 1974–1978.

Colletti, V., Shannon, R., Carner, M., Sacchetto, L., Turazzi, S., Masotto, B., et al. (2007). The first successful case of hearing produced by electrical stimulation of the human midbrain. *Otology and Neurotology*, 28(1), 39–43.

Edgerton, B. J., House, W. F., & Hitselberger, W. (1982). Hearing by cochlear nucleus stimulation in humans. *Annals of Otology, Rhinology Otolaryngology*, 91, 117–124.

Eisenberg, L. S., Johnson, K. C., Martinez, A. S., DesJardin, J. L., Stika, C. J., Dzubak, D., et al. (2008). Comprehensive evaluation of a child with an auditory brainstem implant. *Otology and Neurotology*, 29(2), 251–257.

Evans, D. G. R., Huson, S. M., Neary, W., Blair, V., Newton, V., Strachan, T., et al. (1992). A genetic study of type 2 neurofibromatosis in the United Kingdom: I. Prevalence, mutation rate, fitness, and confirmation of maternal transmission effect on severity. *Journal of Medical Genetics*, 29, 841–846.

Fu, Q.-J., Galvin, J. J., & Wang, X. (2005). Moderate auditory training can improve speech performance of adult cochlear implant users. *Journal of the Acoustical Society of America*, 6(3), 106–111.

Fu, Q.-J., Nogaki, G., & Galvin, J. J., III (2005). Auditory training with spectrally shifted speech: an implication for cochlear implant users' auditory rehabilitation. *Journal of the Association for Research in Otolaryngology*, 6(2), 180–189.

Gantz, B. J., Turner, C. W., & Gefeller, K. E. (2006). Acoustic plus electric speech processing: preliminary results of a multicenter clinical trial of the Iowa/nucleus hybrid implant. *Audiology Neuro-otology*, 11(Suppl. 1), 63–68.

Gibson, W. P., & Sanli, H. (2007). Auditory neuropathy: an update. *Ear and Hearing*, 28(Suppl.), 102S–106S.

Govaerts, P. J., Casselman, J., & Daemers, K. (2003). Cochlear implants in aplasia and hypoplasia of the cochleovestibular nerve. *Otology and Neurotology*, 24, 887–891.

Hitselberger, W., House, W., Edgerton, B., & Whitaker, S. (1984). Cochlear nucleus implant. *Otolaryngology and Head and Neck Surgery*, 92, 52–54.

Laszig, R., Marangos, N., Sollmann, P., Ramsden, R., Fraysse, B., Lenarz, T., et al. (1997). Initial results from the clinical trial of the nucleus 21-channel auditory brain stem implant. *American Journal of Otolaryngology*, 18(6), 160.

Lenarz, M., Lim, H., & Patrick, J. F. (2006a). Electrophysiological validation of a human prototype auditory midbrain implant (AMI) in a guinea pig model. *Journal of the Association for Research in Otolaryngology*, 7, 283–298.

Lenarz, M., Lim, H. H., & Patrick, J. F. (2006b). The auditory midbrain implant: a new auditory prosthesis for neural deafness-concept and device description. *Otology and Neurotology*, 27(6), 838–843.

Lenarz, M., Matthies, C., & Lesinski-Schiedat, A. (2002). Auditory brainstem implant: Part II. Subjective assessment of functional outcome. *Otology and Neurotology*, 23, 694–697.

Lenarz, T., Moshrefi, M., Matthies, M., Frohne, C., Lesinski-Schiedat, A., Illg, A., et al. (2001). Auditory brainstem implant: Part I. Auditory performance and its evolution over time. *Otology and Neurotology*, 22, 823–833.

Lim, H. H., & Anderson, D. J. (2006). Auditory cortical responses to electrical stimulation of the inferior colliculus: implications for an auditory midbrain implant. *Journal of Neurophysiology*, 96, 975–988.

Litovsky, R., Parkinson, A., Arcaroli, J., & Sammeth, C. (2006). Simultaneous bilateral cochlear implantation in adults: a multicenter clinical study. *Ear and Hearing*, 27, 714–731.

Litovsky, R. Y., Parkinson, A., Arcaroli, J., Peters, R., Lake, J., Johnstone, P., et al. (2004). Bilateral cochlear implants in adults and children. *Archives of Otolaryngology—Head and Neck Surgery*, 130(5), 648–655.

Manrique, M., Cevera-Paz, F. J., Huarte, A., & Molina, M. (2004). Advantages of cochlear implantation in prelingual deaf children before 2 years of age when compared to later implantation. *Laryngoscope*, 114, 1462–1469.

McCreery, D. G., Shannon, R. V., Moore, J. K., Chatterjee, M., & Agnew, W. F. (1998). Accessing the tonotopic organization of the ventral cochlear nucleus by intranuclear microstimulation. *IEEE Transactions on Rehabilitation Engineering*, 6(4), 391–399.

Neff, B. A., Wiet, R. M., & Lasak, J. (2007). Cochlear implantation in the neurofibromatosis type 2: long-term follow-up. *Laryngoscope*, 117(6), 1069–1072.

Nevison, B., Laszig, R., Sollmann, W. P., Lenarz, T., Sterkers, O., Ramsden, R., et al. (2002). Results from a European clinical investigation of the nucleus multichannel auditory brainstem implant. *Ear and Hearing*, 23(3), 170–183.

Otto, S. A., Brackmann, D. E., Hitselberger, W. E., Shannon, R. V., & Kuchta, J. (2002). The multichannel auditory brainstem implant update: performance in 60 patients. *Journal of Neurosurgery*, 96, 1063–1071.

Otto, S. R., Shannon, R. V., Brackmann, D. E., Hitselberger, W. E., McCreery, D., Moore, J., et al. (2008). Audiological outcomes with the penetrating electrode auditory brainstem implant. *Otology and Neurotology*, 29(8), 1147–1153.

Robbins, K. M., Koch, D. B., Osberger, M. J., Zimmerman-Phillips, S., & Kishon-Rabin, L. (2004). The effect of age at cochlear implantation on auditory skill development in infants and toddlers. *Archives of Otolaryngology—Head and Neck Surgery*, 130, 570–574.

Shannon, R. V., Fayad, J., Moore, J. K., Lo, W., O'Leary, M., Otto, S., et al. (1993). Auditory brainstem implant. II: Post-surgical issues and performance. *Otolaryngology and Head and Neck Surgery*, 108, 635–643.

Shannon, R. V., Fu, Q.-J., & Galvin, J. (2004). The number of spectral channels required for speech recognition depends on the difficulty of the listening situation. *Acta Oto-Laryngologica, Supplementum*, 552, 50–54.

Shannon, R. V., Zeng, F. G., Kamath, V., Wygonski, J., & Ekelid, M. (1995). Speech recognition with primarily temporal cues. *Science*, *270*, 303–304.

Shannon, R. V., Zeng, F. G., & Wygonski, J. (1998). Speech recognition with altered spectral distribution of envelope cues. *Journal of the Acoustical Society of America*, *104*(4), 2467–2476.

Spahr, A. J., & Dorman, M. F. (2006). Performance of subjects fit with the Advanced Bionics CII and Nucleus 3G cochlear implant devices. *Archives Otolaryngology—Head and Neck Surgery*, *130*, 624–628.

Svirsky, M. A., Robbins, A. M., Kirk, K. I., Pisoni, D. B., & Miyamoto, R. T. (2000). Language development in profoundly deaf children with cochlear implants. *Psychological Science*, *11*(2), 153–158.

Tyler, R. S., Parkinson, A. J., & Woodworth, G. G. (1997). Performance over time of adult patients using the ineraid or nucleus cochlear implant. *Journal of the Acoustical Society of America*, *102*, 508–522.

Zeng, F. G., & Liu, S. (2006). Speech perception in individuals with auditory neuropathy. *Journal of Speech, Language, and Hearing Research*, *49*(2), 367–380.

CHAPTER 23

# Microstimulation of visual cortex to restore vision

Edward J. Tehovnik[1,*], Warren M. Slocum[1], Stelios M. Smirnakis[2] and Andreas S. Tolias[2]

[1]*Department of Brain and Cognitive Sciences, Massachusetts Institute of Technology, Cambridge, MA, USA*
[2]*Department of Neuroscience, Baylor College of Medicine, Houston, TX, USA*

**Abstract:** This review argues that one reason why a functional visuo-cortical prosthetic device has not been developed to restore even minimal vision to blind individuals is because there is no animal model to guide the design and development of such a device. Over the past 8 years we have been conducting electrical microstimulation experiments on alert behaving monkeys with the aim of better understanding how electrical stimulation of the striate cortex (area V1) affects oculo- and skeleto-motor behaviors. Based on this work and upon review of the literature, we arrive at several conclusions: (1) As with the development of the cochlear implant, the development of a visuo-cortical prosthesis can be accelerated by using animals to test the perceptual effects of microstimulating V1 in intact and blind monkeys. (2) Although a saccade-based paradigm is very convenient for studying the effectiveness of delivering stimulation to V1 to elicit saccadic eye movements, it is less ideal for probing the volitional state of monkeys, as they perceive electrically induced phosphenes. (3) Electrical stimulation of V1 can delay visually guided saccades generated to a punctate target positioned in the receptive field of the stimulated neurons. We call the region of visual space affected by the stimulation a *delay field*. The study of delay fields has proven to be an efficient way to study the size and shape of phosphenes generated by stimulation of macaque V1. (4) An alternative approach to ascertain what monkeys see during electrical stimulation of V1 is to have them signal the detection of current with a lever press. Monkeys can readily detect currents of 1–2 μA delivered to V1. In order to evoke featured phosphenes currents of under 5 μA will be necessary. (5) Partially lesioning the retinae of monkeys is superior to completely lesioning the retinae when determining how blindness affects phosphene induction. We finish by proposing a future experimental paradigm designed to determine what monkeys see when stimulation is delivered to V1, by assessing how electrical fields generated through multiple electrodes interact for the production of phosphenes, and by depicting a V1 circuit that could mediate electrically induced phosphenes.

**Keywords:** striate cortex; phosphenes; saccadic eye movements; forelimb movements; electrical microstimulation; rhesus monkeys; cortical visual prosthesis

---

*Corresponding author.
Tel.: +1 617 324 3726; Fax: +1 617 324 3725;
E-mail: tehovnik@mit.edu

DOI: 10.1016/S0079-6123(09)17524-6

## Introduction

Blindness is a condition suffered by over 1 million people in the United States and by over 40 million people throughout the world (Leonard, 2002). As early as 1968, Brindley and Lewin (1968a, 1968b) implanted an array of electrodes onto the surface of striate cortex (i.e., area V1) of a blind patient with the ultimate aim of restoring visual function to the blind. V1 is the first station of the visual pathway that receives an integrated visual signal from the two eyes before relaying this signal to higher cortical areas (Hubel and Wiesel, 1977; Miezin et al., 1981; Trotter et al., 2004). Also this portion of neocortex contains the highest density of cortical neurons (O'Kusky and Colonnier, 1982; Rockel et al., 1980), presumably to ensure that the visual scene can be analyzed at a fraction of a minute of visual angle (Levi et al., 1985). When trains of electrical pulses were delivered through a given electrode situated on the surface of V1, the blind patient reported the presence of a bright punctate spot of light — called a phosphene — whose size and visual-field position varied according to what region of the cortical topography was stimulated (Brindley and Lewin, 1968a). Stimulation near the foveal representation of V1 (coding for the center of gaze) produced a phosphene in the center of visual space, whereas stimulation far from the foveal representation of V1 produced a phosphene in peripheral regions of visual space. Centrally evoked phosphenes tended to be smaller ($<0.5°$ of visual angle) than peripherally evoked phosphenes ($>0.5°$ of visual angle). Additionally, the phosphenes tended to persist for the duration of electrical stimulation.

Since the time of Brindley and Lewin (1968a, 1968b), Dobelle and colleagues (Dobelle and Mladejovsky, 1974; Dobelle et al., 1974, 1976) have continued this line of work with limited success. A major impediment arose because their electrode arrays were situated on the surface of the cortex above the pia. This had two consequences: First, in order to affect the underlying cortical tissue, currents in the milliampere range were needed to evoke phosphenes (Brindley and Lewin, 1968a; Dobelle and Mladejovsky, 1974; Dobelle et al., 1974, 1976; Evans et al., 1979; Girvin et al., 1979; Rushton and Brindley, 1977). The use of such high currents hampered the effectiveness of the electrode arrays due to poor spatial resolution. Electrodes had to be many millimeters apart in order for distinct parts of the visual field to be activated. Furthermore, the use of high currents could activate large portions of the visual field. Second, the high currents could cause pain in patients due to cross-dural excitation (Brindley and Lewin, 1968a; Dobelle, 2000; Rushton and Brindley, 1978).

Schmidt et al. (1996) overcame the shortcomings associated with stimulating the surface of V1 using high currents by implanting an array whose electrodes penetrated the gray matter and resided some 1–2 mm below the cortical surface. Systematic experiments performed on one patient supporting such an implant taught several things: First, currents below $30\,\mu A$ — and as low at $2\,\mu A$ — were now effective at evoking phosphenes. Second, the minimal inter-electrode spacing to resolve two distinct phosphenes was reduced from about 3 mm (using surface stimulation) to about 0.5 mm using depth stimulation. Third, for currents below $10\,\mu A$, colored phosphenes could be evoked suggesting that at such low currents elements coding for specific features could be recruited. For over 10 years now, no additional studies have been forthcoming on the effectiveness of electrical microstimulation for the elicitation of phosphenes from humans using indwelling cortical electrodes. The goal of developing a visual prosthesis for the blind using a cortical implant has not been realized. Efforts to implant electrode arrays in earlier portions of the visual pathway (i.e., retina and retinothalamic pathway) have also not restored even minimal vision to the blind (Zrenner, 2002). Retinal prostheses are not possible in patients with retinal damage involving all retinal layers and stimulation of the optic nerve with a spiral cuff electrode results in the evocation of multiple phosphenes, thereby yielding poor spatial resolution (Veraart et al., 1998, 2003). A thalamic prosthesis is limited by the challenge of having many indwelling electrodes passing deeply into the brain to activate a small region within the thalamus (Cohen, 2007; Pezaris and Reid, 2007) and by the problem of activating fibers of passage that can yield distorted percepts (Marg and Dierssen, 1965).

An often overlooked impediment to the development of an effective cortical prosthesis for the blind is the fact that the topographic map of V1 is anchored to the eyes and not to the head, body, or external world. The position of the visual receptive fields of the neurons composing the map are fixed with respect to the fovea (Daniel and Whitterridge, 1961; Dow et al., 1981; Hubel and Wiesel, 1968, 1974a, b; Schiller et al., 1976a, b; Tehovnik et al., 2005b) and electrical stimulation of V1 neurons evokes saccadic eye movements that terminate in the center of a receptive field irrespective of the starting eye position (Keating et al., 1983, Keating and Gooley, 1988; Schiller, 1972, 1977; Tehovnik et al., 2003b). Furthermore, continued stimulation elicits a succession of saccades each of which exhibits the same size and direction in register with the receptive field of the stimulated neurons at the time of fixation (Keating et al., 1983; Keating and Gooley, 1988; Schiller, 1972, 1977). Therefore it is not surprising that the perceived position of a cortically induced phosphene is anchored to the fovea and thus a phosphene's position changes with eye movements (Brindley and Lewin, 1968a; Schmidt et al., 1996). This issue was best appreciated by Rushton and Brindley (1977) who found that when mapping the position of an evoked phosphene the stimulation could compel a blind subject to direct his gaze in the direction of the phosphene thereby causing an underestimation of the true position of the phosphene with respect to the fovea. This means that the position of the eyes with respect to the external world must be stationary during a bout of V1 stimulation so that an evoked phosphene is perceived as being stationary rather than as streaking across the visual field as would occur during eye movements. Indeed, it has been suggested that the V1 map is refreshed between visual fixations by a projection from the omnipause neurons in the brainstem (Yang et al., 2008).

A consequence of having an eye-centered map within V1 is that the depth location of a phosphene might be expected to change with changes in the vergence angle, which is accompanied by changes in the depth of the fixation. It is well known that when a single V1 site is stimulated in a human (whose eye movements are not under control) the perceived depth location of an evoked phosphene can change between trials (Rushton and Brindley, 1977). When several sites are stimulated simultaneously, the induced phosphenes typically occur along the same depth plane (Dobelle and Mladejovsky, 1974; Dobelle et al., 1974; Schmidt et al., 1996). What this means is that the depth of a phosphene is likely anchored to the plane of fixation and if the plane of fixation changes then the depth of the phosphene also changes. Interestingly, Rushton and Brindley (1977) observed that phosphenes generated in the upper field of a blind patient appeared far, whereas those generated in the lower field appeared near. Whether this difference is related to changes in vergence as gaze is directed to the upper and lower field is not known.

The apparent size of an evoked phosphene should change when the fixation plane changes such that near fixations should produce smaller phosphenes than far fixations. This is borne out by observations based on the study of afterimages as well as on spontaneously induced phosphenes and on phosphenes elicited by transcranial magnetic stimulation (Cowey and Walsh, 2000; Grüsser, 1991; Richards, 1971). It has been observed that the apparent size of an evoked phosphene can change between different test sessions (Rushton and Brindley, 1977). Some of this variability might be due to changes in the plane of fixation across sessions.

A major shortcoming of much of the foregoing work conducted over some 50 years is that investigators had no access to an animal test case by which electrode array effectiveness and safety as well as eye movement control could be optimized before implanting a device in a blind patient. It is well accepted that the visual system of the macaque monkey is comparable to that of the human, especially regarding V1 (Cowey, 1979; Golomb et al., 1985; Jacobs, 1981; Polyak, 1957; Sereno et al., 1995; Tootell et al., 2003). Much of what we know about the physiology of the human visual system is based on the seminal electrophysiological experiments of Hubel and Wiesel

(1977). Their experimental results were derived largely from studies of the macaque visual system.

Primate V1 is a superb region of neocortex for conducting work toward the development of a visual prosthesis for the blind for the following reasons: (1) Representation out to 7° of visual field is located on the cortical surface of macaque V1 (Hubel and Wiesel, 1977); therefore, electrodes can penetrate this region readily to have access to all its layers. This is not true for V2 or V3, for example, whose layers are buried in a sulcus. (2) The highest density of neurons in neocortex (i.e., number of neurons per degree of visual angle) devoted to representing the visual field is found in V1 (Barlow, 1981; O'Kusky and Colonnier, 1982; Rockel et al., 1980; Sereno et al., 1995; Van Essen et al., 1992). This density is over 100 times greater than that found for neurons making up the optic tract or the lateral geniculate nucleus (Barlow, 1981). The greater density of V1 neurons should enhance the spatial resolution of percepts evoked by stimulation in V1 as compared to percepts induced by stimulation of the retino-geniculate pathway. (3) Neurons within V1 have the smallest visual receptive fields within neocortex (Felleman and Van Essen, 1991). If receptive field size were to limit the minimal size of percepts evoked from neocortex, as has been commonly assumed, then stimulating V1 over other cortical areas should elicit a percept of the highest spatial resolution (but see Tehovnik and Slocum, 2007c for arguments as to why receptive field size might not be critical in determining phosphenes size). (4) The microstructure of V1 has been worked out to the greatest detail (Hubel and Wiesel, 1977), which permits the generation of well-developed hypotheses on how stimulation affects tissue to induce percepts (Tehovnik and Slocum, 2007c). (5) V1 has the machinery to generate percepts in depth using stereopsis (Poggio and Fischer, 1977). This is not true for earlier stations along the retinostriate pathway. (6) Percepts generated by stimulations of V1 are not altered by the cognitive state of the subject; this is less true for regions outside of V1 such as the temporal lobes (Penfield and Perot, 1963; Penfield and Rasmussen, 1957). (7) Electrical stimulation of V1 is still effective at evoking phosphenes after V1 has been visually deprived for decades (Brindley, 1972; Dobelle, 2000; Dobelle et al., 1974, 1976; Schmidt et al., 1996). Inducing phosphenes by retino-geniculate stimulation following blindness can be made ineffective due to retinal degeneration (Bartlett et al., 2005; Santos et al., 1997; Stone et al., 1992), which often necessitates the use of excessive currents (e.g., 200–6000 µA) and extremely long pulse durations (e.g., 1–8 ms) to evoke phosphenes (Humayun et al., 1999, 2003; Rizzo et al., 2003).

Some have suggested that testing of prosthetic devices should be done on humans only (e.g., Weiland and Humayan, 2003). We believe that adopting this approach in the absence of animal experimentation will at best significantly delay progress. It is noteworthy that some 10 years of development using behaving animals as test subjects preceded the implantation and testing of a cochlear device in humans that decades later has been used successfully to restore function to the hearing impaired (Clark, 2003, 2006; Clark et al., 1972). So far, the field of visuocortical prosthetics has opted for the reverse approach. As mentioned, Schmidt et al. (1996) tested the effects of stimulating through an array of electrodes implanted in V1 of a blind patient. This work ended due to uncertain complications (Schiller and Tehovnik, 2008). The work was followed 9 years later by the implantation in a monkey with a similar device that had previously been implanted in the blind patient (Bradley et al., 2005). Some 5 months after the electrodes had been implanted in the monkey, the animal became lethargic due to fluid build up around the electrodes. Following recovery, the animal was left with a persistent upward nystagmus and visual-field defects. Subjecting blind humans to cortical implants that have not been perfected in animals should probably be deemed premature at this time. Some hundreds of monkeys may eventually be required to perfect a visual cortical prosthesis for the blind. Getting federal approval to test one patient with an implant may be considered a major achievement given some of the current difficulties with device safety (Bradley et al., 2005; Schmidt et al., 1996). With continued experimentation and testing on animals (and later on humans), however, these difficulties will likely be overcome as they eventually were for the cochlear implant (Clark, 2003).

The following sections describe three behavioral paradigms — electrically evoked saccades, electrically evoked saccadic delays, and the electrically evoked detection response — that have been used to assess what a monkey experiences during microstimulation of V1. We argue that these paradigms can be used on both sighted monkeys and monkeys with visual-field scotomas to determine the size and shape of phosphenes evoked from V1, to assess whether the evoked phosphenes exhibit visual features, and to ascertain what elements within V1 generate phosphenes.

**Microstimulation of V1 elicits saccadic eye movements**

The knowledge that eye movements can be elicited electrically from V1 has been known about for over 100 years (Doty, 1965; Grünbaum and Sherrington, 1901, 1903; Schäfer, 1888; Wagman, 1964; Wagman et al., 1958; Walker and Weaver, 1940). Currents from 2 to 3000 μA have been used to evoke saccadic eye movements from V1 (Keating et al., 1983; Schiller, 1972, 1977; Tehovnik et al., 2003b). This broad range of currents has been attributed to the behavioral state of the animal, which can radically alter the current threshold for evoking saccades from V1 (Tehovnik and Slocum, 2004). While monkeys fixate a visual target for juice reward, currents up to 1500 μA fail to evoke saccades from V1; yet by imposing a gap between the termination of the visual target and onset of electrical stimulation, currents under 100 μA are sufficient for evoking saccades (Tehovnik et al., 2003a). Not only does active fixation affect the current threshold for evoking saccades from V1, but parameters such as time of juice delivery, the ratio of stimulation to non-stimulation trials, and the characteristics of the trial types interleaved with the stimulation trials can greatly affect the current threshold for evoking saccades from V1. For example, the current threshold for evoking saccades on blank trials is lowest when interleaved trials require a monkey to generate saccadic eye movements to a visual target positioned within the receptive field of the stimulated V1 cells, and this threshold increases as the visual target is positioned further away from the receptive field (Tehovnik et al., 2003a).

By imposing a gap between the fixation period and onset of electrical stimulation and by interleaving trials in which a monkey is required to generate a saccade to a target falling in the receptive field of the stimulated neurons, we have been able to study the excitability properties of the neurons in V1 that mediate saccadic eye movements using currents below 30 μA and as low as 2 μA (Tehovnik et al., 2003a, b). Such currents are confined to an ocular dominance column and therefore can be used to disclosed laminar effects (Tehovnik et al., 2006). The lowest current thresholds ($<5\,\mu A$) for evoking saccades occurred when stimulations were delivered to the deepest layers of V1 between 1.5 and 2 mm below the cortical surface. Currents of 30 μA failed to evoke saccades from the surface of V1. The excitability of the neurons mediating the saccades was the least for stimulations of the superficial layers of V1 (0–1 mm below V1; mean chronaxie: 0.24 ms) and the most for stimulations of the deepest layers of V1 (1–2.25 mm below V1; mean chronaxie: 0.17 ms). Also anodal pulses were the most effective at evoking saccades from superficial V1; cathodal pulses were the most effective at evoking saccades from deep V1. This means that in superficial V1 the elements mediating saccades are composed mainly of cell bodies and terminals and that in deep V1 the elements mediating saccades are composed mainly of axons (Fritsch and Hitzig, 1879; Porter, 1963; Ranck, 1975). The shortest latencies for evoking saccades from V1 (i.e., 50 ms or so) occurred for stimulations of the deepest layers (Tehovnik et al., 2003b). It is believed that the saccade signal generated in V1 gains access to the saccade generator in the brainstem by way of corticotectal axons given that lesions of the superior colliculus abolish all saccades evoked electrically from V1 (Keating et al., 1983; Keating and Gooley, 1988; Schiller, 1977).

Some might say that the electrical evocation of saccades from V1 indicates that such stimulation is generating a motor signal within V1. As indicated earlier, the current threshold for evoking saccades from V1 is greatly affected by the behavior state of the animal. Active fixation can increase the current threshold for evoking saccades from V1 by over 40 times, whereas such fixation only increases the current threshold for evoking saccades from the frontal eye fields and superior colliculus by 2 to 3 times (Sparks and Mays, 1983; Tehovnik et al., 1999, 2003a). This difference we believe is related to the fact that unlike V1, the frontal eye fields and superior colliculus have direct access to the saccade generator in the brainstem for generating motor commands.

Recently, Pezaris and Reid (2007) were able to evoke saccadic eye movements from the lateral geniculate nucleus, a structure that is even more removed from the saccade generator in the brainstem than is area V1. To increase the chances of evoking saccades from lateral geniculate nucleus, trials with a visual target falling in the visual receptive field of the stimulated thalamic neurons were interleaved with blank stimulation trials as has been done for V1 stimulation (Tehovnik et al., 2003a). Microampere currents averaging 40 μA were effective at eliciting saccades from the lateral geniculate nucleus on the blank stimulation trials; the latencies of these saccades (i.e., 150 ms or so) were comparable to those of the visually triggered saccades. The comparable latencies were used by Pezaris and Reid (2007) to argue that the electrical saccades were not being evoked per se but were rather being generated volitionally by the monkey in pursuit of a stimulation-evoked phosphene confined to the receptive field of the stimulated thalamic neurons.

The excitability — or chronaxies — of the directly stimulated elements within V1 for the generation of saccades are similar to those described for elements generating phosphenes by V1 stimulation in humans (i.e., between 0.1 and 0.4 ms; Brindley and Lewin, 1968a; Dobelle and Mladejovsky, 1974; Rushton and Brindley, 1978; Tehovnik et al., 2003b). Thus every time a saccadic eye movement is evoked electrically from V1 the monkey is likely in pursuit of a phosphene. To date, there is no evidence that V1 contains motor neurons that fire during saccadic eye movements made into the motor field in the absence of visual stimuli, as is true for the frontal eye fields and superior colliculi (Boch, 1986; Bruce and Goldberg, 1985; Schiller and Koerner, 1971; Sparks et al., 1976; Sparks and Mays, 1980; Wurtz and Goldberg, 1971; Wurtz and Mohler, 1976). Thus in the absence of such motor neurons it is highly unlikely that saccades evoked electrically from V1 are due to the elicitation of a motor command.

It takes a battery of tests to establish that electrical stimulation of macaque V1 indeed generates a phosphene (Tehovnik et al., 2005a). Figure 1 illustrates the results of V1 stimulation under different task conditions. The probability of saccades made into the visual receptive field of the stimulated neurons is plotted for each task condition (Fig. 1, left). Task (a) and (b) were conditions in which a visual target was displayed and saccades were made in response to the target. Stimulation occurred in condition (b) but not condition (a). Not surprising, regardless of stimulation, the percentage of correct saccades was very high (>93%) in both conditions. Task (c) was identical to (b) except that the visual target was absent. Under this condition, the monkey was able to detect a phosphene and make a correct saccade; the percent of correct saccades (those made into the receptive field) was relatively high (66%). In contrast, in conditions (d) and (f) in which visual targets and stimulation were both absent, the percentage of correct saccades was very low. This result rules out the possibility that the monkey was conditioned to make saccades despite the absence of a phosphene or visual target. Finally, in task (e) the juice reward was delivered as soon as the fixation spot was extinguished. Despite the delivery of stimulation, correct saccades were virtually absent (6%). This outcome occurred because the juice reward effectively terminated the monkey's pursuit of phosphenes, ruling out the possibility that the injected signal evoked motor commands to initiate saccades. Similar conclusions were reached in studies in which saccades were replaced by limb movements to register the detection of phosphenes (Bartlett et al., 2005; DeYoe et al., 2005). These observations are in line with the

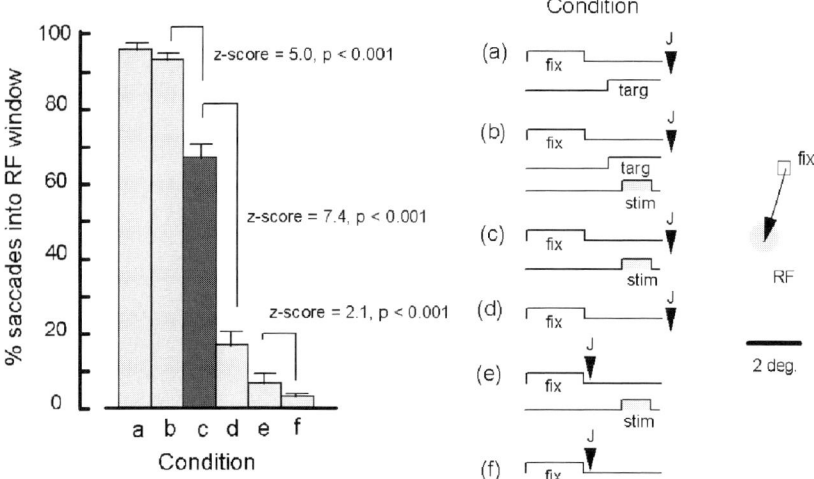

Fig. 1. Left: The percentage of saccade evoked electrically from V1 and made into the receptive field (RF) under six task conditions (right: a–f). Task conditions were randomly selected with the same block of testing. Each condition was tested for 120 trials (20 trials per condition per block for a total of six blocks). The Z-statistic was used to assess the effect of condition on the visually- and electrically evoked saccades terminating in the receptive field of the stimulated V1 cells. The receptive field of the stimulated neurons at the electrode tip was located 2.6° in eccentricity and 265° of meridian, as schematically illustrated on the extreme right in reference to the position of the fixation spot (fix). A visual target (targ) in conditions (a) and (b) was displayed in the location of the receptive field 100 ms following the termination of the fixation spot. Stimulation (stim) was delivered 130 ms after termination of the fixation spot. A drop of apple juice (j) was delivered after the monkey's eyes entered the target window (a–d) or immediately after the termination of the fixation spot (e and f). The depth of stimulation was 1.25 mm below the cortical surface. Stimulation current, pulse duration, pulse frequency, and train duration were 30 μA, 0.2 ms, 200 Hz, and 100 ms, respectively, in the form of anode-first pulses. Adapted from Tehovnik et al. (2005a).

finding that monkeys can detect currents as low as 1–2 μA (DeYoe et al., 2005; Murphey and Maunsell, 2007; Tehovnik et al., 2005a), which is believed to be below the threshold to evoke any non-volitional movements from neocortex (Chen and Tehovnik, 2007).

To establish further that phosphenes are evoked by stimulation of macaque V1, a double-step paradigm has been used (Schiller et al., 2005). Monkeys were trained to generate two saccadic eye movements in succession to two briefly presented targets. Both targets were extinguished well before any eye movements were generated. The first target (T1) was presented several degrees lateral and left of the fixation spot, and the second target (T2) was presented in the receptive field of the stimulated V1 cells under study at the time of initial fixation (Fig. 2, left-top). On some trials, the monkey was required to generate remembered saccades that terminated on target T2′, which was situated outside the receptive field of the stimulated neurons at the time of initial fixation (Fig. 2, left-bottom). Once trained, a monkey had no difficulty performing this task from memory. During stimulation trials (Fig. 2, right), the second target (T2) was replaced by stimulation of the neurons coding for the receptive field location at the time of initial fixation. It was found that even after a saccade was generated to the first target (T1) the monkey factored in this displacement and generated a saccade back to the receptive-field location at the time of initial fixation at stim-2 (Fig. 2, right-top) rather than to stim-2′ (Fig. 2, right-bottom). This suggests that the electrical stimulation of V1 was generating a phosphene anchored to the V1 map such that any eye displacement was compensated. Similar results have been found for stimulations of the lateral geniculate nucleus (Pezaris and Reid, 2007).

It is believed that the double-step test should distinguish target location with respect to sensory

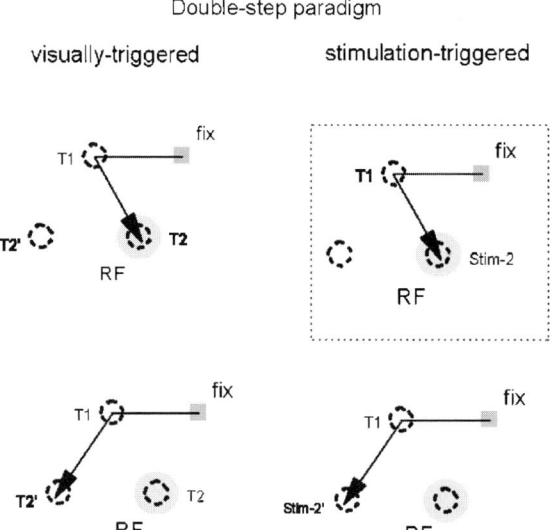

Fig. 2. Double-step paradigm. Monkeys were trained to generate saccadic eye movements to two visual targets flashed in the visual field such that the 2nd target often fell in the visual receptive field of the V1 cells under study during initial fixation (fix, left-top). On some trials the 2nd target was presented outside the visual receptive field at T2′ (left-bottom) or at one of two other locations in the upper left field (not shown). The first target (T1) was flashed for 50 ms and the 2nd target (T2) was flashed for 100 ms. The monkey had 500 ms to acquire the 2nd target in order to obtain a juice reward. On stimulation trials (right), the 2nd target was replaced by a 50-ms train of electrical stimulation. Following stimulation, the monkey always generated the 2nd saccade to the receptive-field location of the stimulated cells at the time of initial fixation (right-top) rather than to a target location outside of the receptive field location (right-bottom). On blank trials in which no stimulation was delivered no saccades were generated to any target locations within the 500-ms choice period. Adapted from Schiller et al. (2005).

space (i.e., proof for phosphene induction) from target location with respect to motor space (i.e., proof for motor induction). In the latter case, any changes due to eye displacement should not be factored in. Therefore the stimulation should generate a motor vector invariant to changes in eye position (Fig. 2, right-bottom, stim-2′). This might lead one to think that stimulations of the frontal eye fields and superior colliculus during the double-step task should produce results that differ from those found for V1 and the lateral geniculate nucleus. Stimulations of the frontal eye fields and superficial superior colliculi have produced results comparable to those reported for V1 and the lateral geniculate nucleus in that the generated eye movements factor in eye displacement (Dassonville et al., 1992; Schlag et al., 1989; Schlag-Rey et al., 1989). Stimulation of the deepest layers of the colliculus produced motor vectors, however, with no correction for eye displacement, presumably due to the stimulating electrode being close to the saccade generator in the brainstem (Fig. 2, right-bottom, stim-2′).

One might suspect that in order to determine what a monkey sees when currents are being delivered to V1, the use of saccadic eye movements as the output metric to make this assessment might be somewhat limited. At the very least, the visual input should be divorced from the motor output by imposing a gap between the stimulation and the response using eye movements (Bradley et al., 2005). This would guarantee that the response was a measure of the monkey's choice rather than an electrically induced motor response. Nevertheless, in the absence of such controls, Schiller et al. (2005) have used saccadic eye movements in combination with microstimulation of area V1 to discern what a monkey sees using a saccade-based, target-choice paradigm (Fig. 3, right). So far the preliminary results are promising and consistent with what one might predict based on the known

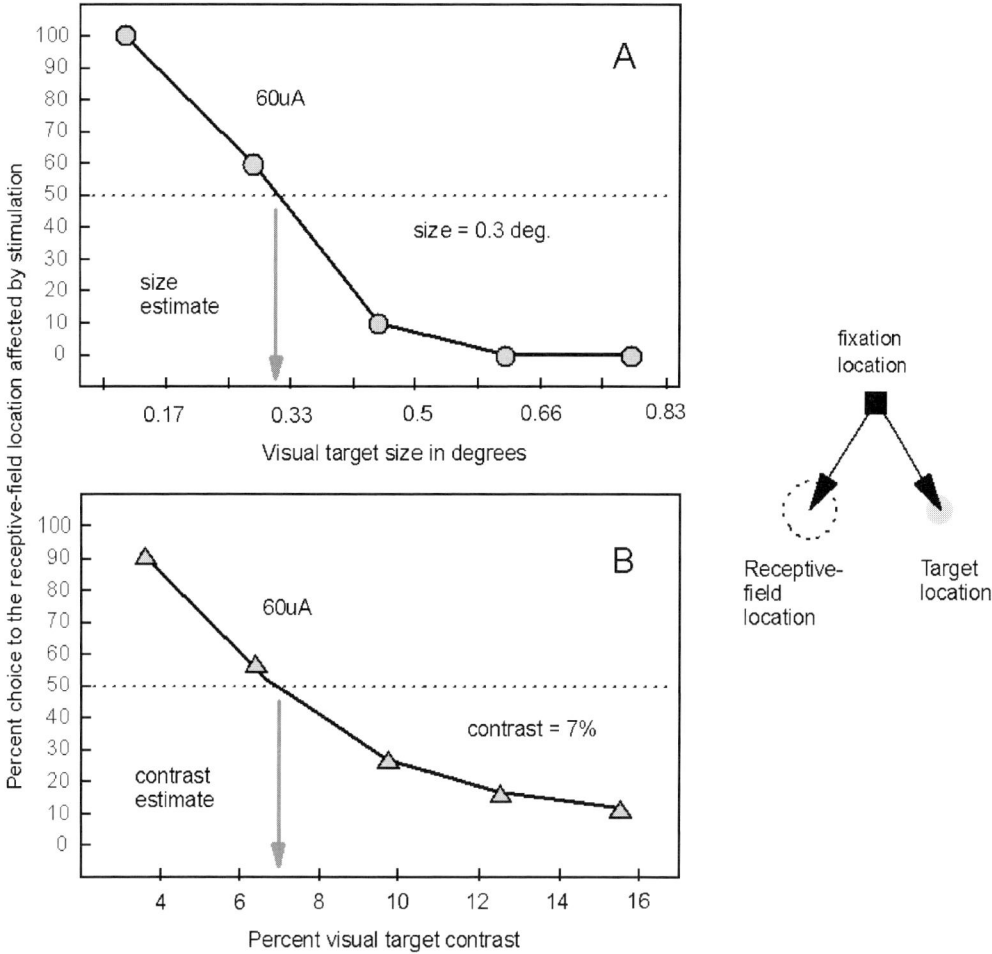

Fig. 3. (A) Percent choice to the receptive-field location affected by stimulation is plotted as a function of the size of visual target (at 10% positive contrast) presented in the right field opposite to that of the receptive field of the directly stimulated V1 cells (right inset). Using 60-μA pulses delivered at 200 Hz embedded in an 80-ms train, the phosphene generated was 0.3° in size as indicated by the 50% crossover point (arrow). (B) Percent choice to the receptive-field location affected by stimulation is plotted as a function of target contrast (at 0.3° in size) presented in the right field opposite to that of the receptive field of the directly stimulated V1 cells (right inset). Using 60-μA pulses delivered at 200 Hz embedded in an 80-ms train, the phosphene generated was 7% in contrast as indicated by the 50% crossover point (arrow). The receptive-field location of the stimulated neurons was 2.8° of eccentricity. At the stimulated V1 location the receptive field is about 0.3° in diameter (Dow et al., 1981; Hubel and Wiesel, 1974b) and is modulated by a stimulus contrast of 2–20% (Albrecht and Hamilton, 1982). Adapted from Schiller et al. (2005).

properties of V1 visual receptive fields. A 60 μA current induced a 0.3° (Fig. 3A), low-contrast stimulus (7%, Fig. 3B) of nondescript polarity that conformed to the size of a visual receptive field of about 0.3° of visual angle.

One challenge in using saccadic eye movements as a metric to assess what a monkey sees when delivering currents to V1 (or any other part of the retino-geniculostriate pathway) is to have the ability to suppress all motor effects that might arise from the elicited effect of stimulation to reach the saccade generator in the brainstem as a motor command (Fig. 4, right). Stimulations delivered to brainstem structures that are connected directly to the eye musculature produce motor responses that are little altered by the

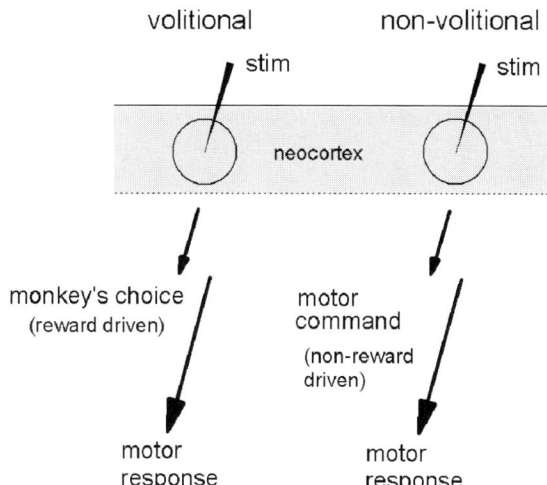

Fig. 4. Two schemes are illustrated for the electrical evocation of motor responses from the neocortex of monkeys. On the right, the stimulation (stim) evokes a motor response that is non-reward driven and that is described as being mediated outside of the animal's volitional state as a motor command. On the left, the stimulation evokes a motor response that is reward driven and is a reflection of a monkey's choice. Using behavioral paradigms that make distinctions between these two extremes is central to making clear-cut interpretations about the effects of electrical microstimulation (Chen and Tehovnik, 2007; Tehovnik and Slocum, 2004).

volitional state of the animal (Tehovnik and Slocum, 2004). As the stimulating electrode is positioned further from the motor output, rigorous behavior controls must be added to any behavioral paradigm to prevent against misattributing an elicited motor response to a volitional act (Chen and Tehovnik, 2007). Suppression of a motor response using reward contingencies (e.g., Fig. 1, condition e) allows one to say that the monkey is choosing a phosphene rather than having its eyes driven electrically to an endpoint as specified by a motor command. Another way is by choosing a behavioral response that is unlikely to be evoked as a purely motor response (Tehovnik and Slocum, 2004). As mentioned, by imposing a gap between the effect of stimulation and the behavioral response (Bradley et al., 2005) or by requiring the animal to generate a behavioral response that differs from the evoked response (Bartlett et al., 2005; DeYoe et al., 2005; Murphey and Maunsell, 2007) breaks the connection between the stimulation and the response so that one may infer that the animal is generating the behavior (Fig. 4, left). The next sections describe behavioral paradigms that have been designed to determine what a monkey sees without evoking saccadic eye movements electrically.

## Microstimulation of V1 delays the execution of visually guided saccades

Not only does electrical microstimulation of V1 evoke saccadic eye movements but it can also disrupt visual processing if delivered at the appropriate time (Morell and Naitoh, 1962; Schiller and Tehovnik, 2001; Slocum and Tehovnik, 2004; Tehovnik and Slocum, 2003; Tehovnik et al., 2002; Ward and Weiskrantz, 1969). If electrical stimulation is delivered during active fixation the subsequent generation of a saccadic eye movement to a visual target situated in the receptive field of the stimulated neurons is delayed (Tehovnik and Slocum, 2005, 2007a–d; Tehovnik et al., 2004, 2005b). We call the region of visual space affected by the stimulation a *delay field*. It is noteworthy that such stimulation also delays forelimb movements, which means that the delay effect is due to a disruption of the visual signal as it is transmitted along the geniculostriate pathway (Tehovnik and Slocum, 2007b).

The size of a delay field is found to vary with the receptive-field eccentricity of the stimulated V1 cells such that the size increases as the site of stimulation within the operculum of V1 is more distant from the foveal representation within V1 (Fig. 5B). Anode-first pulses of 100 μA delivered at 200 Hz were used for these experiments and the depth of stimulation ranged from 0.9 to 2.0 mm below the cortical surface. All stimulations included 100 ms trains commencing at the end of the fixation period immediately before the execution of the visually evoked saccade (Fig. 5A). To map a delay field, a 0.2° target was presented at various locations with respect to the center of the receptive field of the stimulated neurons. Sites coding for a receptive-field eccentricity of 2° exhibited delay fields of 0.14° of visual angle, on average, and sites coding for an eccentricity of 4° exhibited delay fields of 0.35 of visual angle, on average. For receptive-field centers over the range of 1.8°–4.4°, the size of the delay fields varied from a minimum of 0.1° of visual angle to a maximum of 0.55° of visual angle (Fig. 5B).

Since the average receptive-field size of single units in V1 increases with increased distance from the foveal representation (Fig. 5C, I: Hubel and Wiesel, 1974b; Fig. 5C, II: Dow et al., 1981), the size of a delay field might be thought to vary with the size of the visual receptive fields. However, the slope of the function for the size of a delay field across eccentricity differs from that of the function for the size of visual receptive fields (Fig. 5C, I, II). Also at the lowest eccentricities (e.g., 2°), the size of a delay field is less than the size of the receptive fields. Our preferred interpretation (Tehovnik and Slocum, 2007c) is that the size of a delay field varies with the (inverse) retino-cortical magnification factor of V1 (Dow et al., 1981; Hubel and Wiesel, 1974b; Tootell et al., 1988). Note that the concept of the receptive field provides a point-to-point mapping between cortical and visual space (i.e., between a neuron's location in cortex and the center of the neuron's receptive field). By extension, one can map a contiguous region of cortical space to a contiguous region of visual space. One therefore can map the region of current spread (in cortex) to a corresponding region in visual space (the delay field). If one assumes that the distance that current spreads effectively in V1 is invariant for a given current regardless of the location of the stimulation in V1, this model predicts that delay-field size should increase with eccentricity due simply to the retino-cortical magnification factor. If one wishes to use this model to predict delay-field size precisely, one must supply a specific mapping function between cortex and visual space. One such mapping is the complex logarithmic function offered by Schwartz (1994). Using this model, with a realistic model parameter for macaque ($a = 0.3$; Schwartz, 1994), it is possible to calculate the size of a delay field as a function of eccentricity for a given size of current spread. For current equal to 100 μA (which corresponds to activation of V1 tissue in a region with diameter 0.75 mm, see caption of Fig. 5C), anticipated delay-field size as calculated using the Schwartz model is shown (Fig. 5C, III). Note the substantial agreement between the model and the experimental result for delay-field size (Fig. 5C, III and Fig. 5C, delay field at 100 μA). That is, the size of a region of current spread in V1, when mapped to visual space under reasonable assumptions, closely agrees with the size of the corresponding delay field. Certainly, the agreement is better than that between either the Schwartz model or the delay-field data, on the one hand, and receptive-field size on the other hand. Thus the change in delay-field size with the visual-field eccentricity coded by the V1 site stimulated covaries best with the retino-cortical magnification factor.

Five pieces of evidence support the assertion that the delay effect is due to phosphene induction. First, V1 neurons that mediate the delay effect exhibit excitability properties that are similar to those of V1 cells in humans that mediate electrically evoked phosphenes (i.e., chronaxies ranging between 0.1 and 0.4 ms; Brindley and Lewin, 1968a; Dobelle and Mladejovsky, 1974; Rushton and Brindley, 1978; Tehovnik et al., 2004). Second, stimulation of the deepest layers of V1 is the most effective at inducing a delay effect and is the most effective at evoking a phosphene (Bak et al., 1990; Schmidt et al., 1996; Tehovnik et al., 2004). Third,

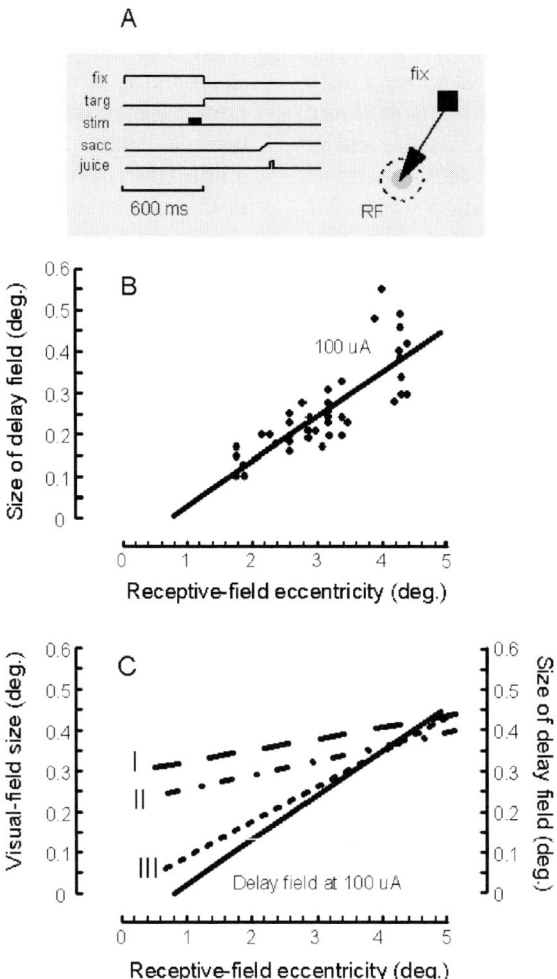

Fig. 5. (A) A delay field was mapped by presenting a 100-ms train of stimulation at the end of the fixation period (fix) immediately before the presentation of the visual target (targ). The animal was given 500 ms to generate a saccadic eye movement (sacc) to the visual target in order to obtain a juice reward (juice). The visual target [0.2 degrees in diameter and at 33% positive contrast (Michaelson)] was presented at various locations with respect to the visual receptive field (RF) of the stimulated V1 cells. (B) Size of the delay field varies with the site of stimulation within the operculum of V1. Size of the delay field is plotted as a function of the eccentricity of the receptive-field center of the V1 cells stimulated. A total of 41 V1 sites (located from 0.9 to 2.0 mm below the cortical surface) were studied. At each site, stimulation was composed of 100-μA anode-first pulses (at 0.2-ms duration) delivered at 200 Hz using a 100-ms train. The solid curve is a regression line representing the data ($r = 0.81$, $n = 41$, $p < 0.01$). (C) The relationship between average receptive-field size (in visual-field coordinates) for macaque V1 and receptive-field eccentricity [(I) Hubel and Wiesel, 1974b; (II) Dow et al., 1981] and the relationship between amount of visual field represented by activation of a 0.75-mm diameter region of macaque V1 and receptive-field eccentricity using the complex logarithmic function of Schwartz (1994) (III) are compared to the relationship between the size of the delay field and receptive-field eccentricity using a 100 μA current (from B). Using the delay-field data of (B), a calculation can be made of how far the 100 μA current was spreading in V1. At an eccentricity of 3°, the size of the visual field affected by the 100 μA current was 0.24° (on average). At a 3° eccentricity, 1000 μm of V1 tissue represents about 0.32° of visual field, which is based on the retino-cortical magnification factor of macaque V1 (Hubel and Wiesel, 1974b; Dow et al., 1981; Tootell et al., 1988). Therefore, 100 μA affects V1 tissue within 375 μm from the electrode tip (i.e., 0.24°/0.32° × 1000 μm/2). A simplified version of the Schwartz (1994) algorithm was used to compute the function (III) in (C) representing the amount of visual field affected by activating a sphere of tissue with radius of 0.375 mm: Visual-field size $= K \times E \times R$, where $K$ is a constant of 0.23°/deg/mm, $E$ (in degrees) the visual-field eccentricity coded by the stimulated cells, and $R$ (in mm) the radial spread of the current, that is, 0.375 mm. Adapted from Tehovnik et al. (2005b).

delay fields tend to be circular in shape and they increase in size with the retino-cortical magnification factor, which concurs with what has been reported for the induction of phosphenes (Brindley and Lewin, 1968a). Fourth, increases in current increase the size of delay fields as well as the size of phosphenes (Dobelle and Mladejovsky, 1974; Rushton and Brindley, 1978; Tehovnik and Slocum, 2007a). Fifth, during migraines associated with the activation of human V1, Grüsser (1995) as well as others (e.g., Airy, 1870; Dahlem et al., 2000; Hadjikhani et al., 2001; Lashley, 1941; Richards, 1971) observed that a temporary scotoma is produced immediately after the evocation of a phosphene (for a review see Grüsser, 1991). The size and shape of the scotoma is the same as that of the phosphene, and both are anchored to the fovea given that their positions shift with changes in gaze or with the tilting of the head. Delay fields, like the visual receptive fields of V1 neurons, are also anchored to the fovea (Tehovnik et al., 2005b). Therefore, every time stimulation is delivered to macaque V1, a phosphene followed by a temporary scotoma of comparable size and shape is presumed to be experienced by the animal.

Given that delay fields and phosphenes exhibit similar properties we believe that the study of delay fields can be used to make inferences about phosphenes evoked electrically from macaque V1. Phosphene size and shape are not limited by the properties of visual receptive fields as has been suggested by some (Maynard, 2001; Warren et al., 2001) but are rather constrained by the retino-cortical magnification factor. The latter can be used to determine the optimal spacing between electrodes for the elicitation of phosphenes. Using currents between 2 and 30 μA delivered between electrodes spaced by 0.5 mm, human subjects never report overlapping phosphenes (Schmidt et al., 1996). Such estimates are consistent with what have been described using delay fields (Fig. 6).

**Microstimulation of V1 elicits a detection response**

Stimulating currents delivered to neocortex (e.g., area V1, extrastriate cortex, parietal and temporal cortex, motor and premotor cortex including the frontal eye fields, etc.) and to subcortical areas (e.g., optic tract, medial and lateral geniculate nuclei, pulvinar, hippocampus, superior colliculus, mesencephalic reticular formation, etc.) can be detected by animals (Bartlett et al., 2005; Doty, 1965; Nielson et al., 1962; Schuckman et al., 1970). Monkeys readily register the detection of electrical stimulation delivered to the brain by producing a lever press in exchange for a food reward, for example. This method has been used to study what monkeys experience perceptually during electrical microstimulation of V1 (Bartlett and Doty, 1980; DeYoe et al., 2005; Murphey and Maunsell, 2007).

A 100-ms train was delivered to V1 such that a monkey was required to depress a lever in response to the stimulation in order to obtain a juice reward (Tehovnik and Slocum, 2009). The deepest layers of V1 were found to be the most sensitive for the elicitation of the detection response. This conclusion is based on four observations. First, the lowest current thresholds for eliciting a detection response occurred when the electrode was positioned from 1 to 2.25 mm below the cortical surface. The thresholds could be as low as 2 μA, but the average minimal threshold was 6 μA. Second, the shortest chronaxies of stimulated V1 neurons mediating the detection response occurred between 1.6 and 2.5 mm below the cortical surface (i.e., ranging from 0.11 to 0.15 ms). Third, the shortest latency for the detection response occurred when the deepest layers of V1 were stimulated. Finally, cathode pulses were more readily detected than anode pulses when delivered to the deepest layers of V1. This suggests that the output axons subserve the detection response (Fritsch and Hitzig, 1879; Porter, 1963; Ranck, 1975). Indeed, transection of the corticofugal fibers from the deepest portions of V1 are known to abolish the detection response (Rutledge and Doty, 1962).

Several studies have been conducted looking at how the current threshold to evoke a behavioral response from V1 varies with cortical depth (Bak et al., 1990; Schmidt et al., 1996; Tehovnik et al., 2003b, 2004). In all cases, the current threshold for evoking saccades, saccadic delays,

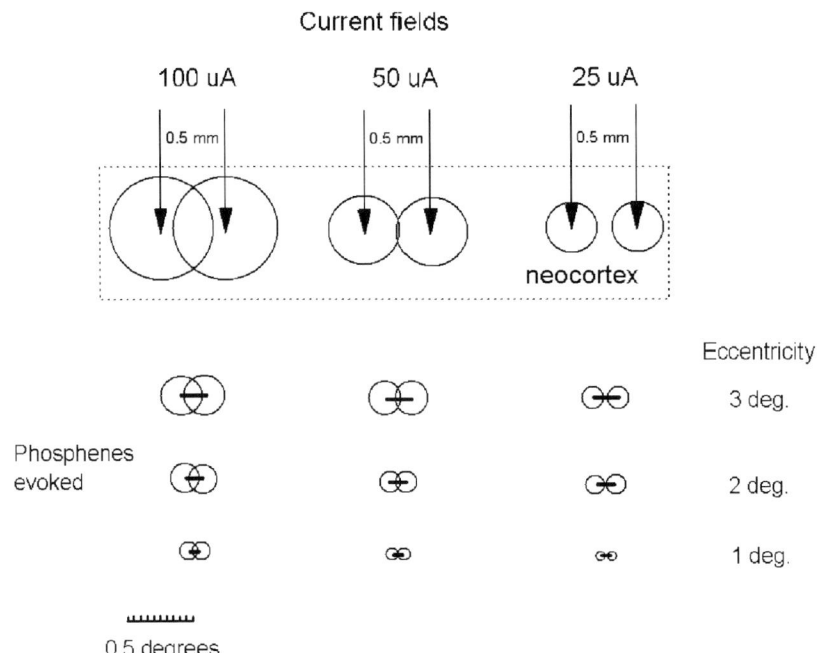

Fig. 6. The size of phosphenes evoked from macaque V1 is illustrated for three levels of current: 25, 50, and 100 μA. The electrodes were spaced by 0.5 mm within the neocortex. Based on delay-field data (Tehovnik and Slocum, 2007d) the current was found to spread by 0.19, 0.27, and 0.39 mm from the electrode tip using the current–distance equation: $I = 675\,\mu\text{A/mm}^2 \times r^2$, where $I$ is the current in microamperes and $r$ the radial spread of current from the electrode tip in millimeters. The current–distance constant of $675\,\mu\text{A/mm}^2$ is derived from experimental data (Tehovnik et al., 2006). Visual-field size was computed using a variation of the Schwartz (1994) algorithm: Visual-field size $= K \times E \times R$, where $K$ is a constant of 0.23°/deg/mm, $E$ (in degrees) the visual-field eccentricity coded by the stimulated cells, and $R$ (in mm) the radial spread of current derived from the current–distance equation. Notice that for currents under 50 μA the current fields are not overlapped; nor are the evoked phosphenes overlapped for the three eccentricities shown.

and phosphenes was the least for stimulations of the deepest layers of V1, which is consistent with what has been observed for V1 stimulation and the detection response (Bartlett and Doty, 1980; Tehovnik and Slocum, 2009). Also the excitability of V1 neurons mediating these responses were found to be comparable, exhibiting chronaxies falling between 0.1 and 0.4 ms (Bartlett et al, 2005; Brindley and Lewin, 1968a; Dobelle and Mladejovsky, 1974; Rushton and Brindley, 1978; Tehovnik et al., 2003b, 2004, 2007). This means that the same V1 neurons mediate saccades, saccadic delays, phosphenes, and the detection response. The chronaxie values indicate that the directly stimulated neurons are composed of pyramidal fibers (Asanuma et al., 1976; Stoney et al., 1968). This conclusion agrees with the observation that many simple and complex cells exhibit a pyramidal morphology whose activation would be expected to generate a phosphene rather than a motor response (Gilbert and Wiesel, 1979; Hubel and Wiesel, 1977).

DeYoe et al. (2005) found that they could evoke a detection response using currents between 1 and 3 μA (at 0.2-ms pulse durations) if delivered to particular laminae within V1 when using high-impedance electrodes (i.e., up to 1.5 MΩ). Of the 16 electrode penetrations made into V1 with tests every 100 μm only two sites exhibited current thresholds below 3 μA. DeYoe et al. concluded from this that the low-threshold sites within V1 are sparsely distributed. We repeated the experiments of DeYoe et al. with slight modifications. First, to assure that our

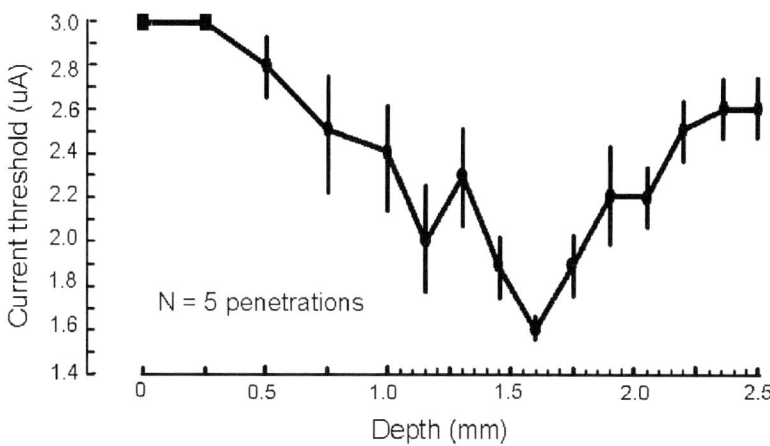

Fig. 7. The current threshold to evoke a detection response on 50% of stimulation trials is plotted as a function of cortical depth. For all experiments ($N = 5$), a 100-ms train composed of anode-first pulses was delivered at 200 Hz. Stimulation was delivered through a platinum–iridium electrode whose impedance value was maintained at 1.8 MΩ. All stimulation occurred 200 ms before the end of the fixation period. The monkey had 800 ms to register a response by depressing a lever following the termination of the fixation spot. Blank trials having no stimulation were interleaved with stimulation trials 50% of the time. The rate of responding on such trials was less than 10%. The square marker indicates that a detection response was not readily evoked from the site using a maximal current of 3 μA. The lowest threshold of 1.6 μA occurred at 1.6 mm below the cortical surface. Standard errors of the mean are shown.

high-impedance electrode (at 1.8 MΩ) was not degraded, the highest current passed through this electrode was 3 μA. We found that if we used currents comparable to the highest currents used by DeYoe et al. (i.e., up to 25 μA) this would melt the electrode tip, drop the impedance to 0.5 MΩ or so, and reduce the effectiveness of the stimulation. Second, to better target the cell bodies with our stimulation, we used anodal pulses instead of cathodal pulses (Fritsch and Hitzig, 1879; Porter, 1963; Ranck, 1975). Using this procedure we found that the lowest current threshold for evoking a detection response occurred some 1.6 mm below the cortical surface (Fig. 7), which is roughly within lamina V of area V1 (Peters and Sethares, 1991) and which is consistent with the original observations of Bartlett and Doty (1980). At this depth, currents of 1.6 μA were effective at evoking a response. Such currents within V1 activate about 47 pyramidal neurons directly [calculation based on 120,000 V1 neurons per mm$^3$ (Peters, 1994); radial spread of 1.6 μA is 0.049 mm [$(I/K)^{1/2}$ = (1.6 μA/ 675 μA/mm$^2$)$^{1/2}$, $I$ = current, $K$ = current–distance constant] (Tehovnik et al., 2006); volume = 4/3π $(0.049 \text{ mm})^3 = 0.00049 \text{ mm}^3$; number of neurons = $0.00049 \text{ mm}^3 \times 120,000$ neurons per mm$^3$ = 59 neurons; pyramidal cells make up 80% of cells in V1 (Peters, 1994): $0.8 \times 59$ neurons = 47 pyramidal neurons]. That activation of such small numbers of pyramidal neurons can be detected by animals has been established using optical stimulation and two-photon microscopy (Huber et al., 2008).

**Can microstimulation of V1 evoke phosphenes exhibiting features?**

Many believe that microstimulation of V1 is limited to the evocation of pixelized vision or a spot of light whose size and brightness can be made to vary by changing the strength of stimulation (e.g., Bartlett et al., 2005; Cha et al., 1992a–c; Chen et al., 2004, 2005a, b, 2006, 2007; DeYoe et al., 2005; Dagnelie et al., 2006, 2007; Fu et al., 2006; Hayes et al., 2003; Schiller et al., 2005; Thompson et al., 2003). This uniformity of perception is believed to arise from activation of common V1 elements no matter where one stimulates in the V1 map (Bartlett et al., 2005; DeYoe et al., 2005). Doty and colleagues have suggested that phosphenes are mediated by excitation of luxotonic neurons, which respond in a sustained fashion to a fixed level of illumination and which are sparsely distributed throughout V1 (Bartlett and Doty, 1974; Kayama et al., 1979; Schiller et al., 1976a). This suggestion, however, goes against the observation that changes in background luminance failed to affect the threshold for detecting currents delivered to V1 (Bartlett et al., 2005) and that most cells in V1 fire transiently to the presentation of visual stimuli confined to their receptive fields (Hubel and Wiesel, 1968; Schiller et al., 1976a, b).

In 2003, Troyk and colleagues proposed that V1 stimulation should evoke featured phosphenes given that cells in V1 code for the stimulus orientation, motion, luminance, and color and that they have convergent inputs from the two eyes to mediate stereoscopic vision (Hubel and Livingston, 1990; Hubel and Wiesel, 1977; Michael, 1981; Poggio and Fischer, 1977; Schiller et al., 1976a, b; Troyk et al., 2003). For current levels greater than 10 µA phosphenes are commonly described as being featureless and circular in shape (Brindley and Lewin, 1968a; Dobelle and Mladejovsky, 1974; Schmidt et al., 1996), whereas for current levels under 10 µA phosphenes are found to exhibit distinct features of red, yellow, or blue (Schmidt et al., 1996). In the latter case such low currents can be confined to sub-portions of a hypercolumn, which contains the neural machinery to represent distinct visual attributes (Hubel and Wiesel, 1977). According to our estimates, to confine current to a group of neurons coding for a common color or line orientation currents of 5 µA or less would be needed (Tehovnik and Slocum, 2007c). The fact that monkeys can reliably detect currents between 1 and 2 µA (Fig. 7) as long as they are delivered to the deepest layers of V1 means that we can now test empirically whether monkeys experience phosphenes exhibiting features. Using such stimulation, monkeys could be trained to match the visual features induced by V1 stimulation with the features coded by the cells at the electrode tip. The precise feature coding of the cells could be ascertained by using conventional electrophysiological methods (Hubel and Wiesel, 1977).

## Effects of blindness on phosphene induction

In intact monkeys, once an animal has been trained to detect stimulation delivered to V1 this response can be transferred immediately to any site within the V1 map of either hemisphere (Bartlett et al., 2005; Doty, 1965, 1969; Schuckman and Battersby, 1966; Schuckman et al., 1970). Such transfers do not occur readily between V1 and extrastriate cortex or between V1 and the lateral geniculate nucleus (Bartlett et al., 2005; Doty, 1969; Schuckman et al., 1970). Furthermore, stimulation of blinded (i.e., retinally deprived) portions of V1 cannot be transferred immediately to intact portions of V1 (Bartlett et al., 2005). This has been interpreted to mean that identical stimulation paradigms generate phosphenes of different quality during blindness and normal vision (Doty, 1965; Schuckman et al., 1970). Indeed it has been found that white phosphenes are often elicited in blind individuals (Brindley, 1972; Brindley and Lewin, 1968a, b; Dobelle et al., 1974), whereas white, black, and colored phosphenes are frequently evoked in sighted persons (Dobelle and Mladejovsky, 1974; Dobelle, 2000; Lee et al., 2000; Penfield and Perot, 1963; Penfield and Rasmussen, 1957; Talalla et al., 1974).

It has also been found that retinal ablation reduces the current threshold for evoking a detection response from V1 (Bartlett et al., 2005), perhaps because V1 now becomes hyperexcitable (Fentress and Doty, 1971; Sakakura and Doty, 1976; Talbot and Marshall, 1941) by virtue of the fact that the gamma-aminobutyric acid (GABA)ergic system is downregulated (Hendry et al., 1990; Jones, 1993; Jones et al., 1994; Rosier et al., 1995). This result agrees with the observation that the threshold to evoke phosphenes from human V1 using transcranial magnetic stimulation is reduced after subjects have been dark adapted (Boroojerdi et al., 2000).

The foregoing observations based on retinal deprivation could be studied systematically using photocoagulation on monkeys. Forming large homonymous retinal lesions in both eyes (corresponding to >1 cm of cortex, Fig. 8) is an effective and long-lasting way of depriving deafferented

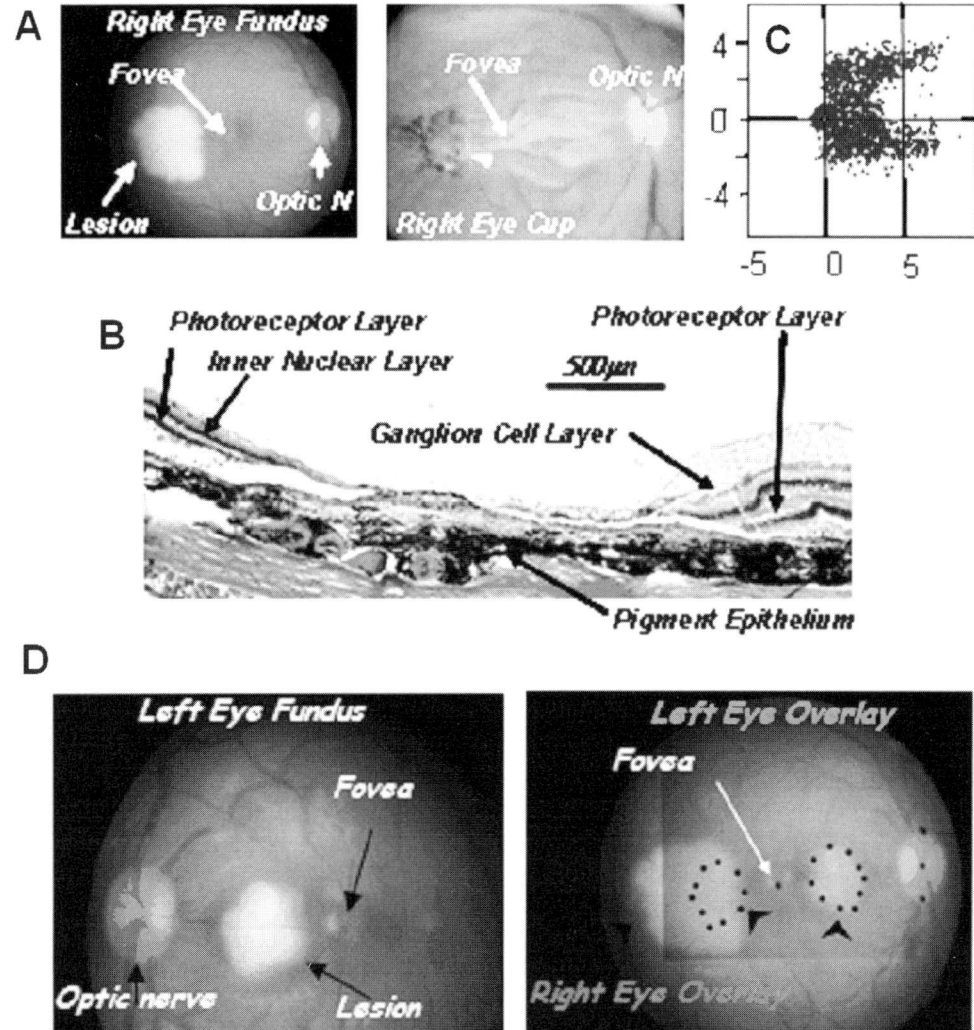

Fig. 8. Retinal lesioning. All lesions were made using a NIDEK GYC-2000 532-nm photocoagulation laser. (A) Left: Picture of the right eye fundus immediately after inducing a laser photocoagulation lesion. The lesion appears white. Right: Picture of the right fundus after extraction and fixation in formaldehyde at the end of the experiment, 9 months later. The monkey had been euthanized and perfused with formaldehyde prior to extraction. Note the hyperpigmented scar (white arrow), which corresponds to the retinal lesion. (B) Fifteen-micrometer thick section through the center of the same lesion, stained with hematoxylin-eosin. Note that all retinal layers are essentially completely destroyed at the center of the lesion. (C) Saccade to visual target task used to map the visual-field scotoma induced in a macaque after a homonymous retinal lesion (D). Adjacent saccade targets were spaced by ~0.5°. Each dot represents a successful saccade to target. Note the absence of saccades outlining the area of the induced scotoma. Scale: Degrees. (D) Illustration of the lesions made in both fundi in order to induce a partly homonymous lesion. Left: Lesion in the left eye fundus. Right: Overlay of the right eye fundus (larger) and the left eye fundus (smaller square) illustrating the lesions (white patches marked by the black arrowheads). The left retina (small square overlay) was mirrored along the vertical axis and scaled to make the optical nerves overlap. Black dots outline the left eye lesion and its homonymous location in the right eye, which lies almost entirely over the right eye lesion resulting in an essentially homonymous left visual field scotoma. Note that lesions in each eye were offset from the fovea such that the foveal representation was spared, thereby allowing the animal to fixate accurately. Adapted from Smirnakis et al. (2005). (See Color Plate 23.8 in color plate section.)

areas of V1 of the capacity for significant visually driven modulation.[1] A photocoagulation laser is used to ablate the photoreceptor and bipolar cell layers of the retina. Two advantages of using this procedure are (I) that by sparing the fovea monkeys can still fixate visual targets, which is a requirement for performing visual tasks and (II) the effect of stimulating deafferented regions of V1 can be compared to stimulation of intact (contralateral) regions of V1 in the same animal under identical experimental conditions. Contrary to this approach, complete blinding can induce a persistent nystagmus that would cause the evoked phosphenes to jerk about in the visual field as the eyes move about (Bartlett et al., 2005; Bradley et al., 2005; Hall and Ciuffreda, 2002; Kömpf and Piper, 1987; Leigh and Zee, 1980), even though such nystagmus is diminished if the blindness occurs in adulthood (Hall and Ciuffreda, 2002; Kömpf and Piper, 1987).

Another way of deafferenting V1 is to entirely destroy the retina of one eye (by way of optic nerve head photocoagulation, for example) while sparing the other eye. Since eye movements are yoked, knowing the position of the intact eye will provide the position of the blinded eye. Using this method the detection threshold, for example, could be determined for the deprived ocular dominance columns versus the non-deprived ocular dominance columns. It is well established that the GABAergic system of the deprived ocular dominance columns is downregulated following monocular deprivation (Hendry et al., 1990; Jones, 1993; Jones et al., 1994). Therefore, stimulation of the deprived columns should result in a lower detection threshold as compared to stimulation of non-deprived columns (Bartlett et al., 2005). Also the detection response evoked by stimulation of deafferented versus non-deafferented columns should not transfer readily given that stimulation of each would be expected to generate qualitatively different phosphenes (Doty, 1965; Schuckman et al., 1970).

It is noteworthy that retinal photocoagulation as a method of inducing a visual-field scotoma has different pathophysiology from naturally occurring ailments that induce blindness (e.g., cataract, glaucoma, macular degeneration, retinitis pigmentosa, diabetic retinopathy, optic nerve atrophy; see Leonard, 2002 for complete list) which have generally different time course and can exhibit defects due to photoreceptor or ganglion cell damage in one or both eyes for central and/or peripheral vision. Nevertheless, since retinal photocoagulation like most conditions of non-cortical blindness deprives area V1 of direct visual input, we expect that the proposed paradigm is going to be a valuable model for studying the basic neurophysiological rules that govern the successful implementation of a visual cortical prosthesis. Performing bilateral homonymous retinal lesions (while sparing the fovea) or disabling one eye entirely by photocoagulating its optic nerve head deprives area V1 of visual input in a controlled fashion, so that the effects of such deprivation on phosphene induction can be studied systematically as the monkey fixates and performs visual tasks. The latter is a central requirement for studying the perceptual characteristics of induced phosphenes.

## Discussion

This review arrives at several conclusions. (1) As with the development of the cochlear implant, the development of a visuo-cortical prosthesis can be

---

[1] It has been reported that 2–6 months following damage of the retinae in adult animals the receptive fields of V1 cells within the region representing the scotoma are altered if corresponding regions of each retina are lesioned such that V1 cells within the scotoma begin to respond to visual stimuli outside of the scotoma (Chino et al., 1992; Darian-Smith and Gilbert, 1995; Gilbert et al., 1990; Gilbert and Wiesel, 1992; Heinen and Skavanski, 1991; Kaas et al., 1990). The extent of reorganization is reported to reach up to but not beyond 5 mm from the border of the lesion (Gilbert et al., 1990). These measurements have recently been challenged by Smirnakis et al. (2005) (as well as Horton and Hocking, 1998) who used fMRI and multiunit electrophysiology measurements to conclude that the extent of reorganization, if present, appears to be <1 mm. In any event, this issue does not affect our discussion here, since we can always adjust the homonymous retinal lesions to create a deafferented region of V1 that lies well beyond the extent of the proposed reorganization (whether this is 0 or 5 mm from the border of the lesion projection zone).

accelerated by using animals to test the perceptual effects of microstimulating V1 in intact and deafferented monkeys. (2) Although a saccade-based paradigm is very convenient in studying the effectiveness of delivering stimulation to V1 to elicit a behavioral response, it is less suitable for probing the volitional state of monkeys as they respond to electrically induced phosphenes. (3) The study of delay fields is an effective way to study the size and shape of phosphenes evoked from macaque V1. (4) An ideal approach to ascertain what a monkey sees when stimulation is delivered to V1 is to have a monkey detect currents delivered to V1 as signaled by a lever press. Monkeys can readily detect currents as low as 1–2 µA. In order to study feature vision currents under 5 µA are necessary. (5) To study the effects of blindness on the ability to generate phosphenes from V1, a macaque model of blindness is proposed. Partially lesioning the retinal input in monkeys while sparing the fovea is superior to completely lesioning this input when assessing the effectiveness of microstimulation delivered to deafferented portions of V1. This allows the deafferented monkey to fixate visual targets and therefore to perform visual tasks. The following section discusses some further issues in using monkeys for the development of a visuo-cortical prosthesis for the blind.

## *Assessing what monkeys sees during microstimulation of V1*

The brightness and size of phosphenes induced from V1 have been studied extensively in human subjects. That the brightness of a phosphenes can be altered by changing parameters of stimulation is well established (Brindley and Lewin, 1968a; Dobelle and Mladejovsky, 1974; Evans et al., 1979; Rushton and Brindley, 1978; Schmidt et al., 1996). The parameters that are the most effective at changing the brightness of phosphenes in both sighted and blind individuals are current and the frequency of pulses. Doubling current increases brightness more than doubling the frequency of pulses, and for frequencies beyond 200 Hz the brightness saturates (Evans et al., 1979). Rushton and Brindley (1978) determined that by manipulating the current level some 12 levels of brightness could be discriminated, which they indicate is far below the levels that can be discriminated by a sighted observer. The main reason for this difference is that phosphenes generated by V1 stimulation do not have access to retinal light and dark adaptive mechanisms which extend to range of ambient light (i.e., by some 10 log units) over which the visual system is responsive. As for phosphene size, Rushton and Brindley (1978) found that increases in current tend to increase the size of a phosphene, whereas Schmidt et al. (1996) found that such increases cause a decrease in size or an increase followed by a decrease (also see Bak et al., 1990). A major difficulty in measuring phosphene size is that size can change with changes in the plane of fixation (Rushton and Brindley, 1977).

By combining the detection of V1 stimulation with the target-choice paradigm (Fig. 9) one can assess the brightness and size of phosphenes evoked from macaque V1. Since the plane of fixation can be controlled in such a paradigm by using eye coils, estimates of phosphene size are precise. Here, a monkey is required to fixate a spot of light on a monitor after which a pair of visual target would be presented. One of the targets of a pair is located within the receptive field of the V1 neurons under study and the other target of a pair is presented in the mirror-opposite hemifield (Fig. 9). In order to obtain a juice reward, a monkey must pull the lever representing the brighter or larger target. When the two targets of a pair are made equivalent, a monkey pulls each lever roughly 50% of the time. During stimulation trials, a train of electrical stimulation is delivered to the V1 cells instead of presenting a visual target within the receptive field of those cells. The visual target in the hemifield opposite the receptive field of the stimulated neurons is manipulated until a point of perceptual equivalency is established between the stimulation-evoked target and the reference target. This point is reached when the monkey generates a lever response to each target location 50% of the time. Brightness and size experiments will need to be conducted as the current and the

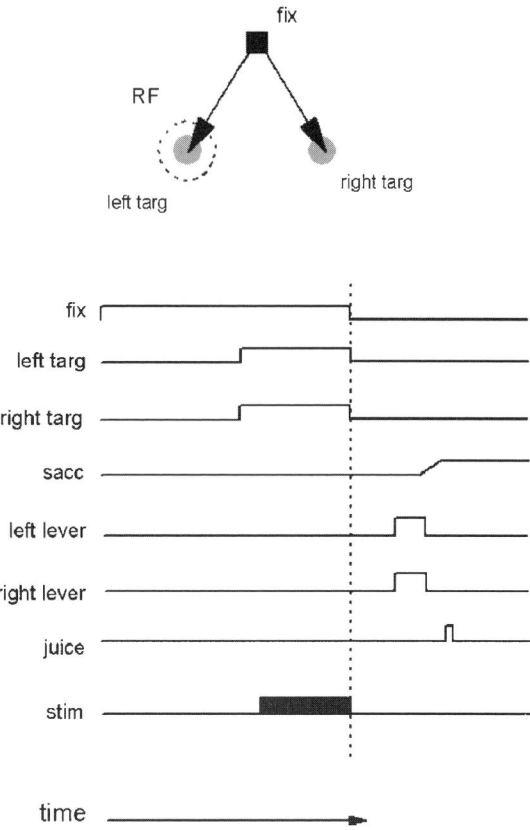

Fig. 9. Detection of V1 stimulation and the target-choice paradigm. Monkeys are required to fixate (fix) a spot of light as targets are presented briefly within each hemifield: the left target (left targ) is situated in the receptive field of the cells under study and the right target (right targ) is positioned in the mirror-opposite hemifield. Following termination of the fixation spot, a monkey must pull the left lever (left lever) if the left target is brighter or larger than the right target and the monkey must pull the right lever (right lever) if the right target is brighter or larger than the left target. A juice reward (juice) is delivered for a correct response. The monkey is free to generate a saccade (sacc) away from the fixation location once the fixation spot has been terminated. On a fraction of trials, electrical stimulation (stim) is delivered to the receptive-field neurons instead of presenting the target (left targ). On stimulation trials all responses are rewarded.

frequency of pulses are manipulated. Also the number of levels of brightness and size that can be discriminated during V1 stimulation will need to be ascertained for monkeys.

Given that blindness is believed to abolish the generation of dark and color phosphenes from V1 while sparing the generation of white phoshenes (Brindley, 1972; Brindley and Lewin, 1968a, b; Dobelle and Mladejovsky, 1974; Dobelle et al., 1974; Dobelle, 2000; Lee et al., 2000; Penfield and Perot, 1963; Penfield and Rasmussen, 1957; Talalla et al., 1974), it would be instructive to use the target-choice paradigm on both blind and sighted monkeys to study this problem. Here the contrast polarity and color of the reference target would need to be varied.

*Generating electric fields through multiple electrodes*

In designing a prosthetic device that will deliver currents through multiple electrodes it is essential to have a clear understanding of how current fields activate neural tissue under different conditions of activation and how these fields behave when currents are delivered serially or simultaneously through each electrode. We have recently used the properties of delay fields to assess how far current spreads in V1 using monopolar stimulation (Tehovnik et al., 2006). This approach has yielded the following formula of current spread ($I = 675\,\mu A/mm^2 \times r^2$, where $I$ is the current in microamperes and $r$ the radial spread of current from the electrode tip in millimeters), which has been used in this review.

Effective current spread is affected by electrode configuration. Current fields become more restricted going from monopolar to bipolar to tripolar stimulation, even though the lowest current threshold for evoking a neural response is with monopolar stimulation (Bierer and Middlebrook, 2002; Chowdhury et al., 2004; Kral et al., 1998). When conducting multipolar stimulation, the anodal pulses at a ground electrode can inhibit axonal firing (Bagshaw and Evans, 1976), which might explain why monopolar stimulation is more effective than multipolar stimulation. Also a current shunt between a closely spaced source and ground electrode can reduce the efficacy of stimulation (Bierer and Middlebrook, 2002; Chowdhury et al., 2004; Kral et al., 1998), but stimulation at the source and ground can also activate tissue at each electrode (Black et al., 1981). Regardless of the number of electrodes passing current, increases in current always results in an increase in the field of activation at the source electrode (Bierer and Middlebrook, 2002, 2004).

Electrically stimulating many sites simultaneously is a requirement for a visuo-cortical prosthesis (Lovell et al., 2005). Simultaneous stimulation between adjacent electrodes can produce summation between the electric fields. For example, delivering currents simultaneously to four electrodes located next to a central electrode doubled the current threshold to evoke a V1 detection response at the central electrode (Bartlett et al., 2005). Some have used an interleaved pulsation strategy to reduce interaction between electrodes (Dobelle et al., 1974, 1976; Wilson et al., 1991), whereas others have optimized the position of the ground electrode to reduce such interaction (Dobelle, 2000). Delivering currents simultaneously through adjacent monopolar electrodes produces more interaction than delivering such current through adjacent bipolar or tripolar electrodes (Bierer and Middlebrook, 2004). Also interaction can be reduced by increasing the interelectrode distance and by delaying pulses delivered through adjacent electrodes by hundreds of microseconds. The optimal delay should be related to the chronaxie of the directly stimulated elements (Bierer and Middlebrook, 2004), which is an estimate of the time constant of neurons (Ranck, 1975).

The size and shape of delay fields as generated by V1 stimulation would be an effective way to measure the size and shape of the electrical fields produced during multipolar stimulation. This could be done for both sequential and concurrent activation of electrodes.

## *Determining what cortical circuit mediates phosphene induction*

The collision method has been instrumental in charting the neural circuits subserving the startle response, lateral head and body movements, and stimulation-induced reward (Gallistel et al., 1981; Yeomans, 1995; Yeomans and Tehovnik, 1988). By positioning one electrode at the somal end of a fibers bundle and another at the terminal end, one can render the bundle ineffective to somal stimulation by delivering electrical pulses to the terminal end within the conduction time of action potentials between the somal and terminal electrodes (Fig. 10A). Behaviorally, collision between the evoked action potentials is realized when the threshold to elicit a behavioral response doubles since the pulses delivered to the somal end of the bundle have been made ineffective (Yeomans and Tehovnik, 1988). This method yields two pieces of information upon collision. First, it proves that the two sites of stimulation are connected by common axons that mediate the same behavior thereby establishing a part of the circuit. Second, this method yields the collision interval between pulses, which can be used to determine the

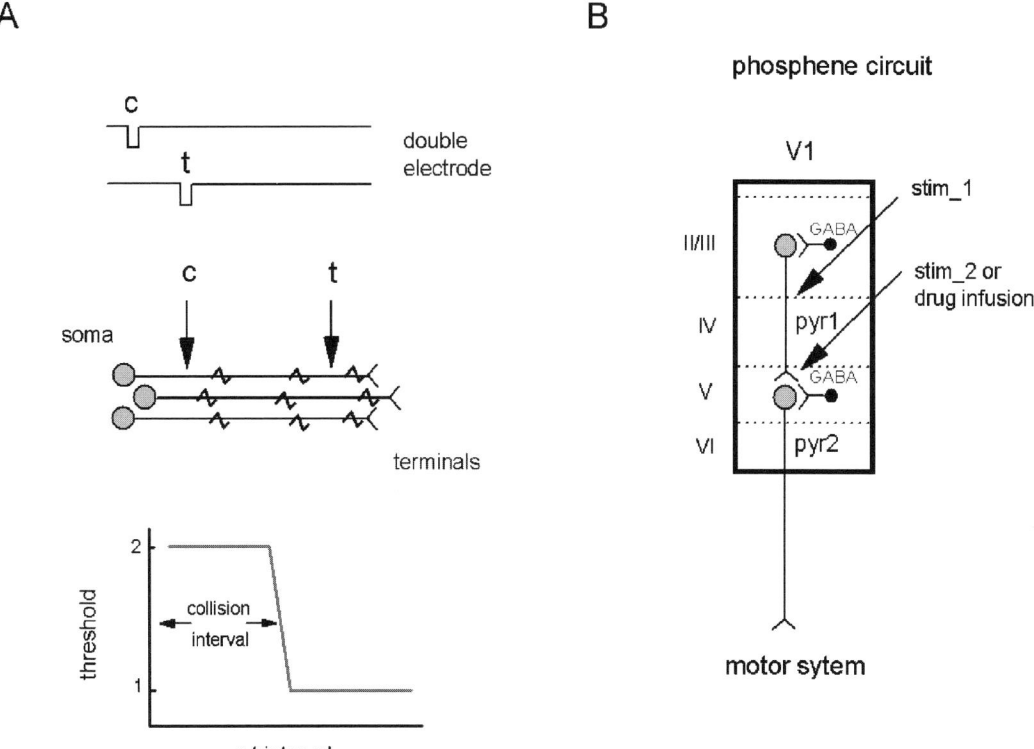

Fig. 10. (A) The collision method. (Top) By first delivering conditioning pulses (*c*) to the somal end of a fiber bundle and then by the delivery test pulses (*t*) to the terminal end of the fiber bundle (middle), the interval at which the threshold to evoke a behavioral response doubles (for reduced *c–t* intervals) marks the collision interval between the action potentials evoked by the *c* and *t* pulses (bottom). When *t* pulses are delivered outside the collision interval the threshold to evoke a behavioral response is reduced since both *c* and *t* pulses now contribute to the elicitation of behavior. Once the collision interval is determined, the conduction time between the somal and terminal electrode can be computed by subtracting the refractory period at the terminal electrode from the collision interval. Refractory periods are ascertained by delivering *c* and *t* pulse through a single electrode (not shown). As with the collision interval, the refractory period is the *c–t* interval at which the threshold doubles. For complete methodological details see Yeomans and Tehovnik (1988). (B) Phosphene circuit. A laminated portion of V1 is shown including laminae II–VI. Vertically aligned pyramidal fibers are proposed to mediate phosphenes, which are extinguished by GABAergic interneurons (GABA) (Tehovnik and Slocum, 2007d). Pyramidal fibers from superficial V1 (pyr1) terminate on pyramidal fibers from deep V1 (pyr2) (Peters and Sethares, 1991). The latter send projections to the motor system. To determine whether vertically aligned pyramidal fibers mediate phosphenes the collision test could be performed by delivering pulses through the top and bottom electrode (i.e., stim_1 and stim_2, respectively). Additionally, pharmacological agents could be infused at the bottom site (stim_2) to study their effects on the threshold to evoke phosphenes from the top electrode (stim_1).

conduction velocities of the directly stimulated axons. This gives information about the speed of signal transmission within the circuit. This method can also be combined with the infusion of various pharmacological agents at the somal or terminal ends of the circuit to study the contribution of neurotransmitters (Kofman and Yeomans, 1988).

We propose that electrical microstimulation of V1 activates vertically aligned pyramidal fibers intrinsic to V1 for the induction of phosphenes, which is then followed by a GABAergic mediated hyperpolarization of these fibers to be reset for future activation (Fig. 10B, Tehovnik and Slocum, 2007d). Pyramidal fibers of superficial V1 connect synaptically to the pyramidal fibers of deep V1 (Peters and Sethares, 1991) whose axons project subcortically to gain access to the motor system. It has been suggested that such vertically aligned modules of pyramidal cells (each spaced by 30 μm or so) mediate a particular stimulus location, orientation, and color (Peters, 1994). Understanding the circuit that is responsible for phosphene induction by using the collision method (Fig. 10A, B) may allow for the development of electrical and pharmacological devices that target specific locations of the circuit to thereby exert more control over the induced percept. In this endeavor, optical stimulation of targeted neurons might be a fruitful approach to study this problem (Han and Boyden, 2007; Huber et al., 2008).

## Conclusions

With continued experimentation on monkeys (as well as humans) we are confident that we will eventually have a cortical device that provides some semblance of vision to the blind. However, three issues need attention before such a device is realized: (1) further development of behavioral paradigms to assess what monkeys see when currents are delivered to V1; (2) an appreciation of how electric fields interact for currents delivered simultaneously through multiple electrodes; and (3) a better understanding of the V1 circuit that mediates phosphenes.

## Abbreviations

V1         striate cortex
GABA       gamma-aminobutyric acid

## Acknowledgment

This work was support by National Eye Institute Grants EY-014884 and EY-08502.

## References

Airy, H. (1870). On a distinct form of transient hemiopsia. *Philosophical Transactions of the Royal Society of London, 160*, 247–264.

Albrecht, D. G., & Hamilton, D. B. (1982). Striate cortex of monkey and cat: contrast response function. *Journal of Neurophysiology, 48*, 217–237.

Asanuma, H., Arnold, A., & Zarzecki, P. (1976). Further study on the execution of pyramidal tract cells by intracortical microstimulation. *Experimental Brain Research, 26*, 443–461.

Bagshaw, E. V., & Evans, M. H. (1976). Measuring of current spread from microelectrodes when stimulating within the nervous system. *Experimental Brain Research, 25*, 391–400.

Bak, M., Girvin, J. P., Hambrecht, F. T., Kufta, C. V., Loeb, G. E., & Schmidt, E. M. (1990). Visual sensations produced by intracortical microstimulation of the human occipital cortex. *Medical & Biological Engineering & Computing, 28*, 257–259.

Barlow, H. B. (1981). Critical limiting factors in the design of the eye and visual cortex. *Proceedings of the Royal Society of London B, 212*, 1–34.

Bartlett, J. R., DeYoe, E. A., Doty, R. W., Lee, B. B., Lewine, J. D., Negrão, N., et al. (2005). Psychophysics of electrical stimulation of striate cortex in macaques. *Journal of Neurophysiology, 94*, 3430–3442.

Bartlett, J. R., & Doty, R. W. (1974). Response of units in striate cortex of squirrel monkeys to visual and electrical stimuli. *Journal of Neurophysiology, 37*, 621–641.

Bartlett, J. R., & Doty, R. W. (1980). An exploration of the ability of macaques to detect microstimulation of striate cortex. *Acta Neurobiologiae Experimentalis (Warszawa), 40*, 713–728.

Bierer, J. A., & Middlebrook, J. C. (2002). Auditory cortical imaging of cochlear-implant stimuli: dependence on electrical configuration. *Journal of Neurophysiology, 87*, 478–492.

Bierer, J. A., & Middlebrook, J. C. (2004). Cortical responses to cochlear implant stimulation: channel interactions. *Journal of the Association for Research in Otolaryngology, 5*, 32–48.

Black, R. C., Clark, G. M., & Patrick, J. F. (1981). Current distribution measurement within the human cochlea. *IEEE Transactions on Biomedical Engineering, 28*, 721–724.

Boch, R. (1986). Behavioral modulation of neuronal activity in monkey striate cortex: excitation in the absence of active fixation. *Experimental Brain Research, 64*, 610–614.

Boroojerdi, B., Bushara, K. O., Corwell, B., Immisch, I., Battaglia, F., Muellbacher, W., et al. (2000). Enhanced excitability of human visual cortex induced by short-term light deprivation. *Cerebral Cortex, 10*, 529–534.

Bradley, D. C., Troyk, P. R., Berg, J. A., Cogan, M., Erickson, R., Kufta, C., et al. (2005). Visuotopic mapping through multichannel stimulating implant in primate V1. *Journal of Neurophysiology, 93*, 1659–1670.

Brindley, G. S. (1972). Sensory effects of electrical stimulation of the visual and paravisual cortex in man. In H. Autrum, R. Jung, W. R. Loewenstein, D. M. MacKay, & H. L. Teuber (Eds.), *Handbook of sensory physiology, central processing of visual information* (pp. 585–594). Heidelberg, Germany: Springer-Verlag Berlin.

Brindley, G. S., & Lewin, W. S. (1968a). The sensations produced by electrical stimulation of visual cortex. *Journal of Physiology (London), 196*, 479–493.

Brindley, G. S., & Lewin, W. S. (1968b). The visual sensation produced by electrical stimulation of the medial occipital cortex. *Journal of Physiology (London)*, 194, 54–55.

Bruce, C. J., & Goldberg, M. E. (1985). Primate frontal eye field. I. Single neurons discharge before saccades. *Journal of Neurophysiology*, 53, 603–635.

Cha, K., Horch, K., & Normann, R. A. (1992a). Simulation of a phosphene-based visual field: visual acuity in a pixelized visual system. *Annals of Biomedical Engineering*, 20, 439–449.

Cha, K., Horch, K., & Normann, R. A. (1992b). Mobility performance with a pixelized visual system. *Vision Research*, 32, 1367–1372.

Cha, K., Horch, K. W., Normann, R. A., & Boman, D. K. (1992c). Reading speed with a pixelized visual system. *Journal of the Optical Society of America*, 9, 673–677.

Chen, L. L., & Tehovnik, E. J. (2007). Cortical control of eye and head movements: integration of movements and percepts. *The European Journal of Neuroscience*, 25, 1253–1264.

Chen, S. C., Hallum, L. E., Lovell, N. H., & Suaning, G. J. (2005a). Visual acuity measurement of prosthetic vision: a virtual-reality simulation study. *Journal of Neural Engineering*, 2, S135–S145.

Chen, S. C., Hallum, L. E., Lovell, N. H., & Suaning, G. J. (2005b). Learning prosthetic vision: a virtual-reality study. *IEEE Transactions on Neural Systems and Rehabilitation Engineering*, 13, 249–255.

Chen, S. C., Hallum, L. E., Lovell, N. H., & Suaning, G. J. (2007). A quantitative analysis of head movement behaviour during visual acuity assessment under prosthetic vision simulation. *Journal of Neural Engineering*, 4, S108–S123.

Chen, S. C., Hallum, L. E., Suaning, G. J., & Lovell, N. H. (2006). Psychophysics of prosthetic vision: I. Visual scanning and visual acuity. In: *Conference Proceedings IEEE Engineering Medicine Biological Society* (vol. 1, pp. 4400–4403).

Chen, S. C., Lovell, N. H., & Suaning, G. J. (2004). Effects on prosthetic vision visual acuity by filtering schemes, filter-off frequency and phosphene matrix: a virtual reality simulation. In: *Conference Proceedings IEEE Engineering Medicine Biological Society* (vol. 6, pp. 4201–4204).

Chino, Y. M., Kaas, J. H., Smith, E. L., III, Langston, A. L., & Cheng, H. (1992). Rapid reorganization of cortical maps in adult cats following restricted deafferentation in retina. *Vision Research*, 32, 789–796.

Chowdhury, V., Morley, J. W., & Cornoneo, M. T. (2004). Surface stimulation of the brain with a prototype array for a visual cortex prosthesis. *Journal of Clinical Neuroscience*, 11, 750–755.

Clark, G. M. (2003). *Cochlear implants: fundamentals and applications*. New York: Springer-Verlag, pp. 1–830.

Clark, G. M. (2006). The multi-channel cochlear implant: the interface between sound and the central nervous system for hearing, speech, and language in deaf people — a personal perspective. *Philosophical Transactions of the Royal Society of London. Series B Biological Sciences*, 361, 791–810.

Clark, G. M., Nathar, J. M., Kranz, H. G., & Maritz, J. S. (1972). A behavioral study on electrical stimulation of the cochlea and central auditory pathways of the cat. *Experimental Neurology*, 36, 350–361.

Cohen, E. D. (2007). Prosthetic interfaces with the visual system: biological issues. *Journal of Neural Engineering*, 4, R14–R31.

Cowey, A. (1979). Cortical maps and visual perception. The Grindley Memorial Lecture. *The Quarterly Journal of Experimental Psychology*, 31, 1–17.

Cowey, A., & Walsh, V. (2000). Magnetically induced phosphenes in sighted, blind and blindsighted observers. *Neuroreport*, 11, 3269–3273.

Daniel, P. M., & Whitterridge, D. (1961). The representation of the visual field on the cerebral cortex in monkeys. *Journal of Physiology (London)*, 159, 203–221.

Dagnelie, G., Barnett, D., Humayun, M. S., & Thompson, R. W., Jr. (2006). Paragraph text reading using a pixelized vision simulator: parameter dependence and task learning in free-viewing condition. *Investigative Ophthalmology & Visual Science*, 47, 1241–1250.

Dagnelie, G., Keane, P., Narla, V., Yang, L., Weiland, J., & Humayun, M. (2007). Real and virtual mobility performance in simulated prosthetic vision. *Journal of Neural Engineering*, 4, S92–S101.

Dahlem, M. A., Engelmann, R., Löwel, S., & Müller, S. C. (2000). Does the migraine aura reflect cortical organization? *The European Journal of Neuroscience*, 12, 767–770.

Darian-Smith, D., & Gilbert, C. D. (1995). Topographic reorganization in the striate cortex of the adult cat and monkey is cortically mediated. *Journal of Neuroscience*, 15, 1631–1647.

Dassonville, P., Schlag, J., & Schlag-Rey, M. (1992). The frontal eye field provides the goal for saccadic eye movement. *Experimental Brain Research*, 89, 300–310.

DeYoe, E. A., Lewine, J., & Doty, R. W. (2005). Laminar variation in threshold for detection of electrical excitation of striate cortex in macaques. *Journal of Neurophysiology*, 94, 3443–3450.

Dobelle, W. H. (2000). Artificial vision for the blind by connecting a television camera to the visual cortex. *ASAIO Journal*, 46, 3–9.

Dobelle, W. H., & Mladejovsky, M. G. (1974). Phosphenes produced by electrical stimulation of human occipital cortex, and their application to the development of a prosthesis for the blind. *Journal of Physiology (London)*, 243, 553–576.

Dobelle, W. H., Mladejovsky, M. G., Evens, J. R., Roberts, T. S., & Girvin, J. P. (1976). 'Braille' reading by a blind volunteer by visual cortex stimulation. *Nature*, 259, 111–112.

Dobelle, W. H., Mladejovsky, M. G., & Girvin, J. P. (1974). Artificial vision for the blind: electrical stimulation of visual cortex offers hope for a functional prosthesis. *Science*, 183, 440–444.

Doty, R. W. (1965). Conditioned reflexes elicited by electrical stimulation of the brain in macaques. *Journal of Neurophysiology*, 28, 623–640.

Doty, R. W. (1969). Electrical stimulation of the brain in behavioral context. *Annual Review of Psychology*, 20, 289–320.

Dow, B. M., Snyder, A. Z., Vautin, R. G., & Bauer, R. (1981). Magnification factor and receptive field size in foveal striate cortex of the monkey. *Experimental Brain Research*, *44*, 213–228.

Evans, J. R., Gordon, J., Abramov, I., Mladejovsky, M. G., & Dobelle, W. H. (1979). Brightness of phosphenes elicited by electrical stimulation of human visual cortex. *Sensory Processes*, *3*, 82–94.

Felleman, D. J., & Van Essen, D. C. (1991). Distribution hierarchical processing in the primate cerebral cortex. *Cerebral Cortex*, *1*, 1–47.

Fentress, J. C., & Doty, R. W. (1971). Effect of tetanization and enucleation upon excitability of visual pathways in squirrel monkeys and cats. *Experimental Neurology*, *30*, 535–554.

Fritsch, G., & Hitzig, E. (1879). Über die electrische Erregbarkeit des Grosshirns. *Archives of Anatomy and Physiology*, *37*, 300–332.

Fu, L., Cai, S., Zhang, H., Hu, G., & Zhang, X. (2006). Psychophysics of reading with a limited number of pixels: towards the rehabilitation of reading ability with visual prosthesis. *Vision Research*, *46*, 1292–1301.

Gallistel, C. R., Shizgal, P., & Yeomans, J. S. (1981). A portrait of the substrate for self-stimulation. *Psychological Review*, *88*, 228–273.

Gilbert, C. D., Hirsch, J. A., & Wiesel, T. N. (1990). Lateral interactions in visual cortex. *Cold Spring Harbor Symposia on Quantitative Biology*, *55*, 663–677.

Gilbert, C. D., & Wiesel, T. N. (1979). Morphological and intracortical projections of functionally characterized neurons in the cat visual cortex. *Nature*, *280*, 120–125.

Gilbert, C. D., & Wiesel, T. N. (1992). Receptive field dynamics in adult primary visual cortex. *Nature*, *356*, 150–152.

Girvin, J. P., Evans, J. R., Dobelle, W. H., Mladejovsky, M. G., Henderson, D. C., Abramov, I., et al. (1979). Electrical stimulation of human visual cortex: the effect of stimulation parameters on phosphene threshold. *Sensory Processes*, *3*, 66–81.

Golomb, B., Andersen, R. A., Nakayama, K., MacLeod, D. I. A., & Wong, A. (1985). Visual thresholds for shearing motion in monkey and man. *Vision Research*, *25*, 813–820.

Grünbaum, A. S. F., & Sherrington, C. S. (1901). Observations on physiology of the cerebral cortex of some of the higher apes. *Proceedings of the Royal Society of London. Series B Biological Sciences*, *69*, 206.

Grünbaum, A. S. F., & Sherrington, C. S. (1903). Observations on physiology of the cerebral cortex of anthropoid apes. *Proceedings of the Royal Society of London. Series B Biological Sciences*, *72*, 152.

Grüsser, O.-J. (1991). Light not illuminating the world: phosphenes and related phenomena. In O.-J. Grüsser & T. Landis (Eds.), *Visual agnosias and other disturbances of visual perception and cognition, vision and visual dysfunction* (pp. 158–178). Boston, MA: CRC.

Grüsser, O.-J. (1995). Migraine phosphenes and the retinocortical magnification factor. *Vision Research*, *35*, 1125–1134.

Hadjikhani, N., Sanchez del Rio, M., Wu, O., Schwartz, D., Bakker, D., Fischl, B., et al. (2001). Mechanisms of migraine aura revealed by functional MRI in human visual cortex. *Proceedings of the National Academy of Sciences of the United States of America*, *98*, 4687–4692.

Hall, E. C., & Ciuffreda, K. J. (2002). Fixational ocular motor control is plastic despite visual deprivation. *Visual Neuroscience*, *19*, 475–481.

Han, X., & Boyden, E. S. (2007). Multiple-color optical activation, silencing, and desynchronization of neural activity with single spike temporal resolution. *PLoS ONE*, *2*(3), e299. doi:10.1371/journal.pone.0000299.

Hayes, J. S., Yin, V. T., Piyathaisere, D., Weiland, J. D., Humayun, M. S., & Dagnelie, G. (2003). Visually guided performance of simple tasks using prosthetic vision. *Artificial Organs*, *27*, 1016–1028.

Heinen, S. J., & Skavenski, A. A. (1991). Recovery of visual responses in foveal V1 neurons following bilateral foveal lesions in adult monkey. *Experimental Brain Research*, *83*, 670–774.

Hendry, S. H., Fuchs, J., DeBlas, A. L., & Jones, E. G. (1990). Distribution and plasticity of immunocytochemically localized GAGA A receptors in adult monkey visual cortex. *Cerebral Cortex*, *3*, 361–372.

Horton, J. C., & Hocking, D. R. (1998). Monocular core zones and binocular border strips in primate striate cortex revealed by the contrasting effects of enucleation, eyelid suture, and retinal laser lesions on cytochrome oxidase activity. *Journal of Neuroscience*, *18*, 5433–5455.

Hubel, D. H., & Livingston, M. S. (1990). Color and contrast sensitivity in the lateral geniculate nucleus and primary visual cortex of the macaque monkey. *Journal of Neuroscience*, *10*, 2223–2237.

Hubel, D. H., & Wiesel, T. N. (1968). Receptive fields and functional architecture of monkey striate cortex. *Journal of Physiology (London)*, *195*, 215–243.

Hubel, D. H., & Wiesel, T. N. (1974a). Sequence regularity and geometry of orientation columns in the monkey striate cortex. *The Journal of Comparative Neurology*, *158*, 267–294.

Hubel, D. H., & Wiesel, T. N. (1974b). Uniformity of monkey striate cortex: a parallel relationship between field size, scatter, and magnification factor. *The Journal of Comparative Neurology*, *158*, 295–306.

Hubel, D. H., & Wiesel, T. N. (1977). Functional architecture of macaque monkey visual cortex. *Proceedings of the Royal Society of London B*, *198*, 1–59.

Huber, D., Petreanu, L., Ghitani, N., Ranade, S., Hromádka, T., Mainen, Z., et al. (2008). Sparse optical microstimulation in barrel cortex drives learned behaviour in freely moving mice. *Nature*, *451*, 61–64.

Humayun, M. S., de Juan, E., Jr., Weiland, J. D., Dagnelie, G., Katona, S., Greenberg, R., et al. (1999). *Vision Research*, *39*, 2569–2576.

Humayun, M. S., Weiland, J. D., Fujii, G. Y., Greenberg, R., Williamson, R., Little, J., et al. (2003). Visual perception in a blind subject with a chronic microelectrode retinal prosthesis. *Vision Research*, *43*, 2573–2581.

Jacobs, G. H. (1981). *Comparative color vision*. New York: Academic Press.

Jones, E. G. (1993). GABAergic neurons and their role in cortical plasticity in primates. *Cerebral Cortex, 3*, 361–372.

Jones, E. G., Hendry, S. H. C., DeFelipe, J., & Benson, D. L. (1994). GABA neurons and their role in activity-dependent plasticity of adult primate visual cortex. In A. Peters & K. S. Rockland (Eds.), *Cerebral cortex* Vol. 10, (pp. 61–140). New York: Plenum Press.

Kaas, J. H., Krubitzer, L. A., Chino, Y. M., Lanston, A. L., Polley, E. H., & Blair, N. (1990). Reorganization of retinotopic cortical maps in adult mammals after lesions of the retina. *Science, 248*, 229–231.

Kayama, Y., Riso, R. R., Bartlett, J. R., & Doty, R. W. (1979). Luxotonic responses of units in macaque striate cortex. *Journal of Neurophysiology, 42*, 1495–1517.

Keating, E. G., & Gooley, S. G. (1988). Disconnection of parietal and occipital access to the saccadic oculomotor system. *Experimental Brain Research, 70*, 385–398.

Keating, E. G., Gooley, S. G., Pratt, S. E., & Kelsey, J. (1983). Removing the superior colliculus silences eye movements normally evoked from stimulation of the parietal and occipital eye fields. *Brain Research, 269*, 145–148.

Kofman, O., & Yeomans, J. S. (1988). Cholinergic antagonists in ventral tegmentum elevates thresholds for lateral hypothalamic and brainstem self-stimulation. *Pharmacology, Biochemistry, and Behavior, 31*, 547–559.

Kömpf, D., & Piper, H-F. (1987). Eye movements and vestibulo-ocular reflex in the blind. *Journal of Neurology, 234*, 337–341.

Kral, A., Hartmann, R., Mortazavi, D., & Klinke, R. (1998). Spatial resolution of cochlear implants: the electrical field and excitation of auditory afferents. *Hearing Research, 121*, 11–28.

Lashley, K. S. (1941). Patterns of cerebral integration indicated by the scotomas of migraine. *Archives of Neurological Psychiatry, 46*, 331–339.

Lee, H. W., Hong, S. B., Seo, D. W., Tae, W. S., & Hong, S. C. (2000). Mapping of functional organization in human visual cortex. *Neurology, 54*, 849–854.

Leigh, R. J., & Zee, D. S. (1980). Eye movements of the blind. *Investigative Ophthalmology & Visual Science, 19*, 328–331.

Leonard, R. (2002). *Statistics on vision impairment: a resource manual*. New York: Arlene R. Gordon Research Institute of Lighthouse International, pp. 1–49.

Levi, D. M., Kein, S. A., & Aitsebaomo, A. P. (1985). Verneir acuity, crowding and cortical magnification. *Vision Research, 25*, 963–977.

Lovell, N. H., Dokos, S., Cheng, E., & Suaning, G. J. (2005). Stimulation of parallel current injection for use in visual prosthesis. In: *Neural Engineering Conference Proceedings. 2nd International IEEE Engineering Medicine Biological Society* (pp. 458–461). March 16–19.

O'Kusky, J., & Colonnier, M. (1982). A laminar analysis of the number of neurons, glia, and synapses in the visual cortex (area 17) of adult macaque monkeys. *The Journal of Comparative Neurology, 210*, 278–290.

Marg, E., & Dierssen, G. (1965). Reported visual percepts from stimulation of the human brain with microelectrodes during therapeutic surgery. *Confined Neurology, 26*, 57–75.

Maynard, E. M. (2001). Visual prostheses. *Annual Review of Biomedical Engineering, 3*, 145–168.

Michael, R. C. (1981). Columnar organization of color cells in monkey's striate cortex. *Journal of Neurophysiology, 46*, 587–604.

Miezin, F. M., Myerson, J., Julesz, B., & Allman, J. M. (1981). Evoked potential to dynamic random-dot correlograms in monkey and man: a test for cyclopean perception. *Vision Research, 21*, 177–179.

Morell, F., & Naitoh, P. (1962). Effect of cortical polarization on a conditioned avoidance response. *Experimental Neurology, 6*, 507–523.

Murphey, D. K., & Maunsell, J. H. R. (2007). Behavioral detection of electrical microstimulation in different cortical visual areas. *Current Biology, 17*, 862–867.

Nielson, H. C., Knight, J. M., & Porter, P. B. (1962). Subcortical conditioning, generalization, and transfer. *Journal of Comparative Physiological Psychology, 55*, 168–173.

Penfield, W., & Perot, P. (1963). The brain's record of auditory and visual experience. *Brain, 86*, 595–696.

Penfield, W., & Rasmussen, T. (1957). *The cerebral cortex of man*. New York: Macmillan Co., pp. 1–248.

Peters, A. (1994). The organization of the primary visual cortex in the macaque. In A. Peters & K. S. Rockland (Eds.), *Cerebral cortex* Vol. 10, (pp. 1–35). New York: Plenum Press.

Peters, A., & Sethares, C. (1991). Organization of pyramidal neurons in area 17 of monkey visual cortex. *The Journal of Comparative Neurology, 306*, 1–23.

Pezaris, J. S., & Reid, R. C. (2007). Demonstration of artificial visual percepts generated through thalamic microstimulation. *Proceedings of the National Academy of Sciences of the United States of America, 104*, 7670–7675.

Poggio, G. F., & Fischer, B. (1977). Binocular interaction and depth sensitivity in striate and prestriate cortex of behaving rhesus monkey. *Journal of Neurophysiology, 40*, 1392–1405.

Polyak, S. (1957). In H. Klüver (Ed.), *The vertebrate visual system*. Chicago, IL: University of Chicago Press.

Porter, R. (1963). Focal stimulation of the hypoglossal neurons in the cat. *Journal of Physiology (London), 169*, 630–640.

Ranck, J. B., Jr. (1975). Which elements are excited in electrical stimulation of mammalian central nervous system: a review. *Brain Research, 98*, 417–440.

Richards, W. (1971). The fortification illusions of migraines. *Scientific American, 224*, 88–95.

Rizzo, J. F., III, Wyatt, J., Loewenstein, J., & Wyatt, J. (2003). Perceptual efficacy of electrical stimulation of human retina with a microelectrode array during short-term surgical trials. *Investigative Ophthalmology & Visual Science, 44*, 5362–5369.

Rockel, A. J., Hiorns, R. W., & Powell, T. P. S. (1980). The basic uniformity in structure of the neocortex. *Brain, 103*, 221–244.

Rosier, A. M., Arckens, L., Demeulemeester, H., Orban, G. A., Eysel, U. T., Wu, Y-J., et al. (1995). Effect of sensory deafferentation on immunoreactivity of GABAergic cells and on GABA receptors in the adult cat visual cortex. *The Journal of Comparative Neurology, 359*, 476–489.

Rushton, D. N., & Brindley, G. S. (1977). Short- and long-term stability of cortical electrical phosphenes. In F. C. Rose (Ed.), *Physiological aspects of clinical neurology* (pp. 123–153). London: Blackwell Scientific Publishing.

Rushton, D. N., & Brindley, G. S. (1978). Properties of cortical electrical phosphenes. In S. J. Cool & E. L. Smith (Eds.), *Frontiers in visual science* (pp. 574–593). New York: Springer-Verlag.

Rutledge, L. T., & Doty, R. W. (1962). Surgical interference with pathways mediating responses conditioned to cortical stimulation. *Experimental Neurology, 6*, 478–492.

Sakakura, H., & Doty, R. W. (1976). EEG of striate cortex in blind monkeys: effects of eye movements and sleep. *Archives Italiennes de Biologie, 114*, 23–48.

Santos, A., Humayun, M. S., de Juan, E., Greenburg, R. J., Marsh, M. J., Klock, I. B., et al. (1997). Preservation of the inner retina in retinitis pigmentosa. *Archives of Ophthalmology, 115*, 511–515.

Schäfer, E. A. (1888). Experiments on the electrical excitation of the cerebral cortex in the monkey. *Brain, 11*, 1–6.

Schiller, P. H. (1972). The role of the monkey superior colliculus in eye movements and vision. *Investigative Ophthalmology & Visual Science, 11*, 451–459.

Schiller, P. H. (1977). The effect of superior colliculus ablation on saccades elicited by cortical stimulation. *Brain Research, 122*, 154–156.

Schiller, P. H., Finlay, B. L., & Volman, S. F. (1976a). Quantitative studies of single-cell properties in monkey striate cortex. I. Spatiotemporal organization of receptive fields. *Journal of Neurophysiology, 39*, 1288–1319.

Schiller, P. H., Finlay, B. L., & Volman, S. F. (1976b). Quantitative studies of single-cell properties in monkey striate cortex. II. Orientation specificity and ocular dominance. *Journal of Neurophysiology, 39*, 1320–1333.

Schiller, P. H., & Koerner, F. (1971). Discharge characteristics of single units in superior colliculus of the alert rhesus monkey. *Journal of Neurophysiology, 34*, 920–935.

Schiller, P. H., & Tehovnik, E. J. (2001). Look and see: how the brain moves your eyes about. *Progress in Brain Research, 134*, 127–142.

Schiller, P. H., & Tehovnik, E. J. (2008). Visual prosthesis. *Perception, 37*, 1529–1559.

Schiller, P. H., Weiner, V. S., & Tehovnik E. J. (2005). Preliminary studies examining the feasibility of a visual prosthetic device. 1. What does a monkey see when area V1 is stimulated electrically. *Society for Neuroscience Abstracts*, 16.1.

Schlag, J., Schlag-Rey, M., & Dassonville, P. (1989). Interactions between natural and electrically evoked saccades. II. At what time is eye position samples as a reference for the location of the target? *Experimental Brain Research, 76*, 548–558.

Schlag-Rey, M., Schlag, J., & Shook, B. (1989). Interactions between natural and electrically evoked saccades. I. Differences between sites carrying retinal error and more error signals in monkey superior colliculus. *Experimental Brain Research, 76*, 537–547.

Schmidt, E. M., Bak, M. J., Hambrecht, F. T., Kufta, C. V., O'Rourke, D. K., & Vallabhanath, P. (1996). Feasibility of a visual prosthesis for the blind based on intracortical microstimulation of the visual cortex. *Brain, 119*, 507–522.

Schuckman, H., & Battersby, W. S. (1966). Frequency specific mechanisms in learning. II. Discriminatory conditioning induced by intracortical stimulation. *Journal of Neurophysiology, 29*, 31–43.

Schuckman, H., Kluger, A., & Frumkes, T. E. (1970). Stimulus generalization within the geniculostriate system of the monkey. *Journal of Comparative and Physiological Psychology, 73*, 494–500.

Schwartz, E. L. (1994). Computational studies of the spatial architecture of primate visual cortex: columns, maps, and protomaps. In A. Peters & K. S. Rockland (Eds.), *Cerebral cortex* Vol. 10, (pp. 359–411). New York: Plenum Press.

Sereno, M. I., Dale, A. M., Reppas, J. B., Kwong, K. K., Belliveau, J. W., Brady, T. J., et al. (1995). Borders of multiple visual areas in humans revealed by functional magnetic resonance imaging. *Science, 268*, 889–893.

Slocum, W. M., & Tehovnik, E. J. (2004). Microstimulation of V1 input layers disrupts the selection and detection of visual targets by monkeys. *The European Journal of Neuroscience, 20*, 1674–1680.

Smirnakis, S. M., Brewer, A. A., Schmid, M. C., Tolias, A. S., Schüz, A., Augath, M., et al. (2005). Lack of long-term cortical reorganization after macaque retinal lesions. *Nature, 435*, 300–307.

Sparks, D. L., Holland, L. R., & Guthrie, B. L. (1976). Size and distribution of movement fields in the monkey superior colliculus. *Brain Research, 113*, 21–34.

Sparks, D. L., & Mays, L. E. (1980). Movement fields of saccade-related burst neurons in the monkey superior colliculus. *Brain Research, 190*, 39–50.

Sparks, D. L., & Mays, L. E. (1983). Spatial localization of saccade targets. I. Compensation for stimulation-induced perturbations in eye position. *Journal of Neurophysiology, 49*, 45–63.

Stone, J. L., Barlow, W. E., Humayun, M. S., de Juan, E., Jr., & Milam, A. H. (1992). Morphometric analysis of macular photoreceptor and ganglion cells in retinas with retinitis pigmentosa. *Archives of Ophthalmology, 110*, 1634–1639.

Stoney, S. D., Jr., Thompson, W. D., & Asanuma, H. (1968). Excitation of pyramidal tract cells by intracortical microstimulation: effective extent of stimulating current. *Journal of Neurophysiology, 31*, 659–669.

Talalla, A., Bullara, L., & Pudenz, R. (1974). Electrical stimulation of the human visual cortex. *Canadian Journal of Neurological Sciences, 1*, 236–238.

Talbot, S. A., & Marshall, W. H. (1941). Physiological studies on neural mechanisms of visual localization and discrimination. *American Journal of Ophthalmology, 24*, 1255–1264.

Tehovnik, E. J., & Slocum, W. M. (2003). Microstimulation of macaque V1 disrupts target selection: effects of stimulation polarity. *Experimental Brain Research, 148*, 233–237.

Tehovnik, E. J., & Slocum, W. M. (2004). Behavioural state affects saccades elicited electrically from neocortex. *Neuroscience & Biobehavioral Reviews, 28*, 13–25.

Tehovnik, E. J., & Slocum, W. M. (2005). Microstimulation of V1 affects the detection of visual targets: manipulation of target contrast. *Experimental Brain Research, 165*, 205–314.

Tehovnik, E. J., & Slocum, W. M. (2007a). Microstimulation of V1 delays visually-guided saccades: a parametric evaluation of delay fields. *Experimental Brain Research, 176*, 413–424.

Tehovnik, E. J., & Slocum, W. M. (2007b). Delaying forelimb responses by microstimulation of macaque V1. *Experimental Brain Research, 178*, 422–426.

Tehovnik, E. J., & Slocum, W. M. (2007c). Phosphene induction by microstimulation of macaque V1. *Brain Research Reviews, 53*, 337–343.

Tehovnik, E. J., & Slocum, W. M. (2007d). What delay fields tell us about striate cortex. *Journal of Neurophysiology, 98*, 559–576.

Tehovnik, E. J., & Slocum, W. M. (2009). Depth-dependent detection of microampere currents delivered to monkey V1. *The European Journal of Neuroscience, 29*, 1477–1489.

Tehovnik, E. J., Slocum, W. M., & Carvey, C. E. (2003a). Behavioural state affects saccadic eye movements evoked by microstimulation of striate cortex. *The European Journal of Neuroscience, 18*, 969–979.

Tehovnik, E. J., Slocum, W. M., Carvey, C. E., & Schiller, P. H. (2005a). Phosphene induction and the generation of saccadic eye movements by striate cortex. *Journal of Neurophysiology, 93*, 1–19.

Tehovnik, E. J., Slocum, W. M., & Schiller, P. H. (1999). Behavioural conditions affecting saccadic eye movements elicited electrically from the frontal lobes of primates. *The European Journal of Neuroscience, 11*, 2431–2443.

Tehovnik, E. J., Slocum, W. M., & Schiller, P. H. (2002). Differential effects of laminar stimulation of V1 cortex on target selection by macaque monkeys. *The European Journal of Neuroscience, 16*, 751–760.

Tehovnik, E. J., Slocum, W. M., & Schiller, P. H. (2003b). Saccadic eye movements evoked by microstimulation of striate cortex. *The European Journal of Neuroscience, 17*, 870–878.

Tehovnik, E. J., Slocum, W. M., & Schiller, P. H. (2004). Microstimulation of V1 delays the execution of visually guided saccades. *The European Journal of Neuroscience, 20*, 264–272.

Tehovnik, E. J., Slocum, W. M., & Schiller, P. H. (2005b). Delaying visually guided saccades by microstimulation of macaque V1: spatial properties of delay fields. *The European Journal of Neuroscience, 22*, 2635–2643.

Tehovnik, E. J., Tolias, A. S., Sultan, F., Slocum, W. M., & Logothetis, N. K. (2006). Direct and indirect activation of cortical neurons by electrical microstimulation. *Journal of Neurophysiology, 96*, 512–521.

Thompson, R. W., Jr., Barnett, G. D., Humayun, M. S., & Dagnelie, G. (2003). Facial recognition using simulated prosthetic pixelized vision. *Investigative Ophthalmology & Visual Science, 44*, 5035–5047.

Tootell, R. B. H., Switkes, E., Silverman, M. S., & Hamilton, S. L. (1988). Functional anatomy of macaque striate cortex. II. Retinotopic organization. *Journal of Neuroscience, 8*, 1531–1568.

Tootell, R. B. H., Tsao, D., & Vanduffel, W. (2003). Neuroimaging weights in: humans meet macaques in "primate" visual cortex. *Journal of Neuroscience, 23*, 3981–3989.

Trotter, Y., Celebrini, S., & Durand, J. B. (2004). Evidence for implication of primate area V1 in neural 3-D spatial localization processing. *Journal of Physiology (Paris), 98*, 125–134.

Troyk, P., Bak, M., Berg, J., Bradley, D., Cogan, S., Erickson, R., et al. (2003). A model for intracortical visual prosthesis research. *Artificial Organs, 27*, 1005–1015.

Van Essen, D. C., Anderson, C. H., & Felleman, D. J. (1992). Information processing in the primate visual system: an integrated systems perspective. *Science, 255*, 419–423.

Veraart, C., Raftopoulos, C., Mortimer, T., Delbeke, J., Pins, D., Michaux, G., et al. (1998). Visual sensations produced by optic nerve stimulation using an implanted self-sizing spiral cuff electrode. *Brain Research, 813*, 181–186.

Veraart, C., Wanet-Defalque, M.-C., Gérard, B., Vanlierde, A., & Delbeke, J. (2003). Pattern recognition with the optic nerve visual prosthesis. *Artificial Organs, 27*, 996–1004.

Wagman, I. H. (1964). Eye movements induced by electrical stimulation of cerebrum in monkeys and their relationship to bodily movements. In M. B. Bender(Ed.), *The oculomotor system* (pp. 18–39). New York: Harper and Row.

Wagman, I. H., Krieger, H. P., & Bender, M. B. (1958). Eye movements elicited by surface and depth stimulation of the occipital lobe of *Macacca mulatta*. *The Journal of Comparative Neurology, 109*, 169–212.

Walker, E. A., & Weaver, T. A. (1940). Ocular movements from the occipital lobe in the monkey. *Journal of Neurophysiology, 3*, 353–357.

Ward, R., & Weiskrantz, L. (1969). Impaired discrimination following polarization of striate cortex. *Experimental Brain Research, 9*, 346–356.

Warren, D. J., Fernandez, E., & Normann, R. A. (2001). High-resolution spatial mapping of cat striate cortex using a 100-microelectrode array. *Neuroscience, 105*, 19–31.

Weiland, J. D., & Humayan, M. S. (2003). Past, present, and future of artificial vision. *Artificial Organs, 27*, 961–962.

Wilson, B. S., Finley, C. C., Lawson, D. T., Wolford, R. D., Eddingston, D. K., & Rabinowitz, W. M. (1991). Better speech recognition with cochlear implants. *Nature, 352,* 236–238.

Wurtz, R. H., & Goldberg, M. E. (1971). Superior colliculus cell responses related to eye movements in awake monkeys. *Science, 171,* 82–84.

Wurtz, R. H., & Mohler, C. W. (1976). Enhancement of visual responses in monkey striate cortex and frontal eye fields. *Journal of Neurophysiology, 39,* 766–772.

Yang, Y., Cao, P., Yang, Y., & Wang, S.-R. (2008). Corollary discharge circuits for saccadic modulation of the pigeon visual system. *Nature Neuroscience, 11,* 595–602.

Yeomans, J. S. (1995). Electrically evoked behaviors: axons and synapses mapped with collision tests. *Behavioural Brain Research, 67,* 121–132.

Yeomans, J. S., & Tehovnik, E. J. (1988). Turning responses evoked by stimulation of visuomotor pathways. *Brain Research Reviews, 13,* 235–259.

Zrenner, E. (2002). Will retinal implants restore vision? *Science, 295,* 1022–1025.

SECTION VI

# Deep Brain Stimulation, FES and TMS

# CHAPTER 24

# Functional neurosurgery for movement disorders: a historical perspective

Alim Louis Benabid*, Stephan Chabardes, Napoleon Torres, Brigitte Piallat, Paul Krack, Valerie Fraix and Pierre Pollak

*CEA Minatec LETI and INSERM, Joseph Fourier University, Grenoble, France*

**Abstract:** Since the 1960s, deep brain stimulation and spinal cord stimulation at low frequency (30 Hz) have been used to treat intractable pain of various origins. For this purpose, specific hardware have been designed, including deep brain electrodes, extensions, and implantable programmable generators (IPGs). In the meantime, movement disorders, and particularly parkinsonian and essential tremors, were treated by electrolytic or mechanic lesions in various targets of the basal ganglia, particularly in the thalamus and in the internal pallidum. The advent in the 1960s of levodopa, as well as the side effects and complications of ablative surgery (e.g., thalamotomy and pallidotomy), has sent functional neurosurgery of movement disorders to oblivion. In 1987, the serendipitous discovery of the effect of high-frequency stimulation (HFS), mimicking lesions, allowed the revival of the surgery of movement disorders by stimulation of the thalamus, which treated tremors with limited morbidity, and adaptable and reversible results. The stability along time of these effects allowed extending it to new targets suggested by basic research in monkeys. The HFS of the subthalamic nucleus (STN) has profoundly challenged the practice of functional surgery as the effect on the triad of dopaminergic symptoms was very significant, allowing to decrease the drug dosage and therefore a decrease of their complications, the levodopa-induced dyskinesias.

In the meantime, based on the results of previous basic research in various fields, HFS has been progressively extended to potentially treat epilepsy and, more recently, psychiatric disorders, such as obsessive–compulsive disorders, Gilles de la Tourette tics, and severe depression. Similarly, suggested by the observation of changes in PET scan, applications have been extended to cluster headaches by stimulation of the posterior hypothalamus and even more recently, to obesity and drug addiction.

In the field of movement disorders, it has become clear that STN stimulation is not efficient on the nondopaminergic symptoms such as freezing of gait. Based on experimental data obtained in MPTP-treated parkinsonian monkeys, the pedunculopontine nucleus has been used as a new target, and as suggested by the animal research results, its use indeed improves walking and stability when stimulation is performed at low frequency (25 Hz). The concept of simultaneous stimulation of multiple targets eventually at low or high frequency, and that of several electrodes in one target, is being accepted to increase the efficiency. This leads to and is being facilitated by the development of new hardware (multiple-channel IPGs, specific electrodes, rechargeable batteries). Still additional efforts are needed at the level of the

---

*Corresponding author.
Tel.: +33680872491;
E-mail: alimlouis@aol.com

stimulation paradigm and in the waveform. The recent development of nanotechnologies allows the design of totally new systems expanding the field of deep brain stimulation. These new techniques will make it possible to not only inhibit or excite deep brain structures to alleviate abnormal symptoms but also open the field for the use of recording cortical activities to drive neuroprostheses through brain–computer interfaces. The new field of compensation of deficits will then become part of the field of functional neurosurgery.

**Keywords:** deep brain stimulation; Parkinson's disease; history; high-frequency stimulation; low-frequency stimulation; subthalamic nucleus; pallidum; thalamus; pedunculopontine nucleus; obsessive–compulsive disorder; epilepsy; dystonia

## Introduction

Parkinson's disease (PD) is due to the nigral degenerescence of dopaminergic neurons, leading to a disorganization of functional circuits in the basal ganglia (BG). This involves primarily the substantia nigra pars compacta (SNc) sending both inhibitory and excitatory dopaminergic projections to the caudate putamen or striatum. From the striatum, two descending parallel pathways project either directly (GABAergic output) to the globus pallidus internus (GPi)/substantia nigra pars reticulata (SNr) complex, or indirectly through (gamma-aminobutyric acid) GABAergic projections to the external pallidum (GPe) and then to the subthalamic nucleus (STN). The STN projects through a glutamatergic pathway to the GPi–SNr complex. This complex is considered to be the final output structure of the BG and projects through a GABAergic projection to the motor thalamus, which itself projects to the cortex. In turn, the cortex sends a strong glutamatergic projection to the STN, which receives also a glutamatergic projection from the centrum medianum–parafascicularis (CM–Pf) complex, and from the pedunculopontine nucleus (PPN).

The decrease of the dopaminergic output of the SNc induces an imbalance between the two output pathways from the striatum to the pallidum. GPi is disinhibited and hyperexcited and, as a consequence, increases its inhibitory output to the thalamus and then to the cortex, which is therefore inhibited. This is the putative pathogenicity of akinesia and rigidity, which are two main symptoms of the parkinsonian triad. The third component, tremor, can be explained as the symptomatic expression of the disorganization of the BG, leading to a periodic or oscillatory behavior.

## Therapy for movement disorders: a random walk around a logical thread

The underlying logic or rationale is that a lesion of the motor system would weaken the motor function, and thereby the motor disturbance and disorders. These lesions can be produced by pyramidotomies or corticectomies, but essentially by lesions in the motor control system (Hassler et al., 1960; Meyers, 1942).

Serendipity has played a role in the discovery of surgical solutions to treat movement disorders. Cooper (1953), during a surgical procedure to perform pyramidotomy, had an accidental choroidal artery injury, which did not create deficits but suppressed the tremor the patient was suffering from. Cooper had the idea to make it to a method and performed it in a certain number of patients. The logical extension from the hypothetical mechanism underlying the beneficial effect of the ligature of the choroidal artery (which was supposed to create an ischemic area in the pallidum) led to the method of pallidotomy, which was then transformed in a thalamotomy when anatomical controls showed that the efficient lesion was actually situated in the thalamus (Cooper, 1956).

The combination of trial and error approach and integrated neurophysiological knowledge led to a semiconsensual pragmatic and science-based definition of the thalamus as a target for parkinsonian symptoms, mainly for tremor.

Then, the results and the quest for improved efficiency of deep brain stimulation (DBS) in PD will justify, in the following period of time:

(1) variations on the aspect of (i) targets (thalamus or pallidum, fibers, or bundles), and (ii) tools [mechanical, chemical (alcohol, wax), physical (electricity, temperature, which is cooling, etc.)],
(2) extensions to other indications, mainly dystonia, chorea, and tics,
(3) ongoing research on anatomy, which should be more functional than nowadays, and must led to the definition of structures and anatomy-based functional paths, and
(4) the establishment of ventralis intermedius nucleus (VIM) as a "consensual target" for tremor, based on intraoperative electrophysiology. VIM, receiving proprioceptive (indirect?) and cerebellar inputs, contains large kinasthetic cells, and is situated within the thalamus, between the motor parts (ventralis oralis anterior and posterior, Voa and Vop, respectively) that are anterior to it, and the sensory parts (ventroposterolateralis, VPL) that are posterior to it. However, it becomes clear at this time that the effects of VIM thalamotomy on other symptoms than tremor are absent or much smaller.

## *Getting serious in front of success in the treatment of motor disturbances: the joint venture of therapy and neuroscience*

At this point in time of history, the only treatment for movement disorders is surgical! It has advantages: it is efficient, immediate, and cheap. It has drawbacks: the effect is decaying, quite often associated to complications, and a cognitive decline might be expected if the procedure is bilateral. Unfortunately, a successful technique becomes overused, depredated by the necessity of speed and the lack of training. Surgeons are too busy treating patients, and they do not have time reporting the results, or they do it poorly! Nevertheless, ablative surgery by lesioning is the dominant therapy: It could be better, "but that's all we have."

The thalamic VIM seems to be the best target for DBS to treat tremor: The effects are better for essential tremor than for PD rest tremor and even better than for dystonia. Tremor is the most improved symptom; akinesia and rigidity are only mildly altered. Considering the methods, electrolytic lesioning is the most used, but the predictability is low and the mechanical lesions (by leucotome) are still used at this time. The bilateral lesions are poorly tolerated, which leads the surgeon to try staged procedures (6 months to 1 year of interval), and sometimes surgeons will even make them asymmetrical! The evaluation of the method is essentially based on expert's opinion rather than on evidence-based medicine. The neurologists send patients to the surgery as a last resort procedure and do not mind publishing midrange results and complications. However, in rare teams, there is a multidisciplinary cooperation: Electrophysiology becomes an intraoperative high-tech approach, scientific data are gathered, new concepts are drawn, and basic science is emulated, eventually providing better understanding, suggesting new targets, and helping in preparing for future evolution.

The forgotten target GPi in the pallidum, for instance, is now revisited (Laitinen et al., 1992). Pallidotomy was initiated as an established procedure by Leksell in the 1950s, but in the post-levodopa era for PD (see below), this method was considered insufficient. However, it becomes the target for DBS as it improves the newly recognized complication of chronic levodopa treatment, dyskinesias that occur 5–7 years after the onset of the therapy, profoundly disabling the patient and disturbing the neurologist. The pallidum will become eventually the target for dystonia, although it was already widely used, described, and documented by Cooper (1956).

## *Lesioning at its peak: functional model of the basal ganglia*

Based on the clinical observations and on animal experiments, a global scheme emerges and will lead to the creation of a functional model of the

BG. Its establishment (Alexander and Crutcher, 1990) sets the stage for strategic surgical thinking, replaces experts' hints and claims, establishes links between VIM, GPi, and STN, and suggests new approaches and targets. Electrophysiology has become a full partner for targeting through microrecording, along with X-rays particularly for ventriculography, while microstimulation, putting function upon an individual anatomical structure, becomes better used (though this requires a higher level of training of the team). But the role of frequency, as one of the important parameters of stimulation used to explore the functional anatomy of the region in which the lesion is intended to be performed, is not yet recognized.

At this time, one sees the neurologists becoming attracted and being even welcomed in the operating room. The new image modalities, such as computed tomography (CT) X-ray scan in 1972 and magnetic resonance imaging (MRI) in 1985, created a real breakthrough and allowed targeting based on visualization of the targets, at least for some of them. However, these improvements, contributing to a much more efficient and safer surgical practice, came too late to rescue the ablative methods, which were suddenly stricken by a deadly blow in the 1960s by an outsider, the pharmacology, so far an unheard and impotent method in the field of treatment for human motor disturbances.

## *Thunderstorms in a blue sky: the levodopa lightning*

Indeed the storm was maturing since the middle of the 20th century. The degenerescence of the substantia nigra (locus niger of Soemmering) neurons had been described in 1919 by Tretiakoff (1919) in his doctoral dissertation. Serendipity strikes back through Arvid Carlsson (Carlsson et al., 1957) who was working on the model of psychosis using reserpine-induced dopamine depletion. He realized that what was considered as a psychotic apathy in the reserpine-treated rabbits was truly a severe akinesia. He then discovered that cerebral dopamine was concentrated in the striatum and that levodopa, a dopamine precursor crossing the blood–brain barrier, could reverse the reserpine-induced akinesia. Ehringer and Hornykiewicz (1960) demonstrated the striatal dopamine depletion in PD patients. This logically calls for the use of levodopa as a substitute therapy (Cotzias et al., 1967). This substitute therapy is truly a miracle for both the patients and their doctors. The levodopa acts on the triad of symptoms: tremor, akinesia, and rigidity. Its effect is immediate. It is reversible and titrable. It does not have complications like surgery had. Surgery has now to disappear, which almost happened.

But levodopa has late side effects, as every gold medal takes time to show its backside. This almost systematic complication, which plagues the therapy after 5–7 years of intense usage in severe PD patients, occurs at the same time as motor fluctuations. This puts the patient in a flip-flop sequence of severe akinesia, acutely replaced by recovered motricity when levodopa is administrated, leaving only a short period of free time before the next occurrence of levodopa-induced involuntary movements, which sometimes may be as invalidating as the parkinsonian symptoms themselves.

These ON–OFF motor fluctuations become a real therapeutic problem, and the neurologists, once again, wanted a solution or an alternative. The solution at this time was pallidotomy, as its relieving effects on dyskinesias are spectacular, allowing the patient to tolerate higher levodopa doses of treatment to improve akinesia and tremor and helping the neurologists in a better and softer management of the drug therapy.

The search for an alternative, on the other hand, is still going on in experimental clinical studies. Intrastriatal neural dopaminergic grafts were tried aimed at replacing the neurons that produce the missing neurotransmitter. This is a brilliant experimental approach, but the treatment is so far ineffective except in specific indications and circumstances, and cannot be proposed to patients as a standard treatment. Stem cells as implants are the next and current perspective, but this methodology brings its own challenges (Morizane et al., 2008).

Gene therapy is the most recent new hope, which provides the possibility to cure the disease by a change in the behavior of the cells by "injecting" genes that can transform the pharmacology of the neurons [such as the glutamic acid decarboxylase (GAD) gene able to transform the glutamatergic neurons of STN in GABAergic neurons; Kaplitt et al., 2007] or induce the local production of growth factors, such as GDNF or nurturin, which have interesting neuroprotective properties for dopaminergic neurons in animal experiments (Marks et al., 2008; also see Chapter 14), or by transfecting striatal neurons with a lentiviral vector bearing a combination of three genes governing the synthesis of dopamine. All these trials are still in a preliminary phase and it is too early to foresee their efficiency.

## Luck and serendipity: high-frequency stimulation mimics ablation

In 1987, during a thalamotomy for essential tremor, using electrical stimulation to probe where a lesion was intended to be placed and to check whether it was not an unwanted functional site (such as the pyramidal tract, lateral to the thalamus, that potentially induce a motor deficit if lesioned, or the ventro-postero-lateral somato-sensory thalamus where a lesion would induce the severe painful Dejerine-Roussy syndrome), we observed that at frequencies higher than 100 Hz the symptoms of the patient, actually a tremor, could be acutely and reversibly modified (Benabid et al., 1987). This effect occurred in a wide range of frequencies between 100 and several thousands of Hertz. It was reversible, adaptable, and associated to a low morbidity, even in bilateral procedures performed in one session. The pragmatic observation of these paradoxical effects of HFS, mimicking the lesion, was the true basis of the development of a totally new method, the mechanism of which is still under investigation. This was what the field of treatment of movement disorders was waiting for.

## How this could work? We don't know but "who cares, it works"

The mechanism (Benabid et al., 2005; Hammond et al., 2008) is clearly not a single one and probably it is a function of several mechanisms, altogether combined with different weights (Fig. 1):

$$\text{Mechanism} = f(A_1 \text{ Mech}_1, \ldots, A_i \text{ Mech}_i, \ldots, A_n \text{ Mech}_n)$$

The question today is to know what are the mechanism variables $\text{Mech}_i$? Are they inhibitory, excitatory, or both? What are the weighing coefficients $A_i$ of these different variables in the equation? Whatever the mechanism could be, it has to be frequency dependent and must mimic lesions as a final result. At the present time, one may try to sketch a generalized model that would gather direct effects of stimulation, such as (i) inhibition of firing of the stimulated neurons (Filali et al., 2004; Welter et al., 2004), (ii) an excitation of the fibers of the neighbor bundles (such as the internal capsule or the fibers of the third nerve or motor common), (iii) the excitation of the outgoing axons (which, as fibers become excited but still behave as if they were inhibited, due to a possible depletion of the neurotransmitters at the synaptic level, can become functionally inefficacious) (Xia et al., 2007), (iv) the retrograde activation of the incoming axons or of axons de passage (which might be responsible for a jamming mechanism blurring the abnormal bursting and sometimes oscillatory firing rate of the various components of the BG), and (v) the resetting (or jamming?) of a network activity (Meissner et al., 2005; Tai et al., 2003). The latter point may also explain why stimulating different nuclei within these BG may lead to almost similar effects on different symptoms.

## What is the current status of the efficiency of high-frequency stimulation in various nuclei in different diseases?

### The thalamic ventral intermedius nucleus (VIM)

Stimulation of the thalamic VIM works essentially on tremor with a long-lasting effect that does not change along time (Benabid et al., 1991, 1996), even not over a period of 20 years (Fig. 2).

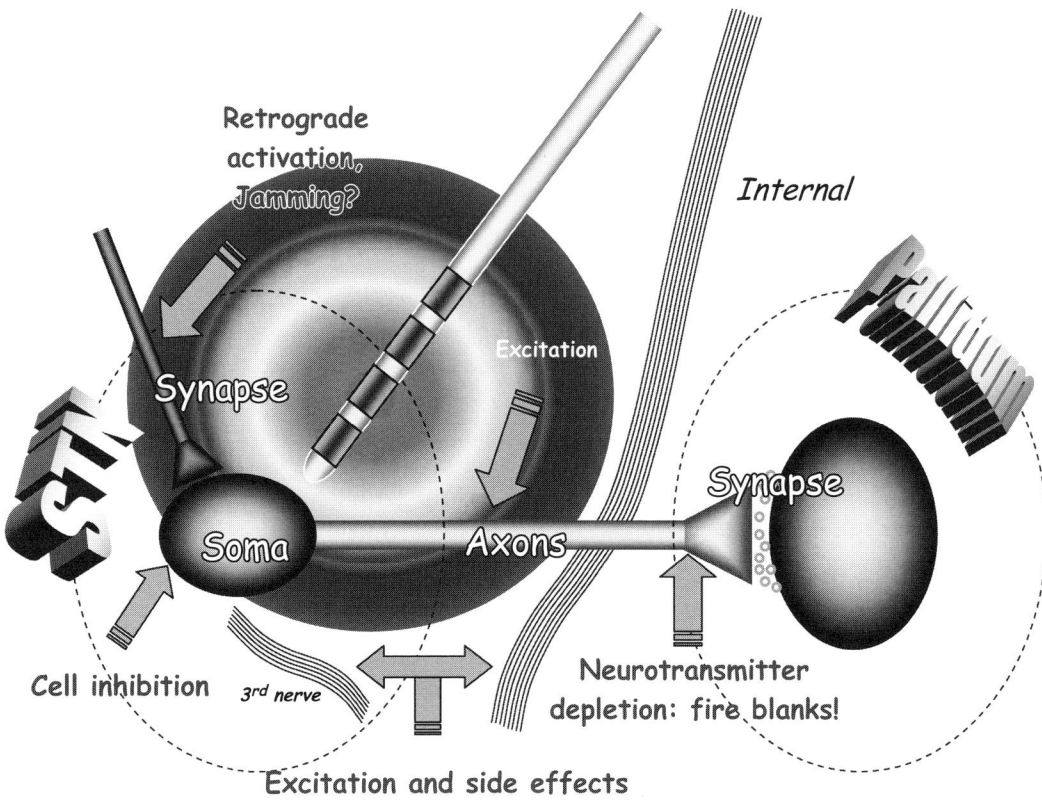

Fig. 1. Schematic representation of the putative mechanisms that contribute to the functional effects.

As tremor in PD is equally improved by STN HFS (Benabid et al., 2009), the indications of VIM stimulation have strongly decreased. They are currently restricted to essential tremor where VIM is still the consensual target, although data are being reported on the efficiency of STN stimulation in such cases. The problem with VIM stimulation in essential tremor (Koller et al., 1999) is that tolerance occurs along time, particularly in patients in whom the intensity of tremor and stimulation is high, and when the patients are stimulated continuously. Tolerance improves after a stimulation holiday when stimulation is resumed at a lower voltage without trying to suppress totally the tremor, and stopping stimulation at night. Another indication for VIM stimulation is writer's cramp or monosymptomatic PD, particularly in the elderly.

*The internal pallidum (GPi)*

Pallidal stimulation, similarly to pallidotomy, is mainly efficient on levodopa-induced dyskinesias (Laitinen et al., 1992). One may observe a progressive decrease in the tendency for PD patients to develop dyskinesias along time. This might be due to a form of neural plasticity, because of the more stable regimen of treatment produced by stimulation as compared to the periodic administration of oral drugs, which creates at the level of the striatal dopaminergic receptors an oscillatory dopamine concentration. These dyskinesias are also decreased during the STN stimulation (Krack et al., 1997; Fraix et al., 2000), but indirectly due to the decrease in drug dosage, which is permitted by the clinical efficiency of STN stimulation. Other indications of pallidal stimulation are currently dystonias

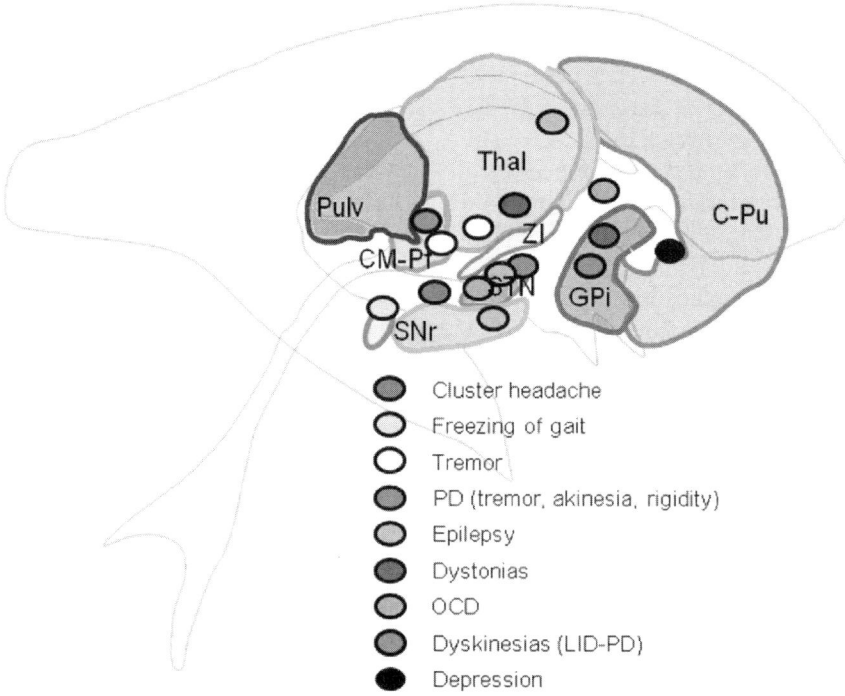

Fig. 2. Locations of all brain regions mentioned as targets for DBS and the related major indications for these targets. Abbreviations: Thal, thalamus; Pulv, pulvenar part of the thalamus; CM–pf, centrum medianum–parafascicularis; STN, subthalamic nucleus; SNr, substantia nigra pars reticulata; Gpi, globus pallidus internus; C–Pu, caudate–putamen complex; ZI, zona interna. (See Color Plate 24.2 in color plate section.)

(Kupsch et al., 2006), although here again recent data might suggest the strong efficiency of STN stimulation.

## The subthalamic nucleus (STN)

This is currently the preferred target because of its efficiency on the triad of dopaminergic symptoms of PD, from akinesia and rigidity to tremor (Fraix et al., 2000; Krack et al., 2003; Funkiewiez et al., 2004; Hamani et al., 2005; Deuschl et al., 2006; Kleiner-Fisman et al., 2006; Benabid et al., 2009). The effects are immediate and reversible; they allow a decrease in drug dosage of about 65%, which is responsible for a comparable decrease in the intensity and duration of the dyskinesias. The consensus indications for STN stimulation are essentially advanced PD with a tendency to operate the patients earlier, as soon as it becomes clear that they will need it, sooner or later. There is no reason to keep them impairing uselessly.

Besides advanced PD, STN HFS has proven to be efficient in epilepsy, dystonia, and obsessive–compulsive disorder where indications are being validated by ongoing clinical studies. The efficiency on multiple system atrophy (MSA) and progressive supranuclear palsy (PSP) does not seem to be significant, although no clinical trial has been specifically designed. The improvements of HFS of STN are persistent along several years of evolution (Krack et al., 2003; Deuschl et al., 2006) on both symptoms of the triad and quality of life. We have currently 15 years of experience in more than 300 bilateral cases of STN stimulation.

As in all surgical procedures, DBS has complications (Blomstedt and Hariz, 2005; Hamani and Lozano, 2006; Benabid et al., 2007; Seijo et al., 2007; Videnovic and Verhagen Metman, 2008). In one retrospective analysis of 526 cases including all targets (138 VIM, 63 GPi, 325 STN), 68% of the patients had no adverse effect of any kind (VIM, 86%; Gpi, 84%; STN, 58%) (Benabid et al., 2007) and 32% had at least one adverse effect (VIM, 15%; GPI, 16%; STN, 43%). The percentage of severe or significant adverse effects was 9% and 8% for VIM- and GPi-stimulated patients, respectively, while it was 22% for STN-stimulated subjects.

The neurocognitive complications are reported in all series with a large variability of occurrence probably depending on the indications and the methodology. In our series, transient postoperative confusion happened in 10% of all cases, temporary postoperative depression in 12%, suicidal attempts in 1.3%, and suicide in 1.2% (1 case in 325 STN patients). Depression and suicidality are multifactorial, including acute changes in treatment, which is a combination of HFS and drugs, and societal issues. All the complications can be minimized by the quality of surgery, the prevention of skin erosion by careful placement of the hardware, the education of patients to avoid and manage irritation of the skin, and optimization of hardware as complications had also clearly decreased over time when the surgical team's learning curve increased. However, in order to better evaluate and eliminate the occurrence of these complications, there is a need for consensus on complications report format, particularly reporting the number of patients without having a complication.

### A newcomer in the club: the pedunculopontine nucleus (PPN)

Following the experimental work of Tipu Aziz (Munro-Davies et al., 1999), patients presenting freezing of gait have been implanted with DBS electrodes (Mazzone et al., 2005; Plaha and Gill, 2005; Stefani et al., 2007). This nucleus is located in the brainstem and difficult to define. However, although the data are still preliminary, it appears that the stimulation of PPN improves PD gait disturbance significantly. It is interesting to note that the efficiency is obtained at low frequency (25 Hz), which suggests that the nucleus has to be excited, coherently with the results of experimental animal data. Whether PPN stimulation might be able to replicate the results of STN stimulation or whether it could be only an additional target to treat the specific symptoms of gait disturbance is not yet clarified.

In our own series of six patients, the coordinates of the electrode are with respect to the posterior commissure (CP) and the midline, −1.84 mm behind CP, −13.31 mm below ACPC level, and 6.8 mm lateral (not published).

### The centrum medianum–parafascicularis (CM–Pf)

The CM–Pf has been shown (Caparros-Lefebvre et al., 1999) to be equivalently efficient as target for stimulation as the thalamic VIM in the treatment of tremor in PD, as well as on levodopa-induced dyskinesias. The reason why dyskinesias are improved by CM–Pf DBS and not by VIM DBS might be related to the fact that GPi sends a heavy projection onto CM–Pf. However, as the other parkinsonian symptoms, bradykinesia and rigidity, are not significantly improved, this does not make CM–Pf a good alternative to GPi or STN DBS.

## Deep brain stimulation: a preferred tool in functional neurosurgery?

DBS has advantages, particularly flexibility, in addition to reversibility and adjustability of the intensity of stimulation (Table 1). The best demonstration is given in PD patients suffering from freezing of gait, where bilateral STN stimulation at 130 Hz is able to control akinesia and rigidity and simultaneous bilateral simulation of the PPN at 25 Hz improves the freezing of gait (Mazzone et al., 2005; Plaha and Gill, 2005; Stefani et al., 2007). This combination of targets at different frequencies is not unique, as there are currently reports of combination of several targets in one patient such as the pallidum and the STN for dystonias. Current protocols in clinical trials are investigating the feasibility and interest of this type of strategy.

Table 1. Pros and cons of deep brain stimulation

|  | Pros | Cons |
|---|---|---|
| Clinical effects | Reversibility of side effects | Voltage-dependent side effects |
| Therapeutic use | Adaptability of parameters to benefits | Necessity for tuning/time |
| Implantation | Relative safety | Cost of hardware and replacement |
|  | Multiple electrodes and targets | Surgical procedure |
| Hardware | Technological evolution | Cosmetic aspects |
|  | Increasing capabilities | Infection and skin erosion |
|  |  | Fracture of leads and extension |
|  |  | Replacement of exhausted battery |

The other advantages of HFS are as follows: it is reversible (permitting intermittent stimulation, close loop, investigation of new targets, reversal of side effects); it is adaptable (fine-tuning, compromise with side effects, optimization of parameters); it is a nonlesioning method with low morbidity (open to new treatment, possibility of replacement of electrodes, multiple electrodes in one target, multiple targets in one patient); and it allows intraoperative investigations (typical firing pattern recognition, specific functional response observation) (Fig. 3 and Table 1).

## *What is the future?*

New targets are being investigated increasingly, suggested often by the results of basic research in neurosciences. Implanting multiple electrodes in one target is also an option as it may lead to a three-dimensional DBS strategy. This would take into account the heterogeneity of most targets, including the smallest ones such as STN, where a more spatially defined delivery of current could help in improving the benefits and at the same time decreasing the side effects.

## *Is STN stimulation neuroprotective?*

The working hypothesis is that HFS might inhibit the hyperactivity of STN neurons and hence decrease the deleterious effect of hyperproduction of glutamate. Experiments in rats as well as in monkeys (Piallat et al., 1996; Nakao et al., 1999; Maesawa et al., 2004; Temel et al., 2006; Wallace et al., 2007) have provided experimental data supporting this hypothesis using either lesioning or HFS of STN. This should be investigated in human patients with clinical trials involving PD at early stages.

## *What are the alternatives to high-frequency stimulation?*

Gene transfer is an option currently investigated by several clinical trials for the treatment of PD (Kaplitt et al., 2007; Marks et al., 2008). Growth factor infusion in the striatum, for instance of GDNF, has been reported but still need validation (Gill et al., 2003). Cortical stimulation is under investigation with interesting preliminary results (Drouot et al., 2004; Canavero and Bonicalzi, 2007; Pagni et al., 2008), but neural grafts of dopaminergic neurons are still the ultimate goal to reach, although this approach is particularly difficult and may find new hopes from the field of stem cell technology. Whatever will come to replace HFS would be, by definition, a better method, which would provide better benefits to the patients.

## What did we learn from the practice of deep brain stimulation in human patients?

## *The pharmacology of STN high frequency stimulation*

From depression to normal mood and then hypomania, there is a continuum that can be traveled along the intensity of any treatment,

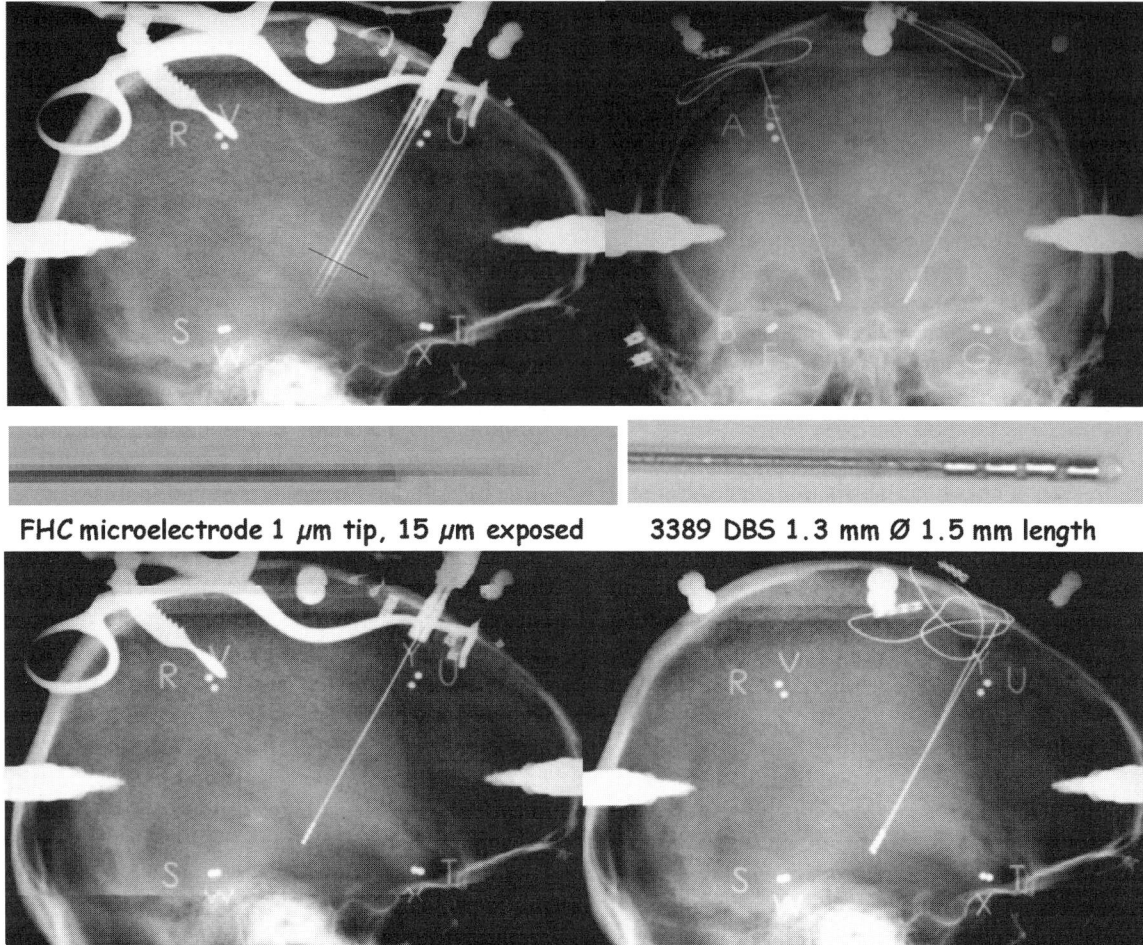

Fig. 3. Shape, size, and dimensions of the electrodes used for target exploration (FHC microelectrodes used as a barrel of five) and chronic stimulation (3389 DBS electrode here bilaterally implanted in the STN for PD).

whether pharmacological or electrical. There is a narrow band called "normality," corresponding to an optimal value of DBS treatment or drug dosage. Below these levels, one may see bradykinesia, doubt, obsessions, depression, sadness (Bejjani et al., 1999), and apathy (Krack et al., 2003; Funkiewiez et al., 2004). For higher values than needed, one may see dyskinesias, overconfidence, hypomania, (Krack et al., 2003; Deuschl et al., 2006; Funkiewiez et al., 2004), laughter (Krack et al., 2001), jokes, and gambling (Herzog et al., 2003; Romito et al., 2002).

*Challenges for the future*

There are many challenges to be addressed in order to improve the efficiency, morbidity, and basic understanding of DBS. This concerns hardware and software (nanotechnologies: miniaturization, multiple electrodes, rechargeability, telemaintenance, and stimulation paradigms with the question: Is the current stimulation the best? Tass and Hauptmann, 2006), mechanisms (the role of low and high frequencies, the involvement of excitation and inhibition), image guidance (MRI deformations), and intraoperative guidance (electrophysiology, local field potentials, target specific markers).

## Conclusions

The saga of functional neurosurgery for movement disorders runs over almost a century. It has been driven by the combined and intertwined forces of therapeutic needs from the pathology, available tools from technology and pharmacology, scientific knowledge from clinical and basic research, anatomy, and neurophysiology. Its course has been shaped by creativity of investigators, positive thinking, luck, and serendipity. The increasing amount of technologies, the improved efficiency of multidisciplinarity, and the harmonious associations of various methods make it increasingly powerful and adaptable to the panel of functional symptoms to alleviate and to the various dysfunctions of the CNS to correct, particularly in the field of movement disorders. The advent of nanotechnologies and the opening of the field of neural prosthesis will broaden the scope of functional neurosurgery even more. There is no doubt that this historical perspective has considered only the very beginning of the saga.

## References

Alexander, G., & Crutcher, M. (1990). Functional architecture of basal ganglia circuits: neural substrates of parallel processing. *Trends in Neurosciences*, *13*, 266–271.

Bejjani, B. P., Damier, P., Arnulf, I., Thivard, L., Bonnet, A. M., Dormont, D., et al. (1999). Transient acute depression induced by high-frequency deep-brain stimulation [see comments]. *New England Journal of Medicine*, *340*, 1476–1480.

Benabid, A. L., Chabardes, S., Mitrofanis, J., & Pollak, P. (2009). Deep brain stimulation of the subthalamic nucleus for the treatment of Parkinson's disease. *Lancet Neurology*, *8*, 67–81.

Benabid, A. L., Chabardes, S., Seigneuret, E., Seigneuret, E., Torres, N., Castana, L., et al. (2007). Surgical complications in 526 DBS patients operated over 20 years. *75th Annual Meeting of the American Association of NeuroSurgeons*; Washington, DC, USA; April. Available at: http://www.aans.org/library/article.aspx?ArticleId = 40839 (November 16, 2008).

Benabid, A. L., Pollak, P., Gervason, C., Hoffmann, D., Gao, D. M., Hommel, M., et al. (1991). Long-term suppression of tremor by chronic stimulation of the ventral intermediate thalamic nucleus. *Lancet*, *337*, 403–406.

Benabid, A. L., Pollak, P., Gao, D. M., Hoffmann, D., Limousin, P., Gay, E., et al. (1996). Chronic electrical stimulation of the ventralis intermedius nucleus of the thalamus as a treatment of movement disorders. *Journal of Neurosurgery*, *84*, 203–214.

Benabid, A. L., Pollak, P., Louveau, A., Henry, S., & Rougemont, J. de. (1987). Combined (thalamotomy and stimulation) stereotactic surgery of the VIM thalamic nucleus for bilateral Parkinson disease. *Applied Neurophysiology*, *50*, 344–346.

Benabid, A. L., Wallace, B., Mitrofanis, J., Xia, R., Piallat, B., Chabardes, S., et al. (2005). A putative generalized model of the effects and mechanism of action of high frequency electrical stimulation of the central nervous system. *Acta Neurologica Belgica*, *105*, 149–157.

Blomstedt, P., & Hariz, M. I. (2005). Hardware-related complications of deep brain stimulation: a ten year experience. *Acta Neurochirurgica*, *147*, 1061–1064.

Canavero, S., & Bonicalzi, V. (2007). Extradural cortical stimulation for movement disorders. *Acta Neurochirurgica. Supplement*, *97*, 223–232.

Caparros-Lefebvre, D., Blond, S., Feltin, M. P., Pollak, P., & Benabid, A. L. (1999). Improvement of levodopa induced dyskinesias by thalamic deep brain stimulation is related to slight variation in electrode placement: possible involvement of the centre median and parafascicularis complex. *Journal of Neurology, Neurosurgery, and Psychiatry*, *67*, 308–314.

Carlsson, A., Lindqvist, M., & Magnusson, T. (1957). 3,4-Dihydroxyphenylalanine and 5-hydroxytryptophanas reserpine antagonists. *Nature*, *180*, 1200.

Cooper, I. S. (1953). Ligation of anterior choroidal artery for involuntary movements-parkinsonism. *Psychiatric Quarterly*, *27*, 317–319.

Cooper, I. S. (1956). An investigation of neurosurgical alleviation of parkinsonism, chorea, athetosis and dystonia. *Annales Medicinae Internae Fenniae*, *45*, 381–392.

Cotzias, G., Van Woert, M., & Schiffer, L. M. (1967). Aromatic amino acids and modification of parkinsonism. *New England Journal of Medicine*, *276*, 374–379.

Deuschl, G., Schade-Brittinger, C., Krack, P., Volkmann, J., Schäfer, H., Bötzel, K., et al. (2006). A randomised trial of deep-brain stimulation for Parkinson's disease. *New England Journal of Medicine*, *355*, 896–908. Erratum in: *New England Journal of Medicine*, *355*, p. 1289

Drouot, X., Oshino, S., Jarraya, B., Besret, L., Kishima, H., Remy, P., et al. (2004). Functional recovery in a primate model of Parkinson's disease following motor cortex stimulation. *Neuron*, *44*, 769–778.

Ehringer, H., & Hornykiewicz, O. (1960). Verteilung von noradrenalin und dopamin (3-hydroxytryptamin) im gehirn des menschen und ihr verhaltenbei erkranktungendes extrapyramidalen systems. *Wiener Klinische Wochenschrift*, 38, 1236–1239.

Filali, M., Hutchison, W. D., Palter, V. N., Lozano, A. M., & Dostrovsky, J. O. (2004). Stimulation-induced inhibition of neuronal firing in human subthalamic nucleus. *Experimental Brain Research*, 156, 274–281.

Fraix, V., Pollak, P., Van Blercom, N., Xie, J., Krack, P., Koudsie, A., et al. (2000). Effect of subthalamic nucleus stimulation on levodopa-induced dyskinesia in Parkinson's disease. *Neurology*, 55, 1921–1923.

Funkiewiez, A., Ardouin, C., Caputo, E., Krack, P., Fraix, V., Klinger, H., et al. (2004). Long term effects of bilateral subthalamic nucleus stimulation on cognitive function, mood, and behaviour in Parkinson's disease. *Journal of Neurology, Neurosurgery, and Psychiatry.*, 75, 834–839.

Gill, S. S., Patel, N. K., Hotton, G. R., O'Sullivan, K., McCarter, R., Bunnage, M., et al. (2003). Direct brain infusion of glial cell line-derived neurotrophic factor in Parkinson disease. *Nature Medicine*, 9, 589–595. Epub 2003 March 31.

Hamani, C., & Lozano, A. M. (2006). Hardware-related complications of deep brain stimulation: a review of te published literature. *Stereotactic and Functional Neurosurgery*, 84, 248–251.

Hamani, C., Richter, E., Schwalb, J. M., & Lozano, A. M. (2005). Bilateral subthalamic nucleus stimulation for Parkinson's disease: a systematic review of the clinical literature. *Neurosurgery*, 56, 1313–1321. Discussion 21-4.

Hammond, C., Ammari, R., Bioulac, B., & Garcia, L. (2008). Latest view on the mechanism of action of deep brain stimulation. *Movement Disorders*, 23(15), 2111–2121.

Hassler, R., Riechert, T., Mundinger, F., Umbach, W., & Gangleberger, J. A. (1960). Physiological observations on stereotaxic operations in extrapyramidal motor disturbances. *Brain*, 83, 337–350.

Herzog, J., Reiff, J., Krack, P., Witt, K., Schrader, B., Muller, D., et al. (2003). Manic episode with psychotic symptoms induced by subthalamic nucleus stimulation in a patient with Parkinson's disease. *Movement Disorders*, 18, 1382–1384.

Kaplitt, M. G., Feigin, A., Tang, C., Fitzsimons, H. L., Mattis, P., Lawlor, P. A., et al. (2007). Safety and tolerability of gene therapy with an adeno-associated virus (AAV) borne GAD gene for Parkinson's disease: an open label, phase I trial. *Lancet*, 369, 2097–2105.

Kleiner-Fisman, G., Herzog, J., Fisman, D. N., Tamma, F., Lyons, K. E., Pahwa, R., et al. (2006). Subthalamic nucleus deep brain stimulation: summary and meta-analysis of outcomes. *Movement Disorders*, 21(Suppl. 14), S290–S304.

Koller, W. C., Lyons, K. E., Wilkinson, S. B., & Pahwa, R. (1999). Efficacy of unilateral deep brain stimulation of the VIM nucleus of the thalamus for essential head tremor. *Movement Disorders*, 14(5), 847–850.

Krack, P., Batir, A., Van Blercom, N., Chabardes, S., Fraix, V., Ardouin, C., et al. (2003). Five-year follow-up of bilateral stimulation of the subthalamic nucleus in advanced Parkinson's disease. *New England Journal of Medicine*, 349, 1925–1934.

Krack, P., Kumar, R., Ardouin, C., Dowsey, P. L., McVicker, J. M., Benabid, A. L., et al. (2001). Mirthful laughter induced by subthalamic nucleus stimulation. *Movement Disorders*, 16, 867–875.

Krack, P., Limousin-Dowsey, P., Benabid, A. L., & Pollak, P. (1997). Chronic stimulation of subthalamic nucleus improves levodopa-induced dyskinesias in Parkinson's disease. *Lancet*, 350, 1676.

Kupsch, A., Benecke, R., Müller, J., Trottenberg, T., Schneider, G. H., Poewe, W., et al. (2006). Deep brain stimulation for Dystonia Study Group: pallidal deep-brain stimulation in primary generalized or segmental dystonia. *New England Journal of Medicine*, 355, 1978–1990.

Laitinen, L. V., Bergenheim, A. T., & Hariz, M. (1992). Leksell's posteroventral pallidotomy in the treatment of Parkinson's disease. *Journal of Neurosurgery*, 76(1), 53–61.

Maesawa, S., Kaneoke, Y., Kajita, Y., Usui, N., Misawa, N., Nakayama, A., et al. (2004). Long-term stimulation of the subthalamic nucleus in hemiparkinsonian rats: neuroprotection of dopaminergic neurons. *Journal of Neurosurgery*, 100, 679–687.

Marks, W. J., Jr., Ostrem, J. L., Verhagen, L., Starr, P. A., Larson, P. S., Bakay, R. A. E., et al. (2008). Safety and tolerability of intraputaminal delivery of CERE-120 (adeno-associated virus serotype 2-neurturin) to patients with idiopathic Parkinson's disease: an open-label, phase I trial. *Lancet Neurology*, 7, 400–408.

Mazzone, P., Lozano, A., Stanzione, P., Galati, S., Scarnati, E., Peppe, A., et al. (2005). Implantation of human pedunculopontine nucleus: a safe and clinically relevant target in Parkinson's disease. *Neuroreport*, 16, 1877–1881.

Meissner, W., Leblois, A., Hansel, D., Bioulac, B., Gross, C. E., Benazzouz, A., et al. (2005). Subthalamic high frequency stimulation resets subthalamic firing and reduces abnormal oscillations. *Brain*, 128, 2372–2382. Epub 2005 August 25.

Meyers, R. (1942). Surgical interruption of the pallido-fugal fibers. Its effect on the syndrome agitans and technical considerations in its applications. *New York State Journal of Medicine*, 42, 317–325.

Morizane, A., Li, J. Y., & Brundin, P. (2008). From bench to bed: the potential of stem cells for the treatment of Parkinson's disease. *Cell and Tissue Research*, 331, 323–336. Epub 2007 November 22.

Munro-Davies, L. E., Winter, J., Aziz, T. Z., & Stein, J. F. (1999). The role of the pedunculopontine region in basal-ganglia mechanisms of akinesia. *Experimental Brain Research*, 129, 511–517.

Nakao, N., Nakai, E., Nakai, K., & Itakura, T. (1999). Ablation of the subthalamic nucleus supports the survival of nigral dopaminergic neurons after nigrostriatal lesions induced by mitochondrial toxin 3-nitropropionic acid. *Annals of Neurology*, 45, 640–651.

Pagni, C. A., Albanese, A., Bentivoglio, A., Broggi, G., Canavero, S., Cioni, B., et al. (2008). The experience of the Italian Study Group of the Italian Neurosurgical Society. *Acta Neurochirurgica. Supplement*, 101, 13–21.

Piallat, B., Benazzouz, A., & Benabid, A. L. (1996). Subthalamic nucleus lesion in rats prevents dopaminergic nigral neuron degeneration after striatal 6-OHDA injection: behavioural and immunohistochemical studies. *Eurupean Journal Neuroscience*, *8*, 1408–1414.

Plaha, P., & Gill, S. S. (2005). Bilateral deep brain stimulation of the pedunculopontine nucleus for Parkinson's disease. *Neuroreport*, *16*, 1883–1887.

Romito, L. M., Raja, M., Daniele, A., Contarino, M. F., Bentivoglio, A. R., Barbier, A., et al. (2002). Transient mania with hypersexuality after surgery for high frequency stimulation of the subthalamic nucleus in Parkinson's disease. *Movement Disorders*, *17*, 1371–1374.

Seijo, F. J., Alvarez-Vega, M. A., Gutierrez, J. C., Fdez-Glez, F., & Lozano, B. (2007). Complications in subthalamic nucleus stimulation surgery for treatment of Parkinson's disease. Review of 272 procedures. *Acta Neurochirurgica*, *149*, 867–875.

Stefani, A., Lozano, A. M., Peppe, A., Stanzione, P., Galati, S., Tropepi, D., et al. (2007). Bilateral deep brain stimulation of the pedunculopontine and subthalamic nuclei in severe Parkinson's disease. *Brain*, *130*, 1596–1607. Epub 2007 January 24.

Tai, C. H., Boraud, T., Bezard, E., Bioulac, B., Gross, C., & Benazzouz, A. (2003). Electrophysiological and metabolic evidence that high-frequency stimulation of the subthalamic nucleus bridles neuronal activity in the subthalamic nucleus and the substantia nigra reticulata. *FASEB Journal*, *17*, 1820–1830.

Tass, P. A., & Hauptmann, C. (2006). Therapeutic modulation of synaptic connectivity with desynchronizing brain stimulation. *International Journal of Psychophysiology*, *64*, 53–61. Epub 2006 September 25.

Temel, Y., Visser-Vandewalle, V., Kaplan, S., Kozan, R., Daemen, M. A., Blokland, A., et al. (2006). Protection of nigral cell death by bilateral subthalamic nucleus stimulation. *Brain Research*, *1120*, 100–105.

Tretiakoff, K. (1919). *Contribution a l'étude de l'anatomie pathologique du Locus Niger de Soemmering avec quelques déductions relatives à la pathogénie des troubles du tonus musculaire et de la maladie de Parkinson*. Thesis, University of Paris.

Videnovic, A., & Verhagen Metman, L. (2008). Deep brain stimulation for Parkinson's disease: prevalence of adverse events and need for standardized reporting. *Movement Disorders*, *23*, 343–349.

Wallace, B. A., Ashkan, K., Heise, C. E., Foote, K. D., Torres, N., Mitrofanis, J., et al. (2007). Survival of midbrain dopaminergic cells after lesion or deep brain stimulation of the subthalamic nucleus in MPTP-treated monkeys. *Brain*, *130*, 2129–2145.

Welter, M. L., Houeto, J. L., Bonnet, A. M., Bejjani, P. B., Mesnage, V., Dormont, D., et al. (2004). Effects of high-frequency stimulation on subthalamic neuronal activity in parkinsonian patients. *Archives of Neurology*, *61*, 89–96.

Xia, R., Berger, F., Piallat, B., & Benabid, A. L. (2007). Alteration of hormone and neurotransmitter production in cultured cells by high and low frequency electrical stimulation. *Acta Neurochirurgica*, *149*, 67–73. Discussion, p. 73; Epub 2006 December 15.

CHAPTER 25

# Recovery of control of posture and locomotion after a spinal cord injury: solutions staring us in the face

Andy J. Fong[7], Roland R. Roy[1,3], Ronaldo M. Ichiyama[4], Igor Lavrov[1], Grégoire Courtine[5], Yury Gerasimenko[1,6], Y.C. Tai[7,8], Joel Burdick[7,8] and V. Reggie Edgerton[1,2,3],*

[1]*Department of Physiological Science, University of California, Los Angeles, Los Angeles, CA, USA*
[2]*Department of Neurobiology, University of California, Los Angeles, Los Angeles, CA, USA*
[3]*Brain Research Institute, University of California, Los Angeles, Los Angeles, CA, USA*
[4]*Institute of Membrane and Systems Biology, University of Leeds, Leeds, UK*
[5]*Neurobiology Department, University of Zurich, Zurich, Switzerland*
[6]*Pavlov Institute of Physiology, St. Petersburg, Russia*
[7]*Division of Engineering, Bioengineering, California Institute of Technology, Pasadena, CA, USA*
[8]*Division of Engineering, Mechanical Engineering Options, California Institute of Technology, Pasadena, CA, USA*

**Abstract:** Over the past 20 years, tremendous advances have been made in the field of spinal cord injury research. Yet, consumed with individual pieces of the puzzle, we have failed as a community to grasp the magnitude of the sum of our findings. Our current knowledge should allow us to improve the lives of patients suffering from spinal cord injury. Advances in multiple areas have provided tools for pursuing effective combination of strategies for recovering stepping and standing after a severe spinal cord injury. Muscle physiology research has provided insight into how to maintain functional muscle properties after a spinal cord injury.

Understanding the role of the spinal networks in processing sensory information that is important for the generation of motor functions has focused research on developing treatments that sharpen the sensitivity of the locomotor circuitry and that carefully manage the presentation of proprioceptive and cutaneous stimuli to favor recovery. Pharmacological facilitation or inhibition of neurotransmitter systems, spinal cord stimulation, and rehabilitative motor training, which all function by modulating the physiological state of the spinal circuitry, have emerged as promising approaches. Early technological developments, such as robotic training systems and high-density electrode arrays for stimulating the spinal cord, can significantly enhance the precision and minimize the invasiveness of treatment after an injury.

Strategies that seek out the complementary effects of combination treatments and that efficiently integrate relevant technical advances in bioengineering represent an untapped potential and are likely to have an immediate impact. Herein, we review key findings in each of these areas of research and present

*Corresponding author.
Tel.: +310 825 1910; Fax: +310 267 2071;
E-mail: vre@ucla.edu

a unified vision for moving forward. Much work remains, but we already have the capability, and more importantly, the responsibility, to help spinal cord injury patients now.

**Keywords:** spinal cord injury; rehabilitation; robotic motor training; pharmacological intervention; skeletal muscle adaptation; proprioception; epidural stimulation; locomotion

**Why does spinal cord injury result in a loss of movement control?**

Movements are defined by the combination of motor pools activated, the level at which they are recruited, and the effectiveness with which the corresponding muscles generate force. The diminished level of movement that follows a spinal cord injury has been attributed generally to an inability to activate motor pools. For most individuals with a spinal cord injury, however, this is less of a factor than typically assumed. Three more salient issues are linked to the impaired ability to recruit appropriate ensembles of motor units in a manner that yields effective movement. First, a significant portion of movement loss is attributable to functional alterations of the spinal circuitry that disrupt the coordination of the motor pools. Second, when a person with a severe, incomplete spinal cord lesion attempts to perform a movement, the level of recruitment is insufficient for some motor pools, while actually exceeding normal levels for others. Finally, it is well known that chronic spinal cord injuries lead to a progressive decline in muscle function. All three of these impairments must be addressed to realize the maximal potential for recovering functional locomotion.

*Aberrant synapse formation leads to inappropriate muscle recruitment and poor coordination*

The loss of most, if not all, descending neural control after a spinal cord injury rapidly triggers adaptation of circuits in the brain and spinal cord. In particular, the neural circuitries responsible for posture and locomotion undergo major reorganization, a process that can continue to evolve for years (Humphrey et al., 2006). These adaptations include the formation of new functional connections. Yet while a large number of new synapses are formed, there is overwhelming evidence that many of these are abnormal connections that misdirect neurons to inappropriate downstream motor networks. The development of such aberrant connections (between the brain and the spinal cord for incomplete injuries, and within the spinal cord circuitry for complete injuries) generally results in poor coordination, unintended movements, and spasticity. For example, when individuals with a severe mid-thoracic spinal cord injury attempt to flex or extend the ankle on one side, often the entire lower limb will flex or extend, or the movement will occur bilaterally (Maegele et al., 2002). Such stimulus-evoked activation of abnormally large numbers of muscles is very common, and may correspond to the widespread synapsing of locomotor network neurons onto multiple nonspecific targets (Calancie et al., 1993). This lack of specificity in synapse formation leads to coactivation of circuits that are not normally activated synchronously, which is a major determinant of step failure.

*Changes in the excitability of the spinal locomotor networks render some synapses hyperexcitable and others hypoexcitable*

In the literature, the spasticity and other functional deficits associated with spinal cord injury are often attributed to hyperexcitability of the spinal circuitry (Nance, 2003). It seems unlikely, however, that this explanation is sufficient to explain the complex cadre of neural changes that accompany spinal cord injury. Hyperexcitability is not always detrimental. In some cases, a higher-than-normal level of recruitment can actually serve as a significant positive adaptation: cooperation between motor pathways, wherein hyperexcitation of one motor pool helps to compensate for hypoexcitation of a related motor pool, can be an important mechanism and strategy for regaining motor function. Furthermore, although there is an increase in the number of aberrant connections after a complete spinal cord injury, some of the spinal circuits are hypoexcitable. Up-regulation of the inhibitory neurotransmitter systems, i.e., the GABAergic and glycinergic systems, depresses the excitability of the spinal circuitry after injury (Edgerton et al., 2001). Strategies targeted at reversing this hypoexcitable state have been very effective. Within minutes, pharmacological treatment with antagonists of these inhibitory neurotransmitters dramatically enhances the locomotor capability of complete spinal cats, taking them from being completely unable to step, to being able to execute successful weight-bearing stepping over a range of speeds (De Leon et al., 1999b). In addition, locomotor training can reverse the depression of the spinal circuits by reducing the number of glycine receptors and the level of $GAD_{67}$ expression (Edgerton et al., 1991, 2001; Tillakaratne et al., 2002). After a spinal cord injury, pathways can be rendered either hyper- or hypoexcitable. Successful rehabilitation requires properly managing the level of excitability, as necessary, of each of the critical locomotor circuits. Excitatory treatments are needed for some circuits, while inhibition is required for others.

## *Progressive deterioration of muscle properties diminishes the ability to generate movements*

The skeletal musculature is highly sensitive to the level of neuromuscular activity, i.e., the levels of activation and loading imposed on the muscles (Edgerton and Roy, 1996; Roy et al., 1991). The chronic decrease in both the activation and loading levels of the muscles below the level of a spinal cord injury results in atrophy, a concomitant loss of force generating potential, and a decrease in fatigue resistance (Castro et al., 1999; Gerrits et al., 2002, 2003; Shields and Dudley-Javoroski, 2006). In other words, the muscles become weak and easy to fatigue in the absence of any countermeasure interventions. The functional consequence of these muscle adaptations is that the individual must recruit a higher percentage of motor units from the muscles involved in performing any given task. Regaining the ability to stand or step most likely will be adversely affected if the muscles are allowed to deteriorate over any prolonged period. A number of rehabilitative strategies have been implemented in attempts to prevent muscle deterioration associated with a chronic decrease in muscle use. The most effective intervention has been the use of electrical stimulation under loaded conditions. With the appropriate use of this countermeasure, it seems feasible that the skeletal musculature can be maintained in a state that will provide the optimum conditions for regaining standing and stepping ability via epidural stimulation and/or pharmacological interventions (see below).

The motor deficits associated with spinal cord injury arise from multiple deficiencies in the neuromuscular system. In this review, we will identify how appropriate interventions involving activity-based therapies can be used to improve posture and locomotion by restoring muscle properties and reinforcing appropriate synaptic connections.

## Initiating, sustaining, and stopping movements: sources of control

### *"Conscious" control*

A common, but incorrect, assumption is that control of movement occurs almost exclusively in the motor cortex. Likewise, it is a misconception that most movements are controlled consciously. On the contrary, there is overwhelming evidence that the details of most movements are performed routinely, with little conscious or voluntary effort. Shik and Orlovsky (1976) proposed the concept of "automaticity" in movement control, suggesting that many movements are executed by parts of the brain and spinal cord that are not commonly associated with "voluntary" or "conscious" control. For over a century, it has been known that even tasks as complex as weight-bearing locomotion can be executed rather effectively in animals after the cerebral cortex is ablated (Grillner, 1981). This evidence of subcortical control of posture and locomotion suggests that there are potential sites in the brainstem and spinal cord that can generate complex movements in response to general stimulating signals without requiring detailed, millisecond-to-millisecond conscious control.

### "Brainstem" control

Shik et al. (1966) convincingly demonstrated in decerebrate cats that a region within the mesencephalon caudal to the pons (now commonly referred to as the mesencephalic locomotor region) can be tonically electrically stimulated to induce stepping. When either the voltage or frequency of stimulation is increased, animals step faster. These authors recognized, however, that the speed of stepping also is regulated strongly by the speed of the treadmill belt. Mori et al. (1991) demonstrated that an awake, sitting cat can be induced to rise and to begin stepping by stimulating chronically implanted electrodes placed in the ventral tegmental field of the caudal pons along its midline. In addition, cats can be induced to stop stepping and to sit on their hindquarters by stimulating the dorsal tegmental field. These results demonstrate that very complex motions can be triggered with relatively non-specific stimulation parameters, but at specific sites within the brainstem.

### "Spinal" control

Multiple sites within the spinal cord can be stimulated to induce or facilitate stepping movements. Stepping can be initiated in cats by providing relatively nonspecific tonic stimulation to a region of the spinal cord referred to as the "locomotor strip" (Kazennikov et al., 1983). This strip is located just lateral to the dorsal boundary of the dorsal horn, lying at a depth of approximately 1–2 mm, and extending from C1 caudally to approximately the L1 spinal cord segment. The capability of the spinal cord to convert rather non-specific stimulating signals into functional motor activity is even more remarkable than in the brainstem. For example, stimulation via epidural electrodes placed anywhere between the T12 and the L6 spinal cord segments can induce locomotor-like and standing-like movements in both cats and rats (Gerasimenko et al., 2002, 2003, 2008; Ichiyama et al., 2005; Kazennikov et al., 1983). Nevertheless, regional differentiation does exist, and it dictates how the spinal cord responds to electrical stimulation. For example, in rats, electrodes placed along the midline of spinal cord segments L2 and S1 seem to be more effective in facilitating stepping compared to electrodes placed at other levels (Ichiyama et al., 2005), whereas stimulation of the L2 and L5 spinal levels is most effective in humans and cats, respectively (Gerasimenko et al., 2002, 2008). In addition, stimulating at spinal cord segments L2 and S1 seem to be more effective in facilitating stepping compared to stimulating at either segmental level alone in the rat (van den Brand et al., 2007).

The ability of epidural stimulation to generate effective stepping is attributed frequently to the activation of neural circuits in the spinal cord responsible for central pattern generation, i.e., circuits that generate coordinated alternating flexor-extensor neuromotor patterns in the absence of supraspinal or sensory modulation. Central pattern generation certainly has an important role in locomotion, but when the injury spares the locomotor circuitry and afferent inputs, sensory information, e.g., proprioceptive, cutaneous, etc., can be equally, if not more, important in shaping the recovered locomotor patterns. Effective execution of weight-bearing stepping in spinal subjects appears to be accomplished when the spinal locomotor circuits are modulated by stepping-associated sensory input. The important point is that the stimulation parameters can be rather nonspecific if the sensory input associated with stepping is available to provide the fine tuning.

### "Sensory" control

The intact spinal cord has a remarkable ability to utilize cutaneous and proprioceptive sensory information to adapt to different environmental conditions during locomotion (Buford and Smith, 1993; Forssberg, 1979). A series of experiments reported over the last few years demonstrate that these capabilities are retained after spinal cord injury (Cai et al., 2006; Cote and Gossard, 2004; Musienko et al., 2007; Timoszyk et al., 2002, 2005). For example, it is well known that complete, low-thoracic spinal cats can modify their stepping kinematics during the swing and stance phases to adapt to changes in treadmill speed, making essentially the same adjustments as intact animals (De Leon et al., 1998). These complete spinal cats can modulate the excitatory levels of appropriate motor pools during different levels of weight bearing, and they can even walk backwards when the treadmill is reversed (Musienko et al., 2007). Essentially the same responses have been observed in human subjects with complete spinal cord injuries when they are partially assisted to walk on a treadmill (Harkema et al., 1997). These experiments demonstrate that it is the combination of the intrinsic ability of the spinal circuitry to execute rhythmic motor patterns with the accessibility to activity-specific sensory information that allows the spinal cord to function with effective automaticity and minimal or no control from the brain. The realization that the locomotor circuits can function independently from brain control opens the door to new paradigms for recovering posture and locomotion in individuals with severe spinal cord injuries.

## Treatment paradigms for preserving muscle function after a spinal cord injury

### Spinal cord injury leads to the degradation of muscle properties

One of the primary effects of a spinal cord injury on the motor system is a loss of mass and function of the muscles below the level of the injury. These effects appear to be muscle type specific. Muscles that function as extensors, i.e., those that are heavily involved in weight support and propulsive functions and show the highest daily activity levels, are the most affected after the injury. For example, in chronic complete spinal cats, the soleus and medial gastrocnemius muscles (primary plantarflexors) show a greater loss in mass and maximum force potential than the tibialis anterior (a primary dorsiflexor) (Roy et al., 1991). In addition, the predominantly slow soleus is affected more than the predominantly fast medial gastrocnemius. Spinal cord injury also has a severe impact on the muscle phenotype, i.e., there is a general shift toward an increase in the percentage of fibers having a faster phenotype within the affected muscles (Talmadge, 2000). This is particularly evident in muscles that normally not only have a high percentage of type I (slow) fibers such as the soleus, but also involve a shift to the fastest phenotypes in normally predominantly fast muscles such as the medial gastrocnemius and tibialis anterior. In effect, the muscles below a spinal cord lesion become smaller, weaker, and more fatigable after the injury. Similar effects generally are observed in human subjects after a spinal cord injury: atrophy, loss of maximum force potential, slow to fast fiber type conversion, and increased fatigability (Burnham et al., 1997; Castro et al., 1999; Gerrits et al., 1999; Shields, 1995).

### Activity-based treatments help to maintain muscle properties

A number of interventions have been attempted to prevent the loss of muscle function associated with spinal cord injury. Several exercise modalities have been used with varying results. For example, we have shown that training complete spinal cord transected cats to step on a treadmill ameliorates, but does not prevent, the loss of mass and force potential of the soleus muscle (Roy et al., 1998), but has a minimal effect on these properties in the medial gastrocnemius and tibialis anterior muscles (Roy et al., 1999). Training the cats to support their weight (stand training), on the other hand, had a positive effect in both the soleus (Roy et al., 1998) and the medial gastrocnemius (Roy et al., 1999). A "passive" cycling exercise regime in spinal rats also had a muscle-specific effect: the total cross sectional area of all fiber types was maintained near control levels in the soleus, but not in the extensor digitorum longus (a predominantly fast synergist of the tibialis anterior for dorsiflexion) (Dupont-Versteegden et al., 1998). This same cycling paradigm in combination with fetal spinal cord tissue implant was more effective in maintaining the mass of the soleus and plantaris (a predominantly fast plantar flexor) muscle than either intervention alone (Dupont-Versteegden et al., 2000), highlighting the importance of combinatory strategies for rehabilitation.

Even limited amounts of daily muscle stimulation can help to maintain muscle properties. The effectiveness of stimulation depends, in part, on how the subjects are trained. We have used the model of spinal cord isolation, which results in neuromuscular inactivity, to examine these issues. The advantage of the spinal cord isolation model is that the baseline level of neuromuscular activity in the muscle is known, i.e., the muscles are virtually inactive (Roy et al., 2007), and thus the effects of imposing known patterns and/or amounts of activity can be determined. For example, we found that the same amount of activation (electrical stimulation through the lateral gastrocnemius-soleus nerve mimicking the EMG pattern observed during treadmill walking) imposed during repeated isometric contractions was more effective in maintaining mass and phenotype of the cat soleus muscle closer to normal than when stimulating with either lengthening or shortening contractions (Roy et al., 2002). Using the rat model of spinal cord isolation, we have initiated studies to determine the minimum number of contractions required to maintain the properties of a muscle. As little as 1 min of brief, high-load isometric contractions per day was sufficient to significantly ameliorate the loss of mass and maximum tetanic tension, and the shift from slower-to-faster phenotypes in the otherwise inactive medial gastrocnemius muscle (Kim et al., 2007). In addition, delivering the same amount of activity during two sessions per day was more effective than one session per day (Fig. 1). Furthermore, a recent study in our laboratory using the same two per day stimulation protocol indicates that a total of 4 min of stimulation per day maintained the mass of the medial gastrocnemius at normal control values. In effect, these data from animal studies indicate that a minimal amount of high-load activity is very effective in maintaining skeletal muscle properties (Kim et al., 2008).

In humans, the most common rehabilitative strategy used to maintain skeletal muscle mass after a spinal cord injury has been functional electrical stimulation, usually administered during cycling activity. Functional electrical stimulation ameliorates muscle loss if the muscles are allowed to produce significant forces during the stimulation. For example, Crameri et al. (2004) highlighted the importance of the loading characteristics in paraplegic subjects: both fiber size and phenotype were maintained closer to control values in the vastus lateralis (a knee extensor) using an isometric loading paradigm as opposed to a dynamic, concentric, minimally loaded paradigm. In addition, this group also showed that an isometric electrical stimulation regime could largely prevent the adverse effects of spinal cord injury on the fiber size and fiber type of paralyzed human muscle if initiated soon after the injury (Crameri et al., 2000). Similarly, Stein et al. (1992) stimulated the tibialis anterior of spinal cord injured subjects for 6 weeks under conditions where the muscle was allowed to shorten (low load) and found an increase in fatigue resistance but no change in tetanic force. The importance of load in maintaining muscle properties also was demonstrated for the wrist extensors of tetraplegic subjects (Hartkopp et al., 2003). A high resistance protocol (30 Hz, maximum load) improved muscle strength and fatigue resistance, whereas a low resistance protocol (15 Hz, 50% of maximum load for the same total work as the high resistance protocol) improved only the fatigue resistance over a 12-week period. A potential complication with functional electrical stimulation, however, is that electrically evoked isometric contractions can cause muscle damage in long-term spinal cord injured patients (Bickel et al., 2004). Whether or not the muscles become more damaged or nonfunctional, however, is probably determined by the specific parameters of the stimulation paradigm.

Functional electrical stimulation also can ameliorate the adaptations in fiber phenotype associated with spinal cord injury. For example, long-term electrical stimulation (1 year) of the vastus lateralis can prevent the shift from slow to fast phenotypes usually associated with spinal

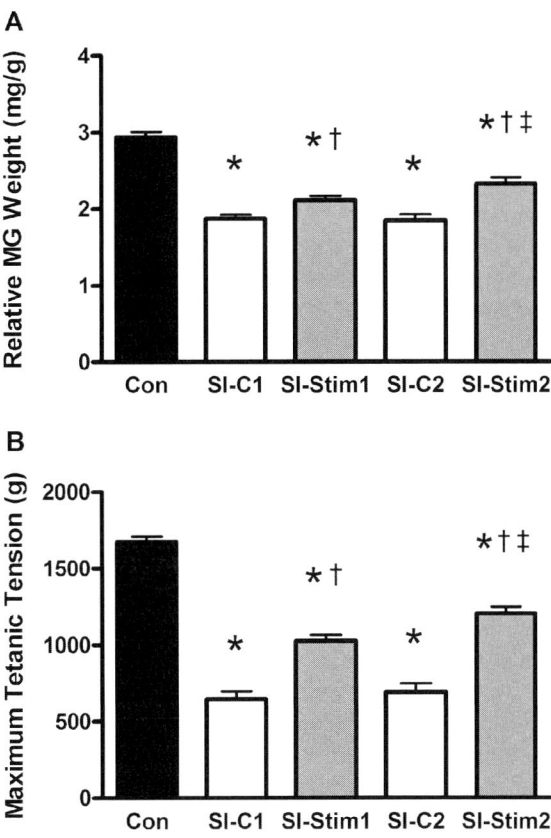

Fig. 1. Brief periods of daily, high-load isometric contraction reduce the loss of muscle function after spinal cord injury. Spinal cord isolated rats (complete spinal cord transections at a mid-thoracic and a high-sacral level, plus dorsal rhizotomy performed between the two transection sites) that were administered one (SI-Stim1) or two (SI-Stim2) bouts of muscle stimulation daily exhibited less atrophy (A, muscle mass normalized to body mass) and a smaller loss of force generation capability (B, maximum tetanic tension) in the stimulated medial gastrocnemius (MG) muscle (gray bars) compared to the non-stimulated contralateral muscle (white bars, SI-C1 and SI-C2) after 30 days of treatment. Stimulation was applied at twice the minimum amplitude necessary to generate a maximum tetanic contraction. The stimulation algorithm applied a 1-s-duration, 100-Hz pulse train that was repeated every 30 s for 5 min. For Stim1, this algorithm was repeated six times over a 1-h period, with 5-min rest periods between repetitions. Stim2 also was stimulated six times, but a 9-h rest period was imposed between the third and fourth cycles. While the daily amount of stimulation was the same for both groups, rats that were stimulated twice/day versus once/day maintained muscle properties that were more similar to those of uninjured controls (black bars). Values are reported as mean ± SEM. *, †, and ‡, significantly different from uninjured control (Con), from non-stimulated contralateral muscle, and from Stim1, respectively. Adapted with permission from Kim et al. (2007).

cord injury (Andersen et al., 1996). Harridge et al. (2002) reported that a fast-to-slow conversion can occur in muscles of human spinal cord injured subjects if the stimulation is intensive enough: they showed an up-regulation of type I myosin heavy chain in the tibialis anterior when stimulated at 10 Hz, 2–6 h/day for 4 weeks. Martin et al. (1992) reported an increase in the oxidative capacity and type I (slow) fibers in the tibialis anterior muscle of spinal cord injured patients after 24 weeks of functional electrical stimulation under no load conditions (the muscles were free to shorten with no external loading of the foot). Other interventions also have been shown to be effective: Stewart et al. (2004) reported that 6 months of body weight support training in incomplete (ASIA C) spinal cord injured subjects resulted in an increase in the mean cross sectional area of type I and IIa fibers, an increase in the percentage of type IIa fibers, a concomitant decrease in type IIa/IIx fibers, and an increase in the oxidative capacity of the vastus lateralis muscle. There also was an improvement in ambulatory capacity and fatigue resistance (time on the treadmill). These data are particularly intriguing and begin to address the important issue of whether the maintenance of muscle mass has a positive effect on the recovery of locomotor ability.

One of the primary deleterious effects of a spinal cord injury in humans is an increase in the fatigability of the muscles. It appears that the decrease in fatigue resistance after an injury is progressive: the soleus muscle of chronic paralyzed subjects ($3.7 \pm 2.05$ years) is more fatigable than in acute paralyzed subjects ($4.6 \pm 1.1$ weeks) (Shields, 1995). Functional electrical stimulation, under either loaded or unloaded conditions, has been effective in restoring or maintaining fatigue resistance. For example, Gerrits et al. (2000) reported an increase in fatigue resistance in the quadriceps muscles of motor-complete spinal cord injured subjects after 6 weeks of functional electrical stimulation cycle ergometry training. Subsequently, they showed that low-frequency stimulation (10 Hz) was more effective than high frequency stimulation (50 Hz) in increasing fatigue resistance while having similar effects in increasing maximum tension capability (20%) over a 12-week period (Gerrits et al., 2002). When initiated within 6 weeks of injury, a unilateral plantar flexion electrical stimulation protocol applied under high-load conditions resulted in the following: compared to the non-stimulated limb, the stimulated limb was less fatigable and produced higher torques under the same testing conditions (Shields and Dudley-Javoroski, 2006). Some possible mechanisms involved in regaining fatigue resistance in muscles after a spinal cord injury include: (1) an increase in the percentage of high oxidative fibers and/or fibers containing the slow isoform of sarco(endo)-plasmic reticulum calcium ATPase (SERCA2) (Talmadge et al., 2002), (2) an improved oxidative capacity of the muscles, e.g., increased succinate dehydrogenase activity (Gerrits et al., 2003), and (3) an increase in fiber size, thus decreasing the number of activated motor units required to perform a given task.

## *Neurotrophic factors help to maintain muscle properties even in the absence of neuromuscular activity*

Thus far we have emphasized the role of neuromuscular activity, i.e., the amount and pattern of loading and activation, in maintaining the homeostatic level of skeletal muscles. It is important to realize, however, that several other factors must be considered. For example, in the absence of neuromuscular activity, the presence of an intact neuromuscular connectivity has a beneficial effect on the muscle properties. After 60 days of inactivity, the relative mass, maximum tetanic tension, specific tension (tension/physiological cross sectional area), and fatigue resistance of the rat soleus is significantly higher in spinal cord isolated (neuromuscular connectivity intact) than in denervated (no neuromuscular connectivity) rats (see Table 1 in Roy et al., 2002). Similarly, the motoneurons associated with the affected muscles are differentially affected: even after prolonged periods of inactivity the motoneurons in spinal cord isolated animals maintain their size and succinate dehydrogenase activity level near control values, whereas axotomized motoneurons have decreased succinate dehydrogenase levels (Chalmers et al., 1992; Roy et al., 2007). The differences in the effects on the muscle properties in these two models of inactivity are most likely related to the presence of activity-independent neurotrophic influences between the muscle and the innervating motoneurons. The recent report by Lee et al. (2007) showing that a long-term peripheral nerve graft in combination with acidic fibroblast growth factor repair (6 months) in complete spinal transected rats was effective in partially restoring the mass and slow phenotype composition of the soleus muscle is consistent with a possible beneficial role of neurotrophic (growth) factors. Spinal cord injury depresses the mRNA and/or protein levels of brain-derived neurotrophic factor and neurotrophin-3 in the spinal cord and/or skeletal muscles, and exercise subsequently elevates these levels to or above control values (Dupont-Versteegden et al., 2004; Gomez-Pinilla et al., 2004; Ying et al., 2005). Thus, these activity-independent neurotrophic factors may be significant contributors to the amelioration of muscle fiber atrophy and phenotype shifts observed with the functional electrical stimulation and exercise countermeasures described above. For injuries such as denervation, where the endogenous supply of neurotrophic factors is lost, exogenous administration may serve as an effective treatment for maintaining muscle function. This may be particularly true for cauda equina lesions where, despite massive denervation, functional electrical stimulation can recover some of the lost muscle mass, even after prolonged periods of injury (Kern et al., 2004).

**Significance of the concept of "physiological state" of the spinal circuitry in relearning to step**

*Phase-dependent modulation of proprioceptive input during stepping*

A clear example of the dynamic ability of the spinal locomotor circuitry to process and adapt to sensory information is the enhanced activation of flexor motor pools in response to a mechanical tripping stimulus. When an obstacle is placed in front of the paw of a spinal cat during the swing phase of stepping, there is enhanced flexion of the tripped limb (Forssberg, 1979). If this same mechanical stimulus is applied during the stance phase, however, there is enhanced excitation of the ipsilateral extensor motor pools. In other words, the same stimulus will cause opposite effects depending on the phase of the step cycle.

The functional importance of phase-dependent modulation of sensory input to the spinal cord has been reinforced by recent experiments demonstrating a very predictable suppression or potentiation of monosynaptic and polysynaptic responses when electrically evoked stimuli are applied to the dorsum of the spinal cord. The amplitudes of these responses are increased during the normal active bursting phases of a given muscle and suppressed during the interburst intervals. In other words during the stance phase, the net effect of sensory input is potentiated in the extensor musculature phase, while during the swing phase the sensory input is potentiated in the flexor musculature. This type of modulation occurs in both uninjured and complete spinal rats, cats, and humans (Gerasimenko et al., 2007; Lavrov et al., 2006, 2008). A similar modulation of responses has been reported in uninjured human subjects during treadmill locomotion (Courtine et al., 2007) (Fig. 2). From these data, it is clear that the spinal circuitry processes sensory information in a strongly cyclic, phase-dependent manner whether or not the spinal cord is injured.

Although the mechanisms for these dynamic responses are unknown, the results reflect a level of "smartness" and decision making capability of the spinal cord circuitry, and provide some insight into how sensory information combines with central pattern generation to generate such remarkably effective locomotion after a spinal cord injury. For instance, these observations make it obvious that phase-dependent processing of proprioceptive input provides a remarkable means of coordinating massive amounts of dynamic sensory information projecting to the motor pools that generate stepping.

*More long-term changes in "state dependence"*

Another kind of state-dependent modulation of the motor output that can be achieved with a longer time constant can be mediated pharmacologically in the spinal animals. Administration of a number of agonists and antagonists of each of the neurotransmitter systems within the spinal circuitry can readily improve or depress locomotor function. The direction and magnitude of this modulation depends, of course, on dosage, but each pharmacological intervention will depend in large part on the functional state of the locomotor circuitry. For example, administration of a modest dose of strychnine, a relatively specific blocker of glycine-mediated inhibition, can have a dramatic effect in facilitating effective weight-bearing stepping within a matter of minutes, whereas the same dosage administered to a spinal animal that has been trained, and that therefore can step well, will have minimal effect (De Leon et al., 1999b). Similar responses have been observed

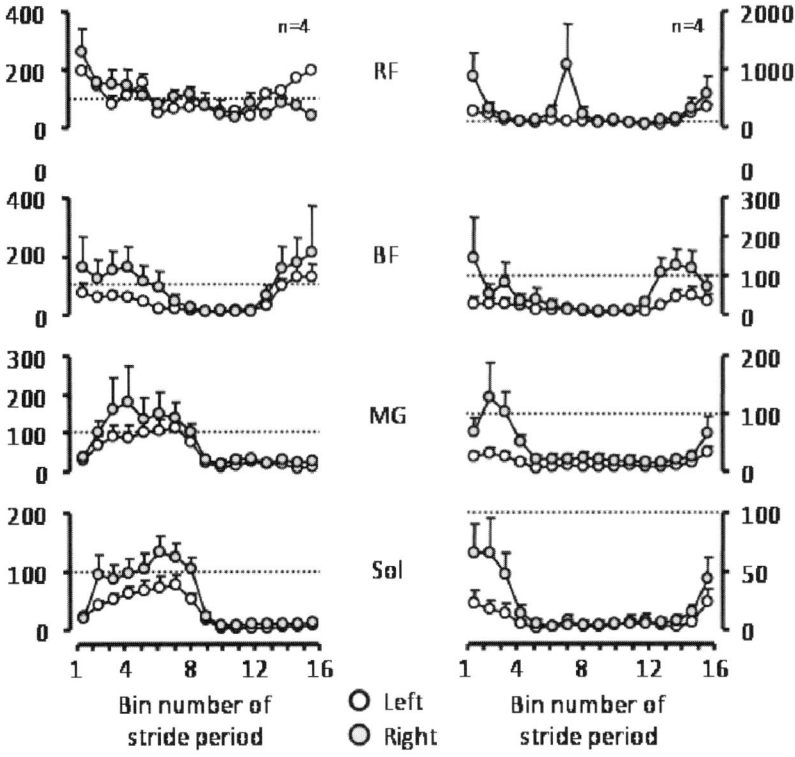

Fig. 2. Monosynaptic muscle responses evoked by spinal cord stimulation are task and phase dependent. Phase-dependent modulation of the multi-segmental monosynaptic response (MMR) amplitude was observed throughout the gait cycle in the leg muscles studied in uninjured subjects. The MMR modulation pattern also was motor-task specific, differing during walking compared to running. Transcutaneous spinal cord stimulation was applied using a AgCl cathode placed on the skin overlying the T11 and T12 spinous processes during walking (left panels: 3.5 km/h) and running (right panels: 8.0 km/h). The resultant MMR responses were recorded bilaterally from selected leg muscles in eight individuals (open and shaded circles depict the left leg and right leg, respectively). Ten step cycles were analyzed for each subject. The data were discretized into 16 time bins corresponding to different periods of the step cycle, beginning with heel strike. Each data point represents the mean ± SD of the MMR amplitude, reported in each individual as a percentage of the MMR amplitude recorded during standing (dashed horizontal line). All evoked potentials are recorded in millivolts. Muscles recorded: RF, rectus femoris; BF, biceps femoris; MG, medial gastrocnemius; Sol, soleus. Adapted with permission from Courtine et al. (2007).

while modulating serotonergic, noradrenergic, and GABAergic systems. These observations provide examples of pharmacological modulation that can be accomplished within a timeframe of minutes. Step training or stand training can change the physiological state of the spinal circuitry that generates stepping over a period of weeks. In general, the efficacy of the spinal pathways can be modulated with time constants ranging from almost instantaneous (mechanical tripping model), to minutes (pharmacological treatments), and even up to weeks or months (step training) (De Leon et al., 1999a).

## Treatment paradigms for restoring locomotor control after a spinal cord injury

As discussed above, two of the fundamental elements for controlling movement are (a) regulating the levels of activation of the appropriate motor pools, and (b) managing how these motor pools are coordinated, i.e., controlling the relative amplitude and timing of activation among muscles. Since it becomes more difficult to control these factors after a spinal cord injury, pharmacological and spinal cord stimulation strategies that increase the excitability of the locomotor circuits, as well as activity-based training techniques that reinstate functional motor pool coordination, can be highly effective in helping subjects regain the ability to step.

## *Pharmacological treatments*

Pharmacological treatments can have an important role in restoring the chemical environment of critical locomotor circuits after a spinal cord injury. Many of the central nervous system neurotransmitters, including the monoamines, are synthesized in isolated regions of the brain (e.g., 5-HT is synthesized in the raphe nucleus) and then transported to the spinal cord. Spinal cord injuries that disrupt the descending flow of neurotransmitters can severely hinder synaptic communication caudal to the lesion, which translates into motor function loss. The damage caused by diminished supraspinal input is aggravated by a significant up-regulation in the inhibitory potential of spinal neurons that mediate locomotion. This causes the locomotor circuitry to become less responsive to excitation from peripheral afferents, which normally provide important proprioceptive triggers that can control many of the details of locomotion.

Despite the disruption of critical neurotransmitter systems, the spinal locomotor circuits retain the capability to respond to sensory-driven presynaptic excitation. For example, serotonergic receptors remain functional after a spinal cord injury and under some circumstances are even upregulated (Fuller et al., 2005; Kim et al., 1999; Otoshi et al., 2009). Since reversing the chemical changes caused by spinal cord injury relates to increasing neurotransmitter supply rather than regenerating lost receptors, pharmacological treatments that supplement the spinal cord with an exogenous supply of neurotransmitter agonists can help regulate synaptic communication and coordinate the activation of stepping-related motor pools. A number of studies have shown that the responsiveness of the locomotor circuitry to sensory input can be readily tuned by "bathing" the lumbosacral spinal cord with various neurotransmitter agonists and antagonists (De Leon et al., 1999b; Edgerton et al., 1997a, b; Kiehn et al., 2008; Rossignol and Barbeau, 1993; Rossignol et al., 1998, 2001). The effectiveness of such blunt pharmacological presentation is quite remarkable.

Low-dose pharmacological treatments, including monoaminergic, glycinergic, and GABAergic agonists, help to supplement the neurotransmitter environment of the post-injury spinal cord and thus can partially restore synaptic communication (Parker, 2005; Rossignol and Barbeau, 1993). Treatments using quipazine (Antri et al., 2002; Barbeau and Rossignol, 1990; Feraboli-Lohnherr et al., 1999; Fong et al., 2005; Guertin, 2004a, b), clonidine (Barbeau and Rossignol, 1991; Chau et al., 1998; Cote et al., 2003), L-DOPA (Barbeau and Rossignol, 1991; Guertin, 2004b; De Mello et al., 2004; Doyle and Roberts, 2004; McEwen and Stehouwer, 2001), strychnine (De Leon et al., 1999b; Edgerton et al., 1997a, b), and/or bicuculline (Edgerton et al., 1997a, b; Robinson and Goldberger, 1986) have been shown to facilitate locomotor recovery. The enhanced synaptic transmission generated by using these drugs potentiates other treatments, including spinal cord stimulation and locomotor training, by lowering the activation threshold of the neurons associated with locomotion. When quipazine, a broad-spectrum serotonin agonist, is combined with step training, it removes the "ceiling" on locomotor recovery that occurs when training is applied alone, resulting in much higher levels of performance in spinal mice (Fong et al., 2005) (Fig. 3). Similarly, treatment with both quipazine and 8-OH-DPAT [8-hydroxy-2-(di-*n*-propylamino)-tetraline], a 5-HT$_{1A/7}$ agonist, increases the stepping performance in spinal rats above and beyond that which can be elicited using spinal cord stimulation or either drug alone (Antri et al., 2005). Cells engineered to secrete serotonin also can enhance locomotion when transplanted into the lumbosacral spinal cord, presumably by making a continuous source of the neurotransmitter available during training (Feraboli-Lohnherr et al., 1997; Gimenez et al., 1998; Ribotta et al., 2000; Hains et al., 2001; Majczynski et al., 2005). When used to facilitate rather than directly generate locomotion, low-dose pharmacological treatments potentially provide an effective solution for enhancing stepping in spinal cord injured subjects.

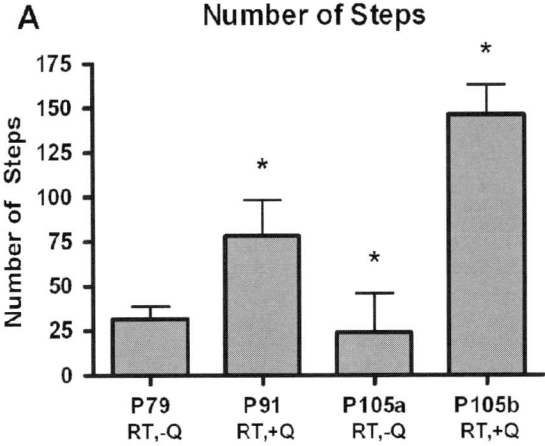
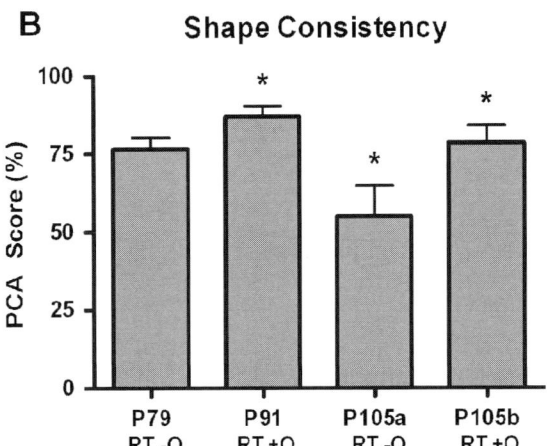
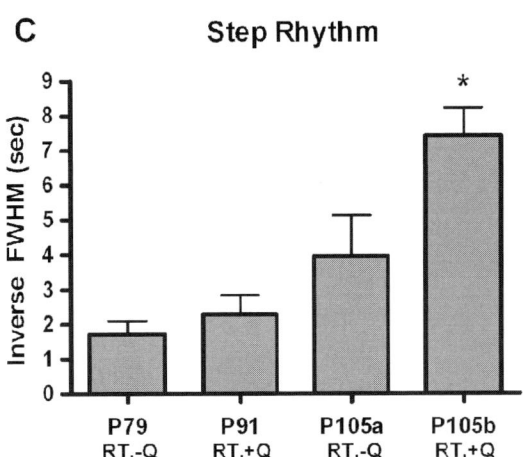

Repeated administration of quipazine also facilitates cellular modification. Immunohistological analysis of the lumbar spinal cord of spinal mice that were given various combinations of quipazine and locomotor training showed that while quipazine had no discernable cellular effect when administered alone, when combined with step training it significantly increased 5-HT$_{2A}$ receptor expression, as well as the levels of AMPA GluR1 and pCREB, which are markers for early and late long-term potentiation, respectively (Otoshi et al., 2005). These results suggest that exogenously administered neurotransmitter agonists may have an important role in facilitating learning and memory during activity-based treatments. If this is indeed the case, spinal cord injury could provide a useful model for examining which neurotransmitter receptors are involved in memory formation.

Advancing current pharmacological paradigms will require development in several key areas including: determining the most effective drug "cocktail" for facilitating various treatments, identifying specific regions of the spinal cord to target, and defining the optimum timeframe for administration. Hochman et al. (2001) and Jordan and Schmidt (2002) have shown that 5-HT$_{1A}$ and 5-HT$_7$ receptors, both of which have

Fig. 3. Pharmacological treatment complements robotic training in enhancing spinal locomotion. In robotically trained spinal mice ($n = 8$), coadministration of quipazine increased the number of steps performed (A) and improved step shape consistency (B), but did not affect step rhythm (C). After an initial period of robotic training, which ended at 79 days postlesion (P79), increases in the number of steps performed and in step shape consistency were observed when quipazine was used to supplement training (P91). This effect was reversed when quipazine was withdrawn, demonstrating that the improvement in locomotion was attributable to the quipazine treatment (P105a). An additional bolus dose of quipazine immediately restored the pharmacologically mediated enhancement (P105b). These results suggest that quipazine and robotic training have complementary effects. Step rhythm, on the other hand, improved steadily throughout the course of robotic training, which is consistent with previous results that suggest that robotic training has a greater effect on step rhythm than quipazine. ∗, Significantly different from P79 (RT, -Q). RT, robotically trained; +Q, treated with quipazine; -Q, not treated with quipazine. Adapted with permission from Fong et al. (2005).

been implicated in the control of stepping, are centered in different regions of the lumbar spinal cord. Ichiyama et al. (2005) has demonstrated that activating each of these pools affects different aspects of stepping. These findings provide examples of the differential concentration of neurotransmitter receptors within the spinal cord, and highlight the need for spatially specific treatments. Furthermore, since the spinal chemical environment is highly dynamic during the first few months after a spinal cord injury, due largely to the antagonism between recovery processes and spreading secondary damage, a different pharmacological intervention will likely be required at each stage of injury progression. Since synaptic communication is essential for all motor functions, pharmacological treatments are a critical area of ongoing spinal cord rehabilitation research.

## *Locomotor training*

It is well established that locomotor training can enhance the recovery of stepping (Edgerton et al., 1997a, b, 2001, 2008) after a spinal cord injury in mice (Cai et al., 2006; Fong et al., 2005), rats (Cha et al., 2007; Timoszyk et al., 2005), cats (Barbeau and Rossignol, 1987; De Leon et al., 1998, 1999a; Edgerton et al., 1991; Lovely et al., 1986, 1990) and, excitingly, even human subjects (Dietz and Harkema, 2004; Harkema et al., 1997; Van De Crommert et al., 1998; Wernig et al., 1998). Engaging the spinal circuitry with sensory input associated with weight-bearing stepping is essential to activating the locomotor circuitry so that effective locomotion can be regained. Using this information, the spinal cord likely performs functional pruning of the many aberrant pathways that form after a spinal cord injury, strengthening those circuits that are relevant to the trained stepping patterns (Ahn et al., 2006; Ichiyama et al., 2008). Traditional manual training involves supporting the subject in a harness over a moving treadmill while a team of therapists/researchers repeatedly guides the legs through a step cycle. In complete injured individuals remarkable levels of recovery can be attained if training is provided persistently over a period of weeks to months.

Experience has helped define a set of critical requirements for effective step training. First, the stepping pattern used to train must have kinematics and kinetic parameters that are stable and appropriate to the training conditions. Second, it is important for training to provide sensory stimuli that closely match normal conditions. The spinal cord circuitry is highly sensitive to proprioceptive and cutaneous inputs: "good" stimuli are processed with exquisite efficiency, whereas "bad" stimuli can lead to failure. Using the same stepping pattern, when spinal rats are trained on an elliptical-like device that maintains continuous contact with the hindpaw, recovery is poorer than on a standard treadmill, where paw contact is broken during swing (Timoszyk et al., 2003). In this example, even seemingly innocuous application of plantar stimulation during stance is enough to detract from the training effect. Third, weight bearing is essential for maximizing recovery. In addition to helping maintain muscle properties (Roy et al., 1991; Stewart et al., 2004), subjects who are challenged to bear increasing amounts of their weight are more likely to achieve better stepping than subjects who are fully supported during training (Edgerton et al., 1991). Finally, although repetitive and consistent application of training paradigms is essential to recovery, there should be a small degree of variability in the parameters that are used to train to prevent locomotor performance from becoming dependent on a single set of stimuli. A controlled amount of variability in training enables subjects to benefit from experiential learning (Cai et al., 2006). This important concept will be discussed in greater detail below.

Several variables affect the extent of recovery that can be attained using step training. Recovery is partially dependent on the severity of the injury and the developmental stage of the subject when it occurs. Training is particularly effective in subjects with an incomplete injury, who typically recover more easily than subjects with a complete injury (Coleman and Geisler, 2004; Marino et al., 1999; Waters et al., 1995). Subjects injured at an early age exhibit better neurologic recovery than those who are older at injury due to the ability to develop alternative neural pathways during the formative period of the central nervous system (Scivoletto et al., 2003). The method of training also affects recovery. While traditional forms of therapist/researcher-assisted training have shown tremendous benefit, there are limitations on the extent of recovery that can be achieved using manual approaches. For example, the size disparity between human hands and small mouse legs makes it difficult to control the hindlimb movements of spinal mice with sufficient consistency and precision to be effective. In a study using spinal mice, after more than a month of training, it was clear that a manual approach was unable to provide efficient training. In contrast, statistically significant improvement was attained when training was carried out using a high-precision robotic system (Fong et al., 2005).

From a research perspective, an inability to replicate training movements day-to-day, or even step-to-step, makes it impossible to evaluate training techniques rigorously. Inconsistency and the lack of precision inherent to manual training were the impetus for developing robotic training systems. Diagrammed in Fig. 4, our robotic step-training system for rodents consists of two motor-driven arms (one for each leg), a weight-support device, and a computer-controlled treadmill. The first robotically assisted training algorithm examined used the robot arms to train a single stepping pattern repetitively during alternate periods for 15 min per day in mice. During the intervening periods, robotic control was turned off to allow the mice to step freely. Compared to the manual method, daily robotic training generated visible and statistically significant improvement within 2 weeks, although the quality of stepping remained poorer than in uninjured control mice (Fong et al., 2005). Using this technique, we observed a plateau in the level of improvement, which we interpreted as saturation in the amount of benefit that the spinal circuitry could extract from a single training pattern. To overcome this, the next set of robotically assisted training algorithms were designed to give subjects the opportunity to experience multiple viable stepping patterns, and even to expose them to occasional failure.

The general idea behind these "assist-as-needed" paradigms was to provide sufficient variability to enable the subjects to learn from several good and bad stepping patterns, yet enough control to prevent catastrophic failure. The results highlight the importance of experiential learning and demonstrate that, while it is important to enforce proper interlimb coordination, mice that are exposed to a continuum of stepping patterns recover stepping more quickly and more robustly than those locked into a fixed trajectory training pattern (Fig. 5): essentially, how you train does matter (Cai et al., 2006).

Technological development has provided a quantum leap in the understanding of how locomotor training produces recovery after a spinal cord injury, and continued advancement will enable novel treatment paradigms to be conceived and tested (Winchester and Querry, 2006).

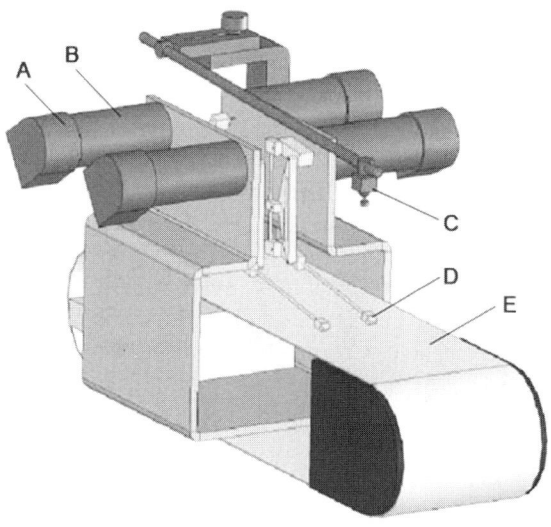

Fig. 4. Diagram of rodent robotic step training and evaluation system. The rodent robotic system consists of the following major components: (A) four optical encoders, (B) four DC motors, (C) a weight-support device, (D) two 5-bar parallelogram linkages, and (E) a motor-driven treadmill. When used in an active mode, the system applies step-training algorithms that are commanded by an external motion controller. In a passive mode, the optical encoders record the trajectories of the legs during free stepping. Robotics thus enables quantitative monitoring of both training and recovery. Adapted with permission from Cai et al. (2005).

Currently, ongoing research is testing the capability of robotic systems to train animals to stand (Bigbee et al., 2007; Liang et al., 2006). Another important study is examining how learning algorithms can utilize robotically sampled stepping data to design optimal training protocols for specific injuries (Cai et al., 2006). This is particularly important as advanced training paradigms make the transition from the laboratory to the clinic, since the wide variability in spinal cord injuries is certain to require customized treatment for each patient. At the same time, robotic training systems for humans are being developed in earnest to assist therapists in overburdened clinics and to give more patients access to the highest standard of care (Aoyagi et al., 2007; Hesse et al., 2003; Ohta et al., 2007; Reinkensmeyer et al., 2006). Beyond spinal cord injury, robotically assisted training has broad applicability to stroke, Parkinson's disease, and many other conditions that involve loss of motor function. The convergence of technology and science has tremendous potential, not yet fully explored, that can rapidly translate into better patient outcomes.

## Spinal cord stimulation

The discovery that spinal cord stimulation can be used in multiple ways to facilitate locomotion has opened new avenues for locomotor rehabilitation. Currently, there are two major strategies for spinal cord stimulation. First, in its more common usage, low-level stimulation is applied to broad areas of the spinal cord to increase the general excitability of the locomotor circuits. Typically, fine-wire (dimensions: $\sim 300 \times 1000\,\mu m$), ball (diameter: $\sim 900\,\mu m$), or spring (dimensions: $1000 \times 3600\,\mu m$) electrodes are placed on the

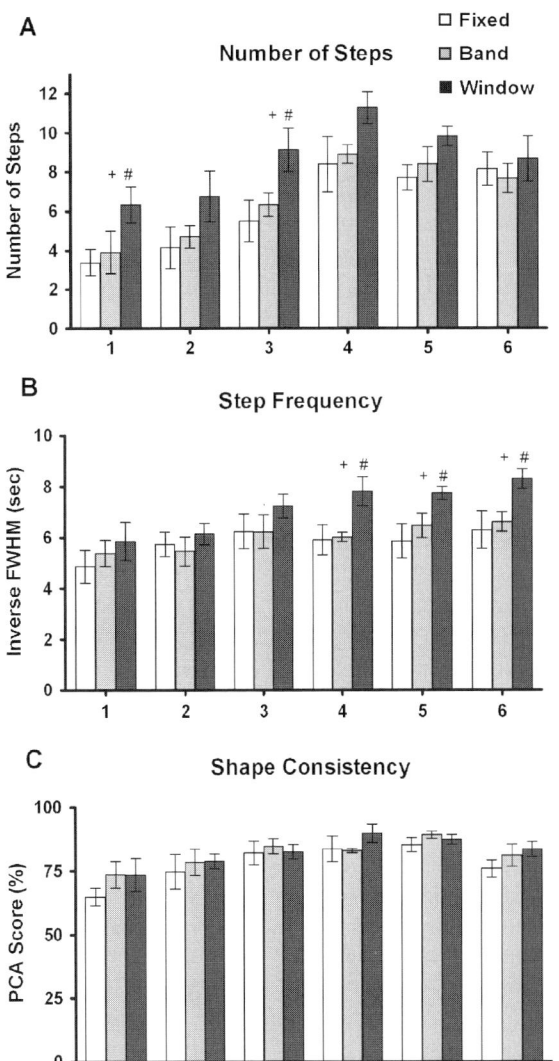

Fig. 5. Variability in robotic step training promotes robust locomotor recovery. After 4 weeks of robotically assisted step training, complete spinal mice that were trained using a "window" control algorithm (black bar) were able to execute more steps (A), and displayed better step rhythm (B), than mice that were trained with either a "band" algorithm (gray bar) or using traditional, continuous-assistance, "fixed" trajectory training (white bar). Contrasted with fixed trajectory training, "window" and "band" training i.e., assist-as-needed paradigms allow the hindlimb to deviate to some degree away from the nominal trained trajectory: the robotics only exert corrective action when the position of the hindlimb moves beyond a set limit, at which point a restoring force is generated (force magnitude encoded by an error-dependent velocity field). "Window" training enforces alternating interlimb coordination, whereas "band" training does not. The data suggest that the additional sensory information provided to the spinal circuitry during "window" training enhances locomotor recovery, but that interlimb coordination should be controlled when training an alternating gait. Values are reported as mean ± SEM. + and #, significantly different from "fixed" and "band" training group, respectively. Adapted with permission from Cai et al. (2006).

epidural surface of the spinal cord, and square wave pulses (amplitude: 1–10 V or 10–200 µA, duration: 100–250 µs, frequency: 1–100 Hz) are applied (Gerasimenko et al., 2003; Ichiyama et al., 2008). Stimulation over this range of parameters is below the threshold for direct motor activation, but is sufficient to facilitate sensory-triggered movements. Used in this manner, the effect of spinal cord stimulation is similar to that of the pharmacological treatments: by lowering the activation threshold of locomotor neurons, spinal cord stimulation makes it easier for proprioceptive and cutaneous signals to enable stepping. The type of motor output elicited by stimulation is affected by both extrinsic and intrinsic parameters. Extrinsically, the effect of stimulation is activity-dependent: during stimulation, complete spinal cats and rats will adapt their stepping pattern, stepping forwards or backwards in accordance with their orientation on the treadmill (Musienko et al., 2007). Intrinsically, stimulation-evoked movement is frequency-dependent in humans, stimulation between 5 and 15 Hz preferentially induces standing, whereas stimulation between 25 and 50 Hz favors stepping movements (Jilge et al., 2004). A second approach to spinal cord stimulation involves direct generation of muscle movement by supra-threshold stimulation of motoneurons. Rather than facilitating locomotor circuits at the sensory or interneuronal levels, penetrating cylindrical electrodes (diameter: 25–30 µm, height: 60–100 µm) are inserted into the ventral horn at sites that are specific for particular muscles (amplitude: 20–300 µA, duration: 200–300 µs, frequency: 1–50 Hz) (Gaunt et al., 2006; Mushahwar et al., 2002, 2004). Given a sufficiently large number of implanted electrodes, this approach bypasses the intrinsic circuitry and allows for external (e.g., computer-driven) control of sequences of muscle movements. The disadvantages of this method are: penetrating electrodes can damage the spinal cord tissue, circumventing the locomotor control circuitry eliminates the benefit of intrinsic synergies that coordinate agonist and antagonist muscles, and real-time control of the many muscles necessary to generate smooth and stable locomotion is computer intensive. In general, an advantage of spinal cord stimulation over direct muscle stimulation is that it typically recruits muscle fibers in a more normal physiological order, i.e., fatigue resistant (slow oxidative) before fast fatiguing (fast glycolytic) fibers, and thus helps to maintain endurance (Bamford et al., 2005). Both epidural stimulation, which facilitates the locomotor circuitry, and intraspinal stimulation, which triggers movements directly, are active areas of research.

Our research in spinal cord stimulation is focused on developing high-density epidural electrode arrays to extend the potential of the epidural stimulation approach (Fig. 6). These arrays consist of platinum electrode contacts and wire lines that are embedded in a parylene-C substrate, and are fabricated using techniques borrowed from semiconductor and microelectromechanical systems processing (Rodger et al., 2007, 2008). The high biocompatibility of the constituent materials makes these arrays well suited for chronic implantation: platinum has a long history of biocompatibility and meets both Tripartite and ISO 10993 standards, while parylene-C is certified as a United States Pharmacopeia Class VI plastic. Parylene-based electrode arrays offer numerous advantages over existing technologies. Parylene arrays allow for stable implantation: formed as thin films ($\sim$ 20 µm thick), they are highly flexible and conform to the spinal cord surface, resisting displacement. Furthermore, the close fit to the spinal cord promotes encapsulating connective tissue growth, which further secures the array and effectively precludes movement. Postimplantation migration is the most common cause of clinical electrode failure (Barolat, 2000; LeDoux and Langford, 1993; North et al., 2005; Renard and North, 2006), in one study forcing 23% of implanted patients to have corrective, follow-up operations (Andersen, 1997). Thus, thin-film parylene arrays provide a critical advancement. Microfabrication processes enable the design of arrays that have novel electrode configurations that can be used to test advanced stimulation algorithms. Using multilayer fabrication techniques, it is now technically possible to build arrays with densities up to 1024 electrodes in a 5 mm × 6 mm area (although, currently, practical application of such high densities is limited by connector and stimulator technologies). Additionally,

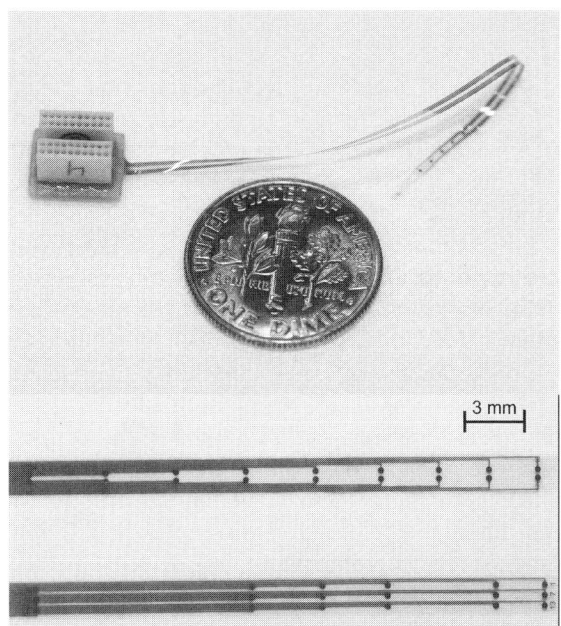

Fig. 6. Photographs of spinal cord electrode arrays. (Top) Photograph of a spinal cord electrode array and 36-pin head connector juxtaposed against a small coin for size comparison. (Bottom) Close-up photograph of the 18-electrode contacts of a 3 × 6 electrode array with physiologically determined rostrocaudal inter-electrode spacing.

electroplating can be used to increase the surface area of each electrode by around 40-fold, which provides two important advantages: it enables higher levels of charge transfer from small electrodes, and it expands the current range over which charge can be transferred capacitively, reducing the occurrence of tissue and electrode damage associated with high current, Faradaic charge transfer (Merrill et al., 2005).

Our electrode arrays provide several key benefits over existing technologies. Access to a large number of small electrodes makes possible selective, high-precision stimulation of focal regions of the spinal cord. Electrode arrays are enabling the identification of specific regions of the spinal cord that are responsible for different components and phases of the step cycle. As our knowledge of the somatotopic organization of the locomotor circuitry increases, the selectivity provided by electrode arrays will make it possible to target stimulation appropriately to address the specific deficiencies of individual injuries. When stimulation is targeted directly at the tissue of interest, less current is needed to generate a desired effect than with bulk stimulation methods. Moreover, by confining the stimulating current to the regions that require it, high-density electrode arrays eliminate unnecessary stimulation of neural tissue, which reduces the potential for long-term damage due to repeated pulsed electrical stimulation. High-density electrode arrays are thus making spinal cord stimulation more effective and safer. Finally, the electrode arrays can be used as a diagnostic tool to measure the properties of spinal cord evoked potentials to assess the locomotor circuitry at different stages of injury and recovery.

While it is remarkable that relatively non-specific stimulation can promote stable, state-dependent treadmill locomotion in the absence of supraspinal control, this represents just the beginning of what can be achieved using spinal cord stimulation. Since the effect of spinal cord stimulation can change dramatically when the electrode site is moved by as little as 200–300 μm (Kazennikov et al., 1983), targeted stimulation approaches can be leveraged to control different components of stepping. In spinal rats, stimulation of the L2 spinal segment induces a general enhancement of the locomotor rhythm, while stimulation of the S1 segment activates the extensor muscles during stance (van den Brand et al., 2007). Stimulation applied to each of the spinal segments between T12 and L6 produces different locomotor effects (Ichiyama et al., 2005). The effects of stimulation also vary in the medial-lateral direction, a finding that provides rationale for pursuing 2-D array designs. As the spatial resolution of the electrode arrays continues to improve, spinal cord stimulation will evolve from providing nonspecific excitation of the spinal circuits to fine tuning very specific aspects of the locomotor pattern. What is most exciting about spinal cord stimulation is that it can be used effectively at both of these levels.

Future stimulation approaches will examine the implementation of biologically inspired stimulation patterns, as well as patterns that involve simultaneous or sequential stimulation of multiple electrode sites. Other applications for multielectrode spinal cord stimulation include management of chronic pain, stroke, and other conditions involving motor function loss. The continued development of electrode array technology is providing unparalleled access to the interneuronal circuitry, and may serve as the best technique for fine-tuning gross motor behaviors after a spinal cord injury.

## Integrating neuroengineering and biological concepts to regain posture and locomotion

Based on the successful recovery that we have attained using muscle stimulation, spinal cord stimulation, pharmacological interventions, and activity-based training, the potential for enhancing locomotor recovery by aggressively pursuing complementary and synergistic strategies is clear and represents a logical direction for translating some of the basic biological concepts to the clinical setting. While it cannot always be assumed that multiple interventions will be complementary (Maier et al., 2009) with careful consideration of their interactive effects, multi-intervention approaches truly are the obvious solutions that are staring us in the face.

We already have observed significant positive interaction when multiple modes of treatment are combined (Fig. 7). Optimal recovery of locomotion requires two important factors: the damaged spinal cord must be provided with adequate information that it can use to relearn to step, but, before that, it must be prepared to receive that information. This explains why the recovery of locomotion using robotically assisted training, which provides information on functional stepping patterns, is significantly enhanced by coadministration of pharmacological agonists that improve synaptic signaling. In mice, e.g., while robotic training restores gross stepping function, pharmacological modulation with quipazine further improves locomotion by facilitating the recovery of movements that are difficult to access with training alone, e.g., activation of the distal extensor muscles during weight-bearing stance (Fong et al., 2005). We also have observed substantial recovery in rats from a combination of locomotor training, two serotonergic drugs, and multiple-site epidural stimulation, and have shown that selective combinations of these treatments lead to very different locomotor effects (van den Brand et al., 2007). The next step is to optimize the combination treatment parameters to maximize the synergies between the constituent interventions. All evidence suggests that engaging complementary approaches may result in the greatest functional gains.

Combinations of paradigms can be effective when each component treatment focuses on repairing a different aspect of motor function loss. Figure 8 depicts the recovery of stepping after spinal cord injury as a multicomponent process. Although this diagram to some degree oversimplifies the complexity of the underlying mechanisms, it provides a didactic representation of how the different treatments interact to promote locomotor recovery. First, muscle stimulation initiated early after injury helps to maintain muscle properties at a functional level. Insufficient muscle tone and/or lack of a normal complement of fatigue-resistant fibers make it more difficult for subsequent treatments to generate appropriate movements, increase the likelihood of injury, and may result in poor endurance. Second, both pharmacological facilitation and spinal cord stimulation can be used to increase the general excitability of the spinal circuits and to strengthen the efficacy of synaptic transmission. Neither low-dose drug treatment nor subthreshold spinal cord stimulation can generate movement independently or produce long-lasting recovery, but they can sufficiently lower the activation threshold of relevant sensory neurons and interneurons to enable even small amounts of stepping-associated stimuli to trigger and sustain locomotion. Third, activity-dependent training helps to reinforce kinematically appropriate stepping patterns. Repetition of these facilitated movements improves muscle recruitment and coordination, and enables the spinal cord to learn to perform new tasks. Over time, the persistent activation of specific spinal pathways results in changes that strengthen those synaptic connections and provide the structural basis for motor learning. The most effective

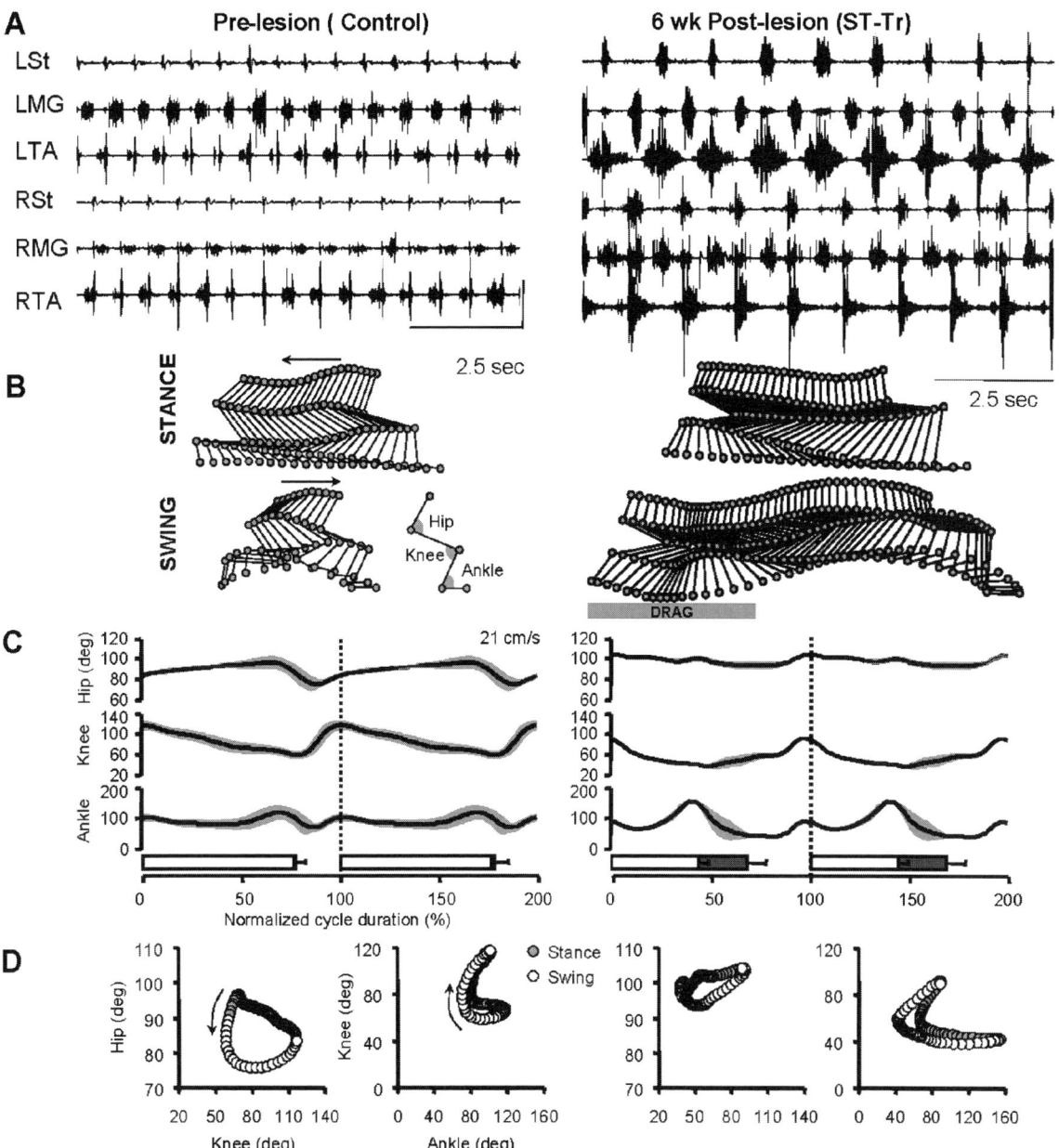

Fig. 7. Bipedal stepping approximating that observed pre-lesion can be recovered after a complete spinal cord transection with the aid of epidural stimulation and pharmacological facilitation. EMG (A) and kinematic (B–D) data are shown for a rat before and 6 weeks after receiving a complete mid-thoracic (∼T9) spinal cord transection while stepping on a treadmill at 21 cm/s. After the transection, the rat was administered quipazine and epidural stimulation. Representative stick diagram decompositions of the left hindlimb movements during the stance and swing phases of gait are shown in B. Mean waveforms of the hip, knee, and ankle joint angle for the left hindlimb are plotted for a normalized gait cycle duration in C. Each trace is an average of 15 (control) and 18 (spinal transected-trained, ST-Tr) successive steps. Horizontal bars at bottom indicate mean value of stance phase (blank) and foot drag duration (shaded). Angle–angle plots showing coupling between hip and knee (left) and knee and ankle (right) from the same data shown in C are shown in D. Filled and empty circles represent stance and swing phases of gait, respectively. Arrows indicate direction along which time is evolving. Shaded portion of the lines in C shows SEM. Adapted with permission from Gerasimenko et al. (2007).

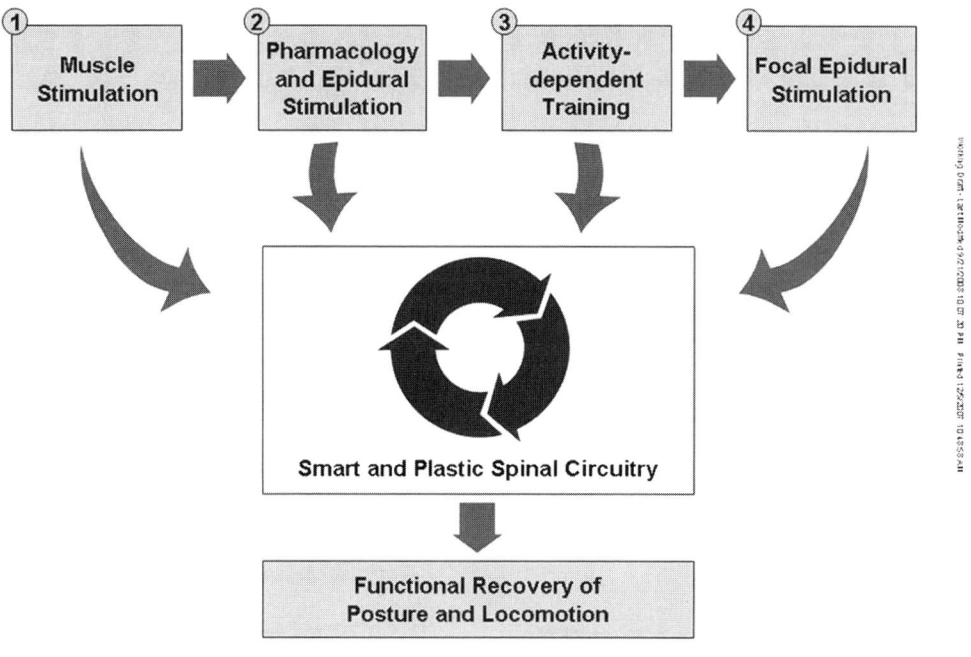

Fig. 8. A multimodal approach to spinal cord rehabilitation. It is becoming increasingly clear that combining multiple treatment paradigms can produce enhanced recovery. This diagram depicts a promising four-step approach to recovering locomotor function: (1) application of muscle stimulation to maintain normal properties of the muscles; (2) use of pharmacological treatments and epidural stimulation to recreate an electrochemical environment conducive to spinal learning; (3) administration of activity-dependent motor training to provide the appropriate cues necessary to teach the spinal cord to walk; and (4) delivery of focal epidural stimulation to refine and facilitate functional stepping patterns. All of these treatments modulate sensory input to the lumbosacral spinal circuitry, which processes the information and uses it to recover functional posture and locomotion.

training techniques promote locomotion that is robust to disturbances by exposing the spinal cord to a range of viable stepping patterns. With minimal invasiveness, rather simple methods of locomotor training can be used to recover gross motor function. Fourth, spinal cord stimulation can be used in a secondary role to fine-tune the locomotor pattern. Since high-density electrode arrays provide access to focal regions of spinal cord tissue, they can be used to target selectively the circuits that control particular components of a movement, and thus to address more specific motor defects. In some cases, targeted stimulation approaches may supplant specialized training techniques, which can be both invasive and resource-intensive.

Continued technological advancement in pharmacological treatment, spinal cord stimulation, and activity-based training offers great potential. Pharmacological therapies will improve with the arrival of sophisticated drug delivery systems that enable treatment of focal regions of the spinal cord. Spinal cord stimulation will continue to progress with electrode array development. Activity-based treatments will advance in conjunction with the development of learning algorithms that will help define optimal training protocols that adapt dynamically with the constantly evolving state of the recovering spinal circuitry. With the aggressive pursuit of the combination therapies "staring us in the face," the expectations for recovery of locomotion are now significantly higher for individuals with spinal cord injury, their family, friends, therapists, and physicians.

## Acknowledgments

The work presented in this paper was supported by the National Institutes of Health Grants NS16333 and NS42291, the Russian Foundation for Basic Research — US Civilian Research and Development Foundation Grant 07-04-91106, the Christopher and Diana Reeve Paralysis Foundation, and the Roman Reed Spinal Cord Injury Research Fund of California.

## References

Ahn, S. N., Guu, J. J., Tobin, A. J., Edgerton, V. R., & Tillakaratne, N. J. (2006). Use of c-fos to identify activity-dependent spinal neurons after stepping in intact adult rats. *Spinal Cord, 44*, 547–559.

Andersen, C. (1997). Complications in spinal cord stimulation for treatment of angina pectoris. Differences in unipolar and multipolar percutaneous inserted electrodes. *Acta Cardiologica, 52*, 325–333.

Andersen, J. L., Mohr, T., Biering-Sorensen, F., Galbo, H., & Kjaer, M. (1996). Myosin heavy chain isoform transformation in single fibres from m. vastus lateralis in spinal cord injured individuals: effects of long-term functional electrical stimulation (FES). *Pflugers Archive, 431*, 513–518.

Antri, M., Barthe, J. Y., Mouffle, C., & Orsal, D. (2005). Long-lasting recovery of locomotor function in chronic spinal rat following chronic combined pharmacological stimulation of serotonergic receptors with 8-OHDPAT and quipazine. *Neuroscience Letters, 384*, 162–167.

Antri, M., Orsal, D., & Barthe, J. Y. (2002). Locomotor recovery in the chronic spinal rat: effects of long-term treatment with a 5-HT2 agonist. *The European Journal of Neurosciences, 16*, 467–476.

Aoyagi, D., Ichinose, W. E., Harkema, S. J., Reinkensmeyer, D. J., & Bobrow, J. E. (2007). A robot and control algorithm that can synchronously assist in naturalistic motion during body-weight-supported gait training following neurologic injury. *IEEE Transactions on Neural Systems and Rehabilitation Engineering, 15*, 387–400.

Bamford, J. A., Putman, C. T., & Mushahwar, V. K. (2005). Intraspinal microstimulation preferentially recruits fatigue-resistant muscle fibres and generates gradual force in rat. *Journal of Physiology, 569*, 873–884.

Barbeau, H., & Rossignol, S. (1987). Recovery of locomotion after chronic spinalization in the adult cat. *Brain Research, 412*, 84–95.

Barbeau, H., & Rossignol, S. (1990). The effects of serotonergic drugs on the locomotor pattern and on cutaneous reflexes of the adult chronic spinal cat. *Brain Research, 514*, 55–67.

Barbeau, H., & Rossignol, S. (1991). Initiation and modulation of the locomotor pattern in the adult chronic spinal cat by noradrenergic, serotonergic and dopaminergic drugs. *Brain Research, 546*, 250–260.

Barolat, G. (2000). Spinal cord stimulation for chronic pain management. *Archives of Medical Research, 31*, 258–262.

Bickel, C. S., Slade, J. M., & Dudley, G. A. (2004). Long-term spinal cord injury increases susceptibility to isometric contraction-induced muscle injury. *European Journal of Applied Physiology, 91*, 308–313.

Bigbee, A. J., Crown, E. D., Ferguson, A. R., Roy, R. R., Tillakaratne, N. J., Grau, J. W., et al. (2007). Two chronic motor training paradigms differentially influence acute instrumental learning in spinally transected rats. *Behavioural Brain Research, 180*, 95–101.

Buford, J. A., & Smith, J. L. (1993). Adaptive control for backward quadrupedal walking. III. Stumbling corrective reactions and cutaneous reflex sensitivity. *Journal of Neurophysiology, 70*, 1102–1114.

Burnham, R., Martin, T., Stein, R., Bell, G., Maclean, I., & Steadward, R. (1997). Skeletal muscle fibre type transformation following spinal cord injury. *Spinal Cord, 35*, 86–91.

Cai, L. L., Fong, A. J., Otoshi, C. K., Liang, Y., Burdick, J. W., Roy, R. R., et al. (2006). Implications of assist-as-needed robotic step training after a complete spinal cord injury on intrinsic strategies of motor learning. *The Journal of Neuroscience, 26*, 10564–10568.

Cai, L. L., Fong, A. J., Otoshi, C. K., Liang, Y. Q., Cham, J. G., Zhong, H., et al. (2005). Effects of consistency vs. variability in robotically controlled training of stepping in adult spinal mice. *Proceedings of International Conference on Rehabilitation Robotics*, 575–579.

Calancie, B., Broton, J. G., Klose, K. J., Traad, M., Difini, J., & Ayyar, D. R. (1993). Evidence that alterations in presynaptic inhibition contribute to segmental hypo- and hyperexcitability after spinal cord injury in man. *Electroencephalography and Clinical Neurophysiology, 89*, 177–186.

Castro, M. J., Apple, D. F., Jr., Staron, R. S., Campos, G. E., & Dudley, G. A. (1999). Influence of complete spinal cord injury on skeletal muscle within 6 mo of injury. *Journal of Applied Physiology, 86*, 350–358.

Cha, J., Heng, C., Reinkensmeyer, D. J., Roy, R. R., Edgerton, V. R., & De Leon, R. D. (2007). Locomotor ability in spinal rats is dependent on the amount of activity imposed on the hindlimbs during treadmill training. *Journal of Neurotrauma, 24*, 1000–1012.

Chalmers, G. R., Roy, R. R., & Edgerton, V. R. (1992). Adaptability of the oxidative capacity of motoneurons. *Brain Research, 570*, 1–10.

Chau, C., Barbeau, H., & Rossignol, S. (1998). Early locomotor training with clonidine in spinal cats. *Journal of Neurophysiology, 79*, 392–409.

Coleman, W. P., & Geisler, F. H. (2004). Injury severity as primary predictor of outcome in acute spinal cord injury:

retrospective results from a large multicenter clinical trial. *The Spine Journal, 4*, 373–378.

Cote, M. P., & Gossard, J. P. (2004). Step training-dependent plasticity in spinal cutaneous pathways. *The Journal of Neuroscience, 24*, 11317–11327.

Cote, M. P., Menard, A., & Gossard, J. P. (2003). Spinal cats on the treadmill: changes in load pathways. *The Journal of Neuroscience, 23*, 2789–2796.

Courtine, G., Harkema, S. J., Dy, C. J., Gerasimenko, Y. P., & Dyhre-Poulsen, P. (2007). Modulation of multisegmental monosynaptic responses in a variety of leg muscles during walking and running in humans. *Journal of Physiology, 582*, 1125–1139.

Crameri, R. M., Cooper, P., Sinclair, P. J., Bryant, G., & Weston, A. (2004). Effect of load during electrical stimulation training in spinal cord injury. *Muscle & Nerve, 29*, 104–111.

Crameri, R. M., Weston, A. R., Rutkowski, S., Middleton, J. W., Davis, G. M., & Sutton, J. R. (2000). Effects of electrical stimulation leg training during the acute phase of spinal cord injury: a pilot study. *European Journal of Applied Physiology, 83*, 409–415.

De Leon, R. D., Hodgson, J. A., Roy, R. R., & Edgerton, V. R. (1998). Locomotor capacity attributable to step training versus spontaneous recovery after spinalization in adult cats. *Journal of Neurophysiology, 79*, 1329–1340.

De Leon, R. D., Hodgson, J. A., Roy, R. R., & Edgerton, V. R. (1999a). Retention of hindlimb stepping ability in adult spinal cats after the cessation of step training. *Journal of Neurophysiology, 81*, 85–94.

De Leon, R. D., Tamaki, H., Hodgson, J. A., Roy, R. R., & Edgerton, V. R. (1999b). Hindlimb locomotor and postural training modulates glycinergic inhibition in the spinal cord of the adult spinal cat. *Journal of Neurophysiology, 82*, 359–369.

De Mello, M. T., Esteves, A. M., & Tufik, S. (2004). Comparison between dopaminergic agents and physical exercise as treatment for periodic limb movements in patients with spinal cord injury. *Spinal Cord, 42*, 218–221.

Dietz, V., & Harkema, S. J. (2004). Locomotor activity in spinal cord-injured persons. *Journal of Applied Physiology, 96*, 1954–1960.

Doyle, L. M., & Roberts, B. L. (2004). Functional recovery and axonal growth following spinal cord transection is accelerated by sustained L-DOPA administration. *European Journal of Neuroscience, 20*, 2008–2014.

Dupont-Versteegden, E. E., Houle, J. D., Dennis, R. A., Zhang, J., Knox, M., Wagoner, G., et al. (2004). Exercise-induced gene expression in soleus muscle is dependent on time after spinal cord injury in rats. *Muscle & Nerve, 29*, 73–81.

Dupont-Versteegden, E. E., Houle, J. D., Gurley, C. M., & Peterson, C. A. (1998). Early changes in muscle fiber size and gene expression in response to spinal cord transection and exercise. *American Journal of Physiology, 275*, C1124–C1133.

Dupont-Versteegden, E. E., Murphy, R. J., Houle, J. D., Gurley, C. M., & Peterson, C. A. (2000). Mechanisms leading to restoration of muscle size with exercise and transplantation after spinal cord injury. *American Journal of Physiology. Cell Physiology, 279*, C1677–C1684.

Edgerton, V. R., Courtine, G., Gerasimenko, Y. P., Lavrov, I., Ichiyama, R. M., Fong, A. J., et al. (2008). Training locomotor networks. *Brain Research. Brain Research Reviews, 57*, 241–254.

Edgerton, V. R., De Guzman, C. P., Gregor, R. J., Roy, R. R., Hodgson, J. A., & Lovely, R. G. (1991). Trainability of the spinal cord to generate hindlimb stepping patterns in adult spinalized cats. In M. Shimamura, S. Grillner, & V. R. Edgerton (Eds.), *Neurobiological basis of human locomotion*. Tokyo: Japan Scientific Societies Press.

Edgerton, V. R., De Leon, R. D., Tillakaratne, N., Recktenwald, M. R., Hodgson, J. A., & Roy, R. R. (1997a). Use-dependent plasticity in spinal stepping and standing. *Advances in Neurology, 72*, 233–247.

Edgerton, V. R., Leon, R. D., Harkema, S. J., Hodgson, J. A., London, N., Reinkensmeyer, D. J., et al. (2001). Retraining the injured spinal cord. *Journal of Physiology, 533*, 15–22.

Edgerton, V. R., & Roy, R. R. (1996). Neuromuscular adaptations for actual and simulated spaceflight. In M. J. Fregly & C. M. Blatteis (Eds.), *Environmental physiology*. Bethesda, MD: American Physiological Society.

Edgerton, V. R., Roy, R. R., De Leon, R., Tillakaratne, N., & Hodgson, J. A. (1997b). Does motor learning occur in the spinal cord? *Neuroscientist, 3*, 294.

Feraboli-Lohnherr, D., Barthe, J. Y., & Orsal, D. (1999). Serotonin-induced activation of the network for locomotion in adult spinal rats. *Journal of Neuroscience Research, 55*, 87–98.

Feraboli-Lohnherr, D., Orsal, D., Yakovleff, A., Gimenez, Y., Ribotta, M., & Privat, A. (1997). Recovery of locomotor activity in the adult chronic spinal rat after sublesional transplantation of embryonic nervous cells: specific role of serotonergic neurons. *Experimental Brain Research, 113*, 443–454.

Fong, A. J., Cai, L. L., Otoshi, C. K., Reinkensmeyer, D. J., Burdick, J. W., Roy, R. R., et al. (2005). Spinal cord-transected mice learn to step in response to quipazine treatment and robotic training. *The Journal of Neuroscience, 25*, 11738–11747.

Forssberg, H. (1979). Stumbling corrective reaction: a phase-dependent compensatory reaction during locomotion. *Journal of Neurophysiology, 42*, 936–953.

Fuller, D. D., Baker-Herman, T. L., Golder, F. J., Doperalski, N. J., Watters, J. J., & Mitchell, G. S. (2005). Cervical spinal cord injury upregulates ventral spinal 5-HT2A receptors. *Journal of Neurotrauma, 22*, 203–213.

Gaunt, R. A., Prochazka, A., Mushahwar, V. K., Guevremont, L., & Ellaway, P. H. (2006). Intraspinal microstimulation excites multisegmental sensory afferents at lower stimulus levels than local alpha-motoneuron responses. *Journal of Neurophysiology, 96*, 2995–3005.

Gerasimenko, Y., Roy, R. R., & Edgerton, V. R. (2008). Epidural stimulation: comparison of the spinal circuits that generate and control locomotion in rats, cats and humans. *Experimental Neurology, 209*, 417–425.

Gerasimenko, Y. P., Avelev, V. D., Nikitin, O. A., & Lavrov, I. A. (2003). Initiation of locomotor activity in spinal cats by epidural stimulation of the spinal cord. *Neuroscience and Behavioral Physiology, 33*, 247–254.

Gerasimenko, Y. P., Ichiyama, R. M., Lavrov, I. A., Courtine, G., Cai, L., Zhong, H., et al. (2007). Epidural spinal cord stimulation plus quipazine administration enable stepping in complete spinal adult rats. *Journal of Neurophysiology, 98*, 2525–2536.

Gerasimenko, Y. P., Makarovskii, A. N., & Nikitin, O. A. (2002). Control of locomotor activity in humans and animals in the absence of supraspinal influences. *Neuroscience and Behavioral Physiology, 32*, 417–423.

Gerrits, H. L., De Haan, A., Hopman, M. T., Van Der Woude, L. H., Jones, D. A., & Sargeant, A. J. (1999). Contractile properties of the quadriceps muscle in individuals with spinal cord injury. *Muscle & Nerve, 22*, 1249–1256.

Gerrits, H. L., De Haan, A., Sargeant, A. J., Dallmeijer, A., & Hopman, M. T. (2000). Altered contractile properties of the quadriceps muscle in people with spinal cord injury following functional electrical stimulated cycle training. *Spinal Cord, 38*, 214–223.

Gerrits, H. L., Hopman, M. T., Offringa, C., Engelen, B. G., Sargeant, A. J., Jones, D. A., et al. (2003). Variability in fibre properties in paralysed human quadriceps muscles and effects of training. *Pflugers Archive, 445*, 734–740.

Gerrits, H. L., Hopman, M. T., Sargeant, A. J., Jones, D. A., & De Haan, A. (2002). Effects of training on contractile properties of paralyzed quadriceps muscle. *Muscle & Nerve, 25*, 559–567.

Gimenez, Y., Ribotta, M., Orsal, D., Feraboli-Lohnherr, D., Privat, A., Provencher, J., et al. (1998). Kinematic analysis of recovered locomotor movements of the hindlimbs in paraplegic rats transplanted with monoaminergic embryonic neurons. *Annals of the New York Academy of Sciences, 860*, 521–523.

Gomez-Pinilla, F., Ying, Z., Roy, R. R., Hodgson, J., & Edgerton, V. R. (2004). Afferent input modulates neurotrophins and synaptic plasticity in the spinal cord. *Journal of Neurophysiology, 92*, 3423–3432.

Grillner, S. (1981). Control of locomotion in bipeds, tetrapods, and fish. In J. M. Brookhart & V. B. Mountcastle (Eds.), *The nervous system: motor control*. Bethesda, MD: American Physiological Society.

Guertin, P. A. (2004a). Role of NMDA receptor activation in serotonin agonist-induced air-stepping in paraplegic mice. *Spinal Cord, 42*, 185–190.

Guertin, P. A. (2004b). Synergistic activation of the central pattern generator for locomotion by l-beta-3,4-dihydroxyphenylalanine and quipazine in adult paraplegic mice. *Neuroscience Letters, 358*, 71–74.

Hains, B. C., Johnson, K. M., Mcadoo, D. J., Eaton, M. J., & Hulsebosch, C. E. (2001). Engraftment of serotonergic precursors enhances locomotor function and attenuates chronic central pain behavior following spinal hemisection injury in the rat. *Experimental Neurology, 171*, 361–378.

Harkema, S. J., Hurley, S. L., Patel, U. K., Requejo, P. S., Dobkin, V. R., & Edgerton, V. R. (1997). Human lumbosacral spinal cord interprets loading during stepping. *Journal of Neurophysiology, 77*, 797–811.

Harridge, S. D., Andersen, J. L., Hartkopp, A., Zhou, S., Biering-Sorensen, F., Sandri, C., et al. (2002). Training by low-frequency stimulation of tibialis anterior in spinal cord-injured men. *Muscle & Nerve, 25*, 685–694.

Hartkopp, A., Harridge, S. D., Mizuno, M., Ratkevicius, A., Quistorff, B., Kjaer, M., et al. (2003). Effect of training on contractile and metabolic properties of wrist extensors in spinal cord-injured individuals. *Muscle & Nerve, 27*, 72–80.

Hesse, S., Schmidt, H., Werner, C., & Bardeleben, A. (2003). Upper and lower extremity robotic devices for rehabilitation and for studying motor control. *Current Opinion in Neurology, 16*, 705–710.

Hochman, S., Garraway, S. M., Machacek, D. W., & Shay, B. L. (2001). 5-HT receptors and the neruomodulatory control of spinal cord functions. In T. C. Cope (Ed.), *Motor neurobiology of the spinal cord*. Boca Raton, FL: CRC Press.

Humphrey, D. R., Mao, H., & Griffith, R. W. (2006). Brain reorganization continues for 25 years after spinal injury: changing patterns of activation in human cerebral cortex during attempts to move. Society of Neuroscience Abstract, Program Number 88.10.

Ichiyama, R. M., Gerasimenko, Y. P., Zhong, H., Roy, R. R., & Edgerton, V. R. (2005). Hindlimb stepping movements in complete spinal rats induced by epidural spinal cord stimulation. *Neuroscience Letters, 383*, 339–344.

Ichiyama, R. M., Courtine, G., Gerasimenko, Y. P., Yang, G. J., van den Brand, R., Lavrov, I. A., et al. (2008). Step training reinforces specific spinal locomotor circuitry in adult spinal rats. *The Journal of Neuroscience, 28*, 7370–7375.

Jilge, B., Minassian, K., Rattay, F., & Dimitrijevic, M. R. (2004). Frequency-dependent selection of alternative spinal pathways with common periodic sensory input. *Biological Cybernetics, 91*, 359–376.

Jordan, L. M., & Schmidt, B. J. (2002). Propriospinal neurons involved in the control of locomotion: potential targets for repair strategies? In L. McKerracher, G. Doucet, & S. Rossignol (Eds.), *Progress in brain research*. Amsterdam: Elsevier Science, B.V.

Kazennikov, O. V., Shik, M. L., & Yakovleva, G. V. (1983). Stepping movements induced in cats by stimulation of the dorsolateral funiculus of the spinal-cord. *Bulletin of experimental biology and medicine, 96*, 1036–1039.

Kern, H., Boncompagni, S., Rossini, K., Mayr, W., Fano, G., Zanin, M. E., et al. (2004). Long-term denervation in humans causes degeneration of both contractile and excitation-contraction coupling apparatus, which is reversible by functional electrical stimulation (FES): a role for myofiber regeneration? *Journal of neuropathology and experimental neurology, 63*, 919–931.

Kiehn, O., Quinlan, K. A., Restrepo, C. E., Lundfald, L., Borgius, L., Talpalar, A. E., et al. (2008). Excitatory components of the mammalian locomotor CPG. *Brain Research. Brain Research Reviews, 57*, 56–63.

Kim, D., Adipudi, V., Shibayama, M., Giszter, S., Tessler, A., Murray, M., et al. (1999). Direct agonists for serotonin receptors enhance locomotor function in rats that received neural transplants after neonatal spinal transection. *The Journal of Neuroscience, 19*, 6213–6224.

Kim, S. J., Roy, R. R., Kim, J. A., Zhong, H., Haddad, F., Baldwin, K. M., et al. (2008). Gene expression during inactivity-induced muscle atrophy: effects of brief bouts of a forceful contraction countermeasure. *Journal of Applied Physiology, 105*, 1246–1254.

Kim, S. J., Roy, R. R., Zhong, H., Suzuki, H., Ambartsumyan, L., Haddad, F., et al. (2007). Electromechanical stimulation ameliorates inactivity-induced adaptations in the medial gastrocnemius of adult rats. *Journal of Applied Physiology, 103*, 195–205.

Lavrov, I., Dy, C. J., Fong, A. J., Gerasimenko, Y., Courtine, G., Zhong, H., et al. (2008). Epidural stimulation induced modulation of spinal locomotor networks in adult spinal rats. *The Journal of Neuroscience, 28*, 6022–6029.

Lavrov, I., Gerasimenko, Y. P., Ichiyama, R. M., Courtine, G., Zhong, H., Roy, R. R., et al. (2006). Plasticity of spinal cord reflexes after a complete transection in adult rats: relationship to stepping ability. *Journal of Neurophysiology, 96*, 1699–1710.

Ledoux, M. S., & Langford, K. H. (1993). Spinal cord stimulation for the failed back syndrome. *Spine, 18*, 191–194.

Lee, Y. S., Lin, C. Y., Caiozzo, V. J., Robertson, R. T., Yu, J., & Lin, V. W. (2007). Repair of spinal cord transection and its effects on muscle mass and myosin heavy chain isoform phenotype. *Journal of Applied Physiology, 103*, 1808–1814.

Liang, Y., Cai, L. L., Burdick, J. W., & Edgerton, V. R. (2006). A robotic training system for studies of post-SCI stand rehabilitation. In: *The First IEEE/RAS-EMBS International Conference on Biomedical Robotics and Biomechatronics, 2006 (BioRob 2006).*

Lovely, R. G., Gregor, R. J., Roy, R. R., & Edgerton, V. R. (1986). Effects of training on the recovery of full-weight-bearing stepping in the adult spinal cat. *Experimental Neurology, 92*, 421–435.

Lovely, R. G., Gregor, R. J., Roy, R. R., & Edgerton, V. R. (1990). Weight-bearing hindlimb stepping in treadmill-exercised adult spinal cats. *Brain Research, 514*, 206–218.

Maegele, M., Muller, S., Wernig, A., Edgerton, V. R., & Harkema, S. J. (2002). Recruitment of spinal motor pools during voluntary movements versus stepping after human spinal cord injury. *Journal of Neurotrauma, 19*, 1217–1229.

Maier, I. C., Ichiyama, R. M., Courtine, G., Schnell, L., Lavrov, I., Edgerton, V. R., et al. (2009). Differential effects of anti-Nogo-A antibody treatment and treadmill training in rats with incomplete spinal cord injury. *Brain* (in press).

Majczynski, H., Maleszak, K., Cabaj, A., & Slawinska, U. (2005). Serotonin-related enhancement of recovery of hind limb motor functions in spinal rats after grafting of embryonic raphe nuclei. *Journal of Neurotrauma, 22*, 590–604.

Marino, R. J., Ditunno, J. F., Jr., Donovan, W. H., & Maynard, F., Jr. (1999). Neurologic recovery after traumatic spinal cord injury: data from the Model Spinal Cord Injury Systems. *Archives of Physical Medicine and Rehabilitation, 80*, 1391–1396.

Martin, T. P., Stein, R. B., Hoeppner, P. H., & Reid, D. C. (1992). Influence of electrical stimulation on the morphological and metabolic properties of paralyzed muscle. *Journal of Applied Physiology, 72*, 1401–1406.

Mcewen, M. L., & Stehouwer, D. J. (2001). Kinematic analyses of air-stepping of neonatal rats after mid-thoracic spinal cord compression. *Journal of Neurotrauma, 18*, 1383–1397.

Merrill, D. R., Bikson, M., & Jefferys, J. G. (2005). Electrical stimulation of excitable tissue: design of efficacious and safe protocols. *Journal of Neuroscience Methods, 141*, 171–198.

Mori, S., Sakamoto, T., & Takakusaki, K. (1991). Interaction of posture and locomotion in cats: its automatic and volitional control aspects. In M. Shimamura, S. Grillner, & V. R. Edgerton (Eds.), *Neurobiological basis of human locomotion*. Tokyo: Japan Scientific Societies Press.

Mushahwar, V. K., Aoyagi, Y., Stein, R. B., & Prochazka, A. (2004). Movements generated by intraspinal microstimulation in the intermediate gray matter of the anesthetized, decerebrate, and spinal cat. *Canadian Journal of Physiology and Pharmacology, 82*, 702–714.

Mushahwar, V. K., Gillard, D. M., Gauthier, M. J., & Prochazka, A. (2002). Intraspinal micro stimulation generates locomotor-like and feedback-controlled movements. *IEEE Transactions on Neural Systems and Rehabilitation Engineering, 10*, 68–81.

Musienko, P. E., Bogacheva, I. N., & Gerasimenko, Y. P. (2007). Significance of peripheral feedback in the generation of stepping movements during epidural stimulation of the spinal cord. *Neuroscience and Behavioral Physiology, 37*, 181–190.

Nance, P. W. (2003). Management of spasticity. In V. W. Lin (Ed.), *Spinal cord medicine: Principles and practice*. New York: Demos Medical Publishing.

North, R. B., Kidd, D. H., Petrucci, L., & Dorsi, M. J. (2005). Spinal cord stimulation electrode design: a prospective, randomized, controlled trial comparing percutaneous with laminectomy electrodes: part II-clinical outcomes. *Neurosurgery, 57*, 990–996.

Ohta, Y., Yano, H., Suzuki, R., Yoshida, M., Kawashima, N., & Nakazawa, K. (2007). A two-degree-of-freedom motor-powered gait orthosis for spinal cord injury patients. *Proceedings of the Institution of Mechanical Engineers. Part H, Journal of Engineering in Medicine, 221*, 629–639.

Otoshi, C. K., Fong, A. J., Cai, L. L., Zhong, H., Roy, R. R., Tillakaratne, N. J. K., et al. (2005). 5-HT receptor distribution after spinal cord transection: effects of chronic serotonergic agonist administration and robotic training. Society of Neuroscience Abstract, Program Number 396.11.

Otoshi, C. K., Walwyn, W. M., Tillakaratne, N. J. K., Zhong, H., Roy, R. R., & Edgerton, V. R. (2009). Distribution and localization of 5-HT(1A) receptors in the rat lumbar

Parker, D. (2005). Pharmacological approaches to functional recovery after spinal injury. *Current Drug Targets. CNS and Neurological Disorders*, *4*, 195–210.

Reinkensmeyer, D. J., Aoyagi, D., Emken, J. L., Galvez, J. A., Ichinose, W., Kerdanyan, G., et al. (2006). Tools for understanding and optimizing robotic gait training. *Journal of Rehabilitation Research and Development*, *43*, 657–670.

Renard, V. M., & North, R. B. (2006). Prevention of percutaneous electrode migration in spinal cord stimulation by a modification of the standard implantation technique. *Journal of Neurosurgery: Spine*, *4*, 300–303.

Ribotta, M. G., Provencher, J., Feraboli-Lohnherr, D., Rossignol, S., Privat, A., & Orsal, D. (2000). Activation of locomotion in adult chronic spinal rats is achieved by transplantation of embryonic raphe cells reinnervating a precise lumbar level. *The Journal of Neuroscience*, *20*, 5144–5152.

Robinson, G. A., & Goldberger, M. E. (1986). The development and recovery of motor function in spinal cats. II. Pharmacological enhancement of recovery. *Experimental Brain Research*, *62*, 387–400.

Rodger, D. C., Fong, A. J., Li, W., Ameri, H., Ahuja, A. K., Gutierrez, C., et al. (2008). Flexible parylene-based multi-electrode array technology for high-density neural stimulation and recording. *Sensors and Actuators B: Chemical*, *132*, 449–460.

Rodger, D. C., Fong, A. J., Li, W., Ameri, H., Lavrov, I., Zhong, H., et al. (2007). High-density flexible parylene-based multielectrode arrays for retinal and spinal cord stimulation. *Solid-State Sensors, Actuators and Microsystems Conference, 2007*. Transducers, pp. 1385–1388.

Rossignol, S., & Barbeau, H. (1993). Pharmacology of locomotion: an account of studies in spinal cats and spinal cord injured subjects. *The Journal of the American Paraplegia Society*, *16*, 190–196.

Rossignol, S., Chau, C., Brustein, E., Giroux, N., Bouyer, L., Barbeau, H., et al. (1998). Pharmacological activation and modulation of the central pattern generator for locomotion in the cat. *Annals of the New York Academy of Sciences*, *860*, 346–359.

Rossignol, S., Giroux, N., Chau, C., Marcoux, J., Brustein, E., & Reader, T. A. (2001). Pharmacological aids to locomotor training after spinal injury in the cat. *Journal of Physiology*, *533*, 65–74.

Roy, R. R., Baldwin, K. M., & Edgerton, V. R. (1991). The plasticity of skeletal muscle: effects of neuromuscular activity. *Exercise and Sport Sciences Reviews*, *19*, 269–312.

Roy, R. R., Talmadge, R. J., Hodgson, J. A., Oishi, Y., Baldwin, K. M., & Edgerton, V. R. (1999). Differential response of fast hindlimb extensor and flexor muscles to exercise in adult spinalized cats. *Muscle & Nerve*, *22*, 230–241.

Roy, R. R., Talmadge, R. J., Hodgson, J. A., Zhong, H., Baldwin, K. M., & Edgerton, V. R. (1998). Training effects on soleus of cats spinal cord transected (T12-13) as adults. *Muscle & Nerve*, *21*, 63–71.

Roy, R. R., Zhong, H., Hodgson, J. A., Grossman, E. J., Siengthai, B., Talmadge, R. J., et al. (2002). Influences of electromechanical events in defining skeletal muscle properties. *Muscle & Nerve*, *26*, 238–251.

Roy, R. R., Zhong, H., Khalili, N., Kim, S. J., Higuchi, N., Monti, R. J., et al. (2007). Is spinal cord isolation a good model of muscle disuse? *Muscle & Nerve*, *35*, 312–321.

Scivoletto, G., Morganti, B., Ditunno, P., Ditunno, J. F., & Molinari, M. (2003). Effects on age on spinal cord lesion patients' rehabilitation. *Spinal Cord*, *41*, 457–464.

Shields, R. K. (1995). Fatigability, relaxation properties, and electromyographic responses of the human paralyzed soleus muscle. *Journal of Neurophysiology*, *73*, 2195–2206.

Shields, R. K., & Dudley-Javoroski, S. (2006). Musculoskeletal plasticity after acute spinal cord injury: effects of long-term neuromuscular electrical stimulation training. *Journal of Neurophysiology*, *95*, 2380–2390.

Shik, M. L., & Orlovsky, G. N. (1976). Neurophysiology of locomotor automatism. *Physiological Reviews*, *56*, 465–501.

Shik, M. L., Severin, F. V., & Orlovskii, G. N. (1966). Control of walking and running by means of electric stimulation of the midbrain. *Biofizika*, *11*, 659–666.

Stein, R. B., Gordon, T., Jefferson, J., Sharfenberger, A., Yang, J. F., De Zepetnek, J. T., et al. (1992). Optimal stimulation of paralyzed muscle after human spinal cord injury. *Journal of Applied Physiology*, *72*, 1393–1400.

Stewart, B. G., Tarnopolsky, M. A., Hicks, A. L., Mccartney, N., Mahoney, D. J., Staron, R. S., et al. (2004). Treadmill training-induced adaptations in muscle phenotype in persons with incomplete spinal cord injury. *Muscle & Nerve*, *30*, 61–68.

Talmadge, R. J. (2000). Myosin heavy chain isoform expression following reduced neuromuscular activity: potential regulatory mechanisms. *Muscle & Nerve*, *23*, 661–679.

Talmadge, R. J., Castro, M. J., Apple, D. F., Jr., & Dudley, G. A. (2002). Phenotypic adaptations in human muscle fibers 6 and 24 wk after spinal cord injury. *Journal of Applied Physiology*, *92*, 147–154.

Tillakaratne, N. J., De Leon, R. D., Hoang, T. X., Roy, R. R., Edgerton, V. R., & Tobin, A. J. (2002). Use-dependent modulation of inhibitory capacity in the feline lumbar spinal cord. *The Journal of Neuroscience*, *22*, 3130–3143.

Timoszyk, W. K., De Leon, R. D., London, N., Joynes, R., Minakata, K., Roy, R. R., et al. (2003). Comparison of virtual and physical treadmill environments for training stepping after spinal cord injury. *Robotica*, *21*, 25–32.

Timoszyk, W. K., De Leon, R. D., London, N., Roy, R. R., Edgerton, V. R., & Reinkensmeyer, D. J. (2002). The rat lumbosacral spinal cord adapts to robotic loading applied during stance. *Journal of Neurophysiology*, *88*, 3108–3117.

Timoszyk, W. K., Nessler, J. A., Acosta, C., Roy, R. R., Edgerton, V. R., Reinkensmeyer, D. J., et al. (2005). Hindlimb loading determines stepping quantity and quality following spinal cord transection. *Brain Research*, *1050*, 180–189.

Van De Crommert, H. W., Mulder, T., & Duysens, J. (1998). Neural control of locomotion: sensory control of the central pattern generator and its relation to treadmill training. *Gait & Posture*, 7, 251–263.

Van Den Brand, R. J., Gerasimenko, Y., Dy, C. J., Ichiyama, R. M., Lavrov, I., Zhong, H., et al. (2007). Epidural stimulation and pharmacological interventions facilitate neurorehabilitation by enabling stepping following spinal cord injury. Society of Neuroscience Abstract, Program Number 75.13.

Waters, R. L., Sie, I., Adkins, R. H., & Yakura, J. S. (1995). Injury pattern effect on motor recovery after traumatic spinal cord injury. *Archives of Physical Medicine and Rehabilitation*, 76, 440–443.

Wernig, A., Nanassy, A., & Muller, S. (1998). Maintenance of locomotor abilities following Laufband (treadmill) therapy in para- and tetraplegic persons: follow-up studies. *Spinal Cord*, 36, 744–749.

Winchester, P., & Querry, R. (2006). Robotic orthoses for body weight-supported treadmill training. *Physical Medicine and Rehabilitation Clinics of North America*, 17, 159–172.

Ying, Z., Roy, R. R., Edgerton, V. R., & Gomez-Pinilla, F. (2005). Exercise restores levels of neurotrophins and synaptic plasticity following spinal cord injury. *Experimental Neurology*, 193, 411–419.

CHAPTER 26

# Deep brain stimulation in obsessive–compulsive disorder

Damiaan Denys* and Mariska Mantione

*Department of Psychiatry, AMC, and the Netherlands Institute for Neuroscience, an Institute of the Royal Netherlands Academy of Arts and Sciences, Amsterdam, The Netherlands*

**Abstract:** The use of deep brain stimulation in psychiatric disorders has received great interest owing to the small risk of the operation, the reversible nature of the technique, and the possibility of optimizing treatment postoperatively. Currently, deep brain stimulation in psychiatry is investigated for obsessive–compulsive disorder, Gilles de la Tourette's syndrome, and major depression. This chapter reviews the application of deep brain stimulation in obsessive–compulsive disorder. Preliminary results suggest that deep brain stimulation in obsessive–compulsive disorder can effectuate a decrease of 40–60% in at least half of the patients. Although various side effects occur, most of these are transitory and linked to specific stimulation parameters which can be changed. Because only a few studies have been performed with a limited number of patients in accordance with varying research protocols, appliance of deep brain stimulation to obsessive–compulsive disorder is still at an experimental stage. The speed of the effect of deep brain stimulation causes fundamental assumptions on the pathophysiology of obsessive–compulsive disorder.

**Keywords:** deep brain stimulation; therapy-refractory obsessive-compulsive disorder; clinical effects; side effects; future

## Introduction

Nowadays psychiatry uses six different electric stimulation techniques in the brain for clinical purposes: electro convulsion therapy (ECT), magnetic convulsion therapy (MCT), transcranial magnetic stimulation (TMS), transcranial direct stimulation, nervus vagus stimulation (NVS), and deep brain stimulation (DBS). DBS, the most invasive technique, has attracted an increasing interest in the treatment of therapy-refractory psychiatric disorders in the past few years. This interest is due to various factors (Wichmann and Delong, 2006). First, the failure of existing treatments for a group of seriously ill, untreatable psychiatric patients. Second, the success of DBS in movement disorders. Third, the small risk of the operation, and the reversible nature of the technique. Fourth, the increasing social awareness of the cost of chronically ill patients. Finally, the research potential of the technique for the understanding of the pathophysiology of psychiatric disorders. This article presents a topical survey of DBS in obsessive–compulsive disorder (OCD).

*Corresponding author.
Tel.: +31 20 8913899; Fax: +31 20 8913898;
E-mail: d.denys@amc.nl

## DBS technique

DBS is a technique whereby one or more electrodes of approximately 1.27 mm diameter are implanted stereotactically on the left and right sides of the brain in a specific brain area (Fig. 1). The precise anatomic position of the electrode is calculated beforehand on the basis of magnetic resonance (MR) and computer tomography (CT) scans. Under the skin, the electrodes are attached via a conductivity cable to a pulse generator, which is inserted into a pocket under the clavicle. The activity of the electrode can be programed externally with a portable appliance communicating with the pulse generator through telemetry. The electrodes have various contact points (mostly four), which can be stimulated separately, as a result of which the anatomic reach of the stimulation area can be adjusted. Frequency, intensity, and pulse width are also programmable. Typical stimulation parameters vary for the frequencies between 2 and 185 Hz, for the current power between 0 and 10 V, and for the pulse widths between 60 and 150 μs. The programing facility has the advantage that, after implantation, the stimulation can be optimized in order to increase the therapeutic effect and to decrease side effects.

## Mechanism of action of DBS

Until now, the exact operating mechanism of DBS is not known. There are two general hypotheses (McIntyre et al., 2004). (1) DBS causes a functional lesion by inhibiting the brain core which is stimulated. This inhibition can be caused by a depolarization blockage of the neurons, by synaptic depression (exhaustion) or by synaptic inhibition via "neuronal jamming," inducing a meaningless activation pattern (McIntyre et al., 2004). (2) DBS activates the neuronal network connected to the brain core which is stimulated. Then stimulation leads to a modulation of the pathological activity in the neuronal network. It is most likely that the therapeutic effects of DBS are caused by a combination of direct and indirect effects dependent on the specific cytoarchitecture of the stimulated brain area. Because the field intensity of the electrode decreases exponentially with distance, neurons are influenced in various ways. The neuronal cell body is probably inhibited in the center of the stimulation area, and the axonal terminals are stimulated on the edge of the stimulation area.

## Psychiatric comorbidity in DBS of neurological disorders

At a rough estimate, more than 35,000 patients with movement disorders have been treated with DBS at present. The clinical experience with therapeutic effects in case of neurological disorders taught us that DBS also influences psychiatric symptoms. Stimulation of the globus pallidus interna (GPi) and the subthalamic nucleus (STN) may lead to panic attacks, hypomania, apathy, increased suicidal inclination, mania, aggression, hallucinations, depression, and pathological gambling (Wichmann and Delong, 2006). After stimulation, hypomania and mania usually occur acutely, whereas depressive disorders begin after some weeks. It is not clear whether the depressive complaints are due to the DBS itself, preceding psychiatric comorbidity, consequences of changing the dopaminergic medication, disillusionment by an unreal expectation from the operation, or estrangement by the abrupt and massive decrease of motorial complaints. No matter how, the impact of DBS on the mental health of patients with movement disorders necessitates psychiatric

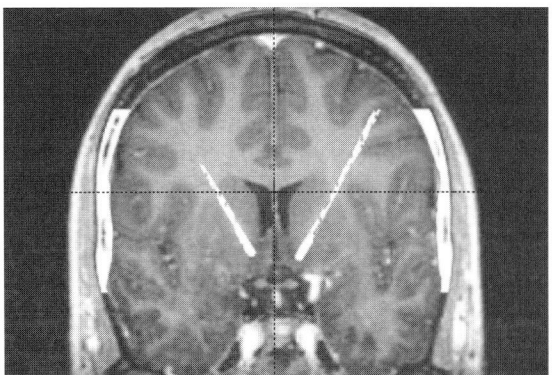

Fig. 1. Coronal section of the brain near the nucleus accumbens with the track of the electrodes on the left and right side.

support, both in the selection process before and after the operation. Both the increases and decreases of psychic complaints are observed with DBS in case of movement disorders. The striking improvement of OCDs in case of Parkinson's disease among others led to applying the technique.

**Obsessive–compulsive disorder**

OCD is a chronic psychiatric disorder characterized by obsessions and compulsions. Obsessions are recurrent, continuous, and disturbing thoughts causing anxiety or unrest, such as the fear to be contaminated, the fear to harm somebody else, and the need for symmetry or perfectionism. Compulsions are actions with a ritual character which are performed to rectify the anxiety caused by the obsession, such as repeated cleaning, washing, checking, classifying, or counting. Approximately 2% of the general population suffers from OCD. Three-quarter of the patients experiences an average decrease of 35% of the complaints with selective serotonin reuptake inhibitors and behavioral therapy. One out of 10 patients cannot be helped with even the best possible treatment and is a candidate for neurosurgical treatment.

DBS is applied to OCD for the first time in psychiatry because this disorder, as one of the few, has been clearly associated with a dysfunction of a neuroanatomic system — a hyperactivity of the corticostriatal circuit. For decades, neurosurgeons have made lesions in the frontal horn of the capsula interna (CI) and the basal ganglia for therapy-refractory OCD by means of classic, ablative neurosurgery (Lipsman et al., 2007). With the first DBS, it is obvious to choose a selective stimulation of the frontal arm of the CI in order to imitate the effect of the capsulotomy.

**Clinical efficacy of DBS in OCD**

Since the introduction of DBS in 1999 as a potential treatment for patients with treatment refractory OCD, several studies about its clinical efficacy have been published (Table 1). At present, approximately 60 patients with OCD have been implanted, and 40 patients have been reported in three double-blind verified studies (Nuttin et al., 1999, 2003; Abelson et al., 2005; Greenberg et al., 2006), six case studies (Mallet et al., 2002; Anderson and Ahmed, 2003; Sturm et al., 2003; Fontaine et al., 2004; Aouizerate et al., 2004, 2005; Jiménez et al., 2007), one study on the acute effects of DBS (Okun et al., 2007), and one recently published study about the combined results of three groups worldwide (Greenberg et al., 2008).

DBS in OCD was initially started in 1998 at the Karolinska Institute in Stockholm where two patients received bilateral implantation in the CI, but the results were never published (Andreewitch, personal communication). In 1999, the Leuven Group continued with the bilateral implantation of electrodes in the CI of four patients (Nuttin et al., 1999) and proceeded with the implantation of another six patients at the same target (Nuttin et al., 2003). These six patients were followed for a period of 21 months. Four patients completed the study and three of them experienced a decrease of complaints of at least 35%. In the double-blind part of the study, where the stimulator is put on and off for a period of time, an average decrease of 40% was observed. After 3 months of continuous stimulation, PET scans resulted in finding three patients with decreased activity in the frontal cortex indicating a decrease of hyperactivity in the corticostriatal circuit. This study from the Leuven Group was followed in 2003 by a study from the German Group where the nucleus accumbens (Nacc) was chosen as a target in four patients (Sturm et al., 2003). In three of the four patients, open stimulation resulted in nearly total recovery from both anxiety and OCD symptoms with follow-up periods of 24–30 months. The lack of effect in the fourth patient appeared to be caused by a displacement of the electrode in the caudoventral direction, thereby missing the target area. In 2005, the Michigan Group reported another study of implantation of electrodes in the CI of four patients. Patients were stimulated in a double-blind way in a phase of four times 3 weeks, with the stimulator on or off, followed by an open phase (Abelson et al., 2005). Only one patient had a

Table 1. Overview of DBS studies in OCD

| Reference | Side | Target | n | Diagnosis | Y-BOCS | Response |
|---|---|---|---|---|---|---|
| Nuttin et al. (1999) | Bilateral | Anterior limbs of internal capsules | 4 | OCD | No Y-BOCS scores were mentioned | In three of four patients, some beneficial results were seen. |
| Nuttin et al. (2003) | Bilateral | Anterior limbs of internal capsules | 6 | OCD | Mean score in the stimulation off condition: $32.3 \pm 3.9$ Mean score in the stimulation on condition: $19.8 \pm 8$ | Of six patients, two patients did not enter the crossover phase. Of the other four patients, three patients had at least a 35% reduction in preoperative Y-BOCS scores. |
| Mallet et al. (2002) | Bilateral | Subthalamic nucleus | 2 | OCD+PD | No Y-BOCS scores were mentioned | Compulsions disappeared and obsessions greatly diminished in two patients. Patient 1 had a 58% improvement on the Y-BOCS, patient 2 had a 64% improvement on the Y-BOCS. |
| Anderson and Ahmed (2003) | Bilateral | Anterior limbs of internal capsules | 1 | OCD | Preoperative score: 34 Postoperative score: 7 | One patient had a reduction of 79%. |
| Sturm et al. (2003) | Unilateral | Accumbens (right) | 4 | OCD | No Y-BOCS scores were mentioned | Nearly total recovery was seen from both anxiety and OCD symptoms in three of four patients. |
| Fontaine et al. (2004) | Bilateral | Subthalamic nucleus | 1 | OCD+PD | Preoperative score: 32 Postoperative: 1 | One patient had a reduction of 97%. |
| Aouizerate et al. (2004) | Bilateral | Accumbens (2)+ caudatus (2) | 1 | OCD+MDD | Preoperative score: 25 Postoperative score at 12 months: 10 Postoperative score at 15 months: 14 Postoperative score at 27 months: 12 | One patient had a reduction of 52%. |

| Study | Laterality | Target | N | Diagnosis | Scores | Outcome |
|---|---|---|---|---|---|---|
| Abelson et al. (2005) | Bilateral | Anterior limbs of internal capsules | 4 | OCD | Mean preoperative score: 32.8<br>Mean postoperative score in open testing: 23 | Double-blind testing: one of four patients had at least 35% reduction of symptoms. One of four patients had a moderate reduction of symptoms (17%).<br>Open stimulation: one of four patients had a 44% reduction of symptoms and one patient had a 73% reduction of symptoms. |
| Greenberg et al. (2006) | Bilateral | Anterior limbs of internal capsules | 10 | OCD | Mean preoperative score: 34.6±0.6<br>Mean postoperative score at 3 months: 25±1.6<br>Mean postoperative score at 36 months: 22.3±2.1 | Of 10 patients, 2 patients did not reach the 36 months follow-up.<br>After 36 months, four of eight patients met or exceeded a 35% reduction in symptoms. |
| Okun et al. (2006) | Bilateral | Anterior limbs of internal capsules | 5 | OCD | No long-term outcomes have been published | |
| Jiménez et al. (2007) | Bilateral | Inferior thalamic peduncle | 1 | OCD+MDD | | Significant decrease in obsessive and compulsive symptoms in one patient. |
| Greenberg et al. (2008) | Bilateral | Anterior limbs of internal capsules | 26 | OCD | Mean preoperative score: 34.0±0.6<br>Mean postoperative score at 3 months: 21.0±1.8<br>Mean postoperative score at 36 months: 20.9±2.4 | Of 26 patients, 16 patients did meet or exceed a 35% reduction at last follow-up. Combined long-term results of 11 patients from Leuven/Antwerp (Nuttin et al., 1999, 2003), 10 patients from the Butler hospital and Cleveland Clinic (Greenberg et al., 2006) and 5 patients from the university of Florida (Okun et al., 2006) |

decrease of more than 35% in the double-blind phase. In the open phase, this patient progressed from severe disability to relatively normal life (73% improvement over baseline) in 8 months. Another patient who showed only a 17% decline in the double-blind phase showed improvement in the open phase, with a final reduction of 44% after completing an intensive behavioral treatment program. In the two patients who were considered responders, PET scans showed decreased activity of the orbitofrontal cortex. An American-Belgian Group described in 2006 the effects of stimulation of the CI and the ventral striatum of 10 OCD patients (Greenberg et al., 2006). Eight of them were followed during 3 years after bilateral implantation. Over the whole group, the complaints decreased with 30% on average. Over 3 years, OCD improved from severe to moderate. Four out of eight patients were considered responders with an average symptom reduction of at least 35%. Recently, the combined data of 26 patients from the Leuven Group, the American-Belgian Group, and a group from the university of Florida were published (Greenberg et al., 2008). The percentage of responders was 28% at 1 month and increased to 61.5% at the last follow-up at 36 months. Patients implanted more recently are better responders due to the gradual change of the target leaving the CI for the ventral striatum. PET imaging in part of these patients found that acute high-frequency DBS increased perfusion in the orbitofrontal cortex, anterior cingulated cortex, striatum, pallidum, and thalamus compared to control conditions.

A number of case studies stated good results with stimulation in other brain targets. Two case reports of implantation in the STN of three patients with OCD and Parkinson's disease were published in 2002 (Mallet et al., 2002) and 2004 (Fontaine et al., 2004). Stimulation showed significant decrease of compulsive symptoms in all the three patients. Furthermore, in 2003, a case study of CI stimulation in one patient was published in which a 79% reduction in symptoms was noticed at 3 months follow-up (Anderson and Ahmed, 2003). At 10 months follow-up, the patient was able to return to work and all compulsions subsided. In 2004 and 2007, two case reports about stimulation in patients with OCD and major depression were published targeting the ventral caudate nucleus/nucleus accumbens (Aouizerate et al., 2004) and inferior thalamic peduncle (Jiménez et al., 2007). In the first study, a marked but delayed reduction of symptoms up to 52% was seen at 15 months follow-up. Likewise, in the second study, stimulation showed a significant reduction of obsessive and compulsive symptoms.

There is a significant difference in time to response between studies (Nuttin et al., 1999, 2003; Mallet et al., 2002; Sturm et al., 2003; Fontaine et al., 2004; Aouizerate et al., 2004; Abelson et al., 2005). Nuttin et al. (1999) and Mallet et al. (2002) reported acute relief of anxiety and obsessive thinking. In the latter studies of Nuttin et al., 2003, reduction of obsessions and compulsions was not reported until the first week after stimulation, a finding that was replicated by Fontaine et al. (2004). Sturm et al. (2003) reported onset of clinical improvement a few days to several weeks after the beginning of the stimulation. In the study from Abelson et al. (2005), beneficial effects were seen within the 3-week period during blinded study, whereas Aouizerate et al. (2005) reported improvement of symptoms merely after 9 months.

In conclusion, 26 of 40 patients who have been reported appear to have at least a 35% reduction of Y-BOCS scores. Therefore, 65% of patients are considered responders. With this in mind, DBS seems to be a promising technique for the treatment of refractory OCD. Three extensive studies will be published shortly: a French multicenter study with 18 patients bilaterally in the STN, a German study with 16 patients unilaterally on the right side in the nucleus accumbens, and a Dutch study with 16 patients bilaterally in the nucleus accumbens. These results will have to be awaited in order to be able to draw definitive conclusions on the effectiveness of DBS with therapy-refractory OCD.

### Side effects of DBS in OCD

Potential complications of DBS can arise: (1) as a result of surgery (procedure related), (2) due to the implanted device (device related), and (3) due to stimulation or cessation of stimulation.

1. Procedure related effects were reported by Greenberg et al. (2006). One patient had a small asymptomatic intracerebral hemorrhage after lead insertion, but this resolved within days after implantation. Another patient had a single intraoperative generalized tonic–clonic seizure after lead implantation. One patient, who had comorbid insulin-dependent diabetes mellitus, developed a superficial surgical wound infection after implantation. Greenberg et al. (2008) reported that 2 of the 26 patients had small intracerebral hemorrhages after lead insertion. Other studies did not describe operative complications.
2. Device related effects were reported by Greenberg et al. (2008) where a break in a stimulating lead or an extension wire requiring a replacement occurred in one patient. Other studies did not describe operative complications.
3. Side effects of stimulation can be divided in acute effects and effects of chronic stimulation which can be separated in effects on mood, cognition, and personality.

## *Acute effects*

Studies of Greenberg et al. (2006) and Okun et al. (2007) found several acute mood effects stimulating the anterior limb of the internal capsule and nucleus accumbens such as transient sadness, anxiety, euphoria, or giddiness. Okun et al. (2007) reported olfactory, gustatory, and motor sensations which were strongly associated with the most ventral lead positions, as well as physiological responses such as autonomic changes, increased breath rate, sweating, nausea, cold sensation, heat sensation, fear, and panic episodes. All these effects reversed when DBS was stopped or parameters were changed.

## *Mood effects*

Most patients with treatment refractory OCD suffer from comorbid major depression. Mood elevation is a favorable but unintended constant side effect of DBS in OCD patients. Patients start to laugh, experience blissful feelings, and describe that they can see the world more bright and clear within seconds after stimulation. Abelson et al. (2005) reported improvement of depression in one out of four patients while stimulating the anterior limbs of internal capsule. Decreased depression scores were also found by Greenberg et al. (2006), Aouizerate et al. (2005), and Jiménez et al. (2007). Greenberg et al. (2006) observed elevated mood in 5 out of 10 patients with increased energy, speech production, and spontaneity of social interactions, but without an increase in behavioral impulsivity.

## *Cognitive changes*

It is at present undecided whether DBS is associated with cognitive decline. Some patients do complain about memory problems and language problems but they are difficult to objectify with neuropsychological tests. Gabriëls et al. (2003), Abelson et al. (2005), Aouizerate et al. (2005), and Greenberg et al. (2006) reported no decline in cognitive and executive functioning. On the contrary, in the latter study, a group analysis revealed significant improvements in recall and prose passages. Acute effects of diminished concentration, 'clouding,' verbal perseveration, and flashbacks were reported by Greenberg et al. (2008), though these side effects ceased after parameters were changed.

## *Personality changes*

Gabriëls et al. (2003) reported no major adverse or harmful personality changes after 1 year of DBS with the Minnesota Multiphasic Personality Inventory (MMPI). Patients or family members did not report of any changes in personality in the study of Abelson et al. (2005).

## *Effects of DBS interruption*

Nuttin et al. (2003) noted severe worsening of mood during the off phase: three out of four patients reported suicidal thoughts. In the study of

Greenberg et al. (2006), six out of nine patients experienced depressed mood during the off phase. Clinical worsening abated to some extent over several days and improved after DBS was restarted. Five of these patients had a worsening in OCS symptoms, although this was less rapid and marked than the change in mood state.

## Conclusion and future

DBS is applied in OCD for approximately 10 years now. From the available literature, some careful conclusions can be drawn.

First of all, preliminary results suggest that DBS in OCD can effectuate a decrease of 40–60% in at least half of the patients. Although various side effects occur, most of these are transitory and linked to specific stimulation parameters which can be changed. Because only a few studies have been performed with a limited number of patients in accordance with varying research protocols, appliance of DBS to psychiatric disorders is still at an experimental stage. There is obviously a need for large double-blind, placebo-controlled studies using uniform treatment strategies before the effect and side effect profiles can be definitively assessed. DBS can only be accepted in clinical practice when agreement is reached on uniform, protocolized treatment algorithms. Even then the technique can only be applied in a specialized third-line setting with sufficient clinical expertise. However, DBS certainly has the potential of becoming preferential treatment for a specific group of seriously ill, therapy-refractory patients because of the small risk of the operation, the reversible nature of the technique, and the possibility to optimize treatment postoperatively.

Second, it is encouraging, from a research point of view, that the various stimulated brain areas, in many cases developed empirically, are still in agreement with recent theoretical findings on the neuroanatomy of OCD. On the other hand, it is surprising that for various disorders the same brain area can be stimulated, and also different brain areas can be stimulated for the same disorder. This observation does not spoil the effect of the causal relation between brain circuits and psychiatric disorders, but does call into question the specificity of that relation. The interest for DBS for various new indications is great by the increasing knowledge of neuroanatomy. Unfortunately, there are no guidelines yet for determining indications of DBS in psychiatry. The following minimum criteria can be an initial impetus (Table 2). A disease qualifies for DBS if (1) there is general agreement on the validity and specificity as an independent psychiatric disorder, (2) there is a clear link with a dysfunctional brain circuit, and (3) the symptoms can be measured objectively. A patient qualifies for DBS if there are (1) serious symptoms and intense suffering, (2) no other available treatment is effective, and (3) after the operation the patient might lead a normal life again by the decrease of the complaints, with the prospect of independence, labor integration, and social development.

Finally, it already appears now from the limited studies that DBS has an immense impact on the individual and the society. For psychiatry, the entirely uncommon quick and overall decrease of complaints is a new phenomenon both for patient and doctor. During the postoperative phase, the patient has to be carefully supported, not onlyto

Table 2. Criteria for DBS in psychiatry

| A disease qualifies for deep brain stimulation if | A patient qualifies for deep brain stimulation if |
|---|---|
| (1) There is general agreement on the validity and specificity as an independent psychiatric disorder | (1) There are serious symptoms and intense suffering |
| (2) There is a clear link with a dysfunctional brain circuit | (2) There are no other available treatment effective |
| (3) The symptoms can be measured objectively | (3) There is reasonable prospect of independence, labor integration, and social development after surgery |

achieve the best possible effect, but also to help him deal with the decrease of the complaints appropriately. DBS can alter the fundamental experience of the patient and face psychiatry with an ethical challenge at various levels. What attitude will the doctor take when a patient feels fine with a certain adjustment but the symptoms have not decreased objectively? What if the partner, who has lived with a sick patient, does not recognize his/her spouse anymore because of the sudden decrease of complaints and wants to break the relationship? In the social sphere it is not yet clear what the social and cultural consequences will be of a technique by which people can manipulate their own brains. For psychiatrists to be able to provide an adequate answer to this, it will have to start the dialog with neurosciences and human sciences. Technological progress in medicine is preeminently represented by DBS. Not only can it alleviate the suffering but also eliminate defects, and even improve achievements. But can man excessively rise above himself with technology?

## References

Abelson, J. L., Curtis, G. C., Sagher, O., Albucher, R. C., Harrigan, M., Taylor, S. F., et al. (2005). Deep brain stimulation for refractory obsessive–compulsive disorder. *Biological Psychiatry, 57,* 510–516.

Anderson, D., & Ahmed, A. (2003). Treatment of patients with intractable obsessive–compulsive disorder with anterior capsular stimulation. *Journal of Neurosurgery, 98,* 1104–1108.

Aouizerate, B., Cuny, E., Martin-Guehl, C., Guehl, D., Amieva, H., Benazzouz, A., et al. (2004). Deep brain stimulation of the ventral caudate nucleus in the treatment of obsessive–compulsive disorder and major depression. *Journal of Neurosurgery, 101,* 682–686.

Aouizerate, B., Martin-Guehl, C., Cuny, E., Guehl, D., Amieva, H., Bernazzouz, A., et al. (2005). Deep brain stimulation for OCD and major depression. *American Journal of Psychiatry, 162,* 2192.

Fontaine, D., Mattei, V., Borg, M., Von Langsdorff, D., Magnie, M., Chanalet, S., et al. (2004). Effect of subthalamic nucleus stimulation on obsessive–compulsive disorder in a patient with Parkinson disease. *Journal of Neurosurgery, 100,* 1084–1086.

Gabriëls, L., Cosyns, P., Nuttin, B., Demeulmeester, H., & Gybels, J. (2003). Deep brain stimulation for treatment-refractory obsessive–compulsive disorder: psychopathological and neuropsychological outcome in three cases. *Acta Psychiatrica Scandinavica, 107,* 275–282.

Greenberg, B. D., Gabriëls, L. A., Malone, D. A., Rezai, A. R., Friehs, G. M., Okun, M. S., et al. (2008). Deep brain stimulation of the ventral internal capsule/ventral striatum for obsessive compulsive disorder worldwide. *Molecular Psychiatry,* 1–16.

Greenberg, B. D., Malone, D. A., Friehs, G. M., Rezai, A. R., Kubu, C. S., Malloy, P. F., et al. (2006). Three-year outcomes in deep brain stimulation for highly resistant obsessive–compulsive disorder. *Neuropsychopharmacology, 31,* 2384–2393.

Jiménez, F., Velasco, F., Salin-Pascual, R., Velasco, M., Nicolini, H., et al. (2007). Neuromodulation of the inferior thalamic peduncle for major depression and obsessive compulsive disorder. *Acta Neurochirurgica Supplementum, 97,* 393–398.

Lipsman, N., Neimat, J. S., & Lozano, A. M. (2007). Deep brain stimulation for treatment-refractory obsessive–compulsive disorder: the search for a valid target. *Neurosurgery, 61*(1), 1–11.

Mallet, L., Mesnage, V., Houeto, J. L., Pelissolo, A., Yelnik, J., Behar, C., et al. (2002). Compulsions, Parkinson's disease, and stimulation. *Lancet, 360,* 1302–1304.

McIntyre, C. C., Savasta, M., Walter, B. L., & Vitek, J. L. (2004). How does deep brain stimulation work? Present understanding and future questions. *Journal of Clinical Neurophysiology, 21,* 40–50.

Nuttin, B., Cosyns, P., Demeulmeester, H., Gybels, J., & Meyerson, B. (1999). Electrical stimulation in anterior limbs of internal capsules in patients with obsessive compulsive disorder. *Lancet, 354,* 1526.

Nuttin, B. J., Gabriels, L. A., Cosyns, P. R., Meyerson, B. A., Andreewitch, S., Sunaert, S. G., et al. (2003). Long-term electrical capsular stimulation in patients with obsessive–compulsive disorder. *Neurosurgery, 52,* 1263–1272.

Okun, M. S., Mann, G., Foote, K. D., Shapira, N. A., Bowers, D., Springer, U., et al. (2007). Deep brain stimulation in the internal capsule and nucleus accumbens region: responses observed during active and sham programming. *Journal of Neurology, Neurosurgery, and Psychiatry, 78,* 310–314.

Sturm, V., Lenartz, D., Koulousakis, A., Treuer, H., Herholz, K., Klein, J. C., et al. (2003). The nucleus accumbens: a target for deep brain stimulation in obsessive–compulsive and anxiety-disorders. *Journal of Chemical Neuroanatomy, 26,* 293–299.

Wichmann, T., & Delong, M. R. (2006). Deep brain stimulation for neurologic and neuropsychiatric disorders. *Neuron, 52,* 197–204.

CHAPTER 27

# The use of repetitive transcranial magnetic stimulation (rTMS) for the treatment of spasticity

Francesco Mori[1,2], Giacomo Koch[1,2], Calogero Foti[3], Giorgio Bernardi[1,2] and Diego Centonze[1,2,*]

[1]*Clinica Neurologica, Dipartimento di Neuroscienze, Università Tor Vergata, Rome, Italy*
[2]*Fondazione Santa Lucia, IRCCS, Rome, Italy*
[3]*Medicina Fisica e Riabilitativa, Dipartimento di Sanità Pubblica e Biologia Cellulare, Università Tor Vergata, Rome, Italy*

**Abstract:** Spasticity is a common disorder in patients with injury of the brain and spinal cord, especially in patients affected by multiple sclerosis (MS). In MS, spasticity is a major cause of long-term disability, it significantly impacts daily activities and quality of life and is only partially influenced by traditional spasmolytic drugs. Transcranial magnetic stimulation (TMS) is a noninvasive tool that can be used to modulate cortical excitability of the leg motor area, inducing remote effects on the excitability of the spinal circuits. The H reflex is a reliable electrophysiologic measure of the stretch reflex, and has been used in previous studies to test the effects of rTMS of the motor cortex on spinal circuitry. Based on these premises, originating from physiological studies in normal subjects, some studies have demonstrated that rTMS of the leg motor cortex can be beneficial in the management of spasticity by enhancing corticospinal tract excitability and reducing H reflex amplitude.

**Keywords:** spasticty; multiple sclerosis; transcranial magnetic stimulation; rTMS; H reflex

## Introduction

Spasticity is a common disorder in patients with injury of the brain and spinal cord. Its prevalence is reported to be around 35% in stroke patients with persistent hemiplegia (Sommerfeld et al., 2004), 65–78% in patients with spinal cord injury (Maynard et al., 1990), and up to 85% in patients with multiple sclerosis (MS) (Rizzo et al., 2004). In this disorder, spasticity is a major cause of long-term disability, and significantly impacts daily activities and quality of life of these subjects.

The great challenge for rehabilitation is to induce active movement in the paretic extremities, even years after injury, and release the spasticity, which is hardly influenced at all by traditional spasmolytic drugs (Dietz, 2001).

Animal studies showed that spinal neuronal circuits present plastic changes following a central motor lesion (Raisman, 1969; Raineteau and Schwab, 2001; Bareyre et al., 2004). This has implications for any rehabilitative therapy that should be directed to take advantage of the plasticity of the central nervous system (CNS), leading to the idea that techniques that are able to

*Corresponding author.
Tel.: 0039 06 7259 6010; Fax: 0039 06 7259 6006;
E-mail: centonze@uniroma2.it

drive CNS plasticity, as for example transcranial magnetic stimulation (TMS), may be of great interest in rehabilitation of spasticity.

**Pathophysiology of spasticity**

Spasticity can be defined as a velocity-dependent increase in muscle tone. This pathological condition is commonly believed to follow the exaggerated activation of the stretch reflex, secondary to the lesion of the upper motoneuron of the corticospinal tract (Young, 1994). The stretch reflexes may therefore be elicited easily in an antagonistic muscle in spastic patients, and this impedes the initiation or execution of agonist movement (el-Abd et al., 1993). Evidence exists, however, that spasticity features are not equivalent in patients with cerebral and spinal cord injury, since increased passive joint resistance in resting limbs in stroke patients has been found to be highly correlated with the activation of the tonic stretch reflex, while this evidence is not so consistent in spinal injury. It is therefore possible that other mechanisms might contribute to hypertonia in spinal cord injury (Woolacott and Burne, 2006).

The stretch reflex arc is composed mainly of a chain of two neurons: a sensory neuron, termed the Ia neuron, is activated in response to the stretch of the muscle spindle, which stimulates, in turn, α motoneurons directed to extrafusal fibers of the same muscle. Thus, as a result of this stimulation, the muscle reacts to the stretch with a contraction. The ventral horn of the spinal cord is the site where Ia afferents connect with α motoneurons. Moreover, the excitability of this circuit is further regulated by a network of spinal interneurons that are responsible for specific phenomena such as the disynaptic reciprocal inhibition, presynaptic Ia inhibition, or autogenic Ib inhibition that have been found to be altered in spasticity (see below).

Several mechanisms have been proposed to explain the origin of stretch reflex hyperexcitability in spasticity. Animal models suggested that spasticity may occur because of changes in supraspinal drive and secondary changes at the cellular level in the spinal cord below the lesion, such as collateral sprouting (Raisman, 1969; Raineteau and Schwab, 2001; Bareyre et al., 2004). These observations, however, have not been verified for human spasticity.

In humans, numerous studies have suggested that a reduction of spinal inhibitory mechanisms is involved in the pathogenesis of spasticity. Different mechanisms have been shown to be altered in spastic patients, indicating that this condition may not be caused by a single mechanism, but rather by an intricate chain of alterations in different interdependent networks. These alterations might comprise:

- Reduced presynaptic inhibition of Ia afferents from muscle spindle terminals of flexor muscles (Nielsen et al., 1995b).
- Reduced disynaptic reciprocal inhibition from Ia afferents of antagonist muscles (Meunier and Pierrot-Deseilligny, 1998; Nielsen et al., 2007).
- Abnormal activity of Ib afferents from Golgi tendon organs (autogenetic Ib inhibition), found to facilitate, instead of inhibiting, α motoneuron activity (Delwaide and Olivier, 1988).
- Impaired recurrent motoneuronal inhibition mediated by Renshaw cells (Katz and Pierrot-Deseilligny, 1982, 1999).

John Eccles demonstrated the existence of presynaptic inhibition of Ia afferents evoked by activation of muscle spindle afferents from flexor muscles in the cat spinal cord. This presynaptic inhibition of Ia terminals, caused by axo-axonal GABA (gamma-aminobutyric acid)-ergic synapses, may substantially modulate the monosynaptic transmission of Ia excitatory effects to motoneurons, since interneurons transmitting this presynaptic inhibition are controlled by descending tracts (Rudomin and Schmidt, 1999). In humans, presynaptic inhibition can be valuated indirectly by measuring the size of heteronymous monosynaptic facilitation of the Soleus H reflex that is induced by stimulation of the femoral nerve

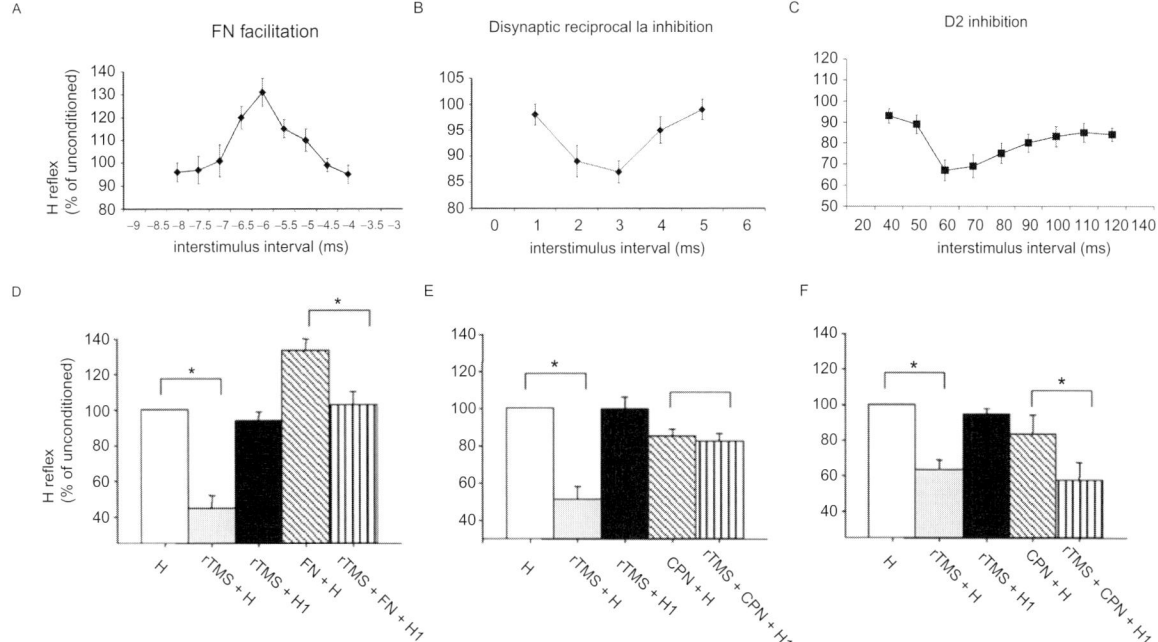

Fig. 1. Effects of conditioning stimulation over different nerves on the SOL H reflex and changes induced by 5 Hz rTMS on presynaptic inhibition and disynaptic reciprocal Ia inhibition on healthy subjects. SOL H-reflex amplitude changes at different interstimulus intervals through (A) femoral nerve facilitation, (B) short-latency depression (disynaptic reciprocal Ia inhibition), and (C) long-latency depression (D2 inhibition). The ordinate shows the size of the conditioned SOL H reflex (as percentage of the control reflex size). The abscissa shows conditioning-test interstimulus intervals. Positive values indicate that the conditioning pulse preceded the control pulse and negative values indicate that the control pulse preceded the conditioning pulse. (D) Following rTMS FN facilitation of the SOL H reflex (conditioning test interval −8 to −4.5 ms) is decreased; (E) short-latency depression of the SOL H reflex (2–3 ms conditioning test interval) is unchanged; (F) long-latency depression of the SOL H reflex (D2 inhibition, 60–80 ms conditioning test interval) is increased. The ordinate shows the size of the conditioned SOL H reflex (as a percentage of the control reflex size), and the abscissa shows different testing conditions. Bars indicate standard errors ($^*p<0.05$). Since rTMS evoked a depression of the H reflex, it was necessary to compensate for this depression by using a second stimulator set at a higher stimulation intensity to evoke the H reflex (H1) in conditions 3 and 5. In this way, the H reflex conditioned by femoral nerve stimulation and peroneal nerve stimulation had the same size with and without rTMS in all experiments; SOL, soleus muscle; CPN, common peroneal nerve; FN, femoral nerve. (Adapted with permission from Perez et al., 2005.)

(Hultborn et al., 1987) (see Fig. 1A). The reflex facilitation, assessed $0.3 \pm 0.4$ ms after its onset, depends only on the size of the conditioning monosynaptic Ia excitatory postsynaptic potential. If presynaptic inhibition of Ia afferents mediating the conditioning Ia volley changes, the size of heteronymous facilitation will change as well: the smaller the ongoing presynaptic inhibition, the larger the reflex facilitation. Using this technique, presynaptic inhibition has been demonstrated to be reduced in spastic patients with MS (Nielsen et al., 1995a, b), or with spinal cord injury, but not with hemiplegic stroke (Faist et al., 1994).

In the lower limbs, a strong candidate for playing a major role in the pathophysiology of spasticity is the reduction of reciprocal inhibition. Disynaptic reciprocal Ia inhibition, in fact, regulates muscle tone of antagonist muscles from ankle dorsiflexors to plantarflexor (Crone and Nielsen, 1994), maintaining relaxed flexor muscles during voluntary movements. Reciprocal inhibition can be tested in humans studying the conditioning effect from the common peroneal nerve (CP) stimulation on the Sol H reflex evoked by posterior tibial nerve (PTN) stimulation. A reciprocal Ia inhibition is observed when the CP stimulation precedes the PTN stimulation by interstimulus intervals of 2–3 ms (early reciprocal Ia inhibition, see Fig. 1B), 10–25 ms (D1 inhibition), and 60–80 ms (D2 inhibition, see Fig. 1C) (Mizuno et al., 1971). In healthy subjects, the interneurons interposed in this pathway are tonically active at rest, thus contributing to maintain low the excitability of soleus motoneurons (Nielsen et al., 1995a). In spastic patients with MS (Crone et al., 1994), stroke, spinal cord injury (Crone et al., 2006), and hereditary spastic paraparesis (Crone et al., 2004), however, it has been found that reciprocal inhibition from ankle dorsiflexors to plantarflexor motoneurons is reduced. In contrast, reciprocal inhibition from ankle plantarflexors to dorsiflexors has been found to be increased in spastic patients (Yanagisawa et al., 1976; Mailis and Ashby, 1990). Dorsiflexor muscles seldom show any spasticity; thus, this finding does not invalidate the hypothesis that reduced reciprocal inhibition may be implicated in the pathogenesis of spasticity in the involved muscles.

Autogenetic Ib inhibition is caused by activation of Ib afferents coming from Golgi tendon organs and mediated by segmental inhibitory interneurons projecting to the motoneurons of the same muscle. It can be easily demonstrated in healthy human subjects by stimulating the branch from the tibial nerve that innervates the medial gastrocnemius muscle and measuring the subsequent depression of the soleus H reflex (Pierrot-Deseilligny et al., 1979). Instead of inhibition, a facilitatory effect was observed on the paretic side in six of the six hemiplegic patients (Delwaide and Olivier, 1988). This facilitation developed in parallel with hyper-reflexia. Studies on the cat spinal cord also showed Ib afferent excitatory pathways. It thus seems likely that changes in the balance between inhibitory and excitatory Ib pathways play an important role in the development of spasticity.

Recurrent inhibition is mediated by Renshaw cells. These cells are located in the ventral horn of the spinal cord, receive excitatory collaterals from the motor axons, and project back to the motoneurons (Baldissera et al., 1981). Recurrent inhibition can be studied in humans by means of the effect of activation of Renshaw cells by a previous reflex discharge on a subsequently evoked test reflex (Pierrot-Deseilligny and Bussel, 1975). In most spastic patients, recurrent inhibition appears to be normal at rest, but during voluntary movement, its modulation is altered compared to healthy subjects (Katz and Pierrot-Deseilligny, 1982, 1999).

Not only the above-described alterations at the presynaptic level of the stretch arc reflex may account for spasticity, but also changes in the excitability of motoneuronal membranes may be involved in the pathogenesis of spasticity. The motoneuronal response to a synaptic input is conditioned by active membrane properties. In animal models, generation of plateau potentials has been described in $\alpha$ motoneurons. After upper motoneuron denervation their elicitation threshold lowers. Voltage-dependent calcium and sodium channels are highly expressed in the motoneuron membrane. The activation of these channels can amplify and prolong the response of motoneurons to synaptic excitation, and if outward currents are diminished or calcium channels are sensitized (for example by serotonergic and noradrenergic inputs), persisting plateau potentials can be generated (Rekling et al., 2000; Powers and Binder, 2001; Heckman et al., 2003; Hultborn et al., 2004; Heckmann et al., 2005). Although experimental data on humans are scarce, it has been reported that excitatory sensory drive to spinal motoneurons does not induce a plateau-like behavior of motor unit activity in patients with spinal cord lesions as easily as in healthy subjects. The authors argued that plateau potentials might play a role to the clinical manifestation of spasticity (Nickolls et al., 2004). Furthermore it has been observed in spinal cord injured patients that motor units require less synaptic drive to be de-recruited, arguing that plateau potentials are activated during the spasm contributing to its occurrence (Gorassini et al., 2004).

## Effects of rTMS on cortical excitability

TMS is a noninvasive tool for brain activation based on the principle of electromagnetic induction. According to this principle, a current passing through a coil placed over the scalp of a subject can induce a magnetic field reaching the surface of the brain with negligible attenuation. This magnetic field is, in turn, able to generate a secondary ionic current that depolarizes the neurons. The standard device for TMS is quite simple and its use is flexible. Accordingly, several parameters can be changed according to the purpose of the stimulation, such as site of application, intensity of stimulation, frequency of stimulation, stimulus duration, and treatment duration.

Modulating the frequency of repetitive TMS (rTMS) pulses makes it possible to achieve remarkably different changes on the excitability of the stimulated area, while prolonged treatments result in enhanced duration of the effects. The 1 Hz frequency is normally used to reduce the excitability of the motor cortex, as demonstrated by a reduction of the size of the motor evoked potentials (MEPs) following its application over the primary motor cortex (Chen et al., 1997). A similar depression was also demonstrated to occur in the visual cortex where the excitability changes can be detected as an increase in the phosphene threshold (Boroojerdi et al., 2000).

On the contrary, the 5 Hz stimulation protocol is used to enhance cortical excitability, as demonstrated by a progressive increase in the size of the MEPs during its application over the motor cortex (Berardelli et al., 1998). Also prolonged 5 Hz rTMS results in increased duration of the effects. In an early study, indeed, it has been convincingly shown that the effects of both 900 and 1800 pulses were still present 5 min after the end of the stimulation, while 20 and 40 min later, only the excitatory effects of 1800 pulses were still detectable (Peinemann et al., 2004). Moreover, TMS has previously been used to reveal remote functional effects between distant interconnected areas (i.e. Civardi et al., 2001; Siebner et al., 2003; Koch et al., 2006, 2007).

## Effects of rTMS on spinal excitability

Corticospinal cells modulate the activity of $\alpha$ and $\gamma$ motoneurons, Ia afferents, and a large group of spinal interneurons. Thus, it is not surprising that the modulation of motor cortex excitability by rTMS results in a significant modulation of spinal excitability. The H reflex is a reliable electrophysiologic measure of the stretch reflex, and has been used in previous studies to test the effects of rTMS of the motor cortex on spinal circuitry.

It has been reported that TMS of the motor cortex exerts a complex modulatory action on the soleus H reflex. Most often the reflex was transiently inhibited but sometimes facilitated by the stimulation, depending on the subject and/or the experimental condition (Nielsen and Petersen, 1995). Single pulses of cortical TMS have also been found to evoke significant and congruent changes in the lower limb, providing increased heteronymous Ia facilitation and decreased D1 inhibition, both of which suggest a decrease in presynaptic inhibition of Ia afferents (Meunier and Pierrot-Deseilligny, 1998). Furthermore, suprathreshold, 5 Hz rTMS delivered to the motor cortex suppressed H reflexes in forearm muscles for a time period of 900 ms while it produced MEPs in hand and forearm muscles, which gradually increased in size (Berardelli et al., 1998).

One-Hz rTMS on the motor cortex has been conversely shown to depress MEP amplitude with negligible (Touge et al., 2001) or enhancing effects on the H reflex (Valero-Cabre et al., 2001). Valero-Cabre et al. (2001) also found that 1 Hz rTMS of the motor cortex did not change the compound motor action potential ($M$) evoked by peripheral nerve stimulation, so that H/$M$ amplitude ratio was increased, arguing that low-frequency rTMS might facilitate monosynaptic spinal cord reflexes by inhibiting the corticospinal projections modulating spinal excitability. More recently, Perez and coworkers investigated whether short trains of 20 pulses at 5 Hz rTMS over the primary motor cortex of the leg area produced adaptations at the spinal level. They found that MEPs in the soleus and tibialis anterior muscles were facilitated by rTMS at rest, while the soleus H reflex was depressed for 1 s. 5 Hz rTMS also increased the size of the long-latency depression of the soleus H reflex evoked by CP nerve stimulation and decreased the femoral nerve facilitation of the soleus H reflex, suggesting that the depression of the H reflex by rTMS can be explained, at least partly, by an increased presynaptic inhibition of soleus Ia afferents, (Perez et al., 2005; see Fig. 1D and F).

## Effects of rTMS in patients with spasticity

Based on the premises originating from physiological studies in normal subjects, some studies have explored the possibility that rTMS of the motor cortex could be beneficial in the management of spasticity by enhancing corticospinal tract excitability and reducing stretch reflex (see overview Table 1).

### MS patients

Nielsen and Sinkjaer evaluated the effect of rTMS on spasticity in 38 patients with MS in a double-blind placebo-controlled study. Patients were treated twice daily for seven consecutive days. Self-score of ease of everyday activities and clinical spasticity score significantly improved after the treatment, and the stretch reflex threshold increased, thus proposing that rTMS has an antispastic effect in these patients (Nielsen et al., 1996). In a subsequent study, these authors also found that high-frequency rTMS (16 stimuli at 25 Hz) over the leg motor area of MS patients with lower limb spasticity induced a decrease of the unconditioned soleus muscle H-reflex amplitude (Nielsen and Sinkjaer, 1997).

The effects of rTMS on spasticity in MS were also evaluated in our laboratory (Centonze et al., 2007). We selected 19 patients with MS and spasticity affecting exclusively or predominantly one lower limb. The rTMS protocol was applied by using a standard device connected to a figure-of-eight coil, and stimuli were delivered over the scalp site corresponding to the leg area of the primary motor cortex contralateral to the affected limb. The effects of rTMS on spasticity were measured both electrophysiologically, by measuring the H reflex, and clinically, by using the modified Ashworth scale (MAS). All the patients underwent three distinct stimulation protocols:

1. A low-frequency rTMS consisting in one train of 900 pluses at 1 Hz. The total duration of the stimulation was 15 min.
2. A high-frequency rTMS consisting of 18 trains of 50 stimuli each, delivered at 5 Hz and separated by a 40 s pause. The total number of the pulses and the duration of the stimulation were 900 and 15 min, respectively, as in the case of the 1 Hz protocol.
3. Sham stimulation. This stimulation was similar to the 5 Hz protocol, but the coil was angled away, so that no current was induced in the brain.

Each patient was exposed to a single session of 1 Hz, 5 Hz, and sham stimulation. The interval between each protocol was one week. Immediately before and immediately after the rTMS session, the amplitude of the H reflex was measured. We found that 5 Hz significantly reduced the H response, according to the idea that this protocol of stimulation inhibited the stretch reflex. The 1 Hz protocol, in contrast, had opposite effects on the H response, and both 1 and 5 Hz effects disappeared after 10 and 20 min. In spite of the effects of rTMS on H reflex, a single session of stimulation did not affect spasticity measured clinically with the MAS applied at three joints of the studied limb. We also evaluated the effects of 5 Hz rTMS on the amplitude of MEP and found that this protocol was indeed able to increase the excitability of the corticospinal tract. In a final set of experiments, we studied the effects of a two-week protocol of 5 Hz rTMS on both H reflex and spasticity. We found that repeated sessions of rTMS did not enhance the size of the effect on H response but remarkably enhanced its duration. Accordingly, a week after the two-week stimulation, it was still possible to record a reduction of the H reflex. Noticeably, two weeks of rTMS at 5 Hz were also able to reduce spasticity in MS patients, and this effect paralleled the reduction of the H response, since it persisted one week after the end of the stimulation (see Fig. 2).

Table 1. Effects of rTMS on spastic patients

| Authors | Study population | Stimulation protocol | Results |
|---|---|---|---|
| Nielsen et al. (1996) | 38 MS spastic patients | 5 Hz rTMS, twice a day for one week | Reduction of stretch reflex threshold and improvement of self-score spasticity scale |
| Nielsen and Sinkjaer (1997) | 11 MS patients with lower limb spasticity induced | One session of 16 stimuli at 25 Hz over the leg motor area and one session of 5 min rTMS | Decrease of the unconditioned soleus muscle H-reflex amplitude |
| Centonze et al. (2007) | 19 patients with MS and spasticity affecting exclusively or predominantly one lower limb | 900 TMS pulses delivered over spastic lower limb motor cortex for two weeks in a double blind controlled study with three groups: 5 Hz (18 trains separated by 40 s pause) versus 1 Hz (continuous) versus Sham | Decrease of H/M ratio, improvement of the MAS, lasting one week after the end of stimulation |
| Valle et al. (2007) | 17 children (mean 9.1 years, SD 3.2) with cerebral palsy and spastic quadriplegia | Five consecutive days of rTMS over APB M1: 1 Hz (1500 pulses, 90% RMT) versus 5 Hz (1500 pulses in 5 × 1 min trains, 2 min intertrain interval, 90% RMT) rTMS versus sham | Improvement of passive range of motion of wrist and elbow dorsal flexion and extension. No significant changes in MAS scores of fingers, wrist, and elbow joints |
| Mälly and Dinya (2008) | 64 patients with chronic stroke (mean age $57.6 \pm 10.8$ years; $10.0 \pm 6.4$ years disease duration). | 1 Hz rTMS at 30% of 2.3 T, 100 stimuli per session, twice a day for a week over both hemispheres together, affected and unaffected, or separately, over hand motor cortex | Improvement of spasticity and of movement and behavior of the paretic extremities with all protocols assessed by motor by the Fugl–Meyer scale |
| Izumi et al. (2008) | Nine adult chronic stroke patients with hemiplegia (mean age 66.3 years) | Four weekly sessions of 100 TMS pulses at 0.1 Hz (stimulation $f$) versus sham, over the motor "hot spot" of extensor digitorum muscles of the paretic hand, during extension and abduction of the paretic thumb and fingers | Improvement of MAS at wrist flexion and MFT (manual function test) immediately after the fourth session and one week later (follow-up was not extended further) |

Based on these data, we argue that the beneficial effects on spasticity observed in that study were possibly related to plastic changes in spinal cord circuits induced by repetitive sessions of rTMS for two weeks presumably through long-term potentiation (LTP)-like mechanisms.

### Cerebral palsy

Similar results were obtained by Valle et al. (2007) on children with cerebral palsy and spastic quadriplegia. These authors found a significant reduction of spasticity after five consecutive days of 5 Hz rTMS stimulation, as indexed by the degree of passive movement. This clinical effect was not evident when using the MAS, although a trend for improvement was seen for elbow movement.

### Poststroke patients

By studying recovery of motor disability and spasticity in severe cases of poststroke patients, Mälly and Dinya (2008) found that spasticity can improve even 10 years after the stroke episode following 1 Hz rTMS at 30% of 2.3 T, 100 stimuli per session, twice a day for a week. The stimulation over both hemispheres together, affected and unaffected, or separately resulted in the release of spasticity during the treatment period. In this study, the authors found that spasticity could be modified by the stimulation of either the affected or the unaffected hemisphere, but the induction of movement could be achieved only by the stimulation of an intact motor pathway and its surrounding area. On the basis of these results, these authors conclude

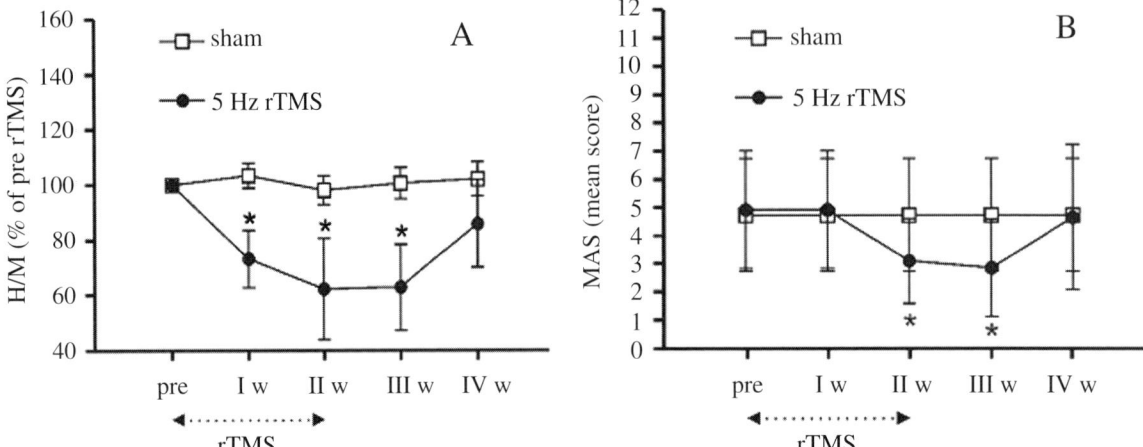

Fig. 2. Effects of a two-week protocol of rTMS on H/M ratio and spasticity in patients with MS. (A) The graph shows that 5 Hz rTMS decreases H/M ratio measured immediately after the first stimulation session (I w), at the end of the two-week stimulation protocol (II w), and one week after the end of stimulation (III w). A decrease of H/M ratio was also evident two weeks after the end of treatment (IV w), although this effect was not significant. Sham stimulation, conversely, did not show any noticeable effect. (B) Clinical improvement in spasticity (measured by the MAS score) paralleled the electrophysiological changes observed immediately after the two-week rTMS protocol (II w) and one week later (III w). * = $p < 0.05$.

that spasticity in one half of the body can be influenced from both sides of the brain. One of the explanations of the effectiveness of this treatment may be that, after stroke, the balance in the counterpart inhibition between the two hemispheres is impaired. Accordingly, an elevated inhibitory drive was demonstrated with TMS from the intact hemisphere to the lesioned hemisphere during the process of generation of a voluntary movement by the paretic hand, which might adversely influence the recovery of a paresis (Murase et al., 2004). This may mean that the development of spasticity includes many neurotransmitter systems from the cortical and subcortical areas of the brain. rTMS may create a homeostatic change in the brain, which contributes to the normalization of muscle tone.

Izumi et al. (2008) studied the effects of rTMS delivered during maximal movement effort of chronic hemiplegic hand after stroke. The stimulation protocol consisted of four weekly sessions of 100 TMS pulses with an interstimulus interval of 10 s, delivered over the motor "hot spot" of extensor digitorum muscles of the paretic hand of nine adult chronic stroke patients with hemyplegia, during extension and abduction of the paretic thumb and fingers.

After stimulation, patients had either reduced wrist flexor spasticity or improved manual performance that persisted one week later. The authors argued that a possible mechanism for improvement in manual function test or amelioration of spasticity is use-dependent plastic change of the motor cortex and spinal cord (Izumi et al., 2008).

## Conclusions

It is generally accepted that locomotion in mammals, including humans, is based on the activity of complex neuronal circuits within the spinal cord (the central pattern generator) (e.g. Dietz, 2003). Afferent information from the periphery (i.e. the limbs) influences the central pattern and, conversely, the central pattern generator selects appropriate afferent information according to the external requirement. Both the central pattern generator and the reflexes that mediate afferent input to the spinal cord are under the control of a supraspinal drive. There is increasing evidence that in central motor diseases, a defective utilization of afferent input, in combination with secondary compensatory processes, is involved in typical movement disorders, such as spasticity.

Reduction of the sensory feedback by the current available antispastic therapy (i.e. diazepam, baclofen, tizanidine) may thus not only reduce spasticity but also inevitably influence the ability of the patients to perform voluntary movements. From this point of view, it is evident that antispastic therapy should be given with care (Nielsen et al., 2007). Along this line, rTMS seems to be a promising tool for the treatment of spasticity in various pathological conditions, such as MS and stroke. Further investigations are needed to test whether different protocols of stimulation, applied with different frequencies (i.e. theta burst stimulation (Huang et al., 2005) or transcranial direct current stimulation (Priori et al., 1998; Nitsche and Paulus, 2000; Fregni and Pascual-Leone, 2007) or in alternative cortical sites, may induce more pronounced and beneficial clinical effects.

## Abbreviations

| | |
|---|---|
| CP | common peroneal nerve |
| FN | femoral nerve |
| LTP | long-term potentiation |
| MAS | modified Ashworth scale |
| MEP | motor evoked potentials |
| MS | multiple sclerosis |
| PTN | posterior tibial nerve |
| rTMS | repetitive transcranial magnetic stimulation |
| Sol | soleus muscle |
| TMS | transcranial magnetic stimulation |

## References

Baldissera, F., Hultborn, H., & Illert, M. (1981). Integration in spinal neuronal systems. In V. B. Brooks (Ed.), *Handbook of physiology, motor control* (pp. 509–595). Baltimore: Williams and Wilkins Publishing Company.

Bareyre, F. M., Kerschensteiner, M., Raineteau, O., Mettenleiter, T. C., Weinmann, O., & Schwab, M. E. (2004). The injured spinal cord spontaneously forms a new intraspinal circuit in adult rats. *Nature Neuroscience*, 7, 269–277.

Berardelli, A., Inghilleri, M., Rothwell, J. C., Romeo, S., Currà, A., Gilio, F., et al. (1998). Facilitation of muscle evoked responses after repetitive cortical stimulation in man. *Experimental Brain Research*, 122(1), 79–84.

Boroojerdi, B., Prager, A., Muellbacher, W., & Cohen, L. G. (2000). Reduction of human visual cortex excitability using 1-Hz transcranial magnetic stimulation. *Neurology*, 54(7), 1529–1531.

Centonze, D., Koch, G., Versace, V., Mori, F., Rossi, S., Brusa, L., et al. (2007). Repetitive transcranial magnetic stimulation of the motor cortex ameliorates spasticity in multiple sclerosis. *Neurology*, 68(13), 1045–1050.

Chen, R., Classen, J., Gerloff, C., Celnik, P., Wassermann, E. M., Hallett, M., et al. (1997). Depression of motor cortex excitability by low-frequency transcranial magnetic stimulation. *Neurology*, 48(5), 1398–1403.

Civardi, C., Cantello, R., Asselman, P., & Rothwell, J. C. (2001). Transcranial magnetic stimulation can be used to test connections to primary motor areas from frontal and medial cortex in humans. *Neuroimage*, 14, 1444–1453.

Crone, C., Johnsen, L. L., Biering-Sorensen, F., & Nielsen, J. B. (2006). Appearance of reciprocal facilitation of ankle extensors from ankle flexors in patients with stroke or spinal cord injury. *Brain*, 126, 495–507.

Crone, C., & Nielsen, J. (1994). Central control of disynaptic reciprocal inhibition in humans. *Acta Physiologica Scandinavica*, 152, 351–363.

Crone, C., Nielsen, J., Petersen, N., Ballegaard, M., & Hultborn, H. (1994). Disynaptic reciprocal inhibition of ankle extensors in spastic patients. *Brain*, 117, 1161–1168.

Crone, C., Petersen, N. T., Nielsen, J. E., Hansen, N. L., & Nielsen, J. B. (2004). Reciprocal inhibition and corticospinal transmission in the arm and leg in patients with autosomal dominant pure spastic paraparesis (ADPSP). *Brain*, 127, 2693–2702.

Delwaide, P. J., & Olivier, E. (1988). Short-latency autogenic inhibition (Ib inhibition) in human spasticity. *Journal of Neurology Neurosurgery and Psychiatry*, 51, 1546–1550.

Dietz, V. (2001). Gait disorder in spasticity and Parkinson's disease. *Advances in Neurology*, 87, 143–154.

Dietz, V. (2003). Spinal cord pattern generators for locomotion. *Clinical Neurophysiology*, 114(8), 1379–1389.

el-Abd, M. A., Ibrahim, I. K., & Dietz, V. (1993). Impaired activation pattern in antagonistic elbow muscles of patients with spastic hemiparesis: Contribution to movement disorder. *Electromyography and Clinical Neurophysiology*, 33, 247–255.

Faist, M., Mazevet, D., Dietz, V., & Pierrot-Deseilligny, E. (1994). A quantitative assessment of presynaptic inhibition of Ia afferents in spastics. Differences in hemiplegics and paraplegics. *Brain*, 117, 1449–1455.

Fregni, F., & Pascual-Leone, A. (2007). Technology insight: noninvasive brain stimulation in neurology-perspectives on the therapeutic potential of rTMS and tDCS. *Natural Clinical Practices in Neurology*, 3, 383–393.

Gorassini, M. A., Knash, M. E., Harvey, P. J., Bennett, D. J., & Yang, J. F. (2004). Role of motoneurons in the generation of muscle spasms after spinal cord injury. *Brain*, 127, 2247–2258.

Heckmann, C. J., Gorassini, M. A., & Bennett, D. J. (2005). Persistent inward currents in motoneuron dendrites: Implications for motor output. *Muscle Nerve*, *31*, 135–156.

Heckman, C. J., Lee, R. H., & Brownstone, R. M. (2003). Hyperexcitable dendrites in motoneurons and their neuromodulatory control during motor behavior. *Trends Neuroscience*, *26*, 688–695.

Huang, Y. Z., Edwards, M. J., Rounis, E., Bhatia, K. P., & Rothwell, J. C. (2005). Theta burst stimulation of the human motor cortex. *Neuron*, *45*(2), 201–206.

Hultborn, H., Brownstone, R. B., Toth, T. I., & Gossard, J. P. (2004). Key mechanisms for setting the input–output gain across the motoneuron pool. *Progress in Brain Research*, *143*, 77–95.

Hultborn, H., Meunier, S., Morin, C., & Pierrot-Deseilligny, E. (1987). Assessing changes in presynaptic inhibition of Ia fibres: a study in man and the cat. *Journal of Physiology*, *389*, 729–756.

Izumi, S., Kondo, T., & Shindo, K. (2008). Transcranial magnetic stimulation synchronized with maximal movement effort of the hemiplegic hand after stroke: a double-blinded controlled pilot study. *Journal of Rehabilitation Medicine*, *40*(1), 49–54.

Katz, R., & Pierrot-Deseilligny, E. (1982). Recurrent inhibition of alpha-motoneurons in patients with upper motor neuron lesions. *Brain*, *105*, 103–124.

Katz, R., & Pierrot-Deseilligny, E. (1999). Recurrent inhibition in humans. *Progress in Neurobiology*, *57*, 325–355.

Koch, G., Fernandez Del Olmo, M., Cheeran, B., Ruge, D., Schippling, S., Caltagirone, C., et al. (2007). Focal stimulation of the posterior parietal cortex increases the excitability of the ipsilateral motor cortex. *Journal of Neuroscience*, *27*, 6815–6822.

Koch, G., Franca, M., Del Olmo, M. F., Cheeran, B., Milton, R., Alvarez Sauco, M., et al. (2006). Time course of functional connectivity between dorsal premotor and contralateral motor cortex during movement selection. *Journal of Neuroscience*, *26*(28), 7452–7459.

Mailis, A., & Ashby, P. (1990). Alterations in group Ia projections to motoneurons following spinal lesions in humans. *Journal of Neurophysiology*, *64*, 637–647.

Mály, J., & Dinya, E. (2008). Recovery of motor disability and spasticity in post-stroke after repetitive transcranial magnetic stimulation (rTMS). *Brain Research Bulletin*, *76*, 388–395.

Maynard, F. M., Karunas, R. S., & Waring, W. P. (1990). Epidemiology of spasticity following traumatic spinal cord injury. *Archives of Physical Medicine and Rehabilitation*, *71*, 566–569.

Meunier, S., & Pierrot-Deseilligny, E. (1998). Cortical control of presynaptic inhibition of Ia afferents in humans. *Experimental Brain Research*, *119*, 415–426.

Mizuno, Y., Tanaka, R., & Yanagisawa, N. (1971). Reciprocal group I inhibition of triceps surae motoneurons in man. *Journal of Neurophysiology*, *34*, 1010–1017.

Murase, N., Duque, J., Mazzocchio, R., & Cohen, L. G. (2004). Influence of interhemispheric interactions on motor function in chronic stroke. *Annals of Neurology*, *55*, 400–409.

Nickolls, P., Collins, D. F., Gorman, R. B., Burke, D., & Gandevia, S. C. (2004). Forces consistent with plateau-like behaviour of spinal neurons evoked in patients with spinal cord injuries. *Brain*, *127*, 660–670.

Nielsen, J., Crone, C., Sinkjaer, T., Toft, E., & Hultborn, H. (1995a). Central control of reciprocal inhibition during fictive dorsiflexion in man. *Experimental Brain Research*, *104*, 99–106.

Nielsen, J., & Petersen, N. (1995). Evidence favouring different descending pathways to soleus motoneurones activated by magnetic brain stimulation in man. *Journal of Physiology*, *486*, 779–788.

Nielsen, J., Petersen, N., & Crone, C. (1995b). Changes in transmission across synapses of Ia afferents in spastic patients. *Brain*, *118*, 995–1004.

Nielsen, J. B., Crone, C., & Hultborn, H. (2007). The spinal pathophysiology of spasticity — from a basic science point of view. *Acta Physiology*, *189*, 171–180.

Nielsen, J. F., & Sinkjaer, T. (1997). Long-lasting depression of soleus motoneurons excitability following repetitive magnetic stimuli of the spinal cord in multiple sclerosis patients. *Multiple Sclerosis*, *3*, 18–30.

Nielsen, J. F., Sinkjaer, T., & Jakobsen, J. (1996). Treatment of spasticity with repetitive magnetic stimulation; a double-blind placebo-controlled study. *Multiple Sclerosis*, *2*(5), 227–232.

Nitsche, M. A., & Paulus, W. (2000). Excitability changes induced in the human motor cortex by weak transcranial direct current stimulation. *Journal of Physiology*, *527*, 633–639.

Peinemann, A., Reimer, B., Löer, C., Quartarone, A., Münchau, A., Conrad, B., et al. (2004). Long-lasting increase in corticospinal excitability after 1800 pulses of subthreshold 5 Hz repetitive TMS to the primary motor cortex. *Clinical Neurophysiology*, *115*, 1519–1526.

Perez, M. A., Lungholt, B. K., & Nielsen, J. B. (2005). Short-term adaptations in spinal cord circuits evoked by repetitive transcranial magnetic stimulation: Possible underlying mechanisms. *Experimental Brain Research*, *162*(2), 202–212.

Pierrot-Deseilligny, E., & Bussel, B. (1975). Evidence for recurrent inhibition by motoneurons in human subjects. *Brain Research*, *88*, 105–108.

Pierrot-Deseilligny, E., Katz, R., & Morin, C. (1979). Evidence of Ib inhibition in human subjects. *Brain Research*, *166*, 176–179.

Powers, R. K., & Binder, M. D. (2001). Input–output functions of mammalian motoneurons. *Review of Physiology Biochemistry and Pharmacology*, *143*, 137–263.

Priori, A., Berardelli, A., Rona, S., Accornero, N., & Manfredi, M. (1998). Polarization of the human motor cortex through the scalp. *Neuroreport*, *9*, 2257–2260.

Raineteau, O., & Schwab, M. E. (2001). Plasticity of motor systems after incomplete spinal cord injury. *Nature Reviews Neuroscience*, *2*, 263–273.

Raisman, G. (1969). Neuronal plasticity in the septal nuclei of the adult rat. *Brain Research*, *14*, 25–48.

Rekling, J. C., Funk, G. D., Bayliss, D. A., Dong, X. W., & Feldman, J. L. (2000). Synaptic control of motoneuronal excitability. *Physiology Reviews, 80*, 767–852.

Rizzo, M. A., Hadjimichael, O. C., Preiningerova, J., & Vollmer, T. L. (2004). Prevalence and treatment of spasticity reported by multiple sclerosis patients. *Multiple Sclerosis, 10*, 589–595.

Rudomin, P., & Schmidt, R. F. (1999). Presynaptic inhibition in the vertebrate spinal cord revisited. *Experimental Brain Research, 129*, 1–37.

Siebner, H. R., Filipovic, S. R., Rowem, J. B., Cordivari, C., Gerschlager, W., Rothwell, J. C., et al. (2003). Patients with focal arm dystonia have increased sensitivity to slow-frequency repetitive TMS of the dorsal premotor cortex. *Brain, 126*, 2710–2725.

Sommerfeld, D. K., Eek, E. U., Svensson, A. K., Holmqvist, L. W., & von Arbin, M. H. (2004). Spasticity after stroke: Its occurrence and association with motor impairments and activity limitations. *Stroke, 35*, 134–139.

Touge, T., Gerschlager, W., Brown, P., & Rothwell, J. C. (2001). Are the after-effects of low-frequency rTMS on motor cortex excitability due to changes in the efficacy of cortical synapses?. *Clinical Neurophysiology, 112*(11), 2138–2145.

Valero-Cabre, A., Oliveri, M., Gangitano, M., & Pascual-Leone, A. (2001). Modulation of spinal cord excitability by subthreshold repetitive transcranial magnetic stimulation of the primary motor cortex in humans. *Neuroreport, 4, 12*(17), 3845–3848.

Valle, A. C., Dionisio, K., Pitskel, N. B., Pascual-Leone, A., Orsati, F., Ferreira, M. J., et al. (2007). Low and high frequency repetitive transcranial magnetic stimulation for the treatment of spasticity. *Developmental Medicine and Child Neurology, 49*(7), 534–538.

Woolacott, A. J., & Burne, J. A. (2006). The tonic stretch reflex and spastic hypertonia after spinal cord injury. *Experimental Brain Research, 174*, 386–396.

Yanagisawa, N., Tanaka, R., & Ito, Z. (1976). Reciprocal Ia inhibition in spastic hemiplegia of man. *Brain, 99*, 555–574.

Young, R. R. (1994). Spasticity: a review. *Neurology, 44*, S12–S20.

# SECTION VII

# Mechanisms of Spontaneous Plasticity and Regeneration

CHAPTER 28

# Calcium signaling and the development of specific neuronal connections

Christian Lohmann*

*Department of Synapse and Network Development, Netherlands Institute for Neuroscience, Amsterdam, The Netherlands*

**Abstract:** During the development of the brain, synaptic connections between nerve cells are being established with remarkable specificity. This is achieved by a series of steps: first, axons grow to their terminal areas. Second, axons and dendrites contact each other and select among potential synaptic partners. Third, after synapses have become functional, the fine-tuning of synaptic connections optimizes emerging networks to perform their specific functions. Here, I summarize the evidence for a central role of intracellular calcium signaling in all three stages of the development of specific synaptic connections. In particular, calcium signaling has the capacity to integrate information from a wide array of extracellular factors that are known to regulate neuronal development, such as molecular cues or neuronal activity. Calcium signaling, in turn, directs structural as well as functional adaptations in individual neurons that underlie the establishment of synaptic specificity. Importantly, evidence is accumulating that errors in calcium-dependent network maturation are associated with neurodevelopmental disorders. Therefore, understanding the role of calcium in setting up brain networks may not only advance our insights into mechanisms of normal brain development, but also help identifying the causes of diseases such as autism or mental retardation.

**Keywords:** axon path finding; dendrite; synapse; neurodevelopmental disorder; synaptic plasticity; calcium

## Introduction

The development of the brain is a highly complex process. Billions of neurons are being connected specifically. Eventually, a neuronal network emerges that determines all aspects of our lives, from sensory experiences to motor coordination, and from feelings and emotions to cognitive functions. Thus, it is not surprising that errors in neurodevelopment can cause severe neurological and psychological disorders.

Brain development proceeds over a number of steps, which are, however, overlapping and not exclusive. First, nerve and glia cells are generated from precursor cells. Many of these cells migrate to their specific locations within the growing nervous system and then start elaborating processes, axons and dendrites. At this stage, developing neurons begin to form connections, and one of their most impressive achievements is to form highly specific connections with the "correct" synaptic partners. In fact, recent experiments and theoretical considerations suggest that connectivity is specific even at the subcellular level in many

*Corresponding author.
Tel.: +31 20 5664943; Fax: +31 20 5666121;
E-mail: c.lohmann@nin.knaw.nl

brain areas (Huang et al., 2007; Poirazi and Mel, 2001).

The formation of specific connections also proceeds over several steps, which, again, are overlapping and nonexclusive. First, axons of projection neurons grow and navigate to their respective target areas in the brain. Second, within each brain area, neurons establish initial contacts, select between potential synaptic partners, and form the first functional synapses. Finally, the synaptic network becomes fine-tuned, a process that involves synapse elimination together with the formation of new synapses. All these steps have to proceed correctly for the mature brain to function properly. As a consequence, perturbations of the wiring process lead to neurodevelopmental disorders, such as autism and mental retardation (Garber, 2007; Geschwind and Levitt, 2007; Rapin and Tuchman, 2008; Zoghbi, 2003).

For the formation of specific connections, neurons rely on a vast array of extracellular signals that regulate or modulate axonal pathfinding, the formation of contacts with specific partners, as well as the fine-tuning of synaptic connectivity. These factors include adhesion molecules, growth factors, diffusible guidance molecules, as well as neuronal activity in the form of synaptic transmission and action potential firing (Cline, 2003; Huang and Reichardt, 2001; Tessier-Lavigne and Goodman, 1996; Washbourne et al., 2004; Yamagata et al., 2003). Despite the number and diversity of these factors, it seems that only a limited number of intracellular pathways integrate this wealth of extracellular information and, in turn, direct the development of synapse structure and function.

In this review, I will argue that particularly intracellular calcium signaling plays a pivotal role in the establishment of specific connectivity at all stages: axonal navigation, initial partner selection, and network fine-tuning. I will summarize the evidence for an integrative role of calcium signaling, in the sense that many developmentally relevant extracellular stimuli trigger characteristic calcium transients in developing neurons. Next, the role of calcium signaling in directing structural as well as functional plasticity in developing neurons will be addressed. Finally, I will describe how calcium-mediated plasticity contributes to establishing specific connections. Calcium signaling endows developing neurons with the capacity to respond to the changing environment with very high spatial (micrometers) and temporal (seconds/minutes) specificity, and is therefore ideal for setting up specific connections in the complex emerging environment of the developing brain.

Many aspects and facts that are assembled here to support the idea that calcium signaling integrates developmentally relevant stimuli to establish specific connections have been previously summarized in a number of excellent reviews (Gomez and Zheng, 2006; Kater et al., 1988; Oertner and Matus, 2005; Wong and Ghosh, 2002; Yuste and Bonhoeffer, 2004; Zheng and Poo, 2007).

**Axonal growth cone navigation**

The first step in establishing specific connectivity — and also historically the first in which the importance of calcium signaling was discovered — is axonal growth cone navigation (Gomez and Spitzer, 2000). It has been clear for some time that calcium levels in the growth cone affect growth cone navigation (Kater et al., 1988). Advances in imaging methods then facilitated the investigation of growth cone regulation by calcium. Gomez, Spitzer, and colleagues (Gomez and Spitzer, 1999; Gomez et al., 2001) were the first to show that growing axons display spontaneous calcium transients in their growth cones in vivo. For example, imaging calcium levels and axonal growth simultaneously in the zebrafish spinal chord demonstrated that spontaneously occurring calcium transients are correlated with a decrease in axonal growth (Gomez and Spitzer, 1999). Interestingly, these calcium transients are modulated by components of the extracellular matrix in the developing brain (Gomez et al., 2001; Fig. 1). The mechanism of such substrate-dependent calcium increases is not yet clear, but there are indications that integrin receptor-mediated activation of calcium channels or mechanical forces may be involved (Gomez et al., 2001).

Besides regulating the advancement of axonal growth, calcium transients exert even finer control

## Substrate induced growth cone calcium transients

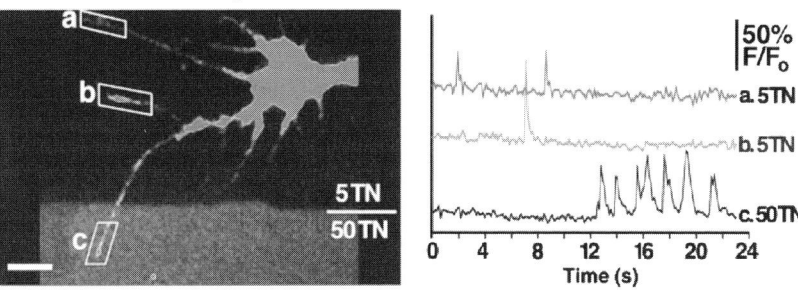

## Calcium triggered growth cone turning

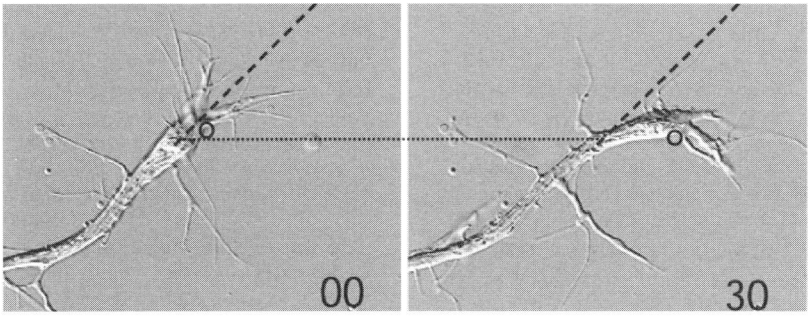

Fig. 1. Calcium signaling in axonal growth cones. Substrate-induced growth cone calcium transients in filopodia: one filopodium (c) contacting high-concentration tenascin substrate (50 μg/ml TN), a component of the extracellular matrix, displays high-frequency calcium signaling. In contrast, filopodia on a low-concentration tenascin substrate generate only few calcium transients. Adapted with permission from Gomez et al. (2001), AAAS. Calcium-triggered growth cone turning: local calcium uncaging within a growth cone (circle) induces a turn toward the side of calcium elevation after 30 minutes. Adapted with permission from Zheng (2000), Macmillan Publishers Ltd. (See Color Plate 28.1 in color plate section.)

over axonal navigation by directing growth cone turns: local calcium transients in individual filopodia or in restricted volumes of the growth cone are associated with growth cone turns (Gomez et al., 2001). Furthermore, uncaging calcium with very high spatial accuracy triggers turning responses in individual growth cones (Gomez et al., 2001; Zheng, 2000; Fig. 1). Importantly, local calcium rises do not elicit stereotypic responses; these also depend on the state of each neuron or growth cone. For example, a local rise on one side of a growth cone may cause a turn to the same or the opposite side, depending on the global calcium levels (Zheng, 2000), an example for the integration of two different forms of calcium signaling. Furthermore, different second messenger signaling mechanisms act upon calcium-regulated growth cone behavior, for example, cyclic nucleotide signaling cascades (Gomez and Zheng, 2006; Song et al., 1997; Zheng, 2000; Zheng and Poo, 2007), illustrating the interdependency of different intracellular signaling cascades in such developmental processes.

Besides adhesion molecules, many other factors that regulate axonal navigation require calcium as a second messenger. These include the neurotrophin brain-derived neurotrophic factor (BDNF; Song et al., 1997) and other diffusible molecules, such as netrin (Hong et al., 2000), and the growth-inhibiting factor myelin-associated glycoprotein (MAG; Henley et al., 2004). Together, these findings show that calcium signaling triggered by diverse factors in the environment of the advancing growth cone determines its behavior depending on the status of the cell. Such mechanisms underlie the establishment of connections on the level of separate brain areas and probably topographic maps in the maturing brain.

## Synaptic partner selection by dendritic filopodia

After axons have reached their target areas, dendrites and axons start forming contacts and subsequently establish functional synapses. Recently, it has been shown that at this stage of network development developing dendrites also generate spontaneous local calcium transients, for example, in hippocampal and retinal neurons (Koizumi et al., 1999; Lohmann et al., 2002, 2005). Similarly to calcium signaling in growth cones, dendritic calcium transients can be stimulated by many factors. For example, treatment with exogenous metabotropic glutamate receptor (mGluR) agonists or caffeine stimulates the generation of these transients (Koizumi et al., 1999). Furthermore, a fraction of local calcium transients in the developing chick retina is dependent on cholinergic signaling (Lohmann et al., 2002), and GABAergic signaling is required for a population of spontaneously occurring local calcium transients in developing hippocampal neurons (Lohmann et al., 2005). Furthermore, local calcium transients can be triggered by synaptic activation in immature hippocampal neurons (Fig. 2; Albantakis and Lohmann, 2009). Somewhat surprisingly, however, the frequency of spontaneously occurring local calcium transients is barely affected by blocking glutamate receptors (Lohmann et al., 2005). This may be due to the fact that glutamatergic synapses are activated in a correlated fashion during ongoing spontaneous activity in early postnatal development (giant depolarizing potentials; Ben-Ari et al., 1989). Correlated synaptic inputs trigger global, but not local, calcium transients in developing neurons. Consistently, global calcium transients are almost completely blocked by glutamate receptor antagonists in hippocampal neurons (Garaschuk et al., 1998). Thus, glutamatergic synaptic signaling contributes mostly to global,

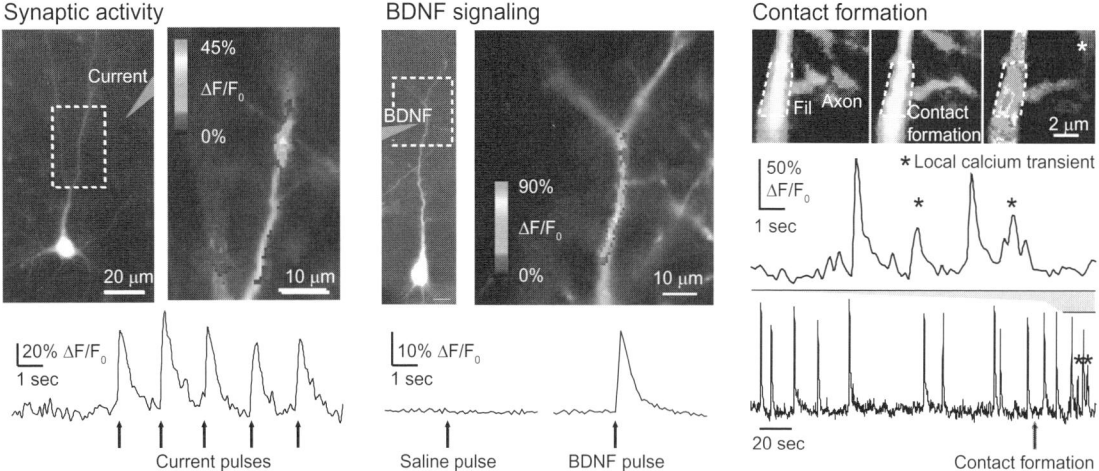

Fig. 2. Various developmentally relevant factors trigger calcium transients in developing hippocampal neurons. Synaptic activity: current stimulation of presynaptic axons leads to reliable dendritic calcium responses. The spatial spread of calcium within the dendrite is shown in pseudo-color. The trace represents the time course of the calcium response. BDNF signaling: focal application of BDNF induces a fast local calcium transient in the apical dendrite of a CA3 pyramidal neuron. Adapted with permission from Lang et al. (2007), the Society for Neuroscience. Contact formation: within less than one minute, local calcium transients are generated in the dendrite at the site where a filopodium has formed a contact with an axon. The expanded view shows the kinetics of the local calcium transients. Calcium transients that are not marked by an asterisk represent ongoing global calcium activity. Adapted with permission from Lohmann and Bonhoeffer (2008), Cell Press. (See Color Plate 28.2 in color plate section.)

but relatively little to local, spontaneous calcium activity in these developing neurons.

Besides neurotransmitters, endogenous BDNF also participates in triggering local calcium transients in hippocampal neurons (Lang et al., 2007), and exogenously applied BDNF induces local calcium transients in their dendrites (Fig. 2; Amaral and Pozzo-Miller, 2007; Lang et al., 2007). Furthermore, local calcium transients can be elicited in dendrites when dendritic filopodia initiate contact with axons (Fig. 2; Lohmann and Bonhoeffer, 2008). Thus, like axons, developing dendrites also have the capacity to integrate a variety of extracellular cues by calcium signaling. In fact, local calcium transients that are elicited by very different factors in axons or dendrites have surprisingly similar spatiotemporal characteristics — for example, transients elicited by extracellular matrix molecules in axonal filopodia (Fig. 1) or transients in dendrites triggered by synaptic activation, BDNF signaling, or contact formation (Fig. 2).

Do fast local calcium transients play a role in the development of dendrites? It became clear that calcium transients can influence the structure of developing dendrites. Blocking local calcium transients in developing chick retinal ganglion cells causes the retraction of their terminal dendrites. Individual dendrites can be rescued from retraction by directed laser uncaging of calcium in these dendrites (Lohmann et al., 2002). These results indicated that calcium can have a stabilizing action on developing dendrites, possibly similar to some axonal growth cones whose motility is negatively regulated by calcium transients (Gomez and Spitzer, 1999). One idea has been that calcium transients may stabilize those dendritic processes that contact appropriate axons; inappropriate connections may fail to trigger calcium transients in dendrites and could thus cause dendritic retraction and, eventually, the elimination of a contact.

To establish contacts with potential presynaptic partners, hippocampal neurons, like many other nerve cells, possess dendritic filopodia, small processes that grow and retract within minutes and even seconds (Dunaevsky and Mason, 2003; Harris, 1999; Jontes and Smith, 2000; Wong and Wong, 2000; Yuste and Bonhoeffer, 2004). Recently, it was shown that dendritic filopodia in addition to establishing contacts with axons have the capacity to select among different types of axons (Fig. 3; Lohmann and Bonhoeffer, 2008). Specifically, filopodia on the apical dendrites of neonatal CA3 pyramidal neurons frequently establish contacts with all types of axons. Many of these contacts are broken up within one or two minutes after their formation. Some contacts, however, become stabilized and may establish functional synapses. Importantly, only if the axon population comprises excitatory axons, contacts become stabilized; with GABAergic axons, filopodia never form stable contacts, demonstrating a role for filopodial motility in selecting among potential synaptic partners.

The growth of filopodia is correlated with calcium signaling in the dendrite, and contact formation triggers dendritic calcium transients. Furthermore, calcium activity is particularly prominent at sites where filopodia stabilize contacts later on, but significantly lower at sites where contacts get eliminated after a few minutes (Fig. 3). Consistently, calcium activity does not rise when dendritic filopodia establish contacts with GABAergic axons, which never become stabilized. These observations suggest that calcium signaling regulates the selection of specific synaptic partners by selectively stabilizing "appropriate" contacts. "Inappropriate" contacts, which are not protected by high levels of local calcium signaling, will be eliminated. Indeed, experimental manipulation of local calcium signaling with high spatiotemporal precision proved that calcium transients can stabilize individual filopodia. Uncaging of calcium, for example, stabilizes filopodia, whereas rapid pharmacological blockade of local calcium signaling by fast superfusion increases their motility (Lohmann et al., 2005).

These observations support the idea that calcium signaling in dendrites helps integrating diverse forms of extracellular signaling to implement specific connectivity. While axon navigation most likely supports connecting brain

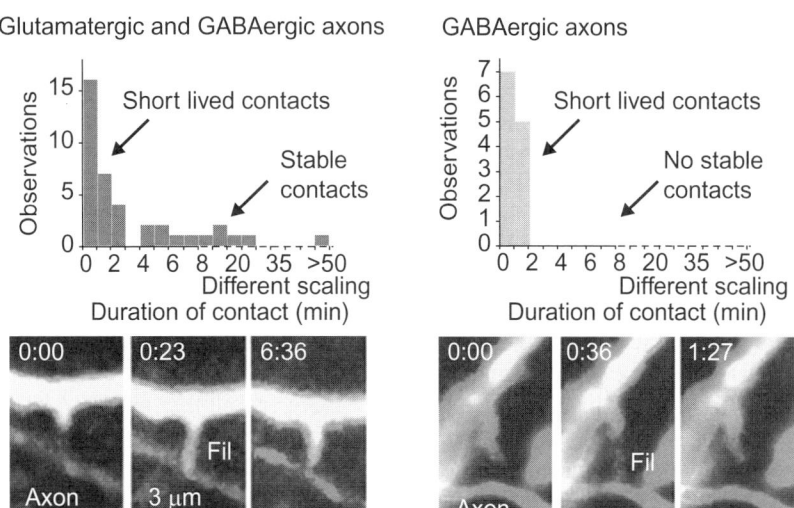

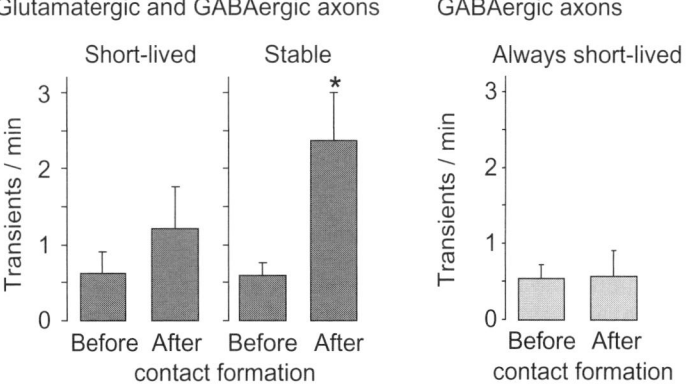

Fig. 3. Role of calcium signaling in synaptic partner selection. Differential stabilization of synaptogenic contacts: contacts with a mixed population of axons are frequently short lived, but sometimes become stabilized. Contacts with GABAergic axons never become stabilized. Adapted with permission from Lohmann and Bonhoeffer (2008), Cell Press. Calcium signaling and contact stabilization: in dendrites where contacts become stabilized, there is a significant increase in calcium signaling after contact formation. In short-lived contacts, calcium signaling is moderate, in particular at contacts with GABAergic axons. Adapted with permission from Lohmann and Bonhoeffer (2008), Cell Press.

areas and establishing topographically organized projections, dendritic calcium signaling contributes to selecting specific partners on the level of single neurons and may also achieve specificity on a subcellular level.

## Refinement of synaptic connections

A high degree of specificity can be achieved by the developmental mechanisms described so far. Nevertheless, networks become even more refined by processes that adjust the strength of individual synapses. Synapses undergo increases and decreases in transmission efficacy. In particular, long-term potentiation (LTP) and depression (LTD) play prominent roles in adjusting networks to changing environments. Synaptic plasticity at earlier stages involves similar, however, oftentimes molecularly and physiologically different mechanisms compared to the mature nervous system (Constantine-Paton and Cline, 1998).

It has been clear for quite some time that influx of calcium at the synapse mediates synaptic plasticity in adult as well as developing neurons (Malenka et al., 1988). Despite this long-standing appreciation of the importance of calcium signaling for synaptic plasticity, it is virtually unknown what the properties of calcium transients are that determine whether a synapse becomes potentiated or depressed (Malenka and Bear, 2004). Some models suggest that moderate increases in calcium may activate primarily phosphatases (e.g., calcineurin and protein phosphatase-1) that in turn facilitate synaptic depression (Mansuy and Shenolikar, 2006). In contrast, the activation of kinases (e.g., calcium/calmodulin-dependent protein kinase II, CaMKII) by high-amplitude calcium transients may favor potentiation (Lisman et al., 2002). This is in fact an interesting parallel to the regulation of attractive vs. repulsive axon guidance by calcium: larger calcium transients can activate CaMKII and induce turns toward the side of calcium elevation, whereas smaller calcium increases activate the phosphatases calcineurin and phosphatase-1 and trigger repulsive turns (Wen et al., 2004; Zheng and Poo, 2007).

A first series of experiments that correlated certain types of calcium transients with synaptic potentiation or depression finds that this relationship may be more complex than previously expected (Nevian and Sakmann, 2006). The lesson from these experiments seems to be that the timing of calcium transients and the involvement of mGluRs are more important than the absolute amplitudes of calcium transients (Nevian and Sakmann, 2006).

Thus, while molecular mechanisms of synaptic plasticity have been investigated in great detail, the exact immediate determinants of synaptic changes are still unclear, at least in as far as they are related to specific forms of calcium signaling. Future experiments, involving highly time-resolved imaging, may help to address these issues and to investigate how certain patterns of calcium dynamics may underlie the generation of specificity by the fine-tuning of synaptic weights.

## Conclusion

Kater et al. (1988) proposed that "Calcium may ... act as a common integrator of environmental cues that influence neurite outgrowth and synaptogenesis, and in this way may play a key role in the establishment and modulation of brain circuitry." Today, we have learned that many more environmental cues than known at that time are integrated by calcium signaling and contribute to the development of the nervous system. Furthermore, this idea has been adapted to mechanisms beyond growth cone navigation (which was the main interest of Kater et al.), such as the selection of synaptic partners by dendritic filopodia and the fine-tuning of neuronal networks by synaptic plasticity (Fig. 4).

Despite this progress, many questions are unsolved. Most importantly, direct evidence that perturbed calcium-dependent axon guidance, synaptic partner selection, or the fine-tuning of synapses leads to changed connectivity and aberrant behavior in animals has not been obtained so far. Interestingly, however, a large number of mutations that have been implicated in autism spectrum disorders, a group of neurodevelopmental diseases, are related to intracellular calcium signaling (summarized in Krey and Dolmetsch, 2007). Together with the idea that neurodevelopmental disorders are the consequence of errors in the wiring of the brain (Garber, 2007; Geschwind and Levitt, 2007; Rapin and Tuchman, 2008; Zoghbi, 2003), this fact suggests a central role of calcium signaling in establishing specific connections that are required for the proper functioning of

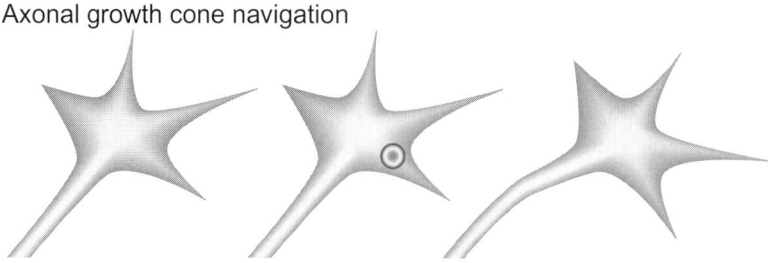

Axonal growth cone navigation

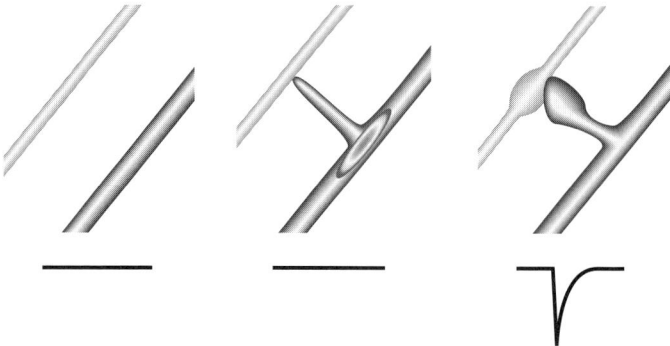

Synaptic partner selection and onset of synaptic function

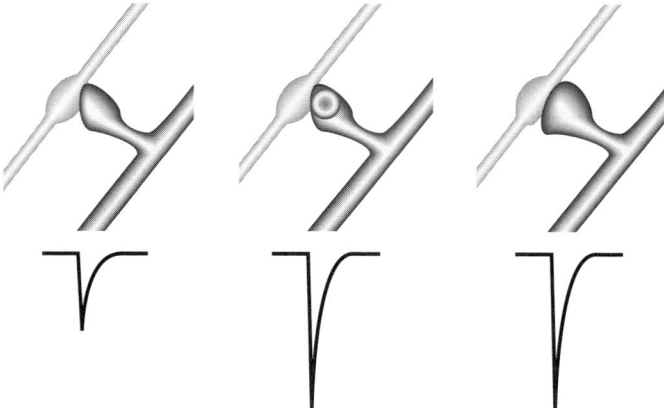

Fine-tuning of synaptic strength and size

Fig. 4. Calcium signaling mediates structural and functional maturation at three stages of network development. Local calcium transients (disk) steer growth cones for the formation of topographic transregional connections. Nonsynaptic calcium transients in dendrites upon contact formation may determine synaptic partner selection and precede the onset of synaptic function. Calcium dynamics underlie the fine-tuning of neuronal networks through adaptations of synaptic strength and size.

the brain. In addition, these observations may guide an alternative route for testing the importance of calcium-based wiring mechanisms, for example, by investigating whether these processes are specifically altered in mouse models of neurodevelopmental diseases.

The observation that calcium transients caused by various extracellular signaling factors are similar (Figs. 1 and 2) raises an important question: are these transients identical or different with respect to the downstream signaling mechanisms? Do they depend on a certain type of calcium channel or release of calcium from internal stores? Furthermore, what is the logic of integration of different types of calcium transients (e.g., local vs. global transients), in particular in synaptic partner selection and synaptic fine-tuning? We now know that calcium signaling facilitates precise connectivity with high temporal and spatial precision. Therefore, to address these questions will require high-resolution probing and manipulating neuronal calcium signaling. Recent technological developments in labeling and imaging neurons as well as manipulating neuronal activity with optical means will greatly assist this task.

## Acknowledgments

I thank Friederike Siegel, Thomas Kleindienst, and Christiaan Levelt for helpful comments on the manuscript.

## References

Albantakis, L., and Lohmann, C. (2009). A simple method for quantitative calcium imaging in unperturbed developing neurons. (Submitted).

Amaral, M. D., & Pozzo-Miller, L. (2007). BDNF induces calcium elevations associated with IBDNF, a nonselective cationic current mediated by TRPC channels. *Journal of Neurophysiology*, 98, 2476–2482.

Ben-Ari, Y., Cherubini, E., Corradetti, R., & Galarsa, J.-L. (1989). Giant synaptic potentials in immature rat CA3 hippocampal neurones. *The Journal of Physiology (London)*, 416, 303–325.

Cline, H. (2003). Sperry and Hebb: oil and vinegar? *Trends in Neurosciences*, 26, 655–661.

Constantine-Paton, M., & Cline, H. T. (1998). LTP and activity-dependent synaptogenesis: the more alike they are, the more different they become. *Current Opinion in Neurobiology*, 8, 139–148.

Dunaevsky, A., & Mason, C. A. (2003). Spine motility: a means towards an end? *Trends in Neurosciences*, 26, 155–160.

Garaschuk, O., Hanse, E., & Konnerth, A. (1998). Developmental profile and synaptic origin of early network oscillations in the CA1 region of rat neonatal hippocampus. *The Journal of Physiology (London)*, 507, 219–236.

Garber, K. (2007). Neuroscience. Autism's cause may reside in abnormalities at the synapse. *Science*, 317, 190–191.

Geschwind, D. H., & Levitt, P. (2007). Autism spectrum disorders: developmental disconnection syndromes. *Current Opinion in Neurobiology*, 17, 103–111.

Gomez, T. M., Robles, E., Poo, M., & Spitzer, N. C. (2001). Filopodial calcium transients promote substrate-dependent growth cone turning. *Science*, 291, 1983–1987.

Gomez, T. M., & Spitzer, N. C. (1999). In vivo regulation of axon extension and pathfinding by growth-cone calcium transients. *Nature*, 397, 350–355.

Gomez, T. M., & Spitzer, N. C. (2000). Regulation of growth cone behavior by calcium: new dynamics to earlier perspectives. *Journal of Neurobiology*, 44, 174–183.

Gomez, T. M., & Zheng, J. Q. (2006). The molecular basis for calcium-dependent axon pathfinding. *Nature Reviews. Neuroscience*, 7, 115–125.

Harris, K. M. (1999). Structure, development, and plasticity of dendritic spines. *Current Opinion in Neurobiology*, 9, 343–348.

Henley, J. R., Huang, K. H., Wang, D., & Poo, M. M. (2004). Calcium mediates bidirectional growth cone turning induced by myelin-associated glycoprotein. *Neuron*, 44, 909–916.

Hong, K., Nishiyama, M., Henley, J., Tessier-Lavigne, M., & Poo, M. (2000). Calcium signalling in the guidance of nerve growth by netrin-1. *Nature*, 403, 93–98.

Huang, E. J., & Reichardt, L. F. (2001). Neurotrophins: roles in neuronal development and function. *Annual Review of Neuroscience*, 24, 677–736.

Huang, Z. J., Di Cristo, G., & Ango, F. (2007). Development of GABA innervation in the cerebral and cerebellar cortices. *Nature Reviews. Neuroscience*, 8, 673–686.

Jontes, J. D., & Smith, S. J. (2000). Filopodia, spines, and the generation of synaptic diversity. *Neuron*, 27, 11–14.

Kater, S. B., Mattson, M. P., Cohan, C., & Connor, J. (1988). Calcium regulation of the neuronal growth cone. *Trends in Neurosciences*, 11, 315–321.

Koizumi, S., Bootman, M. D., Bobanovic, L. K., Schell, M. J., Berridge, M. J., & Lipp, P. (1999). Characterization of elementary $Ca^{2+}$ release signals in NGF-differentiated PC12 cells and hippocampal neurons. *Neuron*, 22, 125–137.

Krey, J. F., & Dolmetsch, R. E. (2007). Molecular mechanisms of autism: a possible role for $Ca^{2+}$ signaling. *Current Opinion in Neurobiology*, 17, 112–119.

Lang, S. B., Stein, V., Bonhoeffer, T., & Lohmann, C. (2007). Endogenous brain-derived neurotrophic factor triggers fast calcium transients at synapses in developing dendrites. *The Journal of Neuroscience*, 27, 1097–1105.

Lisman, J., Schulman, H., & Cline, H. (2002). The molecular basis of CaMKII function in synaptic and behavioural memory. *Nature Reviews. Neuroscience*, 3, 175–190.

Lohmann, C., & Bonhoeffer, T. (2008). A role for local calcium signaling in rapid synaptic partner selection by dendritic filopodia. *Neuron*, 59, 253–260.

Lohmann, C., Finski, A., & Bonhoeffer, T. (2005). Local calcium transients regulate the spontaneous motility of dendritic filopodia. *Nature Neuroscience*, 8, 305–312.

Lohmann, C., Myhr, K. L., & Wong, R. O. (2002). Transmitter-evoked local calcium release stabilizes developing dendrites. *Nature, 418*, 177–181.

Malenka, R. C., & Bear, M. F. (2004). LTP and LTD: an embarrassment of riches. *Neuron, 44*, 5–21.

Malenka, R. C., Kauer, J. A., Zucker, R. S., & Nicoll, R. A. (1988). Postsynaptic calcium is sufficient for potentiation of hippocampal synaptic transmission. *Science, 242*, 81–84.

Mansuy, I. M., & Shenolikar, S. (2006). Protein serine/threonine phosphatases in neuronal plasticity and disorders of learning and memory. *Trends in Neurosciences, 29*, 679–686.

Nevian, T., & Sakmann, B. (2006). Spine $Ca^{2+}$ signaling in spike-timing-dependent plasticity. *The Journal of Neuroscience, 26*, 11001–11013.

Oertner, T. G., & Matus, A. (2005). Calcium regulation of actin dynamics in dendritic spines. *Cell Calcium, 37*, 477–482.

Poirazi, P., & Mel, B. W. (2001). Impact of active dendrites and structural plasticity on the memory capacity of neural tissue. *Neuron, 29*, 779–796.

Rapin, I., & Tuchman, R. F. (2008). What is new in autism? *Current Opinion in Neurology, 21*, 143–149.

Song, H. J., Ming, G. L., & Poo, M. M. (1997). cAMP-induced switching in turning direction of nerve growth cones. *Nature, 388*, 275–279.

Tessier-Lavigne, M., & Goodman, C. S. (1996). The molecular biology of axon guidance. *Science, 274*, 1123–1133.

Washbourne, P., Dityatev, A., Scheiffele, P., Biederer, T., Weiner, J. A., Christopherson, K. S., et al. (2004). Cell adhesion molecules in synapse formation. *The Journal of Neuroscience, 24*, 9244–9249.

Wen, Z., Guirland, C., Ming, G. L., & Zheng, J. Q. (2004). A CaMKII/calcineurin switch controls the direction of Ca(2+)-dependent growth cone guidance. *Neuron, 43*, 835–846.

Wong, R. O., & Ghosh, A. (2002). Activity-dependent regulation of dendritic growth and patterning. *Nature Reviews. Neuroscience, 3*, 803–812.

Wong, W. T., & Wong, R. O. (2000). Rapid dendritic movements during synapse formation and rearrangement. *Current Opinion in Neurobiology, 10*, 118–124.

Yamagata, M., Sanes, J. R., & Weiner, J. A. (2003). Synaptic adhesion molecules. *Current Opinion in Cell Biology, 15*, 621–632.

Yuste, R., & Bonhoeffer, T. (2004). Genesis of dendritic spines: insights from ultrastructural and imaging studies. *Nature Reviews. Neuroscience, 5*, 24–34.

Zheng, J. Q. (2000). Turning of nerve growth cones induced by localized increases in intracellular calcium ions. *Nature, 403*, 89–93.

Zheng, J. Q., & Poo, M. M. (2007). Calcium signaling in neuronal motility. *Annual Review of Cell and Developmental Biology, 23*, 375–404.

Zoghbi, H. Y. (2003). Postnatal neurodevelopmental disorders: meeting at the synapse?. *Science, 302*, 826–830.

CHAPTER 29

# Remyelination in multiple sclerosis

Gabrièle Piaton[1,*], Anna Williams[4], Danielle Seilhean[2,3] and Catherine Lubetzki[1,3]

[1]UMRS, Inserm 975, CR-Icm, Paris, France
[2]Laboratoire de Neuropathologie Escourolle, Paris, France
[3]Faculté de médecine, Université Pierre & Marie Curie, Paris, France
[4]MS Centre, Centre for Inflammation Research, QMRI, Edinburgh, UK

**Abstract:** Remyelination in multiple sclerosis is in most cases insufficient, leading to irreversible disability. Different and nonexclusive factors account for this repair deficit. Local inhibitors of the differentiation of oligodendrocyte progenitor cells (OPCs) might play a role, as well as axonal factors impairing the wrapping process. Alternatively, a defect in the recruitment of OPCs toward the demyelinated area may be involved in lesions with oligodendroglial depopulation. Deciphering the mechanisms underlying myelin repair success or failure should open new avenues for designing strategies aimed at favoring endogenous remyelination.

**Keywords:** multiple sclerosis; remyelination; oligodendrocyte progenitor cell; guidance molecules; semaphorins

## Introduction

Multiple sclerosis (MS) is an inflammatory demyelinating disease of the central nervous system (CNS), affecting more than 350,000 persons in Europe. It is considered as the second cause of acquired disability in the young adult, after trauma.

Demyelinated lesions of MS, disseminated throughout the CNS, may remyelinate. Neuropathological demonstration of remyelination in MS lesions was first reported at the ultrastructural level in the 1960s (Perier and Gregoire, 1965). Remyelinated lesions correspond to the classical "shadow plaques," described by Charcot (1868) which are only faintly colored by a myelin stain.

These shadow plaques are composed of fields of thinly myelinated axons that are similar in appearance to remyelinating axons observed in numerous experimental models of demyelination (Lassmann, 1983; Ludwin, 1987). CNS remyelination first reported in chronic lesions has since been observed in active lesions, occurring in association with ongoing demyelination (Prineas et al., 1993; Raine et al., 1981; Raine and Wu, 1993).

In addition to its role in the restoration of a rapid, saltatory conduction of nerve impulses, which was demonstrated in experimental models, recent experimental studies have ascertained that prompt remyelination is capable of preventing axonal degeneration (Irvine and Blakemore, 2008). However, in contrast with experimental models of demyelination in which remyelination is almost complete, the capacity for repair in MS is insufficient and decreases with time, resulting in neurological dysfunction.

*Corresponding author.
Tel.: 00 33 1 42 16 21 59; Fax: 00 33 1 45 84 80 08;
E-mail: gabiepiaton@gmail.com

Currently available therapies in MS limit CNS inflammation, but have not demonstrated significant effects on long-term disability. Therefore, improvement of myelin repair is a major goal in MS research. In this context, understanding why remyelination fails is crucial for devising effective methods in order to enhance it.

## Capacity for remyelination: quantitative data

Although remyelination has long been considered as being very limited, especially in the late or progressive phases of the disease, recent neuropathological studies have disputed this. A large study, performed by the group of Hans Lassmann, has analyzed remyelination in autopsies of MS cases (Patrikios et al., 2006). The extent of remyelination was variable between cases. In 20% of the patients, remyelination was extensive with 60–96% of the global lesion area remyelinated. In contrast, 67% of the cases were characterized by a low percentage of repair, below 25% of the plaque area. This study also suggested that repair capacity may be related to the location of the demyelinated area, as cortical lesions were often well remyelinated, and peri-ventricular lesions were poorly remyelinated. Extensive repair was not limited to the relapsing and remitting stage of the disease, but was also detected in both primary and secondary progressive disease. In addition, extensive remyelination was detected despite a long disease course, as shown in another study on two cases of long-lasting disease, with an analysis of 168 white matter lesions, of which 22% were shadow plaques, 73% were partially remyelinated, and only 5% were completely demyelinated (Patani et al., 2007). Results from these studies demonstrate that remyelination in MS may be more extensive than previously thought, but there might also be a relative intraindividual homogeneity of myelin repair capacity, contrasting with a large interindividual heterogeneity.

## What are the mechanisms underlying the success or failure of myelin repair in MS?

This is one of the key questions in MS neurobiology. It is clear that these mechanisms are many and not exclusive, and may vary between different types of lesions. Demyelinated lesions differ in their oligodendroglial cell content. Some demyelinated lesions are characterized by an oligodendroglial depopulation (Fig. 1D, E). In contrast, in other lesions, oligodendrocytes and/or oligodendrocyte progenitor cells (OPCs) persist within the

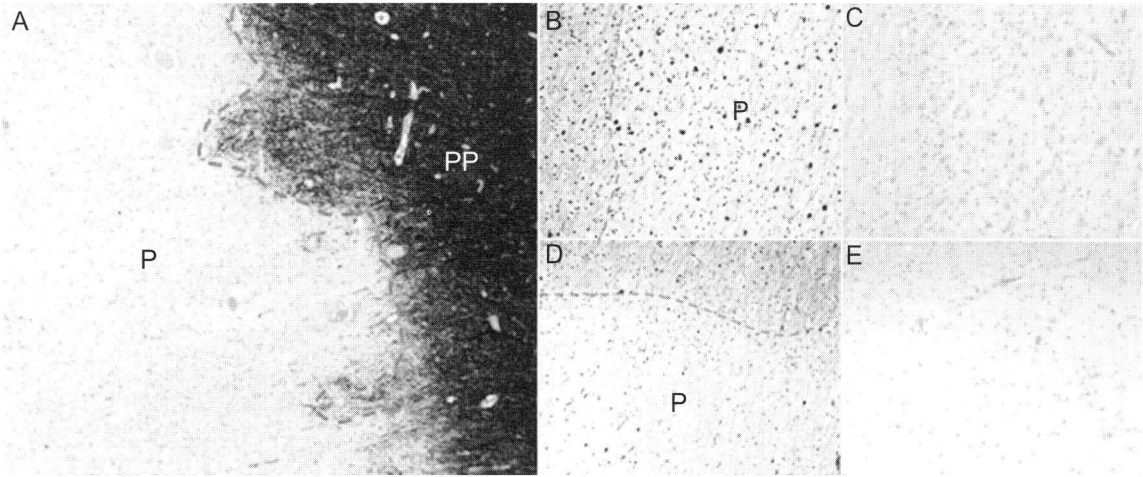

Fig. 1. (A) Immunolabeling with an anti-MBP (myelin basic protein) antibody showing the sharp demarcation between the plaque (P) and the periplaque (PP) with bundles of remyelination from the border. (B–E) Two different demyelinated MS lesions, stained with Bodian silver impregnation associated with Luxol Fast Blue (B, D) and with hematoxylin-eosin (C, E). Cell bodies suggestive of oligodendroglial cells persist in one lesion (B, C), whereas the other lesion (D, E) is characterized by a cell depopulation. Scale: (A) 1 cm = 400 μm, (B–E) 1 cm = 100 μm.

demyelinated area (Fig. 1B, C). The potential mechanisms leading to either success or failure of remyelination might differ between these different lesions.

## Demyelinated MS lesions with persistence of cells of the oligodendrocyte lineage

Some chronically demyelinated MS lesions contain numerous cells of the oligodendrocyte lineage (Chang et al., 2000, 2002; Wolswijk, 1998). These cells express different markers of immature oligodendroglial cells (staining with O4 antibody, expression of the proteoglycan NG2), and also express markers characteristic of more differentiated cells such as proteolipid protein (PLP) or myelin–oligodendrocyte glycoprotein (MOG). However, these oligodendroglial cells appear to have been unable to wrap around axons and form a new myelin sheath. It is assumed that these cells are derived from OPCs, either resident or having migrated in, and which have failed to mature/develop correctly into myelinating oligodendrocytes, but it is possible that they were mature oligodendrocytes which are attempting to dedifferentiate.

Local inhibitors of oligodendroglial cell differentiation may be present within these chronic lesions. Different inhibitory cues have been reported recently, expressed either by glial cells or neurons. These inhibitors target the differentiation or maturation step, which transforms immature oligodendroglial cells into mature, myelinating cells. Other inhibitors may act directly on the wrapping step of the myelination process. These molecules have been studied in vitro, in cultures of cells of the oligodendroglial lineage, as well as in vivo, in experimental models of demyelination. In addition, some have been detected in MS brain tissue.

### Inhibitors of differentiation or maturation of OPCs

*The Notch/Jagged inhibitory pathway*

The arrest of OPCs in their maturation program may be linked to an activation of the Notch/Jagged pathway. During development, the expression of the ligand protein Jagged1 by the axon inhibits oligodendrocyte differentiation through its binding to the receptor protein Notch1, located on OPCs. Then Jagged1 is down-regulated by axons, allowing oligodendrocyte differentiation and maturation (Wang et al., 1998). Activation of this pathway has been hypothesized to be involved in the maturation defect of OPCs within MS plaques. Indeed, in chronically demyelinated lesions populated with oligodendroglial cells, expression of Jagged1 by reactive astrocytes of the plaque has been detected, and thus could interact with Notch1 protein on OPCs, and explain, at least in part, the differentiation arrest. In contrast, within remyelinated areas, astrocytes do not express Jagged (John et al., 2002). However, the importance of the reactivation of this developmental pathway in MS is unclear. Conditional ablation of Notch1 in OPCs in transgenic mice which were then treated with cuprizone yielded no significant differences in remyelination between these and control mice (Stidworthy et al., 2004). Conversely, it was recently shown in experimental autoimmune encephalomyelitis (EAE) that blocking Notch pathway activation through inhibition of gamma-secretase promotes remyelination (Jurynczyk et al., 2008).

*LINGO-1, an oligodendroglial inhibitor of oligodendrocyte differentiation*

Nogo receptor-interacting protein (LINGO-1), selectively expressed in oligodendrocytes and neurons in the brain and spinal cord, has been identified as an in vitro and in vivo negative regulator of oligodendrocyte differentiation and myelination. Attenuation of LINGO-1 function in OPCs promotes oligodendroglial differentiation, with an increase in the number of mature oligodendrocytes in vitro and increased synthesis of myelin proteins. This promyelinating effect has been shown to be mediated through increased Fyn phosphorylation and down-regulation of RhoA-GTP (Mi et al., 2005; Park et al., 2005). Recently, these results have been confirmed in vivo in EAE. In this model, it was shown that loss of LINGO-1 function by *Lingo1* gene knockout or by treatment with an antibody antagonist leads to increased remyelination (Mi et al., 2007).

## Inhibitors of the myelination process

### PSA-NCAM

In addition to glial inhibitors, axonal signals expressed or reexpressed at the surface of denuded axons might also impair the remyelination process. This is the case for the polysialylated acid form of the neural cell adhesion molecule (PSA-NCAM) which has been shown to negatively regulate myelination during normal development, and to be absent from axons in the adult brain, except in areas known to exhibit permanent plasticity and neurogenesis (Charles et al., 2000). In demyelinated MS plaques, reexpression of PSA-NCAM on a percentage of denuded axons has been detected, whereas PSA-NCAM was never detected on remyelinated internodes within partially remyelinated plaques (Charles et al., 2002).

### Electrical activity

Other axonal dysfunction, besides expression of inhibitory molecules, may affect remyelination. During development, electrical activity along axons has been shown to play a crucial role in the initiation of the myelination process (Demerens et al., 1996; Stevens et al., 2002). Therefore, it could be speculated that modifications of action potential or conduction block along demyelinated internodes may impair the positive signal necessary for remyelination to proceed. This neuron–glial signal provides a molecular mechanism for promoting oligodendrocyte development and myelination, and may prove to be a promising strategy to improve remyelination. However, drugs such as lamotrigine and flecainide which block sodium channels and reduce action potentials have been shown in in vitro and in vivo models of MS to promote neuronal protection and survival which is a prerequisite for remyelination [reviewed in (Smith, 2007)]. Perhaps, what is needed is a cessation of electrical activity initially to preserve axons while debris is cleared and harmful inflammation subsides, followed by reestablishment of action potentials, encouraging oligodendroglial cell maturation and remyelination.

## Lesions with oligodendroglial depopulation

Demyelinated MS lesions with oligodendroglial depopulation are frequently detected, despite the presence of OPCs expressing platelet derived growth factor alpha receptor (PDGF$\alpha$R) or NG2 in the white matter distant from the plaque. One hypothesis is that there is a defect in the recruitment of adult OPCs toward the demyelinated area through dysregulation of oligodendroglial guidance cues.

### Different mechanisms of OPCs guidance

OPCs migration has mostly been studied during development. It involves two types of mechanisms: contact mediated and chemokinetic migration, which involve guidance cues imparted to the cell by the surrounding environment to help migratory choices. These signals can be soluble, matrix bound, or displayed at the cell surface, and may exist in concentration gradients. Different developmental studies suggest that glial precursors can use similar molecular cues to axons. Some of these cues are reexpressed under demyelinating conditions, and so are likely to influence myelin repair through their attractive or repulsive chemotactic properties on adult OPCs. We will briefly review the main cues that have been implicated in OPCs developmental migration, with special emphasis on class 3 semaphorins. In relation to our recent results, we will then discuss the potential involvement of these latter cues in influencing remyelination in MS.

### Contact-mediated cues

Adhesion molecules and components of the ECM have been implicated in OPCs migration during development from the germinal zones where they are generated to their final destination. Inhibitory adhesion molecules include tenascin-C (Garcion et al., 2001; Kiernan et al., 1996; Sobel, 2005), ephrins (Prestoz et al., 2004), and N-cadherins (Schnadelbach et al., 2000). Among the pro-migratory adhesion molecules, integrins have been reported to act on OPCs migration in vitro, the αVβ1 integrin being the major player (Milner et al., 1996). Attachment of the PSA residue to NCAM enhances cell migratory potential (Barral-Moran et al., 2003; Vitry et al., 2001; Wang et al., 1994), and is necessary for OPCs migration before being down-regulated at the axonal surface at the onset of myelination (Charles et al., 2000; Fewou et al., 2007; Franceschini et al., 2004; Jakovcevski et al., 2007). Recently, anosmin-1 expressed by both migrating OPCs and axons in the embryonic optic nerve has been shown to be involved in oligodendrocyte adhesion and migration (Bribian et al., 2008).

*Secreted molecules*

*Growth factors.* The first growth factors involved in OPCs migration were PDGF and basic fibroblast growth factor (bFGF, also known as FGF-2) in the 1990s. Adult OPCs express receptors for FGF-2 and PDGF both constitutively and during the proliferative response to demyelination (Maeda et al., 2001; Redwine and Armstrong, 1998; Redwine et al., 1997). PDGF, in addition to its mitogenic role, also promotes OPCs motility and is chemoattractive for cultured progenitors while inhibiting their premature differentiation (Noble et al., 1988). FGF-2 has also been suggested to be crucial for OPCs motility (Osterhout et al., 1997). Both factors act in a nonadditive manner via different signaling pathways, with PDGF having a stronger influence (Milner et al., 1997).

In addition, epidermal growth factor (EGF) (Aguirre et al., 2005; Fricker-Gates et al., 2000; Ivkovic et al., 2008) and hepatocyte growth factor (HGF) (Yan and Rivkees, 2002) have also been shown to be involved in OPCs migration.

Interactions clearly exist between contact-mediated and soluble growth factor guidance mechanisms. For example, PSA-NCAM selectively enhances migration toward PDGF but not toward FGF-2 (Zhang et al., 2004), and the effect of tenascin-C on OPCs proliferation is mediated by an integrin and interacts with the PDGF-stimulated mitogenic pathway (Garcion et al., 2001).

*Chemokines.* Chemokines are chemotactic cytokines classically involved in leukocyte activation and chemotaxis; some of them have also been shown to influence OPCs migration.

Neonatal OPCs express CXCR4, the receptor for the chemokine CXCL12, and they show a migratory response to CXCL12 in vitro. In vivo, CXCR4−/− mice have a strong reduction of the number of PDGFαR-positive OPCs in the embryonic spinal cord (Dziembowska et al., 2005). These results suggesting that CXCR4 signaling is involved in OPCs survival and outward migration during development are controversial, as this migratory influence has not been confirmed in other experimental conditions (Maysami et al., 2006).

Among the other chemokine receptors, CXCR1, CXCR2, and CXCR3 are expressed by oligodendroglial cells both in normal adult human tissue and in culture. Their respective ligands are CXCL8, CXCL1, and CXCL10. CXCL1/CXCR2 signaling is known to have a biological effect on OPCs migration. CXCL1 is a clear stop signal for migrating OPCs in the spinal cord (Tsai et al., 2002), whereas the influence of other chemokines remains to be evaluated.

These different results suggest that, in addition to their immunological effect, glial-secreted chemokines may influence OPCs recruitment.

*Neurotransmitters.* OPCs express most neurotransmitter receptors, and this expression decreases during oligodendrocyte differentiation, suggesting a specific role in progenitor cells. Activation of AMPA (Gudz et al., 2006), GABA (Luyt et al., 2007), and adenosine (Agresti et al., 2005; Othman et al., 2003) receptors may influence the migratory capacity of OPCs.

In summary, these secreted molecules may influence OPCs migration either independently from their main function (as for neurotransmitters) or in addition to their alternative influence on OPCs proliferation or differentiation (as for growth factors).

*Netrin-1 and Class 3 semaphorins.* Other secreted molecules have been reported mainly as guidance cues during development. To date, the chemotropic molecules studied in detail are netrins and secreted class 3 semaphorins. Netrins and semaphorins consist of two families of proteins that were first identified as regulating the migration of neurons and axonal growth cones during development. They can act as bifunctional signals, being chemoattractive for some neurons and chemorepellent for others. As neurons and OPCs share common sites of origin in the embryonic neural tube and furthermore develop in close timing, the influence of these neuronal cues on OPCs behavior is of particular interest.

Netrin-1 can function either as a chemoattractant or a chemorepellent depending on the cellular context and receptor complexes: in neuronal cells, chemoattraction is mediated by deleted in colorectal cancer (DCC) or neogenin receptors (Keino-Masu et al., 1996), whereas chemorepulsion is mediated by expression of both DCC and Unc5 receptors together or Unc5 receptors alone (Hong et al., 1999; Keleman and Dickson, 2001).

In oligodendroglial cells, netrin-1 has been shown to act mainly as a chemorepellent. In the developing spinal cord, netrin-1 is expressed by floor plate cells as OPCs migrate away from the ventral midline, and both *dcc* and *unc5h1* receptors are expressed by migrating OPCs. In vivo absence of netrin-1 or DCC results in fewer OPCs migrating from the ventral to dorsal embryonic spinal cord (Jarjour et al., 2003), and an overall reduction in the number of oligodendroglial lineage cells (Tsai et al., 2006). Blocking netrin-1 signaling in embryonic spinal cord slices also inhibits OPCs migration from the ventricular zone (Tsai et al., 2003). These results indicate that netrin-1 signaling is not only necessary for initial dispersal and ventral to dorsal OPCs migration but also for their subsequent development. In the neonatal rat optic nerve, netrin-1 expressed in the optic chiasma also acts as a chemorepellent for migrating OPCs (Sugimoto et al., 2001). Adult spinal cord neural progenitor cells seeded ex vivo onto organotypic slice preparation of both intact embryonic and injured adult spinal cord are repelled by netrin-1 from the floor plate and the injury core, respectively (Petit et al., 2007). Netrin-1 may therefore act as an inhibiting factor for OPCs recruitment toward demyelinated lesions, but further description of netrin-1 and receptor expression and functional studies in demyelinating models are needed to confirm this hypothesis.

Semaphorins constitute a large and highly conserved family of molecular signals that were initially identified through their role in axon guidance (Kolodkin et al., 1993; Luo et al., 1993) and later implicated in a range of functions from cell guidance to regulation of immune function, angiogenesis and pathologies as neurological diseases and cancer (Casazza et al., 2007; Mann et al., 2007; Neufeld et al., 2007; Suzuki et al., 2008). They include both membrane-associated and secreted proteins, and bind to two classes of receptors: neuropilin co-receptors and plexins, the latter being transducing units (Zhou et al., 2008) (Fig. 2).

Sugimoto et al. (2001) first provided evidence that class 3 semaphorins and Netrin-1 are involved in OPCs migration in the developing optic nerve. They analyzed OPCs migration in the neonatal rat optic nerve and found that netrin-1 and semaphorin 3A are expressed in the optic chiasma and that they are chemorepellent for migrating OPCs (Sugimoto et al., 2001). Another study confirmed that migrating OPCs in the optic nerve express semaphorin receptors (neuropilin-1 and -2) and that semaphorin 3F expressed in the retina has a dual role, with both a chemoattractive and a mitogenic influence on embryonic OPCs (Spassky et al., 2002). Finally, it was shown that both neonatal OPCs and adult oligodendrocytes express NP-1, and different NP-2 isoforms with a down-regulation of this expression as differentiation takes place (Cohen et al., 2003; Ricard et al., 2001).

In addition to their influence on OPCs migration, semaphorin 3A has been shown to induce OPCs and oligodendrocyte processes retraction, through an intracellular signaling pathway involving Rho kinase (Ricard et al., 2000, 2001). The effect of semaphorin 3F on OPCs and oligodendrocyte processes may be different and needs further elucidation.

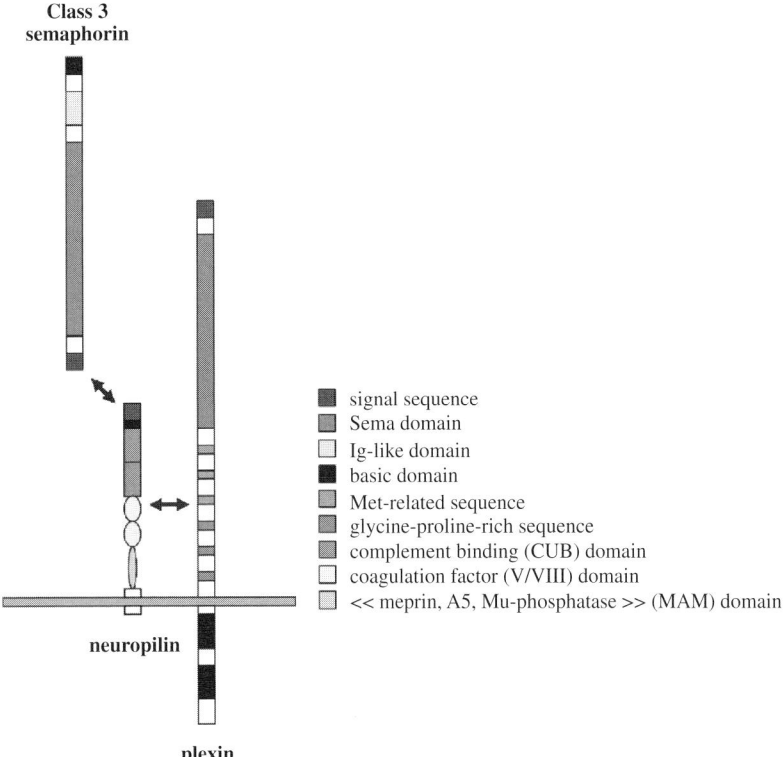

Fig. 2. Class 3 semaphorins can signal through a neuropilin/plexin A receptor complex which can be modulated by cell adhesion molecules (not shown here). Adapted from Pasterkamp and Verhaagen (2006).

## Class 3 semaphorins in CNS pathology

### CNS injury

Semaphorin expression in the adult CNS is related to structural plasticity and neural regeneration (de Wit and Verhaagen, 2003). Dysregulation of class 3 semaphorins and neuropilins has been reported in different models of adult CNS injury, either traumatic or ischemic.

An increased expression has been detected in different cell types: fibroblast-like cells of the neural scar (De Winter et al., 2002; Nitzan et al., 2006; Pasterkamp et al., 1999), neurons (Fujita et al., 2001; Hashimoto et al., 2004; Lindholm et al., 2004), and activated microglial cells (Agudo et al., 2005; Fujita et al., 2001). Absence of semaphorin 3A after injury in the neonatal spinal cord is associated with strong axonal regeneration, in contrast to injury in the adult spinal cord (Pasterkamp et al., 1999), and in human neuromata, where fibroblasts surrounding nerve fibers produce increased levels of semaphorin 3A which inhibits neurite extension in vitro (Tannemaat et al., 2007). These data suggest that in the injured adult CNS, semaphorin 3A is likely to inhibit axonal regeneration and branching and participate in the glial-fibrotic scar. Selective in vivo inhibition of semaphorin 3A enhances axon regeneration, Schwann cell-mediated myelination, cell survival and angiogenesis in the lesion site after spinal cord transection in rats, resulting in better functional recovery (Kaneko et al., 2006), confirming the functional inhibitory role of semaphorin 3A. However, these results were not reproduced in another model of spinal cord injury in the mouse (Mire et al., 2008).

These data, together with the reported influence of semaphorins 3A and 3F on OPCs migration during development, have prompted us to analyze the expression of class 3 semaphorins after CNS demyelination.

*MS and experimental demyelination*

Given their involvement in OPCs guidance and in the response to injury shown by different groups, we hypothesized that, by reiterating developmental mechanisms, semaphorins 3A and 3F could represent repulsive or attractive signals (respectively) for adult OPCs in MS brain, and that their relative production (and that of their receptors) could influence OPCs recruitment to demyelinated lesions, and hence remyelination.

We first analyzed semaphorin and neuropilin mRNA expression in MS brain tissue. In control tissue, semaphorins were only expressed in neurons of the cortical gray matter, with no expression in the normal appearing white matter. In MS tissue, both *semaphorins 3A* and *3F* were expressed within and around active demyelinated white matter lesions. This expression was mostly by astroglial and microglial cells. In addition, within these active lesions, a subset of oligodendroglial cells, detected by Olig2 expression, expressed *NP-1* and *NP-2* receptors. The relative levels of expression of *semaphorins 3A* and *3F* was correlated with the inflammatory status of the lesion, with the more inflammatory lesions having higher levels of both semaphorins compared to less active lesions, but also expressing more *semaphorin 3F* (attractive) than *3A* (repulsive). In contrast, no *semaphorin 3A* or *3F* expression was detected within or around chronic, noninflammatory white matter lesions.

In addition to this expression in the white matter, both semaphorin levels were overexpressed in cortical neurons (i.e., distant from a demyelinated white matter plaque), suggesting that this expression may be secondary to the axonal demyelinating insult.

These local and distant effects were then reproduced in a rat model of demyelination in which we induced a spinal cord demyelinating lesion targeting the corticospinal tract of the spinal cord by lysophosphatidylcholine injection. Demyelination was associated with *semaphorins 3A* and *3F* expression at the level of the lesion, mostly by glial cells. In addition, *semaphorins 3A* and *3F* expression was also detected in the cortical neurons of the contralateral motor cortex corresponding to the demyelinated lesion. Although no clear difference in the percentage of cells expressing *semaphorin 3A* or *3F* was detected in cortical neurons, the percentage of cells expressing *semaphorin 3F* in the spinal cord was higher compared to *semaphorin 3A*. This latter result showing an increased attractive signal for OPCs after spinal cord demyelination is in agreement with the rapid and extensive spontaneous remyelination in this model (Williams et al., 2007).

## Hypothesis and perspectives

OPCs express a variety of receptors to different factors that have been shown to affect their developmental migration to different extents, allowing multiple and adaptive responses to environmental cues. This could be related to the existence of different OPCs subpopulations or to the modulation of their response to one factor by changing expression of specific receptors.

Adult OPCs are likely to be responsive to the same signals, which have been shown to influence OPCs migration during development, and which are reexpressed after injury or demyelination. These signals may alter adult OPCs migration and guidance to the lesion sites, and the balance between attractive and repulsive cues expressed in a same lesion could dictate the remyelination efficiency. However, the functional migratory influence of these guidance cues in the adult demyelinated CNS has to be ascertained. This might be approached by taking advantage of specific conditional knockout animals in which oligodendroglial expression of guidance cue receptors are ablated in the adult CNS. These studies will help to assess the potential role of these cues in OPCs recruitment toward demyelinated lesions, and whether the rate of recruitment and total OPCs numbers within the lesion influence remyelination outcome. If this is the case, then molecules able to guide adult OPCs migration in the injured brain may represent powerful therapeutic targets to increase adult OPCs migration to lesion sites, allowing these cells to colonize and remyelinate these lesions, eventually in association with pro-differentiation factors. Finally, it is important to stress that the different molecules involved in OPCs guidance can interact in complex ways, and that their expression in the demyelinated CNS might be influenced by the inflammatory process. A better understanding of these multiple interactions will be crucial for a translational strategy.

## Conclusion

Recent years have yielded important results concerning the occurrence and extent of the myelin repair process after demyelination. The different and nonexclusive mechanisms supporting either failure or success of the remyelination are being deciphered currently, and this should open promising avenues, complementary to anti-inflammatory strategies, aimed at favoring remyelination, which in turn will favor axonal protection and prevention of disability in MS.

## Acknowledgments

Catherine Lubetzki thanks INSERM and the French MS society ARSEP for their support. Anna Williams is funded by The Wellcome Trust.

## References

Agresti, C., Meomartini, M. E., Amadio, S., Ambrosini, E., Volonte, C., Aloisi, F., et al. (2005). ATP regulates oligodendrocyte progenitor migration, proliferation, and differentiation: involvement of metabotropic P2 receptors. *Brain Research. Brain Research Reviews*, 48(2), 157–165.

Agudo, M., Robinson, M., Cafferty, W., Bradbury, E. J., Kilkenny, C., Hunt, S. P., et al. (2005). Regulation of neuropilin 1 by spinal cord injury in adult rats. *Molecular and Cellular Neurosciences*, 28(3), 475–484.

Aguirre, A., Rizvi, T. A., Ratner, N., & Gallo, V. (2005). Overexpression of the epidermal growth factor receptor confers migratory properties to nonmigratory postnatal neural progenitors. *Journal of Neuroscience*, 25(48), 11092–11106.

Barral-Moran, M. J., Calaora, V., Vutskits, L., Wang, C., Zhang, H., Durbec, P., et al. (2003). Oligodendrocyte progenitor migration in response to injury of glial monolayers requires the polysialic neural cell-adhesion molecule. *Journal of Neuroscience Research*, 72(6), 679–690.

Bribian, A., Esteban, P. F., Clemente, D., Soussi-Yanicostas, N., Thomas, J. L., Zalc, B., et al. (2008). A novel role for anosmin-1 in the adhesion and migration of oligodendrocyte precursors. *Developmental Neurobiology*, 68(13), 1503–1516.

Casazza, A., Fazzari, P., & Tamagnone, L. (2007). Semaphorin signals in cell adhesion and cell migration: functional role and molecular mechanisms. *Advances in Experimental Medicine and Biology*, 600, 90–108.

Chang, A., Nishiyama, A., Peterson, J., Prineas, J., & Trapp, B. D. (2000). NG2-positive oligodendrocyte progenitor cells in adult human brain and multiple sclerosis lesions. *Journal of Neuroscience*, 20(17), 6404–6412.

Chang, A., Tourtellotte, W. W., Rudick, R., & Trapp, B. D. (2002). Premyelinating oligodendrocytes in chronic lesions of multiple sclerosis. *The New England Journal of Medicine*, 346(3), 165–173.

Charcot, J. M. (1868). Histologie de la sclérose en plaques. *Gazette des Hôpitaux de Paris*, 41, 554–566.

Charles, P., Hernandez, M. P., Stankoff, B., Aigrot, M. S., Colin, C., Rougon, G., et al. (2000). Negative regulation of central nervous system myelination by polysialylated-neural cell adhesion molecule. *Proceedings of the National Academy of Sciences of the United States of America*, 97(13), 7585–7590.

Charles, P., Reynolds, R., Seilhean, D., Rougon, G., Aigrot, M. S., Niezgoda, A., et al. (2002). Re-expression of PSA-NCAM by demyelinated axons: an inhibitor of remyelination in multiple sclerosis? *Brain*, 125(Pt. 9), 1972–1979.

Cohen, R. I., Rottkamp, D. M., Maric, D., Barker, J. L., & Hudson, L. D. (2003). A role for semaphorins and neuropilins in oligodendrocyte guidance. *Journal of Neurochemistry*, 85(5), 1262–1278.

Demerens, C., Stankoff, B., Logak, M., Anglade, P., Allinquant, B., Couraud, F., et al. (1996). Induction of myelination in the central nervous system by electrical activity. *Proceedings of the National Academy of Sciences of the United States of America*, 93(18), 9887–9892.

De Winter, F., Oudega, M., Lankhorst, A. J., Hamers, F. P., Blits, B., Ruitenberg, M. J., et al. (2002). Injury-induced class 3 semaphorin expression in the rat spinal cord. *Experimental Neurology*, 175(1), 61–75.

De Wit, J., & Verhaagen, J. (2003). Role of semaphorins in the adult nervous system. *Progress in Neurobiology*, 71(2–3), 249–267.

Dziembowska, M., Tham, T. N., Lau, P., Vitry, S., Lazarini, F., & Dubois-Dalcq, M. (2005). A role for CXCR4 signaling in survival and migration of neural and oligodendrocyte precursors. *Glia*, 50(3), 258–269.

Fewou, S. N., Ramakrishnan, H., Bussow, H., Gieselmann, V., & Eckhardt, M. (2007). Down-regulation of polysialic acid is required for efficient myelin formation. *The Journal of Biological Chemistry*, 282(22), 16700–16711.

Franceschini, I., Vitry, S., Padilla, F., Casanova, P., Tham, T. N., Fukuda, M., et al. (2004). Migrating and myelinating potential of neural precursors engineered to overexpress PSA-NCAM. *Molecular and Cellular Neurosciences, 27*(2), 151–162.

Fricker-Gates, R. A., Winkler, C., Kirik, D., Rosenblad, C., Carpenter, M. K., & Bjorklund, A. (2000). EGF infusion stimulates the proliferation and migration of embryonic progenitor cells transplanted in the adult rat striatum. *Experimental Neurology, 165*(2), 237–247.

Fujita, H., Zhang, B., Sato, K., Tanaka, J., & Sakanaka, M. (2001). Expressions of neuropilin-1, neuropilin-2 and semaphorin 3A mRNA in the rat brain after middle cerebral artery occlusion. *Brain Research, 914*(1–2), 1–14.

Garcion, E., Faissner, A., & ffrench-Constant, C. (2001). Knockout mice reveal a contribution of the extracellular matrix molecule tenascin-C to neural precursor proliferation and migration. *Development, 128*(13), 2485–2496.

Gudz, T. I., Komuro, H., & Macklin, W. B. (2006). Glutamate stimulates oligodendrocyte progenitor migration mediated via an alphav integrin/myelin proteolipid protein complex. *Journal of Neuroscience, 26*(9), 2458–2466.

Hashimoto, M., Ino, H., Koda, M., Murakami, M., Yoshinaga, K., Yamazaki, M., et al. (2004). Regulation of semaphorin 3A expression in neurons of the rat spinal cord and cerebral cortex after transection injury. *Acta Neuropathologica, 107*(3), 250–256.

Hong, K., Hinck, L., Nishiyama, M., Poo, M. M., Tessier-Lavigne, M., & Stein, E. (1999). A ligand-gated association between cytoplasmic domains of UNC5 and DCC family receptors converts netrin-induced growth cone attraction to repulsion. *Cell, 97*(7), 927–941.

Irvine, K. A., & Blakemore, W. F. (2008). Remyelination protects axons from demyelination-associated axon degeneration. *Brain, 131*(Pt. 6), 1464–1477.

Ivkovic, S., Canoll, P., & Goldman, J. E. (2008). Constitutive EGFR signaling in oligodendrocyte progenitors leads to diffuse hyperplasia in postnatal white matter. *Journal of Neuroscience, 28*(4), 914–922.

Jakovcevski, I., Mo, Z., & Zecevic, N. (2007). Down-regulation of the axonal polysialic acid-neural cell adhesion molecule expression coincides with the onset of myelination in the human fetal forebrain. *Neuroscience, 149*(2), 328–337.

Jarjour, A. A., Manitt, C., Moore, S. W., Thompson, K. M., Yuh, S. J., & Kennedy, T. E. (2003). Netrin-1 is a chemorepellent for oligodendrocyte precursor cells in the embryonic spinal cord. *Journal of Neuroscience, 23*(9), 3735–3744.

John, G. R., Shankar, S. L., Shafit-Zagardo, B., Massimi, A., Lee, S. C., Raine, C. S., et al. (2002). Multiple sclerosis: re-expression of a developmental pathway that restricts oligodendrocyte maturation. *Nature Medicine, 8*(10), 1115–1121.

Jurynczyk, M., Jurewicz, A., Bielecki, B., Raine, C. S., & Selmaj, K. (2008). Overcoming failure to repair demyelination in EAE: gamma-secretase inhibition of Notch signaling. *Journal of Neurological Sciences, 265*(1-2), 5–11.

Kaneko, S., Iwanami, A., Nakamura, M., Kishino, A., Kikuchi, K., Shibata, S., et al. (2006). A selective Sema3A inhibitor enhances regenerative responses and functional recovery of the injured spinal cord. *Nature Medicine, 12*(12), 1380–1389.

Keino-Masu, K., Masu, M., Hinck, L., Leonardo, E. D., Chan, S. S., Culotti, J. G., et al. (1996). Deleted in colorectal cancer (DCC) encodes a netrin receptor. *Cell, 87*(2), 175–185.

Keleman, K., & Dickson, B. J. (2001). Short- and long-range repulsion by the Drosophila Unc5 netrin receptor. *Neuron, 32*(4), 605–617.

Kiernan, B. W., Gotz, B., Faissner, A., & ffrench-Constant, C. (1996). Tenascin-C inhibits oligodendrocyte precursor cell migration by both adhesion-dependent and adhesion-independent mechanisms. *Molecular and Cellular Neurosciences, 7*(4), 322–335.

Kolodkin, A. L., Matthes, D. J., & Goodman, C. S. (1993). The semaphorin genes encode a family of transmembrane and secreted growth cone guidance molecules. *Cell, 75*(7), 1389–1399.

Lassmann, H. (1983). Comparative neuropathology of chronic experimental allergic encephalomyelitis and multiple sclerosis. *Schriftenreihe Neurologie, 25*, 1–135.

Lindholm, T., Skold, M. K., Suneson, A., Carlstedt, T., Cullheim, S., & Risling, M. (2004). Semaphorin and neuropilin expression in motoneurons after intraspinal motoneuron axotomy. *Neuroreport, 15*(4), 649–654.

Ludwin, S. K. (1987). Remyelination in demyelinating diseases of the central nervous system. *Critical Reviews in Neurobiology, 3*(1), 1–28.

Luo, Y., Raible, D., & Raper, J. A. (1993). Collapsin: a protein in brain that induces the collapse and paralysis of neuronal growth cones. *Cell, 75*(2), 217–227.

Luyt, K., Slade, T. P., Dorward, J. J., Durant, C. F., Wu, Y., Shigemoto, R., et al. (2007). Developing oligodendrocytes express functional GABA(B) receptors that stimulate cell proliferation and migration. *Journal of Neurochemistry, 100*(3), 822–840.

Maeda, Y., Solanky, M., Menonna, J., Chapin, J., Li, W., & Dowling, P. (2001). Platelet-derived growth factor-alpha receptor-positive oligodendroglia are frequent in multiple sclerosis lesions. *Annals of Neurology, 49*(6), 776–785.

Mann, F., Chauvet, S., & Rougon, G. (2007). Semaphorins in development and adult brain: implication for neurological diseases. *Progress in Neurobiology, 82*(2), 57–79.

Maysami, S., Nguyen, D., Zobel, F., Pitz, C., Heine, S., Hopfner, M., et al. (2006). Modulation of rat oligodendrocyte precursor cells by the chemokine CXCL12. *Neuroreport, 17*(11), 1187–1190.

Mi, S., Hu, B., Hahm, K., Luo, Y., Kam Hui, E. S., Yuan, Q., et al. (2007). LINGO-1 antagonist promotes spinal cord remyelination and axonal integrity in MOG-induced experimental autoimmune encephalomyelitis. *Nature Medicine, 13*(10), 1228–1233.

Mi, S., Miller, R. H., Lee, X., Scott, M. L., Shulag-Morskaya, S., Shao, Z., et al. (2005). LINGO-1 negatively regulates myelination by oligodendrocytes. *Nature Neuroscience, 8*(6), 745–751.

Milner, R., Anderson, H. J., Rippon, R. F., McKay, J. S., Franklin, R. J., Marchionni, M. A., et al. (1997). Contrasting effects of mitogenic growth factors on oligodendrocyte precursor cell migration. *Glia, 19*(1), 85–90.

Milner, R., Edwards, G., Streuli, C., & Ffrench-Constant, C. (1996). A role in migration for the alpha V beta 1 integrin expressed on oligodendrocyte precursors. *Journal of Neuroscience, 16*(22), 7240–7252.

Mire, E., Thomasset, N., Jakeman, L. B., & Rougon, G. (2008). Modulating Sema3A signal with a L1 mimetic peptide is not sufficient to promote motor recovery and axon regeneration after spinal cord injury. *Molecular and Cellular Neurosciences, 37*(2), 222–235.

Neufeld, G., Lange, T., Varshavsky, A., & Kessler, O. (2007). Semaphorin signaling in vascular and tumor biology. *Advances in Experimental Medicine and Biology, 600*, 118–131.

Nitzan, A., Kermer, P., Shirvan, A., Bahr, M., Barzilai, A., & Solomon, A. S. (2006). Examination of cellular and molecular events associated with optic nerve axotomy. *Glia, 54*(6), 545–556.

Noble, M., Murray, K., Stroobant, P., Waterfield, M. D., & Riddle, P. (1988). Platelet-derived growth factor promotes division and motility and inhibits premature differentiation of the oligodendrocyte/type-2 astrocyte progenitor cell. *Nature, 333*(6173), 560–562.

Osterhout, D. J., Ebner, S., Xu, J., Ornitz, D. M., Zazanis, G. A., & McKinnon, R. D. (1997). Transplanted oligodendrocyte progenitor cells expressing a dominant-negative FGF receptor transgene fail to migrate in vivo. *Journal of Neuroscience, 17*(23), 9122–9132.

Othman, T., Yan, H., & Rivkees, S. A. (2003). Oligodendrocytes express functional A1 adenosine receptors that stimulate cellular migration. *Glia, 44*(2), 166–172.

Park, J. B., Yiu, G., Kaneko, S., Wang, J., Chang, J., He, X. L., et al. (2005). A TNF receptor family member, TROY, is a coreceptor with Nogo receptor in mediating the inhibitory activity of myelin inhibitors. *Neuron, 45*(3), 345–351.

Pasterkamp, R. J., Giger, R. J., Ruitenberg, M. J., Holtmaat, A. J., De Wit, J., De Winter, F., et al. (1999). Expression of the gene encoding the chemorepellent semaphorin III is induced in the fibroblast component of neural scar tissue formed following injuries of adult but not neonatal CNS. *Molecular and Cellular Neurosciences, 13*(2), 143–166.

Pasterkamp, R. J., & Verhaagen, J. (2006). Semaphorins in axon regeneration: developmental guidance molecules gone wrong? *Philosophical Transactions of the Royal Society of London. Series B, Biological Sciences, 361*(1473), 1499–1511.

Patani, R., Balaratnam, M., Vora, A., & Reynolds, R. (2007). Remyelination can be extensive in multiple sclerosis despite a long disease course. *Neuropathology and Applied Neurobiology, 33*(3), 277–287.

Patrikios, P., Stadelmann, C., Kutzelnigg, A., Rauschka, H., Schmidbauer, M., Laursen, H., et al. (2006). Remyelination is extensive in a subset of multiple sclerosis patients. *Brain, 129*(Pt. 12), 3165–3172.

Perier, O., & Gregoire, A. (1965). Electron microscopic features of multiple sclerosis lesions. *Brain, 88*(5), 937–952.

Petit, A., Sellers, D. L., Liebl, D. J., Tessier-Lavigne, M., Kennedy, T. E., & Horner, P. J. (2007). Adult spinal cord progenitor cells are repelled by netrin-1 in the embryonic and injured adult spinal cord. *Proceedings of the National Academy of Sciences of the United States of America, 104*(45), 17837–17842.

Prestoz, L., Chatzopoulou, E., Lemkine, G., Spassky, N., Lebras, B., Kagawa, T., et al. (2004). Control of axonophilic migration of oligodendrocyte precursor cells by Eph-ephrin interaction. *Neuron Glia Biology, 1*(1), 73–83.

Prineas, J. W., Barnard, R. O., Kwon, E. E., Sharer, L. R., & Cho, E. S. (1993). Multiple sclerosis: remyelination of nascent lesions. *Annals of Neurology, 33*(2), 137–151.

Raine, C. S., Scheinberg, L., & Waltz, J. M. (1981). Multiple sclerosis. Oligodendrocyte survival and proliferation in an active established lesion. *Laboratory Investigation, 45*(6), 534–546.

Raine, C. S., & Wu, E. (1993). Multiple sclerosis: remyelination in acute lesions. *Journal of Neuropathology and Experimental Neurology, 52*(3), 199–204.

Redwine, J. M., & Armstrong, R. C. (1998). In vivo proliferation of oligodendrocyte progenitors expressing PDGF-alphaR during early remyelination. *Journal of Neurobiology, 37*(3), 413–428.

Redwine, J. M., Blinder, K. L., & Armstrong, R. C. (1997). In situ expression of fibroblast growth factor receptors by oligodendrocyte progenitors and oligodendrocytes in adult mouse central nervous system. *Journal of Neuroscience Research, 50*(2), 229–237.

Ricard, D., Rogemond, V., Charrier, E., Aguera, M., Bagnard, D., Belin, M. F., et al. (2001). Isolation and expression pattern of human Unc-33-like phosphoprotein 6/collapsin response mediator protein 5 (Ulip6/CRMP5): coexistence with Ulip2/CRMP2 in Sema3a- sensitive oligodendrocytes. *Journal of Neuroscience, 21*(18), 7203–7214.

Ricard, D., Stankoff, B., Bagnard, D., Aguera, M., Rogemond, V., Antoine, J. C., et al. (2000). Differential expression of collapsin response mediator proteins (CRMP/ULIP) in subsets of oligodendrocytes in the postnatal rodent brain. *Molecular and Cellular Neurosciences, 16*(4), 324–337.

Schnadelbach, O., Blaschuk, O. W., Symonds, M., Gour, B. J., Doherty, P., & Fawcett, J. W. (2000). N-cadherin influences migration of oligodendrocytes on astrocyte monolayers. *Molecular and Cellular Neurosciences, 15*(3), 288–302.

Smith, K. J. (2007). Sodium channels and multiple sclerosis: roles in symptom production, damage and therapy. *Brain Pathology, 17*(2), 230–242.

Sobel, R. A. (2005). Ephrin A receptors and ligands in lesions and normal-appearing white matter in multiple sclerosis. *Brain Pathology, 15*(1), 35–45.

Spassky, N., de Castro, F., Le Bras, B., Heydon, K., Queraud-LeSaux, F., Bloch-Gallego, E., et al. (2002). Directional guidance of oligodendroglial migration by class 3 semaphorins and netrin-1. *Journal of Neuroscience, 22*(14), 5992–6004.

Stevens, B., Porta, S., Haak, L. L., Gallo, V., & Fields, R. D. (2002). Adenosine: a neuron-glial transmitter promoting myelination in the CNS in response to action potentials. *Neuron*, *36*(5), 855–868.

Stidworthy, M. F., Genoud, S., Li, W. W., Leone, D. P., Mantei, N., Suter, U., et al. (2004). Notch1 and Jagged1 are expressed after CNS demyelination, but are not a major rate-determining factor during remyelination. *Brain*, *127*(Pt. 9), 1928–1941.

Sugimoto, Y., Taniguchi, M., Yagi, T., Akagi, Y., Nojyo, Y., & Tamamaki, N. (2001). Guidance of glial precursor cell migration by secreted cues in the developing optic nerve. *Development*, *128*(17), 3321–3330.

Suzuki, K., Kumanogoh, A., & Kikutani, H. (2008). Semaphorins and their receptors in immune cell interactions. *Nature Immunology*, *9*(1), 17–23.

Tannemaat, M. R., Korecka, J., Ehlert, E. M., Mason, M. R., van Duinen, S. G., Boer, G. J., et al. (2007). Human neuroma contains increased levels of semaphorin 3A, which surrounds nerve fibers and reduces neurite extension in vitro. *Journal of Neuroscience*, *27*(52), 14260–14264.

Tsai, H. H., Frost, E., To, V., Robinson, S., Ffrench-Constant, C., Geertman, R., et al. (2002). The chemokine receptor CXCR2 controls positioning of oligodendrocyte precursors in developing spinal cord by arresting their migration. *Cell*, *110*(3), 373–383.

Tsai, H. H., Macklin, W. B., & Miller, R. H. (2006). Netrin-1 is required for the normal development of spinal cord oligodendrocytes. *Journal of Neuroscience*, *26*(7), 1913–1922.

Tsai, H. H., Tessier-Lavigne, M., & Miller, R. H. (2003). Netrin 1 mediates spinal cord oligodendrocyte precursor dispersal. *Development*, *130*(10), 2095–2105.

Vitry, S., Avellana-Adalid, V., Lachapelle, F., & Evercooren, A. B. (2001). Migration and multipotentiality of PSA-NCAM+ neural precursors transplanted in the developing brain. *Molecular and Cellular Neurosciences*, *17*(6), 983–1000.

Wang, C., Rougon, G., & Kiss, J. Z. (1994). Requirement of polysialic acid for the migration of the O-2A glial progenitor cell from neurohypophyseal explants. *Journal of Neuroscience*, *14*(7), 4446–4457.

Wang, S., Sdrulla, A. D., diSibio, G., Bush, G., Nofziger, D., Hicks, C., et al. (1998). Notch receptor activation inhibits oligodendrocyte differentiation. *Neuron*, *21*(1), 63–75.

Williams, A., Piaton, G., Aigrot, M. S., Belhadi, A., Theaudin, M., Petermann, F., et al. (2007). Semaphorin 3A and 3F: key players in myelin repair in multiple sclerosis? *Brain*, *130*(Pt. 10), 2554–2565.

Wolswijk, G. (1998). Chronic stage multiple sclerosis lesions contain a relatively quiescent population of oligodendrocyte precursor cells. *Journal of Neuroscience*, *18*(2), 601–609.

Yan, H., & Rivkees, S. A. (2002). Hepatocyte growth factor stimulates the proliferation and migration of oligodendrocyte precursor cells. *Journal of Neuroscience Research*, *69*(5), 597–606.

Zhang, H., Vutskits, L., Calaora, V., Durbec, P., & Kiss, J. Z. (2004). A role for the polysialic acid-neural cell adhesion molecule in PDGF-induced chemotaxis of oligodendrocyte precursor cells. *Journal of Cell Science*, *117*(Pt. 1), 93–103.

Zhou, Y., Gunput, R. A., & Pasterkamp, R. J. (2008). Semaphorin signaling: progress made and promises ahead. *Trends in Biochemical Sciences*, *33*(4), 161–170.

CHAPTER 30

# Magnetic resonance techniques to quantify tissue damage, tissue repair, and functional cortical reorganization in multiple sclerosis

M. Filippi[*] and F. Agosta

*Neuroimaging Research Unit, Institute of Experimental Neurology, Division of Neuroscience, Scientific Institute and University Ospedale San Raffaele, Milan, Italy*

**Abstract:** A dramatic paradigm shift is taking place in our understanding of the pathophysiology of multiple sclerosis (MS). An important contribution to such a shift has been made possible by the advances in magnetic resonance imaging (MRI) technology, which allows structural damage to be quantified in the brains of patients with MS and to be followed over the course of the disease.

Modern quantitative MR techniques have reshaped the picture of MS, leading to the definition of the so-called "axonal hypothesis" (i.e., changes in axonal metabolism, morphology, or density are important determinants of functional impairment in MS). Metrics derived from magnetization transfer and diffusion-weighted MRI enable us to quantify the extent of structural changes occurring within T2-visible lesions and normal-appearing tissues (including gray matter), with increased pathological specificity over conventional MRI to irreversible tissue damage; proton MR spectroscopy adds valuable pieces of information on the biochemical nature of such changes. Finally, functional MRI can provide new insights into the role of cortical adaptive changes in limiting the clinical consequences of MS-related irreversible structural damage.

Our current understanding of the pathophysiology of MS is that this is not only a disease of the white matter, characterized by focal inflammatory lesions, but also a disease involving more subtle and diffuse damage throughout the white and gray matter. The inflammatory and neurodegenerative components of the disease process are present from the earliest observable phases of the disease, but appear to be, at least partially, dissociated. In addition, recovery and repair play an important role in the genesis of the clinical manifestations of the disease, involving both structural changes and plastic reorganization of the cortex. This new picture of MS has important implications in the context of treatment options, since it suggests that agents that protect against neurodegeneration or promote tissue repair may have an important role to play alongside agents acting on the inflammatory component of the disease

**Keywords:** magnetic resonance imaging; multiple sclerosis; tissue damage; disability; functional reorganization

---

[*]Corresponding author.
Tel.: +39 02 26433032; Fax: +39 02 26435972;
E-mail: filippi.massimo@hsr.it

## Introduction

Over the past decade, conventional and modern structural magnetic resonance imaging (MRI) techniques have been extensively used to study patients with multiple sclerosis (MS) with the ultimate goal to increase the understanding of the mechanisms responsible for the accumulation of irreversible disability (Rovaris et al., 2005b; De Stefano et al., 2007; Filippi and Agosta, 2007; Rocca and Filippi, 2007). Despite this, the magnitude of the correlation between structural MRI and clinical findings remains suboptimal. Among the reasons for such a discrepancy, the limited ability of conventional MRI (cMRI) to grade the extent of tissue injury as well as the variable effectiveness of reparative and recovery mechanisms following central nervous system (CNS) damage has been suggested to play a role.

In the last 10–15 years, we have witnessed an unprecedented development and application of new strategies to obtain hidden pieces of information from cMRI images (Filippi and Rocca, 2007) as well as of newer MRI-based techniques to quantify the extent and define the nature of focal and diffuse abnormalities associated with MS (Rovaris et al., 2005b; De Stefano et al., 2007; Filippi and Agosta, 2007). cMRI has been used to measure atrophy of the brain tissue (whole or segmented in white matter [WM] and gray matter [GM]) as a tool to grade the extent of neurodegeneration related to MS (Miller et al., 2002).

Magnetization transfer (MT) MRI allows the calculation of an index, the MT ratio (MTR), which when reduced indicates a diminished capacity of the protons bound to the brain tissue matrix to exchange magnetization with the surrounding "free" water, and provides an accurate estimate of the extent of tissue disruption (Filippi and Agosta, 2007). Diffusion tensor (DT) MRI enables the random diffusional motion of water molecules to be measured, thus providing metrics, such as mean diffusivity (MD) and fractional anisotropy (FA), which provide estimates of the size and geometry of water-filled spaces (Rovaris et al., 2005b). Proton MR spectroscopy ($^1$H-MRS) can complement structural MRI in the assessment of patients with MS by simultaneously defining several chemical correlates of the pathological changes occurring in the brain (De Stefano et al., 2007). Water-suppressed, proton MR spectra of the human brain at long echo times reveal four major resonances from (a) choline-containing phospholipids (Cho), (b) creatine and phosphocreatine (Cr), (c) N-acetyl-aspartate (NAA), and (d) lactate (Lac) methyl group. NAA is a marker of axonal integrity, whereas Cho and Lac are considered as chemical correlates of acute inflammatory or demyelinating changes. $^1$H-MRS studies with shorter echo times can detect additional metabolites, such as lipids and myoinositol (mI), which are also regarded as markers of ongoing myelin damage. More recently, functional MRI (fMRI) has also been used to study MS in an attempt to measure the ability of the MS brain to respond to tissue injury (Rocca and Filippi, 2007). fMRI is a noninvasive technique to define abnormal patterns of brain activation caused by injury or disease. The fMRI signal depends on blood oxygenation level-dependent (BOLD) alterations in the deoxyhemoglobin concentration associated with the increased metabolic activity in activated neural tissue, which in turn alters the transverse magnetization relaxation time.

This chapter summarizes the main results obtained from the use of conventional and modern quantitative MR-based techniques (namely, MT MRI, DT MRI, and $^1$H-MRS) for the assessment of WM and GM pathology in patients with MS. It also reviews the main contributions to the understanding of MS pathobiology gained by the use of fMRI with different paradigms of stimulation.

## Imaging brain atrophy

Atrophy may occur very early in MS, even at the clinically isolated syndrome (CIS) stage (Dalton et al., 2002; Filippi et al., 2004b). Brain atrophy is also present in the early stages of primary progressive MS (PPMS), affecting both WM and GM (Sastre-Garriga et al., 2004). Moreover, cross-sectional and longitudinal studies have shown a moderate correlation between brain atrophy and the clinical manifestations of MS (for review, see Miller et al., 2002).

A recent large-scale study of WM and GM atrophies involving 597 patients and 104 controls (Tedeschi et al., 2005) reported that global WM and GM fractions were reduced in patients compared to controls. In addition, both WM and GM atrophy were significantly more severe in secondary progressive MS (SPMS) than in relapsing-remitting MS (RRMS) patients. WM and GM atrophies were also significantly related to the Expanded Disability Status Scale (EDSS) score and age at onset, suggesting that the younger the age at disease onset, the worse the brain atrophy. In this study, GM atrophy was found to be the most significant MRI-derived variable associated with clinical disability. Numerous pieces of evidence indicate that GM volume decrease may well be a correlate of MS-related impairment (De Stefano et al., 2003; Amato et al., 2004; Sanfilipo et al., 2005). A study by Sanfilipo et al. (2005) showed that MS patients had lower GM and total parenchymal volumes, but that their WM volume was not significantly different from that of controls. GM atrophy was related to clinical status as well as T1 and T2 lesion loads.

Several studies have also shown that neuropsychological dysfunction is more closely associated with measures of brain atrophy than with those of the lesion burden (Zivadinov et al., 2001; Benedict et al., 2004). Changes of brain parenchymal volume have been shown to predict cognitive impairment over two years in patients with early RRMS (Zivadinov et al., 2001). In addition, in patients with early RRMS, cortical atrophy was found only in cognitively impaired patients and was significantly correlated with a poorer performance on tests of verbal memory, attention/concentration, and verbal fluency (Amato et al., 2004). Recent studies suggest that regional atrophy measurement may be more important than macroscopic lesion burden and whole brain atrophy in predicting selective patterns of MS-associated cognitive dysfunction (Locatelli et al., 2004; Benedict et al., 2005).

In MS, brain atrophy develops in different structures in the different clinical disease phenotypes (Pagani et al., 2005b). Thus, ventricular enlargement is predominant in RRMS, whereas cortical atrophy seems to be more important in the progressive forms of the disease. Sophisticated atrophy measurement techniques have also been used to improve the correlations between measures of GM loss and clinical findings (Sailer et al., 2003). For example, Sailer et al. (2003) evaluated cortical thickness in the entire brain, and its relationship with disability, disease duration, T2-hyperintense, and T1-hypointense lesion volumes in a group of 20 patients with MS, and found that the mean cortical thickness was reduced in MS patients compared with controls. In addition, patients with long-standing disease or severe disability had focal thinning of the primary motor cortex area. Recently, using voxel-based morphometry, GM loss in the thalamus has been demonstrated not only in patients with PPMS (Sepulcre et al., 2006) but also in those with pediatric MS (Mesaros et al., 2008).

Finally, thanks to recent advances in image postprocessing techniques, the measurement of atrophy in selected pathways, such as the corpus callosum and the corticospinal tract (CST) (Pagani et al., 2005a), may also represent a novel strategy to improve the in vivo monitoring of MS damage in clinically eloquent WM regions.

## Intrinsic lesion damage

Although cMRI has a great sensitivity in detecting the presence and extent of macroscopic lesions in MS, it lacks specificity toward the heterogeneous pathological substrates of these lesions, which range from edema to demyelination, remyelination, gliosis, and axonal loss, as demonstrated by several pathological studies (Allen and McKeown, 1979; Ferguson et al., 1997; Trapp et al., 1998; Evangelou et al., 2000).

In chronic lesions that appear hyperintense on T2-weighted scans, quantitative MRI studies have shown variable degrees of MTR, FA, and NAA reductions, and MD increase (De Stefano et al., 1995; Filippi et al., 1999; Filippi and Inglese, 2001). All these values vary dramatically across individual lesions, but are typically more pronounced in lesions that are hypointense on T1-weighted images and in patients with the most disabling courses of the disease (Falini et al., 1998; Filippi et al., 1999; van Walderveen et al., 1999; Filippi and Inglese, 2001). The variability of MTR, MD, FA, and NAA values seen in MS lesions also suggests that different proportions of lesions with different degrees of structural changes might contribute to the evolution of the disease. This concept is supported by a three-year follow-up study (Rocca et al., 1999) showing that newly formed lesions from SPMS patients have more severe MTR deterioration than those from mildly disabling RRMS patients.

New enhancing lesions have different MTR values, according to their size, modality, and duration of enhancement. In particular, MTR is higher in homogeneously enhancing lesions than in ring-enhancing lesions (Silver et al., 1998); in lesions enhancing on a single scan than in those enhancing on two or more serial scans (Filippi et al., 1998a); and in lesions enhancing after the injection of a triple dose of gadolinium than in those enhancing after the injection of a standard dose (Filippi et al., 1998c). DT MRI characteristics of enhancing lesions are less well defined. While FA values are consistently lower in enhancing than in nonenhancing lesions (Werring et al., 1999; Rovaris et al., 2005b), conflicting results have been achieved when comparing MD between these two lesion populations. While some studies reported higher MD values in nonenhancing than in enhancing lesions (Werring et al., 1999; Droogan et al., 1999), others, based on larger samples of patients and lesions, did not report any significant difference between these two lesion populations (Filippi et al., 2000b). The heterogeneity of enhancing lesions has also been underlined by the demonstration that water diffusivity is markedly increased in ring-enhancing lesions when compared to homogeneously enhancing lesions or in the nonenhancing portions of enhancing lesions when compared with enhancing portions (Roychowdhury et al., 2000).

$^1$H-MRS of acute MS lesions at both short and long echo times reveals increases in Cho and Lac resonance intensities (Davie et al., 1994; De Stefano et al., 1995). In large, acute demyelinating lesions, decreases in Cr can also be seen (De Stefano et al., 1995). Short-echo time spectra can detect transient increases in visible lipids, released during myelin breakdown and mI (Narayana et al., 1998). All these changes are usually associated with a decrease in NAA. After the acute phase and over a period of days to weeks, there is a progressive reduction of raised Lac resonance intensities to normal levels. Resonance intensities of Cr also return to normal within a few days. Cho, lipid, and mI resonance intensities return to normal over months. The signal intensity of NAA may remain decreased or show partial recovery, starting soon after the acute phase and lasting for several months (Davie et al., 1994; De Stefano et al., 1995).

## Imaging "diffuse" normal-appearing brain tissue (NABT) and normal-appearing WM (NAWM) damage

A reduction of MTR values is already detectable in the normal-appearing WM (NAWM) prior to T2-visible lesion appearance (Filippi et al., 1998b; Laule et al., 2003), in NAWM areas adjacent to focal T2-weighted lesions, particularly in progressive MS patients (Filippi et al., 1995), and in patients with MS and no T2-visible WM abnormalities (Filippi et al., 1999). Normal-appearing brain tissue (NABT) MTR histogram-derived measures are different and evolve at a different pace in the major MS clinical phenotypes (Filippi et al., 2000c; Tortorella et al., 2000). A multivariate analysis of several cMRI- and MT MRI-derived variables found that average NABT MTR can be more strongly associated with cognitive impairment in MS patients than the extent of T2-visible lesions and their intrinsic tissue damage (Filippi et al., 2000d). Reduced NABT MTR has also been found in asymptomatic relatives of patients with MS and in patients at presentation with CIS (Miller et al., 2005). Although not confirmed by other studies, the extent of NABT changes in patients at presentation with CIS has been found to be an independent predictor of subsequent evolution to clinically definite MS (Miller et al., 2005). A significant prognostic value of NAWM MTR abnormalities for the medium-term evolution of disability in patients with established MS has also been independently reported by longitudinal studies (Santos et al., 2002; Khaleeli et al., 2007).

In recent studies, voxel-based analysis has been applied to brain MTR maps to better assess the severity of tissue damage in specific WM regions (Audoin et al., 2004; Ranjeva et al., 2005). Patients with CIS showed significantly lower MTR values than healthy controls in multiple WM regions, including the corpus callosum, the occipitofrontal fascicles, the external capsule, and the optic radiations (Audoin et al., 2004; Ranjeva et al., 2005). Significant correlations between regional MTR values and the Multiple Sclerosis Functional Composite (MSFC) scores were found for the right superior longitudinal fasciculus, the right frontal WM, the splenium, and the genu of the corpus callosum. MTR values in the right superior longitudinal fasciculus and in the splenium of the corpus callosum were also correlated with the patients' performance at the paced auditory serial addition task (PASAT) test (Ranjeva et al., 2005). These results suggest a potential role for voxel-based analysis of brain MTR data to achieve a better understanding of the relationship between the location of NAWM damage and its functional impact in patients with MS.

Similarly to what was found using MT MRI, numerous DT MRI studies have consistently shown the presence of diffusion abnormalities in the NAWM of patients with MS, which can precede the development of T2-visible lesions by several weeks (for a review, see Rovaris et al., 2005b). DT MRI studies showed that NAWM abnormalities in MS are widespread, but tend to be more severe in sites where MRI-visible lesions are usually located and in periplaque regions. A recent study, using histogram-based analysis of segmented diffusion maps of the brain, reported a significant increase in MD and decrease in FA values in CIS patients with paraclinical evidence of disease dissemination in space, although the changes did not predict the short-term occurrence of cMRI disease activity (Gallo et al., 2005). Cross-sectional studies did not find significant correlations between the average NAWM diffusivity or anisotropy values and the severity of MS neurological disability, as assessed by the EDSS score, nor significant differences between MS clinical phenotypes as regards the severity of overall DT MRI changes in the NAWM (for a review, see Rovaris et al., 2005b). However, when the DT MRI characteristics of clinically eloquent NAWM regions were studied, a significant relationship with patients' EDSS scores was found for the corpus callosum and internal capsule FA values (Filippi et al., 2001) and for the cerebral peduncle MD and FA values (Ciccarelli et al., 2001). In addition, in a preliminary DT MRI study of cognitive impairment in patients with RRMS (Rovaris et al., 2002b), moderate correlations were found between several NAWM histogram-derived quantities and neuropsychological test scores. A different spatial distribution of NAWM damage between patients with benign MS (BMS) and those with RRMS has been demonstrated, which was in apparent contrast with the between-group similarity of the overall extent of WM structural changes (Ceccarelli et al., 2008).

DT MRI tractography is a promising technique for in vivo segmentation of the major WM tract fiber bundles in the brain WM (Mori et al., 2002). In MS patients, diffusivity and anisotropy along the CST correlate with clinical outcome measures of locomotor disability, such as the EDSS score or the pyramidal functional system score at this scale, more than T2 lesion burden and the overall brain extent of diffusivity changes of the brain (Wilson et al., 2003; Lin et al., 2005). Moreover, a DT MRI tractography study of patients with CIS and motor impairment showed that these patients had increased MD and T2 lesion volume in the CST compared to patients without pyramidal motor symptoms (Pagani et al., 2005a). In MS patients, apparent diffusion coefficient (ADC) values of the corpus callosum are associated with the level of cognitive performance (Lin et al., 2005). A DT MRI tractography study in patients with optic neuritis showed reduced connectivity values in both left and right optic radiations compared with controls, suggesting the occurrence of mechanisms of trans-synaptic degeneration secondary to optic nerve (ON) damage (Ciccarelli et al., 2005). DT MRI tractography also provided a method to identify NAWM fibers at risk for degeneration because they intersect focal T2-visible lesions (Simon et al., 2006).

Regional increases in the concentrations of Cho and mobile lipids, as well as decreases in NAA concentrations, can precede the appearance of cMRI-visible WM lesions by several months (De Stefano et al., 2007). For several months after the acute stage of WM lesions, the signal intensity of NAA may remain decreased or just show a partial recovery. Chronic lesions have higher NAA signal intensities in patients with BMS than those in patients with SPMS, indicating a more efficient tissue integrity recovery in patients with less severe disability (De Stefano et al., 2007). Decreases in NAA concentrations are also well-known to occur in the NAWM of MS patients (Caramanos et al., 2005). Reversible changes in NAA concentration in the NAWM of the hemisphere contralateral to large, acute demyelinating lesions have been shown (De Stefano et al., 1999), suggesting that part of these diffuse changes in the NAWM may be related to a sublethal damage of axons passing through inflammatory lesions. In the NAWM of patients with CIS (Fernando et al., 2004) and established MS (Vrenken et al., 2005), an increase in mI levels has also been detected, confirming the hypothesis that increased glial cell activity is an important feature of "occult" WM damage together with axonal damage. A significant correlation of $^1$H-MRS findings with neurological disability or selective motor impairment has been shown by several studies (De Stefano et al., 2007), but these findings have not been confirmed by others (Vrenken et al., 2005). In a recent study (Sastre-Garriga et al., 2005), the EDSS score was found to correlate with mI and glutamate-glutamine concentrations in the NAWM, suggesting that $^1$H-MRS markers of NAWM pathological features other than axonal damage may also have a functional relevance in MS.

## Imaging "diffuse" GM damage

Although MS has been classically regarded as a WM disease of the CNS, pathological studies (Kidd et al., 1999; Peterson et al., 2001) have shown the presence of MS-related damage in the GM of MS patients. Recently, a multislab 3-D double inversion recovery (DIR) sequence has been developed to improve the detection of GM lesions in MS (Geurts et al., 2005). As a result of the increased contrast between GM and WM, DIR images were able to detect more intracortical lesions than standard MRI techniques (Geurts et al., 2005). Higher contrast between lesion and its surroundings also resulted in an improved distinction between juxta-cortical and mixed WM–GM lesions on DIR images (Geurts et al., 2005). In a large cohort of MS patients, intracortical lesions were detected in 58% of cases: 36% of patients with CIS, 64% of patients with RRMS, and 73% of patients with SPMS (Calabrese et al., 2007).

It is likely that GM damage in MS is not limited to focal lesions, but might also cause "diffuse" tissue changes, for instance, through retrograde and trans-synaptic degeneration. Modern quantitative MR-based techniques have the potential to provide accurate estimates of "overall" (focal, and "diffuse") GM abnormalities, and might therefore contribute to a more complete picture of GM damage associated with MS.

Several studies have demonstrated reduced MTR values in the brain GM (which was considered to be normal on conventional MR scans) from patients with different MS phenotypes (Cercignani et al., 2001; Ge et al., 2001; Dehmeshki et al., 2003), including those at the earliest clinical stage of MS (Fernando et al., 2005; Davies et al., 2005a; Ramio-Torrenta et al., 2006). GM abnormalities increase with disease duration, since they were found to be more pronounced in patients with PPMS or SPMS (Rovaris et al., 2001). In a recent, large, multicenter study of PPMS patients, greater MT MRI-detectable GM damage was found in patients who required walking aids than in those who did not (Rovaris et al., 2008). The distribution of MTR abnormalities in the various brain GM structures still needs to be defined. In a voxel-based MTR study (Audoin et al., 2005), a regional pattern of brain MTR decrease, more evident in the basal ganglia, was found in patients with early

MS. On the contrary, in another study (Davies et al., 2005b) that assessed thalamic MTR in early RRMS patients, no significant difference was observed at baseline between patients and controls. After one or two years, however, the mean thalamic MTR became significantly lower in patients.

GM MTR changes have been found to correlate with clinical disability (Ge et al., 2001; Dehmeshki et al., 2003; Ramio-Torrenta et al., 2006; Oreja-Guevara et al., 2006) and cognitive impairment (Rovaris et al., 2000b; Ranjeva et al., 2005), whereas no correlation with fatigue emerged (Codella et al., 2002). A voxel-based MTR study (Ranjeva et al., 2005) showed that regional MTR values of several cortical areas are correlated significantly with the MSFC and PASAT scores. GM MTR was also found to be an independent predictor of subsequent accumulation of disability in patients with MS followed up for eight years (Agosta et al., 2006).

In line with MT MRI findings, DT MRI confirmed the presence of GM damage in MS (Cercignani et al., 2001; Bozzali et al., 2002; Rovaris et al., 2002b; Oreja-Guevara et al., 2005; Pulizzi et al., 2007) and showed that the extent of such damage differs among the various disease phenotypes, being more severe in patients with SPMS (Cercignani et al., 2001; Bozzali et al., 2002; Pulizzi et al., 2007). An increased diffusivity in the thalami of MS patients has also been found, which was again more pronounced in SPMS than in RRMS patients (Fabiano et al., 2003). More intriguingly, DT MRI has been shown to be sensitive to the evolution of MS damage over short time periods. Longitudinal studies (Rovaris et al., 2002a, 2005a, 2006; Oreja-Guevara et al., 2005) have demonstrated a worsening of GM damage over time in patients with RRMS (Oreja-Guevara et al., 2005), SPMS, and PPMS (Rovaris et al., 2002a, 2006). Several studies suggested that DT MRI might be more sensitive to the accrual of GM damage than to that of NAWM (Rovaris et al., 2005a). A moderate correlation between MD of the GM and the degree of cognitive impairment has been detected in mildly disabled RRMS patients (Rovaris et al., 2002b). GM diffusivity was also found to predict the accumulation of disability over a five-year period in patients with PPMS (Rovaris et al., 2006).

Using $^1$H-MRS, several studies have found metabolite abnormalities, including reduced concentrations of NAA and Cho, and increased concentrations of mI, in the cortical GM (Sharma et al., 2001; Chard et al., 2002) and subcortical tissue (Adalsteinsson et al., 2003; Cifelli et al., 2002; Inglese et al., 2004) from MS patients. This was shown to occur also in early RRMS (Sharma et al., 2001), and in patients with CIS suggestive of MS (Kapeller et al., 2002). This disagrees, at least partially, with the results of another study (Sijens et al., 2006) where significant decreases in Cho, creatine, and NAA concentrations were found in the GM of patients with the progressive forms of the disease, but not in the GM of those with RRMS. NAA reduction has also been demonstrated in the thalamus of SPMS (Adalsteinsson et al., 2003) and RRMS patients (Cifelli et al., 2002; Inglese et al., 2004). In addition, a reduced concentration of glutamate and glutamine in the cortical GM of patients with PPMS has also been measured (Sastre-Garriga et al., 2005), which was significantly correlated with the EDSS score.

**Cortical reorganization**

Brain plasticity is a well-known feature of the human brain, which is likely to have several different substrates (including increased axonal expression of sodium channels, synaptic changes, increased recruitment of parallel existing pathways or "latent" connections, and reorganization of distant sites), and which might have a major adaptive role in limiting the functional consequences of axonal loss in MS.

An altered brain pattern of movement-associated cortical activations, characterized by an increased recruitment of the contralateral primary sensorimotor cortex (SMC) during the performance of simple tasks (Rocca et al., 2003a, b, c; Filippi et al., 2004a) and by the recruitment of additional "classical" and "higher-order" sensorimotor areas during the performance of more complex tasks (Filippi et al., 2004a), has been demonstrated in patients with CIS suggestive of

MS. The clinical and cMRI follow-up of these patients has shown that, at disease onset, CIS patients with a subsequent evolution to clinically definite MS tend to recruit a more widespread sensorimotor network than those without short-term disease evolution (Rocca et al., 2005b).

An increased recruitment of several sensorimotor areas, mainly located in the cerebral hemisphere ipsilateral to the limb that performed the task, has also been demonstrated in patients with early RRMS and a previous episode of hemiparesis (Pantano et al., 2002a). In patients with similar characteristics, but who presented with an episode of optic neuritis, this increased recruitment involved sensorimotor areas that were mainly located in the contralateral cerebral hemisphere (Pantano et al., 2002b).

In patients with established MS and an RR course, functional cortical changes, mainly characterized by an increased recruitment of "classical" motor areas, including the primary SMC, the supplementary motor area (SMA), and the secondary sensorimotor cortex (SII), have been shown during the performance of simple motor (Rocca et al., 2002a) and visuo-motor integration tasks (Cerasa et al., 2006). Movement-associated cortical changes, characterized by the activation of highly specialized cortical areas, have also been described in patients with SPMS (Rocca et al., 2003a) and in patients with PPMS (Rocca et al., 2002b; Filippi et al., 2002).

The concept that movement-associated cortical reorganization varies across patients at different stages of the disease has been shown by an fMRI study of patients with different disease phenotypes (Rocca et al., 2005a). The study suggested that early in the course of the disease, more areas typically devoted to motor tasks (such as the primary SMC) are recruited, and then a bilateral activation of these regions is seen, and late in the course of the disease, areas that healthy people recruit to perform novel or complex tasks are activated, perhaps in an attempt to limit the functional consequences of accumulating tissue damage.

Recent fMRI studies have suggested that functional cortical changes might have an adaptive role also in limiting MS-related cognitive impairment (Audoin et al., 2003, 2005; Staffen et al., 2002; Mainero et al., 2004; Hillary et al., 2003; Wishart et al., 2004; Li et al., 2004; Cader et al., 2006; Rocca et al., 2009). Several cognitive domains have been investigated in MS patients with fMRI. Working memory has been the most extensively studied by means of the PASAT or the paced visual serial addition task (PVSAT) (which also involves sustained attention, information processing speed, and simple calculation), the n-back task, or a task adapted from the Sternberg paradigm. Additional cognitive domains including attention and planning have been interrogated, too.

In CIS patients, an altered pattern of cortical activations has been described during the performance of the PASAT (Audoin et al., 2003, 2005). During the performance of the PVSAT, RRMS patients with intact task performance had an increased activation of several regions located in the frontal and parietal lobes, bilaterally, compared to healthy volunteers, suggesting the presence of functional compensatory mechanisms (Staffen et al., 2002). An increased recruitment of several cortical areas during the performance of a simple cognitive task has also been shown in patients with RRMS and mild clinical disability (Mainero et al., 2004). An increased activation of regions exclusively located in the right cerebral hemisphere (mainly in the frontal and temporal lobes) has also been found in MS patients when testing rehearsal within working memory (Hillary et al., 2003). The degree of right hemisphere recruitment was strongly related to patient neuropsychological performance. In patients with RRMS and no cognitive deficits, during an n-back task, a reduced activation of the "core" areas of the working memory circuitry (including prefrontal and parietal regions) and an increased activation of other regions within and beyond the typical working memory circuitry (including areas in the frontal, parietal, temporal, and occipital lobes) have been found (Wishart et al., 2004). This shift of activation was most prominent with increased working memory demands. These findings suggest that, as shown for motor and visual tasks, dynamic cognitive-associated changes of brain activation patterns can occur in RRMS patients. Other studies that also investigated working memory performance in MS patients demonstrated (1) an increased recruitment of regions related to sensorimotor functions and anterior attentional/executive components of the working memory system in patients compared to healthy controls and (2) a reduced recruitment of several regions in the right cerebellar hemisphere in patients compared with healthy individuals (Li et al., 2004), thus suggesting that the cerebellum might play a role in the working memory impairment of MS.

Recently, a study in RRMS patients (Cader et al., 2006) investigated working memory with an n-back task and functional connectivity analysis. Compared to controls, patients had relatively reduced activations of the superior frontal and anterior cingulate gyri. Patients also showed a variable but substantially smaller increase in activation than healthy controls with greater task complexity, suggesting a reduced functional reserve for cognition relevant to memory in MS patients. The functional connectivity analysis revealed increased correlations between right dorsolateral prefrontal and superior frontal/anterior cingulate activations in controls, and increased correlations between activations in the right and left prefrontal cortices in patients, indicating that altered interhemispheric interactions between dorsal and lateral prefrontal regions may yet be an additional adaptive mechanism distinct from recruitment of novel processing regions.

More significant activations of several areas of the cognitive network involved in the performance of the Stroop test have also been demonstrated in a group of 15 cognitively preserved patients with BMS when compared to 19 healthy controls (Rocca et al., 2009). BMS patients also showed an increased connectivity of several cortical areas of the sensorimotor network with the right inferior frontal gyrus (IFG) and the right cerebellum, as well as a decreased connectivity between some areas and the anterior cingulate cortex. These results suggest an altered interhemispheric balance in favor of the right hemisphere in BMS patients in comparison with healthy controls, when performing cognitive tasks.

## Imaging the spinal cord

The cord is a clinically eloquent region whose damage has the potential to significantly influence the functional outcome of MS. MRI-detectable lesions can be found in up to approximately 90% of patients with established disease and are more frequently located in the cervical and the thoracic cord (Lycklama à Nijeholt et al., 2003; Agosta and Filippi, 2007). Cord lesions are usually located peripherically, rarely exceed two vertebral segments in length, and occupy less than half of the cord cross-sectional area. Acute lesions are often associated with cord swelling, whereas chronic lesions are isointense on T1-weighted images. Enhancing lesions are less frequently seen in the cord than in the brain. Cord imaging can be particularly helpful in the diagnosis of MS since cord lesions are now accepted to meet the international panel criteria for disease dissemination in space (Agosta and Filippi, 2007). It is also useful in those cases where brain MRI is normal or equivocal and in patients more than 50 years old or with nonspecific T2 abnormalities of the brain because, contrary to what happens for the brain, cord lesions develop rarely with aging per se (Agosta et al., 2007b).

Significant reduction of cervical cord size can be observed in the early phase of MS (Brex et al., 2001). The severity of cord atrophy is, however, more pronounced in the progressive forms of MS (Rovaris et al., 2001). Progressive cord and brain atrophies have been observed over a five-year period in PPMS, but the lack of correlation between the two suggests that independent processes may be contributing to progressive tissue loss in the two regions (Ingle et al., 2003).

In addition to atrophy measurements, reliable MT MRI can be obtained from the spinal cord. The use of MTR histogram analysis has allowed to obtain a more global picture of cord pathology in patients with MS (Agosta and Filippi, 2007). Histogram analysis has demonstrated that cord MTR histogram metrics in patients with CIS, RRMS, and early onset MS are similar to those of healthy individuals (Mezzapesa et al., 2004; Rovaris et al., 2004). On the contrary, cord MTR metrics are markedly reduced in patients with SPMS and PPMS in a similar manner (Bozzali et al., 1999;

Filippi et al., 2000a; Rovaris et al., 2001). The average cervical cord MTR is lower in MS patients with locomotor disability than in those without (Bozzali et al., 1999). In PPMS, a model including cervical cord area and MTR histogram peak height was significantly, albeit modestly, associated with the level of disability (Filippi et al., 2000a). In MS, only a moderate correlation has been found between the average brain MTR and cervical cord MTR, suggesting that cervical cord damage in MS is not a mere reflection of brain pathology (Rovaris et al., 2000a). Interestingly, the extent of cervical cord damage (measured using MT MRI) has been found to be strictly associated with the extent of movement-associated cortical activations (measured using fMRI) in patients with cord demyelination (Rocca et al., 2003c, 2006).

With increasing technical advances, it has also become possible to study cord MS pathology using diffusion-weighted MRI. We used histogram analysis to assess water molecular diffusivity of the cervical cord from patients with RRMS or SPMS and found reduced average cord FA in MS patients compared to controls (Valsasina et al., 2005). In MS, the reduction of cord FA was correlated with the degree of disability. Altered MD and FA cord histogram-derived metrics have also been found in patients with PPMS (Agosta et al., 2005). Conventional and DT MRI of the cervical cord was recently obtained from 42 MS patients at baseline and after a mean follow-up of 2.4 years (Agosta et al., 2007a). In MS patients, the cervical cord cross-sectional area and FA decreased, and cervical cord MD increased during follow-up. The baseline cord cross-sectional area and FA correlated with increase in disability at follow-up.

Pathological studies demonstrated that extensive demyelination can occur in the spinal cord GM of MS patients (Lycklama à Nijeholt et al., 2001), while a recent postmortem study showed that spinal cord atrophy in MS is almost exclusively secondary to WM volume loss (Gilmore et al., 2005). Consistently with pathological data, $^1$H-MRS studies have recently described a reduction of NAA in the cervical cord of MS patients (Kendi et al., 2004; Ciccarelli et al., 2007). Moreover, patients with RRMS demonstrated lower cervical cord GM average MTR compared to healthy controls (Agosta et al., 2007c). Interestingly, GM average MTR also correlated with the degree of disability, thus suggesting that cervical cord GM is not spared by MS pathology and such a damage is an additional factor contributing to the disability of these patients.

fMRI has been recently applied to the assessment of functional changes in the spinal cord in MS. Among the reasons for the paucity of fMRI investigations in the human spinal cord is the significant technical challenge associated with the detection and measurement of activation changes in such a small structure, subjected to periodic movements. Although BOLD fMRI of the spinal cord has provided reliable results in healthy subjects, several studies put forward the concept that the exploitation of the so-called signal enhancement by extravascular protons (SEEP) effect rather than the "classical" BOLD effect might be more suitable for the assessment of spinal cord activity (Stroman, 2005). While the BOLD effect depends on the local MR signal change due to variation in blood oxygenation, the SEEP effect is proposed to arise from a local change in fluid balance that may result from changes in perfusion pressure, production of extracellular fluid, cellular swelling, and maintenance of ion and neurotransmitter concentrations at sites of neuronal activity. Following a tactile stimulation of the right palm, a significant task-related mean signal change in the entire cervical cord has been detected in 12 right-handed healthy subjects (Agosta et al., 2009). In this study, cord activity was higher in the right than in the left cervical cord, and a significant heterogeneity in frequency of fMRI activity between cord levels was also observed, with the highest frequencies of fMRI activity detected at C6 and C7.

Two studies have interrogated cervical cord neuronal activity during a proprioceptive (Agosta et al., 2008a) and a tactile (Agosta et al., 2008a) stimulation of the right upper limb in patients with relapsing MS. In the first study (Agosta et al., 2008b), MS patients had a higher average fMRI signal change of the overall cord, as well as higher average fMRI signal changes in the anterior section of the right cord at C5 and the left cord at C5–C6 in comparison to controls. In MS patients, the overall cord average signal change correlated significantly with quantitative MRI measures of cord and brain tissue damage, suggesting an adaptive role of such an abnormal recruitment. In the second study (Agosta et al., 2008b), MS patients and healthy controls were scanned during a tactile stimulation of the right palm. MS patients had a 20% higher cord fMRI signal change than healthy controls. In addition, MS patients also showed a different topographical distribution of fMRI signal changes at the level of the different portions of the cervical cord in comparison with controls, mainly characterized by an activation of regions located in the anterior and left portions of the cord. Such an over-recruitment of the ipsilateral posterior cervical cord associated with a reduced functional lateralization suggests an abnormal function of the spinal relay interneurons in MS patients.

## Imaging the optic nerve

MRI of the ON presents a number of technical difficulties (Vinogradov et al., 2005). First, the ON is small and requires high spatial resolution. Second, the fat, bone, and CSF that surround the ON can produce artifacts that degrade image quality. However, with the development of dedicated MR coils and fast imaging techniques, the study of this structure has been considerably improved. In a study of 17 patients with an initial episode of unilateral optic neuritis, using short-echo fast fluid-attenuated inversion recovery, the mean cross-sectional area of the intraorbital portion of the ON was lower in diseased eyes than in fellow eyes and in healthy controls' eyes (Hickman et al., 2001). However, there was no correlation of atrophy with final visual function, which was good in most patients. By contrast, a subsequent study in patients with more severe visual deficits showed that visual acuity was associated with the degree of ON atrophy (Hickman et al., 2002). Similar findings have also been reported by others (Inglese et al., 2002; Hickman et al., 2004b). The lack of an association between early atrophy and subsequent visual outcome may be due to the redundancy of function and plasticity (Werring et al., 2000). However, as atrophy evolves over time, the subsequent loss of tissue may have functional importance.

MTR can be measured in the ON (Thorpe et al., 1995; Inglese et al., 2002; Hickman et al., 2004a; Melzi et al., 2007). MTR of the ON correlates with the visual-evoked potentials (VEP) P100 latency (Thorpe et al., 1995) and with the degree of visual function recovery after an acute episode of optic neuritis (Inglese et al., 2002). It may also be used to study longitudinally the ON, although, surprisingly, the time course of the MTR change (a progressive decline reaching a nadir after eight months) does not seem to match that of the clinical recovery (Hickman et al., 2004a). More recently, 11 patients with a first episode of acute optic neuritis were evaluated, using conventional and MT MRI at baseline and after 3 and 12 months. At the onset of acute optic neuritis, MTR values in the affected ON were significantly higher than those of the healthy ON, thus suggesting the presence of inflammatory cellular infiltrates due to the breakdown of the blood–ON barrier. Then, during follow-up, MTR values of affected ON progressively decreased over time, without a subsequent increase, suggesting a progression of ON damage despite the early visual recovery.

More recently, full DT measurements from the ON have been obtained using high-resolution fat- and CSF-suppressed zonal oblique multisection echoplanar imaging sequences (Hickman et al., 2005; Trip et al., 2006). For example, one study found that the mean ON ADC was significantly higher following optic neuritis than in either the contralateral or healthy control eyes (Hickman et al., 2005). Moreover, the ADC was strongly correlated with both visual acuity and VEP parameters.

## Conclusions

The extensive application of modern MR-based techniques to the assessment of brain and cord pathology in patients with MS has considerably improved our understanding of MS pathophysiology and has provided new objective metrics that might be useful to monitor disease evolution, either in natural history studies or in treatment trials. From the large body of available literature, it is clear, however, that none of these quantitative techniques, taken in isolation, is able to provide a complete picture of the complexity of the MS process. There are several pieces of evidence indicating that a multiparametric approach, combining aggregates of different MR quantities, might improve our ability to monitor the disease. Such an approach should include not only the assessment of brain damage but also that of cord pathology, as suggested by a recent study in which, putting together brain and cord measures reflecting the severity of MS-related abnormalities, it was possible to explain approximately 50% of the variance of patients' disability. Finally, in the evaluation of the relationship between clinical and MRI markers of disease severity and evolution, the presence and efficacy of functional cortical changes should also be considered.

## References

Adalsteinsson, E., Langer-Gould, A., Homer, R. J., Rao, A., Sullivan, E. V., Lima, C. A., et al. (2003). Gray matter N-acetyl aspartate deficits in secondary progressive but not relapsing-remitting multiple sclerosis. *American Journal of Neuroradiology*, 24, 1941–1945.

Agosta, F., Absinta, M., Sormani, M. P., Ghezzi, A., Bertolotto, A., Montanari, E., et al. (2007a). In vivo assessment of cervical cord damage in MS patients: a longitudinal diffusion tensor MRI study. *Brain*, 130, 2211–2219.

Agosta, F., Benedetti, B., Rocca, M. A., Valsasina, P., Rovaris, M., Comi, G., et al. (2005). Quantification of cervical cord pathology in primary progressive MS using diffusion tensor MRI. *Neurology*, 64, 631–635.

Agosta, F., & Filippi, M. (2007). MRI of spinal cord in multiple sclerosis. *Journal of Neuroimaging*, 17(Suppl. 1), 46S–49S.

Agosta, F., Laganà, M., Valsasina, P., Sala, S., Dall'Occhio, L., Sormani, M. P., et al. (2007b). Evidence for cervical cord tissue disorganisation with aging by diffusion tensor MRI. *NeuroImage*, 36, 728–735.

Agosta, F., Pagani, E., Caputo, D., & Filippi, M. (2007c). Associations between cervical cord gray matter damage and disability in patients with multiple sclerosis. *Archives of Neurology*, 64, 1302–1305.

Agosta, F., Rovaris, M., Pagani, E., Sormani, M. P., Comi, G., & Filippi, M. (2006). Magnetization transfer MRI metrics predict the accumulation of disability 8 years later in patients with multiple sclerosis. *Brain*, 129, 2620–2627.

Agosta, F., Valsasina, P., Caputo, D., Rocca, M. A., & Filippi, M. (2009). Tactile-associated fMRI recruitment of the cervical cord in healthy subjects. *Human Brain Mapping*, 30, 340–345.

Agosta, F., Valsasina, P., Caputo, D., Stroman, P. W., & Filippi, M. (2008a). Tactile-associated recruitment of cervical cord is altered in patients with multiple sclerosis. *NeuroImage*, 39, 1542–1548.

Agosta, F., Valsasina, P., Rocca, M. A., Caputo, D., Sala, S., Judica, E., et al. (2008b). Evidence for enhanced functional recruitment of cervical cord in relapsing multiple sclerosis. *Magnetic Resonance in Medicine*, 59, 1035–1042.

Allen, I. V., & McKeown, S. R. (1979). A histological, histochemical and biochemical study of the macroscopically normal white matter in multiple sclerosis. *Journal of the Neurological Sciences*, 41, 81–91.

Amato, M. P., Bartolozzi, M. L., Zipoli, V., Portaccio, E., Mortella, M., Guidi, L., et al. (2004). Neocortical volume decrease in relapsing-remitting MS patients with mild cognitive impairment. *Neurology*, 63, 89–93.

Audoin, B., Au Duong, M. V., Ranjeva, J. P., Ibarrola, D., Malikova, I., Confort-Gouny, S., et al. (2005). Magnetic resonance study of the influence of tissue damage and cortical reorganization on PASAT performance at the earliest stage of multiple sclerosis. *Human Brain Mapping*, 24, 216–228.

Audoin, B., Ibarrola, D., Ranjeva, J. P., Confort-Gouny, S., Malikova, I., Ali-Chérif, A., et al. (2003). Compensatory cortical activation observed by fMRI during a cognitive task at the earliest stage of MS. *Human Brain Mapping*, 20, 51–58.

Audoin, B., Ranjeva, J. P., Au Duong, M. V., Ibarrola, D., Malikova, I., Confort-Gouny, S., et al. (2004). Voxel-based analysis of MTR images: a method to locate grey matter abnormalities in patients at the earliest stage of multiple sclerosis. *Journal of Magnetic Resonance Imaging*, 20, 765–771.

Benedict, R. H., Weinstock-Guttman, B., Fishman, I., Sharma, J., Tjoa, C. W., & Bakshi, R. (2004). Prediction of neuropsychological impairment in multiple sclerosis: comparison of conventional magnetic resonance imaging measures of atrophy and lesion burden. *Archives of Neurology*, 61, 226–230.

Benedict, R. H., Zivadinov, R., Carone, D. A., Weinstock-Guttman, B., Gaines, J., Maggiore, C., et al. (2005). Regional lobar atrophy predicts memory impairment in multiple sclerosis. *American Journal of Neuroradiology*, 26, 1824–1831.

Bozzali, M., Cercignani, M., Sormani, M. P., Comi, G., & Filippi, M. (2002). Quantification of brain grey matter damage in different MS phenotypes by use of diffusion tensor MR imaging. *American Journal of Neuroradiology*, 23, 985–988.

Bozzali, M., Rocca, M. A., Iannucci, G., Pereira, C., Comi, G., & Filippi, M. (1999). Magnetization-transfer histogram analysis of the cervical cord in patients with multiple sclerosis. *American Journal of Neuroradiology, 20*, 1803–1808.

Brex, P. A., Leary, S. M., O'Riordan, J. I., Miszkiel, K. A., Plant, G. T., Thompson, A. J., et al. (2001). Measurement of spinal cord area in clinically isolated syndromes suggestive of multiple sclerosis. *Journal of Neurology, Neurosurgery, and Psychiatry, 70*, 544–547.

Cader, S., Cifelli, A., Abu-Omar, Y., Palace, J., & Matthews, P. M. (2006). Reduced brain functional reserve and altered functional connectivity in patients with multiple sclerosis. *Brain, 129*, 527–537.

Calabrese, M., De Stefano, N., Atzori, M., Bernardi, V., Mattisi, I., Barachino, L., et al. (2007). Detection of cortical inflammatory lesions by double inversion recovery magnetic resonance imaging in patients with multiple sclerosis. *Archives of Neurology, 64*, 1416–1422.

Caramanos, Z., Narayanan, S., & Arnold, D. L. (2005). $^1$H-MRS quantification of tNA and tCr in patients with multiple sclerosis: a meta-analytic review. *Brain, 128*, 2483–2506.

Ceccarelli, A., Rocca, M. A., Pagani, E., Ghezzi, A., Capra, R., Falini, A., et al. (2008). The topographical distribution of tissue injury in benign MS: a 3T multiparametric MRI study. *NeuroImage, 39*, 1499–1509.

Cerasa, A., Fera, F., Gioia, M. C., Liguori, M., Passamonti, L., Nicoletti, G., et al. (2006). Adaptive cortical changes and the functional correlates of visuo-motor integration in relapsing-remitting multiple sclerosis. *Brain Research Bulletin, 69*, 597–605.

Cercignani, M., Bozzali, M., Iannucci, G., Comi, G., & Filippi, M. (2001). Magnetisation transfer ratio and mean diffusivity of normal-appearing white and gray matter from patients with multiple sclerosis. *Journal of Neurology, Neurosurgery, and Psychiatry, 70*, 311–317.

Chard, D. T., Griffin, C. M., McLean, M. A., Kapeller, P., Kapoor, R., Thompson, A. J., et al. (2002). Brain metabolite changes in cortical gray and normal-appearing white matter in clinically early relapsing-remitting multiple sclerosis. *Brain, 125*, 2342–2352.

Ciccarelli, O., Toosy, A. T., Hickman, S. J., Parker, G. J., Wheeler-Kingshott, C. A., Miller, D. H., et al. (2005). Optic radiation changes after optic neuritis detected by tractography-based group mapping. *Human Brain Mapping, 25*, 308–316.

Ciccarelli, O., Werring, D. J., Wheeler-Kingshott, C. A., Barker, G. J., Parker, G. J., Thompson, A. J., et al. (2001). Investigation of MS normal-appearing brain using diffusion tensor MRI with clinical correlations. *Neurology, 56*, 926–933.

Ciccarelli, O., Wheeler-Kingshott, C. A., McLean, M. A., Cercignani, M., Wimpey, K., Miller, D. H., et al. (2007). Spinal cord spectroscopy and diffusion-based tractography to assess acute disability in multiple sclerosis. *Brain, 130*, 2220–2231.

Cifelli, A., Arridge, M., Jezzard, P., Esiri, M. M., Palace, J., & Matthews, P. M. (2002). Thalamic neurodegeneration in multiple sclerosis. *Annals of Neurology, 52*, 650–653.

Codella, M., Rocca, M. A., Colombo, B., Rossi, P., Comi, G., & Filippi, M. (2002). A preliminary study of magnetization transfer and diffusion tensor MRI of multiple sclerosis patients with fatigue. *Journal of Neurology, 249*, 535–537.

Dalton, C. M., Brex, P. A., Jenkins, R., Fox, N. C., Miszkiel, K. A., Crum, W. R., et al. (2002). Progressive ventricular enlargement in patients with clinically isolated syndromes is associated with the early development of multiple sclerosis. *Journal of Neurology, Neurosurgery, and Psychiatry, 73*, 141–147.

Davie, C. A., Hawkins, C. P., Barker, G. J., Brennan, A., Tofts, P. S., Miller, D. H., et al. (1994). Serial proton magnetic resonance spectroscopy in acute multiple sclerosis lesions. *Brain, 117*, 49–58.

Davies, G. R., Altmann, D. R., Hadjiprocopis, A., Rashid, W., Chard, D. T., Griffin, C. M., et al. (2005a). Increasing normal-appearing gray and white matter magnetisation transfer ratio abnormality in early relapsing-remitting multiple sclerosis. *Journal of Neurology, 252*, 1037–1044.

Davies, G. R., Altmann, D. R., Rashid, W., Chard, D. T., Griffin, C. M., Barker, G. J., et al. (2005b). Emergence of thalamic magnetization transfer ratio abnormality in early relapsing-remitting multiple sclerosis. *Multiple Sclerosis, 11*, 276–281.

Dehmeshki, J., Chard, D. T., Leary, S. M., Watt, H. C., Silver, N. C., Tofts, P. S., et al. (2003). The normal appearing gray matter in primary progressive multiple sclerosis: a magnetisation transfer imaging study. *Journal of Neurology, 250*, 67–74.

De Stefano, N., Filippi, M., Miller, D., Pouwels, P. J., Rovira, A., Gass, A., et al. (2007). Guidelines for using proton MR spectroscopy in multicenter clinical MS studies. *Neurology, 69*, 1942–1952.

De Stefano, N., Matthews, P. M., Antel, J. P., Preul, M., Francis, G., & Arnold, D. L. (1995). Chemical pathology of acute demyelinating lesions and its correlation with disability. *Annals of Neurology, 38*, 901–909.

De Stefano, N., Matthews, P. M., Filippi, M., Agosta, F., De Luca, M., Bartolozzi, M. L., et al. (2003). Evidence of early cortical atrophy in MS: relevance to white matter changes and disability. *Neurology, 60*, 1157–1162.

De Stefano, N., Narayanan, S., Matthews, P. M., Francis, G. S., Antel, J. P., & Arnold, D. L. (1999). In vivo evidence for axonal dysfunction remote from focal cerebral demyelination of the type seen in multiple sclerosis. *Brain, 122*, 1933–1939.

Droogan, A. G., Clark, C. A., Werring, D. J., Barker, G. J., McDonald, W. I., & Miller, D. H. (1999). Comparison of multiple sclerosis clinical subgroups using navigated spin echo diffusion-weighted imaging. *Magnetic Resonance Imaging, 17*, 653–661.

Evangelou, N., Esiri, M. M., Smith, S., Palace, J., & Matthews, P. M. (2000). Quantitative pathological evidence for axonal loss in normal appearing white matter in multiple sclerosis. *Annals of Neurology, 47*, 391–395.

Fabiano, A. J., Sharma, J., Weinstock-Guttman, B., Munschauer, F. E., III, Benedict, R. H., Zivadinov, R., et al.

(2003). Thalamic involvement in multiple sclerosis: a diffusion-weighted magnetic resonance imaging study. *Journal of Neuroimaging, 13*, 307–314.

Falini, A., Calabrese, G., Filippi, M., Origgi, D., Lipari, S., Colombo, B., et al. (1998). Benign versus secondary progressive multiple sclerosis: the potential role of $^1$H MR spectroscopy in defining the nature of disability. *American Journal of Neuroradiology, 19*, 223–229.

Ferguson, B., Matyszak, M. K., Esiri, M. M., & Perry, V. H. (1997). Axonal damage in acute multiple sclerosis lesions. *Brain, 120*, 393–399.

Fernando, K. T., McLean, M. A., Chard, D. T., MacManus, D. G., Dalton, C. M., Miszkiel, K. A., et al. (2004). Elevated white matter myo-inositol in clinically isolated syndromes suggestive of multiple sclerosis. *Brain, 127*, 1361–1369.

Fernando, K. T., Tozer, D. J., Miszkiel, K. A., MacManus, D. G., Dalton, C. M., Miszkiel, K. A., et al. (2005). Magnetization transfer histograms in clinically isolated syndromes suggestive of multiple sclerosis. *Brain, 128*, 2911–2925.

Filippi, M., & Agosta, F. (2007). Magnetization transfer MRI in multiple sclerosis. *Journal of Neuroimaging, 17*(Suppl. 1), 22S–26S.

Filippi, M., Bozzali, M., Horsfield, M. A., Rocca, M. A., Sormani, M. P., Iannucci, G., et al. (2000a). A conventional and magnetization transfer MRI study of the cervical cord in patients with MS. *Neurology, 54*, 207–213.

Filippi, M., Campi, A., Dousset, V., Baratti, C., Martinelli, V., Canal, N., et al. (1995). A magnetization transfer imaging study of normal-appearing white matter in multiple sclerosis. *Neurology, 45*, 478–482.

Filippi, M., Cercignani, M., Inglese, M., Horsfield, M. A., & Comi, G. (2001). Diffusion tensor magnetic resonance imaging in multiple sclerosis. *Neurology, 56*, 304–311.

Filippi, M., Iannucci, G., Cercignani, M., Rocca, M. A., Pratesi, A., & Comi, G. (2000b). A quantitative study of water diffusion in multiple sclerosis lesions and normal-appearing white matter using echo-planar imaging. *Archives of Neurology, 57*, 1017–1021.

Filippi, M., & Inglese, M. (2001). Overview of diffusion-weighted magnetic resonance studies in multiple sclerosis. *Journal of the Neurological Sciences, 186*(Suppl. 1), S37–S43.

Filippi, M., Inglese, M., Rovaris, M., Sormani, M. P., Horsfield, M. A., Iannucci, G., et al. (2000c). Magnetization transfer imaging to monitor the evolution of MS: a 1-year follow-up study. *Neurology, 55*, 940–946.

Filippi, M., & Rocca, M. A. (2007). Conventional MRI in multiple sclerosis. *Journal of Neuroimaging, 17*(Suppl. 1), 3S–9S.

Filippi, M., Rocca, M. A., & Comi, G. (1998a). Magnetization transfer ratios of multiple sclerosis lesions with variable durations of enhancement. *Journal of the Neurological Sciences, 159*, 162–165.

Filippi, M., Rocca, M. A., Falini, A., Caputo, D., Ghezzi, A., Colombo, B., et al. (2002). Correlations between structural CNS damage and functional MRI changes in primary progressive MS. *NeuroImage, 15*, 537–546.

Filippi, M., Rocca, M. A., Martino, G., Horsfield, M. A., & Comi, G. (1998b). Magnetization transfer changes in the normal appearing white matter precede the appearance of enhancing lesions in patients with multiple sclerosis. *Annals of Neurology, 43*, 809–814.

Filippi, M., Rocca, M. A., Mezzapesa, D. M., Ghezzi, A., Falini, A., Martinelli, V., et al. (2004a). Simple and complex movement-associated functional MRI changes in patients at presentation with clinically isolated syndromes suggestive of MS. *Human Brain Mapping, 21*, 108–117.

Filippi, M., Rocca, M. A., Minicucci, L., Martinelli, V., Ghezzi, A., Bergamaschi, R., et al. (1999). Magnetization transfer imaging of patients with definite MS and negative conventional MRI. *Neurology, 52*, 845–848.

Filippi, M., Rocca, M. A., Rizzo, G., Horsfield, M. A., Rovaris, M., Minicucci, L., et al. (1998c). Magnetization transfer ratios in multiple sclerosis lesions enhancing after different doses of gadolinium. *Neurology, 50*, 1289–1293.

Filippi, M., Rovaris, M., Inglese, M., Barkhof, F., De Stefano, N., Smith, S., et al. (2004b). Interferon beta-1a for brain tissue loss in patients at presentation with syndromes suggestive of multiple sclerosis: a randomised, double-blind, placebo-controlled trial. *Lancet, 364*, 1489–1496.

Filippi, M., Tortorella, C., Rovaris, M., Bozzali, M., Possa, F., Sormani, M. P., et al. (2000d). Changes in the normal appearing brain tissue and cognitive impairment in multiple sclerosis. *Journal of Neurology, Neurosurgery, and Psychiatry, 68*, 157–161.

Gallo, A., Rovaris, M., Riva, R., Ghezzi, A., Benedetti, B., Martinelli, V., et al. (2005). Diffusion tensor MRI detects normal-appearing white matter damage unrelated to short-term disease activity in patients at the earliest clinical stage of multiple sclerosis. *Archives of Neurology, 62*, 803–808.

Ge, Y., Grossman, R. I., Udupa, J. K., Babb, J. S., Kolson, D. L., & McGowan, J. C. (2001). Magnetization transfer ratio histogram analysis of gray matter in relapsing-remitting multiple sclerosis. *American Journal of Neuroradiology, 22*, 470–475.

Geurts, J. J., Pouwels, P. J., Uitdehaag, B. M., Polman, C. H., Barkhof, F., & Castelijns, J. A. (2005). Intracortical lesions in multiple sclerosis: improved detection with 3D double inversion-recovery MR imaging. *Radiology, 236*, 254–260.

Gilmore, C. P., DeLuca, G. C., Bo, L., Owens, T., Lowe, J., Esiri, M. M., et al. (2005). Spinal cord atrophy in multiple sclerosis caused by white matter volume loss. *Archives of Neurology, 62*, 1859–1862.

Hickman, S. J., Brex, P. A., Brierley, C. M., Silver, N. C., Barker, G. J., Scolding, N. J., et al. (2001). Detection of optic nerve atrophy following a single episode of unilateral optic neuritis by MRI using a fat-saturated short-echo fast FLAIR sequence. *Neuroradiology, 43*, 123–128.

Hickman, S. J., Brierley, C. M., Brex, P. A., MacManus, D. G., Scolding, N. J., Compston, D. A., et al. (2002). Continuing optic nerve atrophy following optic neuritis: a serial MRI study. *Multiple Sclerosis, 8*, 339–342.

Hickman, S. J., Toosy, A. T., Jones, S. J., Altmann, D. R., Miszkiel, K. A., MacManus, D. G., et al. (2004a). Serial magnetization transfer imaging in acute optic neuritis. *Brain*, *127*, 692–700.

Hickman, S. J., Toosy, A. T., Jones, S. J., Altmann, D. R., Miszkiel, K. A., MacManus, D. G., et al. (2004b). A serial magnetic resonance imaging study following optic nerve mean area in acute optic neuritis. *Brain*, *127*, 2498–2505.

Hickman, S. J., Wheeler-Kingshott, C. A., Jones, S. J., Miszkiel, K. A., Barker, G. J., Plant, G. T., et al. (2005). Optic nerve diffusion measurement from diffusion-weighted imaging in optic neuritis. *American Journal of Neuroradiology*, *26*, 951–956.

Hillary, F. G., Chiaravalloti, N. D., Ricker, J. H., Steffener, J., Bly, B. M., Lange, G., et al. (2003). An investigation of working memory rehearsal in multiple sclerosis using fMRI. *Journal of Clinical and Experimental Neuropsychology*, *25*, 965–978.

Ingle, G. T., Stevenson, V. L., Miller, D. H., & Thompson, A. J. (2003). Primary progressive multiple sclerosis: a 5-year clinical and MR study. *Brain*, *126*, 2528–2536.

Inglese, M., Ghezzi, A., Bianchi, S., Gerevini, S., Sormani, M. P., Martinelli, V., et al. (2002). Irreversible disability and tissue loss in multiple sclerosis: a conventional and magnetization transfer magnetic resonance imaging study of the optic nerves. *Archives of Neurology*, *59*, 250–255.

Inglese, M., Liu, S., Babb, J. S., Mannon, L. J., Grossman, R. I., & Gonen, O. (2004). Three-dimensional proton spectroscopy of deep gray matter nuclei in relapsing-remitting MS. *Neurology*, *63*, 170–172.

Kapeller, P., Brex, P. A., Chard, D. T., Dalton, C., Griffin, C. M., McLean, M. A., et al. (2002). Quantitative $^1$H MRS imaging 14 years after presenting with a clinically isolated syndrome suggestive of multiple sclerosis. *Multiple Sclerosis*, *8*, 207–210.

Kendi, A. T., Tan, F. U., Kendi, M., Yilmaz, S., Huvaj, S., & Tellioğlu, S. (2004). MR spectroscopy of cervical spinal cord in patients with multiple sclerosis. *Neuroradiology*, *46*, 764–769.

Kidd, D., Barkhof, F., McConnell, R., Algra, P. R., Allen, I. V., & Revesz, T. (1999). Cortical lesions in multiple sclerosis. *Brain*, *122*, 17–26.

Khaleeli, Z., Sastre-Garrga, J., Ciccarelli, O., Miller, D. H., & Thompson, A. J. (2007). Magnetisation transfer ratio in the normal appearing white matter predicts progression of disability over 1 year in early primary progressive multiple sclerosis. *Journal of Neurology, Neurosurgery, and Psychiatry*, *78*, 1076–1082.

Laule, C., Vavasour, I. M., Whittall, K. P., Oger, J., Paty, D. W., Li, D. K., et al. (2003). Evolution of focal and diffuse magnetisation transfer abnormalities in multiple sclerosis. *Journal of Neurology*, *250*, 924–931.

Li, Y., Chiaravalloti, N. D., Hillary, F. G., Deluca, J., Liu, W. C., Kalnin, A. J., et al. (2004). Differential cerebellar activation on functional magnetic resonance imaging during working memory performance in persons with multiple sclerosis. *Archives of Physical Medicine and Rehabilitation*, *85*, 635–639.

Lin, X., Tench, C. R., Morgan, P. S., Niepel, G., & Constantinescu, C. S. (2005). 'Importance sampling' in MS: use of diffusion tensor tractography to quantify pathology related to specific impairment. *Journal of the Neurological Sciences*, *237*, 13–19.

Locatelli, L., Zivadinov, R., Grop, A., & Zorzon, M. (2004). Frontal parenchymal atrophy measures in multiple sclerosis. *Multiple Sclerosis*, *10*, 562–568.

Lycklama à Nijeholt, G., Thompson, A., Filippi, M., Miller, D., Polman, C., Fazekas, F., et al. (2003). Spinal-cord MRI in multiple sclerosis. *Lancet Neurology*, *2*, 555–562.

Lycklama à Nijeholt, G. J., Bergers, E., Kamphorst, W., Bot, J., Nicolay, K., Castelijns, J. A., et al. (2001). Post-mortem high-resolution MRI of the spinal cord in multiple sclerosis: a correlative study with conventional MRI, histopathology and clinical phenotype. *Brain*, *124*, 154–166.

Mainero, C., Caramia, F., Pozzilli, C., Pisani, A., Pestalozza, I., Borriello, G., et al. (2004). fMRI evidence of brain reorganization during attention and memory tasks in multiple sclerosis. *NeuroImage*, *21*, 858–867.

Melzi, L., Rocca, A., Bianchi Marzoli, S., Falini, A., Pezzulli, P., Ghezzi, A., et al. (2007). A longitudinal conventional and magnetization transfer MRI study of optic neuritis. *Multiple Sclerosis*, *13*, 265–268.

Mesaros, S., Rocca, M. A., Absinta, M., Ghezzi, A., Milani, N., Moiola, L., et al. (2008). Evidence of thalamic gray matter loss in pediatric multiple sclerosis. *Neurology*, *70*, 1107–1112.

Mezzapesa, D. M., Rocca, M. A., Falini, A., Rodegher, M. E., Ghezzi, A., Comi, G., et al. (2004). A preliminary diffusion tensor and magnetization transfer magnetic resonance imaging study of early-onset multiple sclerosis. *Archives of Neurology*, *61*, 366–368.

Miller, D., Barkhof, F., Montalban, X., Thompson, A., & Filippi, M. (2005). Clinically isolated syndromes suggestive of multiple sclerosis, part 2: non-conventional MRI, recovery processes, and management. *Lancet Neurology*, *4*, 341–348.

Miller, D. H., Barkhof, F., Frank, J. A., Parker, G. J., & Thompson, A. J. (2002). Measurement of atrophy in multiple sclerosis: pathological basis, methodological aspects and clinical relevance. *Brain*, *125*, 1676–1695.

Mori, S., Kaufmann, W. E., Davatzikos, C., Stieltjes, B., Amodei, L., Fredericksen, K., et al. (2002). Imaging cortical association tracts in the human brain using diffusion-tensor-based axonal tracking. *Magnetic Resonance in Medicine*, *47*, 215–223.

Narayana, P. A., Doyle, T. J., Lai, D., & Wolinsky, J. S. (1998). Serial proton magnetic resonance spectroscopic imaging, contrast-enhanced magnetic resonance imaging, and quantitative lesion volumetry in multiple sclerosis. *Annals of Neurology*, *43*, 56–71.

Oreja-Guevara, C., Charil, A., Caputo, D., Cavarretta, R., Sormani, M. P., & Filippi, M. (2006). MT MRI reflects clinical changes over 18 months in relapsing-remitting MS patients. *Archives of Neurology*, *63*, 736–740.

Oreja-Guevara, C., Rovaris, M., Iannucci, G., Valsasina, P., Caputo, D., Cavarretta, R., et al. (2005). Progressive grey matter damage in patients with relapsing-remitting multiple sclerosis: a longitudinal diffusion tensor magnetic resonance imaging study. *Archives of Neurology, 62*, 578–584.

Pagani, E., Filippi, M., Rocca, M. A., & Horsfield, M. A. (2005a). A method for obtaining tract-specific diffusion tensor MRI measurements in the presence of disease: application to patients with clinically isolated syndromes suggestive of multiple sclerosis. *NeuroImage, 26*, 258–265.

Pagani, E., Rocca, M. A., Gallo, A., Rovaris, M., Martinelli, V., Comi, G., et al. (2005b). Regional brain atrophy evolves differently in patients with multiple sclerosis according to clinical phenotype. *American Journal of Neuroradiology, 26*, 341–346.

Pantano, P., Iannetti, G. D., Caramia, F., Mainero, C., Di Legge, S., Bozzao, L., et al. (2002a). Cortical motor reorganization after a single clinical attack of multiple sclerosis. *Brain, 125*, 1607–1615.

Pantano, P., Mainero, C., Iannetti, G. D., Caramia, F., Di Legge, S., Piattella, M. C., et al. (2002b). Contribution of corticospinal tract damage to cortical motor reorganization after a single clinical attack of multiple sclerosis. *NeuroImage, 17*, 1837–1843.

Peterson, J. W., Bö, L., Mork, S., Chang, A., & Trapp, B. D. (2001). Transected neurites, apoptotic neurons, and reduced inflammation in cortical multiple sclerosis lesions. *Annals of Neurology, 50*, 389–400.

Pulizzi, A., Rovaris, M., Judica, E., Sormani, M. P., Martinelli, V., Comi, G., et al. (2007). Determinants of disability in multiple sclerosis at various disease stages: a multiparametric magnetic resonance study. *Archives of Neurology, 64*, 1163–1168.

Ramio-Torrenta, L., Sastre-Garriga, J., Ingle, G. T., Davies, G. R., Ameen, V., Miller, D. H., et al. (2006). Abnormalities in normal appearing tissues in early primary progressive multiple sclerosis and their relation to disability: a tissue specific magnetisation transfer study. *Journal of Neurology, Neurosurgery, and Psychiatry, 77*, 40–45.

Ranjeva, J. P., Audoin, B., Au Duong, M. V., Ibarrola, D., Confort-Gouny, S., Malikova, I., et al. (2005). Local tissue damage assessed with statistical mapping analysis of brain magnetization transfer ratio: relationship with functional status of patients in the earliest stage of multiple sclerosis. *American Journal of Neuroradiology, 26*, 119–127.

Rocca, M. A., Agosta, F., Martinelli, V., Falini, A., Comi, G., & Filippi, M. (2006). The level of spinal cord involvement influences the pattern of movement-associated cortical recruitment in patients with isolated myelitis. *NeuroImage, 30*, 879–884.

Rocca, M. A., Colombo, B., Falini, A., Ghezzi, A., Martinelli, V., Scotti, G., et al. (2005a). Cortical adaptation in patients with MS: a cross-sectional functional MRI study of disease phenotypes. *Lancet Neurology, 4*, 618–626.

Rocca, M. A., Falini, A., Colombo, B., Scotti, G., Comi, G., & Filippi, M. (2002a). Adaptive functional changes in the cerebral cortex of patients with non-disabling MS correlate with the extent of brain structural damage. *Annals of Neurology, 51*, 330–339.

Rocca, M. A., & Filippi, M. (2007). Functional MRI in multiple sclerosis. *Journal of Neuroimaging, 17*(Suppl. 1), 36S–41S.

Rocca, M. A., Gavazzi, C., Mezzapesa, D. M., Ghezzi, A., Falini, A., Scotti, G., et al. (2003a). A functional magnetic resonance imaging study of patients with secondary progressive multiple sclerosis. *NeuroImage, 19*, 1770–1777.

Rocca, M. A., Mastronardo, G., Rodegher, M., Comi, G., & Filippi, M. (1999). Long-term changes of magnetization transfer-derived measures from patients with relapsing-remitting and secondary progressive multiple sclerosis. *American Journal of Neuroradiology, 20*, 821–827.

Rocca, M. A., Matthews, P. M., Caputo, D., Ghezzi, A., Falini, A., Scotti, G., et al. (2002b). Evidence for widespread movement-associated functional MRI changes in patients with PPMS. *Neurology, 58*, 866–872.

Rocca, M. A., Mezzapesa, D. M., Falini, A., Ghezzi, A., Martinelli, V., Scotti, G., et al. (2003b). Evidence for axonal pathology and adaptive cortical reorganisation in patients at presentation with clinically isolated syndromes suggestive of MS. *NeuroImage, 18*, 847–855.

Rocca, M. A., Mezzapesa, D. M., Ghezzi, A., Falini, A., Agosta, F., Martinelli, V., et al. (2003c). Cord damage elicits brain functional reorganization after a single episode of myelitis. *Neurology, 61*, 1078–1085.

Rocca, M. A., Mezzapesa, D. M., Ghezzi, A., Falini, A., Martinelli, V., Scotti, G., et al. (2005b). A widespread pattern of cortical activations in patients at presentation with clinically isolated symptoms is associated with evolution to definite multiple sclerosis. *American Journal of Neuroradiology, 26*, 1136–1139.

Rocca, M. A., Valsasina, P., Ceccarelli, A., Absinta, M., Ghezzi, A., Riccitelli, G., et al. (2009). Structural and functional MRI correlates of Stroop control in benign MS. *Human Brain Mapping, 30*, 276–290.

Rovaris, M., Bozzali, M., Iannucci, G., Ghezzi, A., Caputo, D., Montanari, E., et al. (2002a). Assessment of normal-appearing white and grey matter in patients with primary progressive multiple sclerosis. *Archives of Neurology, 59*, 1406–1412.

Rovaris, M., Bozzali, M., Santuccio, G., Ghezzi, A., Caputo, D., Montanari, E., et al. (2001). In vivo assessment of the brain and cervical cord pathology of patients with primary progressive multiple sclerosis. *Brain, 124*, 2540–2549.

Rovaris, M., Bozzali, M., Santuccio, G., Iannucci, G., Sormani, M. P., Colombo, B., et al. (2000a). Relative contributions of brain and cervical cord pathology to multiple sclerosis disability: a study with magnetisation transfer ratio histogram analysis. *Journal of Neurology, Neurosurgery, and Psychiatry, 69*, 723–727.

Rovaris, M., Filippi, M., Minicucci, L., Iannucci, G., Santuccio, G., Possa, F., et al. (2000b). Cortical/subcortical disease burden and cognitive impairment in patients with multiple sclerosis. *American Journal of Neuroradiology, 21*, 402–408.

Rovaris, M., Gallo, A., Riva, R., Ghezzi, A., Bozzali, M., Benedetti, B., et al. (2004). An MT MRI study of the cervical cord in clinically isolated syndromes suggestive of MS. *Neurology, 63*, 584–585.

Rovaris, M., Gallo, A., Valsasina, P., Benedetti, B., Ghezzi, A., Montanari, E., et al. (2005a). Short-term accrual of gray matter pathology in patients with progressive multiple sclerosis: an in vivo study using diffusion tensor MRI. *NeuroImage, 24*, 1139–1146.

Rovaris, M., Gass, A., Bammer, R., Hickman, S. J., Ciccarelli, O., Miller, D. H., et al. (2005b). Diffusion MRI in multiple sclerosis. *Neurology, 65*, 1526–1532.

Rovaris, M., Iannucci, G., Falautano, M., Possa, F., Martinelli, V., Comi, G., et al. (2002b). Cognitive dysfunction in patients with mildly disabling relapsing-remitting multiple sclerosis: an exploratory study with diffusion tensor MR imaging. *Journal of the Neurological Sciences, 195*, 103–109.

Rovaris, M., Judica, E., Gallo, A., Sormani, M. P., Benedetti, B., Caputo, D., et al. (2006). Grey matter damage predicts the evolution of primary progressive multiple sclerosis at 5 years. *Brain, 129*, 2628–2634.

Rovaris, M., Judica, E., Sastre-Garriga, J., Rovira, A., Sormani, M. P., Benedetti, B., et al. (2008). Large-scale, multicentre, quantitative MRI study of brain and cord damage in primary progressive multiple sclerosis. *Multiple Sclerosis, 14*, 455–464.

Roychowdhury, S., Maldijan, J. A., & Grossman, R. I. (2000). Multiple sclerosis: comparison of trace apparent diffusion coefficients with MR enhancement pattern of lesions. *American Journal of Neuroradiology, 21*, 869–874.

Sailer, M., Fischl, B., Salat, D., Tempelmann, C., Schonfeld, M. A., Busa, E., et al. (2003). Focal thinning of the cerebral cortex in multiple sclerosis. *Brain, 126*, 1734–1744.

Sanfilipo, M. P., Benedict, R. H., Sharma, J., Weinstock-Guttman, B., & Bakshi, R. (2005). The relationship between whole brain volume and disability in multiple sclerosis: a comparison of normalized gray vs. white matter with misclassification correction. *NeuroImage, 26*, 1068–1077.

Santos, A. C., Narayanan, S., De Stefano, N., Tartaglia, M. C., Francis, S. J., Arnaoutelis, R., et al. (2002). Magnetization transfer can predict clinical evolution in patients with multiple sclerosis. *Journal of Neurology, 249*, 662–668.

Sastre-Garriga, J., Ingle, G. T., Chard, D. T., Ramió-Torrentà, L., McLean, M. A., Miller, D. H., et al. (2005). Metabolite changes in normal-appearing gray and white matter are linked with disability in early primary progressive multiple sclerosis. *Archives of Neurology, 62*, 569–573.

Sastre-Garriga, J., Ingle, G. T., Chard, D. T., Ramio-Torrenta, L., Miller, D. H., & Thompson, A. J. (2004). Grey and white matter atrophy in early clinical stages of primary progressive multiple sclerosis. *NeuroImage, 22*, 353–359.

Sepulcre, J., Sastre-Garriga, J., Cercignani, M., Ingle, G. T., Miller, D. H., & Thompson, A. J. (2006). Regional gray matter atrophy in early primary progressive multiple sclerosis: a voxel-based morphometry study. *Archives of Neurology, 63*, 1175–1180.

Sharma, R., Narayana, P. A., & Wolinsky, J. S. (2001). Grey matter abnormalities in multiple sclerosis: proton magnetic resonance spectroscopic imaging. *Multiple Sclerosis, 7*, 221–226.

Sijens, P. E., Mostert, J. P., Oudkerk, M., & De Keyser, J. (2006). $^1$H MR spectroscopy of the brain in multiple sclerosis subtypes with analysis of the metabolite concentrations in gray and white matter: initial findings. *European Radiology, 16*, 489–495.

Silver, N. C., Lai, M., Symms, M. R., Barker, G. J., McDonald, W. I., & Miller, D. H. (1998). Serial magnetization transfer imaging to characterize the early evolution of new MS lesions. *Neurology, 51*, 758–764.

Simon, J. H., Zhang, S., Laidlaw, D. H., Miller, D. E., Brown, M., Corboy, J., et al. (2006). Identification of fibers at risk for degeneration by diffusion tractography in patients at high risk for MS after a clinically isolated syndrome. *Journal of Magnetic Resonance Imaging, 24*, 983–988.

Staffen, W., Mair, A., Zauner, H., Unterrainer, J., Niederhofer, H., Kutzelnigg, A., et al. (2002). Cognitive function and fMRI in patients with multiple sclerosis: evidence for compensatory cortical activation during an attention task. *Brain, 125*, 1275–1282.

Stroman, P. W. (2005). Magnetic resonance imaging of neuronal function in the spinal cord: spinal FMRI. *Clinical Medicine & Research, 3*, 146–156. Review.

Tedeschi, G., Lavorgna, L., Russo, P., Prinster, A., Minacci, D., Savettieri, G., et al. (2005). Brain atrophy and lesion load in a large population of patients with multiple sclerosis. *Neurology, 65*, 280–285.

Thorpe, J. W., Barker, G. J., Jones, S. J., Moseley, I., Losseff, N., MacManus, D. G., et al. (1995). Magnetisation transfer ratios and transverse magnetisation decay curves in optic neuritis: correlation with clinical findings and electrophysiology. *Journal of Neurology, Neurosurgery, and Psychiatry, 58*, 487–492.

Tortorella, C., Viti, B., Bozzali, M., Sormani, M. P., Rizzo, G., Gilardi, M. C., et al. (2000). A magnetization transfer histogram study of normal appearing brain tissue in multiple sclerosis. *Neurology, 54*, 186–193.

Trapp, B. D., Peterson, J., Ransohoff, R. M., Rudick, R., Mörk, S., & Bö, L. (1998). Axonal transection in the lesions of multiple sclerosis. *The New England Journal of Medicine, 338*, 278–285.

Trip, S. A., Wheeler-Kingshott, C., Jones, S. J., Li, W. Y., Barker, G. J., Thompson, A. J., et al. (2006). Optic nerve diffusion tensor imaging in optic neuritis. *NeuroImage, 30*, 498–505.

Valsasina, P., Rocca, M. A., Agosta, F., Benedetti, B., Horsfield, M. A., Gallo, A., et al. (2005). Mean diffusivity and fractional anisotropy histogram analysis of the cervical cord in MS patients. *NeuroImage, 26*, 822–828.

van Walderveen, M. A., Barkhof, F., Pouwels, P. J., van Schijndel, R. A., Polman, C. H., & Castelijns, J. A. (1999). Neuronal damage in T1-hypointense multiple sclerosis lesions demonstrated in vivo using proton magnetic resonance spectroscopy. *Annals of Neurology, 46*, 79–87.

Vinogradov, E., Degenhardt, A., Smith, D., Marquis, R., Vartanian, T. K., Kinkel, P., et al. (2005). High-resolution anatomic, diffusion tensor, and magnetization transfer magnetic resonance imaging of the optic chiasm at 3 T. *Journal of Magnetic Resonance Imaging, 22*, 302–306.

Vrenken, H., Barkhof, F., Uitdehaag, B. M., Castelijns, J. A., Polman, C. H., & Pouwels, P. J. (2005). MR spectroscopic evidence for glial increase but not for neuro-axonal damage in MS normal-appearing white matter. *Magnetic Resonance in Medicine, 53*, 256–266.

Werring, D. J., Bullmore, E. T., Toosy, A. T., Miller, D. H., Barker, G. J., MacManus, D. G., et al. (2000). Recovery from optic neuritis is associated with a change in the distribution of cerebral response to visual stimulation: a functional magnetic resonance imaging study. *Journal of Neurology, Neurosurgery, and Psychiatry, 68*, 441–449.

Werring, D. J., Clark, C. A., Barker, G. J., Thompson, A. J., & Miller, D. H. (1999). Diffusion tensor imaging of lesions and normal-appearing white matter in multiple sclerosis. *Neurology, 52*, 1626–1632.

Wilson, M., Tench, C. R., Morgan, P. S., & Blumhardt, L. D. (2003). Pyramidal tract mapping by diffusion tensor magnetic resonance imaging in multiple sclerosis: improving correlations with disability. *Journal of Neurology, Neurosurgery, and Psychiatry, 74*, 203–207.

Wishart, H. A., Saykin, A. J., McDonald, B. C., Mamourian, A. C., Flashman, L. A., Schuschu, K. R., et al. (2004). Brain activation patterns associated with working memory in relapsing-remitting MS. *Neurology, 62*, 234–238.

Zivadinov, R., Sepcic, J., Nasuelli, D., De Masi, R., Bragadin, L. M., Tommasi, M. A., et al. (2001). A longitudinal study of brain atrophy and cognitive disturbances in the early phase of relapsing-remitting multiple sclerosis. *Journal of Neurology, Neurosurgery, and Psychiatry, 70*, 773–780.

*J. Verhaagen et al. (Eds.)*
*Progress in Brain Research*, Vol. 175
ISSN 0079-6123
Copyright © 2009 Elsevier B.V. All rights reserved

CHAPTER 31

# MRI of neuronal network structure, function, and plasticity

Henning U. Voss[1],* and Nicholas D. Schiff[2]

[1]*Citigroup Biomedical Imaging Center, Weill Cornell Medical College, New York, NY, USA*
[2]*Department of Neurology and Neuroscience, Weill Cornell Medical College, New York, NY, USA*

**Abstract:** We review two complementary MRI imaging modalities to characterize structure and function of neuronal networks in the human brain, and their application to subjects with severe brain injury.

The structural imaging modality, diffusion tensor imaging, is based on imaging the diffusion of water protons in the brain parenchyma. From the diffusion tensor, several quantities characterizing fiber structure in the brain can be derived. The principal direction of the diffusion tensor has been found to depend on the fiber direction of myelinated axons. It can be used for white matter fiber tracking. The anisotropy (or directional dependence) of diffusion has been shown to be sensitive to developmental as well as white matter changes during training and recovery from brain injury.

The functional MRI imaging modality, resting state fMRI, concerns the functional connectivity of neuronal networks rather than their anatomical structure. Subjects undergo a conventional fMRI imaging protocol without performing specific tasks. Various resting state network patterns can be computed by algorithms that reveal correlations in the fMRI signal. Often, thalamic structures are involved, suggesting that resting state fMRI could reflect global brain network functionality. Clinical applications of resting state fMRI have been reported, in particular relating signal abnormalities to neurodegenerative processes. To better understand to which degree resting state patterns reflect neuronal network function, we are comparing network patterns of normal subjects with those having severe brain lesions in a small pilot study.

**Keywords:** MRI; neuronal networks; diffusion tensor imaging (DTI); functional MRI (fMRI); resting state; brain injury; axonal regrowth

## Introduction

The impressive advancement of magnetic resonance imaging (MRI) in the past three decades provides us with sensitive tools to obtain noninvasive, unprecedented characterizations of the human brain in vivo. As highly versatile machines that can be programmed to yield an ever-expanding variety of different image contrasts, MRI scanners now allow us to ask specific questions about structural and functional connectivity of the brain. In this contribution, we review two complementary MRI imaging modalities with a potential to characterize therapeutic response or recovery after brain injury.

*Corresponding author.
Tel.: +1 212 746 5216; Fax: +1 212 746 6681;
E-mail: hev2006@med.cornell.edu

The first imaging modality, diffusion tensor imaging (DTI), is based on imaging the strength and direction of thermal diffusion of water protons in the brain parenchyma. From the diffusion tensor, several quantities can be derived that indirectly characterize fiber structure in the brain, particularly in the white matter. Direction of diffusion has been found to depend on the predominant fiber direction of myelinated axons and can be used for white matter fiber tracking. Here we review applications of DTI based on the anisotropy (or directional dependence) and strength of diffusion, with emphasis on quantitative measures that sensitively reflect white matter changes during recovery from brain injury, as well as during development.

The second imaging modality, resting state fMRI, is based on the blood–oxygen-level-dependent (BOLD) effect in gray matter during rest and concerns the functional connectivity of neuronal networks rather than their anatomical structure. In resting state fMRI, subjects undergo a conventional fMRI imaging protocol without performing specific tasks. The resulting signal contains various resting state network patterns, which are obtained by algorithms that reveal correlations in the signal. The neurophysiological origins of resting state BOLD fluctuations are not completely understood; neither is the potential use of this measurement for imaging therapeutic response nor its recovery after brain injury. However, during the past 3 years, several clinical applications of resting state fMRI have been reported. Here we review, in particular, applications that relate signal abnormalities to neurodegenerative processes. In addition, we explore the possibility to use this assessment of network integrity in patients with brain injuries. We describe a pilot study with two patients with multiple focal brain lesions, and discuss the potential of resting state fMRI as a tool to assess disruptions and recovery of neuronal networks after brain injury.

Both DTI and resting state fMRI are reviewed with most recent applications to brain injury. Rather than providing a comprehensive review over the techniques, we will focus on the future potential of these techniques as clinical research tools.

## Diffusion tensor imaging

### Technique and applications

DTI is an imaging modality capable of rendering the local direction of fibers and characterizing microscopic tissue properties (Basser et al., 1994; Coremans et al., 1994; Basser and Pierpaoli, 1996). Particularly in the brain parenchyma, diffusion of water parallel to myelinated white matter is larger than that perpendicular to the fiber tracts, where diffusion is hindered by myelin sheaths (Beaulieu and Allen, 1994). Thus, diffusion of water in white matter depends on the direction, i.e., it is anisotropic. In contrast, diffusion in a glass of water or in the cerebrospinal fluid is largely direction-independent, i.e., isotropic. The diffusion constant and other parameters describing diffusion depend on microscopic tissue properties such as density, diameter, degree of myelination, and the geometry of fibers, and influence the MRI signal (LeBihan et al., 1986; Douek et al., 1991; Basser, 1995; Alexander et al., 2007; Assaf and Pasternak, 2008; Mukherjee et al., 2008a). The apparent diffusion of water ("apparent" since physiological processes that are not strictly diffusive in a physical sense may contribute; Neil, 1997) is described, in particular, by the apparent diffusion coefficient (ADC), the degree of anisotropy (e.g., expressed by the fractional anisotropy, FA) and the predominant direction of diffusion. A sketch of the DTI methodology and definitions of these quantities is provided in Fig. 1.

Recently, DTI has attracted much interest in both clinical neurology and basic neurobiological research for its ability to render white matter fiber structures, which cannot be seen in conventional magnetic resonance imaging. For these "fiber tracking" methods, see, for example, Basser (1995); Mori and van Zijl (2002); Mori et al. (2005). The accuracy of DTI-based fiber tracts compared to true neuronal fiber tracts has been continuously improved to the point where, for some structures, DTI fiber tracking approaches or even outperforms the accuracy of anatomical atlases (Douek et al., 1991; Holodny et al., 2005; Ino et al., 2007; Kim et al., 2008; Park et al., 2008).

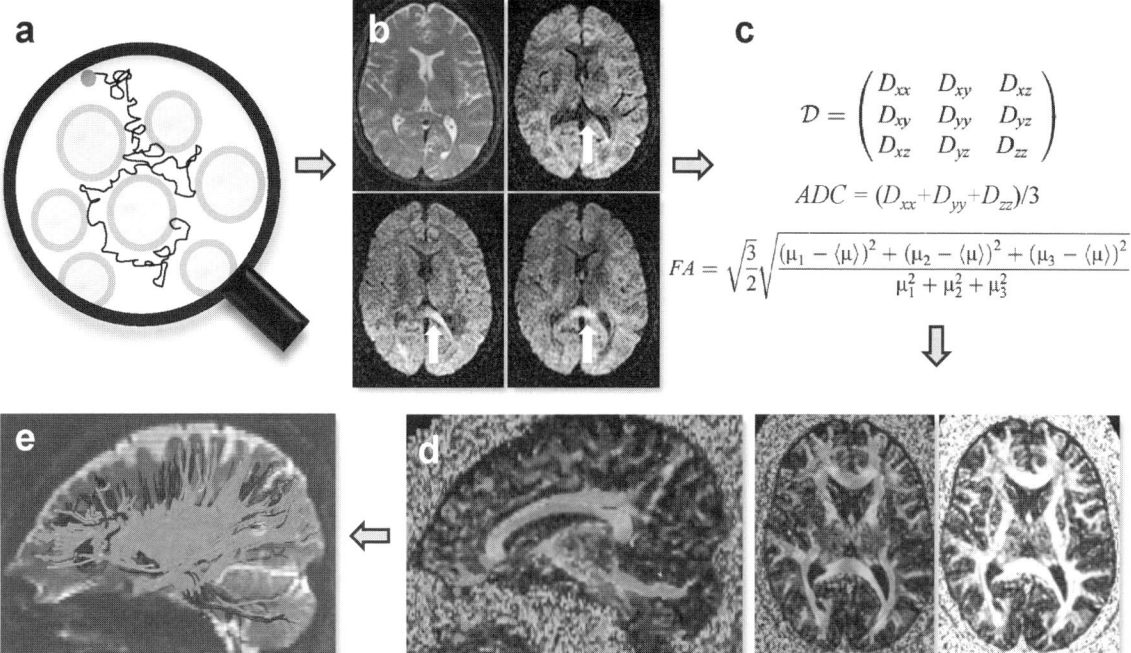

Fig. 1. From microscopic diffusion to cerebral fiber tract reconstructions. (a) Microscopic thermal diffusion of water protons in the interstitial space between myelinated axons is not isotropic but restricted in the directions perpendicular to the fibers. It is less restricted in the direction parallel to the fibers, which would be out-of-plane here. Schematically sketched is a possible diffusive trajectory of a water proton. There is also diffusion of water within the fibers and across the myelin sheaths, which provides a usually smaller contribution to the average diffusion properties of the tissue. (b) A T2 weighted image and three representative axial MRI diffusion weighted images of the brain of a normal subject. The three diffusion weighted images have been acquired with diffusion gradients probing diffusion along different directions. It is evident how the different gradient directions determine image intensity, for example in the splenium of the corpus callosum (arrows). (c) From the T2 and diffusion weighted images, for each voxel a series of mathematical operations is performed. First, a diffusion tensor D is computed by an overdetermined least squares algorithm. In this case, one T2 weighted and 55 diffusion weighted images have been acquired, from which the six independent components of D were estimated. The tensor is diagonalized to compute a coordinate system that is aligned with the diffusion ellipsoid, a three-dimensional shape describing the preferred directions of diffusion in each voxel. The principal axis of the diffusion ellipsoid defines the estimate of the main fiber direction. The lengths of the three ellipsoid axes, or the eigenvalues μ of the matrix D, define useful properties characterizing the microscopic organization of the tissue. In particular, the apparent diffusion constant (ADC) is an estimate of the diffusion strength, and the fractional anisotropy (FA) quantifies the anisotropy of diffusion. (d) A sagittal FA color map, an axial FA color map, and an axial FA intensity map. FA color maps demonstrate the predominant white matter fiber direction with different colors indicating different directions. Color is composed of red, blue, and green, indexing whether anisotropy is most pronounced from left-to-right, superior-to-inferior, or anterior-to-posterior, respectively. The intensity of the color is proportional to FA: it is large for the hindered diffusion of water, such as in case of myelinated axons, and small for more isotropic regions in the brain, such as gray matter or cerebrospinal fluid. (e) The fiber orientation information of the diffusion tensor eigenvectors and the anisotropy values provide the basis for estimating fiber tracts in the brain. In this example, two seed regions in the internal capsule were placed in the right and the left hemispheres. Fiber estimates originating from these seed regions are rendered in green and blue, respectively. (See Color Plate 31.1 in color plate section.)

DTI provides a subject-specific rendering of fiber tracts, which have a certain intersubject variability independent of variability caused by brain disease, injury, or tumors (Liu et al., 2001; Jellison et al., 2004; Sundgren et al., 2004; Narayana et al., 2007; Smith et al., 2007). Fiber tracking algorithms usually depend on a number of MRI and analysis parameters, and their results may not be easily comparable between different sites and subjects, although recent progress has been obtained here, too (Schwartzman et al., 2005). Figure 1e shows a fiber "tractogram."

In anisotropy maps (Pajevic and Pierpaoli, 1999; Jellison et al., 2004), diffusion anisotropy measures such as FA are used as a proxy for white matter fiber integrity. Figure 1d provides examples of color-coded and gray-scale anisotropy maps. Whereas most conventional MRI contrast mechanisms such as T1, T2, and proton density weighting do not provide quantitative values, diffusion anisotropy maps are derived directly from the quantitative measurements of diffusion in space and thus give us a quantitative means of measuring microscopic tissue properties. In principle, results obtained on different MRI scanners should be comparable. However, due to different scanner hardware, imaging protocols, and data processing algorithms, in practice, most often a direct comparison of numbers obtained on different MRI scanners is not advisable, and corrective measures may be required (Croxson et al., 2005; Johansen-Berg et al., 2005; Farrell et al., 2007; Klein et al., 2007; Landman et al., 2007; Niogi et al., 2007). Fiber tractography and quantitative measures can be combined to quantify diffusion along specific white matter fiber tracts (Jones et al., 2005; Smith et al., 2007). For technical aspects of DTI, and computation of indexes such as FA and ADC, as well as more recent technical advances surpassing the Gaussian (tensor) model of diffusion, see Tournier et al. (2004); Tuch (2004); Alexander (2005); Anderson (2005); Campbell et al. (2005); Hagmann et al. (2006); Kingsley (2006a, b); Minati and Weglarz (2007); Mukherjee et al. (2008b). A recent review of pitfalls of DTI is given by Koch and Norris (2006).

### *DTI and recovery from brain injury*

There are at least two motivations for the application of DTI in studies of axonal reorganization after brain injury: (i) The sensitivity of DTI with respect to the degree of myelination and axonal density suggests its potential utility as imaging modality to look at axonal reorganization and growth, although still requiring confirmation by accompanying findings of more direct studies using experimental methods (Pons et al., 1991; Napieralski et al., 1996; Florence et al., 1998; Jain et al., 2000; Carmichael and Chesselet, 2002; Chklovskii et al., 2004; Dancause et al., 2005), or MR spectroscopy (Danielsen et al., 2003). (ii) DTI-based measures such as FA and ADC have proven to provide a sensitive *longitudinal* measure of white and gray matter changes in the developing and adult brain (McKinstry et al., 2002; Partridge et al., 2004; Bengtsson et al., 2005; deIpolyi et al., 2005; Hermoye et al., 2006; Mukherjee and McKinstry, 2006; Mac Donald et al., 2007; Wozniak and Lim, 2006).

We performed a longitudinal case study of a patient who spontaneously recovered reliable expressive language after 19 years in a minimally conscious state, caused by traumatic brain injury (Voss et al., 2006). Comparison of white matter integrity in the patient with 20 control subjects using ADC and FA identified widespread altered diffusivity and decreased anisotropy. These findings remained unchanged over an 18-month interval between two MRI scans. In addition, we identified large, bilateral regions of posterior white matter with significantly increased anisotropy as measured by FA, which reduced over the same time interval. In contrast, notable increases in anisotropy within the midline cerebellar white matter at the second time point correlated with marked clinical improvements in the patient's motor functions (recovery of limited use of lower extremities and left upper extremity, and improvement in articulation). The finding of changes in the cerebellar white matter was also correlated with an increase in resting metabolism measured by PET in this region. We conjectured that axonal regrowth may account for these observations and provide a biological mechanism for the late recovery of motor function over the interval between DTI measurements. These correlations of altered white matter structures as determined by DTI with clinical improvements suggest the potential clinical utility of longitudinal tracking of patients with brain injuries using DTI.

A more thorough prospective longitudinal DTI study of recovery from severe traumatic brain injury examined the correlation of DTI measurements with clinical outcome variables in 23 subjects (Sidaros et al., 2008). In this study, the authors looked in detail at the three principal diffusion tensor components, which enabled them to study the origin of changes, specifically, whether they are more likely to represent axonal injury or demyelination. They found that at the first time point, at 5–11 weeks post-trauma, FA showed widespread reduction in white matter compared to controls. Patients with unfavorable 1-year outcome (measured by the Glasgow Outcome Scale) tended to deviate more from control DTI values than patients with favorable outcome. At the same time, FA in the cerebral peduncle correlated with outcome score at about a year later. After a post-trauma time interval of 9–15 months, follow-up scans showed that FA in patients had increased in the internal capsule and the centrum semiovale. Primarily in patients with favorable outcome, FA reached normal levels or was even higher than normal in these regions. No significant DTI parameter changes over time were found in 14 matched controls during the same time interval. Of note, these increases in FA were restricted to the direction parallel to axonal fibers, an observation interpreted by the investigators as potentially consistent with axonal recovery. The prospective cohort design of this study adds significant evidence in favor of the future use of this technique as a biomarker for prediction of outcomes from traumatic brain injury. Taken together with the longitudinal single subject study of Voss et al. (2006), these findings corroborate the observation of longitudinal changes in DTI during recovery from severe brain injury and add weight to the interpretation that such changes reflect a slow variable of structural change in the brain.

Other studies have shown correlations between recovery scores from severe brain injury and anisotropy values (Han et al., 2007; Skoglund et al., 2008). Ramu et al. (2008) demonstrated significant fiber tract reorganization and regeneration in different parts of the brain in response to spinal cord injury in rats, using histology and DTI. Another study in rats with spinal cord injury investigating the significance of DTI to study neuronal recovery mechanisms, but this time in proximity of the injury, was performed by Deo et al. (2006). The authors found region-specific recovery patterns in the spinal cord white matter, including areas with sustained recovery.

Due to its sensitivity to pathological changes of white matter, DTI has found other widespread clinical applications. For recent reviews of other specific applications, see Lim and Helpern (2002); Sundgren et al. (2004); Johansen-Berg and Behrens (2006); Assaf and Pasternak (2008). DTI sequences are now available from all major clinical MRI scanner manufacturers, along with automated basic postprocessing software, such that they can easily be added to conventional clinical imaging protocols.

# Resting state fMRI

## Technique and applications

In conventional fMRI, fluctuations in the baseline BOLD signal during rest are usually discarded as noise, although they exhibit a rich (but not apparent) spatiotemporal dynamics, as demonstrated by several groups (Biswal et al., 1995; Greicius et al., 2003; Fox et al., 2005; Fox and Raichle, 2007). The most prominent resting state pattern, the "default mode" was originally discovered in resting metabolism positron emission tomography (PET) data (Raichle et al., 2001) and then readily hypothesized to reflect the existence of an organized, baseline default mode of brain function. Interestingly, the default mode network also shows up as the most prominent resting state pattern in BOLD-fMRI data. In addition, the default mode has a direct correlate in cognitive processing. In MRI as well as in PET data, one observes deactivation of areas belonging to the default mode network during attention-demanding tasks or goal-directed behavior (Raichle et al., 2001; Singh and Fawcett, 2008).

Besides the default mode network, several other resting state networks have been identified more recently (Esposito et al., 2005; De Luca et al., 2006; Mantini et al., 2007). In particular, Mantini et al. found six different functionally meaningful resting state patterns. Figure 2 provides examples for four of them. These six main resting state networks are as follows.

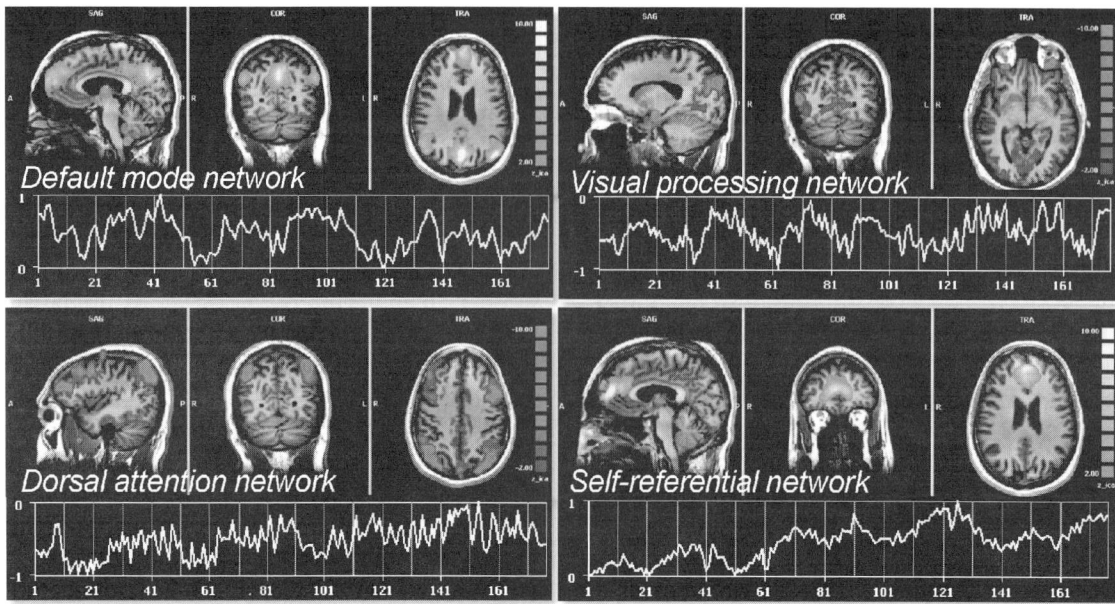

Fig. 2. Four resting state networks in a normal subject. Also shown are corresponding time courses of the independent component coefficients. (Colors quantify the $z$-value of correlation of the time series with the corresponding ICA coefficient. Either positive or negative $z$-values are shown.) (See Color Plate 31.2 in color plate section.)

RSN 1: default mode network, including the posterior cingulate and precuneus, medial prefrontal cortex, dorsal lateral prefrontal cortex, and inferior parietal cortex. RSN 2: dorsal attention network, including the intraparietal sulci, areas at the intersection of precentral and superior frontal sulcus, ventral precentral, and middle frontal gyrus. RSN 3: visual processing network, including retinotopic occipital cortex and temporal–occipital regions. RSN 4: auditory–phonological network, the superior temporal cortices. RSN 5: sensory–motor network, including the precentral, postcentral, and medial frontal gyri, the primary sensory–motor cortices, and the supplementary motor area. RSN 6: self-referential network, including the medial–ventral prefrontal cortex, the pregenual anterior cingulate, the hypothalamus, and the cerebellum. These patterns have considerable overlap with main neuronal networks that have been earlier identified in conventional fMRI studies. (However, the chosen names do not necessarily refer to the exact functions of the resting state networks, which are still putative.) A milestone finding with respect to functional and anatomical significance of resting state networks is the recent demonstration that the oculomotor, the visual, and the somatomotor systems (as determined by corresponding conventional fMRI experiments and fiber staining in anesthetized monkeys) are also identified by BOLD resting state patterns (Vincent et al., 2007).

Recently, it was shown that there is a direct correspondence between fluctuations in EEG oscillations and resting state BOLD baseline fluctuations (Mantini et al., 2007). Correlation of these signals is potentially of considerable relevance as persistent low-frequency oscillations observable in the EEG can provide valuable clinical information about brain function (Niedermeyer and da Silva, 2004; Nunez and Srinivasan, 2005). In further support of the view that the fMRI resting state signal is correlated with electrical activity, Lu et al. (2007) found that in anesthetized rats low-frequency electrical activity underlies the neural correlates of resting state connectivity. This finding is consistent with the suggestion of several authors that resting state BOLD fluctuations are representative of synchronous spontaneous fluctuations of large-scale networks, which occur at low frequencies (von Stein and Sarnthein, 2000; Schnitzler and Gross, 2005). Furthermore, the fMRI resting state signal differentially registers to specific electrical oscillatory frequency band activity, suggesting that fMRI may be able to distinguish the ongoing from the evoked activity of the brain (Fox and Raichle, 2007). Additional studies have related EEG power fluctuations, particularly within the alpha-band, to resting state patterns (Goldman et al., 2002; Leopold et al., 2003; Laufs et al., 2003; Feige et al., 2005).

These findings make it clear that "resting state" cannot be understood as a complete rest of the brain; it is inevitable that the brain constantly engages in some activity during wakefulness. Rather, a resting state experiment should be seen as an unconstrained experiment as opposed to experiments in which subjects perform specific tasks under defined timing constraints (Lowe et al., 1998). Task-dependent fMRI experiments probe the activity of the brain in certain, usually well-defined areas, whereas resting state activations result from spontaneous, intrinsic fluctuations. The approach of looking at intrinsic fluctuations resembles a methodology from the science of complex systems; for example, by virtue of natural laws such as the fluctuation–dissipation theorem, it is possible to test for the functionality of complex biological and physical systems by analyzing their intrinsic fluctuations (Lauk et al., 1998; Crooker et al., 2004; Rossberg et al., 2004). Resting state fMRI is compatible with the view that the brain largely operates intrinsically, with sensory information only modulating system operations (Burton et al., 2004).

Typical resting state spatial patterns have been computed by "blind" methods such as independent component analysis (ICA) (Formisano et al., 2001; Esposito et al., 2002). The resulting patterns can be classified further as attributable to various sources, such as the BOLD effect, and also to artifacts caused by the experimental procedures or motion. ICA separates spatial components by making the assumption that the spatial network patterns are statistically independent in their time course. In the alternative region of interest (ROI) based approach (Greicius et al., 2003), seed ROIs are selected and signals in them are compared to the signal in other parts of the brain. Statistical measures can be derived to understand the correlation structure of these signals in relation to the seed ROI. Both methods have been shown to be able to define very similar resting state correlation structures (Greicius and Menon, 2004).

## Resting state fMRI and potential applications to brain injury

Within the last 3 years, fMRI resting state network analysis has been applied to a variety of diseases and conditions, such as Alzheimer's Disease (AD) (He et al., 2007; Rombouts et al., 2007; Sorg et al., 2007; Wang et al., 2007), the aging brain (Andrews-Hanna et al., 2007; Wu et al., 2007), blindness (Burton et al., 2004), schizophrenia (Jafri et al., 2008; Liu et al., 2008), attention-deficit/hyperactivity disorder (Zhu et al., 2008; Cao et al., 2006), and epilepsy (Waites et al., 2006; Laufs et al., 2007). In these studies, interesting abnormalities have been found in the resting state network patterns. Recent evidence points toward the possibility of extracting information about functionality and connectivity from resting state fMRI fluctuations (Hampson et al., 2002; Greicius et al., 2003; Seeley et al., 2007). The latter property makes resting state fMRI of particular interest for clinical applications. Often, thalamic structures are involved (De Luca et al., 2006), suggesting that resting state fMRI could develop into a technique to assess global brain network functionality. Conventional fMRI approaches that probe specific functions one-by-one by letting the subjects perform specific tasks are technically demanding, and for sufficiently complex tasks, they are usually difficult to perform in the clinical setting. Although such method can be very useful for specific assessments of function (e.g., Owen et al., 2005; Schiff et al., 2005; Giacino et al., 2006), this approach has significant limitations for providing a global view of connectivity in the brain.

The ability to image network integrity or disruptions may be particularly relevant for assessments of severely injured brain. For example, in the minimally conscious state (MCS) some patients exhibit behavioral evidence of the recovery of a minimal dynamic network architecture required to organize behavioral sets and to respond to sensory stimuli, not seen in the vegetative state (VS) patients (Giacino et al., 2002). Assessment of functional network activation using fMRI in such patients suggests preservation of large-scale networks that must retain anatomical connectivity (Schiff et al., 2005). While it is expected that normal subjects show similar location and intensity of networks, we hypothesize that subjects with severe focal brain injury may show distortions of the normal patterns. If so, resting state fMRI could eventually provide functional and diagnostic information. However, more experiments and tools are needed in order to better understand the neurophysiological origins of resting state patterns, and to explore the potential utility of resting state fMRI, a diagnostic tool in clinical imaging of the human brain.

## Initial results

As part of a study characterizing brain function following severe injuries caused by vascular occlusions, we performed resting state fMRI on two subjects with severe multifocal brain injuries. We compared these two studies to those obtained from four normal control subjects without known neurological deficits. Informed consent was obtained from all subjects (or their legally authorized representatives) and the studies were approved by the Institutional Review Board of Weill Cornell Medical College. Subjects were instructed to close their eyes and think of nothing in particular.

Patient 1 (female, 24 years) suffered multifocal brain lesions consisting of infarctions in the brainstem (ventral pons and tegmental midbrain) and thalamus secondary to basilar artery thrombosis, 18 months prior to MRI. Lesions can be seen in the anterior intralaminar nuclei (primarily CL and Pc) bilaterally and in the right LGN (Fig. 3A). The destructive effects of the strokes

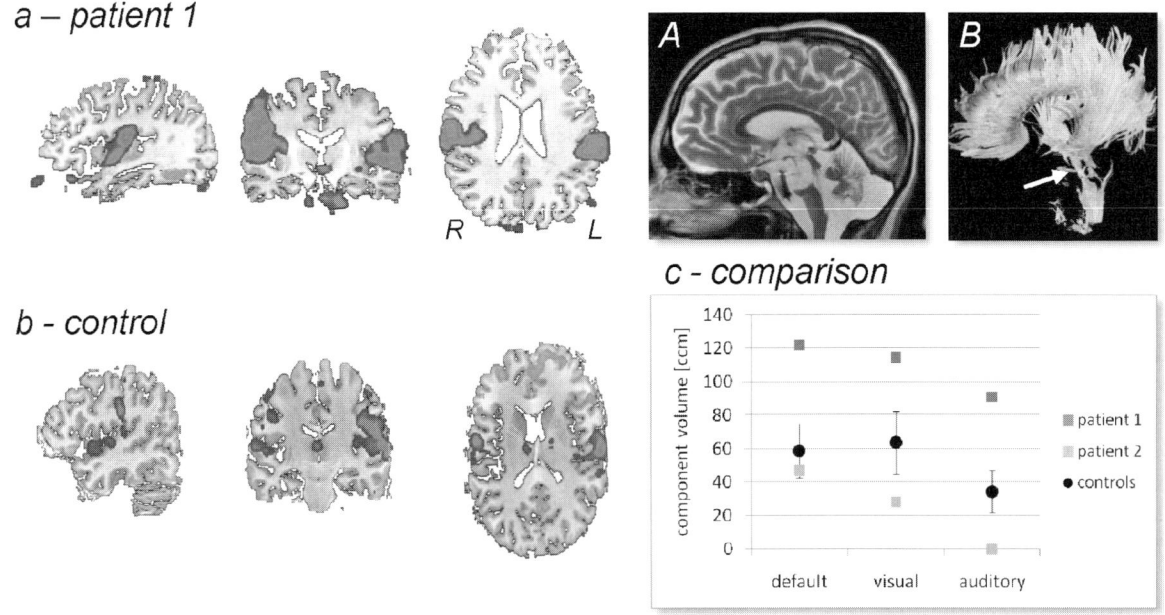

Fig. 3. Auditory–phonological resting state network of (a) Patient 1 and (b) a control subject, as well as (c) a comparison of network volumes between both patients and four control subjects. An anatomical characterization of Patient 1 by means of (A) T1 weighted anatomical imaging and (B) estimates of white matter fiber tracts using DTI. The arrow points to the pontine lesions visible here as disrupted white matter fiber tracts. (See Color Plate 31.3 in color plate section.)

are also visible in DTI-based white matter fiber tract reconstructions (Fig. 3B). Resting state fluorodeoxyglucose (FDG) PET data showed notable metabolic downregulation of left hemisphere frontotemporal cortical regions (not shown). Patient 2 (female, 56 years) suffered multifocal injuries bilaterally within the anterior and middle cerebral artery distributions following severe cerebral vasospasm induced by anterior communicating artery rupture 32 months prior to MRI.

Resting state networks were identified by independent component analysis (BrainVoyagerQX). The volumes of the RSNs were defined by voxels with $z > 2$ in clusters $> 1$ cm$^3$. Resting metabolism (FDG-PET) of patients was qualitatively compared to volume and connectivity (measured as the maximum $z$-value) of RSN clusters. In all control subjects, we could identify the default mode (RSN 1), visual processing (RSN 3), auditory–phonological (RSN 4), and self-referential (RSN 6) networks.

In Patient 1, we could identify RSN 1, 3, and 4 (Fig. 3a). These RSNs had a significantly larger volume than in the control subjects (Fig. 3c). However, due to the small size of the control group this finding had to be considered preliminary. Metabolic rates (FDG-PET) showed left frontotemporal downregulation, which is traceable to the central thalamic lesion (Van der Werf et al., 1999), and loss of calcarine metabolism on the right, ipsilateral to the LGN infarct. In the fMRI data, the parietal regions of RSN 1 (default) appeared stronger with respect to both volume and connectivity, on the right side. RSN 3 (visual) was smaller on the right side, but had a comparable connectivity. RSN 4 (audio) appeared to be of similar volume and connectivity in both hemispheres. The relatively stronger resting state signal in the right parietal regions of the default mode network may correlate with the metabolic downregulation of the left hemisphere. Similarly, the relatively smaller visual processing network on the right side may correlate with the ipsilateral LGN lesion.

In Patient 2, we could identify the anterior part of RSN 1, as well as RSN 3, 5, and 6. The volume of these networks was reduced but not significantly different from the control subjects (Fig. 3c). The posterior areas of RSN 1 were missing. This observation was correlated with particularly low metabolism in the corresponding regions, and extensive loss of midline cortical regions.

In conclusion, preliminary studies of alterations in resting state patterns in subjects with brain lesions show correlations with lesion patterns and resting metabolism PET. Further development will be required to assess the potential role of resting state fMRI as an additional tool in evaluating patients with severe brain injuries.

## Discussion

We have reviewed two complementary MRI modalities, DTI and resting state fMRI, to image neuronal structure and function. Whereas DTI has become a mature tool to study various diseases, resting state fMRI is still in its infancy and requires more research to fully understand the observations. However, our two case studies of patients with severe brain lesions provide hints that resting state fMRI may have the potential to become an important tool to assess large-scale connectivity and functional integrity of neuronal networks.

Whether the discussed DTI or resting state fMRI methods, or their combination with other imaging modalities can be used as reliable predictors of outcome after brain lesions remains a challenging problem warranting further research (Calixto et al., 2007; Galanaud et al., 2007; Laureys and Boly, 2007; Schiff, 2007; Rousseau et al., 2008). In particular, more longitudinal, multimodal imaging studies (such as Naganawa et al., 2004; Ewing-Cobbs et al., 2006; Weiss et al., 2007) are required to obtain a clear picture of the predictive value of white matter imaging.

## Abbreviations

| | |
|---|---|
| AD | Alzheimer's disease |
| BOLD | blood–oxygen-level-dependent |
| DTI | diffusion tensor imaging |
| EEG | electroencephalography |
| FDG | fluorodeoxyglucose |

| | |
|---|---|
| fMRI | functional MRI |
| MCS | minimally conscious state |
| MRI | magnetic resonance imaging |
| PET | positron emission tomography |
| RSN | resting state network |

**Acknowledgments**

We acknowledge funding from NIH-NICHD, the Dana Foundation, the James S. McDonnell Foundation (NDS), and the Cervical Spine Research Society (HUV).

**References**

Alexander, A. L., Lee, J. E., Lazar, M., & Field, A. S. (2007). Diffusion tensor imaging of the brain. *Neurotherapeutics*, 4, 316–329.

Alexander, D. C. (2005). Multiple-fiber reconstruction algorithms for diffusion MRI. *Annals of the New York Academy of Sciences*, 1064, 113.

Anderson, A. W. (2005). Measurement of fiber orientation distributions using high angular resolution diffusion imaging. *Magnetic Resonance in Medicine*, 54, 1194–1206.

Andrews-Hanna, J. R., Snyder, A. Z., Vincent, J. L., Lustig, C., Head, D., Raichle, M. E., et al. (2007). Disruption of large-scale brain systems in advanced aging. *Neuron*, 56, 924–935.

Assaf, Y., & Pasternak, O. (2008). Diffusion tensor imaging (DTI)-based white matter mapping in brain research: a review. *Journal of Molecular Neuroscience*, 34, 51–61.

Basser, P. J. (1995). Inferring microstructural features and the physiological state of tissues from diffusion-weighted images. *NMR in Biomedicine*, 8, 333–344.

Basser, P. J., Mattiello, J., & LeBihan, D. (1994). MR diffusion tensor spectroscopy and imaging. *Biophysical Journal*, 66, 259–267.

Basser, P. J., & Pierpaoli, C. (1996). Microstructural and physiological features of tissues elucidated by quantitative-diffusion-tensor MRI. *Journal of Magnetic Resonance. Series B*, 111, 209–219.

Beaulieu, C., & Allen, P. S. (1994). Determinants of anisotropic water diffusion in nerves. *Magnetic Resonance in Medicine*, 31, 394–400.

Bengtsson, S. L., Nagy, Z., Skare, S., Forsman, L., Forssberg, H., & Ullen, F. (2005). Extensive piano practicing has regionally specific effects on white matter development. *Nature Neuroscience*, 8, 1148–1150.

Biswal, B., Yetkin, F. Z., Haughton, V. M., & Hyde, J. S. (1995). Functional connectivity in the motor cortex of resting human brain using echo-planar MRI. *Magnetic Resonance in Medicine*, 34, 537–541.

Burton, H., Snyder, A. Z., & Raichle, M. E. (2004). Default brain functionality in blind people. *Proceedings of the National Academy of Sciences of the United States of America*, 101, 15500–15505.

Calixto, M., Rafael, R., & Julius, K. (2007). Will diffusion tensor imaging assess rewiring in PVS' brains? *Neurology India*, 55, 88.

Campbell, J. S. W., Siddiqi, K., Rymar, V. V., Sadikot, A. F., & Pike, G. B. (2005). Flow-based fiber tracking with diffusion tensor and q-ball data: validation and comparison to principal diffusion direction techniques. *Neuroimage*, 27, 725–736.

Cao, Q. J., Zang, Y. F., Sun, L., Sui, M. Q., Long, X. Y., Zou, Q. H., & Wang, Y. F. (2006). Abnormal neural activity in children with attention deficit hyperactivity disorder: a resting-state functional magnetic resonance imaging study. *Neuroreport*, 17, 1033–1036.

Carmichael, S. T., & Chesselet, M. F. (2002). Synchronous neuronal activity is a signal for axonal sprouting after cortical lesions in the adult. *The Journal of Neuroscience*, 22, 6062–6070.

Chklovskii, D. B., Mel, B. W., & Svoboda, K. (2004). Cortical rewiring and information storage. *Nature*, 431, 782–788.

Coremans, J., Luypaert, R., Verhelle, F., Stadnik, T., & Osteaux, M. (1994). A method for myelin fiber orientation mapping using diffusion-weighted MR-images. *Magnetic Resonance Imaging*, 12, 443–454.

Crooker, S. A., Rickel, D. G., Balatsky, A. V., & Smith, D. L. (2004). Spectroscopy of spontaneous spin noise as a probe of spin dynamics and magnetic resonance. *Nature*, 431, 49–52.

Croxson, P. L., Johansen-Berg, H., Behrens, T. E. J., Robson, M. D., Pinsk, M. A., Gross, C. G., et al. (2005). Quantitative investigation of connections of the prefrontal cortex in the human and macaque using probabilistic diffusion tractography. *The Journal of Neuroscience*, 25, 8854–8866.

Dancause, N., Barbay, S., Frost, S. B., Plautz, E. J., Chen, D. F., Zoubina, E. V., et al. (2005). Extensive cortical rewiring after brain injury. *The Journal of Neuroscience*, 25, 10167–10179.

Danielsen, E. R., Christensen, P. B., rlien-Soborg, P., & Thomsen, C. (2003). Axonal recovery after severe traumatic brain injury demonstrated in vivo by 1H MR spectroscopy. *Neuroradiology*, 45, 722–724.

deIpolyi, A. R., Mukherjee, P., Gill, K., Henry, R. G., Patridge, S. C., Veeraraghavan, S., et al. (2005). Comparing microstructural and macrostructural development of the cerebral cortex in premature newborns: diffusion tensor imaging versus cortical gyration. *Neuroimage*, 27, 579–586.

De Luca, M., Beckmann, C. F., De Stefano, N., Matthews, P. M., & Smith, S. M. (2006). fMRI resting state networks define distinct modes of long-distance interactions in the human brain. *Neuroimage*, 29, 1359–1367.

Deo, A. A., Grill, R. J., Hasan, K. M., & Narayana, P. A. (2006). In vivo serial diffusion tensor imaging of experimental spinal cord injury. *Journal of Neuroscience Research*, 83, 801–810.

Douek, P., Turner, R., Pekar, J., Patronas, N., & LeBihan, D. (1991). MR color mapping of myelin fiber orientation. *Journal of Computer Assisted Tomography*, 15, 923–929.

Esposito, F., Formisano, E., Seifritz, E., Goebel, R., Morrone, R., Tedeschi, G., et al. (2002). Spatial independent component analysis of functional MRI time-series: to what extent do results depend on the algorithm used? *Human Brain Mapping*, 16, 146–157.

Esposito, F., Scarabino, T., Hyvarinen, A., Himberg, J., Formisano, E., Comani, S., et al. (2005). Independent component analysis of fMRI group studies by self-organizing clustering. *Neuroimage*, 25, 193–205.

Ewing-Cobbs, L., Hasan, K. M., Prasad, M. R., Kramer, L., & Bachevalier, J. (2006). Corpus callosum diffusion anisotropy correlates with neuropsychological outcomes in twins disconcordant for traumatic brain injury. *American Journal of Neuroradiology*, 27, 879–881.

Farrell, J. A. D., Landman, B. A., Jones, C. K., Smith, S. A., Prince, J. L., van Zijl, P. C. M., et al. (2007). Effects of signal-to-noise ratio on the accuracy and reproducibility of diffusion tensor imaging-derived fractional anisotropy, mean diffusivity, and principal eigenvector measurements at 1.5 T. *Journal of Magnetic Resonance Imaging*, 26, 756–767.

Feige, B., Scheffler, K., Esposito, F., Di Salle, F., Hennig, J., & Seifritz, E. (2005). Cortical and subcortical correlates of electroencephalographic alpha rhythm modulation. *Journal of Neurophysiology*, 93, 2864–2872.

Florence, S. L., Taub, H. B., & Kaas, J. H. (1998). Large-scale sprouting of cortical connections after peripheral injury in adult macaque monkeys. *Science*, 282, 1117–1121.

Formisano, E., Esposito, F., Di Salle, F., & Goebel, R. (2001). Cortex-based independent component analysis of fMRI time-series. *Neuroimage*, 13, S119.

Fox, M. D., & Raichle, M. E. (2007). Spontaneous fluctuations in brain activity observed with functional magnetic resonance imaging. *Nature Reviews. Neuroscience*, 8, 700–711.

Fox, M. D., Snyder, A. Z., Vincent, J. L., Corbetta, M., Van Essen, D. C., & Raichle, M. E. (2005). The human brain is intrinsically organized into dynamic, anticorrelated functional networks. *Proceedings of the National Academy of Sciences of the United States of America*, 102, 9673–9678.

Galanaud, D., Naccache, L., & Puybasset, L. (2007). Exploring impaired consciousness: the MRI approach. *Current Opinion in Neurology*, 20, 627–631.

Giacino, J. T., Ashwal, S., Childs, N., Cranford, R., Jennett, B., Katz, D. I., et al. (2002). The minimally conscious state — definition and diagnostic criteria. *Neurology*, 58, 349–353.

Giacino, J. T., Hirsch, J., Schiff, N., & Laureys, S. (2006). Functional neuroimaging applications for assessment and rehabilitation planning in patients with disorders of consciousness. *Archives of Physical Medicine and Rehabilitation*, 87, S67–S76.

Goldman, R. I., Stern, J. M., Engel, J., & Cohen, M. S. (2002). Simultaneous EEG and fMRI of the alpha rhythm. *Neuroreport*, 13, 2487–2492.

Greicius, M. D., Krasnow, B., Reiss, A. L., & Menon, V. (2003). Functional connectivity in the resting brain: a network analysis of the default mode hypothesis. *Proceedings of the National Academy of Sciences of the United States of America*, 100, 253–258.

Greicius, M. D., & Menon, V. (2004). Default-mode activity during a passive sensory task: uncoupled from deactivation but impacting activation. *Journal of Cognitive Neuroscience*, 16, 1484–1492.

Hagmann, P., Jonasson, L., Maeder, P., Thiran, J. P., Wedeen, V. J., & Meuli, R. (2006). Understanding diffusion MR imaging techniques: from scalar diffusion-weighted imaging to diffusion tensor imaging and beyond. *Radiographics*, 26, S205–U219.

Hampson, M., Peterson, B. S., Skudlarski, P., Gatenby, J. C., & Gore, J. C. (2002). Detection of functional connectivity using temporal correlations in MR images. *Human Brain Mapping*, 15, 247–262.

Han, B. S., Kim, S. H., Kim, O. L., Cho, S. H., Kim, Y. H., & Jang, S. H. (2007). Recovery of corticospinal tract with diffuse axonal injury: a diffusion tensor image study. *Neurorehabilitation*, 22, 151–155.

He, Y., Wang, L., Zang, Y. F., Tian, L. X., Zhang, X. Q., Li, K. C., et al. (2007). Regional coherence changes in the early stages of Alzheimer's disease: a combined structural and resting-state functional MRI study. *Neuroimage*, 35, 488–500.

Hermoye, L., Saint-Maitin, C., Cosnard, G., Lee, S. K., Kim, J., Nassogne, M. C., et al. (2006). Pediatric diffusion tensor imaging: normal database and observation of the white matter maturation in early childhood. *Neuroimage*, 29, 493–504.

Holodny, A. I., Gor, D. M., Watts, R., Gutin, P. H., & Ulug, A. M. (2005). Diffusion-tensor MR tractography of somatotopic organization of corticospinal tracts in the internal capsule: initial anatomic results in contradistinction to prior reports. *Radiology*, 234, 649–653.

Ino, T., Nakai, R., Azuma, T., Yamamoto, T., Tsutsumi, S., & Fukuyama, H. (2007). Somatotopy of corticospinal tract in the internal capsule shown by functional MRI and diffusion tensor images. *Neuroreport*, 18, 665–668.

Jafri, M. J., Pearlson, G. D., Stevens, M., & Calhoun, V. D. (2008). A method for functional network connectivity among spatially independent resting-state components in schizophrenia. *Neuroimage*, 39, 1666–1681.

Jain, N., Florence, S. L., Qi, H. X., & Kaas, J. H. (2000). Growth of new brainstem connections in adult monkeys with massive sensory loss. *Proceedings of the National Academy of Sciences of the United States of America*, 97, 5546–5550.

Jellison, B. J., Field, A. S., Medow, J., Lazar, M., Salamat, M. S., & Alexander, A. L. (2004). Diffusion tensor imaging of cerebral white matter: a pictorial review of physics, fiber tract anatomy, and tumor imaging patterns. *American Journal of Neuroradiology*, 25, 356–369.

Johansen-Berg, H., & Behrens, T. E. J. (2006). Just pretty pictures? What diffusion tractography can add in clinical neuroscience. *Current Opinion in Neurology*, 19, 379–385.

Johansen-Berg, H., Behrens, T. E. J., Sillery, E., Ciccarelli, O., Thompson, A. J., Smith, S. M., et al. (2005). Functional-anatomical validation and individual variation of diffusion tractography-based segmentation of the human thalamus. *Cerebral Cortex*, 15, 31–39.

Jones, D. K., Travis, A. R., Eden, G., Pierpaoli, C., & Basser, P. J. (2005). PASTA: pointwise assessment of streamline tractography attributes. *Magnetic Resonance in Medicine, 53*, 1462–1467.

Kim, Y. H., Hong, J. H., Park, C. H., Hua, N., Bickart, K. C., Byun, W. M., et al. (2008). Corticospinal tract location in internal capsule of human brain: diffusion tensor tractography and functional MRI study. *Neuroreport, 19*, 817–820.

Kingsley, P. B. (2006a). Introduction to diffusion tensor imaging mathematics: part I. Tensors, rotations, and eigenvectors. *Concepts in Magnetic Resonance Part A, 28A*, 101–122.

Kingsley, P. B. (2006b). Introduction to diffusion tensor imaging mathematics: part II. Anisotropy, diffusion-weighting factors, and gradient encoding schemes. *Concepts in Magnetic Resonance Part A, 28A*, 123–154.

Klein, J. C., Behrens, T. E. J., Robson, M. D., Mackay, C. E., Higham, D. J., & Johansen-Berg, H. (2007). Connectivity-based parcellation of human cortex using diffusion MRI: establishing reproducibility, validity and observer independence in BA 44/45 and SMA/pre-SMA. *Neuroimage, 34*, 204–211.

Koch, M. A., & Norris, D. G. (2006). Artifacts and pitfalls in diffusion MR imaging. In J. Gillard, A. Waldman, & P. Barker (Eds.), *Clinical MR neuroimaging – Diffusion, perfusion and spectroscopy* (pp. 99–108). Cambridge: Cambridge University Press.

Landman, B. A., Farrell, J. A. D., Jones, C. K., Smith, S. A., Prince, J. L., & Mori, S. (2007). Effects of diffusion weighting schemes on the reproducibility of DTI-derived fractional anisotropy, mean diffusivity, and principal eigenvector measurements at 1.5 T. *Neuroimage, 36*, 1123–1138.

Laufs, H., Hamandi, K., Salek-Haddadi, A., Kleinschmidt, A. K., Duncan, J. S., & Lemieux, L. (2007). Temporal lobe interictal epileptic discharges affect cerebral activity in "default mode" brain regions. *Human Brain Mapping, 28*, 1023–1032.

Laufs, H., Krakow, K., Sterzer, P., Eger, E., Beyerle, A., Salek-Haddadi, A., et al. (2003). Electroencephalographic signatures of attentional and cognitive default modes in spontaneous brain activity fluctuations at rest. *Proceedings of the National Academy of Sciences of the United States of America, 100*, 11053–11058.

Lauk, M., Chow, C. C., Pavlik, A. E., & Collins, J. J. (1998). Human balance out of equilibrium: nonequilibrium statistical mechanics in posture control. *Physical Review Letters, 80*, 413–416.

Laureys, S., & Boly, M. (2007). What is it like to be vegetative or minimally conscious? *Current Opinion in Neurology, 20*, 609–613.

LeBihan, D., Breton, E., Lallemand, D., Grenier, P., Cabanis, E., & Lavaljeantet, M. (1986). MR imaging of intravoxel incoherent motions – application to diffusion and perfusion in neurologic disorders. *Radiology, 161*, 401–407.

Leopold, D. A., Murayama, Y., & Logothetis, N. K. (2003). Very slow activity fluctuations in monkey visual cortex: implications for functional brain imaging. *Cerebral Cortex, 13*, 422–433.

Lim, K. O., & Helpern, J. A. (2002). Neuropsychiatric applications of DTI – a review. *NMR in Biomedicine, 15*, 587–593.

Liu, Y., Liang, M., Zhou, Y., He, Y., Hao, Y. H., Song, M., et al. (2008). Disrupted small-world networks in schizophrenia. *Brain, 131*, 945–961.

Liu, W. C., Ollenschleger, M., Kalnin, A., & Holodny, A. (2001). Two directional method to identify white matter tracts for patients with brain tumors. *Neuroimage, 13*, S1213.

Lowe, M. J., Mock, B. J., & Sorenson, J. A. (1998). Functional connectivity in single and multislice echoplanar imaging using resting-state fluctuations. *Neuroimage, 7*, 119–132.

Lu, H., Zuo, Y., Gu, H., Waltz, J. A., Zhan, W., Scholl, C. A., et al. (2007). Synchronized delta oscillations correlate with the resting-state functional MRI signal. *Proceedings of the National Academy of Sciences of the United States of America, 104*, 18265–18269.

Mac Donald, C., Dikranian, K., Bayly, P., Holtzman, D., & Brody, D. (2007). Diffusion tensor imaging reliably detects experimental traumatic axonal injury and indicates approximate time of injury. *Journal of Neurotrauma, 24*, 1248.

Mantini, D., Perrucci, M. G., Del Gratta, C., Romani, G. L., & Corbetta, M. (2007). Electrophysiological signatures of resting state networks in the human brain. *Proceedings of the National Academy of Sciences of the United States of America, 104*, 13170–13175.

McKinstry, R. C., Mathur, A., Miller, J. H., Ozcan, A., Snyder, A. Z., Schefft, G. L., et al. (2002). Radial organization of developing preterm human cerebral cortex revealed by non-invasive water diffusion anisotropy MRI. *Cerebral Cortex, 12*, 1237–1243.

Minati, L., & Weglarz, W. P. (2007). Physical foundations, models, and methods of diffusion magnetic resonance imaging of the brain: a review. *Concepts in Magnetic Resonance Part A, 30A*, 278–307.

Mori, S., & van Zijl, P. C. M. (2002). Fiber tracking: principles and strategies – a technical review. *NMR in Biomedicine, 15*, 468–480.

Mori, S., Wakana, S., Nagae-Poetscher, L. M., & van Zijl, P. C. M. (2005). *MRI atlas of human white matter*. Amsterdam: Elsevier.

Mukherjee, P., Berman, J. I., Chung, S. W., Hess, C. P., & Henry, R. G. (2008a). Diffusion tensor MR imaging and fiber tractography: theoretic underpinnings. *American Journal of Neuroradiology, 29*, 632–641.

Mukherjee, P., Chung, S. W., Berman, J. I., Hess, C. P., & Henry, R. G. (2008b). Diffusion tensor MR imaging and fiber tractography: technical considerations. *American Journal of Neuroradiology, 29*, 843–852.

Mukherjee, P., & McKinstry, R. C. (2006). Diffusion tensor imaging and tractography of human brain development. *Neuroimaging Clinics of North America, 16*, 19.

Naganawa, S., Sato, C., Ishihra, S., Kumada, H., Ishigaki, T., Miura, S., et al. (2004). Serial evaluation of diffusion tensor brain fiber tracking in a patient with severe diffuse axonal injury. *American Journal of Neuroradiology, 25*, 1553–1556.

Napieralski, J. A., Butler, A. K., & Chesselet, M. F. (1996). Anatomical and functional evidence for lesion-specific sprouting of corticostriatal input in the adult rat. *The Journal of Comparative Neurology, 373*, 484–497.

Narayana, A., Chang, J., Thakur, S., Huang, W., Karimi, S., Hou, B., et al. (2007). Use of MR spectroscopy and functional imaging in the treatment planning of gliomas. *The British Journal of Radiology, 80*, 347–354.

Neil, J. J. (1997). Measurement of water motion (apparent diffusion) in biological systems. *Concepts in Magnetic Resonance, 9*, 385–401.

Niedermeyer, E., & da Silva, F. L. (Eds). (2004). *Electroencephalography: Basic principles, clinical applications, and related fields*. Philadelphia: Lippincott Williams & Wilkins.

Niogi, S. N., Mukherjee, P., & McCandliss, B. D. (2007). Diffusion tensor imaging segmentation of white matter structures using a Reproducible Objective Quantification Scheme (ROQS). *Neuroimage, 35*, 166–174.

Nunez, P. L., & Srinivasan, R. (2005). *Electric fields of the brain: The neurophysics of EEG*. Oxford: Oxford University Press.

Owen, A. M., Coleman, M. R., Menon, D. K., Johnsrude, I. S., Rodd, J. M., Davis, M. H., et al. (2005). Residual auditory function in persistent vegetative state: a combined PET and fMRI study. *Neuropsychological Rehabilitation, 15*, 290–306.

Pajevic, S., & Pierpaoli, C. (1999). Color schemes to represent the orientation of anisotropic tissues from diffusion tensor data: application to white matter fiber tract mapping in the human brain. *Magnetic Resonance in Medicine, 42*, 526–540.

Park, J. K., Kim, B. S., Choi, G., Kim, S. H., Choi, J. C., & Khang, H. (2008). Evaluation of the somatotopic organization of corticospinal tracts in the internal capsule and cerebral peduncle: results of diffusion-tensor MR tractography. *Korean Journal of Radiology, 9*, 191–195.

Partridge, S. C., Mukherjee, P., Henry, R. G., Miller, S. P., Berman, J. I., Jin, H., et al. (2004). Diffusion tensor imaging: serial quantitation of white matter tract maturity in premature newborns. *Neuroimage, 22*, 1302–1314.

Pons, T. P., Garraghty, P. E., Ommaya, A. K., Kaas, J. H., Taub, E., & Mishkin, M. (1991). Massive cortical reorganization after sensory deafferentation in adult macaques. *Science, 252*, 1857–1860.

Raichle, M. E., MacLeod, A. M., Snyder, A. Z., Powers, W. J., Gusnard, D. A., & Shulman, G. L. (2001). A default mode of brain function. *Proceedings of the National Academy of Sciences of the United States of America, 98*, 676–682.

Ramu, J., Herrera, J., Grill, R., Bockhorst, T., & Narayana, P. (2008). Brain fiber tract plasticity in experimental spinal cord injury: diffusion tensor imaging. *Experimental Neurology, 212*, 100–107.

Rombouts, S. A. R. B., Scheltens, P., Kuijer, J. P. A., & Barkhof, F. (2007). Whole brain analysis of T2* weighted baseline FMRI signal in dementia. *Human Brain Mapping, 28*, 1313–1317.

Rossberg, A. G., Bartholome, K., Voss, H. U., & Timmer, J. (2004). Phase synchronization from noisy univariate signals. *Physical Review Letters, 93*, 154103.

Rousseau, M. C., Confort-Gouny, S., Catala, A., Graperon, J., Blaya, J., Soulier, E., et al. (2008). A MRS-MRI-fMRI exploration of the brain. Impact of long-lasting persistent vegetative state. *Brain Injury, 22*, 123–134.

Schiff, N. D. (2007). Bringing neuroimaging tools closer to diagnostic use in the severely injured brain. *Brain, 130*, 2482–2483.

Schiff, N. D., Rodriguez-Moreno, D., Kamal, A., Kim, K. H. S., Giacino, J. T., Plum, F., et al. (2005). fMRI reveals large-scale network activation in minimally conscious patients. *Neurology, 64*, 514–523.

Schnitzler, A., & Gross, J. (2005). Normal and pathological oscillatory communication in the brain. *Nature Reviews. Neuroscience, 6*, 285–296.

Schwartzman, A., Dougherty, R. F., & Taylor, J. E. (2005). Cross-subject comparison of principal diffusion direction maps. *Magnetic Resonance in Medicine, 53*, 1423–1431.

Seeley, W. W., Menon, V., Schatzberg, A. F., Keller, J., Glover, G. H., Kenna, H., et al. (2007). Dissociable intrinsic connectivity networks for salience processing and executive control. *The Journal of Neuroscience, 27*, 2349–2356.

Sidaros, A., Engberg, A., Sidaros, K., Liptrot, M. G., Herning, M., Petersen, P., et al. (2008). Diffusion tensor imaging during recovery from severe traumatic brain injury and relation to clinical outcome: a longitudinal study. *Brain, 131*, 559–572.

Singh, K. D., & Fawcett, I. P. (2008). Transient and linearly graded deactivation of the human default-mode network by a visual detection task. *Neuroimage, 41*, 100–112.

Skoglund, T. S., Nilsson, D., Ljungberg, M., Jonsson, L., & Rydenhag, B. (2008). Long-term follow-up of a patient with traumatic brain injury using diffusion tensor imaging. *Acta Radiologica, 49*, 98–100.

Smith, S. M., Johansen-Berg, H., Jenkinson, M., Rueckert, D., Nichols, T. E., Miller, K. L., et al. (2007). Acquisition and voxelwise analysis of multi-subject diffusion data with Tract-Based Spatial Statistics. *Nature Protocols, 2*, 499–503.

Sorg, C., Riedl, V., Muhlau, M., Calhoun, V. D., Eichele, T., Laer, L., et al. (2007). Selective changes of resting-state networks in individuals at risk for Alzheimer's disease. *Proceedings of the National Academy of Sciences of the United States of America, 104*, 18760–18765.

Sundgren, P. C., Dong, Q., Gomez-Hassan, D., Mukherji, S. K., Maly, P., & Welsh, R. (2004). Diffusion tensor imaging of the brain: review of clinical applications. *Neuroradiology, 46*, 339–350.

Tournier, J. D., Calamante, F., Gadian, D. G., & Connelly, A. (2004). Direct estimation of the fiber orientation density function from diffusion-weighted MRI data using spherical deconvolution. *Neuroimage, 23*, 1176–1185.

Tuch, D. S. (2004). Q-ball imaging. *Magnetic Resonance in Medicine, 52*, 1358–1372.

Van der Werf, Y. D., Weerts, J. G. E., Jolles, J., Witter, M. P., Lindeboom, J., & Scheltens, P. (1999). Neuropsychological correlates of a right unilateral lacunar thalamic infarction. *Journal of Neurology, Neurosurgery, and Psychiatry, 66*, 36–42.

Vincent, J. L., Patel, G. H., Fox, M. D., Snyder, A. Z., Baker, J. T., Van Essen, D. C., et al. (2007). Intrinsic functional architecture in the anaesthetized monkey brain. *Nature, 447*, 83–86.

von Stein, A., & Sarnthein, J. (2000). Different frequencies for different scales of cortical integration: from local gamma to long range alpha/theta synchronization. *International Journal of Psychophysiology, 38*, 301–313.

Voss, H. U., Ulug, A. M., Dyke, J. P., Watts, R., Kobylarz, E. J., McCandliss, B. D., et al. (2006). Possible axonal regrowth in late recovery from the minimally conscious state. *The Journal of Clinical Investigation, 116*, 2005–2011.

Waites, A. B., Briellmann, R. S., Saling, M. M., Abbott, D. F., & Jackson, G. D. (2006). Functional connectivity networks are disrupted in left temporal lobe epilepsy. *Annals of Neurology, 59*, 335–343.

Wang, K., Liang, M., Wang, L., Tian, L. X., Zhang, X. Q., Li, K. C., et al. (2007). Altered functional connectivity in early Alzheimer's disease: a resting-state fMRI study. *Human Brain Mapping, 28*, 967–978.

Weiss, N., Galanaud, D., Carpentier, A., Naccache, L., & Puybasset, L. (2007). Clinical review: prognostic value of magnetic resonance imaging in acute brain injury and coma. *Critical Care, 11*, 230–241.

Wozniak, J. R., & Lim, K. O. (2006). Advances in white matter imaging: a review of in vivo magnetic resonance methodologies and their applicability to the study of development and aging. *Neuroscience and Biobehavioral Reviews, 30*, 762–774.

Wu, T., Zang, Y. F., Wang, L., Long, X. Y., Li, K. C., & Chan, P. (2007). Normal aging decreases regional homogeneity of the motor areas in the resting state. *Neuroscience Letters, 423*, 189–193.

Zhu, C. Z., Zang, Y. F., Cao, Q. J., Yan, C. G., He, Y., Jiang, T. Z., et al. (2008). Fisher discriminative analysis of resting-state brain function for attention-deficit/hyperactivity disorder. *Neuroimage, 40*, 110–120.

# CHAPTER 32

# The eighteenth C.U. Ariëns Kappers lecture: an introduction

Joost Verhaagen and Dick F. Swaab

*Laboratory for Neuroregeneration, Netherlands Institute for Neuroscience, an Institute of the Royal Academy of Arts and Sciences, Amsterdam, The Netherlands*

**C.U. Ariëns Kappers and the Central Institute for Brain Research in Amsterdam**

Modern brain research in the Netherlands dates back to the International Association of Academies in Paris, France, in 1901, where, on the initiative of the anatomist Wilhelm His, it was proposed to place research of the nervous system on an international footing. The resulting "Brain Committee" — with big international names, among which were three Nobel Laureates, namely, Camillo Golgi from Italy, Santiago Ramón y Cajal from Spain, and Charles Scott Sherrington from England — proceeded to launch a daring scheme that entailed "…organizing a network of institutions throughout the civilized world, dedicated to the study of the structure and functions of the central organ and collaborating according to a well thought out scheme" because they felt that "the time was not far distant when brain anatomists would be forced to divide the millions of brain cells among themselves in the same way that astronomers had been obliged to divide the millions of stars into various groups."

The Brain Committee decided to establish nine "interacademic institutes" in eight countries, with each of these institutes representing one of the then existing disciplines of brain research. One of these institutes was the Central Institute for Brain Research in Amsterdam, which opened on June 8, 1909, with C.U. Ariëns Kappers, who had spent the previous two years working as a team leader at the institute of the famous comparative neuroanatomist Ludwig Edinger in Frankfurt, Germany, as its first director.

The Central Institute for Brain Research was the first institute established in the Netherlands with the express purpose of carrying out pure scientific research, as is stated, not without pride, in speeches held during the opening of the Central Institute for Brain Research.

From the start, the institute was internationally oriented and boasted many foreign guest researchers, with Kappers himself spending time as visiting professor at many foreign universities, such as Beirut, and Beijing Union Medical College, where he stayed during 1923–1924.

In those days, the whole of China only counted 2000 medical students. Kappers had to teach brain anatomy in Beijing, and for this purpose, he had 50 brains from the Wilhelmina Gasthuis Hospital in Amsterdam wrapped up and shipped to China. He was also expected to teach histology of every organ of the human body, a subject he had not thought about for 15 years. He used his time on board the ship from Marseille to Shanghai to brush up his knowledge. The porch of the laboratory in Beijing where Kappers worked and was photographed among his students, the dissecting room behind it, and the roofs with the green glazed roofing tiles that he loved have remained intact to this day, incorporated in a huge new hospital complex.

While Kappers was staying in China, at the Beijing Union Medical College, this institute organized archeological excavations in the caves near Beijing, with a team headed by Professor Davidson Black. These excavations resulted in the famous finds of a large number of fossils of "Peking man," *Sinanthropus pekinensis*, who lived in the Dragon Bone Hill area some 40,000 years ago.

Back in the Netherlands, in 1924, Kappers made an attempt to put Black in touch with Eugene Dubois, the Dutch physician who, in 1891, had found the fossils of "Java man," *Pithecanthropus erectus*, in the then Dutch colony of Indonesia, so that they could compare these two vital evolutionary finds. Unfortunately, Dubois was so paranoid about someone taking off with his results that this meeting never came about; Kappers and Black arrived at Teylers Museum in Haarlem, where the fossils were kept in a safe, only to learn that Dubois had left for his country house. He had left strict instructions with the laboratory assistant to show them only the plaster casts of the fossils, but they already had those. At present, the original Java man fossils may be seen in a glass display cabinet at the National Museum of Natural History, Naturalis, in Leiden. Not long ago, it was discovered that the Peking man fossils might have been aboard the Awa Maru, a Japanese Red Cross relief ship that was torpedoed by an American submarine, the USS Queenfish, in the Taiwan Strait on April 1, 1945. Two thousand people perished, and only one survived.

Kappers became an extraordinary professor at the University of Amsterdam in 1929. Black died in 1934, at the age of 50, in Beijing, when the excavations were still in full swing.

In 1941, after the hectic day of the attack on Pearl Harbor, the US Marines packed the many Peking man fossils in crates and put them on a train to get them (and themselves) out of China. However, they were intercepted by the Japanese, and the US Marines were made prisoners of war. The crates were left beside the track and have not been seen since. The emperor of Japan declared he does not have them in his possession. The only things that remain in the Anatomy Department of the Beijing Union Medical College are photographs of the excavations, a cast of a skull, and a model of Peking man that shows a remarkable likeness to modern man. As far as we know, the Chinese have so far not been successful in salvaging the Awa Maru.

In 1921, the Central Institute for Brain Research even drew royal attention; Her Majesty the Queen Mother Emma paid a visit. Because Kappers wanted a decent red carpet for the stairwell, he borrowed his parents' stair carpet, which he rolled up and brought to the institute — "not without some difficulty" — on the handlebars of his bicycle. He could not have acted more Dutch if he tried.

On June 5, 1918, Kappers took the initiative to publish a supplement of the journal The Psychiatric and Neurological Journals, with the visionary title *Neurotherapy*. The first issue appeared one year later. Until 1918, *Neurotherapy* contained papers in Dutch, French, German, and English about the regeneration of nerve fibers. However, it was not until recently that research into neurotherapy gained momentum. It was the topic of our 25th International Summer School of Brain Research in 2008.

Kappers' work has been taken down in his two-volume handbook, in German, *Die vergleichende Anatomie des Nervensystems der Wirbeltiere und des Menschen*, with him as the sole author. The two volumes were published in 1920 and 1921, and appeared later in three volumes in English, with Crosby and Huber as coauthors. These three volumes are still in use today.

He also gained fame with his "neurobiotaxic theory," in which he introduced the idea that the connections between structures in development are determined by simultaneous electrical activities of these structures during development. From his correspondence with Cajal about this subject, it emerges that Cajal was of the contrary opinion that the outgrowth of nerve fibers during development is determined by chemical attraction. To date, both aspects still play a role in the way we think about developmental processes in the brain.

Professor Kappers received many prizes. He was awarded honorary doctorates in Glasgow, Dublin, Chicago, and Yale. He did not get married until he was 60, when he lived in Villa Betty, a stately mansion between Overtoom and Vondelpark in Amsterdam, and let himself be chauffeured in a Rolls Royce. He died in 1946, at the age of 68, while he was in the last stage of writing his memoirs, published in 2001, aptly titled *C.U. Ariëns Kappers, Reiziger in Breinen (Traveller in Brains)*. The book paints an excellent picture of how science was done by "gentlemen" in the early days of the 20th century.

In 1987, in honor of the first director of the institute, the then director Dick Swaab instigated the Ariëns Kappers lecture and an accompanying medal was struck. Pasko Rakic was the first to deliver the lecture and receive the medal in 1987. Since then, 17 leading neuroscientists have received the medal, all for their outstanding contributions to the field. For the 25th International Summer School of Brain Research of the Netherlands Institute for Neuroscience, in 2008, we were very fortunate that James Fawcett, director of the Cambridge Centre for Brain Repair, accepted our invitation to deliver the Ariëns Kappers lecture. He received the accompanying medal for his outstanding contributions to the field of restorative neuroscience and neurology. In his lecture, James Fawcett highlighted his work on the inhibitory role of the extracellular matrix on regenerating axons in the brain and spinal cord. This topic is indeed entirely within the scope of the journal *Neurotherapy*, which Professor Kappers instigated 90 years earlier.

# CHAPTER 33

# Molecular control of brain plasticity and repair

James Fawcett*

*Cambridge University Centre for Brain Repair, Department of Clinical Neurosciences, Robinson Way, Cambridge, UK*

**Abstract:** Recovery of function after damage to the CNS is limited due to the absence of axon regeneration and relatively low levels of plasticity. Plasticity in the CNS can be reactivated in the adult CNS by treatment with chondroitinase ABC, which removes glycosaminoglycan (GAG) chains from chondroitin sulfate proteoglycans (CSPGs). Plasticity in the adult CNS is restricted by perineuronal nets (PNNs) around many neuronal cell bodies and dendrites, which appear at the closure of critical periods and contain several inhibitory CSPGs. Formation of these structures and the turning off of plasticity is triggered by impulse activity in neurons. Expression of a link protein by neurons is the event that triggers the formation of PNNs. Treatment with chondroitinase removes PNNs and other inhibitory influences in the damaged spinal cord and promotes sprouting of new connections. However, promoting plasticity by itself does not necessarily bring back useful behavior; this only happens when useful connections are stabilized and inappropriate connections removed, driven by behavior. Thus after rodent spinal cord injury, combining a daily rehabilitation treatment for skilled paw function with chondroitinase produces much greater recovery than either treatment alone. The rehabilitation must be specific for the behavior that is to be enhanced because non-specific rehabilitation improves locomotor behavior but not skilled paw function. Plasticity-enhancing treatments may therefore open up a window of opportunity for successful rehabilitation.

**Keywords:** spinal cord; plasticity; axon regeneration; proteoglycan; rehabilitation; behavioral therapy; perineuronal net; extracellular matrix

## Mechanisms of recovery after CNS damage

After damage to the CNS, there is loss of function due to structural damage, inflammation, edema, and compression. Edema usually resolves rapidly, but inflammatory processes may continue for weeks, with release of cytokines, nitric oxide, and free radicals leading to conduction failure, demyelination, and other dysfunctions. Eventually, the patient is left with the consequences of structural damage and the loss of function that results from it. There is considerable spontaneous recovery from CNS lesions, which may continue for up to a year, although the most rapid improvement occurs during the first three months (Fawcett et al., 2007). Recovery over the first three weeks is probably mainly due to resolution of edema and inflammation. After that, plasticity and remyelination are the major players. Greater degrees of recovery of function could only occur if the damaged structures could be replenished by cell replacement and axonal regeneration, neither of which occurs spontaneously in the mammalian CNS.

*Corresponding author.
Tel.: +441223331160; Fax: +441223331174;
E-mail: jf108@cam.ac.uk

## Plasticity in recovery of function

The word "plasticity" is used as regards recovery of function in the damaged CNS to mean any process that leads to recreation of functional circuits, excluding long-distance axon regeneration. Short-distance sprouting above and below lesions leading to formation of new connections and alteration in the strength of existing connections are the processes involved. These changes can allow signals to bypass areas of damage through newly created circuits, and can also reassign areas of the CNS to new functions. This is seen most dramatically after cortical or subcortical stroke in human patients, where fMRI studies show that functions that were performed by the damaged area can become widely distributed throughout the brain, including the contralateral side. The more severe the damage, the more widely is the function redistributed (Ward, 2005; Rossini et al., 2003). The functional utility of the more widespread redistributions of function is debated: it is possible to silence some of these cortical regions by using transcranial magnetic stimulation (TMS), and most studies report improved function after silencing brain regions far from the area of damage, but there are also some contrary reports (Boggio et al., 2006). As function improves and rehabilitation continues, activity becomes more focused, usually in the perilesional regions close to the brain area that originally performed those functions. It is this perilesional plasticity and focusing of function that is associated with good functional recovery after cortical lesions (Ward, 2005). After spinal cord injury, the anatomical and physiological rearrangements needed to restore function are different, because the injury mainly affects long axonal tracts. Bypassing an injury therefore requires that the activity is diverted into preserved axons via interneuron circuits. In rat spinal cord injuries, the occurrence of this type of spontaneous plasticity has been demonstrated anatomically. The damaged corticospinal tract was shown to produce sprouts above the lesion that are connected to interneurons, which in turn made contact with motoneuron dendrites (Bareyre et al., 2004), and other descending pathways also sprout new connections (Ballermann and Fouad, 2006). Rearrangements have also been seen in the cortex after spinal cord injury (Fouad and Tse, 2008; Ramanathan et al., 2006). Direct evidence that plastic changes are responsible for recovery of function after CNS damage is sparse. There is an excellent correlation between the occurrence and time course of observed plasticity and recovery of function, but no direct way of preventing the process to see whether recovery is prevented. The most direct way in which the effects of cortical plasticity can be assessed is the use of repetitive TMS to inactivate cortical regions that have taken on new functions. The results of these studies are complex, but inactivation of cortical regions ipsilateral to the original damage and close to the lesion can lead to a significant loss of function, while inactivation of areas in the contralateral hemisphere that are active during a behavior can actually improve function (Nair et al., 2007; Boggio et al., 2006). Clearly, perilesional plastic changes are more likely to be beneficial than more widespread changes. Other evidence for the importance of plasticity in recovery comes from factors that modulate it. Developmental events affect the degree to which plastic changes can occur; there are also pharmacological interventions that directly promote plasticity, and these, as described in the following text, have the predicted effects on recovery of function after CNS damage.

## Plasticity decreases with age

The most studied example of CNS plasticity is the ocular dominance shift that occurs after disadvantaging one eye, leading to the other eye winning more space on the visual cortex. This form of plasticity is developmentally controlled, with a critical period for plasticity after which visual deprivation produces little change in cortical connections (Berardi et al., 2003). Where it has been studied, most other parts of the CNS appear to show similar critical periods, with a peak of plasticity shortly after birth terminating at around 35 days in rats, and 5 years in humans. There are several examples in which this critical period has clinical significance. The correction of amblyopia in human patients with a squint or cataract must be performed before five years if vision is to be restored. Peripheral nerve repair leads to regeneration of damaged axons back to muscle and skin, but there is no specific axon guidance, so reinnervation is almost random. In young patients the CNS is sufficiently plastic to allow some compensation for these incorrect connections, while in adults the functional results of peripheral nerve repair, even where there is good reinnervation, can be very poor (Lundborg, 2000). Many other functions show changes at around the same time, for instance, the acquisition of new language becomes much more difficult after about five years of age, and the new language occupies a new cortical area rather than being superimposed.

## Methods of restoring plasticity in the adult CNS

It might seem logical that the best methods for reactivating plasticity in the CNS would be to undo the events that lead to the closure of the critical periods. This is not exactly how things have worked out. It has been clear for some time that the ending of the critical period for ocular dominance plasticity is associated with changes in the level of GABAergic inhibition and with the maturation of GABAergic interneurons (Hensch, 2004). Unfortunately, any large systemic blocking of GABAergic inhibition leads to a lowered threshold for seizures, but transgenic manipulations that partly disable the inhibitory mechanisms can influence the critical period (Hensch, 2004). The most effective methods of restoring plasticity came from research intended to promote axon regeneration after CNS damage.

## Chondroitinase and CSPGs

Chondroitin sulfate proteoglycans (CSPGs) are upregulated in glial scar tissue and are inhibitory to axon regeneration, mainly through the action of their sulfated glycosaminoglycan (GAG) chains (Galtrey and Fawcett, 2007). CSPGs are generic barrier-forming molecules that guide axons in development and probably act as infection barriers when upregulated in injuries. Bacteria therefore evolved several forms of the enzyme chondroitinase, which digests the GAG chains to disaccharides, in the process removing much of the inhibitory activity of the CSPGs. In an attempt to overcome the inhibition of axon regeneration due to the glial scar, chondroitinase was injected into the brain and spinal cord, where it promoted axon regeneration, but in the spinal cord, it also produced a very marked and rapid improvement in behavior, faster and more profound than would be expected from the axon regeneration that had been observed (Bradbury et al., 2002; Moon et al., 2001). The pattern of recovery was consistent with the possibility that chondroitinase was promoting plasticity, so the next step was to see whether the enzyme would reactivate plasticity in a pure plasticity model, namely, ocular dominance shift. It was found that application of chondroitinase to the visual cortex of adult rats would reactivate ocular dominance plasticity, confirming the action of the enzyme on plasticity (Pizzorusso et al., 2002). Another pure plasticity model that has been investigated recently is compensation for the inaccurate connections made by peripheral nerves when they regenerate to reinnervate muscles and skin. Repair of the median and ulnar nerves leaves the rat forelimb weak and clumsy, as shown by skilled paw-reaching and grip strength tests, and if the nerves are crossed over to produce even more inaccurate reinnervation, there is almost no recovery of skilled paw function. Animals were given a month to achieve such spontaneous recovery as they were capable of, and then chondroitinase was injected into the spinal cord to reactivate plasticity. The result was an improvement in skilled paw-reaching ability and grip strength (Galtrey et al., 2007). In another plasticity model, a lesion of the dorsal columns in the cervical cord was made to partly denervate the cuneate sensory nucleus in the medulla. This is normally followed by limited sprouting of the remaining axons, leading to incomplete reinnervation of the nucleus. Application of chondroitinase increased this sprouting, leading to electrically functional connections throughout the nucleus (Massey et al., 2006).

Recently, chondroitinase has been used to restore breathing ability in animals with unilateral lesions of the descending axons controlling the diaphragm. Further experiments with the enzyme where the plasticity-inducing effect has been combined with rehabilitation, or in which the enzyme has been used to promote axon regeneration, are described later.

The plasticity-promoting effect of chondroitinase is well established, which begs the question: why should digesting the GAG chains of CSPGs in the CNS promote plasticity, or indeed how might CSPG GAG chains be instrumental in turning off plasticity in the first place? The CSPGs form part of the CNS extracellular matrix (ECM), which surrounds all the cells, synapses, and processes in the brain and spinal cord. The matrix can be divided into three main compartments: the free-floating loosely attached matrix that is found throughout the CNS, the matrix molecules that are attached to the cell surface, and the condensed cartilage-like matrix found around some neurons as perineuronal nets (PNNs) (Deepa et al., 2006). Examining developmental changes in CSPGs that might be responsible for terminating critical periods of plasticity, there are two main changes. The first is a developmental change in the sulfation pattern of the CSPG GAG chains, with the predominant form in embryos being sulfated at the 6 position of *N*-acetyl galactosamine, while in postnatal CNS, the majority is 4 sulfated (Properzi et al., 2005; Kitagawa et al., 1997). However, this shift happens at the end of embryogenesis, long before the critical period. Around the end of the critical periods, the obvious change is the beginning of formation of PNNs, which starts in rat cortex around day 30, or around day 18 in the spinal cord, in both cases corresponding with the reduction in plasticity (Bruckner et al., 2000; Pizzorusso et al., 2002). Moreover, it is possible to delay the end of the critical period in the visual cortex by rearing animals in darkness, and this prevents the formation of PNNs in the visual areas until animals are placed back in the light, but other brain regions are unaffected (Pizzorusso et al., 2002). Because PNNs surround synapses and dendrites and contain molecules that inhibit process growth, it has been assumed that they are the matrix structures that control plasticity. However, only a subset of cortical neurons are surrounded by PNNs, particularly the GABAergic interneurons and some pyramidal neurons, so it is not clear why restricting synaptic turnover with these ECM structures should affect the plasticity of the whole cortex. In the ventral spinal cord, most neurons and their dendrites are surrounded by dense PNNs, so a potential mechanism is easier to visualize. A test of the hypothesis that PNNs are responsible for terminating plasticity after critical periods will have to wait for transgenic animals in which the PNNs are absent (see the following text). However, animals lacking the PNN component tenascin-R have partly formed PNNs, and show greater than normal plasticity and recovery after CNS and facial nerve damage (Apostolova et al., 2006; Guntinas-Lichius et al., 2005). Chondroitinase digests the CSPGs in PNNs, but does not completely eliminate the structures. Moreover, chondroitinase has effects on all three compartments of the ECM, so the fact that chondroitinase promotes plasticity does not automatically implicate the PNNs.

The events that lead to the formation of PNNs are now understood. PNNs contain a mixture of CSPGs, tenascin-R, hyaluronan, and link proteins (Carulli et al., 2006). Hyaluronan chains provide the backbone of an ECM, and most of the CSPGs have hyaluronan-binding domains and tenascin-binding domains. This basic set of components is similar to the ECM in other tissues, except that PNNs lack collagen. Where there is a stable condensed matrix, as in cartilage, the binding of CSPGs to hyaluronan is stabilized by link proteins. It is therefore not surprising to find that in the CNS, the link proteins are present in PNNs, but not in other matrix compartments (Galtrey et al., 2007, 2008; Bekku et al., 2003; Oohashi et al., 2002). The molecules of the PNNs are not all made by the neurons that they surround. Tenascin-R, versican, phosphacan, and neurocan are all made by neighboring glial cells, and must therefore be captured into PNNs (Galtrey et al., 2008; Carulli et al., 2006). To form a pericellular matrix, there must be pericellular hyaluronan, which can be achieved either by cells expressing a hyaluronan receptor or the enzyme hyaluronan synthase. Every neuron with a PNN expresses one form of hyaluronan synthase. Also made specifically by the neurons with PNNs are link proteins. Aggrecan and neurocan are made by CNS neurons, although slightly more widely expressed than just PNN neurons. Which of these molecules initiates PNN formation? The molecule responsible should be upregulated at just the time that PNN formation begins. Hyaluronan synthase and tenascin-R expression begins earlier, but link protein and aggrecan expression corresponds closely with the timing of PNN formation (Galtrey et al., 2008; Carulli et al., 2007). We have recently obtained mice that lack expression of one of the link proteins, Crtl1, in the CNS. These animals lack PNNs around dendrites, but there are still vestigial aggrecan-containing PNNs around cell bodies. It is probable that the event that initiates PNN formation is link protein production, with some participation from aggrecan. Also, tenascin-R is necessary for full development of the structures because animals lacking tenascin-R have partly formed PNNs (Bruckner et al., 2000).

The way in which PNNs might affect synaptic behavior is not known. The fact that chondroitinase treatment promotes plasticity suggests that the GAG chains of the CSPGs are important in the control of synapses. In general, proteoglycan GAG chains can act in three ways. They can mask the active sites of other molecules, such as laminin, and can bind and present other active molecules to their receptors. It may also be that they can have direct effects on receptors. The possibility that CSPG GAG chains present active molecules to synapses by concentrating them in PNNs and presenting them to receptors has started to be examined. The binding properties of GAGs depend on their pattern of sulfation, so the sulfation pattern of CSPG GAGs from the three CNS matrix compartments was compared. GAGs from PNNs have more 6 and 4,6 sulfation than the other compartments, so they might be expected to have different binding properties (Deepa et al., 2006). This has been confirmed by ELISA studies showing that several molecules, including brain-derived neurotrophic factor (BDNF), bind more strongly to PNN GAG than to other forms. Most interestingly, one of the molecules that binds selectively to PNN GAG is semaphorin 3A (Vo et al., 2007). This molecule and the other semaphorin 3s have well-established roles in axon guidance during development. They and their receptors continue to be expressed widely in the adult CNS, but their function is largely unknown, although semaphorin 3A has been shown to have direct effects on synaptic behavior (Bouzioukh et al., 2006). Immunohistochemistry of brain sections shows, as expected from the biochemistry, that semaphorin 3A is localized to PNNs, and that treatment with chondroitinase removes it (Vo et al., 2007). Therefore, it seems probable that part of the effect of PNNs on plasticity is through the binding of semaphorins. However, there are probably other active molecules presented in the same way, so the effects may be complex. It is also possible that the CSPGs in PNNs affect access to matrix molecules by integrins on synaptic structures.

**NogoA and plasticity**

Many of the phenomona described previously for the control of plasticity by CSPGs also apply to NogoA. Thus, animals lacking the Nogo receptor show continuing ocular dominance plasticity into adulthood (McGee et al., 2005). Blocking NogoA with antibodies or interfering with the receptor using peptides or knockdowns can all lead to enhanced sprouting, formation of new connections, and recovery of function after damage (Buchli and Schwab, 2005). Although the effects of NogoA manipulations are rather similar to those of chondroitinase, there is no obvious relationship between their biology except for the fact that the inhibitory influences of both types of molecule signal via the Rho pathway. It is not clear why interference with NogoA, which is mainly present on myelin, should have effects on plasticity.

**Inosine**

Inosine was identified as a factor responsible for the regeneration of fish retinal axons. It is now known to stimulate the activity of the enzyme

Mst3b, which is on a neurotrophin signaling pathway in neurons (Irwin et al., 2006). It would therefore be expected to exert its effects by increasing the ability of neurons to grow new processes and overcome inhibitory environments. When given to rats after spinal cord injury, it promotes both axon regeneration and recovery of function (Benowitz et al., 1999). Particularly marked in these experiments were local sprouting of axons across the midline in the brain stem and spinal cord. Inosine has been used in both focal stroke and focal traumatic brain injury models, where there was a very useful return of function in treated animals, combined with corticospinal tract sprouting (Chen et al., 2002; Smith et al., 2007).

## Promoting plasticity opens a window of opportunity for rehabilitation

High levels of plasticity by themselves would not be expected to have behavioral consequences. Increased rates of sprouting and synapse formation, if they are random, will not be useful to the animal. In order to produce useful behavioral outcomes, the CNS must select useful new connections and remove inappropriate ones. During the later stages of development, there is activity-dependent refinement of exuberant connections in which successful connections are strengthened and inappropriate connections repressed. The requirements following CNS damage are, first, that the capacity to make new connections is restored and, second, that a process similar to the developmental refinement of connections selects the appropriate connections to stabilize. From the experiments of Merzenich and his colleagues on remapping of the primate cortex after injury, it became clear that cortical plasticity has to be driven by behavioral reinforcement, and that the animal has to be motivationally involved in the task (Buonomano and Merzenich, 1998). Similar issues apply to recovery from stroke and other CNS lesions in human patients. Rehabilitation can bring back considerable amounts of useful function, but in order to achieve recovery, long periods of repetitive engagement in the rehabilitation task have to be endured. Rehabilitation aims to use the plasticity of the human CNS to bring back useful function, but of course, in the elderly, the CNS is really not particularly plastic. There is a major shutdown of plasticity in humans at the end of the critical periods at around five years, and many of the measures that indicate levels of plasticity also show an additional decline with ageing. What could be achieved if the CNS could be made more plastic during the period of rehabilitation? Experiments to test this idea are in progress in our laboratory and also that of Martin Schwab. Our experiments have focused on plasticity in the corticospinal system. The human spinal cord is particularly specialized in that many functions are controlled by the corticospinal tract, so recovery of function in this pathway is relevant to human spinal cord repair. In the rodent, the task that measures corticospinal function most clearly is skilled paw function, which is severely affected by lesions of this pathway (Piecharka et al., 2005). We have therefore made lesions of the dorsal columns of the cord, including the main corticospinal tract, but leaving the much smaller ventral and lateral tracts intact. After these lesions, animals recover hardly any skilled paw-reaching function. Promoting plasticity with chondroitinase makes only a small improvement (Garcia-Alias et al., 2008). We therefore designed a rehabilitation task, in which animals spend an hour a day reaching into wells for seeds. In order to find out whether general exercise and environmental enrichment are responsible for any recovery of function due to rehabilitation, we also designed an environmental enrichment cage in which animals climb, run, and feed but do not use skilled paw function. Specific paw-reaching rehabilitation produced a small improvement in function compared to no treatment, but a large effect came when this form of rehabilitation was combined with chondroitinase; in other words, promotion of plasticity enabled rehabilitation to be effective. The surprise was when we examined the results from animals with general environmental enrichment, because these animals had completely lost the ability to perform skilled paw reaching, although their ability at walking over narrow beams and ladders was improved.

This suggests that in situations where there may be a limited number of new connections that can be formed, different behaviors may compete for available connections, and learning success in one behavior may disadvantage others. Rehabilitation enhanced by plasticity treatments must therefore be very carefully designed to give the range of behaviors that would be useful to a patient. Experience in Schwab's laboratory has also shown that plasticity-enhanced rehabilitation can be a two-edged sword, in that adding anti-NogoA antibodies to stepping training can make the behavior worse rather than better (Edgerton et al., 2008).

**Do the treatments that restore plasticity also promote axon regeneration?**

The three treatments highlighted previously as promoters of plasticity are also somewhat effective at promoting axon regeneration in the damaged CNS. Some features of plasticity are similar to long-distance axon regeneration, namely, local sprouting of processes, while others such as changes in synaptic strength are different. It is not clear at present which aspects of plasticity are enhanced by chondroitinase, anti-NogoA, and inosine, nor is it known in detail which connections are formed and changed. However, not all treatments that promote axon regeneration will necessarily promote plasticity. For instance, grafting of permissive glial cells into CNS lesions to provide a regeneration bridge is unlikely to affect plasticity. Some of the treatments that specifically affect axon growth mechanisms may also enhance regeneration without affecting plasticity. There are also likely to be plasticity treatments that do not affect axon regeneration, particularly those that operate at the level of the synapse. Overall, it seems probable that combinations of several treatments will be optimal for treating CNS damage.

**How might therapies be combined to treat spinal cord injury?**

There are now several treatments that show promise for treating spinal cord injury and other forms of CNS damage. They are divided roughly into five categories: (1) removal of inhibition in the environment, (2) enhancing the intrinsic ability of process growth, (3) providing cellular bridges for axon growth, (4) actions at synapses, and (5) replacing lost neurons and glia. It is logical to think that interventions from the different categories could have additive effects. It is also possible that more similar treatments, for instance, anti-NogoA and chondroitinase, could be additive, but the combination experiments have not been done. The combination of treatments that might be helpful will differ from lesion type to lesion type, and the treatment that brings back one form of functional recovery may not work for another. Another important variable is how long after the lesion the treatment is applied. For many conditions, there are advantages in patient care and also in clinical trial design from postponing the beginning of treatment for weeks or even months. We are getting near to the time when a systematic examination of combinations will be worthwhile. However, before doing this, it makes sense to understand fully each individual treatment and what it can achieve. It will also be important to standardize on a few animal models so that results between groups are more nearly comparable.

**References**

Apostolova, I., Irintchev, A., & Schachner, M. (2006). Tenascin-R restricts posttraumatic remodeling of motoneuron innervation and functional recovery after spinal cord injury in adult mice. *The Journal of Neuroscience*, 26, 7849–7859.

Ballermann, M., & Fouad, K. (2006). Spontaneous locomotor recovery in spinal cord injured rats is accompanied by anatomical plasticity of reticulospinal fibers. *The European Journal of Neuroscience*, 23, 1988–1996.

Bareyre, F. M., Kerschensteiner, M., Raineteau, O., Mettenleiter, T. C., Weinmann, O., & Schwab, M. E. (2004). The injured spinal cord spontaneously forms a new intraspinal circuit in adult rats. *Nature Neuroscience*, 7, 269–277.

Bekku, Y., Su, W. D., Hirakawa, S., Fassler, R., Ohtsuka, A., Kang, J. S., et al. (2003). Molecular cloning of Bral2, a novel brain-specific link protein, and immunohistochemical colocalization with brevican in perineuronal nets. *Molecular and Cellular Neurosciences*, 24, 148–159.

Benowitz, L. I., Goldberg, D. E., Madsen, J. R., Soni, D., & Irwin, N. (1999). Inosine stimulates extensive axon collateral growth in the rat corticospinal tract after injury. *Proceedings* of the National Academy of Sciences of the United States of America, 96, 13486–13490.

Berardi, N., Pizzorusso, T., Ratto, G. M., & Maffei, L. (2003). Molecular basis of plasticity in the visual cortex. Trends in Neurosciences, 26, 369–378.

Boggio, P. S., Alonso-Alonso, M., Mansur, C. G., Rigonatti, S. P., Schlaug, G., Pascual-Leone, A., et al. (2006). Hand function improvement with low-frequency repetitive transcranial magnetic stimulation of the unaffected hemisphere in a severe case of stroke. American Journal of Physical Medicine & Rehabilitation, 85, 927–930.

Bouzioukh, F., Daoudal, G., Falk, J., Debanne, D., Rougon, G., & Castellani, V. (2006). Semaphorin3A regulates synaptic function of differentiated hippocampal neurons. The European Journal of Neuroscience, 23, 2247–2254.

Bradbury, E. J., Moon, L. D. F., Popat, R. J., King, V. R., Bennett, G. S., Patel, P. N., et al. (2002). Chondroitinase ABC promotes axon regeneration and functional recovery following spinal cord injury. Nature, 416, 636–640.

Bruckner, G., Grosche, J., Schmidt, S., Hartig, W., Margolis, R. U., Delpech, B., et al. (2000). Postnatal development of perineuronal nets in wild-type mice and in a mutant deficient in tenascin-R. The Journal of Comparative Neurology, 428, 616–629.

Buchli, A. D., & Schwab, M. E. (2005). Inhibition of Nogo: a key strategy to increase regeneration, plasticity and functional recovery of the lesioned central nervous system. Annals of Medicine, 37, 556–567.

Buonomano, D. V., & Merzenich, M. M. (1998). Cortical plasticity: from synapses to maps. Annual Review of Neuroscience, 21, 149–186.

Carulli, D., Deepa, S. S., & Fawcett, J. W. (2007). Upregulation of aggrecan, link protein 1 and hyaluronan synthases during formation of perineuronal nets in the rat cerebellum. The Journal of Comparative Neurology, 501, 83–94.

Carulli, D., Rhodes, K. E., Brown, D. J., Bonnert, T. P., Pollack, S. J., Oliver, K., et al. (2006). The composition of perineuronal nets in the adult rat cerebellum and the cellular origin of their components. The Journal of Comparative Neurology, 494, 559–577.

Chen, P., Goldberg, D. E., Kolb, B., Lanser, M., & Benowitz, L. I. (2002). Inosine induces axonal rewiring and improves behavioral outcome after stroke. Proceedings of the National Academy of Sciences of the United States of America, 99, 9031–9036.

Deepa, S. S., Carulli, D., Galtrey, C., Rhodes, K., Fukuda, J., Mikami, T., et al. (2006). Composition of perineuronal net extracellular matrix in rat brain: a different disaccharide composition for the net-associated proteoglycans. The Journal of Biological Chemistry, 281, 17789–17800.

Edgerton, V. R., Courtine, G., Gerasimenko, Y. P., Lavrov, I., Ichiyama, R. M., Fong, A. J., et al. (2008). Training locomotor networks. Brain Research Reviews, 57, 241–254.

Fawcett, J. W., Curt, A., Steeves, J. D., Coleman, W. P., Tuszynski, M. H., Lammertse, D., et al. (2007). Guidelines for the conduct of clinical trials for spinal cord injury as developed by the ICCP panel: spontaneous recovery after spinal cord injury and statistical power needed for therapeutic clinical trials. Spinal Cord, 45, 190–205.

Fouad, K., & Tse, A. (2008). Adaptive changes in the injured spinal cord and their role in promoting functional recovery. Neurological Research, 30, 17–27.

Galtrey, C. M., Asher, R. A., Nothias, F., & Fawcett, J. W. (2007). Promoting plasticity in the spinal cord with chondroitinase improves functional recovery after peripheral nerve repair. Brain, 130, 926–939.

Galtrey, C. M., & Fawcett, J. W. (2007). The role of chondroitin sulfate proteoglycans in regeneration and plasticity in the central nervous system. Brain Research Reviews, 54, 1–18.

Galtrey, C. M., Kwok, J. C., Carulli, D., Rhodes, K. E., & Fawcett, J. W. (2008). Distribution and synthesis of extracellular matrix proteoglycans, hyaluronan, link proteins and tenascin-R in the rat spinal cord. The European Journal of Neuroscience, 27, 1373–1390.

Garcia-Alias, G., Lin, R., Akrimi, S. F., Story, D., Bradbury, E. J., & Fawcett, J. W. (2008). Therapeutic time window for the application of chondroitinase ABC after spinal cord injury. Experimental Neurology, 210, 331–338.

Guntinas-Lichius, O., Angelov, D. N., Morellini, F., Lenzen, M., Skouras, E., Schachner, M., et al. (2005). Opposite impacts of tenascin-C and tenascin-R deficiency in mice on the functional outcome of facial nerve repair. The European Journal of Neuroscience, 22, 2171–2179.

Hensch, T. K. (2004). Critical period regulation. Annual Review of Neuroscience, 27, 549–579.

Irwin, N., Li, Y. M., O'Toole, J. E., & Benowitz, L. I. (2006). Mst3b, a purine-sensitive Ste20-like protein kinase, regulates axon outgrowth. Proceedings of the National Academy of Sciences of the United States of America, 103, 18320–18325.

Kitagawa, H., Tsutsumi, K., Tone, Y., & Sugahara, K. (1997). Developmental regulation of the sulfation profile of chondroitin sulfate chains in the chicken embryo brain. The Journal of Biological Chemistry, 272, 31377–31381.

Lundborg, G. (2000). Brain plasticity and hand surgery: an overview. Journal of Hand Surgery (Edinburgh, Lothian), 25, 242–252.

Massey, J. M., Hubscher, C. H., Wagoner, M. R., Decker, J. A., Amps, J., Silver, J., et al. (2006). Chondroitinase ABC digestion of the perineuronal net promotes functional collateral sprouting in the cuneate nucleus after cervical spinal cord injury. The Journal of Neuroscience, 26, 4406–4414.

McGee, A. W., Yang, Y., Fischer, Q. S., Daw, N. W., & Strittmatter, S. M. (2005). Experience-driven plasticity of visual cortex limited by myelin and Nogo receptor. Science, 309, 2222–2226.

Moon, L. D. F., Asher, R. A., Rhodes, K. E., & Fawcett, J. W. (2001). Regeneration of CNS axons back to their original target following treatment of adult rat brain with chondroitinase ABC. Nature Neuroscience, 4, 465–466.

Nair, D. G., Hutchinson, S., Fregni, F., Alexander, M., Pascual-Leone, A., & Schlaug, G. (2007). Imaging correlates of motor recovery from cerebral infarction and their physiological significance in well-recovered patients. *NeuroImage, 34*, 253–263.

Oohashi, T., Hirakawa, S., Bekku, Y., Rauch, U., Zimmermann, D. R., Su, W. D., et al. (2002). Bral1, a brain-specific link protein, colocalizing with the versican V2 isoform at the nodes of Ranvier in developing and adult mouse central nervous systems. *Molecular and Cellular Neurosciences, 19*, 43–57.

Piecharka, D. M., Kleim, J. A., & Whishaw, I. Q. (2005). Limits on recovery in the corticospinal tract of the rat: partial lesions impair skilled reaching and the topographic representation of the forelimb in motor cortex. *Brain Research Bulletin, 66*, 203–211.

Pizzorusso, T., Medini, P., Berardi, N., Chierzi, S., Fawcett, J. W., & Maffei, L. (2002). Reactivation of ocular dominance plasticity in the adult visual cortex with chondroitinase ABC. *Science, 298*, 1248–1251.

Properzi, F., Carulli, D., Asher, R. A., Muir, E., Camargo, L. M., van Kuppevelt, T. H., et al. (2005). Chondroitin 6-sulphate synthesis is up-regulated in injured CNS, induced by injury-related cytokines and enhanced in axon-growth inhibitory glia. *The European Journal of Neuroscience, 21*, 378–390.

Ramanathan, D., Conner, J. M., & Tuszynski, M. H. (2006). A form of motor cortical plasticity that correlates with recovery of function after brain injury. *Proceedings of the National Academy of Sciences of the United States of America, 103*, 11370–11375.

Rossini, P. M., Calautti, C., Pauri, F., & Baron, J. C. (2003). Post-stroke plastic reorganisation in the adult brain. *Lancet Neurology, 2*, 493–502.

Smith, J. M., Lunga, P., Story, D., Harris, N., Le, B. J., James, M. F., et al. (2007). Inosine promotes recovery of skilled motor function in a model of focal brain injury. *Brain, 130*, 915–925.

Vo, T., Carulli, D., Ehlert, E. M. E., Kwok, J. C., Asher, R. A., Fawcett, et al. (2007). The chemorepulsive axon guidance protein Semaphorin 3A is a constituent of perineuronal nets in the adult rodent brain. *Society for Neuroscience Abstract* (585.1).

Ward, N. S. (2005). Neural plasticity and recovery of function. *Progress in Brain Research, 150*, 527–535.

# Subject Index

AAV. *See* Adeno-associated viral (AAV)
AAV2-NTN gene delivery system, 211–214
AAV serotype-2 (AAV2), 151, 153–156, 163–170
ABI. *See* Auditory brainstem implants (ABI)
Aβ immunogens, for active vaccination, 88–90
Aβ immunotherapy
   clinical trials in humans, 85, 87
   mechanisms of, 87–88
   preclinical trials in
      in monkeys, 85–86
      nonhuman primates, 85–86
      rodents, 84–85
Accessory olfactory system, 34
ACE. *See* Advanced combination encoder (ACE)
Acute motor axonal neuropathy (AMAN), 109, 111
Adeno-associated viral (AAV) vector-mediated gene therapy, GDNF and, 209–210
Adeno-associated viral (AAV) vectors, 73–74, 151–158, 163–170, 178, 193–194
   controlled dissemination in primate brain, 163–170
   dendritic morphology of RGCs transduced with, 156–158
   mediated transduction
      of retinal ganglion cells, 152–154
      and tropic properties in retina, 154–156
   receptors, 167
   serotypes, 151, 153–156, 163–170, 178
   trafficking in brain, 164–170
      axonal projections, 165–167
      clinical implications, 168–170
      perivascular pump, 165–166
      real-time MRI, 167–169
Adeno-associated virus, uses in gene therapy, 188
Adenoviral (AD) vector-mediated gene therapy, GDNF and, 204–205
Adenoviral (AdV) vectors, 152–153
Adjuvant, 83, 85, 88–90

Advanced combination encoder (ACE), 335
Advanced glycation end products (AGEs), 260
AES. *See* Auditory and electric stimulation (AES)
Age-related CNS disorders, NSCs in treatment for, 45–47
Age-related macular degeneration (AMD), 23, 254, 317, 318
   VEGF antagonists in
      bevacizumab (Avastin®), 263
      pegaptanib sodium (Macugen®), 262
      Ranibizumab (Lucentis®), 262–263
AGEs. *See* Advanced glycation end products (AGEs)
Aging, and NSCs, 45–47
Allopregnanolone (AP), 223–225, 232. *See also* Progesterone (PROG)
ALS. *See* Amyotrophic lateral sclerosis (ALS)
Alzheimer's disease (AD), 34, 140, 180, 191–192
   complement activation in, 101–102
   immunogens for vaccine for, 83–91
   olfactory vector hypothesis, 34
   RMS in, 39–40
Amacrine cells, 23–24, 26–28, 155
AMD. *See* Age-related macular degeneration (AMD)
AMI. *See* Auditory midbrain implant (AMI)
Amyotrophic lateral sclerosis (ALS), 189, 300, 302
   clinical symptoms and treatment, 189
   complement activation in, 102–103
   LV vector-mediated gene therapy in, 189–190
AN. *See* Auditory neuropathy (AN)/auditory dissynchrony disorder
Anaphylatoxins, 97
ANCHOR study, 262, 264

Angiogenesis, VEGF antagonists and
    choroidal neovascularization (CNV), AMD
        and, 254
      bevacizumab (Avastin®), 263
      pegaptanib sodium (Macugen®), 262
      Ranibizumab (Lucentis®), 262–263
    mechanisms of, 254–257
    ocular, 253–254
      VEGF in, role of, 256, 257–259
    preretinal, 254
      diabetic macular edema (DME), 260–262
      diabetic retinopathy (DR), 259–260
Animal PD models
  GDNF and
    gene therapy, 204–210
    infusions of, 203
    injections of, 202–203
  NTN and
    administration, 211
    gene therapy, 211–213
Ankle
  dorsiflexors, 432
  plantarflexors, 432
Anodal pulses, 351
Anosmia, 34
Anosmin-1, 457
Anti-MBP (myelin basic protein) antibody,
    immunolabeling with, 454
AN1792 vaccine trial, 85, 87, 91
AP. *See* Allopregnanolone (AP)
Apoptosis hypothesis, 189
Apparent diffusion coefficient (ADC) values, 469
AREDS clinical trial
  RD disease and, 319
Artificial silicone retina (ASR), 322
  timeline for progress of, 329
Artificial vision device (AVD), 325
ASR. *See* Artificial silicone retina (ASR)
AstraZeneca, 222
Astrocytes cells, 10–11, 23, 25–26, 38, 44, 97, 100,
    102–103, 125, 127–128, 139–140
  TLRs in, 142–146
Auditory and electric stimulation (AES), 334
Auditory brainstem implants (ABI), 334–336
  in children, 341–342
  CN and, 337
  microelectrode array and, 335
  NF2, patients, 337, 338
  in nontumor patients, 337–339
  with penetrating microelectrodes (PABI), 336
  physiological damages, analysis of, 339–341
  role of learning, 342
  surface-electrodes and, 336
  VS removal and, 340
Auditory midbrain implant (AMI), 341
Auditory neuropathy (AN)/auditory dissynchrony
    disorder, 340
Auditory–phonological resting state network, 490
Auditory training programs, 342
Autogenetic Ib inhibition, 432
AVD. *See* Artificial vision device (AVD)
Axonal growth cone navigation, 444–446
Axonal projections, 165–167

Bapineuzimab, 87
Basal ganglia (BG), 380
  functional model of, 381–382
Basic fibroblast growth factor (bFGF), 167, 457
Basilar membrane, 333
Basso Beattie Bresnahan (BBB) scale, 278
BBB. *See* Blood-brain barrier (BBB)
BBB scale. *See* Basso Beattie Bresnahan (BBB)
    scale
BCAO. *See* Bilateral common carotid artery
    occlusion (BCAO)
B cells, 38, 83–84, 88, 126–127
BCIs. *See* Brain-computer interfaces (BCIs)
BDNF. *See* Brain-derived neurotrophic factor
    (BDNF)
Benign MS (BMS), 469
Bevacizumab (Avastin®), 256, 263
BG. *See* Basal ganglia (BG)
Bilateral cochlear implant. *See* Cochlear implant
    (CI)
Bilateral common carotid artery occlusion
    (BCAO), 224
Biocompatibility, 299, 304, 408
Bipedal stepping, 411
2,2′-bipyridine-5,5′-dicarboxylic acid (BPY-DCA),
    276
BIV (bovine immunodeficiency virus) vector, 188
Blindness, 348
  effects of, on phosphene, 362–364
Blood–brain barrier (BBB), 220, 227, 231
Blood–ON barrier, 475
Blood oxygenation level-dependent (BOLD), 466

BPY-DCA. *See* 2,2′-bipyridine-5,5′-dicarboxylic acid (BPY-DCA)
Brain. *See also* Primate brain
　AAV vectors trafficking in
　　axonal projections, 165–167
　　clinical implications, 168–170
　　perivascular pump, 165–166
　　real-time MRI, 167–169
　optokinetic testing and, 326
　parenchyma, 484
　plasticity, 471
　regions, targets for DBS, 385
　synapse formation, 394
　tumors and NSCs, 46–49
Brain–computer interfaces (BCIs), 297
　approaches, comparison of, 309–310
　class I, extracorporeal electrodes for, 300–301
　classification of, 300–301
　class II, epicortical electrodes for, 302–305
　class III, intracortical electrodes for, 306–309
Brain-derived neurotrophic factor (BDNF), 38, 71, 73, 144, 151–154, 156–158, 174, 178, 180, 183, 189, 244, 445
　effects of, on muscle properties, 400–401
BrainGate™ system, 306
Brainstem implants. *See* Auditory brainstem implants (ABI)
Brainstem movement control, 396
Bruch's membrane, 24, 254, 262

CAD. *See* Computer-aided design (CAD) file
Cajal–Retzius cells, 37
Calcineurin, 449
Calcitonin gene-related peptide (CGRP)-positive neurons, 272
Calcium/calmodulin-dependent protein kinase II (CaMKII), 449
Calcium signaling, 444, 450
　in axonal growth cones, 444–446
　in synaptic partner selection, 446–448
Calcium transients, dendritic, 446–448
CAMP. *See* Cyclic adenosine monophosphate (cAMP)
Cancer, 58
　initiating stem-like cells, 48
　and NSCs, 47–49
Candy store effect, 179–180
Cannula, 164, 167, 170

Cathodal pulses, 351
[$^{11}$C]carfentanil, 289
CCK. *See* Cholecystokinin (CCK) antagonist
CD4+ T cells, 239
Celebrex, 228
Cells
　Muller, 323
　photoreceptor. *See* Photoreceptor
　RPE, 318
　sorting, 59, 61, 63–64
　stem. *See* Stem cells
　transplantation
　　sources for retinal, 6–10
　　strategies for retinal repair, 3–15
Cell therapy
　limitations, 54, 57
　　fetal tissue source, 54
　　incomplete recovery, 57
　　therapeutic outcome variability, 54, 57
　for Parkinson's disease, 54–57
Center for Neural Communication Technologies, 308
Central nervous system (CNS), 227, 269
　complement in, 97–103, 113–115
　　Alzheimer's disease, 96, 101–102, 140
　　amyotrophic lateral sclerosis, 102–103
　　CNS disease, 101–103, 113–115
　　CNS injury, 99, 101, 113–115
　　expression in brain, 100
　　Huntington's disease, 102–103
　　local synthesis, 97–100
　　neurodegenerative disease, 102–103
　　Parkinson's disease, 103
　　Pick's disease, 102
　　therapeutics in injury and disease, 113–115
　injury, lesion scarring after. *See* Lesion scar, after CNS injury
　plasticity and
　　morphogenesis, 37
　　neurogenesis, 43
　　remyelination, 453
　role of TLRs in, 139–146
Central pattern generation
　role of, in locomotion, 396
Centrum medianum–parafascicularis (CM–Pf) complex, 380
　efficiency of HFS in, 386
CERE-120. *See* AAV2-NTN gene delivery system

Cerebral palsy, 435
Ceregene Inc., 211, 213
CGRP-positive neurons. *See* Calcitonin gene-related peptide (CGRP)-positive neurons
Charcot–Marie–Tooth (CMT) disease
  complement activation in, 109, 112–113
Chemokines, 457
Children
  ABI in, 341–342
  CI in, 341
Chimeric peripheral nerve grafts
  LV use to genetically modify Schwann cells in, 154
Cho and mobile lipids, 470
Cholecystokinin (CCK) antagonist, 286
Choroidal neovascularization (CNV)
  bevacizumab (Avastin®), 263
  pegaptanib sodium (Macugen®), 262
  Ranibizumab (Lucentis®), 262–263
Chronic arterial hypertension, stroke and, 230
CI. *See* Cochlear implant (CI)
Ciliary neurotrophic factor (CNTF), 144, 151–154, 156–158, 174, 181, 194, 319
CIS. *See* Continuous interleaved sampling (CIS)
Class 3 semaphorins, 458
  in CNS pathology
    CNS injury, 459–460
    MS and experimental demyelination, 460
Clinically isolated syndrome (CIS), 466–467
CM–Pf complex. *See* Centrum medianum–parafascicularis (CM–Pf) complex
CN. *See* Cochlear nuclei (CN)
CNS. *See* Central nervous system (CNS)
CNS damage, mechanisms of recovery after, 501
  chondroitinase, 503–505
  chondroitin sulfate proteoglycans (CSPGs), 503–505
  plasticity in promoting axon regeneration, 507
  plasticity in recovery function, 502
    inosine, 505–506
    methods of restoring, 503
    NogoA and, 505
  plasticity in rehabilitation, 506–507
CNTF. *See* Ciliary neurotrophic factor (CNTF)
CNV. *See* Choroidal neovascularization (CNV)

Cochlear implant (CI), 334, 336, 340
  in children, 341
  devices, 335
  role of learning, 342
Cochlear nuclei (CN), 333, 335, 336, 341
  ABI and, 337
  and damage by tumor, 339
Cochrane Review Manager software, 228
Collagen type IV (Coll IV) network, 270, 274
Collateral sprouting, 430
Collision method, 367–369
Coll IV. *See* Collagen type IV (Coll IV) network
Combination therapy, stroke and, 230–231. *See also* Stroke, progesterone and
Common peroneal nerve (CP) stimulation, 432
Complement (C)
  activation and regulation of system, 96–97
  activation in
    Alzheimer's disease, 101–102
    amyotrophic lateral sclerosis, 102–103
    Charcot–Marie–Tooth disease, 109, 112–113
    CNS disease, 101–103
    CNS injury, 99, 101
    Creutzfelt–Jakob disease, 103
    experimental autoimmune neuritis, 103, 109–110
    Guillain–Barré syndrome, 109–112
    HMSNs, 109, 112–113
    Huntington's disease, 102–103
    neurofibromatosis, 109
    neuroma, 109
    Parkinson's disease, 103
    Pick's disease, 102
    PNS disease, 109–113
    PNS injury, 104–106
  in central nervous system, 97–103, 113–115
  expression in
    brain, 100
    healthy human sciatic nerve, 104
  local synthesis in
    CNS, 97–99
    PNS, 103–104
  peripheral nervous system, 103–115
  post-traumatic activation effect, 106–109
  regulation in nerve injury and disease, 113–115
  role in inflammation, 97
  system, 95–97
  therapeutics in CNS and PNS injury, 113–115

Computer-aided design (CAD) file, 304–305
Computerized tomography (CT) scans, 227
Conditioned response, defined, 285. *See also* Placebo response
Connective tissue growth factor (CTGF), 255
Conscious movement control, 395–396
Continuous interleaved sampling (CIS), 335
Convection-enhanced delivery (CED), 164–170
Cord
  atrophy, 473
  demyelination, 474
  MTR metrics, 473–474
  swelling, 473
Corin, 68–69
Cortical reorganization, 471–473
Cortical space, 357
Cortical stimulation, 387
Corticectomies, 380
Corticospinal cells, 433
Corticospinal tract (CST), 467
Corticosteroids After Significant Head Injury (CRASH) trial, 220
COX-2 inhibitors, 228
[$^{11}$C]raclopride, $D_2$-$D_3$ dopamine receptor agonist, 289
CRASH trial. *See* Corticosteroids After Significant Head Injury (CRASH) trial
Creutzfelt–Jakob disease, complement activation in, 103
CSF-suppressed zonal oblique multisection echoplanar imaging sequences, 475
CTGF. *See* Connective tissue growth factor (CTGF)
CT scan. *See* Computerized tomography (CT) scan
C.U. Ariëns Kappers and the Central Institute for Brain Research, 497–499
CXCL12, 457
CXCR4, 457
  signaling, 457
Cyclic adenosine monophosphate (cAMP), 276
Cyclosporine A (CsA), 127
Cystoid macular edema (CME), 14

Data processing algorithms, 486
DBS. *See* Deep brain stimulation (DBS)

Deep brain stimulation (DBS), 419–420
  efficiency of, in PD, 381
  efficiency of HFS in, 386–387
  mechanism of action of, 420
  of neurological disorders, psychiatric comorbidity in, 420–421
  obsessive–compulsive disorder (OCD) and, 421
    clinical efficacy in, 421–424
    side effects, 424–426
  pros and cons of, 387
  in psychiatry, criteria for, 426
  side effects in OCD, 424
    acute effects, 425
    cognitive changes, 425
    DBS interruption, effects of, 425–426
    mood effects, 425
    personality changes, 425
  targets for, brain regions, 385
Defective axonal transport hypothesis, 189
Dehydrotestosterone (DHT), 244
Delayed-type hypersensitivity (DTH), 245, 246
Delay field, 356
  phosphenes, 359
  size of, 357, 358
Deleted in colorectal cancer (DCC), 458
Demyelinated MS lesions, 454, 455–461
  hypothesis and perspectives, 460–461
  with oligodendroglial depopulation, 456–458
Dendritic calcium transients, 446–448
Dendritic filopodia, 446–448
Depression, placebo effects and, 289–290
Developing midbrain, mDA neurons from, 61–65
DHT. *See* Dehydrotestosterone (DHT)
Diabetic macular edema (DME), 260–262
Diabetic retinopathy (DR), 254, 259–260
Diffusion tensor (DT) MRI, 466
Diffusion tensor imaging (DTI), 326
  and recovery from brain injury, 486–487
  technique and applications, 484–486
Dimethylsulfoxide (DMSO), 223
D1 inhibition, 433
Disynaptic reciprocal Ia inhibition, 431–432
Dizocilpine (MK-801), 227
DLPFC. *See* Dorsolateral prefrontal cortex (DLPFC)
DME. *See* Diabetic macular edema (DME)
DMSO. *See* Dimethylsulfoxide (DMSO)
Dopamine, placebo effects and, 289

Dopamine neurons. *See* Midbrain dopamine (mDA) neurons
Dorsal root ganglia (DRG), 270
  axons, 130
Dorsiflexor muscles, 432
Dorsolateral prefrontal cortex (DLPFC), 287
DR. *See* Diabetic retinopathy (DR)
DRG. *See* Dorsal root ganglia (DRG)
DTH. *See* Delayed-type hypersensitivity (DTH)
DTI. *See* Diffusion tensor imaging (DTI)
DT MRI tractography, 469
Dyskinesias
  STN stimulation and, 384

EAE. *See* Experimental autoimmune encephalomyelitis (EAE)
ECM. *See* Extracellular matrix (ECM) network
ECoG. *See* Electrocorticogram (ECoG)
ECT. *See* Electro convulsion therapy (ECT)
Edema, 125
EDSS scores, 469
EEG. *See* Electroencephalogram (EEG)
EEP. *See* Elicit electrically evoked potential (EEP)
EIAV (equine infectious anemia virus) vector, 188
EIAV vector. *See* Equine infectious anaemia virus (EIAV) vector
Electrical activity during myelination, 456
Electric fields
  generation, from multiple electrodes, 366–367
Electro convulsion therapy (ECT), 419
Electrocorticogram (ECoG), 300, 309
Electrode arrays, 408, 409
Electrodes
  generating electric fields from multiple, 366–367
  shape, size, and dimensions of, 388
Electroencephalogram (EEG), 299–300, 309
Electrophysiological assessment of ABI, 341
Electrophysiology, 381, 382
Electroretinography (ERG), 325
Elicit electrically evoked potential (EEP), 321
Embryonic stem (ES) cells, 6, 9–10, 12, 14, 23, 28, 58–61
  retinal cell generation from, 29
Endogenous BDNF, 447
Epicortical electrode array, for BCI class II, 302–305
  laser-structured, 305
  micromachined, 304

Epidermal growth factor (EGF), 7, 457
Epidural stimulation, 396
Equine infectious anaemia virus (EIAV) vector, 209
ERG. *See* Electroretinography (ERG)
ERs. *See* Estrogen receptors (ERs)
Estrogen, multiple sclerosis (MS) and
  immunomodulatory properties of, 242–244
  neuroprotective properties of, 244–245
  as treatment, 245–246
Estrogen receptors (ERs), 242–245
European Conformity (CE), 298
European NeuroProbes, 308
Excitotoxicity hypothesis, 189
Exogenous neurotrophic factors, 179–180
Expanded Disability Status Scale (EDSS) score, 467
Experimental autoimmune encephalomyelitis (EAE), 103, 129, 239, 242, 244, 245
Experimental autoimmune neuritis (EAN)
  complement activation in, 103, 109–110, 112–114
Extracellular matrix (ECM) network, 270
Extracorporeal electrodes, for BCI class I, 300–301
Eye disease, VEGF and, 257
Eye movements, saccadic. *See* Saccadic eye movements

FDA. *See* Food and Drug Administration (FDA)
Fetal retinal sheet transplants, 3–4
FGF-2, 457
Fibers, pyramidal, 368
Fiber tracking algorithms, 485–486
"Fiber tracking" methods, 484–485
Fiber tractogram, 486
Fiber tractography, 486
Fibroblast growth factor-2 (FGF2), 7
Fibrous scar
  formation, therapies to prevent, 273–274
  molecular composition of, 271–272
  morphology of, 271
Filopodia, 445
  dendritic, 446–448
FIV (feline immunodeficiency virus) vector, 188
Flavivirus, uses in gene therapy, 188
Flecainide, 456
Fluoro-dopa ($^{18}$F-dopa) uptake, 204, 209

FMRI. *See* Functional magnetic resonance imaging (fMRI)
Food and Drug Administration (FDA), 222, 228, 263, 298
Foveal stimulation, 348
Fractional anisotropy (FA), 466
Freehand System, 298
Freund's adjuvant, 85
Frontal eye fields
  stimulation of, 354
Functional connectivity analysis, 473
Functional electrical stimulation, 398, 400
Functional magnetic resonance imaging (fMRI), 300, 466, 474

GA. *See* Geographic atrophy (GA)
GABA. *See* Gamma-amino butyric acid (GABA)
GABAergic signaling, 446
(GABA)ergic system. *See* Gamma-aminobutyric acid (GABA)ergic system
GABA (gamma-aminobutyric acid)-ergic synapses, 430
GAD. *See* Glutamic acid decarboxylase (GAD)
Gamma-amino butyric acid (GABA), 220, 231
Gamma-aminobutyric acid (GABA)ergic system, 362, 364
Gaussian (tensor) model of diffusion, 486
GDNF. *See* Glial cell line-derived neurotrophic factor (GDNF)
GDNF family ligands (GFLs), 202
GDNF family receptor alpha (GFRα), 202
Gender gap, in MS, 240–241
Gene therapy, 382–383
  clinical prospects and challenges, 195
    gene transfer system, 188
  for neurodegenerative diseases based on LV vectors, 187–195
  principles of, 187–188
  and transplantation in retinofugal pathway, 151–158
  vectors used in, 187–188
Gene therapy, in animal PD models
  GDNF and
    adeno-associated viral vector-mediated, 209–210
    adenoviral vector-mediated, 204–205
    lentiviral vector-mediated, 205–209
  NTN and, 211–213

Gene transfer, 387
  system, 187–188
Geniculate nucleus. *See* Lateral geniculate nucleus
Geographic atrophy (GA), 323
GFAP. *See* Glial fibrillary acidic protein (GFAP)
GFLs. *See* GDNF family ligands (GFLs)
GFRα. *See* GDNF family receptor alpha (GFRα)
GFRα-1 receptor, 210
GFRα-2 receptor, 210
Glasgow Outcome Scale, 487
Glial cell-derived neurotrophic factor (GDNF), 73–74, 152, 154, 165, 167, 169, 174–175, 178–181, 190, 193–195
Glial cell line-derived neurotrophic factor (GDNF)
  animal PD models
    gene therapy in, 204–210
    infusions in, 203
    injections in, 202–203
  clinical studies with, 210
  PD human subjects, experimental administration in, 203–204
Glial cells, 443
Glial fibrillary acidic protein (GFAP), 270
Globus pallidus internus (GPi), 380, 420
  efficiency of HFS in, 384–385
  pallidal stimulation and, 384–385
Glucocorticoids, 220
Glutamate excitotoxicity, 125
Glutamic acid decarboxylase (GAD), 383
GPi. *See* Globus pallidus internus (GPi)
G-protein, 242
Graft composition, and relevance for functional impact, 65–69
Gray matter atrophy, in MS, 240
Green fluorescent protein (GFP), 28, 55–57, 59, 61–64, 66–68, 70–71, 73–74, 169, 178
Growth-associated protein 43 (GAP43), 204
Growth factors (GF), 255
Guillain–Barré syndrome (GBS), complement activation in, 109–113

Handshake phenomenon, 73
Healthy human sciatic nerve, complement expression in, 104
Hearing
  ABI. *See* Auditory brainstem implants (ABI)
  CI, 334

Hematopoietic stem cells (HSCs), 45, 47
Heparan sulfate proteoglycan (HSPG) receptor, 167
Hepatocyte growth factor (HGF), 457
Hereditary motor and sensory neuropathy (HMSN), complement activation in, 109, 112–113
Herpes simplex virus, uses in gene therapy, 188
HFS. See High-frequency stimulation (HFS)
High-frequency stimulation (HFS), 379, 383
  alternatives to, 387
  efficiency of
    in CM–Pf, 386
    in DBS, 386–387
    in GPi, 384–385
    in PPN, 386
    in STN, 385–386
    in VIM, 383–384
  pharmacology of STN, 387–388
Hippocampal neurons, 446
HIV. See Human immunodeficiency virus (HIV)
HIV (human immunodeficiency virus) vector, 188
$^1$H-MRS, 471
HMSN. See Hereditary motor and sensory neuropathy (HMSN)
Human embryonic stem cells, 29
Human immunodeficiency virus (HIV), 209
Human leukocyte antigens (HLA), 14
Human RMS, 35–36
Huntington's disease (HD), 189
  clinical symptoms, 193
  complement activation in, 102–103
  LV vector-mediated gene therapy in, 194–195
  modeling, 193–194
  RMS in, 39–40
6-hydroxydopamine (6-OHDA)
  lesion model, 202–203, 205, 209–210, 211
  neurotoxin, 55, 68, 70–71, 74, 192–193
Hyperexcitability, 394–395
Hyperglycemia, DR and, 259–260
Hypothermia, for global ischemia, 220

IAM. See Internal auditory meatus (IAM)
IC. See Inferior colliculus (IC)
IL-10, 242
Imaging, 341
  brain atrophy, 466–467
  "diffuse" GM damage, 470–471

Immunohistological analysis of lumbar spinal cord, 404
Immunolabeling with an anti-MBP (myelin basic protein) antibody, 454
Implantable programmable generator (IPG), 379
Induced pluripotent stem (iPS) cells, 14–15, 58
Inferior colliculus (IC), 333
Inferior frontal gyrus (IFG), 473
Inflammation versus neurodegeneration, in MS, 239–240
Infliximab, 128
In-plane electrode arrangement, 308. See also Brain–computer interfaces (BCIs)
Internal auditory meatus (IAM), 339
International Stroke Conference (2007), 222
Intracortical electrodes, for BCI class III, 306–309
Intra-nigral grafting, nigro-striatal pathway reconstruction through, 69–74
Intrinsic lesion damage, 467–468
Iris neovascularization, 261–262
Iron chelators, scar suppression with
  axonal regeneration following, 276
  functional recovery following, 277–278
  neuroprotective effect of, 276–277
  principle of, 274–276
Ischemia, 125, 129

"Java man," 498

Lamotrigine, 456
Laser-evoked potentials (LEPs), 286, 303
Laser treatment, for DR, 260
Lateral geniculate nucleus, 352, 353
LCA. See Leber congenital amaurosis (LCA)
Learning Retina Implant system, 322
Leber congenital amaurosis (LCA), 318
Lentiviral (LV) vector-mediated gene therapy, GDNF and, 205–209
Lentiviral (LV) vectors, 151–152, 154, 173, 177–178, 180
  based gene therapy in
    ALS, 189–190
    clinical prospects and challenges, 188–189
    gene transfer system, 187–188
    Huntington's disease, 194–195
    neurodegeneration, 188–189
    neurodegenerative diseases, 187–195

Parkinson's disease, 191–193
  spinal muscular atrophy, 190–191
 clinical application to injured PN, 177–178
 overexpression of neurotrophic factors by, 178–179
 use in chimeric PN grafts, 154
Lentiviral (LV) vectors encoding nerve growth factor (LVNGF), 178–179
Lentivirus
 uses in gene therapy, 188
LEPs. See Laser-evoked potentials (LEPs)
Lesions, 380
Lesion scar, after CNS injury, 270
 fibrous scar
  formation, therapies to prevent, 273–274
  molecular composition of, 271–272
  morphology of, 271
 formation
  axonal growth responses to, differences in, 272–273
 iron chelators, suppression with
  axonal regeneration following, 276
  functional recovery following, 277–278
  neuroprotective effect of, 276–277
  principle of, 274–276
Leukocytes, 125–127, 129
Levodopa (L-DOPA), 382–383
 dopamine precursor, 54, 57, 68
Lewy bodies, 191–192
Limb movements, 352
LINGO-1. See Nogo receptor-interacting protein (LINGO-1)
Locomotion
 central pattern generation in, 396
 factors for optimal recovery of, 410
Locomotor circuits
 treatment for, 402
  locomotor training, 405–407
  pharmacological, 403–405
  spinal cord stimulation, 407–410
Locomotor strip, 396
Locomotor training, 395, 405–407
Long-term depression (LTD), 449
Long-term functional recovery, progesterone treatment for stroke and, 229
Long-term potentiation (LTP), 449
Long-term potentiation (LTP)-like mechanisms, 435

Lou Gehrig's disease. See Amyotrophic lateral sclerosis (ALS)
Low vision
 measuring visual function in, 324–326
Lumbar spinal cord
 immunohistological analysis of, 404
LV. See Lentiviral (LV) vectors
Lysosomal storage disorders (LSD), 164

Macaque V1
 size of phosphenes evoked from, 360
 stimulation of, 353
Macrophages, role in nerve regeneration and SCI, 129–130
Macular degenerations. See Age-related macular degeneration (AMD)
Magnesium
 as noncompetitive NMDA receptor blocker, 227
Magnesium sulfate trial, 220
Magnetencephalography (MEG), 300
Magnetic convulsion therapy (MCT), 419
Magnetic resonance imaging (MRI), 224
 AAV vectors trafficking in brain by, 167–169
Magnetization transfer (MT) MRI, 466
Magnetization transfer ratio (MTR), 466
 change, 475
 histogram analysis, 473
Mammalian retina, 5
MARINA study, 262, 264
MCA. See Middle cerebral artery (MCA)
MCAO. See MCA occlusion (MCAO)
MCA occlusion (MCAO), 222–223
MCT. See Magnetic convulsion therapy (MCT)
Mean diffusivity (MD), 466
Medial gastrocnemius muscle, 397
MEG. See Magnetencephalography (MEG)
Meningoencephalitis, 83, 85
Mesencephalic locomotor, 396
Mesenchymal stem cells (MSC), 45
Methotrexate, 127
1-Methyl-4-phenyl-1,2,3,6-tetrahydropyridine (MPTP)-lesioned rhesus monkeys, 203, 205
Microelectrode array, ABI and, 335
Microfabrication, 408
Microglia, 125, 128–129, 141–143, 146
 TLRs in, 141–143, 146

Microstimulation
    of V1
        delays execution of visually guided saccades, 356–359
        elicits detection response, 359–361
        elicits saccadic eye movements, 351–356
Midbrain dopamine (mDA) neurons, 53
    anatomical division, 58
    cell therapy for Parkinson's disease and, 54–57
    classification, 57–58
    differentiation states, 63
    graft composition functional impact, 65–69
    isolation from
        developing midbrain, 61–65
        stem cell-derived populations, 58–61
    neurogenesis, 54
    nigro-striatal pathway through intra-nigral grafting, 69–74
    progenitors, 59–62, 64–65, 68, 74
    transplantable, 58–61
Middle cerebral artery (MCA), 222
Miller Fisher syndrome (MFS), 110
Minnesota Multiphasic Personality Inventory (MMPI), 425
Minocycline, 128
Mitochondrial dysfunction hypothesis, 189
MMPI. See Minnesota Multiphasic Personality Inventory (MMPI)
MMR. See Multi-segmental monosynaptic response (MMR)
Modified Ashworth scale (MAS), 434
Monkeys
    detection of V1 stimulation and, 365–366
    double-step paradigm and, 353–354
    neocortex of, electrical evocation of motor responses from, 356
    saccadic eye movements and, 353
Mood elevation, 425. See also Obsessive-compulsive disorder (OCD), DBS in
Morris water maze (MWM) memory test, 89–90
Motoneuronal inhibition, 430
Motoneuron axons pruning, 177
α motoneurons, 430, 432
Motor disturbances
    treatment of, 381
Motor evoked potentials (MEPs), 433
Motor neuron disease (MND). See Amyotrophic lateral sclerosis (ALS)

Motor responses
    electrical evocation of, from neocortex of monkeys, 356
    suppression of, 356
Motor system, 380
Mouse embryonic stem cells, 29
Movement control
    brainstem, 396
    conscious, 395–396
    sensory, 396–397
    spinal, 396
Movement disorders
    basal ganglia. See Basal ganglia (BG)
    levodopa, 382–383
    putative mechanisms. See Putative mechanisms
    surgical treatment of
        advantages, 381
        drawbacks, 381
    therapy for, 380–381
MPTP (1-methyl-4-phenyl-1,2,3,6-tetrahydropyrindine) neurotoxin, 192
MRI. See Magnetic resonance imaging (MRI)
MS. See Multiple sclerosis (MS)
MSA. See Multiple system atrophy (MSA)
Muller cells, 323
Muller glia, 323
Muller glia cell, 23–25
    source of new neurons, 25–27
Muller stem-like (MS) cells, 6, 9, 12
Multiple Sclerosis Functional Composite (MSFC) scores, 469
Multiple sclerosis (MS), 101, 129, 140, 453
    patients, 434–435
Multiple sclerosis (MS), sex hormones and
    estrogen
        immunomodulatory properties of, 242–244
        neuroprotective properties of, 244–245
        as treatment, 245–246
    inflammation versus neurodegeneration in, 239–240
    rationale for
        gender gap, 240–241
        pregnancy, protective effects of, 241–242
    testosterone
        immunomodulatory properties of, 242
        neuroprotective properties of, 244
        treatment, 245
Multiple system atrophy (MSA), 385

Multipolar stimulation, 367
Multi-segmental monosynaptic response (MMR)
 phase-dependent modulation of, 402
Myelin-associated glycoprotein (MAG), 445
Myelination process, inhibitors of
 electrical activity, 456
 PSA-NCAM, 456

N-acetyl-aspartate (NAA), 466
NAWM. See Normal-appearing white matter
 (NAWM)
Neocortex
 of monkeys, electrical evocation of motor
  responses from, 356
Neogenin receptors, 458
Neonatal CA3, 447
Neovascular glaucoma, 261–262
Nerve cells, 443
Nervous system. See also Central nervous
  system (CNS); Peripheral nervous
  system (PNS)
 innate immunity in, 95–115
Nervus vagus stimulation (NVS), 419
Netrin-1, 458
Neural cell adhesion molecule (NCAM), 37–39
Neural-Colony Forming Cell Assay (N-CFCA)
  and NSCs, 44–46, 48
Neural retina, therapeutic strategies to restore, 3–6
Neural stem cells (NSCs), 7–9
 aging and, 45–47
 brain tumors and, 46–49
 cancer and, 47–49
 discovery, 43–44
 enumeration, 44–46, 48
 identification, 44, 46, 48
 mathematical model, 44–45, 48
 neural-colony forming cell assay and, 44–46, 48
 neurosphere assay and, 44, 46, 48
 therapeutic potential, 45–49
 in treatment for age-related CNS disorders,
  45–47
Neural transplantation, 61
"Neurobiotaxic theory", 498
Neuroblast migration
 in olfactory bulb, 39
 in RMS, 36–37
Neurocognitive, complications, 386
NeuroControl, 298

Neurodegeneration, 33–34, 83, 99, 102, 114–115,
  125. See also Neurodegenerative diseases
 in CNS, 33, 139–146
 complement (C) and. See Complement (C)
 olfactory vector hypothesis, 34
 role of TLRs in CNS, 139–146
Neurodegenerative diseases, 47, 95, 103
 LV vectors based gene therapy in, 187–195
  amyotrophic lateral sclerosis, 189–190
  gene transfer system, 187–188
  Huntington's disease, 194–195
  Parkinson's disease, 191–193
  spinal muscular atrophy, 190–191
 olfactory vector hypothesis, 34
Neurodevelopmental disorders, 444
Neurofibromatosis, complement activation in, 109
Neurofibromatosis type 2 (NF2), 334–335
 ABI patients, 337, 338
Neuroinflammation
 cellular and molecular approaches to reduce,
  127–129
 consequences on neuronal plasticity and
  regeneration, 129–131
 diseases, 101
 growth promotion by, 130–131
 immunosuppression, 126–127
 manipulation to improve SCI recovery, 126–129
 in pathogenesis of secondary injury, 125–126
 role of macrophages in, 129–130
 in spinal cord injury, 125–131
Neuroma, complement activation in, 109
Neuromuscular system
 brain-derived neurotrophic factors and, 400–401
 spinal cord injury and, 395
 spinal cord isolation model and, 398
 treatment for activity-based, 397–400
"Neuronal jamming," 420
Neuronal plasticity and regeneration
 consequences of neuroinflammation on,
  129–131
NeuroNexus Technologies, 308
Neuron–glial signal, 456
Neurons
 density of, 350
 Muller glia cell source of, 25–27
 thalamic, 352
 TLRs in, 145–146
 V1, 360

Neuropathies, 95–96, 109–113, 115, 142
Neuroprostheses, in research and clinical applications, 297–299. See also Brain–computer interfaces (BCIs)
Neurosphere assay (NSA), and NSCs, 44, 46, 48
Neurotherapy, 498, 499
Neurotransmitters, 395, 403, 457
Neurotrophic agents, 330
Neurotrophic factors
  exogenous, 179–180
  LV vector-mediated overexpression, 178–179
  motoneuron survival-enhancing properties, 181
  Parkinson's disease (PD) and, 201–202
  precise time- and location-specific application, 179–180
  role in PN repair, 173–175, 177
Neurturin (NTN), 210
  animal PD models
    administration in, 211
    gene therapy in, 211–213
  human PD subjects, experimental gene therapy in, 213–214
NF2. See Neurofibromatosis type 2 (NF2)
NG2, 456
Nigro-striatal pathway reconstruction
  through intra-nigral grafting, 69–74
NMDA receptor. See N-methyl-D-aspartic acid (NMDA) receptor
N-methyl-D-aspartic acid (NMDA) receptor, 227, 231
Nogo receptor-interacting protein (LINGO-1), 455–456
Nonhuman primates (NHP), 83–84, 90, 163–164, 192
  Aβ immunotherapy preclinical trials in, 85–86
Nonneuronal cells toxicity, 189
Nontumor (NT), patients
  ABI in, 337–339
  open-set sentence recognition, results of, 338
Normal appearing brain tissue (NABT), 468–470
Normal-appearing white matter (NAWM), 240
  damage, 468–470
Normality, 388
Notch/Jagged inhibitory pathway, 455
NSC. See Neural stem cells (NSCs)
NTN. See Neurturin (NTN)
NTN mRNA, 211
Nucleus 22 ABI system, 335

Nucleus 24 ABI system, 335
NVS. See Nervus vagus stimulation (NVS)
NXY-059, free radical spin trap drug, 222, 227, 228

Obsessive–compulsive disorder (OCD), DBS in, 419
  clinical efficacy of, 421–424
  side effects, 424
    acute effects, 425
    cognitive changes, 425
    DBS interruption, effects of, 425–426
    mood effects, 425
    personality changes, 425
OCD. See Obsessive-compulsive disorder (OCD)
OCT. See Optical coherence tomography (OCT)
Ocular angiogenesis, 253–254
  VEGF in, role of, 256
  and eye disease, 257
6-OHDA. See 6-hydroxydopamine (6-OHDA) lesion model
Olfaction, purpose and importance, 33–35
Olfactory bulb (OB), 34–37
  neuroblast migration in, 39
  neuron genesis in, 43–44
Olfactory system, 33–36
Olfactory vector hypothesis, 34
Oligodendrocyte progenitor cells (OPCs)
  inhibitors of differentiation or maturation of
    LINGO-1, 455–456
    the Notch/Jagged inhibitory pathway, 455
  migration
    contact-mediated cues, 456–457
    secreted molecules, 457–458
Oligodendrocytes, 454–455
  TLRs, 140
Oligodendroglial cells, 458
Open-set sentence recognition, results of, 338
Optical coherence tomography (OCT), 262–263, 320
  measurement, 327
Optic nerve (ON)
  damage, 469–470
  imaging, 475
  injury, 151–154
Optic neuritis, 472, 475
Optobionics, 322
Optokinetic testing, of visual acuity, 326

Orbitofrontal cortex (OrbC), 286
Ossification, cochlear, 339–340
Oxidative capacity
 of vastus lateralis muscle, 400
Oxidative stress hypothesis, 189

PABI. See Penetrating microelectrodes, ABI with (PABI)
Paced auditory serial-addition task (PASAT), 245
Paced visual serial addition task (PVSAT), 472
Pain, placebo effects and, 286–287
Pallidal stimulation, 384
Pallidotomy, 381
Panretinal laser treatment, 254
Paralysis, 190
Parkinson's disease (PD), 163–164, 180, 189, 244, 380
 cell therapy for, 54–57
 clinical symptoms, 191–192
 complement activation in, 103
 efficiency of DBS in, 381
 levodopa, 382–383
 LV vector-mediated gene therapy in, 192–193
 modeling, 192
 and motor performance, placebo effects and, 287–289
 olfactory vector hypothesis, 34
 RMS in, 39–40
Parkinson's disease (PD), trophic factors therapy in
 glial cell line-derived neurotrophic factor (GDNF), 202
  animal PD models. See Animal PD models
  clinical studies with, 210
  PD human subjects, experimental administration in, 203–204
 and neurotrophic factors, 201–202
 neurturin, 210
  animal PD models, administration in, 211
  animal PD models, gene therapy in, 211–213
  human PD subjects, experimental gene therapy in, 213–214
Parylene arrays, 408
PASAT. See Paced auditory serial-addition task (PASAT)
PBMC. See Peripheral blood mononuclear cell (PBMC)
PD. See Parkinson's disease (PD)

PDGF-BB. See Platelet-derived growth factor (PDGF-BB)
PDR. See Proliferative diabetic retinopathy (PDR)
PDT. See Photodynamic therapy (PDT)
Pedunculopontine nucleus (PPN), 380
 efficiency of HFS in, 386
Pegaptanib sodium (Macugen®), 257, 262
"Peking man", 498
Penetrating microelectrodes, ABI with (PABI), 336
Performance at the paced auditory serial addition task (PASAT) test, 469, 472
Periaqueductal gray (PAG), 286
Peripheral blood mononuclear cell (PBMC), 245, 246
Peripheral nerve (PN)
 grafts, 151, 153–154, 156–158
 injury and regeneration
  assessment of injury, 173, 176
  axonal outgrowth rate, 173, 175–176
  axonotmetic injuries, 175–176
  challenges, 175–177
  clinical outcome, 174
  exogenous neurotrophic factors application, 179–180
  LN vector overexpression of neurotrophic factors for, 178–179
  miniaturization of surgical and diagnostic tools, 183–184
  misrouting of regenerating axons and, 173, 177, 181–182
  molecular mechanisms, 174–175
  motoneuron survival, 173, 177, 181
  neuroma-in-continuity, 175–176
  neurotmetic injuries, 175–176
  post-traumatic complement activation effect on, 106–108
  routing problem, 182–183
  scar formation, 173, 175–176
  sensory nerve grafts and regeneration of motoneurons, 173, 177
  viral vectors clinical application to, 177–178
 repair, 173–184
Peripheral nervous system (PNS), 269
 complement in, 103–115, 109–110
  Charcot–Marie–Tooth disease, 109, 112–113
  experimental autoimmune neuritis, 109–110
  expression in healthy human sciatic nerve, 104

Guillain–Barré syndrome, 109–112
  HMSNs, 109, 112–113
  local synthesis in, 103–104
  neurofibromatosis, 109
  neuroma, 109
  peripheral nerve regeneration, 106–109
  PNS disease, 109–115
  PNS injury, 104–106, 113–115
  therapeutics in injury and disease, 113–115
  Wallerian degeneration, 105–107
Perivascular pump, 165–166
Perivasculature, 165
Permanent stroke (pMCAO), 224, 226
Pharmacological treatments
  of locomotor circuits, 403–405
Phase-dependent modulation, 401
Phenotypes, 398
Phosphatases, 449
Phosphenes, 324, 348, 349, 433
  brightness of, 365
  delay fields and, 359
  effects of blindness on, 362–364
  scotoma and, 359
  size of, evoked from macaque V1, 360
Photocoagulation laser, 363, 364
Photodynamic therapy (PDT), 262
Photoreceptor cells, 3, 5–8, 9, 11–13, 15, 23–28, 155
  degenerations, 318
  stem cells, transplantation of, 319–320
Pick's disease, complement activation in, 102
Pigmented epithelial cells, 23
Pioneer axons concept, 71
*Pithecanthropus erectus,* 498
Placebo response
  clinical studies, 290
  defined, 283
  neurological disorders and
    depression, 289–290
    pain, 286–287
    Parkinson's disease and motor performance, 287–289
  reflex or cognitive, 285–286
  sugar pill, evolution of, 284–285
  two different meanings of, 284
Placenta growth factor (PlGF), 255
Platelet derived growth factor alpha receptor (PDGFaR), 456

Platelet-derived growth factor (PDGF-BB), 245
PlGF. *See* Placenta growth factor (PlGF)
PMCAO. *See* Permanent stroke (pMCAO)
PNS. *See* Peripheral nervous system (PNS)
Poliovirus, 34
Polyglutamine disorders. *See* Huntington's disease (HD)
Polysialylated neural cell adhesion molecule (PSA-NCAM), 37–39, 59
  immunostaining for, 36
Population study, progesterone treatment for stroke and, 229–230
Posterior tibial nerve (PTN) stimulation, 432
Posterior ventral cochlear nuclei (PVCN), 336
Poststroke patients, 435–436
Poxvirus, uses in gene therapy, 188
PPN. *See* Pedunculopontine nucleus (PPN)
Preferential motor reinnervation, 177, 183
Pregnancy, protective effects of, MS and, 241–242
Pregnancy in Multiple Sclerosis (PRIMS) Group, 241
Preretinal angiogenesis, 254
  diabetic macular edema (DME), 260–262
  diabetic retinopathy, 259–260
Presynaptic inhibition, 431
  of Ia afferents, 433
Primate brain. *See also* Brain
  AAV vectors controlled dissemination in, 163–170
Primate embryonic stem cells, 29
PRIMS Group. *See* Pregnancy in Multiple Sclerosis (PRIMS) Group
PROG. *See* Progesterone (PROG)
Progenitor cell, 4–6, 8–9, 14, 24–25, 27–28, 35–39, 44–45, 47–48, 59, 61, 64
Progenitor migration, 38
Progesterone for Traumatic Brain Injury — Experimental Clinical Treatment (Pro-TECT), 221
Progesterone (PROG), 219
  neuroprotective mechanisms, 231–232
  stroke and
    and drugs for, 227
    human translation, improving animal study design for, 227–231
    preclinical studies, 222–227
    problem, scope of, 222

TBI and, neuroprotective effects of, 221
    sex-related differences in, 221
    as therapeutic candidate, 220
    as treatment for, 221–222
Progressive supranuclear palsy (PSP), 385
Proliferative diabetic retinopathy (PDR), 259
Proliferative retinopathies, defined, 254
Prolyl 4-hydroxylase (P4H), 274
Prosthesis
    retinal, 348
    thalamic, 348
Prosthetic devices, electron, 320–323
    clinical trial and testing of, 326–328
Protein kinase C (PKC-β), 259
Protein misfolding and aggregation hypothesis, 189
Protein phosphatase-1, 449
Proton MR spectroscopy ($^{1}$H-MRS), 466
PSA-NCAM, 456, 457
Psoriasis, 241
PSP. See Progressive supranuclear palsy (PSP)
Pulses
    anodal, 351
    cathodal, 351
Putative mechanisms, 383
    schematic representation of, 384
PVCN. See Posterior ventral cochlear nuclei (PVCN)
Pyramidal fibers, 368
Pyramidotomies, 380

QOL. See Quality of life (QOL)
Quality of life (QOL), 317, 319
Quiescent cells, 6
Quipazine, 403

RA. See Rheumatoid arthritis (RA)
Rabies-G virus, 188
RACC. See Rostral anterior cingulate cortex (rACC)
Randomized controlled trials (RCTs), 260, 261
Ranibizumab (Lucentis®), 257, 262–263
RCTs. See Randomized controlled trials (RCTs)
RD. See Retinal degeneration (RD) disease
Real-time MRI, AAV vectors trafficking in brain by, 167–169

Receptive field (RF), 353
    locations affected by stimulation, 353
    visual, 357, 358
Reciprocal inhibition, 431–432
Recombinant lentiviral (rLV) vectors, 205
Recurrent inhibition, 432
Relapsing-remitting MS (RRMS), 246, 467
Remyelination in multiple sclerosis (MS), 453–454
    capacity for, 454
    mechanisms behind, 454–455
Renshaw cells, 430
    activation of, 432
Repetitive TMS (rTMS)
    effect on cortical excitability, 433
    effect on patients with spasticity
        cerebral palsy, 435
        MS patients, 434–435
        poststroke patients, 435–436
    effect on spinal excitability, 433–434
Reporter mice, in neural transplantation studies, 57, 61, 63, 67, 74
Resting state fMRI
    and potential applications to brain injury, 489–490
        initial results, 490
    technique and applications, 487–489
Retina
    cell
        generation from embryonic stem cells, 29
        integration, 10–14
        replacement strategies, 27–29
        sources for transplantation, 6–10
        therapy, 14–15
        transplantation strategies, 3–15
    ciliary marginal zone, 24–25
    degeneration, 3, 9, 11–13, 15, 23–24, 29
    developmental biology, 24
    endogenous repair mechanisms, 25–27
    repair
        cell replacement strategies, 27–29
        endogenous repair mechanisms, 25–27
        strategies for, 23–29
    therapeutic strategies to restore, 3–6
    transplanted cell integration, 10–14
Retina encoder, 322
Retinal degeneration (RD) disease, 317
    AREDS clinical trial and, 319
    overview of, 318–319

and prosthesis implantation, 323–324
  therapies, 319–320
Retinal dystrophies, 15
Retinal electronic prosthesis
  components of, 321
  implantation of, RD and, 323–324
  morphological and neuronal bases, for implantation of, 323
Retinal ganglion cells (RGCs), 23, 151–158
  AAV-mediated transduction of, 152–154
  dendritic morphology, 156–158
  transduced with AAV encoding different genes, 156–158
Retinal lesions, 362, 363
Retinal pigmented epithelial (RPE) cells, 3, 5, 13, 23–24, 318
Retinal pigmented epithelial (RPE) layer, 254
Retinal prosthesis, 348
Retinal stem-like (RS) cells, 6–9, 12, 25, 28
Retinal transplantation
  barriers to, 11–12
  cell integration in, 10–14
  cell sources for, 6–10
  immune rejection, 13–14
Retinitis pigmentosa (RP), 3, 14–15, 24, 317, 318
  X-linked, 327
Retino-cortical magnification factor, 357, 359
Retinofugal pathway, transplantation in, 151–158
Retinopathy, VEGF-induced, in experimental animals, 257–259
Ret receptor tyrosine kinase, 202
Retromastoid area, 335
Retrovirus, uses in gene therapy, 188
RF. See Receptive field (RF)
RGC. See Retinal ganglion cells (RGCs)
Rheumatoid arthritis (RA), 241
Riluzole, 189
RMS. See Rostral migratory stream (RMS)
Rodent RMS, 35
Rodent robotic step training, 406
  variability in, 407
Rostral anterior cingulate cortex (rACC), 286–287
Rostral migratory stream (RMS)
  brain disorders and, 39–40
  in human, 35–36
  neuroblast migration in, 36–37
  pathway, 35–36
  in rodent, 35

RP. See Retinitis pigmentosa (RP)
RPE cells. See Retinal pigmented epithelial (RPE) cells
RPE layer. See Retinal pigmented epithelial (RPE) layer
RRMS. See Relapsing-remitting MS (RRMS)

Saccadic eye movements
  lateral geniculate nucleus and, 352
  microstimulation of V1 elicits, 351–356
  monkeys and, 353
SAILOR study, 264
SAINT III. See Stroke Acute Ischemic NXY-059 Treatment Trial (SAINT III)
Schwann cells, 70, 103–113, 154, 174–178, 182–183
  tumors, 335
SCI. See Spinal cord injury (SCI)
Scotoma, 359
  phosphene and, 359
Secondary-progressive MS (SPMS), 246
Second Sight Medical Products (SSMP), 317, 320
Sensorimotor cortex (SMC), 471–472
Sensory movement control, 396–397
Serendipity, 380
Serial analysis of gene expression (SAGE), 103–104
Serotonergic neurons, 59, 65–66, 68
Sex hormones, multiple sclerosis (MS) and
  estrogen
    immunomodulatory properties of, 242–244
    neuroprotective properties of, 244–245
    treatment, 245–246
  inflammation versus neurodegeneration in, 239–240
  rationale for
    gender gap, 240–241
    pregnancy, protective effects of, 241–242
  testosterone
    immunomodulatory properties of, 242
    neuroprotective properties of, 244
    treatment, 245
Sex-related differences, in brain injury, 221
"Shadow plaques", 453
Sham stimulation, 434
Signal enhancement by extravascular protons (SEEP), 474
Sinanthropus pekinensis, 498
siRNA. See Small interfering RNA (siRNA)

SIV (simian immunodeficiency virus) vector, 188
Skeletal muscle stem cells, 45
Small interfering RNA (siRNA), 257
Small nuclear ribonucleoprotein (snRNP) biogenesis, 191
SNc. See Substantia nigra pars compacta (SNc)
SNr. See Substantia nigra pars reticulata (SNr) complex
Soleus H reflex, 430–431, 433–434
Soleus muscle, 397
Sol H reflex, 432
Spasticity
   effects of rTMS on cortical excitability, 433
   effects of rTMS on spinal excitability, 433–434
SPEAK. See Spectral peak coding (SPEAK)
Spectral peak coding (SPEAK), 335
Spinal circuitry
   phase-dependent modulation, 401
   physiological state of, 401–402
state-dependent modulation, 401–402
Spinal cord
   electrode arrays, 408, 409
   imaging, 473–475
   isolation model, 398
   rehabilitation, multimodal approach to, 412
   stimulation, 407–410, 412
Spinal cord injury (SCI), 99, 114, 269
   cellular and molecular approaches, 127–129
   deleterious effects of, in humans, 400
   hyperexcitability, 394–395
   immunosuppression and, 126–127
   locomotor circuits treatment after. See Locomotor circuits
   macrophages relevance to, 129–130
   motor pools and, 394
   movement control and. See Movement control
   movement loss and, 394
   multi-modal strategy for treating, 412
   neuroinflammation manipulation to recovery from, 126–129
   neuromuscular system and, 395
   pathology, 125–126
   phenotypes, associated with, 398
   promotion by neuroinflammation, 130–131
   synapse formation and, 394
Spinal movement control, 396

Spinal muscular atrophy (SMA), 189
   clinical manifestations and genetics of, 190–191
   LV vector-mediated gene therapy in, 191
SPMS. See Secondary-progressive MS (SPMS)
SSMP. See Second Sight Medical Products (SSMP)
STAIR. See Stroke Therapy Academic Industry Roundtable (STAIR)
Stargardt's disease, 318
Stem cells, 3, 9, 14, 24–25, 37, 39–40, 43–44, 53, 57–58, 69, 382. See also Neural stem cells (NSCs)
   transplantation of, 319–320
Stepping
   bipedal, 411
   locomotor training and, 405
   phase-dependent modulation and, 401
   rodent robotic step training and, 406
   stance phase of, 401
   swing phase of, 401
STN. See Subthalamic nucleus (STN)
STN-DBS. See Subthalamic nucleusdeep brain stimulation (STN-DBS)
Stretch reflex arc, 430
Striate cortex (V1), 347
   deepest layers of, 359
   detection of, stimulation, 365–366
   electrical stimulation of, 350
   microstimulation of
      delays execution of visually guided saccades, 356–359
      elicits detection response, 359–361
      elicits saccadic eye movements, 351–356
   microstructure of, 350
   neurons, 360
   pyramidal fibers of superficial, 368
   saccade evoked electrically from, 352, 353
   stimulation of macaque, 353
   topographic map of, 349
Stroke, progesterone and
   drugs for treatment and, 227
   human translation, improving animal study design for, 227
      combination therapy, 230–231
      long-term functional recovery, 229
      physiological monitoring, 228
      population, study, 229–230

systematic preclinical dose–response studies, 228
treatment window, 228–229
preclinical studies with, 222–227
problem, scope of, 222
thrombolysis therapy and, 222
Stroke Acute Ischemic NXY-059 Treatment Trial (SAINT III), 228
Stroke Therapy Academic Industry Roundtable (STAIR), 222, 227–228, 229
Stroop test, 473
Subretinal neovascularization. See Choroidal neovascularization (CNV)
Substantia nigra, 382
Substantia nigra pars compacta (SNc), 380
Substantia nigra pars reticulata (SNr) complex, 380
Subthalamic nucleusdeep brain stimulation (STN-DBS), 287
Subthalamic nucleus (STN), 379, 380, 420, 424
efficiency of HFS in, 385–386
pharmacology of, HFS, 387–388
stimulation, 384–385
Sugar pill, evolution of, 284–285
Superior colliculus
stimulation of, 354
Superoxide dismutase 1 (SOD1), 103, 189–190
Supplementary motor area (SMA), 472
Surface-electrodes, 301, 341
ABI and, 336
Survival of motor neuron (SMN) gene, 190–191
Synapse
formation of, spinal cord injury and, 394
Synaptic plasticity, 449
α-synuclein-positive Lewy bodies, 209

TBI. See Traumatic brain injury (TBI)
TBI-induced coagulopathy, 232
T cells, 13, 83–84, 88, 90–91, 112, 127, 145
Temporary MCAO (tMCAO), 223–224
Tenascin-C, 457
Testosterones, multiple sclerosis (MS) and
immunomodulatory properties of, 242
neuroprotective properties of, 244
treatment, 245
TGFβ1. See Transforming growth factor β1 (TGFβ1)
Thalamic neurons, 352

Thalamic prosthesis, 348
Thalamotomy, 383
Thalamus, 348, 381
Thrombolysis therapy, stroke and, 222
Tissue plasminogen activators (tPAs), 222, 229, 232
TLR. See Toll-like receptors (TLRs)
TMCAO. See Temporary MCAO (tMCAO)
TMS. See Transcranial magnetic stimulation (TMS)
TNFα. See Tumor necrosis factor α (TNFα)
Toll-like receptors (TLRs), 129–130
in astrocytes, 142–146
in CNS, 139–146
features, 140–141
in microglia, 141–143, 146
in neurons, 145–146
Tonotopic regions, 335
TPAs. See Tissue plasminogen activators (tPAs)
Transcranial magnetic stimulation (TMS), 419
Transforming growth factor β1 (TGFβ1), 245
Transgenic mice, 61–62, 84–85, 87–88, 99, 101, 113, 151, 153, 189–190.95
Transplantable mDA neurons, 58–61
Transplantation, 25, 27–29, 54–55, 57–59, 61–66, 68–69, 73–74, 129, 151–153
barriers to retinal, 11–12
cell integration in, 10–14
cell sources for, 6–10
immune rejection, 13–14
in retinofugal pathway, 151–158
Transplanted cell integration, 10–14
Traumatic brain injury (TBI), 99, 114
progesterone in, neuroprotective effects of, 220
human trials evaluating, as treatment for, 221–222
as therapeutic candidate, 221
sex-related differences in, 221
Tremor, 380, 381
thalamotomy for, 383
VIM stimulation in, 384
Tumor angiogenesis factor, 256
Tumor-initiating cells (TICs), 47
Tumor necrosis factor α (TNFα), 242
Tyrosine hydroxylase (TH)-positive nigral neurons, 202, 203, 205, 208–209, 210, 211

Unc5 receptors, 458
Unified Parkinson's Disease Rating Scale (UPDRS) scores, 203, 204, 214, 287–288
UPDRS scores. See Unified Parkinson's Disease Rating Scale (UPDRS) scores
Utah electrode array, 320
Utah system, 306

V1. See Striate cortex (V1)
Vaccinia virus, uses in gene therapy, 188
Vascular endothelial growth factor (VEGF) antagonists
 in AMD and choroidal neovascularization (CNV), 254
  bevacizumab (Avastin®), 263
  pegaptanib sodium (Macugen®), 262
  Ranibizumab (Lucentis®), 262–263
 angiogenesis and. See Angiogenesis, VEGF antagonists and
 clinical use of, 257–259
 neuroprotective role of, 256
 in pathology outside eye, 255–256
 preretinal neovascularization and
  diabetic macular edema (DME), 260–262
  diabetic retinopathy, 259–260
 side effects of ocular use of, 263–264
 as therapeutic agents, 256–257
Vascular permeability factor (VPF), 257
Vastus lateralis muscle
 oxidative capacity of, 400
α Vβ1 integrin, 457
VEGF-A, 255, 257, 260, 262
VEGF antagonists. See Vascular endothelial growth factor (VEGF) antagonists
VEGF family, 255
VEGFR. See VEGF receptors (VEGFR)
VEGFR-2, 260
VEGF receptors (VEGFR), 255

VEGF (vascular endothelial growth factor), 189–190
Ventralis intermedius nucleus (VIM), thalamic, 381
 efficiency of HFS in, 383–384
Ventral mesencephalon (VM), 53–55, 57, 61–71, 73
Ventriculo-olfactory neurogenic system (VONS), 36
Vergence angle, 349
Vestibular schwannoma (VS), 340
VIM. See Ventralis intermedius nucleus (VIM), thalamic
Vioxx, 228
Viral vectors. See also Adeno-associated viral (AAV) vectors
 clinical application in injured PN, 177–178
 mediated application of GDNF, 181
Visual-evoked potentials (VEP), 475
Visual functions
 measurement of, 325t
Visuo-cortical prosthesis, 367
Voluntary control, 395–396
VPF. See Vascular Permeability Factor (VPF)
VS. See Vestibular schwannoma (VS)
VSV-G (vesicular stomatitis virus), 188

Wallerian degeneration in white matter, 240
Wallerian degeneration (WD), 104–108, 129
Working memory, 472
 circuitry, 472, 473
Wound healing, mechanisms of, 254–257

X-linked, RP, 327

Y-BOCS scores, 424

Zea Longa test, 223
Zymosan, 130
Zymosan-activated macrophages (ZAMs), 130

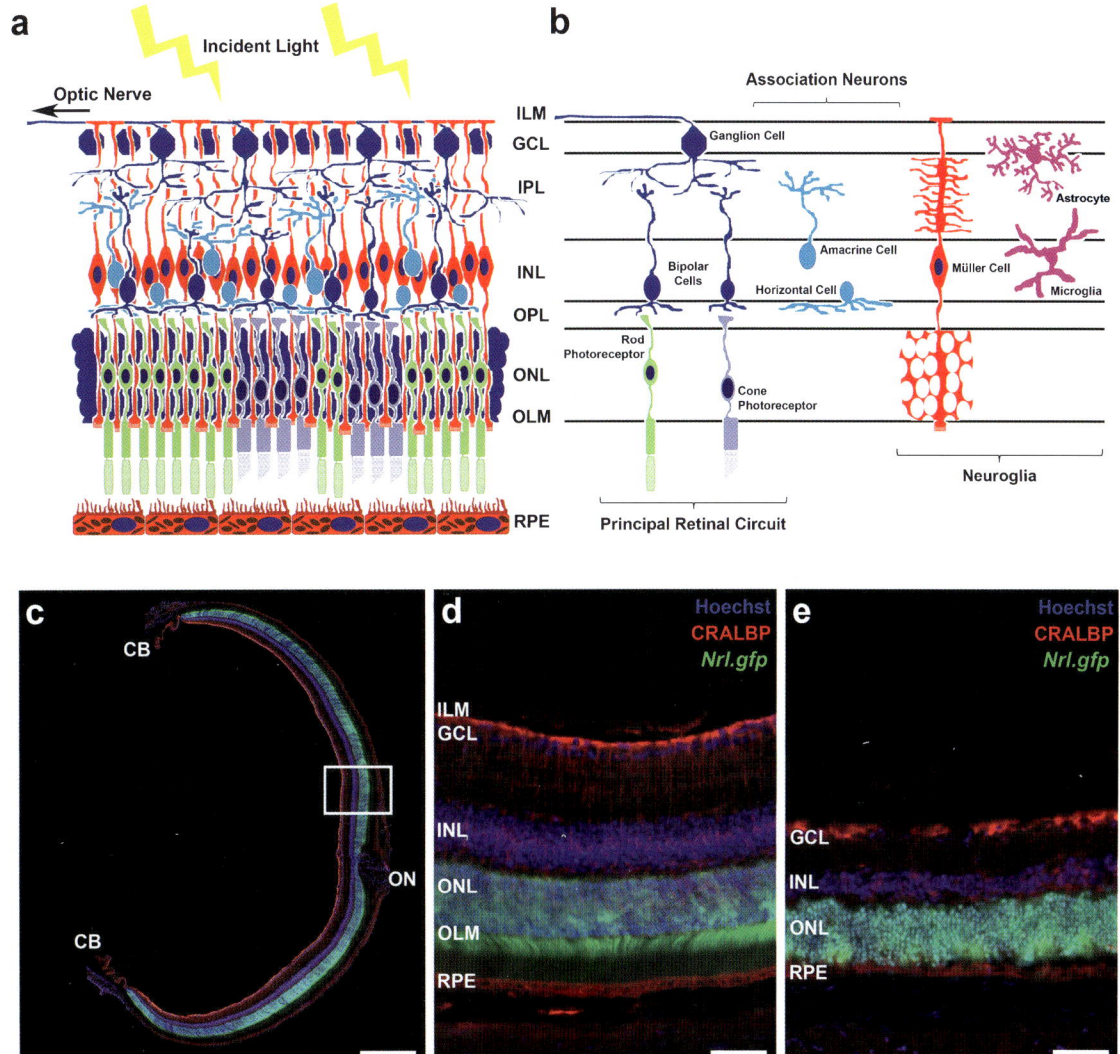

Plate 1.1. The mammalian retina. (a) A schematic diagram illustrating the layers of the mammalian retina (green rod and purple cone photoreceptors; red Müller cells and RPE; blue nuclei). (b) A schematic diagram illustrating the position of the various cell types present in the adult neural retina. These cells are subdivided into (i) the principal retinal circuit, (ii) the association neurons, and (iii) the neuroglia. (c) A sagittal retinal section from an *Nrl.gfp* (green; rod photoreceptors) mouse. Scale bar, 200 μm. (d) A single fluorescence image of an adult *Nrl.gfp* retinal section stained for CRALBP (red), a protein present in Müller cells and the RPE. Scale bar, 40 μm. (e) A single fluorescence image of a degenerating retinal section stained for CRALBP (red), demonstrating the disorganization and loss of photoreceptor cells (*Nrl.gfp*; green). Scale bar, 40 μm. Nuclei were counterstained with Hoechst 33342 (blue). CB, ciliary body; ON, optic nerve; ILM, inner limiting membrane; GCL, ganglion cell layer; IPL, inner plexiform layer; INL, inner nuclear layer; OPL, outer plexiform layer; ONL, outer nuclear layer; OLM, outer limiting membrane; RPE, retinal pigment epithelium.

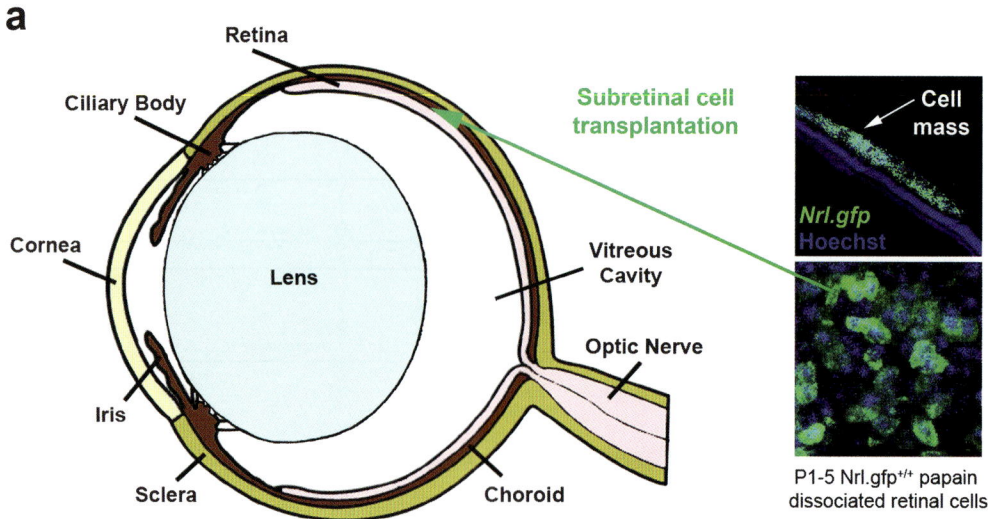
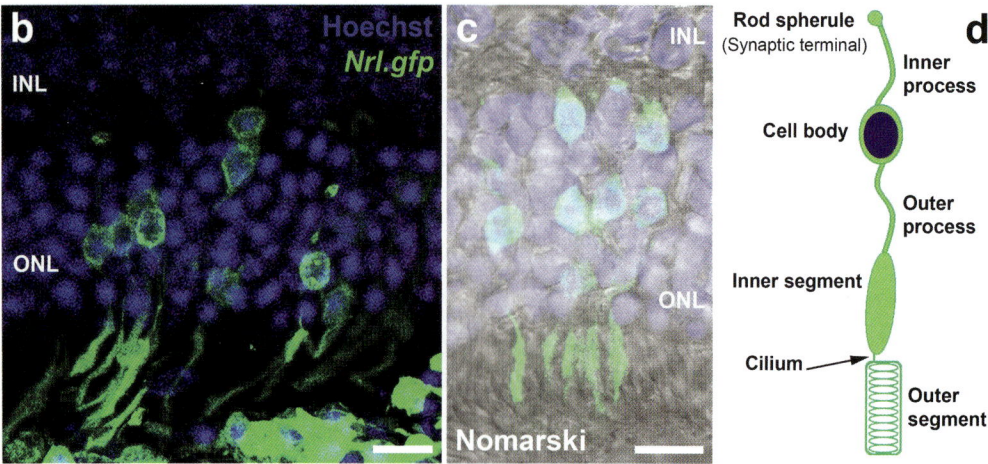

Plate 1.2. Photoreceptor precursor cell transplantation into the adult eye. (a) A schematic diagram of a mouse eye illustrating the subretinal transplantation of *Nrl.gfp* precursor cells (green) and the resulting cell mass (inserts). (b) A confocal image of integrated *Nrl.gfp* rod photoreceptors, 21 days after transplantation to an adult recipient. (c) A Nomarski confocal image of integrated *Nrl.gfp* rod photoreceptors. (d) A schematic representation of the structure of a rod photoreceptor. Nuclei were counterstained with Hoechst 33342 (blue). Scale bars, 20 μm. INL, inner nuclear layer; ONL, outer nuclear layer.

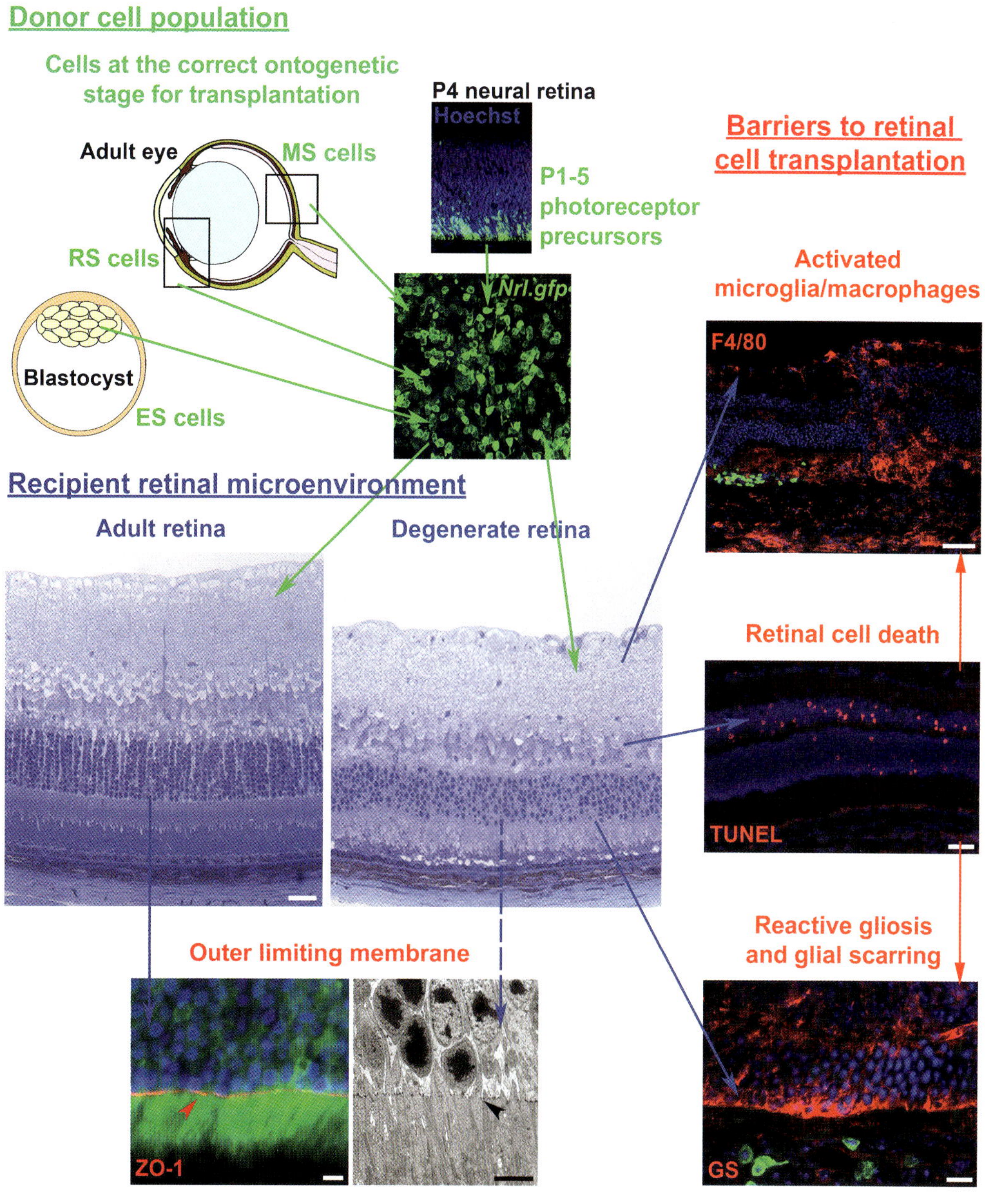

Plate 1.3. A summary of retinal cell transplantation strategies. A diagram to summarize the various retinal cell transplantation strategies and the related barriers that may limit transplanted photoreceptor cell integration in the adult and degenerate neural retina, as discussed in the main text. The donor cell population (top; green) can be derived from a variety of cell sources, but must be differentiated to the correct ontogenetic stage (postmitotic rod precursors, *Nrl.gfp*; green) prior to transplantation to enable photoreceptor cell integration into the host adult retina (MacLaren et al., 2006). The recipient retinal microenvironment (middle; blue) may also limit photoreceptor cell integration if the relevant barriers are not modulated at the time of transplantation. Scale bar, 50 μm. The relevant barriers to retinal cell transplantation and integration (right; red) are indicated. The outer limiting membrane (indicated by the red or black arrow head) forms a barrier to increased cell integration in the adult retina and in some models of retinal degeneration. Scale bars, 10 μm and 5 μm. Other barriers, present predominantly in the degenerate retina, include retinal cell death and the resulting activated microglia/macrophages and reactive gliosis/glial scarring. Scale bars, 50, 100, and 20 μm, respectively. Nuclei were counterstained with Hoechst 33342 (blue). ES cells, embryonic stem cells; GS, glutamine synthetase; MS cells, Müller stem-like cells; RS cells, retinal stem-like cells; ZO-1, zonula occludens-1.

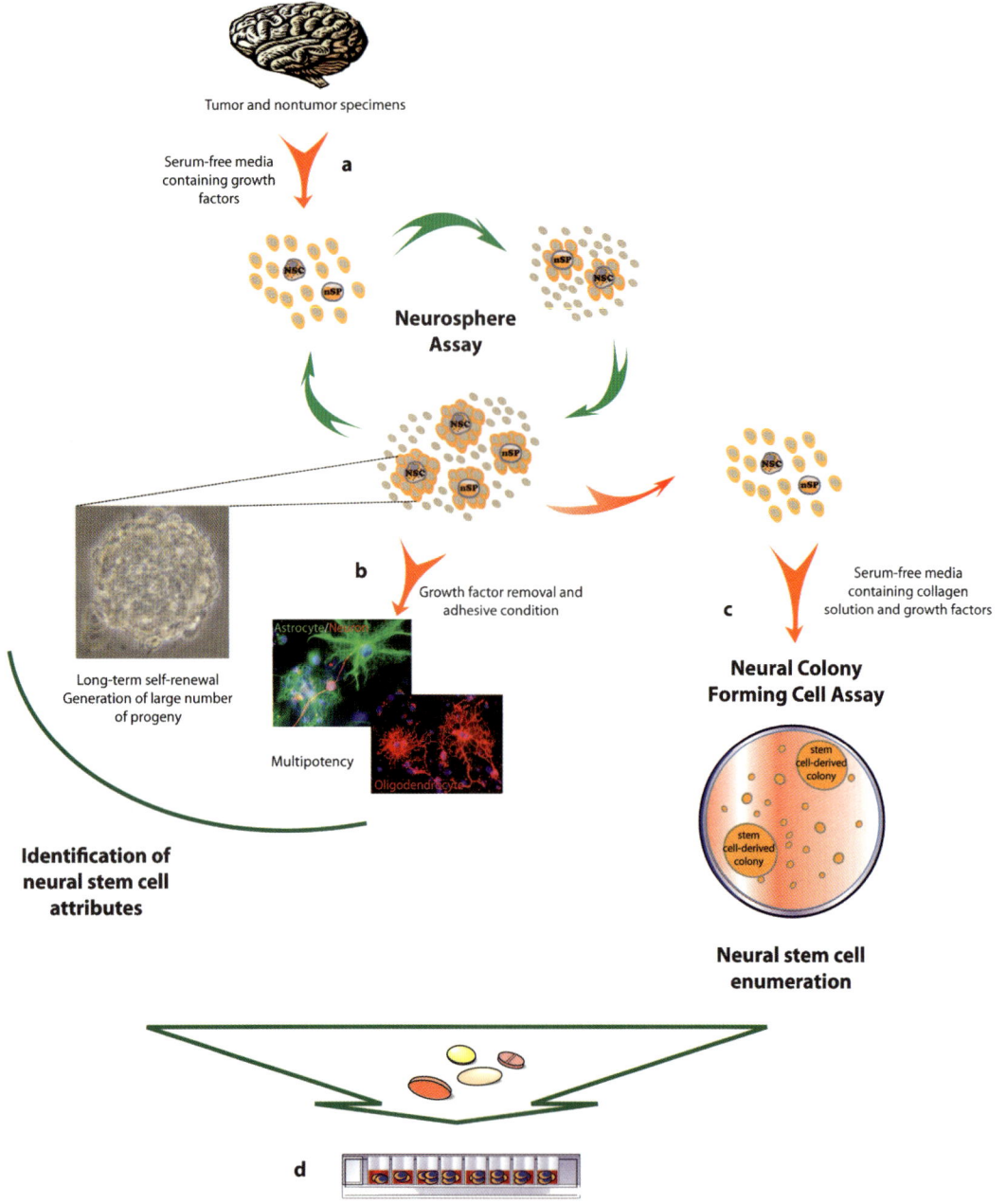

Plate 4.1. (a) Tumor and nontumor brain tissue samples are harvested in single cell suspension and transfer in culture into the neurosphere assay (NSA) consisting of a serum-free media supplemented with growth factors. In these conditions, growth factor-responsive cells survive and proliferate to generate clonal cell aggregates called neurospheres. The neurospheres can be dissociated and replated to form secondary neurospheres. This process can be indefinitely repeated, resulting in an exponential increase in the number of cells and spheres that are generated, describing long-term self-renewal and proliferation properties. (b) Plating the cells or neurospheres on adhesive substrate coupled with growth factor withdrawal creates differentiative conditions leading to the generation of neurons, astrocytes, and oligodendrocytes (nuclei stained with DAPI, blue), demonstrating multipotency characteristic. Therefore, the NSA allows identification of stem cells characteristics. (c) Cells from primary tissue or from the NSA are dissociated into single cell suspension and transfer in the neural colony forming cell assay (N-CFCA) that consists of a serum-free media containing collagen solution and supplemented with growth factors. After 3 weeks of culture, the growth factor-responsive cells generate different ranges of colonies size (i.e. diameter). Quantification of the number of large colonies (>2 mm in diameter, exhibiting stem cell attributes) allows enumeration of actual neural stem cells. (d) The NSA and the N-CFCA represent meaningful tools to design stem cell-based treatment in the setting of age-related diseases and to devise rational drug discovery approaches to cure brain tumors.

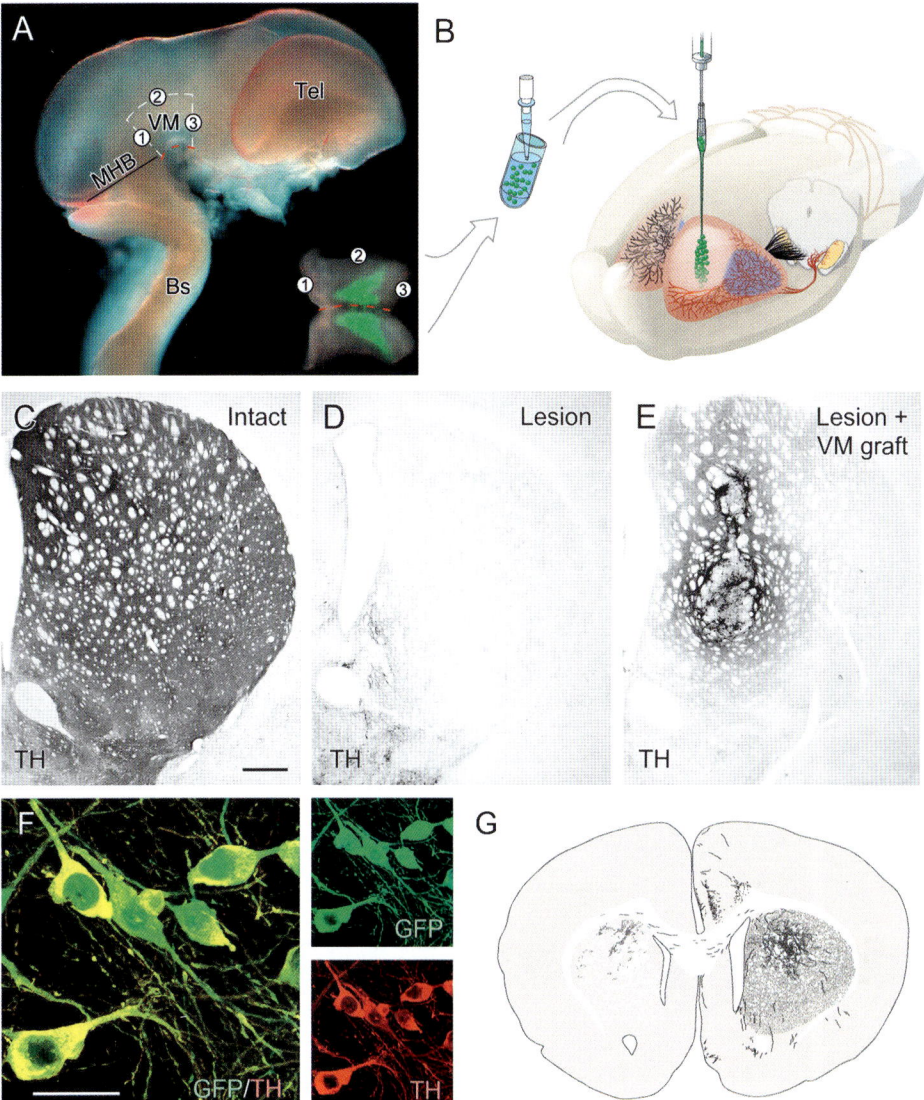

Plate 5.1. Cell therapy for Parkinson's disease. (A) The developing mouse brain at embryonic day 12.5. The dashed lines indicate the approximate region of ventral mesencephalon (VM) dissected in order to generate cell preparations for grafting. The inset shows a piece of VM tissue dissected from a mouse in which all midbrain dopamine neurons express GFP (Pitx3-GFP mouse). The numbers indicated are for orientation relative to the intact brain. The red dashed line marks the midline. (B) A schematic overview of a typical transplantation procedure, whereby the dissected VM is prepared as a single cell suspension (through trypsin digestion and mechanical dissociation), and then cells are microinjected into the host brain. Image used here shows placement into the striatum. (C–E) TH immunohistochemistry in coronal sections through the adult rat brain. The dark staining of the striatum in the intact animal (C) represents the dense terminal network of TH-positive fibers originating from mDA neuronal projections. Lesioning of the mDA neurons through injection of 6-hydroxydopamine removes this TH-positive afferent innervation of the striatum (D). Panel (E) illustrates a 6-OHDA-lesioned animal 6 weeks after grafting of $1.0 \times 10^5$ E12.5 mouse VM cells into the striatum. The graft itself can be seen as a discrete teardrop-shaped deposit of darkly stained TH-positive cells, while the dark gray area surrounding the graft represents the new TH-positive innervation of the host striatum provided by the grafted mDA neurons. (F) Grafting of VM tissue from donor mice in which the dopamine neurons express GFP allows for the unequivocal identification of mDA neurons and their associated fiber outgrowth in the host brain. The images shown are from an intra-striatal graft of cells prepared from VM dissected from the TH-GFP mouse. (G) Schematic tracing of GFP immunoreactivity 6 weeks after grafting of E12.5 TH-GFP VM cells into the striatum of a neonatal rat reveals graft-specific patterns of mDA neuronal fiber outgrowth. Abbreviations: Bs, brainstem; MHB, mid-hindbrain boundary; Tel, telencephalon. Scale bars: C, 500 μm; F, 30 μm.

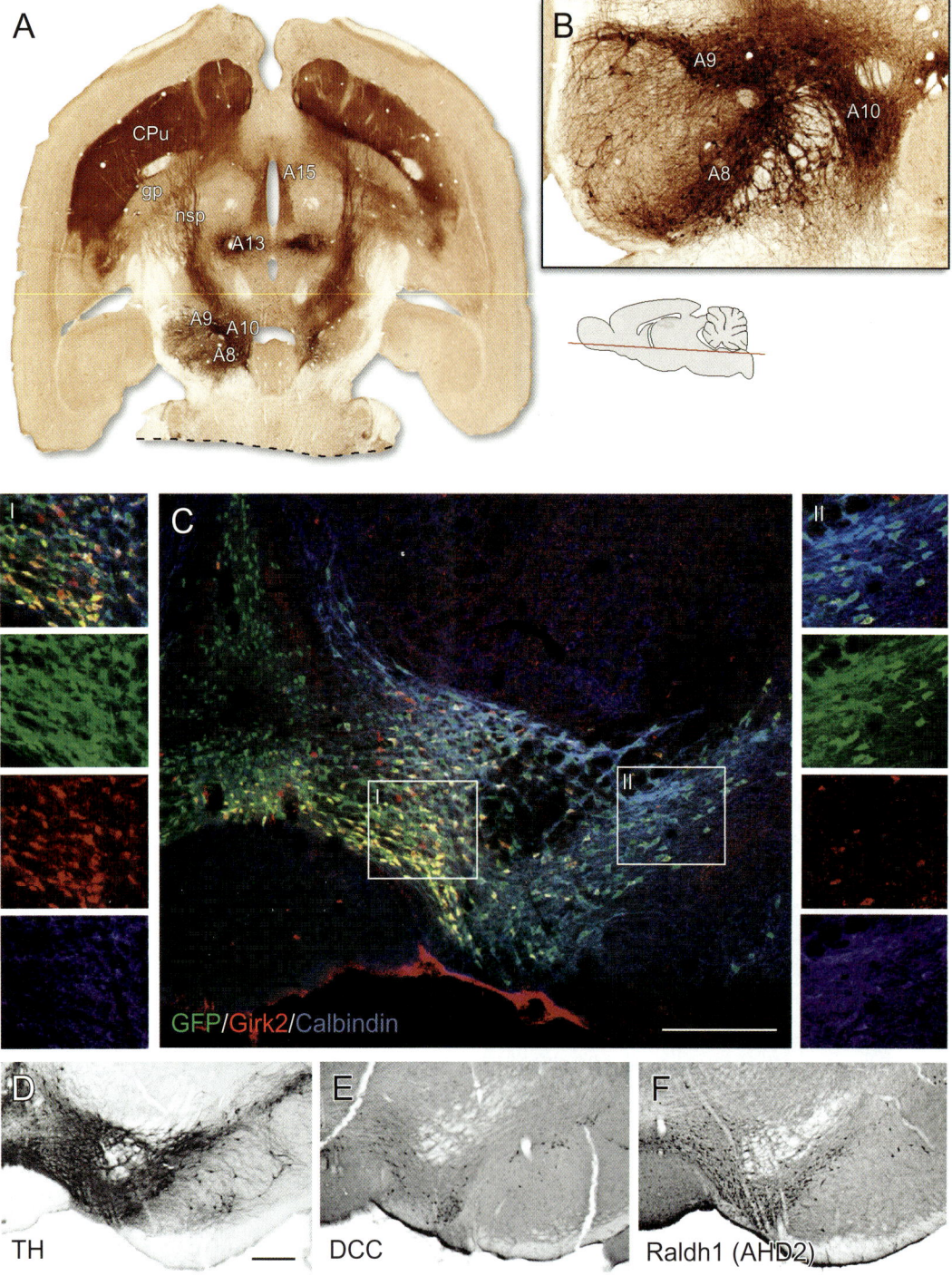

Plate 5.2. Basic neuroanatomical features of the midbrain dopamine neuron projection system. (A) Immunohistochemistry for tyrosine hydroxylase in a horizontal section through the adult mouse brain shows the major midbrain dopaminergic cell groups (A8, A9, and A10) and their efferent projection patterns. The approximate section plane (red line) is indicated in the parasagittal diagram (note, this animal has received a partial 6-OHDA lesion on the right-hand side of the brain, reflected by a notable loss of TH-positive cell bodies on that side). (B) The spatial distribution of the A8, A9, and A10 cell groups shown at greater magnification. (C) Immunohistochemistry for GFP (green), Girk2 (blue), and calbindin (red) in a coronal section through the adult mouse midbrain. In this animal, GFP is expressed under control of the regulatory elements for Pitx3 and, therefore, in all dopamine neurons. Girk2 and calbindin broadly identify A9 and A10 dopamine neurons, respectively. The boxed areas show in greater detail cells in the VTA (I) and the substantia nigra pars compacta (II). Other proteins, including DCC (E) and Raldh1/AHD2 (F) identify ventral subsets of dopamine neurons in the midbrain. Tyrosine hydroxylase expression (D) is shown in an adjacent section as a point of reference. Abbreviations: CPu, caudate putamen unit; DCC, deleted in colorectal cancer; gp, globus pallidus; nsp, nigro-striatal pathway; TH, tyrosine hydroxylase. Scale bar: C, 200 μm (panels A, B, D–F, courtesy: S. Grealish).

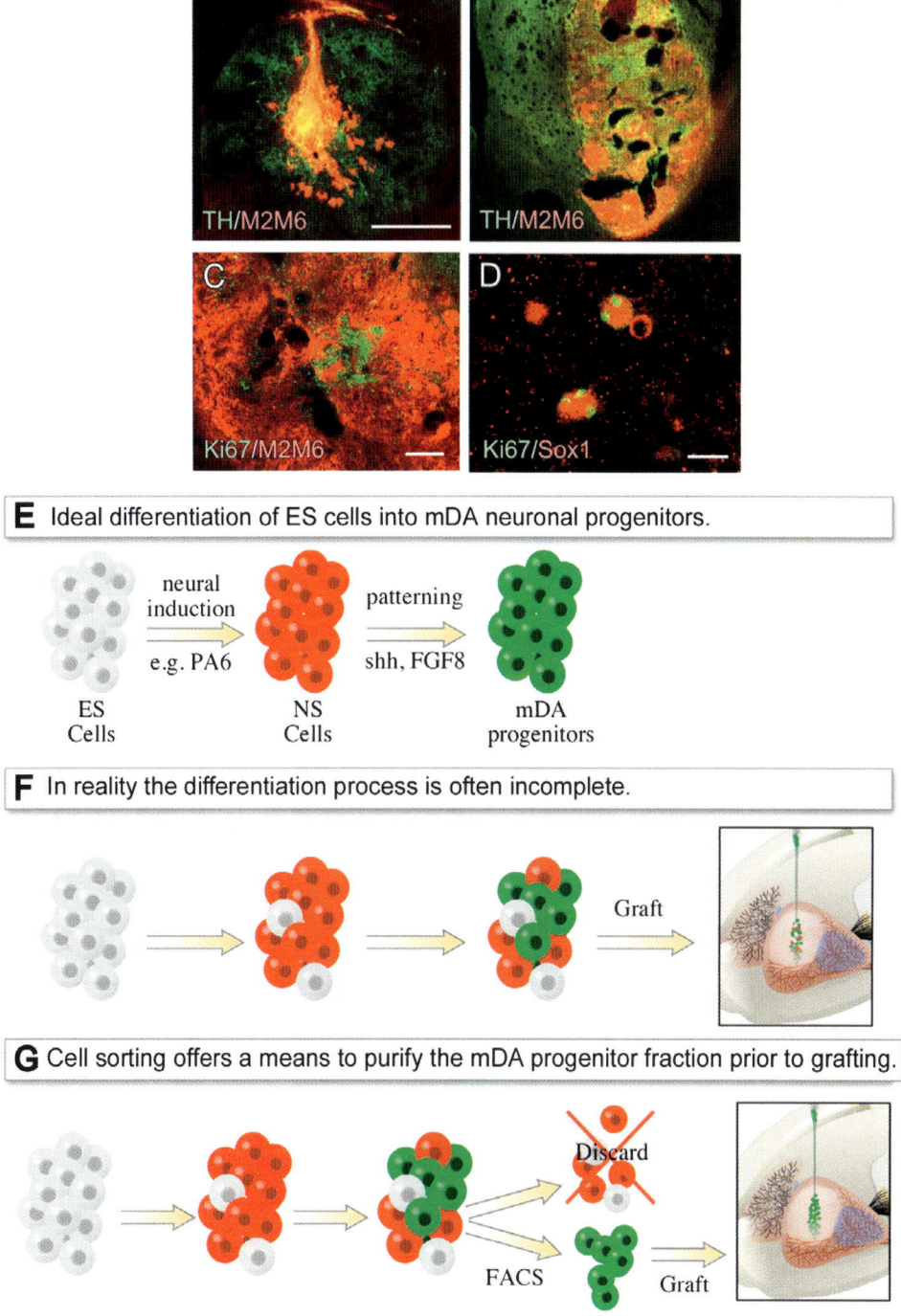

Plate 5.3. Dopamine neuron-containing grafts derived from embryonic stem cells can give rise to tumors. (A, B) Grafts of mouse cells placed in the rat brain can be detected using antibodies against the mouse-specific antigens M2 and M6. Panels A and B illustrate the gross morphology of intra-striatal grafts derived from the same number ($1.0 \times 10^5$) of either E12.5 fetal VM cells (A) or partially differentiated ES cells (B), 6 weeks after grafting into neonatal rat hosts. Note the dramatically larger size of the ES cell-derived graft, along with pockets of necrosis (black) throughout the graft core. Both grafts also contain large numbers of TH-positive mDA neurons (green), which innervate the host striatum. (C) The ES cell-derived grafts contain a population of actively dividing (Ki67-positive, green) cells even 6 weeks after transplantation. (D) Many of the Ki67-positive cells are Sox1-positive and thus are likely to be primitive neural precursors. (E–G) Schematic representation of basic features of the differentiation procedures used to generate transplantable mDA neurons, and how a cell-sorting strategy might be used to avoid tumor formation after grafting. Abbreviations: FACS, fluorescence-activated cell sorting; FGF8, fibroblast growth factor 8; NS, neural stem; shh, sonic hedgehog. Scale bars: A–B, 1 mm; C, 200 μm; D, 20 μm.

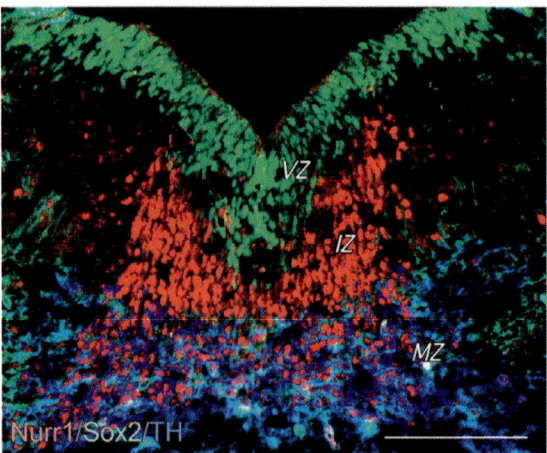

Plate 5.4. Differentiation states of midbrain dopamine neuronal progenitors in the embryonic mouse brain. Immunohistochemistry for Sox2 (green), Nurr1 (red), and TH (blue) in a coronal section through the E12.5 mouse midbrain illustrates the distribution of neural progenitors in distinct states of differentiation. The ventricular zone (VZ) contains actively dividing Sox2-positive precursors. At this late stage of mDA neurogenesis (E12.5) most of the mDA progenitors have already exited the cell cycle and only very few of the VZ precursors will give rise to mDA neurons. Most of the transplantable mDA progenitors at E12.5 reside in the intermediate zone (IZ) as Nurr1-positive, post-mitotic neuroblasts. As these progenitors continue to differentiate, they move into the mantle zone (MZ) and begin to express TH. Very few of these TH-positive mDA neurons are able to survive the transplantation procedure. Scale bar: 200 μm.

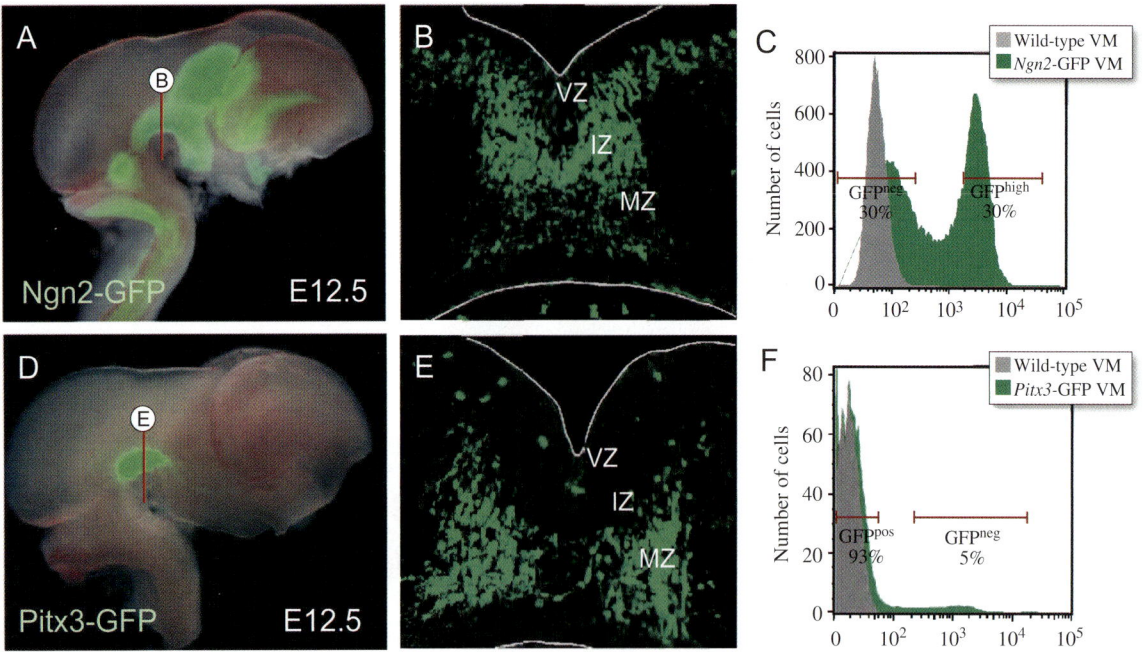

Plate 5.5. Reporter mice can be used to isolate distinct progenitor populations from the embryonic midbrain. GFP fluorescence superimposed over brightfield photographs shows the regional distribution of GFP-expressing cells in the embryonic (E12.5) brains of the Ngn2-GFP (A) and Pitx3-GFP (D) mice. Immunohistochemistry for GFP in coronal sections through the midbrain of these mice illustrates the local distribution of GFP-positive cells. (B) In Ngn2-GFP mice, GFP expression identifies a population of newly post-mitotic neuroblasts in the intermediate zone (IZ). (E) In the Pitx3-GFP mice, the GFP-positive cells represent young mDA neurons in the mantle zone (MZ; the approximate section plane for the coronal images is indicated by the red line in A and D). These GFP-positive cell fractions can be selectively isolated from dissected VM tissue pieces through fluorescence-activated cell sorting (FACS). (C) FACS analysis of VM tissue pieces from E12.5 Ngn2-GFP mice identifies a distinct subpopulation of highly GFP-positive cells, which represents approximately 30% of all viable cells. (F) FACS analysis of E12.5 Pitx3-GFP VM identifies a subset of GFP-positive cells representing around 5% of the viable cell population. In order to establish the threshold for specific detection of GFP-positive cells, the gate settings on the FACS apparatus are determined using cell suspensions prepared from wild-type littermates (gray in C and F). Panel A is a modified reproduction from Thompson et al. (2006).

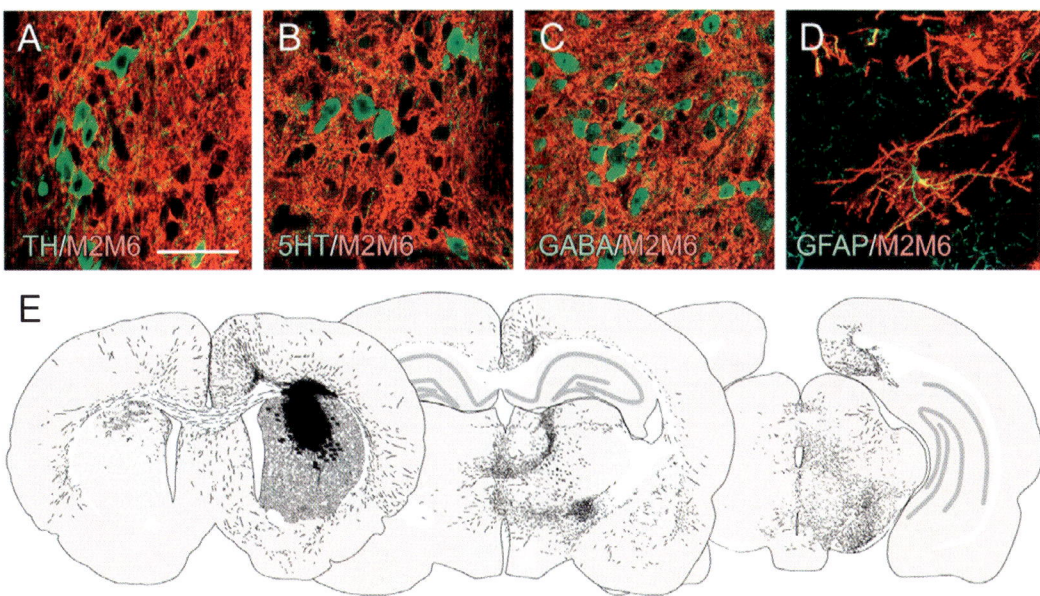

Plate 5.6. Non-dopaminergic cell types in VM grafts. Grafts of VM will contain neurons corresponding to a variety of neurochemical phenotypes, and also various kinds of glial cells. Six weeks after transplantation of E12.5 mouse VM cells into the striatum of neonatal rats, immunohistochemistry for the mouse-specific M2 and M6 proteins allows for clear identification of the grafted cells. In addition to TH-positive mDA neurons (A), the grafts will also contain 5HT-positive serotonergic neurons (B), and a large number of γ-aminobutyric acid (GABA) containing neurons (C). The grafts also contain various glial subtypes, including those that are immunoreactive for glial fibrillary acidic protein (GFAP). The M2M6 antigens are expressed throughout the neuritic processes of certain classes of neurons, allowing for identification of patterns of fiber outgrowth. A schematic representation of immunohistochemistry for M2 and M6 in coronal sections 6 weeks after grafting of E12.5 mouse VM cells into the striatum of a neonatal rat, illustrates the pattern of graft-derived fiber outgrowth in the host brain. Double labeling of M2M6 and TH (not shown), indicates that while the vast majority of striatal M2M6-positive innervation is dopaminergic, most of the fibers found in structures outside the striatum are non-dopaminergic. Scale bar: A–D, 50 μm (images shown here are modified reproductions from Thompson et al., 2008).

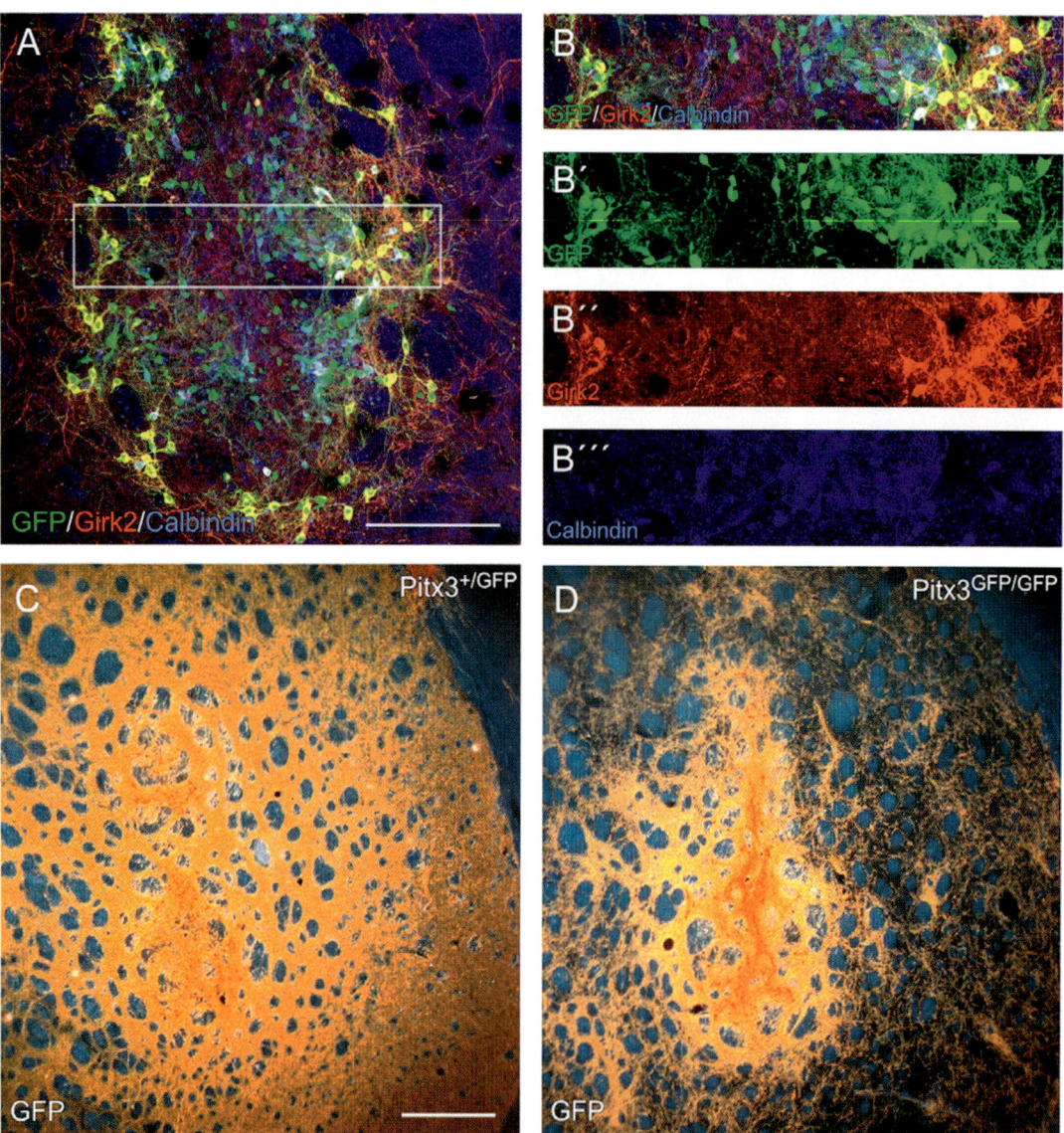

Plate 5.7. Contribution of different midbrain dopamine neuronal subtypes in ventral mesencephalic grafts. The potassium channel protein, Girk2, and the calcium-binding protein, calbindin, can be used to broadly identify mDA neurons of the A9 and A10 subtype, respectively, in VM grafts. (A) Immunohistochemistry for Girk2 (red) and calbindin (blue) in coronal sections of a mouse that received an intra-striatal graft of E12.5 VM cells from Pitx3-GFP donor mice, reveals that the A9, Girk2-positive/GFP-positive neurons (yellow) are distributed throughout the periphery of the graft, while the A10, calbindin-positive/GFP-positive cells (aqua) are clustered mainly in the center of the graft. The boxed area, spanning peripheral and central aspects of the graft, is shown in higher magnification (B) and as individual color channels (B′–B′′′). The knock-in design of the Pitx3-GFP reporter mice means that VM cell suspensions can be prepared from mice either heterozygous (Pitx3$^{wt/GFP}$) or homozygous (Pitx3$^{GFP/GFP}$) for GFP, and therefore null for Pitx3 in the latter case. Darkfield images of immunohistochemistry for GFP 12 weeks after grafting of either Pitx3$^{wt/GFP}$ VM (C) or Pitx3$^{GFP/GFP}$ VM (D) shows that the Pitx3$^{GFP/GFP}$ grafts, which have a markedly reduced proportion of A9 neurons (not shown), also display a significantly reduced capacity to provide dopaminergic innervation of the host striatum. Scale bars: A, 200 μm; C, D, 500 μm.

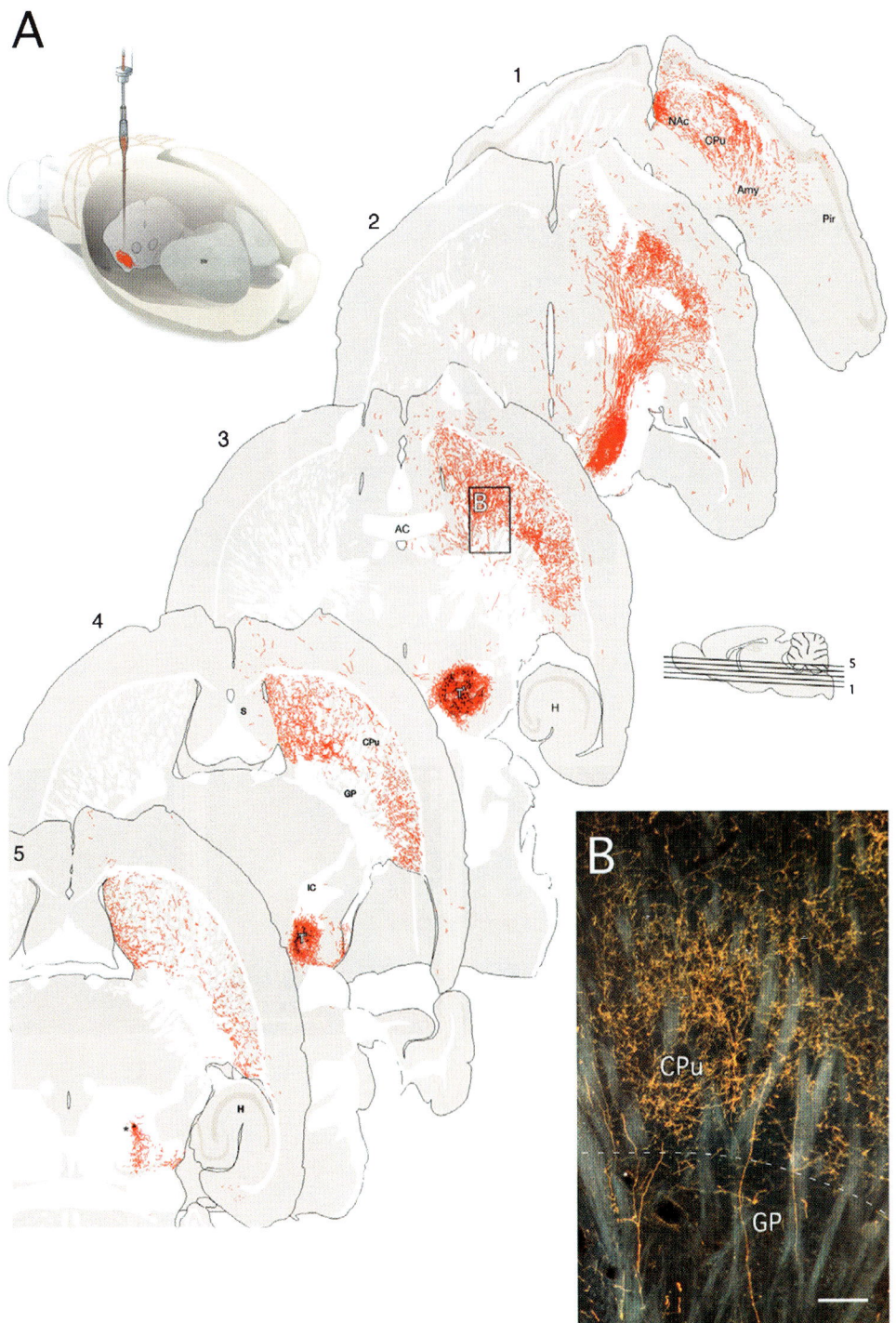

Plate 5.8. Reconstruction of the nigro-striatal pathway through intra-nigral grafting. By using donor tissue from mice in which GFP expression is driven by the TH promoter, detection of GFP in the resulting grafts allows for highly sensitive and unambiguous characterization of graft-derived dopaminergic fiber patterns in the host brain. (A) Schematic reproduction of GFP immunoreactivity in horizontal sections 16 weeks after grafting of $1.5 \times 10^5$ E12.5 TH-GFP VM cells into the substantia nigra of a mouse that had previously received partial lesioning of the intrinsic dopamine neuron projection system. The approximate dorso-negativentral levels of the horizontal sections (1–5) are indicated in the parasagittal diagram. A schematic representation of the whole mouse brain illustrates the targeting of the substantia nigra using a micro-transplantation approach to inject the VM cells. (B) A darkfield photograph of immunohistochemistry for GFP shows the pattern of GFP-positive fibers coursing through the globus pallidus and forming a ramified terminal network in the striatum (from boxed area on Section 3 in panel A). The dashed line approximates the striato-palladial border. Abbreviations: AC, anterior commissure; Amy, amygdala; CPu, caudate putamen; GP, globus pallidus; H, hippocampus; IC, internal capsule; NAc, nucleus accumbens; Pir, piriform cortex; S, septum; T, transplant. Scale bar: B, 200 μm.

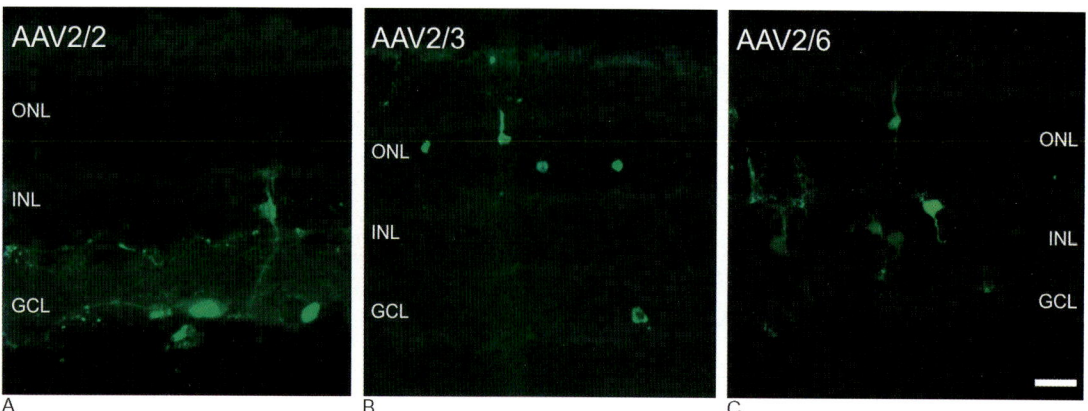

Plate 10.1. Fluorescence photomicrographs of retinal sections from eyes that had been injected intravitreally with different AAV serotypes expressing GFP, 10 weeks post-injection. (A, B, and C) show GFP positive cells in retina from eyes injected with AAV 2/2, AAV 2/3, or AAV 2/6 respectively. Note differences in the distribution and morphology of transduced cells. GCL, retinal ganglion cell layer; INL, inner nuclear layer; ONL, outer nuclear layer. Scale bar for all figures = 20 μm.

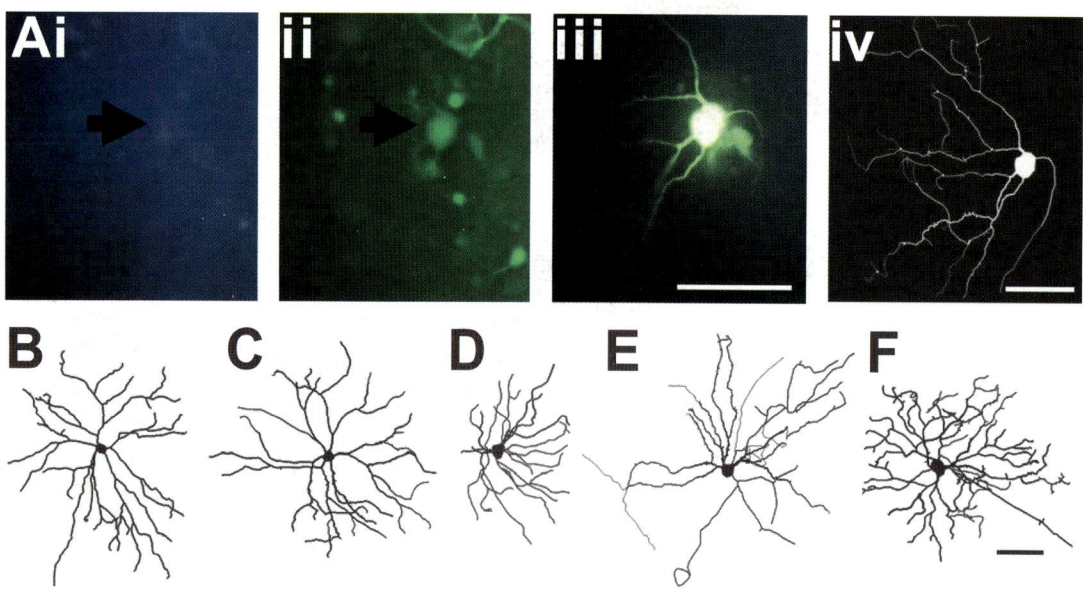

Plate 10.2. (A(i)–A(iv)) Retinal ganglion cell from an AAV–CNTF–GFP-injected retina. The cell is identified by fluorogold labeling (A(i)), is transduced because it expresses GFP (A(ii)), and is injected with Lucifer yellow (A(iii)). The cell is traced using Neurolucida software for analysis. (B–F) traces of Type 1 RGCs from Saline (B), AAV–GFP (C), AAV–CNTF–GFP (D,E), and AAV–BDNF–GFP (F) injected eyes. Scale bar for all figures = 100 μm.

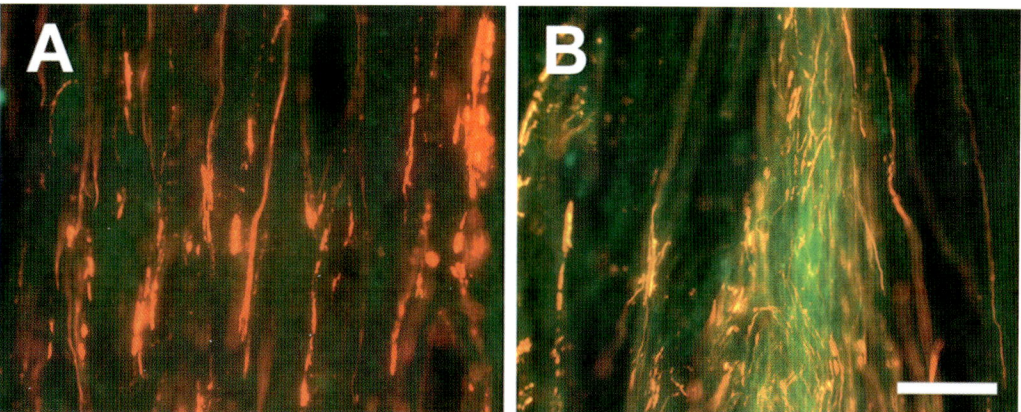

Plate 12.2. The candy store effect. Locally increasing the concentration of neurotrophic factor (e.g., through viral vector-mediated overexpression) at the site of peripheral nerve injury does not improve regeneration. In this longitudinal section of a peripheral nerve after experimental lesioning and subsequent repair, high levels of transgenic GDNF expression at the site of repair (green: GDNF) induce a neuroma-like "candy store" of coiled motoneuron axons (red: choline acetyl transferase). Long-distance outgrowth of neurites towards their targets is thus impaired.

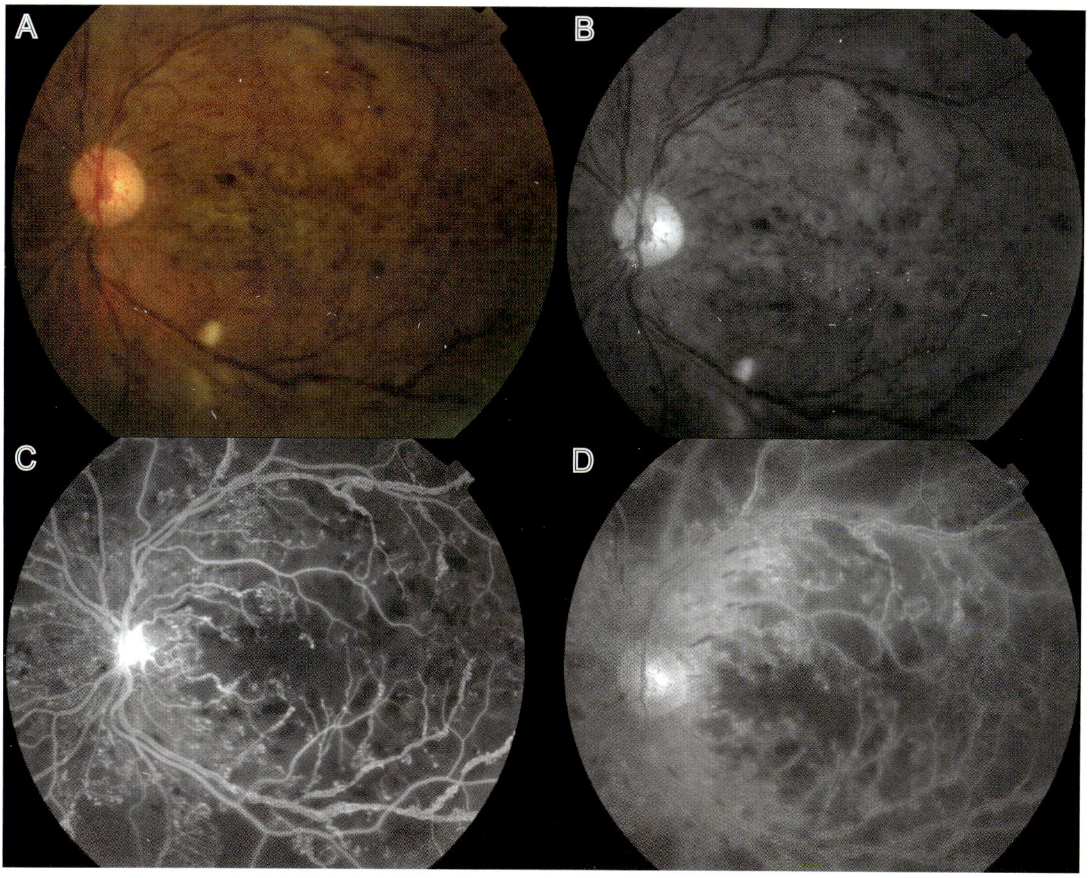

Plate 17.2. Fundus photograph (A), red-free photograph (B), early-phase (C), and late-phase (D) fluorescein angiographic imaging of a patient with severe proliferative diabetic retinopathy (PDR). Note the extensive intraretinal hemorrhages (A, B); a hyperfluorescent lesion on the optic nerve, which is suggestive of a neovascularization (C); extensive venous changes (C, D) and vascular leakage (D). The dark aspect of the retina in the early phase (C) represents widespread capillary nonperfusion and, thus, ischemia.

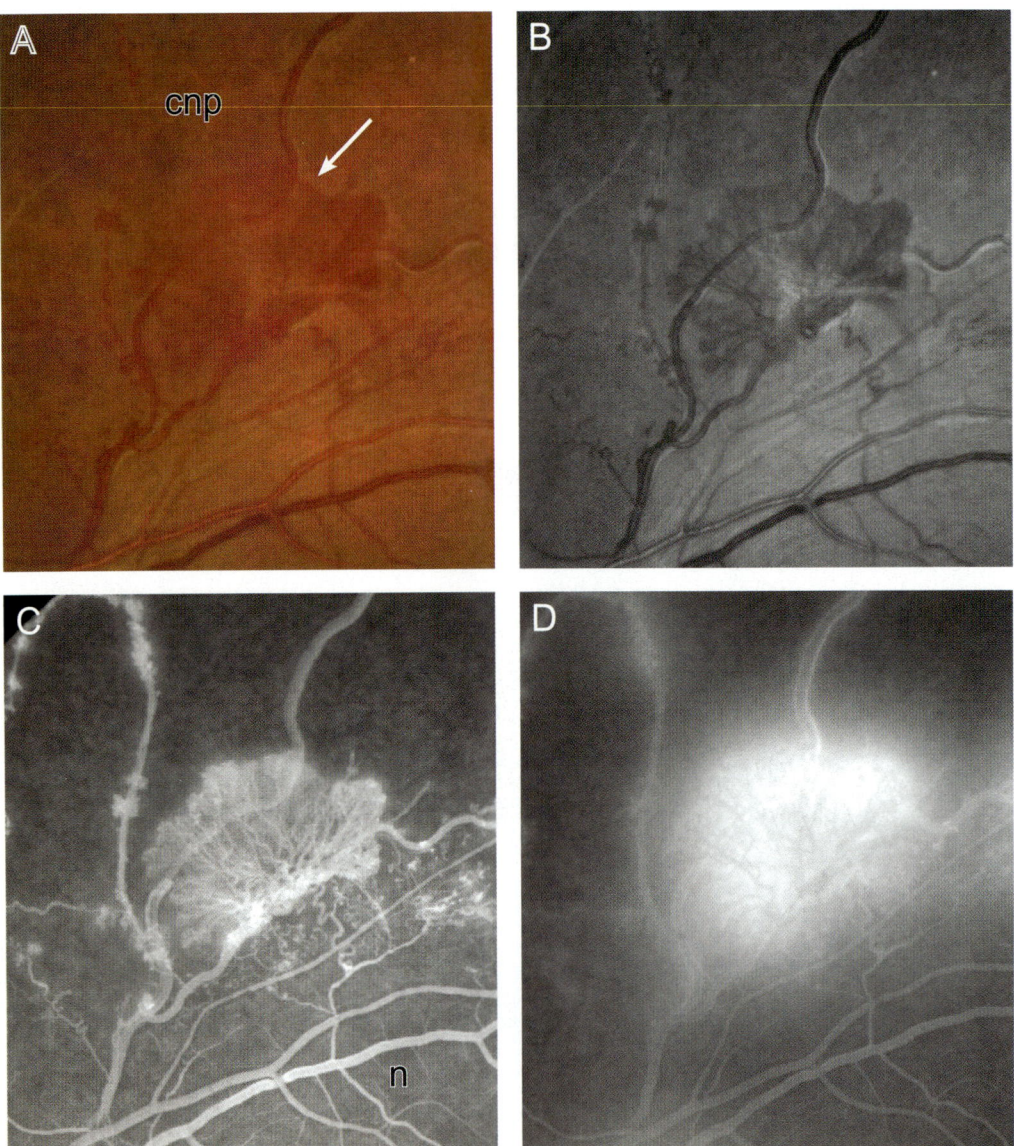

Plate 17.3. Fundus photograph (A), red-free photograph (B), early-phase (C), and late-phase (D) fluorescein angiographic imaging of a patient with preretinal neovascularization after a branch retinal vein occlusion. Note the large neovascularization (white arrow) adjacent to an extensive area of capillary nonperfusion (cnp).

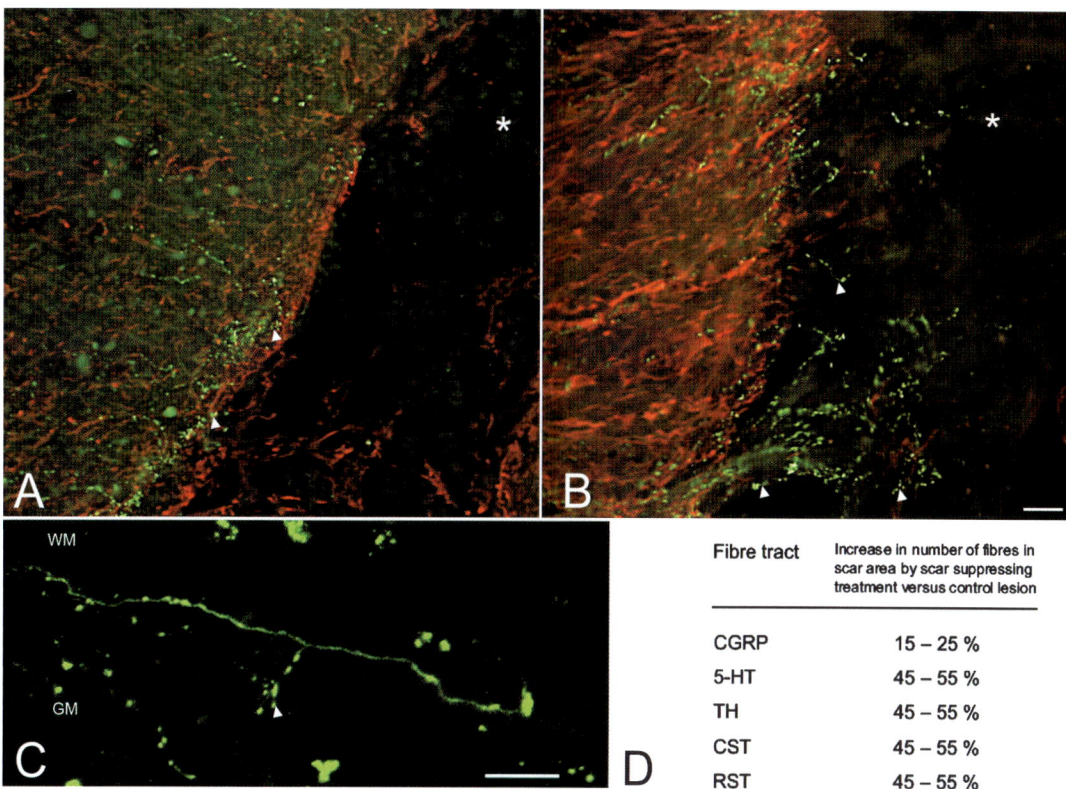

Plate 18.2. Axonal regeneration of different fiber tracts after dorsal hemisection of the rat spinal cord at thoracic level 8 and scar suppressing treatment. Immunohistological double staining of the lesion site at 5 weeks postinjury with antibodies against serotonin (green) and GFAP (red) of a control lesioned rat (A) and a rat receiving the scar suppressing treatment. (B) In contrast to the control, numerous labeled axons enter the GFAP-negative collagenous scar after treatment in B. (C) Anterogradely BDA-labeled CST axon regenerating beyond the lesion site in the distal spinal cord at 4 months after injury and scar suppressing treatment (modified from Klapka et al., 2005). (D) Quantification of axon fragments of different fiber populations detected in the lesion center. The increase in fiber number within the collagenous scar area following the scar suppressing treatment versus treatment with buffer (control) is shown (see also Schiwy et al., 2008). Arrowheads in A and B: serotonergic axons. Arrowhead in C: varicosities on regenerated fiber. Asterisks in A and B: collagenous scar area. GM, gray matter; WM, white matter. Magnification bars: 50 μm.

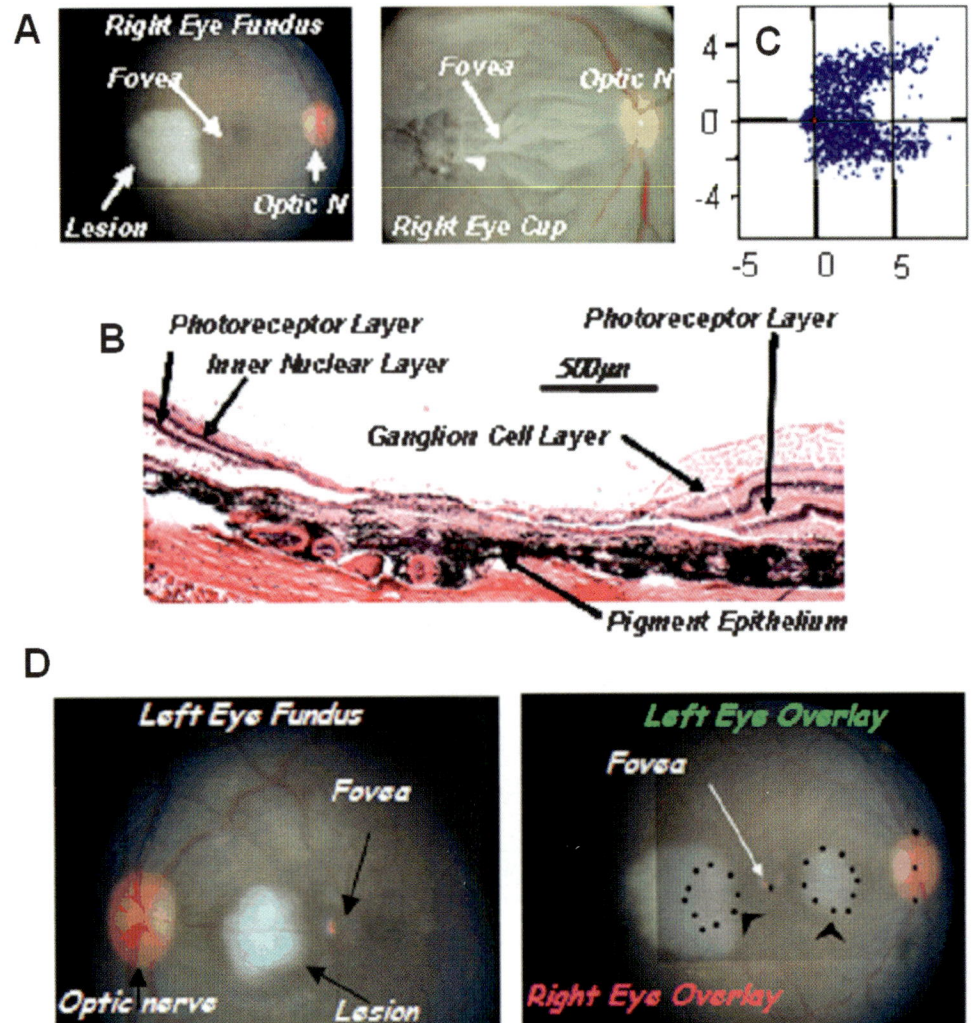

Plate 23.8. Retinal lesioning. All lesions were made using a NIDEK GYC-2000 532-nm photocoagulation laser. (A) Left: Picture of the right eye fundus immediately after inducing a laser photocoagulation lesion. The lesion appears white. Right: Picture of the right fundus after extraction and fixation in formaldehyde at the end of the experiment, 9 months later. The monkey had been euthanized and perfused with formaldehyde prior to extraction. Note the hyperpigmented scar (white arrow), which corresponds to the retinal lesion. (B) Fifteen-micrometer thick section through the center of the same lesion, stained with hematoxylin-eosin. Note that all retinal layers are essentially completely destroyed at the center of the lesion. (C) Saccade to visual target task used to map the visual-field scotoma induced in a macaque after a homonymous retinal lesion (D). Adjacent saccade targets were spaced by $\sim 0.5°$. Each dot represents a successful saccade to target. Note the absence of saccades outlining the area of the induced scotoma. Scale: Degrees. (D) Illustration of the lesions made in both fundi in order to induce a partly homonymous lesion. Left: Lesion in the left eye fundus. Right: Overlay of the right eye fundus (larger) and the left eye fundus (smaller square) illustrating the lesions (white patches marked by the black arrowheads). The left retina (small square overlay) was mirrored along the vertical axis and scaled to make the optical nerves overlap. Black dots outline the left eye lesion and its homonymous location in the right eye, which lies almost entirely over the right eye lesion resulting in an essentially homonymous left visual field scotoma. Note that lesions in each eye were offset from the fovea such that the foveal representation was spared, thereby allowing the animal to fixate accurately. Adapted from Smirnakis et al. (2005).

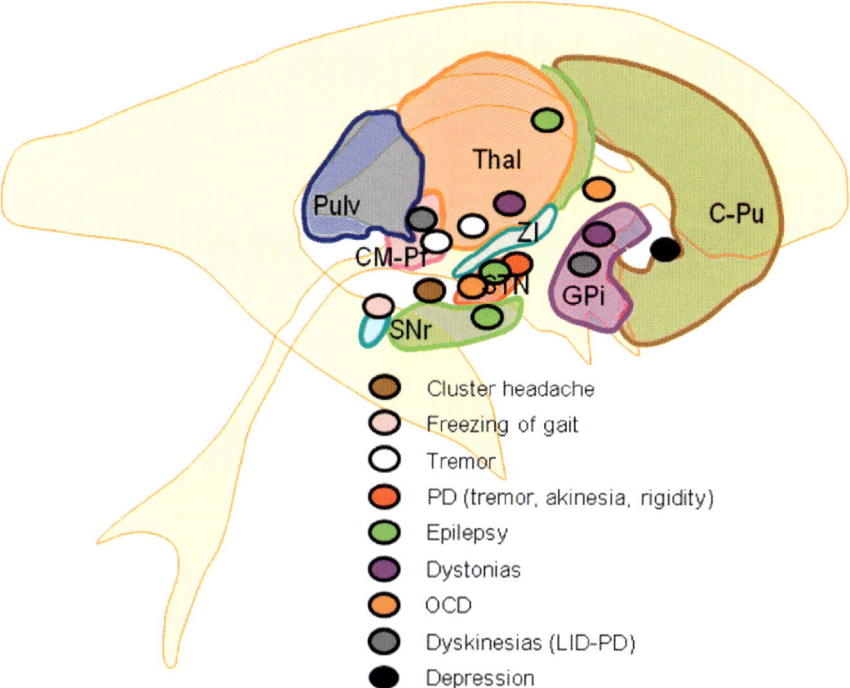

Plate 24.2. Locations of all brain regions mentioned as targets for DBS and the related major indications for these targets. Abbreviations: Thal, thalamus; Pulv, pulvenar part of the thalamus; CM–pf, centrum medianum–parafascicularis; STN, subthalamic nucleus; SNr, substantia nigra pars reticulata; Gpi, globus pallidus internus; C–Pu, caudate–putamen complex; ZI, zona interna.

## Substrate induced growth cone calcium transients

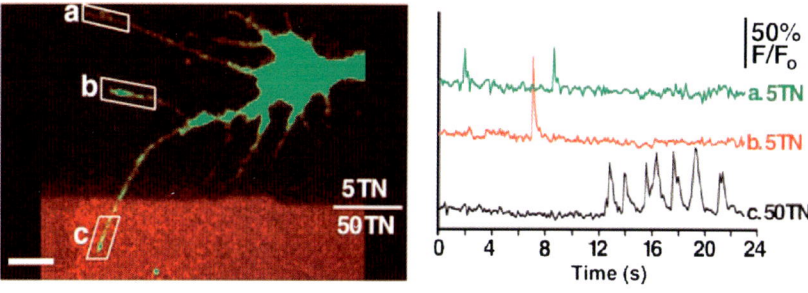

## Calcium triggered growth cone turning

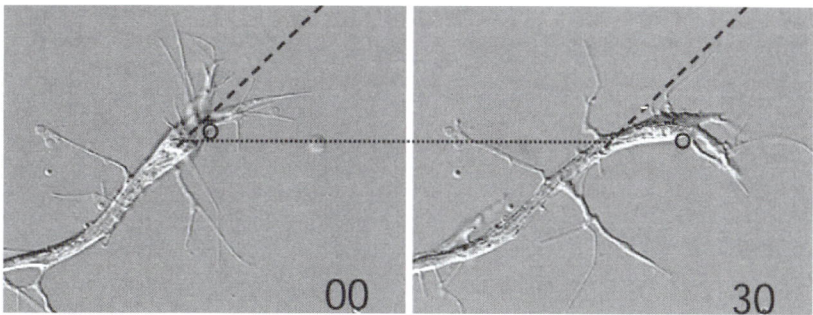

Plate 28.1. Calcium signaling in axonal growth cones. Substrate-induced growth cone calcium transients in filopodia: one filopodium (c) contacting high-concentration tenascin substrate (50 μg/ml TN), a component of the extracellular matrix, displays high-frequency calcium signaling. In contrast, filopodia on a low-concentration tenascin substrate generate only few calcium transients. Adapted with permission from Gomez et al. (2001), AAAS. Calcium-triggered growth cone turning: local calcium uncaging within a growth cone (circle) induces a turn toward the side of calcium elevation after 30 minutes. Adapted with permission from Zheng (2000), Macmillan Publishers Ltd.

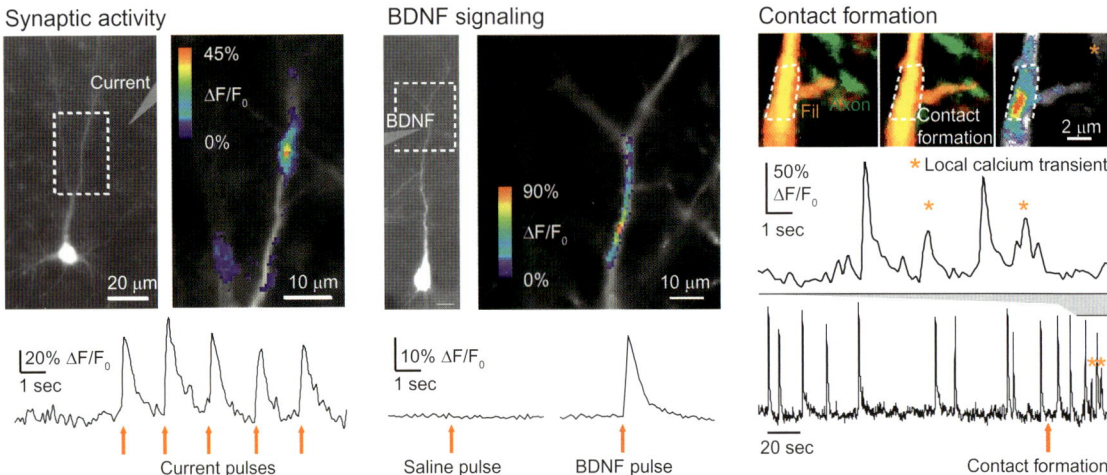

Plate 28.2. Various developmentally relevant factors trigger calcium transients in developing hippocampal neurons. Synaptic activity: current stimulation of presynaptic axons leads to reliable dendritic calcium responses. The spatial spread of calcium within the dendrite is shown in pseudo-color. The trace represents the time course of the calcium response. BDNF signaling: focal application of BDNF induces a fast local calcium transient in the apical dendrite of a CA3 pyramidal neuron. Adapted with permission from Lang et al. (2007), the Society for Neuroscience. Contact formation: within less than one minute, local calcium transients are generated in the dendrite at the site where a filopodium has formed a contact with an axon. The expanded view shows the kinetics of the local calcium transients. Calcium transients that are not marked by an asterisk represent ongoing global calcium activity. Adapted with permission from Lohmann and Bonhoeffer (2008), Cell Press.

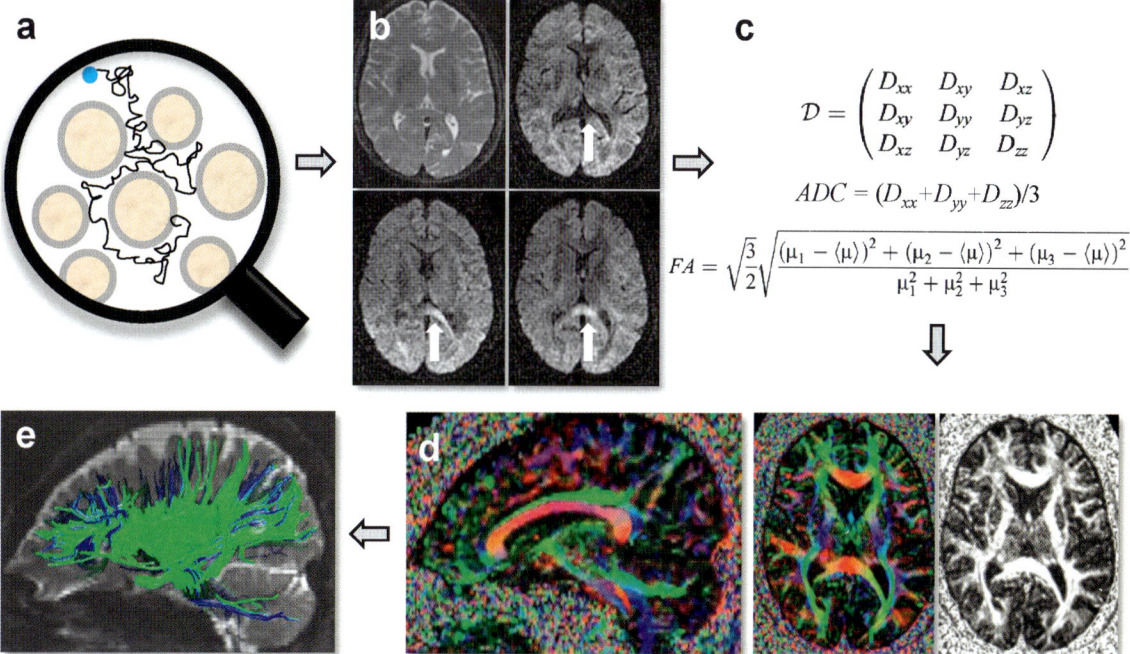

Plate 31.1. From microscopic diffusion to cerebral fiber tract reconstructions. (a) Microscopic thermal diffusion of water protons in the interstitial space between myelinated axons is not isotropic but restricted in the directions perpendicular to the fibers. It is less restricted in the direction parallel to the fibers, which would be out-of-plane here. Schematically sketched is a possible diffusive trajectory of a water proton. There is also diffusion of water within the fibers and across the myelin sheaths, which provides a usually smaller contribution to the average diffusion properties of the tissue. (b) A T2 weighted image and three representative axial MRI diffusion weighted images of the brain of a normal subject. The three diffusion weighted images have been acquired with diffusion gradients probing diffusion along different directions. It is evident how the different gradient directions determine image intensity, for example in the splenium of the corpus callosum (arrows). (c) From the T2 and diffusion weighted images, for each voxel a series of mathematical operations is performed. First, a diffusion tensor D is computed by an overdetermined least squares algorithm. In this case, one T2 weighted and 55 diffusion weighted images have been acquired, from which the six independent components of D were estimated. The tensor is diagonalized to compute a coordinate system that is aligned with the diffusion ellipsoid, a three-dimensional shape describing the preferred directions of diffusion in each voxel. The principal axis of the diffusion ellipsoid defines the estimate of the main fiber direction. The lengths of the three ellipsoid axes, or the eigenvalues μ of the matrix D, define useful properties characterizing the microscopic organization of the tissue. In particular, the apparent diffusion constant (ADC) is an estimate of the diffusion strength, and the fractional anisotropy (FA) quantifies the anisotropy of diffusion. (d) A sagittal FA color map, an axial FA color map, and an axial FA intensity map. FA color maps demonstrate the predominant white matter fiber direction with different colors indicating different directions. Color is composed of red, blue, and green, indexing whether anisotropy is most pronounced from left-to-right, superior-to-inferior, or anterior-to-posterior, respectively. The intensity of the color is proportional to FA: it is large for the hindered diffusion of water, such as in case of myelinated axons, and small for more isotropic regions in the brain, such as gray matter or cerebrospinal fluid. (e) The fiber orientation information of the diffusion tensor eigenvectors and the anisotropy values provide the basis for estimating fiber tracts in the brain. In this example, two seed regions in the internal capsule were placed in the right and the left hemispheres. Fiber estimates originating from these seed regions are rendered in green and blue, respectively.

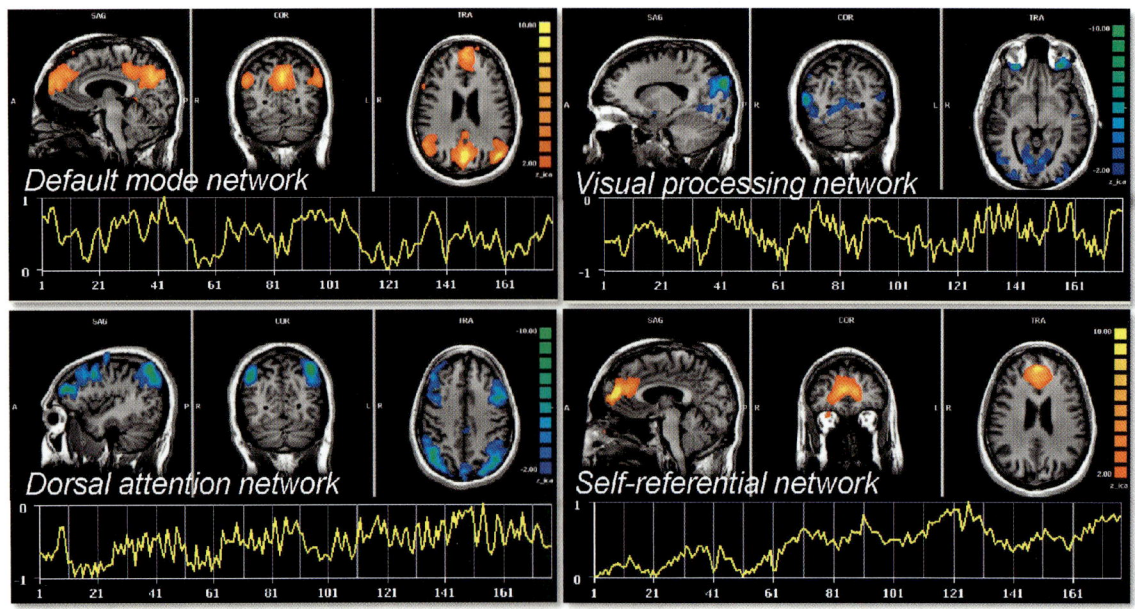

Plate 31.2. Four resting state networks in a normal subject. Also shown are corresponding time courses of the independent component coefficients. (Colors quantify the $z$-value of correlation of the time series with the corresponding ICA coefficient. Either positive or negative $z$-values are shown.)

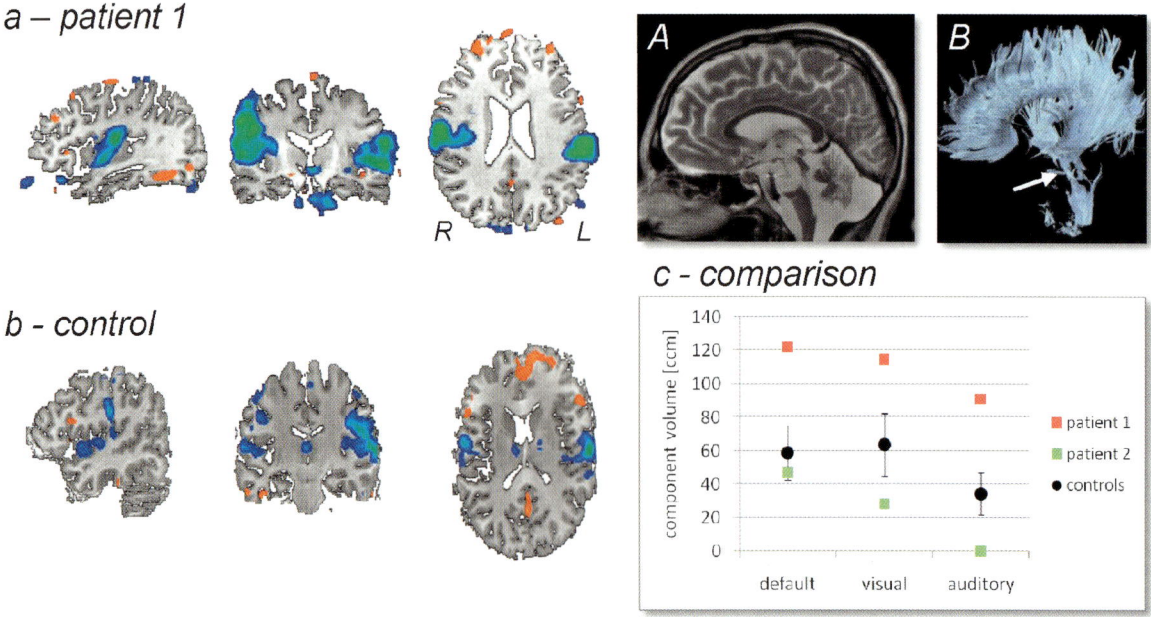

Plate 31.3. Auditory–phonological resting state network of (a) Patient 1 and (b) a control subject, as well as (c) a comparison of network volumes between both patients and four control subjects. An anatomical characterization of Patient 1 by means of (A) T1 weighted anatomical imaging and (B) estimates of white matter fiber tracts using DTI. The arrow points to the pontine lesions visible here as disrupted white matter fiber tracts.